CLASSICAL MECHANICS

SECOND EDITION

Herbert Goldstein

Columbia University

ADDISON-WESLEY PUBLISHING COMPANY

Reading, Massachusetts · Menlo Park, California
London · Amsterdam · Don Mills, Ontario · Sydney

This book is in the
Addison-Wesley Series in Physics

Library of Congress Cataloging in Publication Data

Goldstein, Herbert, 1922–
    Classical mechanics.

    Bibliography: p.
    Includes index.
    1. Mechanics, Analytic. I. Title.
QA805.G6 1980   531   79-23456
ISBN 0-201-02918-9

Second printing, July 1981

ISBN 0-201-02918-9
    FGHIJ-MA-8987654

To the Memory
of
LOUIS J. KLEIN
Teacher and Friend

CONTENTS

CHAPTER 8   THE HAMILTON EQUATIONS OF MOTION   339

CHAPTER 9   CANONICAL TRANSFORMATIONS   378

CHAPTER 10   HAMILTON-JACOBI THEORY   438

CHAPTER 11   CANONICAL PERTURBATION THEORY   499

PREFACE TO THE SECOND EDITION

The prospect of a second edition of *Classical Mechanics,* almost thirty years after initial publication, has given rise to two nearly contradictory sets of reactions. On the one hand it is claimed that the adjective "classical" implies the field is complete, closed, far outside the mainstream of physics research. Further, the first edition has been paid the compliment of continuous use as a text since it first appeared. Why then the need for a second edition? The contrary reaction has been that a second edition is long overdue. More important than changes in the subject matter (which have been considerable) has been the revolution in the attitude towards classical mechanics in relation to other areas of science and technology. When it appeared, the first edition was part of a movement breaking with older ways of teaching physics. But what were bold new ventures in 1950 are the commonplaces of today, exhibiting to the present generation a slightly musty and old-fashioned air. Radical changes need to be made in the presentation of classical mechanics.

In preparing this second edition I have attempted to steer a course somewhere between these two attitudes. I have tried to retain, as much as possible, the advantages of the first edition (as I perceive them) while taking some account of the developments in the subject itself, its position in the curriculum, and its applications to other fields. What has emerged is a thorough-going revision of the first edition. Hardly a page of the text has been left untouched. The changes have been of various kinds:

Errors (some egregious) that I have caught, or which have been pointed out to me, have of course been corrected. It is hoped that not too many new ones have been introduced in the revised material.

The chapter on small oscillations has been moved from its former position as the penultimate chapter and placed immediately after Chapter 5 on rigid body motion. This location seems more appropriate to the usual way mechanics courses are now being given. Some material relating to the Hamiltonian formulation has therefore had to be removed and inserted later in (the present) Chapter 8.

A new chapter on perturbation theory has been added (Chapter 11). The last chapter, on continuous systems and fields, has been greatly expanded, in keeping with the implicit promise made in the Preface to the first edition.

New sections have been added throughout the book, ranging from one in Chapter 3 on Bertrand's theorem for the central-force potentials giving rise to closed orbits, to the final section of Chapter 12 on Noether's theorem. For the most part these sections contain completely new material.

In various sections arguments and proofs have been replaced by new ones that seem simpler and more understandable, e.g., the proof of Euler's theorem in Chapter 4. Occasionally, a line of reasoning presented in the first edition has been supplemented by a different way of looking at the problem. The most important example is the introduction of the symplectic approach to canonical transformations, in parallel with the older technique of generating functions. Again, while the original convention for the Euler angles has been retained, alternate conventions, including the one common in quantum mechanics, are mentioned and detailed formulas are given in an appendix.

As part of the fruits of long experience in teaching courses based on the book, the body of exercises at the end of each chapter has been expanded by more than a factor of two and a half. The bibliography has undergone similar expansion, reflecting the appearance of many valuable texts and monographs in the years since the first edition. In deference to—but not in agreement with—the present neglect of foreign languages in graduate education in the United States, references to foreign-language books have been kept down to a minimum.

The choices of topics retained and of the new material added reflect to some degree my personal opinions and interests, and the reader might prefer a different selection. While it would require too much space (and be too boring) to discuss the motivating reasons relative to each topic, comment should be made on some general principles governing my decisions. The question of the choice of mathematical techniques to be employed is a vexing one. The first edition attempted to act as a vehicle for introducing mathematical tools of wide usefulness that might be unfamiliar to the student. In the present edition the attitude is more one of caution. It is much more likely now than it was 30 years ago that the student will come to mechanics with a thorough background in matrix manipulation. The section on matrix properties in Chapter 4 has nonetheless been retained, and even expanded, so as to provide a convenient reference of needed formulas and techniques. The cognoscenti can, if they wish, simply skip the section. On the other hand, very little in the way of newer mathematical tools has been introduced. Elementary properties of group theory are given scattered mention throughout the book. Brief attention is paid in Chapters 6 and 7 to the manipulation of tensors in non-Euclidean spaces. Otherwise, the mathematical level in this edition is pretty much the same as in the first. It is more than adequate for the physics content of the book, and alternate means exist in the curriculum for acquiring the mathematics needed in other branches of physics. In particular the "new mathematics" of theoretical physics has been deliberately excluded. No mention is made of manifolds or diffeomorphisms, of tangent fibre bundles or invariant torii. There are certain highly specialized areas of classical mechanics where the powerful tools of global analysis and differential topology are useful, probably essential. However, it is not clear to me that they contribute to the understanding of the physics of classical mechanics at the level sought in this edition. To introduce these mathematical concepts, and their applications, would swell the book beyond bursting, and serve, probably, only to obscure the physics. Theoretical physics, current trends to the contrary, is not merely mathematics.

In line with this attitude, the complex Minkowski space has been retained for most of the discussion of special relativity in order to simplify the mathematics. The bases for this decision (which it is realized goes against the present fashion) are given in detail on pages 292–293.

It is certainly true that classical mechanics today is far from being a closed subject. The last three decades have seen an efflorescence of new developments in classical mechanics, the tackling of new problems, and the application of the techniques of classical mechanics to far-flung reaches of physics and chemistry. It would clearly not be possible to include discussions of all of these developments here. The reasons are varied. Space limitations are obviously important. Also, popular fads of current research often prove ephemeral and have a short lifetime. And some applications require too extensive a background in other fields, such as solid-state physics or physical chemistry. The selection made here represents something of a personal compromise. Applications that allow simple descriptions and provide new insights are included in some detail. Others are only briefly mentioned, with enough references to enable the student to follow up his awakened curiosity. In some instances I have tried to describe the current state of research in a field almost entirely in words, without mathematics, to provide the student with an overall view to guide further exploration. One area omitted deserves special mention—nonlinear oscillation and associated stability questions. The importance of the field is unquestioned, but it was felt that an adequate treatment deserves a book to itself.

With all the restrictions and careful selection, the book has grown to a size probably too large to be covered in a single course. A number of sections have been written so that they may be omitted without affecting later developments and have been so marked. It was felt however that there was little need to mark special "tracks" through the book. Individual instructors, familiar with their own special needs, are better equipped to pick and choose what they feel should be included in the courses they give.

I am grateful to many individuals who have contributed to my education in classical mechanics over the past thirty years. To my colleagues Professors Frank L. DiMaggio, Richard W. Longman, and Dean Peter W. Likins I am indebted for many valuable comments and discussions. My thanks go to Sir Edward Bullard for correcting a serious error in the first edition, especially for the gentle and gracious way he did so. Professor Boris Garfinkel of Yale University very kindly read and commented on several of the chapters and did his best to initiate me into the mysteries of celestial mechanics. Over the years I have been the grateful recipient of valuable corrections and suggestions from many friends and strangers, among whom particular mention should be made of Drs. Eric Ericsen (of Oslo University), K. Kalikstein, J. Neuberger, A. Radkowsky, and Mr. W. S. Pajes. Their contributions have certainly enriched the book, but of course I alone am responsible for errors and misinterpretations. I should like to add a collective acknowledgment and thanks to the authors of papers on classical mechanics that have appeared during the last three decades in the *American Journal of Physics,* whose pages I hope I have perused with profit.

The staff at Addison-Wesley have been uniformly helpful and encouraging. I want especially to thank Mrs. Laura R. Finney for her patience with what must have seemed a never-ending process, and Mrs. Marion Howe for her gentle but persistent cooperation in the fight to achieve an acceptable printed page.

To my father, Harry Goldstein ל״ז, I owe more than words can describe for his lifelong devotion and guidance. But I wish at least now to do what he would not permit in his lifetime—to acknowledge the assistance of his incisive criticism and careful editing in the preparation of the first edition. I can only hope that the present edition still reflects something of his insistence on lucid and concise writing.

I wish to dedicate this edition to those I treasure above all else on this earth, and who have given meaning to my life—to my wife, Channa, and our children, Penina Perl, Aaron Meir, and Shoshanna.

And above all I want to register the thanks and acknowledgment of my heart, in the words of Daniel (2:23):

לך אלה אבהתי מהודא ומשבח אנה
די חכמתא וגבורתא יהבת לי

Kew Gardens Hills, New York   HERBERT GOLDSTEIN
January 1980

PREFACE TO THE FIRST EDITION

An advanced course in classical mechanics has long been a time-honored part of the graduate physics curriculum. The present-day function of such a course, however, might well be questioned. It introduces no new physical concepts to the graduate student. It does not lead him directly into current physics research. Nor does it aid him, to any appreciable extent, in solving the practical mechanics problems he encounters in the laboratory.

Despite this arraignment, classical mechanics remains an indispensable part of the physicist's education. It has a twofold role in preparing the student for the study of modern physics. First, classical mechanics, in one or another of its advanced formulations, serves as the springboard for the various branches of modern physics. Thus, the technique of action-angle variables is needed for the older quantum mechanics, the Hamilton-Jacobi equation and the principle of least action provide the transition to wave mechanics, while Poisson brackets and canonical transformations are invaluable in formulating the newer quantum mechanics. Secondly, classical mechanics affords the student an opportunity to master many of the mathematical techniques necessary for quantum mechanics while still working in terms of the familiar concepts of classical physics.

Of course, with these objectives in mind, the traditional treatment of the subject, which was in large measure fixed some fifty years ago, is no longer adequate. The present book is an attempt at an exposition of classical mechanics which does fulfill the new requirements. Those formulations which are of importance for modern physics have received emphasis, and mathematical techniques usually associated with quantum mechanics have been introduced wherever they result in increased elegance and compactness. For example, the discussion of central force motion has been broadened to include the kinematics of scattering and the classical solution of scattering problems. Considerable space has been devoted to canonical transformations, Poisson bracket formulations, Hamilton-Jacobi theory, and action-angle variables. An introduction has been provided to the variational principle formulation of continuous systems and fields. As an illustration of the application of new mathematical techniques, rigid body rotations are treated from the standpoint of matrix transformations. The familiar Euler's theorem on the motion of a rigid body can then be presented in terms of the eigenvalue problem for an orthogonal matrix. As a consequence, such diverse topics as the inertia tensor, Lorentz transformations in Minkowski space, and resonant frequencies of small oscillations become capable of a unified mathematical treatment. Also, by this technique it becomes possible to include at an early stage the difficult concepts of reflection operations and pseudotensor quantities, so important in modern quantum mechanics. A further advantage of matrix methods is that "spinors" can be introduced in connection with the properties of Cayley-Klein parameters.

Several additional departures have been unhesitatingly made. All too often, special relativity receives no connected development except as part of a highly specialized course which also covers general relativity. However, its vital importance in modern physics requires that the student be exposed to special relativity at an early stage in his education. Accordingly, Chapter 6 has been devoted to the subject. Another innovation has been the inclusion of velocity-dependent forces. Historically, classical mechanics developed with the emphasis on static forces dependent on position only, such as gravitational forces. On the other hand, the velocity-dependent electromagnetic force is constantly encountered in modern physics. To enable the student to handle such forces as early as possible, velocity-dependent potentials have been included in the structure of mechanics from the outset, and have been consistently developed throughout the text.

Still another new element has been the treatment of the mechanics of continuous systems and fields in Chapter 11, and some comment on the choice of material is in order. Strictly interpreted, the subject could include all of elasticity, hydrodynamics, and acoustics, but these topics lie outside the prescribed scope of the book, and adequate treatises have been written for most of them. In contrast, no connected account is available on the classical foundations of the variational principle formulation of continuous systems, despite its growing importance in the field theory of elementary particles. The theory of fields can be carried to considerable length and complexity before it is necessary to introduce quantization. For example, it is perfectly feasible to discuss the stress-energy tensor, microscopic equations of continuity, momentum space representations, etc., entirely within the domain of classical physics. It was felt, however, that an adequate discussion of these subjects would require a sophistication beyond what could naturally be expected of the student. Hence it was decided, for this edition at least, to limit Chapter 11 to an elementary description of the Lagrangian and Hamiltonian formulation of fields.

The course for which this text is designed normally carries with it a prerequisite of an intermediate course in mechanics. For both the inadequately prepared graduate student (an all too frequent occurrence) and the ambitious senior who desires to omit the intermediate step, an effort was made to keep the book self-contained. Much of Chapters 1 and 3 is therefore devoted to material usually covered in the preliminary courses.

With few exceptions, no more mathematical background is required of the student than the customary undergraduate courses in advanced calculus and vector analysis. Hence considerable space is given to developing the more complicated mathematical tools as they are needed. An elementary acquaintance with Maxwell's equations and their simpler consequences is necessary for understanding the sections on electromagnetic forces. Most entering graduate students have had at least one term's exposure to modern physics, and frequent advantage has been taken of this circumstance to indicate briefly the relation between a classical development and its quantum continuation.

A large store of exercises is available in the literature on mechanics, easily accessible to all, and there consequently seemed little point to reproducing an extensive collection of such problems. The exercises appended to each chapter therefore have been limited, in the main, to those which serve as extensions of the text, illustrating some particular point or proving variant theorems. Pedantic museum pieces have been studiously avoided.

The question of notation is always a vexing one. It is impossible to achieve a completely consistent and unambiguous system of notation that is not at the same time impracticable and cumbersome. The customary convention has been followed of indicating vectors by bold face Roman letters. In addition, matrix quantities of whatever rank, and tensors other than vectors, are designated by bold face sans serif characters, thus: **A**. An index of symbols is appended at the end of the book, listing the initial appearance of each meaning of the important symbols. Minor characters, appearing only once, are not included.

References have been listed at the end of each chapter, for elaboration of the material discussed or for treatment of points not touched on. The evaluations accompanying these references are purely personal, of course, but it was felt necessary to provide the student with some guide to the bewildering maze of literature on mechanics. These references, along with many more, are also listed at the end of the book. The list is not intended to be in any way complete, many of the older books being deliberately omitted. By and large, the list contains the references used in writing this book, and must therefore serve also as an acknowledgement of my debt to these sources.

The present text has evolved from a course of lectures on classical mechanics that I gave at Harvard University, and I am grateful to Professor J. H. Van Vleck, then Chairman of the Physics Department, for many personal and official encouragements. To Professor J. Schwinger, and other colleagues I am indebted for many valuable suggestions. I also wish to record my deep gratitude to the students in my courses, whose favorable reaction and active interest provided the continuing impetus for this work.

תושלב״ע

Cambridge, Mass.   HERBERT GOLDSTEIN
March 1950

Newt's laws

1) $\dfrac{d\mathbf{v}}{dt} = 0$

2) $F = \dfrac{d\mathbf{p}}{dt}$

3) $f_{12} = -f_{21}$

1) $L - const$

2) $N = \dfrac{dL}{dt}$

CHAPTER 1
Survey of the Elementary Principles

The·motion of material bodies formed the subject of some of the earliest researches pursued by the pioneers of physics. From their efforts there has evolved a vast field known as analytical mechanics or dynamics, or simply, mechanics. In the present century the term "classical mechanics" has come into wide use to denote this branch of physics in contradistinction to the newer physical theories, especially quantum mechanics. We shall follow this usage, interpreting the name to include the type of mechanics arising out of the special theory of relativity. It is the purpose of this book to develop the structure of classical mechanics and to outline some of its applications of present-day interest in pure physics.

Basic to any presentation of mechanics are a number of fundamental physical concepts, such as space, time, simultaneity, mass, and force. In discussing the special theory of relativity the notions of simultaneity and of time and length scales will be examined briefly. For the most part, however, these concepts will not be analyzed critically here; rather, they will be assumed as undefined terms whose meanings are familiar to the reader.

1–1 MECHANICS OF A PARTICLE

Let **r** be the radius vector of a particle from some given origin and **v** its vector velocity:

$$\mathbf{v} = \frac{d\mathbf{r}}{dt}. \tag{1–1}$$

The *linear momentum* **p** of the particle is defined as the product of the particle mass and its velocity:

$$\mathbf{p} = m\mathbf{v}. \tag{1–2}$$

In consequence of interactions with external objects and fields the particle may experience forces of various types, e.g., gravitational or electrodynamic; the vector sum of these forces exerted on the particle is the total force **F**. The mechanics of the particle is contained in *Newton's Second Law of Motion*, which

1

states that there exist frames of reference in which the motion of the particle is described by the differential equation

$$\mathbf{F} = \frac{d\mathbf{p}}{dt},$$ (1–3)

or

$$\mathbf{F} = \frac{d}{dt}(m\mathbf{v}).$$ (1–4)

In most instances the mass of the particle is constant and Eq. (1–3) reduces to

$$\mathbf{F} = m\frac{d\mathbf{v}}{dt} = m\mathbf{a},$$ (1–5)

where **a** is the vector acceleration of the particle defined by

$$\mathbf{a} = \frac{d^2\mathbf{r}}{dt^2}.$$ (1–6)

The equation of motion is thus a differential equation of second order, assuming **F** does not depend on higher order derivatives.

A reference frame in which Eq. (1–3) is valid is called an *inertial* or *Galilean system*. Even within classical mechanics the notion of an inertial system is something of an idealization. In practice, however, it is usually feasible to set up a coordinate system that comes as close to the desired properties as may be required. For many purposes a reference frame fixed in the Earth (the "laboratory system") is a sufficient approximation to an inertial system, while for some astronomical purposes it may be necessary to construct an inertial system by reference to the most distant galaxies.

Many of the important conclusions of mechanics can be expressed in the form of conservation theorems, which indicate under what conditions various mechanical quantities are constant in time. Equation (1–1) directly furnishes the first of these, the

Conservation Theorem for the Linear Momentum of a Particle: If the total force, **F**, *is zero then* $\dot{\mathbf{p}} = 0$ *and the linear momentum*, **p**, *is conserved*.

The angular momentum of the particle about point *O*, denoted by **L**, is defined as

$$\mathbf{L} = \mathbf{r} \times \mathbf{p},$$ (1–7)

where **r** is the radius vector from *O* to the particle. Notice that the order of the factors is important. We now define the *moment of force* or *torque* about *O* as

$$\mathbf{N} = \mathbf{r} \times \mathbf{F}.$$ (1–8)

Si/ rxf = 0

in S₂/ r'xf = (r + r₀)xf = r₀xf ≠ 0   frame dependent.

The equation analogous to (1–3) for $\mathbf{N}$ is obtained by forming the cross product of $\mathbf{r}$ with Eq. (1–4):

$$\mathbf{r} \times \mathbf{F} = \mathbf{N} = \mathbf{r} \times \frac{d}{dt}(m\mathbf{v}). \qquad (1\text{–}9)$$

Equation (1–9) can be written in a different form by using the vector identity:

$$\frac{d}{dt}(\mathbf{r} \times m\mathbf{v}) = \underline{\mathbf{v} \times m\mathbf{v}} + \mathbf{r} \times \frac{d}{dt}(m\mathbf{v}), \qquad (1\text{–}10)$$

where the first term on the right obviously vanishes. In consequence of this identity Eq. (1–9) takes the form

Analogy to N's law

$$\mathbf{N} = \frac{d}{dt}(\mathbf{r} \times m\mathbf{v}) = \frac{d\mathbf{L}}{dt}. \qquad (1\text{–}11)$$

Note that both $\mathbf{N}$ and $\mathbf{L}$ depend upon the point O, about which the moments are taken.

As was the case for Eq. (1–3), the torque equation, (1–11), also yields an immediate conservation theorem, this time the

Conservation Theorem for the Angular Momentum of a Particle: If the total torque, $\mathbf{N}$, *is zero then* $\dot{\mathbf{L}} = 0$, *and the angular momentum* $\mathbf{L}$ *is conserved.*

if >0 KE increases.

Next consider the work done by the external force $\mathbf{F}$ upon the particle in going from point 1 to point 2. By definition this work is

$$W_{12} = \int_1^2 \mathbf{F} \cdot d\mathbf{s}. \qquad F = F(x, y, z) \quad (1\text{–}12)$$

For constant mass (as will be assumed from now on unless otherwise specified), the integral in Eq. (1–12) reduces to

ds = v dt

$$\int \mathbf{F} \cdot d\mathbf{s} = m \int \frac{d\mathbf{v}}{dt} \cdot \mathbf{v}\, dt = \frac{m}{2} \int \frac{d}{dt}(v^2)\, dt,$$

and therefore

$$W_{12} = \frac{m}{2}(v_2^2 - v_1^2). \qquad (1\text{–}13)$$

The scalar quantity $mv^2/2$ is called the kinetic energy of the particle and is denoted by T, so that the work done is equal to the change in the kinetic energy:

$$W_{12} = T_2 - T_1. \qquad (1\text{–}14)$$

If the force field is such that the work W_{12} is the same for any physically possible path between points 1 and 2, then the force (and the system) is said to be *conservative.* An alternative description of a conservative system is obtained by

positive work increases T

imagining the particle being taken from point 1 to point 2 by one possible path and then being returned to point 1 by another path. The independence of W_{12} on the particular path implies that the work done around such a closed circuit is zero, that is:

conservative force.

$$\oint \mathbf{F} \cdot d\mathbf{s} = 0. \tag{1–15}$$

$W_s = \int \vec{F} \cdot d\vec{s}$

Physically it is clear that a system cannot be conservative if friction or other dissipation forces are present, because $\mathbf{F} \cdot d\mathbf{s}$ due to friction is always positive and the integral cannot vanish.

$= \int (\vec{\nabla} \times \vec{F}) \cdot d\vec{A}$

By a well-known theorem of vector analysis* a necessary and sufficient condition that the W_{12} be independent of the physical path taken by the particle is that $\mathbf{F}$ be the gradient of some scalar function of position:

if $\vec{F} = -\vec{\nabla}\phi$ *then* $\vec{\nabla} \times \vec{F} = 0$
$\therefore W_s = 0$

$$\mathbf{F} = -\nabla V(\mathbf{r}), \tag{1–16}$$

conservative
force $\leftrightarrow \vec{\nabla}\phi$

where V is called the *potential*, or *potential energy*. The existence of V can be inferred intuitively by a simple argument. If W_{12} is independent of the path of integration between the end points 1 and 2, it should be possible to express W_{12} as the change in a quantity that depends only upon the positions of the end points. This quantity may be designated by $-V$, so that for a differential path length we have the relation

$V = -\phi$

$F = -\nabla V$

$$\mathbf{F} \cdot d\mathbf{s} = -dV$$

or

$$F_s = -\frac{\partial V}{\partial s},$$

which is equivalent to Eq. (1–16). Note that in Eq. (1–16) we can add to V any quantity constant in space, without affecting the results. Hence, *the zero level of V is arbitrary.*

For a conservative system the work done by the forces is

$$W_{12} = V_1 - V_2. \quad = -(V_2 - V_1) \tag{1–17}$$

Combining Eq. (1–17) with Eq. (1–14) we have the result

$$\boxed{T_1 + V_1 = T_2 + V_2,} \tag{1–18}$$

which states in symbols the

Energy Conservation Theorem for a Particle: If the forces acting on a particle are conservative, then the total energy of the particle, $T + V$, is conserved.

---

* See, for example, W. Kaplan, *Advanced Calculus*, 2d ed. (Reading, Massachusetts: Addison-Wesley, 1973), p. 311, p. 347.

The force applied to a particle may in some circumstances be given by the gradient of a scalar function that depends explicitly on both the position of the particle and the time. However, the work done on the particle when it travels a distance ds,

$$\mathbf{F} \cdot d\mathbf{s} = -\frac{\partial V}{\partial s} ds,$$

is then no longer the total change in $-V$ during the displacement, since V also changes explicitly with time as the particle moves. Hence the work done as the particle goes from point 1 to point 2 is no longer the difference in the function V between those points. While a total energy $T + V$ may still be defined, it is not conserved during the course of the particle's motion.

1-2 MECHANICS OF A SYSTEM OF PARTICLES

In generalizing the ideas of the previous section to systems of many particles, we must distinguish between the *external forces* acting on the particles due to sources outside the system, and *internal forces* on, say, some particle i due to all other particles in the system. Thus the equation of motion (Newton's Second Law) for the ith particle is to be written

$$\sum_j \mathbf{F}_{ji} + \mathbf{F}_i^{(e)} = \dot{\mathbf{p}}_i, \qquad (1\text{-}19)$$

where $\mathbf{F}_i^{(e)}$ stands for an external force, and $\mathbf{F}_{ji}$ is the internal force on the ith particle due to the jth particle ($\mathbf{F}_{ii}$, naturally, is zero). We shall assume that the $\mathbf{F}_{ij}$ (like the $\mathbf{F}_i^{(e)}$) obey Newton's third law of motion in its original form: that the forces two particles exert on each other are equal and opposite. This assumption (which does not hold for all types of forces) is sometimes referred to as the *weak law of action and reaction.*

Summed over all particles Eq. (1-19) takes the form

$$\frac{d^2}{dt^2} \sum_i m_i \mathbf{r}_i = \sum_i \mathbf{F}_i^{(e)} + \sum_{\substack{i,j \\ i \neq j}} \mathbf{F}_{ji}. \qquad (1\text{-}20)$$

The first sum on the right is simply the total external force $\mathbf{F}^{(e)}$, while the second term vanishes, since the law of action and reaction states that each pair $\mathbf{F}_{ij} + \mathbf{F}_{ji}$ is zero. To reduce the left-hand side, we define a vector $\mathbf{R}$ as the average of the radii vectors of the particles, weighted in proportion to their mass:

$$\mathbf{R} = \frac{\sum m_i \mathbf{r}_i}{\sum m_i} = \frac{\sum m_i \mathbf{r}_i}{M}. \qquad (1\text{-}21)$$

The vector $\mathbf{R}$ defines a point known as the *center of mass*, or more loosely as the center of gravity, of the system (cf. Fig. 1-1). With this definition (1-20) reduces to

$$M \frac{d^2 \mathbf{R}}{dt^2} = \sum_i \mathbf{F}_i^{(e)} \equiv \mathbf{F}^{(e)}, \qquad (1\text{-}22)$$

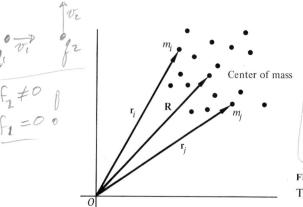

FIGURE 1-1
The center of mass of a system of particles.

which states that the center of mass moves as if the total external force were acting on the entire mass of the system concentrated at the center of mass. Purely internal forces, if they obey Newton's third law, therefore have no effect on the motion of the center of mass. An oft-quoted example is the motion of an exploding shell—the center of mass of the fragments traveling as if the shell were still in a single piece (neglecting air resistance). The same principle is involved in jet and rocket propulsion. In order that the motion of the center of mass be unaffected, the ejection of the exhaust gases at high velocity must be counterbalanced by the forward motion of the vehicle.

By Eq. (1–21) the total linear momentum of the system,

$$\mathbf{P} = \Sigma\, m_i \frac{d\mathbf{r}_i}{dt} = M\frac{d\mathbf{R}}{dt}, \qquad (1\text{–}23)$$

is the total mass of the system times the velocity of the center of mass. Consequently, the equation of motion for the center of mass, (1–23), can be restated as the

Conservation Theorem for the Linear Momentum of a System of Particles: If the total external force is zero, the total linear momentum is conserved.

We obtain the total angular momentum of the system by forming the cross product $\mathbf{r}_i \times \mathbf{p}_i$ and summing over i. If this operation is performed in Eq. (1–19) there results, with the aid of the identity, Eq. (1–10),

$$\sum_i (\mathbf{r}_i \times \dot{\mathbf{p}}_i) = \sum_i \frac{d}{dt}(\mathbf{r}_i \times \mathbf{p}_i) = \dot{\mathbf{L}} = \sum_i \mathbf{r}_i \times \mathbf{F}_i^{(e)} + \sum_{\substack{i,j \\ i \neq j}} \mathbf{r}_i \times \mathbf{F}_{ji}. \qquad (1\text{–}24)$$

The last term on the right in (1–24) can be considered a sum of the pairs of the form

$$\mathbf{r}_i \times \mathbf{F}_{ji} + \mathbf{r}_j \times \mathbf{F}_{ij} = (\mathbf{r}_i - \mathbf{r}_j) \times \mathbf{F}_{ji}, \qquad (1\text{–}25)$$

using the equality of action and reaction. But $\mathbf{r}_i - \mathbf{r}_j$ is identical with the vector $\mathbf{r}_{ij}$ from j to i (cf. Fig. 1–2), so that the right-hand side of Eq. (1–25) can be written as

$$\mathbf{r}_{ij} \times \mathbf{F}_{ji}. \quad = 0 \; ! \quad \text{by third law}$$

If the internal forces between two particles in addition to being equal and opposite also lie along the line joining the particles—a condition known as the *strong law of action and reaction*—then all of these cross products vanish. The sum over pairs is zero under this assumption and Eq. (1–24) may be written in the form

$$\frac{d\mathbf{L}}{dt} = \mathbf{N}^{(e)}. \quad = \sum_i \mathbf{r}_i \times \mathbf{f}_i^{(e)} \tag{1–26}$$

The time derivative of the total angular momentum is thus equal to the moment of the external force about the given point. Corresponding to Eq. (1–26) is the

*Conservation Theorem for Total Angular Momentum: **L** is constant in time if the applied (external) torque is zero.*

(It is perhaps worthwhile to emphasize that this is a *vector* theorem, that is, L_z will be conserved if $N_z^{(e)}$ is zero, even if $N_x^{(e)}$ and $N_y^{(e)}$ are not zero.)

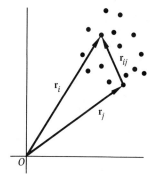

FIGURE 1–2
The vector $\mathbf{r}_{ij}$ between the ith and jth particles.

Note that the conservation of linear momentum in the absence of applied forces assumes that the weak law of action and reaction is valid for the internal forces. The conservation of the total angular momentum of the system in the absence of applied torques requires the validity of the strong law of action and reaction—that the internal forces in addition be *central*. Many of the familiar physical forces, such as that of gravity, satisfy the strong form of the law. But it is possible to find forces for which action and reaction are equal even though the forces are not central (see below). In a system involving moving charges, the forces between the charges predicted by the Biot–Savart law indeed may violate both forms of the action and reaction law.* Equations (1–23) and (1–26), and their corresponding conservation theorems, are not applicable in such cases, at least in the form here given. Usually it is then possible to find some generalization of **P** or **L** that is conserved. Thus, in an isolated system of moving charges it is the sum of the mechanical angular momentum and the electromagnetic "angular momentum" of the field that is conserved.

Equation (1–23) states that the total linear momentum of the system is the same as if the entire mass were concentrated at the center of mass and moving with it. The analogous theorem for angular momentum is more complicated. With the origin O as reference point the total angular momentum of the system is

$$\mathbf{L} = \sum_i \mathbf{r}_i \times \mathbf{p}_i.$$

Let **R** be the radius vector from O to the center of mass, and let $\mathbf{r}_i'$ be the radius vector from the center of mass to the ith particle. Then we have (cf. Fig. 1–3)

$$\mathbf{r}_i = \mathbf{r}_i' + \mathbf{R} \tag{1–27}$$

and

$$\mathbf{v}_i = \mathbf{v}_i' + \mathbf{v},$$

where

$$\mathbf{v} = \frac{d\mathbf{R}}{dt}$$

is the velocity of the center of mass relative to O, and

$$\mathbf{v}' = \frac{d\mathbf{r}'}{dt}$$

is the velocity of the ith particle relative to the center of mass of the system. Using Eq. (1–27), the total angular momentum takes on the form

$$\mathbf{L} = \sum_i \mathbf{R} \times m_i \mathbf{v} + \sum_i \mathbf{r}_i' \times m_i \mathbf{v}_i' + \left(\sum_i m_i \mathbf{r}_i' \right) \times \mathbf{v} + \mathbf{R} \times \frac{d}{dt} \sum_i m_i \mathbf{r}_i'.$$

The last two terms in this expression vanish, for both contain the factor $\sum m_i \mathbf{r}_i'$, which, it will be recognized, defines the radius vector of the center of mass in the very coordinate system whose origin is the center of mass and is therefore a null vector. Rewriting the remaining terms, the total angular momentum about O is

$$\mathbf{L} = \mathbf{R} \times M\mathbf{v} + \sum_i \mathbf{r}_i' \times \mathbf{p}_i'. \tag{1–28}$$

L about O L about CM

---

*If two charges are moving uniformly with parallel velocity vectors that are not perpendicular to the line joining the charges, then the mutual forces are equal and opposite but do not lie along the vector between the charges. Consider, further, two charges moving (instantaneously) so as to "cross the T," i.e., one charge moving directly at the other, which in turn is moving at right angles to the first. Then the second charge exerts a nonvanishing force on the first, without experiencing any reaction force at all.

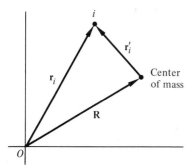

FIGURE 1-3
The vectors involved in the shift of reference point for the angular momentum.

In words, Eq. (1–28) says that the total angular momentum about a point O is the angular momentum of the system concentrated at the center of mass, plus the angular momentum of motion about the center of mass. The form of Eq. (1–28) emphasizes that in general $\mathbf{L}$ depends on the origin O, through the vector $\mathbf{R}$. Only if the center of mass is at rest with respect to O will the angular momentum be independent of the point of reference. In this case the first term in (1–28) vanishes, and $\mathbf{L}$ always reduces to the angular momentum taken about the center of mass.

Finally, let us consider the energy equation. As in the case of a single particle, we calculate the work done by all forces in moving the system from an initial configuration 1, to a final configuration 2:

$$W_{12} = \sum_i \int_1^2 \mathbf{F}_i \cdot d\mathbf{s}_i = \sum_i \int_1^2 \mathbf{F}_i^{(e)} \cdot d\mathbf{s}_i + \sum_{\substack{i,j \\ i \neq j}} \int_1^2 \mathbf{F}_{ji} \cdot d\mathbf{s}_i. \qquad (1\text{–}29)$$

Again, the equations of motion can be used to reduce the integrals to

$$\sum_i \int_1^2 \mathbf{F}_i \cdot d\mathbf{s}_i = \sum_i \int_1^2 m_i \dot{\mathbf{v}}_i \cdot \mathbf{v}_i \, dt = \sum_i \int_1^2 d(\tfrac{1}{2} m_i v_i^2).$$

Hence the work done can still be written as the difference of the final and initial kinetic energies:

$$W_{12} = T_2 - T_1,$$

where T, the total kinetic energy of the system, is

$$T = \frac{1}{2} \sum_i m_i v_i^2. \qquad (1\text{–}30)$$

Making use of the transformations to center-of-mass coordinates, given in Eq. (1–27), we may write T also as

$$T = \frac{1}{2} \sum_i m_i (\mathbf{v} + \mathbf{v}_i') \cdot (\mathbf{v} + \mathbf{v}_i')$$

$$= \frac{1}{2} \sum_i m_i v^2 + \frac{1}{2} \sum_i m_i v_i'^2 + \mathbf{v} \cdot \frac{d}{dt} \left(\sum_i m_i \mathbf{r}_i' \right),$$

and by the reasoning already employed in calculating the angular momentum, the last term vanishes, leaving

$$T = \frac{1}{2}Mv^2 + \frac{1}{2}\sum_i m_i v_i'^2. \tag{1-31}$$

The kinetic energy, like the angular momentum, thus also consists of two parts: the kinetic energy obtained if all the mass were concentrated at the center of mass, plus the kinetic energy of motion about the center of mass.

Consider now the right-hand side of Eq. (1–29). In the special case that the external forces are derivable in terms of the gradient of a potential, the first term can be written as

$$\sum_i \int_1^2 \mathbf{F}_i^{(e)} \cdot d\mathbf{s}_i = -\sum_i \int_1^2 \nabla_i V_i \cdot d\mathbf{s}_i = -\sum_i V_i \Big|_1^2,$$

where the subscript i on the del operator indicates that the derivatives are with respect to the components of $\mathbf{r}_i$. If the internal forces are also conservative, then the mutual forces between the ith and jth particles, $\mathbf{F}_{ij}$ and $\mathbf{F}_{ji}$, can be obtained from a potential function V_{ij}. To satisfy the strong law of action and reaction, V_{ij} can be a function only of the distance between the particles:

$$V_{ij} = V_{ij}(|\mathbf{r}_i - \mathbf{r}_j|). \tag{1-32}$$

The two forces are then automatically equal and opposite,

$$\mathbf{F}_{ji} = -\nabla_i V_{ij} = +\nabla_j V_{ij} = -\mathbf{F}_{ij}, \tag{1-33}$$

and lie along the line joining the two particles,

$$\nabla V_{ij}(|\mathbf{r}_i - \mathbf{r}_j|) = (\mathbf{r}_i - \mathbf{r}_j)f, \tag{1-34}$$

where f is some scalar function. If V_{ij} were also a function of the difference of some other pair of vectors associated with the particles, such as their velocities or (to step into the domain of modern physics) their intrinsic "spin" angular momenta, then the forces would still be equal and opposite, but would not necessarily lie along the direction between the particles.

When the forces are all conservative the second term in Eq. (1–29) can be rewritten as a sum over *pairs* of particles, the terms for each pair being of the form

$$-\int_1^2 (\nabla_i V_{ij} \cdot d\mathbf{s}_i + \nabla_j V_{ij} \cdot d\mathbf{s}_j).$$

If the difference vector $\mathbf{r}_i - \mathbf{r}_j$ be denoted by $\mathbf{r}_{ij}$, and if ∇_{ij} stands for the gradient with respect to $\mathbf{r}_{ij}$, then

$$\nabla_i V_{ij} = \nabla_{ij} V_{ij} = -\nabla_j V_{ij},$$

and

$$d\mathbf{s}_i - d\mathbf{s}_j = d\mathbf{r}_i - d\mathbf{r}_j = d\mathbf{r}_{ij},$$

so that the term for the ij pair has the form

$$- \int \nabla_{ij} V_{ij} \cdot d\mathbf{r}_{ij}.$$

The total work arising from internal forces then reduces to

$$-\frac{1}{2} \sum_{\substack{i,j \\ i \neq j}} \int_1^2 \nabla_{ij} V_{ij} \cdot d\mathbf{r}_{ij} = -\frac{1}{2} \sum_{\substack{i,j \\ i \neq j}} V_{ij} \bigg|_1^2. \tag{1–35}$$

The factor $\frac{1}{2}$ appears in Eq. (1–35) because in summing over *both* i and j each member of a given pair is included twice, first in the i summation and then in the j summation.

From these considerations it is clear that if the external and internal forces are both derivable from potentials it is possible to define a *total potential energy*, V, of the system,

$$V = \sum_i V_i + \frac{1}{2} \sum_{\substack{i,j \\ i \neq j}} V_{ij}, \tag{1–36}$$

such that the total energy $T + V$ is conserved, the analog of the conservation theorem (1–17) for a single particle.

The second term on the right in Eq. (1–36) will be called the internal potential energy of the system. In general, it need not be zero and, more important, it may vary as the system changes with time. Only for the particular class of systems known as *rigid bodies* will the internal potential always be constant. Formally, a rigid body can be defined as a system of particles in which the distances r_{ij} are fixed and cannot vary with time. In such case the vectors $d\mathbf{r}_{ij}$ can only be perpendicular to the corresponding $\mathbf{r}_{ij}$, and therefore to the $\mathbf{F}_{ij}$. Therefore, in a rigid body the *internal forces do no work*, and the internal potential must remain constant. Since the total potential is in any case uncertain to within an additive constant, an unvarying internal potential can be completely disregarded in discussing the motion of the system.

1–3 CONSTRAINTS

From the previous sections one might obtain the impression that all problems in mechanics have been reduced to solving the set of differential equations (1–19):

$$m_i \ddot{\mathbf{r}}_i = \mathbf{F}_i^{(e)} + \sum_j \mathbf{F}_{ji}.$$

One merely substitutes the various forces acting upon the particles of the system, turns the mathematical crank, and grinds out the answers! Even from a purely physical standpoint, however, this view is an oversimplification. For example, it may be necessary to take into account the *constraints* that limit the motion of the system. We have already met one type of system involving constraints, namely rigid bodies, where the constraints on the motions of the particles keep the distances r_{ij} unchanged. Other examples of constrained systems can easily be furnished. The beads of an abacus are constrained to one-dimensional motion by the supporting wires. Gas molecules within a container are constrained by the walls of the vessel to move only *inside* the container. A particle placed on the surface of a solid sphere is subject to the constraint that it can move only on the surface or in the region exterior to the sphere.

Constraints may be classified in various ways and we shall use the following system. If the conditions of constraint can be expressed as equations connecting the coordinates of the particles (and possibly the time) having the form

$$f(\mathbf{r}_1, \mathbf{r}_2, \mathbf{r}_3, \ldots, t) = 0, \tag{1–37}$$

then the constraints are said to be *holonomic*. Perhaps the simplest example of holonomic constraints is the rigid body, where the constraints are expressed by equations of the form

$$(\mathbf{r}_i - \mathbf{r}_j)^2 - c_{ij}^2 = 0.$$

A particle constrained to move along any curve or on a given surface is another obvious example of a holonomic constraint, with the equations defining the curve or surface acting as the equations of constraint.

Constraints not expressible in this fashion are called nonholonomic. The walls of a gas container constitute a nonholonomic constraint. The constraint involved in the example of a particle placed on the surface of a sphere is also nonholonomic, for it can be expressed as an inequality

$$r^2 - a^2 \geqq 0$$

(where a is the radius of the sphere), which is not in the form of (1–37). Thus, in a gravitational field a particle placed on the top of the sphere will slide down the surface part of the way but will eventually fall off.

Constraints are further classified according as the equations of constraint contain the time as an explicit variable (rheonomous) or are not explicitly dependent on time (scleronomous). A bead sliding on a rigid curved wire fixed in space is obviously subject to a scleronomous constraint; if the wire is moving in some prescribed fashion the constraint is rheonomous. Note that if the wire moves, say, as a reaction to the bead's motion, then the time dependence of the constraint enters in the equation of constraint only through the coordinates of the curved wire (which are now part of the system coordinates). The overall constraint is then scleronomous.*

---

* The terminology of constraints can be quite elaborate (cf. Kilmister and Reeve, *Rational Mechanics* (New York: American Elsevier, 1966). Of some use is the distinction between *bilateral* constraints, where the equations of constraint are equalities, and *unilateral* constraints, which involve inequalities. This choice of nomenclature derives from the example of motion constrained relative to a surface. In the bilateral constraint of motion on a surface, the force of constraint may be in either direction normal to the surface. A particle moving on or outside the surface of a sphere is subject to a unilateral constraint where any force of constraint acts in only one direction, along the outward normal.

Constraints introduce two types of difficulties in the solution of mechanical problems. First, the coordinates r_i are no longer all independent, since they are connected by the equations of constraint; hence the equations of motion (1–19) are not all independent. Second, the forces of constraint, e.g., the force that the wire exerts on the bead (or the wall on the gas particle) is not furnished a priori. They are among the unknowns of the problem and must be obtained from the solution we seek. Indeed, imposing constraints on the system is simply another method of stating that there are forces present in the problem that cannot be specified directly but are known rather in terms of their effect on the motion of the system.

In the case of holonomic constraints, the first difficulty is solved by the introduction of *generalized coordinates*. So far we have been thinking implicitly in terms of Cartesian coordinates. A system of N particles, free from constraints, has $3N$ independent coordinates or *degrees of freedom*. If there exist holonomic constraints, expressed in k equations in the form (1–37), then we may use these equations to eliminate k of the $3N$ coordinates, and we are left with $3N - k$ independent coordinates, and the system is said to have $3N - k$ degrees of freedom. This elimination of the dependent coordinates can be expressed in another way, by the introduction of new, $3N - k$, independent variables $q_1, q_2, \ldots, q_{3N-k}$ in terms of which the old coordinates $r_1, r_2, \ldots, r_N$ are expressed by equations of the form

$$\mathbf{r} = \mathbf{r}_1(q_1, q_2, \ldots, q_{3N-k}, t)$$

$$\tag{1–38}$$

$$\mathbf{r}_N = \mathbf{r}_N(q_1, q_2, \ldots, q_{3N-k}, t)$$

containing the constraints in them implicitly. These are *transformation* equations from the set of $(\mathbf{r}_l)$ variables to the (q_l) set, or alternatively Eqs. (1–38) can be considered as parametric representations of the $(\mathbf{r}_l)$ variables. It is always assumed that one can also transform back from the (q_l) set to the $(\mathbf{r}_l)$ set, i.e., that Eqs. (1–38) combined with the k equations of constraint can be inverted to obtain any q_i as a function of the $(\mathbf{r}_l)$ variable and time.

Usually the generalized coordinates, q_l, unlike the Cartesian coordinates, will not divide into convenient groups of three that can be associated together to form vectors. Thus, in the case of a particle constrained to move *on* the surface of a sphere, the two angles expressing position on the sphere, say latitude and longitude, are obvious possible generalized coordinates. Or, in the example of a double pendulum moving in a plane (two particles connected by an inextensible light rod and suspended by a similar rod fastened to one of the particles), satisfactory generalized coordinates are the two angles θ_1, θ_2. (Cf. Fig. 1–4.) Generalized coordinates, in the sense of coordinates other than Cartesian, are often useful in systems without constraints. Thus, in the problem of a particle moving in an external central force field ($V = V(r)$) there is no constraint

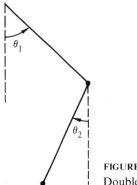

FIGURE 1–4
Double pendulum.

involved, but it is clearly more convenient to use spherical polar coordinates than Cartesian coordinates. One must not, however, think of generalized coordinates in terms of conventional orthogonal position coordinates. All sorts of quantities may be impressed to serve as generalized coordinates. Thus the amplitudes in a Fourier expansion of $\mathbf{r}_j$ may be used as generalized coordinates, or we may find it convenient to employ quantities with the dimensions of energy or angular momentum.

If the constraint is nonholonomic the equations expressing the constraint cannot be used to eliminate the dependent coordinates. An oft-quoted example of a nonholonomic constraint is that of an object rolling on a rough surface without slipping. The coordinates used to describe the system will generally involve angular coordinates to specify the orientation of the body, plus a set of coordinates describing the location of the point of contact on the surface. The constraint of "rolling" connects these two sets of coordinates; they are not independent. A change in the position of the point of contact inevitably means a change in its orientation. Yet we cannot reduce the number of coordinates, for the "rolling" condition is not expressible as an equation between the coordinates, in the manner of (1–37). Rather, it is a condition on the *velocities* (i.e., the point of contact is stationary), a differential condition that can be given in an integrated form only *after* the problem is solved.

A simple case will illustrate the point. Consider a disk rolling on the horizontal xy plane constrained to move so that the plane of the disk is always vertical. The coordinates used to describe the motion might be the x, y coordinates of the center of the disk, an angle of rotation ϕ about the axis of the disk, and an angle θ between the axis of the disk and, say, the x axis (cf. Fig. 1–5). As a result of the constraint the velocity of the center of the disk, $\mathbf{v}$, has a magnitude proportional to $\dot{\phi}$,

$$v = a\dot{\phi},$$

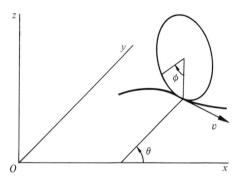

FIGURE 1–5
Vertical disk rolling on a horizontal plane.

where a is the radius of the disk, and its direction is perpendicular to the axis of the disk:

$$\dot{x} = v \sin \theta,$$

$$\dot{y} = -v \cos \theta.$$

Combining these conditions, we have two *differential* equations of constraint:

$$dx - a \sin \theta \, d\phi = 0,$$
$$dy + a \cos \theta \, d\phi = 0.$$

$$(1–39)$$

Neither of Eqs. (1–39) can be integrated without in fact solving the problem; that is, one cannot find an integrating factor $f(x, y, \theta, \phi)$ that will turn either of the equations into perfect differentials (cf. Exercise 7).* Hence the constraints cannot be reduced to the form of Eq. (1–37) and are therefore nonholonomic. Physically we can see that there can be no direct functional relation between ϕ and the other coordinates x, y, and θ by noting that at any point on its path the disk can be made to roll around in a circle tangent to the path and of arbitrary radius. At the end of the process, x, y, and θ have been returned to their original values, but ϕ has changed by an amount depending on the radius of the circle.

Nonintegrable *differential* constraints of the form of Eqs. (1–39) are of course not the only type of nonholonomic constraints. The constraint conditions may involve higher order derivatives, or may appear in the form of inequalities, as we have seen.

Partly because the dependent coordinates can be eliminated, problems involving holonomic constraints are always amenable to a formal solution. But there is no general way of attacking nonholonomic examples. True, if the constraint is nonintegrable, the differential equations of constraint can be introduced into the problem along with the differential equations of motion, and

---

* In principle, an integrating factor can always be found for a first-order differential equation of constraint in systems involving only two coordinates (cf. Kaplan, *Advanced Calculus*, 2d ed., p. 531) and such constraints are therefore holonomic. A familiar example is the two-dimensional motion of a circle rolling on an inclined plane.

the dependent equations eliminated, in effect, by the method of Lagrange multipliers. We shall return to this method at a later point. However, the more vicious cases of nonholonomic constraint must be tackled individually, and consequently in the development of the more formal aspects of classical mechanics it is almost invariably assumed that any constraint, if present, is holonomic. This restriction does not greatly limit the applicability of the theory, despite the fact that many of the constraints encountered in everyday life are nonholonomic. The reason is that the entire concept of constraints imposed in the system through the medium of wires or surfaces or walls is particularly appropriate only in macroscopic or large-scale problems. But the physicist today is primarily interested in atomic problems. On this scale all objects, both in and out of the system, consist alike of molecules, atoms or smaller particles, exerting definite forces, and the notion of constraint becomes artificial and rarely appears. Constraints are then used only as mathematical idealizations to the actual physical case or as classical approximations to a quantum-mechanical property, e.g., rigid body rotations for "spin." Such constraints are always holonomic and fit smoothly into the framework of the theory.

To surmount the second difficulty, namely, that the forces of constraint are unknown a priori, we should like to so formulate the mechanics that the forces of constraint disappear. We need then deal only with the known applied forces. A hint as to the procedure to be followed is provided by the fact that in a particular system with constraints, i.e., a rigid body, the work done by internal forces (which are here the forces of constraint) vanishes. We shall follow up this clue in the ensuing sections and generalize the ideas contained in it.

1–4 D'ALEMBERT'S PRINCIPLE AND LAGRANGE'S EQUATIONS

A virtual (infinitesimal) displacement of a system refers to a change in the configuration of the system as the result of any arbitrary infinitesimal change of the coordinates $\delta \mathbf{r}_i$, *consistent with the forces and constraints imposed on the system at the given instant t*. The displacement is called virtual to distinguish it from an actual displacement of the system occurring in a time interval dt, during which the forces and constraints may be changing. Suppose the system is in equilibrium, i.e., the total force on each particle vanishes, $\mathbf{F}_i = 0$. Then clearly the dot product $\mathbf{F}_i \cdot \delta \mathbf{r}_i$, which is the virtual work of the force $\mathbf{F}_i$ in the displacement $\delta \mathbf{r}_i$, also vanishes. The sum of these vanishing products over all particles must likewise be zero:

$$\sum_i \mathbf{F}_i \cdot \delta \mathbf{r}_i = 0. \tag{1–40}$$

As yet nothing has been said that has any new physical content. Decompose $\mathbf{F}_i$ into the applied force, $\mathbf{F}_i^{(a)}$, and the force of constraint, $\mathbf{f}_i$,

$$\mathbf{F}_i = \mathbf{F}_i^{(a)} + \mathbf{f}_i, \tag{1–41}$$

so that Eq. (1–40) becomes

$$\sum_i \mathbf{F}_i^{(a)} \cdot \delta \mathbf{r}_i + \sum_i \mathbf{f}_i \cdot \delta \mathbf{r}_i = 0. \tag{1–42}$$

We now restrict ourselves to systems for which *the net virtual work of the forces of constraint is zero.* We have seen that this condition holds true for rigid bodies and it is valid for a large number of other constraints. Thus, if a particle is constrained to move on a surface, the force of constraint is perpendicular to the surface, while the virtual displacement must be tangent to it, and hence the virtual work vanishes. This is no longer true if sliding friction forces are present, and we must exclude such systems from our formulation. The restriction is not unduly hampering, since the friction is essentially a macroscopic phenomenon. On the other hand, the forces of rolling friction do not violate this condition, since the forces act on a point that is momentarily at rest and can do no work in an infinitesimal displacement consistent with the rolling constraint. Note that if a particle is constrained to a surface that is itself moving in time, the force of constraint is instantaneously perpendicular to the surface and the work during a virtual displacement is still zero even though the work during an actual displacement in the time dt does not necessarily vanish.

We therefore have as the condition for equilibrium of a system that the virtual work of the *applied forces* vanishes:

$$\sum_i \mathbf{F}_i^{(a)} \cdot \delta \mathbf{r}_i = 0. \tag{1–43}$$

Equation (1–43) is often called the *principle of virtual work.* Notice that the coefficients of $\delta \mathbf{r}_i$ can no longer be set equal to zero, i.e., in general $\mathbf{F}_i^{(a)} \neq 0$, since the $\delta \mathbf{r}_i$ are not completely independent but are connected by the constraints. In order to equate the coefficients to zero, one must transform the principle into a form involving the virtual displacements of the q_i, which are independent. Equation (1–43) satisfies our needs in that it does not contain the $\mathbf{f}_i$, but it deals only with statics; we want a condition involving the general motion of the system.

To obtain such a principle we use a device first thought of by James Bernoulli and developed by D'Alembert. The equation of motion,

$$\mathbf{F}_i = \dot{\mathbf{p}}_i,$$

can be written as

$$\mathbf{F}_i - \dot{\mathbf{p}}_i = 0,$$

which states that the particles in the system will be in equilibrium under a force equal to the actual force plus a "reversed effective force" $-\dot{\mathbf{p}}_i$. Instead of (1–40) we can immediately write

$$\sum_i (\mathbf{F}_i - \dot{\mathbf{p}}_i) \cdot \delta \mathbf{r}_i = 0, \tag{1–44}$$

and, making the same resolution into applied forces and forces of constraint there results

$$\sum_i (\mathbf{F}_i^{(a)} - \dot{\mathbf{p}}_i) \cdot \delta \mathbf{r}_i + \sum_i \mathbf{f}_i \cdot \delta \mathbf{r}_i = 0.$$

We again restrict ourselves to systems for which the virtual work of the forces of constraint vanishes and therefore obtain

$$\sum_i (\mathbf{F}_i^{(a)} - \dot{\mathbf{p}}_i) \cdot \delta \mathbf{r}_i = 0, \qquad (1\text{–}45)$$

which is often called *D'Alembert's principle*. We have achieved our aim, in that the forces of constraint no longer appear, and the superscript $^{(a)}$ can now be dropped without ambiguity. It is still not in a useful form to furnish equations of motion for the system. We must now transform the principle into an expression involving virtual displacements of the generalized coordinates, which are then independent of each other (for holonomic constraints), so that the coefficients of the δq_i can be set separately equal to zero.

The translation from $\mathbf{r}_i$ to q_j language starts from the transformation equations (1–38),

$(- 1 \to 3N - k$

$$\mathbf{r}_i = \mathbf{r}_i(q_1, q_2, \ldots, q_n, t)$$

(assuming n independent coordinates), and is carried out by means of the usual "chain rules" of the calculus of partial differentiation.* Thus $\mathbf{v}_i$ is expressed in terms of the $\dot{q}_k$ by the formula

$$\mathbf{v}_i \equiv \frac{d\mathbf{r}_i}{dt} = \sum_k \frac{\partial \mathbf{r}_i}{\partial q_k} \dot{q}_k + \frac{\partial \mathbf{r}_i}{\partial t}. \qquad (1\text{–}46)$$

Similarly, the arbitrary virtual displacement $\delta \mathbf{r}_i$ can be connected with the virtual displacements δq_i by

$$\delta \mathbf{r}_i = \sum_j \frac{\partial \mathbf{r}_i}{\partial q_j} \delta q_j. \qquad (1\text{–}47)$$

Notice that no variation of time, δt, is involved here, since a virtual displacement by definition considers only displacements of the coordinates. (Only then is the virtual displacement perpendicular to the force of constraint if the constraint itself is changing in time.)

In terms of the generalized coordinates the virtual work of the $\mathbf{F}_i$ becomes

$$\sum_i \mathbf{F}_i \cdot \delta \mathbf{r}_i = \sum_{i,j} \mathbf{F}_i \cdot \frac{\partial \mathbf{r}_i}{\partial q_j} \delta q_j$$

$$= \sum_j Q_j \delta q_j, \qquad (1\text{–}48)$$

---

* See, for example, Kaplan, *Advanced Calculus*, 2d ed., p. 135.

where the Q_j are called the components of the *generalized force*, defined as

Generalized force

$$Q_j = \sum_i \mathbf{F}_i \cdot \frac{\partial \mathbf{r}_i}{\partial q_j}. \tag{1-49}$$

Note that just as the q's need not have the dimensions of length, so the Q's do not necessarily have the dimensions of force, but $Q_j \delta q_j$ must always have the dimensions of work.

We turn now to the other term involved in Eq. (1–45), which may be written as

$$= \sum_i \dot{\mathbf{p}}_i \cdot \delta \mathbf{r}_i = \sum_i m_i \ddot{\mathbf{r}}_i \cdot \delta \mathbf{r}_i.$$

Expressing $\delta \mathbf{r}_i$ by (1–47), this becomes

$$\sum_{i,j} m_i \ddot{\mathbf{r}}_i \cdot \frac{\partial \mathbf{r}_i}{\partial q_j} \delta q_j.$$

Consider now the relation

$$\sum_i m_i \ddot{\mathbf{r}}_i \cdot \frac{\partial \mathbf{r}_i}{\partial q_j} = \sum_i \left\{ \frac{d}{dt} \left(m_i \dot{\mathbf{r}}_i \cdot \frac{\partial \mathbf{r}_i}{\partial q_j} \right) - m_i \dot{\mathbf{r}}_i \cdot \frac{d}{dt} \left(\frac{\partial \mathbf{r}_i}{\partial q_j} \right) \right\}. \tag{1-50}$$

In the last term of Eq. (1–50) we can interchange the differentiation with respect to t and q_j, for, in analogy to (1–46),

$$\frac{\partial}{\partial q_j}\left(\frac{d\mathbf{r}_i}{dt}\right) = \frac{d}{dt}\left(\frac{\partial \mathbf{r}_i}{\partial q_j}\right) = \sum_k \frac{\partial^2 \mathbf{r}_i}{\partial q_j \partial q_k} \dot{q}_k + \frac{\partial^2 \mathbf{r}_i}{\partial q_j \partial t} = \frac{\partial \dot{\mathbf{r}}_i}{\partial q_j},$$

$$= \frac{\partial \mathbf{v}_i}{\partial q_j},$$

by Eq. (1–46). Further, we also see from Eq. (1–46) that $r_i \neq f(\dot{q})$

$$\frac{\partial \left(\frac{\partial r_i}{\partial t} \right)}{\partial \left(\frac{\partial q_j}{\partial t} \right)} = \frac{\partial \mathbf{v}_i}{\partial \dot{q}_j} = \frac{\partial \mathbf{r}_i}{\partial q_j}. \tag{1-51}$$

Substitution of these changes in (1–50) leads to the result that

$$\sum_i m_i \ddot{\mathbf{r}}_i \cdot \frac{\partial \mathbf{r}_i}{\partial q_j} = \sum_i \left\{ \frac{d}{dt}\left(m_i \mathbf{v}_i \cdot \frac{\partial \mathbf{v}_i}{\partial \dot{q}_j} \right) - m_i \mathbf{v}_i \cdot \frac{\partial \mathbf{v}_i}{\partial q_j} \right\},$$

and the second term on the left-hand side of Eq. (1–45) can be expanded into

$$\sum_j \left\{ \frac{d}{dt}\left(\frac{\partial}{\partial \dot{q}_j} \left(\sum_i \frac{1}{2} m_i v_i^2 \right) \right) - \frac{\partial}{\partial q_j}\left(\sum_i \frac{1}{2} m_i v_i^2 \right) \right\} \delta q_j.$$

Identifying $\sum_i \frac{1}{2} m_i v_i^2$ with the system kinetic energy T, D'Alembert's principle becomes

$$\sum_i (\dot{\mathbf{p}} - \mathbf{F}) \cdot \delta \mathbf{r}_i = 0$$

$$\sum_j \left[\left\{ \frac{d}{dt}\left(\frac{\partial T}{\partial \dot{q}_j} \right) - \frac{\partial T}{\partial q_j} \right\} - Q_j \right] \delta q_j = 0. \tag{1-52}$$

must be 0

Note that in a system of Cartesian coordinates the partial derivative of T with respect to q_j vanishes. Thus, speaking in the language of differential geometry, this term arises from the curvature of the coordinates q_j. In polar coordinates, for example, it is in the partial derivative of T with respect to an angle coordinate that the centripetal acceleration term appears.

Up to this point no restriction has been made on the nature of the constraints other than that they be workless in a virtual displacement. The variables q_j can be any set of coordinates used to describe the motion of the system. If, however, the constraints are holonomic, then it is possible to find sets of independent coordinates q_j that contain the constraint conditions implicitly in the transformation equations (1–38). Any virtual displacement δq_j is then independent of δq_k and therefore the only way for (1–52) to hold is for the separate coefficients to vanish:

$$\frac{d}{dt}\left(\frac{\partial T}{\partial \dot{q}_j}\right) - \frac{\partial T}{\partial q_j} = Q_j. \qquad (1\text{--}53)$$

There are n such equations in all.

The Eqs. (1–53) are often referred to as Lagrange's equations, but this designation is frequently reserved for the form of Eqs. (1–53) when the forces are derivable from a scalar potential function V:

$$\mathbf{F}_i = -\nabla_i V.$$

In this case the generalized forces can be written as

$$Q_j = \sum_i \mathbf{F}_i \cdot \frac{\partial \mathbf{r}_i}{\partial q_j} = -\sum_i \nabla_i V \cdot \frac{\partial \mathbf{r}_i}{\partial q_j},$$

which is exactly the same expression for the partial derivative of a function $-V(\mathbf{r}_1, \mathbf{r}_2, \ldots, \mathbf{r}_N, t)$ with respect to q_j:

$$Q_j = -\frac{\partial V}{\partial q_j}. \qquad (1\text{--}54)$$

Equations (1–53) can then be rewritten as

$$\frac{d}{dt}\left(\frac{\partial T}{\partial \dot{q}_j}\right) - \frac{\partial(T-V)}{\partial q_j} = 0. \qquad (1\text{--}55)$$

The equations of motion in the form (1–55) are not necessarily restricted to conservative systems; only if V is not an explicit function of time is the system conservative (cf. p. 5). As here defined the potential V, however, does not depend on the generalized velocities. Hence one can include a term in V in the partial derivative with respect to $\dot{q}_j$:

$$\frac{d}{dt}\left(\frac{\partial(T-V)}{\partial \dot{q}_j}\right) - \frac{\partial(T-V)}{\partial q_j} = 0.$$

Or, defining a new function, the *Lagrangian L*, as

$$L = T - V, \qquad (1\text{--}56)$$

the Eqs. (1–53) become

$$\frac{d}{dt}\left(\frac{\partial L}{\partial \dot{q}_j}\right) - \frac{\partial L}{\partial q_j} = 0. \tag{1–57}$$

Unless otherwise specified we shall mean Eqs. (1–57) whenever the term "Lagrange's equations" is used.

It should be noted that for a particular set of equations of motion there is no unique choice of Lagrangian such that Eqs. (1–53) lead to the equations of motion in the given generalized coordinates. Thus, in Exercise 14 it is shown that if $L(q, \dot{q}, t)$ is an appropriate Lagrangian and $F(q, t)$ is *any* differentiable function of the generalized coordinates and time, then

$$L'(q, \dot{q}, t) = L(q, \dot{q}, t) + \frac{dF}{dt}$$

is a Lagrangian also resulting in the same equations of motion. It is also often possible to find alternate Lagrangians beside those constructed by this prescription (see Exercise 18). While Eq. (1–56) is always a suitable way to construct a Lagrangian for a conservative system, it does not provide the *only* Lagrangian suitable for the given system.

1–5 VELOCITY-DEPENDENT POTENTIALS AND THE DISSIPATION FUNCTION

Lagrange's equations can be put in the form (1–57) even if there is no potential function, V, in the usual sense, providing the generalized forces are obtained from a function $U(q_j, \dot{q}_j)$ by the prescription

$$Q_j = -\frac{\partial U}{\partial q_j} + \frac{d}{dt}\left(\frac{\partial U}{\partial \dot{q}_j}\right). \tag{1–58}$$

In such case Eqs. (1–57) still follow from Eqs. (1–53) with the Lagrangian given by

$$L = T - U. \tag{1–59}$$

U may be called a "generalized potential," or "velocity-dependent potential."[*] The possibility of using such a "potential" is not academic; it applies to one very

---

[*] The history of the designation given to such a potential is curious. Apparently spurred by Weber's early (and erroneous) classical electrodynamics, which postulated velocity-dependent forces, the German mathematician E. Schering seems to have been the first to attempt seriously to include such forces in the framework of mechanics, cf. Gött. Abh. **18**, 3 (1873). The first edition of Whittaker's *Analytical Dynamics* (1904) thus refers to the potential as "Schering's potential function," but the name apparently did not stick, for the title was dropped in later editions. More recently Morgenstern and Szabó (*Vorlesungen über Theoretische Mechanik*, 1961) have used the name "Schering potential" for the specific velocity-dependent potential that gives the Coriolis force in a rotating coordinate system. We shall preferably use the name "generalized potential," including within this designation also the ordinary potential energy, a function of position only.

important type of force field, namely, the electromagnetic forces on moving charges. Considering its importance, a diversion on this subject is well worthwhile.

In Gaussian units the Maxwell equations are

$$\nabla \times \mathbf{E} + \frac{1}{c}\frac{\partial \mathbf{B}}{\partial t} = 0, \qquad\qquad \nabla \cdot \mathbf{D} = 4\pi\rho,$$

$$\nabla \times \mathbf{H} - \frac{1}{c}\frac{\partial \mathbf{D}}{\partial t} = \frac{4\pi\mathbf{j}}{c}, \qquad\qquad \nabla \cdot \mathbf{B} = 0. \qquad (1\text{-}60)$$

The force on a charge q is not given entirely by the electric force

$$\mathbf{F} = q\mathbf{E} = -q\nabla\phi,$$

so that the system is not conservative in this sense. Instead, the complete force is

$$\mathbf{F} = q\left\{\mathbf{E} + \frac{1}{c}(\mathbf{v} \times \mathbf{B})\right\}. \qquad (1\text{-}61)$$

$\mathbf{E}$ is not the gradient of a scalar function since $\nabla \times \mathbf{E} \neq 0$, but from $\nabla \cdot \mathbf{B} = 0$ it follows that $\mathbf{B}$ can be represented as the curl of a vector,

$$\mathbf{B} = \nabla \times \mathbf{A}, \qquad (1\text{-}62)$$

where $\mathbf{A}$ is called the magnetic vector potential. Then the curl $\mathbf{E}$ equation becomes

$$\nabla \times \mathbf{E} + \frac{1}{c}\frac{\partial}{\partial t}(\nabla \times \mathbf{A}) = \nabla \times \left(\mathbf{E} + \frac{1}{c}\frac{\partial \mathbf{A}}{\partial t}\right) = 0.$$

Hence we can set

$$\mathbf{E} + \frac{1}{c}\frac{\partial \mathbf{A}}{\partial t} = -\nabla\phi$$

or

$$\mathbf{E} = -\nabla\phi - \frac{1}{c}\frac{\partial \mathbf{A}}{\partial t}. \qquad 1\text{-}63)$$

In terms of the potentials ϕ and $\mathbf{A}$, the so-called Lorentz force (1-61) becomes

$$\mathbf{F} = q\left\{-\nabla\phi - \frac{1}{c}\frac{\partial \mathbf{A}}{\partial t} + \frac{1}{c}(\mathbf{v} \times (\nabla \times \mathbf{A}))\right\}. \qquad (1\text{-}64)$$

The terms of Eq. (1-64) can be rewritten in a more convenient form. As an example consider the x component

$$(\nabla\phi)_x = \frac{\partial\phi}{\partial x}$$

and

$$(\mathbf{v} \times (\nabla \times \mathbf{A}))_x = v_y \left(\frac{\partial A_y}{\partial x} - \frac{\partial A_x}{\partial y} \right) - v_z \left(\frac{\partial A_x}{\partial z} - \frac{\partial A_z}{\partial x} \right)$$

$$= v_y \frac{\partial A_y}{\partial x} + v_z \frac{\partial A_z}{\partial x} + v_x \frac{\partial A_x}{\partial x} - v_y \frac{\partial A_x}{\partial y} - v_z \frac{\partial A_x}{\partial z} - v_x \frac{\partial A_x}{\partial x},$$

where we have added and subtracted the term

$$v_x \frac{\partial A_x}{\partial x}.$$

Now, the total time derivative of A_x is

$$\frac{dA_x}{dt} = \frac{\partial A_x}{\partial t} + \left(v_x \frac{\partial A_x}{\partial x} + v_y \frac{\partial A_x}{\partial y} + v_z \frac{\partial A_x}{\partial z} \right),$$

where the first term arises from the explicit variation of A_x with time, and the second term results from the motion of the particle with time, which changes the spatial point at which A_x is evaluated. The x component of $\mathbf{v} \times \nabla \times \mathbf{A}$ can therefore be written as

$$(\mathbf{v} \times (\nabla \times \mathbf{A}))_x = \frac{\partial}{\partial x} (\mathbf{v} \cdot \mathbf{A}) - \frac{dA_x}{dt} + \frac{\partial A_x}{\partial t}.$$

With these substitutions, (1–64) becomes

$$F_x = q \left\{ -\frac{\partial}{\partial x} \left(\phi - \frac{1}{c} \mathbf{v} \cdot \mathbf{A} \right) - \frac{1}{c} \frac{d}{dt} \left(\frac{\partial}{\partial v_x} (\mathbf{A} \cdot \mathbf{v}) \right) \right\}. \tag{1–64'}$$

Since the scalar potential is independent of velocity, this expression is equivalent to

$$F_x = -\frac{\partial U}{\partial x} + \frac{d}{dt} \frac{\partial U}{\partial v_x},$$

where

$$U = q\phi - \frac{q}{c} \mathbf{A} \cdot \mathbf{v}. \tag{1–65}$$

U is a generalized potential in the sense of Eq. (1–58), and the Lagrangian for a charged particle in an electromagnetic field can be written

$$L = T - q\phi + \frac{q}{c} \mathbf{A} \cdot \mathbf{v}. \tag{1–66}$$

It should be noted that if not all the forces acting on the system are derivable from a potential, then Lagrange's equations can always be written in the form

$$\frac{d}{dt} \left(\frac{\partial L}{\partial \dot{q}_j} \right) - \frac{\partial L}{\partial q_j} = Q_j,$$

where L contains the potential of the conservative forces as before, and Q_j represents the forces *not* arising from a potential. Such a situation often occurs when there are frictional forces present. It frequently happens that the frictional force is proportional to the velocity of the particle, so that its x component has the form

$$F_{fx} = -k_x v_x.$$

Frictional forces of this type may be derived in terms of a function $\mathcal{F}$, known as *Rayleigh's dissipation function*, and defined as

$$\mathcal{F} = \frac{1}{2} \sum_i (k_x v_{ix}^2 + k_y v_{iy}^2 + k_z v_{iz}^2), \tag{1-67}$$

where the summation is over the particles of the system. From this definition it is clear that

$$F_{fx} = -\frac{\partial \mathcal{F}}{\partial v_x},$$

or, symbolically,

$$\mathbf{F}_f = -\nabla_v \mathcal{F}. \tag{1-68}$$

One can also give a physical interpretation to the dissipation function. The work done *by* the system *against* friction is

$$dW_f = -\mathbf{F}_f \cdot d\mathbf{r} = -\mathbf{F}_f \cdot \mathbf{v}\, dt = (k_x v_x^2 + k_y v_y^2 + k_z v_z^2)\, dt.$$

Hence $2\mathcal{F}$ is the rate of energy dissipation due to friction. The component of the generalized force resulting from the force of friction is then given by

$$Q_j = \sum_i \mathbf{F}_{if} \cdot \frac{\partial \mathbf{r}_i}{\partial q_j} = -\sum \nabla_v \mathcal{F} \cdot \frac{\partial \mathbf{r}_i}{\partial q_i}$$

$$= -\sum \nabla_v \mathcal{F} \cdot \frac{\partial \dot{\mathbf{r}}_i}{\partial \dot{q}_j}, \qquad \text{by (1.51),}$$

$$= -\frac{\partial \mathcal{F}}{\partial \dot{q}_j}. \tag{1-69}$$

The Lagrange equations now become

$$\frac{d}{dt}\left(\frac{\partial L}{\partial \dot{q}_j}\right) - \frac{\partial L}{\partial q_j} + \frac{\partial \mathcal{F}}{\partial \dot{q}_j} = 0, \tag{1-70}$$

so that two scalar functions, L and $\mathcal{F}$, must be specified to obtain the equations of motion.

1–6 SIMPLE APPLICATIONS OF THE LAGRANGIAN FORMULATION

The previous sections show that for systems where one can define a Lagrangian, i.e., holonomic systems with applied forces derivable from an ordinary or generalized potential and workless constraints, we have a very convenient way of setting up the equations of motion. We were led to the Lagrangian formulation by the desire to eliminate the forces of constraint from the equations of motion, and in achieving this goal we have obtained many other benefits. In setting up the original form of the equations of motion, Eqs. (1–19), it is necessary to work with many *vector* forces and accelerations. With the Lagrangian method one has only to deal with two *scalar* functions, T and V, which greatly simplifies the problem. A straightforward routine procedure can now be established for all problems of mechanics to which the Lagrangian formulation is applicable. One has only to write T and V in generalized coordinates, form L from them, and substitute in (1–57) to obtain the equations of motion. The needed transformation of T and V from Cartesian coordinates to generalized coordinates is obtained by applying the transformation equations (1–38) and (1–46). Thus T is given in general by

$$T = \sum_i \frac{1}{2} m_i v_i^2 = \sum_i \frac{1}{2} m_i \left(\sum_j \frac{\partial \mathbf{r}_i}{\partial q_j} \dot{q}_j + \frac{\partial \mathbf{r}_i}{\partial t} \right)^2.$$

It is clear that on carrying out the expansion, the expression for T in generalized coordinates will have the form

$$T = M_0 + \sum_j M_j \dot{q}_j + \frac{1}{2} \sum_{j,k} M_{jk} \dot{q}_j \dot{q}_k \tag{1-71}$$

where M_0, M_j, M_{jk} are definite functions of the $\mathbf{r}$'s and t and hence of the q's and t. In fact, a comparison shows that

$$M_0 = \sum_i \frac{1}{2} m_i \left(\frac{\partial \mathbf{r}_i}{\partial t} \right)^2,$$

$$M_j = \sum_i m_i \frac{\partial \mathbf{r}_i}{\partial t} \cdot \frac{\partial \mathbf{r}_i}{\partial q_j}, \tag{1-72}$$

and

$$M_{jk} = \sum_i m_i \frac{\partial \mathbf{r}_i}{\partial q_j} \cdot \frac{\partial \mathbf{r}_i}{\partial q_k}.$$

Thus the kinetic energy of a system can always be written as the sum of three homogeneous functions of the generalized velocities,

$$T = T_0 + T_1 + T_2, \tag{1-73}$$

where T_0 is independent of the generalized velocities, T_1 is linear in the velocities, and T_2 is quadratic in the velocities. If the transformation equations do not contain the time explicitly, as may occur when the constraints are independent of time (scleronomous), then only the last term in Eq. (1–71) is nonvanishing and T is always a homogeneous quadratic form in the generalized velocities.

Let us now consider simple examples of this procedure:

1. Single particle in space
   a. Cartesian coordinates
   b. Plane polar coordinates

2. Atwood's machine

3. Time-dependent constraint—bead sliding on rotating wire

1. (a) *Motion of one particle: using Cartesian coordinates.* The generalized forces needed in Eq. (1–53) are obviously F_x, F_y, and F_z. Then

$$T = \tfrac{1}{2}m(\dot{x}^2 + \dot{y}^2 + \dot{z}^2),$$

$$\frac{\partial T}{\partial x} = \frac{\partial T}{\partial y} = \frac{\partial T}{\partial z} = 0,$$

$$\frac{\partial T}{\partial \dot{x}} = m\dot{x}, \qquad \frac{\partial T}{\partial \dot{y}} = m\dot{y}, \qquad \frac{\partial T}{\partial \dot{z}} = m\dot{z},$$

and the equations of motion are

$$\frac{d}{dt}(m\dot{x}) = F_x, \qquad \frac{d}{dt}(m\dot{y}) = F_y, \qquad \frac{d}{dt}(m\dot{z}) = F_z. \tag{1–74}$$

We are thus led back to the original Newton's equations of motion.

(b) *Motion of one particle: using plane polar coordinates.* Here we must express T in terms of $\dot{r}$ and $\dot{\theta}$. The equations of transformation, i.e., the Eqs. (1–38), in this case are simply

$$x = r\cos\theta,$$

$$y = r\sin\theta.$$

In analogy to (1–46), the velocities are given by

$$\dot{x} = \dot{r}\cos\theta - r\dot{\theta}\sin\theta,$$

$$\dot{y} = \dot{r}\sin\theta + r\dot{\theta}\cos\theta.$$

The kinetic energy $T = \tfrac{1}{2}m(\dot{x}^2 + \dot{y}^2)$ then reduces formally to

$$T = \tfrac{1}{2}m(\dot{r}^2 + (r\dot{\theta})^2). \tag{1–75}$$

An alternative derivation of Eq. (1–75) is obtained by recognizing that the plane polar components of the velocity are $\dot{r}$ along $\mathbf{r}$, and $r\dot{\theta}$ along the direction perpendicular to r, denoted by the unit vector $\mathbf{n}$. Hence the square of the velocity expressed in polar coordinates is simply $\dot{r}^2 + (r\dot{\theta})^2$. The components of the generalized force can be obtained from the definition, Eq. (1–49),

$$Q_r = \mathbf{F} \cdot \frac{\partial \mathbf{r}}{\partial r} = \mathbf{F} \cdot \frac{\mathbf{r}}{r} = F_r,$$

$$Q_\theta = \mathbf{F} \cdot \frac{\partial \mathbf{r}}{\partial \theta} = \mathbf{F} \cdot r\mathbf{n} = rF_\theta,$$

since the derivative of $\mathbf{r}$ with respect to θ is, by definition of a derivative, a vector in the direction of $\mathbf{n}$ (cf. Fig. 1–6). There are two generalized coordinates, and therefore two Lagrange equations. The derivatives occurring in the r equation are

$$\frac{\partial T}{\partial r} = mr\dot{\theta}^2, \qquad \frac{\partial T}{\partial \dot{r}} = m\dot{r}, \qquad \frac{d}{dt}\left(\frac{\partial T}{\partial \dot{r}}\right) = m\ddot{r},$$

and the equation itself is

$$m\ddot{r} - mr\dot{\theta}^2 = F_r,$$

the second term being the centripetal acceleration term. For the θ equation we have the derivatives

$$\frac{\partial T}{\partial \theta} = 0, \qquad \frac{\partial T}{\partial \dot{\theta}} = mr^2\dot{\theta}, \qquad \frac{d}{dt}(mr^2\dot{\theta}) = mr^2\ddot{\theta} + 2mr\dot{r}\dot{\theta},$$

so that the equation becomes

$$\frac{d}{dt}(mr^2\dot{\theta}) = mr^2\ddot{\theta} + 2mr\dot{r}\dot{\theta} = rF_\theta.$$

Note that the left is just the time derivative of the angular momentum, and the right is exactly the applied torque, so that we have simply rederived the torque equation (1–26).

2. *The Atwood's machine*—an example of a conservative system with holonomic, scleronomous constraint (the pulley is assumed frictionless and

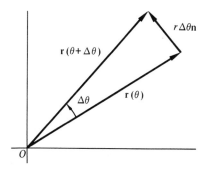

FIGURE 1–6
Derivative of r with respect to θ.

massless). Clearly there is only one independent coordinate x, the position of the other weight being determined by the constraint that the length of the rope between them is l. The potential energy is

$$V = -M_1 g x - M_2 g(l - x),$$

while the kinetic energy is

$$T = \tfrac{1}{2}(M_1 + M_2)\dot{x}^2.$$

Combining the two, the Lagrangian has the form

$$L = T - V = \tfrac{1}{2}(M_1 + M_2)\dot{x}^2 + M_1 g x + M_2 g(l - x).$$

There is only one equation of motion, involving the derivatives

$$\frac{\partial L}{\partial x} = (M_1 - M_2)g,$$

$$\frac{\partial L}{\partial \dot{x}} = (M_1 + M_2)\dot{x},$$

so that we have

$$(M_1 + M_2)\ddot{x} = (M_1 - M_2)g,$$

or

$$\ddot{x} = \frac{M_1 - M_2}{M_1 + M_2}g,$$

which is the familiar result obtained by more elementary means. This trivial problem emphasizes that the forces of constraint—here the tension in the rope—appear nowhere in the Lagrangian formulation. By the same token neither can the tension in the rope be found directly by the Lagrangian method.

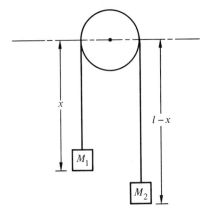

FIGURE 1–7
Atwood's machine.

3. *A bead sliding on a uniformly rotating wire in a force-free space.* The wire is straight, and is rotated uniformly about some fixed axis perpendicular to the wire. This example has been chosen as a simple illustration of a constraint being time dependent, with the transformation equations therefore containing the time explicitly:

$$x = r \cos \omega t,$$
$$y = r \sin \omega t.$$

ω = angular velocity of rotation.

While one could then find T (here the same as L) by the same procedure used to obtain (1–71), it is simpler to take over (1–75) directly, expressing the constraint by the relation $\dot{\theta} = \omega$:

$$T = \tfrac{1}{2}m(\dot{r}^2 + r^2\omega^2).$$

Notice that T is not a homogeneous quadratic function of the generalized velocities, since there is now an additional term not involving $\dot{r}$. The equation of motion is then

$$m\ddot{r} - mr\omega^2 = 0$$

or

$$\ddot{r} = r\omega^2,$$

which is the well-known result that the bead moves outward because of the centripetal acceleration. Again, the method cannot furnish the force of constraint keeping the bead on the wire.

SUGGESTED REFERENCES*

K. R. Symon, *Mechanics.* An excellent and unusually detailed intermediate textbook on mechanics that can be used with much profit as a preliminary, and often as a supplement, to the present book. For the material in this chapter note especially the discussion on conservation theorems and variable mass systems (such as rockets) in Chapter 4, and the extensive consideration of constraints in Chapter 9.

W. Hauser, *Introduction to the Principles of Mechanics.* Another intermediate text with many fresh and original viewpoints. Note especially the discussion of the Lagrange equations of motion, in the form of Eq. (1–53), as Newton's equations of motion in a curvilinear space of the generalized coordinates

C. W. Kilmister and J. E. Reeve, *Rational Mechanics.* Mechanics at an intermediate level presented with great mathematical thoroughness and designed for students of mathematics. Particularly noteworthy is the extensive discussion of constraints and a simplified attempt at an axiomatic formulation of the basic concepts of mechanics.

C. Lanczos, *The Variational Principles of Mechanics.* Of much wider content than the title implies, this book is in fact a survey of all mechanics with emphasis on the bases of the various formulations. Contains many insightful historical notes.

* For convenience the references at the end of each chapter are listed only by short title. Full bibliographical description will be found in the bibliography at the end of the book.

W. F. OSGOOD, *Mechanics*. Long since out-of-print, this book is still well worth hunting up. The first five chapters form an elementary introduction into the subject that is delightfully flavored by the author's long pedagogic experience. In this regard the reader's attention is directed especially to page 102!

E. MACH, *The Science of Mechanics*. A classic analysis and criticism of the fundamental concepts of classical mechanics. In its earlier editions this book did much to clear the way philosophically for relativity theory.

R. B. LINDSAY AND H. MARGENAU, *Foundations of Physics*. Chapter 3 contains a clear discussion of the foundations of classical mechanics. This book, together with Mach's work, can serve as an excellent point of departure for further reading on the nature of the basic ideas involved in mechanics.

P. W. BRIDGMAN, "Significance of the Mach Principle," *American Journal of Physics* **29**, 32–36, Jan. 1961. This article discusses why and to what extent the frame of the "fixed" stars can be taken as operationally defining an inertial system.

C. TRUESDELL, *Essays in the History of Mechanics*. The style is highly personalized and forceful, brooking no gainsaying. But these essays are perfused by an intense historical sense and supported by deep scholarship. Of particular interest here is the chapter "Whence the law of the moment of momentum?" which describes the historical development of the various conservation theorems for angular momentum. Much the same view, in more mathematical detail and with emphasis on continuous systems, is given in Section 196 of the next reference.

C. TRUESDELL AND R. A. TOUPIN, *The Classical Field Theories*, a book 567 pages long that forms part of Vol. 3/1, *Encyclopedia of Physics*. Incidentally, the Bibliography P, p. 788f, of this latter work, provides a set of references (to 1957) on the formal axiomatic treatment of mechanics.

E. T. WHITTAKER, *Analytical Dynamics*. A well-known treatise that presents an exhaustive treatment of analytical mechanics from the older viewpoints. The development is marked, regrettably, by an apparent dislike of diagrams (of which there are only four in the entire book) and of vector notation, and by a fondness for the type of pedantic mechanics problems made famous by the Cambridge Tripos examinations. It remains, however, a practically unique source for the discussion of many specialized topics. For the present chapter reference should be made principally to Chapter II, especially Section 31, which discusses velocity-dependent potentials. Sections 92–94 of Chapter VIII are concerned with the dissipation function.

LORD RAYLEIGH, *The Theory of Sound*. The dissipation function is introduced in Chapter IV, Vol. I of this classic treatise.

EXERCISES

1. A nucleus, originally at rest, decays radioactively by emitting an electron of momentum 1.73 MeV/c, and at right angles to the direction of the electron a neutrino with momentum 1.00 MeV/c. (The MeV (million electron volt) is a unit of energy, used in modern physics, equal to 1.60×10^{-6} erg. Correspondingly, MeV/c is a unit of linear momentum equal to 5.34×10^{-17} gm-cm/sec.) In what direction does the nucleus recoil? What is its momentum in MeV/c? If the mass of the residual nucleus is 3.90×10^{-22} gm, what is its kinetic energy, in electron volts?

2. The *escape velocity* of a particle on the earth is the minimum velocity required at the surface of the earth in order that the particle can escape from the earth's gravitational field. Neglecting the resistance of the atmosphere, the system is conservative. From the conservation theorem for potential plus kinetic energy show that the escape velocity for the earth, ignoring the presence of the moon, is 6.95 mi/sec.

3. Rockets are propelled by the momentum reaction of the exhaust gases expelled from the tail. Since these gases arise from the reaction of the fuels carried in the rocket the mass of the rocket is not constant, but decreases as the fuel is expended. Show that the equation of motion for a rocket projected vertically upward in a uniform gravitational field, neglecting atmospheric resistance, is

$$m\frac{dv}{dt} = -v'\frac{dm}{dt} - mg,$$

where m is the mass of the rocket and v' is the velocity of the escaping gases relative to the rocket. Integrate this equation to obtain v as a function of m, assuming a constant time rate of loss of mass. Show, for a rocket starting initially from rest, with v' equal to 6800 ft/sec and a mass loss per second equal to 1/60th of the initial mass, that in order to reach the escape velocity the ratio of the weight of the fuel to the weight of the empty rocket must be almost 300!

4. Show that for a single particle with constant mass the equation of motion implies the following differential equation for the kinetic energy:

$$\frac{dT}{dt} = \mathbf{F} \cdot \mathbf{v},$$

while if the mass varies with time the corresponding equation is

$$\frac{d(mT)}{dt} = \mathbf{F} \cdot \mathbf{p}.$$

5. Prove that the magnitude R of the position vector for the center of mass from an arbitrary origin is given by the equation

$$M^2 R^2 = M\sum_i m_i r_i^2 - \frac{1}{2}\sum_{i,j} m_i m_j r_{ij}^2.$$

6. Suppose a system of two particles is known to obey the equations of motion, Eqs. (1–22) and (1–26). Then from the equations of the motion of the individual particles show that the internal forces between particles satisfy both the weak and the strong laws of action and reaction. The argument may be generalized to a system with arbitrary number of particles, thus proving the converse of the arguments leading to Eqs. (1–22) and (1–26).

7. The equations of constraint for the rolling disk, Eqs. (1–39), are special cases of general linear differential equations of constraint of the form

$$\sum_{i=1}^{n} g_i(x_1,\dots,x_n)\,dx_i = 0.$$

A constraint condition of this type is holonomic only if an integrating function $f(x_1, \ldots, x_n)$ can be found that turns it into an exact differential. Clearly the function must be such that

$$\frac{\partial(f g_i)}{\partial x_j} = \frac{\partial(f g_j)}{\partial x_i}$$

for all $i \neq j$. Show that no such integrating factor can be found for either of Eqs. (1–39).

8. Two wheels of radius a are mounted on the ends of a common axle of length b such that the wheels rotate independently. The whole combination rolls without slipping on a plane. Show that there are two nonholonomic equations of constraint,

$$\cos \theta \, dx + \sin \theta \, dy = 0,$$

$$\sin \theta \, dx - \cos \theta \, dy = a(d\phi + d\phi'),$$

(where θ, ϕ, and ϕ' have meanings similar to those in the problem of a single vertical disc, and (x, y) are the coordinates of a point on the axle midway between the two wheels) and one holonomic equation of constraint,

$$\theta = C - \frac{a}{b}(\phi - \phi'),$$

where C is a constant.

9. A particle moves in the x–y plane under the constraint that its velocity vector is always directed towards a point on the x axis whose abscissa is some given function of time $f(t)$. Show that for $f(t)$ differentiable, but otherwise arbitrary, the constraint is nonholonomic.

10. Two points of mass m are joined by a rigid weightless rod of length l, the center of which is constrained to move on a circle of radius a. Set up the kinetic energy in generalized coordinates.

11. Show that Lagrange's equations in the form of Eq. (1–53) can also be written as

$$\frac{\partial \dot{T}}{\partial \dot{q}_j} - 2\frac{\partial T}{\partial q_j} = Q_j.$$

These are sometimes known as the *Nielsen* form of the Lagrange equations.

12. A point particle moves in space under the influence of a force derivable from a generalized potential of the form

$$U(\mathbf{r}, \mathbf{v}) = V(r) + \boldsymbol{\sigma} \cdot \mathbf{L},$$

where $\mathbf{r}$ is the radius vector from a fixed point, $\mathbf{L}$ is the angular momentum about that point, and $\boldsymbol{\sigma}$ is a fixed vector in space.

a) Find the components of the force on the particle in both Cartesian and spherical polar coordinates, on the basis of Eq. (1–58).
b) Show that the components in the two coordinate systems are related to each other as in Eq. (1–49).
c) Obtain the equations of motion in spherical polar coordinates.

13. A particle moves in a plane under the influence of a force, acting toward a center of force, whose magnitude is

$$F = \frac{1}{r^2}\left(1 - \frac{\dot{r}^2 - 2\ddot{r}r}{c^2}\right),$$

where r is the distance of the particle to the center of force. Find the generalized potential that will result in such a force, and from that the Lagrangian for the motion in a plane. (The expression for F represents the force between two charges in Weber's electrodynamics.)

14. If L is a Lagrangian for a system of n degrees of freedom satisfying Lagrange's equations, show by direct substitution that

$$L' = L + \frac{dF(q_1, \ldots, q_n, t)}{dt}$$

also satisfies Lagrange's equations where F is any arbitrary, but differentiable, function of its arguments.

15. Let $q_1, \ldots, q_n$ be a set of independent generalized coordinates for a system of n degrees of freedom, with a Lagrangian $L(q, \dot{q}, t)$. Suppose we transform to another set of independent coordinates $s_1, \ldots, s_n$ by means of transformation equations

$$q_i = q_i(s_1, \ldots, s_n, t), \qquad i = 1, \ldots, n.$$

(Such a transformation is called a *point transformation*.) Show that if the Lagrangian function is expressed as a function of s_j, $\dot{s}_j$, and t through the equations of transformation, then L satisfies Lagrange's equations with respect to the s coordinates:

$$\frac{d}{dt}\left(\frac{\partial L}{\partial \dot{s}_j}\right) - \frac{\partial L}{\partial s_j} = 0.$$

In other words, the form of Lagrange's equations is invariant under a point transformation.

16. A Lagrangian for a particular physical system can be written as

$$L' = \frac{m}{2}(a\dot{x}^2 + 2b\dot{x}\dot{y} + c\dot{y}^2) - \frac{K}{2}(ax^2 + 2bxy + cy^2),$$

where a, b, and c are arbitrary constants but subject to the condition that $b^2 - ac \neq 0$. What are the equations of motion? Examine particularly the two cases $a = 0 = c$ and $b = 0$, $c = -a$. What is the physical system described by the above Lagrangian? Show that the usual Lagrangian for this system as defined by Eq. (1–56) is related to L' by a point transformation (cf. Exercise 15 above). What is the significance of the condition on the value of $b^2 - ac$?

17. Obtain the Lagrange equations of motion for a spherical pendulum, i.e., a mass point suspended by a rigid weightless rod.

18. A particle of mass m moves in one dimension such that it has the Lagrangian

$$L = \frac{m^2\dot{x}^4}{12} + m\dot{x}^2 V(x) - V^2(x),$$

where V is some differentiable function of x. Find the equation of motion for $x(t)$ and describe the physical nature of the system on the basis of this equation.

19. Two mass points of mass m_1 and m_2 are connected by a string passing through a hole in a smooth table so that m_1 rests on the table surface and m_2 hangs suspended. Assuming m_2 moves only in a vertical line, what are the generalized coordinates for the system? Write down the Lagrange equations for the system and, if possible, discuss the physical significance any of them might have. Reduce the problem to a single second-order differential equation and obtain a first integral of the equation. What is its physical significance? (Consider the motion only so long as neither m_1 nor m_2 passes through the hole.)

20. Obtain the Lagrangian and equations of motion for the double pendulum illustrated in Fig. 1–4, where the lengths of the pendula are l_1 and l_2 with corresponding masses m_1 and m_2.

21. The electromagnetic field is invariant under a gauge transformation of the scalar and vector potential given by

$$\mathbf{A} \rightarrow \mathbf{A} + \nabla\psi(\mathbf{r}, \mathbf{t}),$$

$$\phi \rightarrow \phi - \frac{1}{c}\frac{\partial\psi}{\partial t},$$

where ψ is arbitrary (but differentiable). What effect does this gauge transformation have on the Lagrangian of a particle moving in the electromagnetic field? Is the motion affected?

22. Obtain the equation of motion for a particle falling vertically under the influence of gravity when frictional forces obtainable from a dissipation function $\frac{1}{2}kv^2$ are present. Integrate the equation to obtain the velocity as a function of time and show that the maximum possible velocity for fall from rest is $v = mg/k$.

CHAPTER 2
Variational Principles and Lagrange's Equations

2–1 HAMILTON'S PRINCIPLE

The derivation of Lagrange's equations presented in the previous chapter has started from a consideration of the instantaneous state of the system and small virtual displacements about the instantaneous state, i.e., from a "differential principle" such as D'Alembert's principle. It is also possible to obtain Lagrange's equations from a principle that considers the entire motion of the system between times t_1 and t_2, and small virtual variations of the entire motion from the actual motion. A principle of this nature is known as an "integral principle."

Before presenting the integral principle, the meaning attached to the phrase "motion of the system between times t_1 and t_2" must first be stated in more precise language. The instantaneous configuration of a system is described by the values of the n generalized coordinates $q_1 \ldots q_n$, and corresponds to a particular point in a Cartesian hyperspace where the q's form the n coordinate axes. This n-dimensional space is therefore known as configuration space. As time goes on the state of the system changes, and the system point moves in configuration space tracing out a curve, described as "the path of motion of the system." The "motion of the system," as used above, then refers to the motion of the system point along this path in *configuration space*. Time can be considered formally as a parameter of the curve; to each point on the path there is associated one or more values of the time. It must be emphasized that configuration space has no necessary connection with the physical three-dimensional space, just as the generalized coordinates are not necessarily position coordinates. The path of motion in configuration space will not have any necessary resemblance to the path in space of any actual particle; each point on the path represents the *entire* system configuration at some given instant of time.

The integral *Hamilton's Principle* describes the motion of those mechanical systems for which all forces (except the forces of constraint) are derivable from a generalized scalar potential that may be a function of the coordinates, velocities, and time. Borrowing from the terminology devised by C. Lanczos, such systems will be denoted as *monogenic*.* Where the potential is an explicit function of

---

* C. Lanczos, *The Variational Principles of Mechanics*, 4th ed. (Toronto: U. of Toronto Press, 1970), p. 30. The term indicates all forces are generated from a single function.

position coordinates only, then a monogenic system is also conservative (cf. Section 1–2). For monogenic systems, Hamilton's principle can be stated as *The motion of the system from time t_1 to time t_2 is such that the line integral*

$$I = \int_{t_1}^{t_2} L\,dt,\tag{2–1}$$

where $L = T - V$, has a stationary value for the correct path of the motion.*

That is, out of all possible paths by which the system point could travel from its position at time t_1 to its position at time t_2, it will actually travel along that path for which the value of the integral (2–1) is stationary. By the term "stationary value" for a line integral we mean that the integral along the given path has the same value to within first-order infinitesimals as that along all neighboring paths (i.e., those that differ from it by infinitesimal displacements). (Cf. Figure 2–1.) The notion of a stationary value for a line integral thus corresponds in ordinary function theory to the vanishing of the first derivative.

We can summarize Hamilton's principle by saying that the motion is such that the *variation* of the line integral I for fixed t_1 and t_2 is zero:

$$\delta I = \delta \int_{t_1}^{t_2} L(q_1,\dots,q_n,\dot{q}_1,\dots,\dot{q}_n,t)\,dt = 0.\tag{2–2}$$

Where the system constraints are holonomic, Hamilton's principle, Eq. (2–2) is both a necessary and sufficient condition for Lagrange's equations, Eqs. (1–57). Thus, it can be shown that Hamilton's principle follows directly from Lagrange's equations (cf. Whittaker's *Analytical Dynamics*, 4th ed., p. 245). Instead we shall prove the converse, namely, that Lagrange's equations follow from Hamilton's principle, as being the more important theorem. That Hamilton's principle is a sufficient condition for deriving the equations of motion enables us to construct

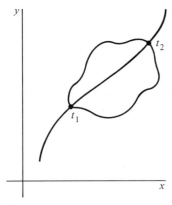

FIGURE 2–1
Path of the system point in configuration space.

---

* The quantity I is referred to as the *action* or *action integral*.

the mechanics of monogenic systems from Hamilton's principle as the basic postulate rather than Newton's laws of motion. Such a formulation has advantages; for example, since the integral I is obviously invariant to the system of generalized coordinates used to express L, the equations of motion must always have the Lagrangian form no matter how the generalized coordinates are transformed. More important, the formulation in terms of a variational principle is the route that must be followed when we try to describe apparently nonmechanical systems in the mathematical clothes of classical mechanics, as in the theory of fields.

2–2 SOME TECHNIQUES OF THE CALCULUS OF VARIATIONS

Before demonstrating that Lagrange's equations do follow from (2–2), a digression must first be made on the methods of the calculus of variations, for one of the chief problems of this calculus is to find the curve for which some given line integral has a stationary value.

Consider first the problem in an essentially one-dimensional form: we have a function $f(y, \dot{y}, x)$ defined on a path $y = y(x)$ between two values x_1 and x_2, where $\dot{y}$ is the derivative of y with respect to x. We wish to find a particular path $y(x)$ such that the line integral J of the function f between x_1 and x_2,

$$\dot{y} \equiv \frac{dy}{dx},$$

$$J = \int_{x_1}^{x_2} f(y, \dot{y}, x)\, dx, \tag{2–3}$$

has a stationary value relative to paths differing infinitesimally from the correct function $y(x)$. The variable x here plays the role of the parameter t, and we consider only such varied paths for which $y(x_1) = y_1$, $y(x_2) = y_2$. (Cf. Fig. 2–2.)

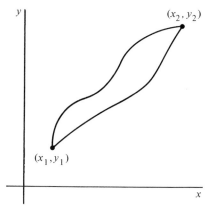

FIGURE **2–2**
Varied paths in the one-dimensional extremum problem.

Note that the diagram does *not* represent configuration space. In the one-dimensional configuration space both the correct and varied paths are the segment of the straight line connecting y_1 and y_2; the paths differ only in the functional relation between y and x.)

We put the problem in a form that enables us to use the familiar apparatus of the differential calculus for finding the stationary points of a function. Since J must have a stationary value for the correct path relative to *any* neighboring path, the variation must be zero relative to *some* particular set of neighboring paths labeled by an infinitesimal parameter α. Such a set of paths might be denoted by $y(x, \alpha)$, with $y(x, 0)$ representing the correct path. For example, if we select any function $\eta(x)$ that vanishes at $x = x_1$ and $x = x_2$, then a possible set of varied paths is given by

$$y(x, \alpha) = y(x, 0) + \alpha\eta(x). \tag{2-4}$$

For simplicity, it is assumed that both the correct path $y(x)$ and the auxiliary function $\eta(x)$ are well-behaved functions—continuous and nonsingular between x_1 and x_2, with continuous first and second derivatives in the same interval. For any such parametric family of curves, J in Eq. (2–3) is also a function of α:

$$J(\alpha) = \int_{x_1}^{x_2} f(y(x, \alpha), \dot{y}(x, \alpha), x)\, dx, \tag{2-5}$$

and the condition for obtaining a stationary point is the familiar one that

$$\left(\frac{dJ}{d\alpha}\right)_{\alpha=0} = 0. \tag{2-6}$$

By the usual methods of differentiating under the integral sign one finds that

$$\frac{\partial x}{\partial \alpha} = 0$$

$$\frac{dJ}{d\alpha} = \int_{x_1}^{x_2} \left\{ \frac{\partial f}{\partial y}\frac{\partial y}{\partial \alpha} + \frac{\partial f}{\partial \dot{y}}\frac{\partial \dot{y}}{\partial \alpha} \right\} dx. \tag{2-7}$$

Consider the second of these integrals:

$$\int_{x_1}^{x_2} \frac{\partial f}{\partial \dot{y}}\frac{\partial \dot{y}}{\partial \alpha}\, dx = \int_{x_1}^{x_2} \frac{\partial f}{\partial \dot{y}}\frac{\partial^2 y}{\partial x\, \partial \alpha}\, dx.$$

Integrating by parts the integral becomes

$$\int_{x_1}^{x_2} \frac{\partial f}{\partial \dot{y}}\frac{\partial^2 y}{\partial x\, \partial \alpha}\, dx = \frac{\partial f}{\partial \dot{y}}\frac{\partial y}{\partial \alpha}\bigg|_{x_1}^{x_2} - \int_{x_1}^{x_2} \frac{d}{dx}\left(\frac{\partial f}{\partial \dot{y}}\right)\frac{\partial y}{\partial \alpha}\, dx. \tag{2-8}$$

The conditions on all the varied curves are that they pass through the points $(x_1, y_1), (x_2, y_2)$, and hence the partial derivative of y with respect to α at x_1 and x_2 must vanish. Therefore the first term of (2–8) vanishes and Eq. (2–7) reduces to

$$\frac{dJ}{d\alpha} = \int_{x_1}^{x_2} \left(\frac{\partial f}{\partial y} - \frac{d}{dx}\frac{\partial f}{\partial \dot{y}}\right)\frac{\partial y}{\partial \alpha}\, dx.$$

The condition for a stationary value, Eq. (2–6), is therefore equivalent to the equation

$$\int_{x_1}^{x_2} \left(\frac{\partial f}{\partial y} - \frac{d}{dx}\frac{\partial f}{\partial \dot{y}} \right) \left(\frac{\partial y}{\partial \alpha} \right)_0 dx = 0. \tag{2–9}$$

Now, the partial derivative of y with respect to α occurring in Eq. (2–9) is a function of x that is arbitrary except for continuity and end point conditions. For example, for the particular parametric family of varied paths given by Eq. (2–4), it is the arbitrary function $\eta(x)$. One can therefore apply to Eq. (2–9) the so-called "fundamental lemma" of the calculus of variations, which says if

$$\int_{x_1}^{x_2} M(x)\eta(x)\,dx = 0 \tag{2–10}$$

for all arbitrary functions $\eta(x)$ continuous through the second derivative, then $M(x)$ must identically vanish in the interval (x_1, x_2). While a formal mathematical proof of the lemma may be found in the texts on the calculus of variations cited in the references, the validity of the lemma is easily seen intuitively. One can imagine constructing a function η that is positive in the immediate vicinity of any chosen point in the interval and zero everywhere else. Equation (2–10) can then hold only if $M(x)$ vanishes at that (arbitrarily) chosen point, which shows M must be zero throughout the interval. From Eq. (2–9) and the fundamental lemma it therefore follows that J can have a stationary value only if

$$\frac{\partial f}{\partial y} - \frac{d}{dx}\left(\frac{\partial f}{\partial \dot{y}}\right) = 0. \tag{2–11}$$

The differential quantity

$$\left(\frac{\partial y}{\partial \alpha}\right)_0 d\alpha \equiv \delta y \tag{2–12}$$

represents the infinitesimal departure of the varied path from the correct path $y(x)$ at the point x and thus corresponds to the virtual displacement introduced in Chapter 1 (hence the notation δy). Similarly the infinitesimal variation of J about the correct path can be designated

$$\left(\frac{dJ}{d\alpha}\right)_0 d\alpha \equiv \delta J. \tag{2–13}$$

The assertion that J is stationary for the correct path can thus be written

$$\delta J = \int_{x_1}^{x_2} \left\{ \frac{\partial f}{\partial y} - \frac{d}{dx}\frac{\partial f}{\partial \dot{y}} \right\} \delta y\,dx = 0,$$

requiring that $y(x)$ satisfy the differential equation (2–11). The δ-notation, introduced through equations (2–12) and (2–13), may be used as a convenient shorthand for treating the variation of integrals, remembering always that it stands for the manipulation of parametric families of varied paths such as Eq. (2–4).

Some simple examples of the application of Eq. (2–11) (which clearly resembles a Lagrange equation) may now be considered:

1. *Shortest distance between two points in a plane.* An element of arc length in a plane is

$$ds = \sqrt{dx^2 + dy^2}$$

and the total length of any curve going between points 1 and 2 is

$$I = \int_1^2 ds = \int_{x_1}^{x_2} \sqrt{1 + \left(\frac{dy}{dx}\right)^2}\, dx.$$

The condition that the curve be the shortest path is that I be a minimum. This is an example of the extremum problem as expressed by Eq. (2–3), with

$$f = \sqrt{1 + \dot{y}^2}.$$

Substituting in (2–11) with

$$\frac{\partial f}{\partial y} = 0, \qquad \frac{\partial f}{\partial \dot{y}} = \frac{\dot{y}}{\sqrt{1 + \dot{y}^2}},$$

we have

$$\frac{d}{dx}\left(\frac{\dot{y}}{\sqrt{1 + \dot{y}^2}}\right) = 0$$

or

$$\frac{\dot{y}}{\sqrt{1 + \dot{y}^2}} = c,$$

where c is constant. This solution can be valid only if

$$\dot{y} = a$$

where a is a constant related to c by

$$a = \frac{c}{\sqrt{1 - c^2}}.$$

But this is clearly the equation of a straight line,

$$y = ax + b,$$

where b is another constant of integration. Strictly speaking, the straight line has only been proved to be an extremum path, but for this problem it is obviously also a minimum. The constants of integration, a and b, are determined by the condition that the curve pass through the two end points, (x_1, y_1), (x_2, y_2).

In a similar fashion one could obtain the shortest distance between two points on a sphere, by setting up the arc length on the surface of the sphere in terms of the angle coordinates of position on the sphere. In general, curves that give the shortest distance between two points on a given surface are called the *geodesics* of the surface.

2. *Minimum surface of revolution.* Suppose we form a surface of revolution by taking some curve passing between two fixed end points (x_1, y_1) and (x_2, y_2) and revolving it about the y axis (cf. Fig. 2–3). The problem then is to find that curve for which the surface area is a minimum. The area of a strip of the surface is $2\pi x \, ds = 2\pi x\sqrt{1 + \dot{y}^2} \, dx$, and the total area is

$$2\pi \int_1^2 x\sqrt{1 + \dot{y}^2} \, dx.$$

The extremum of this integral is again given by (2–11) where

$$f = x\sqrt{1 + \dot{y}^2}$$

and

$$\frac{\partial f}{\partial y} = 0, \qquad \frac{\partial f}{\partial \dot{y}} = \frac{x\dot{y}}{\sqrt{1 + \dot{y}^2}}.$$

Equation (2–11) becomes in this case

$$\frac{d}{dx}\left(\frac{x\dot{y}}{\sqrt{1 + \dot{y}^2}}\right) = 0$$

or

$$\frac{x\dot{y}}{\sqrt{1 + \dot{y}^2}} = a,$$

where a is some constant of integration clearly smaller than the minimum value of x. Squaring the above equation and factoring terms we have

$$\dot{y}^2(x^2 - a^2) = a^2$$

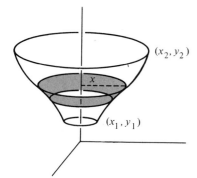

(x_2, y_2)

x

(x_1, y_1)

FIGURE 2–3
Minimum surface of revolution.

or solving,

$$\frac{dy}{dx} = \frac{a}{\sqrt{x^2 - a^2}}.$$

The general solution of this differential equation, in light of the nature of a, is

$$y = a \int \frac{dx}{\sqrt{x^2 - a^2}} + b = a \operatorname{arc\,cosh} \frac{x}{a} + b$$

or

$$x = a \cosh \frac{y - b}{a},$$

which is the equation of a catenary. Again the two constants of integration, a and b, are determined in principle by the requirements that the curve pass through the two given end points. It must be pointed out, however, that when examined in detail the nature of the solution turns out to be a good deal more complicated than these considerations suggest. For some pairs of end points, unique constants of integration a and b can indeed be found. But for other end points, *two* catenary solutions result, while in still other regions no possible values can be found for a and b. Further, it should be remembered that Eq. (2–11) represents a condition for finding curves $y(x)$, continuous through the second derivatives, that render the integral stationary. The catenary solutions therefore do not always represent minimum values, but may give "points of inflexion." For certain combinations of end points, the absolute minimum in the surface of revolution is provided by a curve composed of straight line segments—from the first end point parallel to the x axis until the y axis is reached, then along the y axis until the point $(0, y_2)$ and then out in a straight line to the second end point. Such a curve has discontinuous first derivatives and one could not expect to find it as a solution to Eq. (2–11). This example is valuable in emphasizing the restrictions that surround the derivation and meaning of the stationary condition. Further details will be found in some of the exercises and in the cited texts on the calculus of variations.

3. *The brachistochrone problem.* This well-known problem is to find the curve joining two points, along which a particle falling from rest under the influence of gravity travels from the higher to the lower point in the least time.

If v is the speed along the curve, then the time required to fall an arc length ds is ds/v, and the problem is to find a minimum of the integral

$$t_{12} = \int_1^2 \frac{ds}{v}.$$

If y is measured down from the initial point of release the conservation theorem for the energy of the particle can be written as

$$\tfrac{1}{2}mv^2 = mgy$$

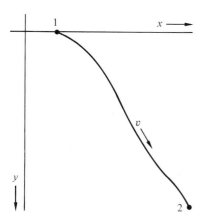

FIGURE 2–4
The brachistochrone problem.

or

$$v = \sqrt{2gy}.$$

Then the expression for t_{12} becomes

$$t_{12} = \int_1^2 \frac{\sqrt{1 + \dot{y}^2}}{\sqrt{2gy}} \, dx,$$

and f is identified as

$$f = \sqrt{\frac{1 + \dot{y}^2}{2gy}}.$$

The integration of Eq. (2–11) with this form for f is straightforward and will be left as one of the exercises for this chapter. The brachistochrone problem is famous in the history of mathematics, for it was the analysis of this problem by John Bernoulli that led to the formal foundation of the calculus of variations.

2–3 DERIVATION OF LAGRANGE'S EQUATIONS FROM HAMILTON'S PRINCIPLE

The fundamental problem of the calculus of variations is easily generalized to the case where f is a function of many independent variables y_i, and their derivatives $\dot{y}_i$. (Of course, all these quantities are considered as functions of the parametric variable x.) Then a variation of the integral J,

$$\delta J = \delta \int_1^2 f(y_1(x); y_2(x), \ldots, \dot{y}_1(x); \dot{y}_2(x), \ldots, x) \, dx, \qquad (2\text{–}14)$$

is obtained, as before, by considering J as a function of a parameter α that labels a possible set of curves $y_i(x, \alpha)$. Thus we may introduce α by setting

$$y_1(x, \alpha) = y_1(x, 0) + \alpha \eta_1(x),$$

$$y_2(x, \alpha) = y_2(x, 0) + \alpha \eta_2(x),$$

$$\vdots \qquad \vdots \qquad \vdots$$

$$(2\text{–}15)$$

where $y_1(x, 0)$, $y_2(x, 0)$, etc., are the solutions of the extremum problem (to be obtained) and η_1, η_2, etc., are independent functions of x that vanish at the end points and that are continuous through the second derivative, but otherwise are completely arbitrary. The calculation proceeds as before. The variation of J is given in terms of

$$\frac{\partial J}{\partial \alpha} d\alpha = \int_1^2 \sum_i \left(\frac{\partial f}{\partial y_i} \frac{\partial y_i}{\partial \alpha} d\alpha + \frac{\partial f}{\partial \dot{y}_i} \frac{\partial \dot{y}_i}{\partial \alpha} d\alpha \right) dx. \qquad (2\text{–}16)$$

$$\frac{\partial x}{\partial \alpha} = 0$$

Again we integrate by parts the integral involved in the second sum of Eq. (2–16):

$$\int_1^2 \frac{\partial f}{\partial \dot{y}_i} \frac{\partial^2 y_i}{\partial \alpha \, \partial x} dx = \frac{\partial f}{\partial \dot{y}_i} \frac{\partial y_i}{\partial \alpha} \Big|_1^2 - \int_1^2 \frac{\partial y_i}{\partial \alpha} \frac{d}{dx} \left(\frac{\partial f}{\partial \dot{y}_i} \right) dx,$$

where the first term vanishes because all curves pass through the fixed end points. Substituting in (2–16), δJ becomes

$$\delta J = \int_1^2 \sum_i \left(\frac{\partial f}{\partial y_i} - \frac{d}{dx} \frac{\partial f}{\partial \dot{y}_i} \right) \delta y_i \, dx, \qquad (2\text{–}17)$$

where, in analogy with (2–12), the variation δy_i is

$$\delta y_i = \left(\frac{\partial y_i}{\partial \alpha} \right)_0 d\alpha.$$

Since the y variables are independent, the variations δy_i are independent (e.g., the functions $\eta_i(x)$ will be independent of each other). Hence, by an obvious extension of the fundamental lemma (cf. above Eq. 2–10), the condition that δJ is zero requires that the coefficients of the δy_i separately vanish:

$$\frac{\partial f}{\partial y_i} - \frac{d}{dx} \frac{\partial f}{\partial \dot{y}_i} = 0, \qquad i = 1, 2, \dots, n. \qquad (2\text{–}18)$$

Equations (2–18) represent the appropriate generalization of (2–11) to several variables and are known as the *Euler–Lagrange Differential Equations*. Their solutions represent curves for which the variation of an integral of the form given in (2–14) vanishes. Further generalizations of the fundamental variational problem are easily possible. Thus one can take f as a function of higher derivatives $\ddot{y}$, $\dddot{y}$, etc., leading to equations different from (2–18). Or it can be extended to cases where there are several parameters x_j and the integral is then multiple, with f also involving as variables derivatives of y_i with respect to each of the parameters x_j. Finally, it is possible to consider variations in which the end points are *not* held fixed.

For present purposes what we have derived here suffices, for the integral in Hamilton's principle,

$$I = \int_1^2 L(q_i, \dot{q}_i, t) \, dt \tag{2-19}$$

has just the form stipulated in (2–14) with the transformations

$$x \rightarrow t$$

$$y_i \rightarrow q_i$$

$$f(y_i, \dot{y}_i, x) \rightarrow L(q_i, \dot{q}_i, t).$$

In deriving Eqs. (2–18) it was assumed that the y_i variables are independent. The corresponding condition in connection with Hamilton's principle is that the generalized coordinates q_i be independent, which requires that the constraints be holonomic. The Euler–Lagrange equations corresponding to the integral I then become the Lagrange equations of motion,

$$\frac{d}{dt}\frac{\partial L}{\partial \dot{q}_i} - \frac{\partial L}{\partial q_i} = 0, \qquad i = 1, 2, \ldots, n.$$

and we have accomplished our original aim, to show that Lagrange's equations follow from Hamilton's principle—for monogenic systems with holonomic constraints.

2–4 EXTENSION OF HAMILTON'S PRINCIPLE TO NONHOLONOMIC SYSTEMS

It is possible to extend Hamilton's principle, at least in a formal sense, to cover certain types of nonholonomic systems. In deriving Lagrange's equations from either Hamilton's or D'Alembert's principle, the requirement of holonomic constraints does not appear until the last step, when the variations q_i are considered as independent of each other. With nonholonomic systems the generalized coordinates are not independent of each other, and it is not possible to reduce them further by means of equations of constraint of the form $f(q_1, q_2, \ldots, q_n, t) = 0$. Hence it is no longer true that the q_i's are all independent.

Another difference that must be considered in treating the variational principle is the manner in which the varied paths are constructed. In the discussion of Section 2–2 it was pointed out that δy (or δq) represents a virtual displacement from a point on the actual path to some point on the neighboring varied path. But, with independent coordinates it is the final varied path that is significant, and not how it is constructed. When the coordinates are not independent, but subject to constraint relations, it becomes important whether the varied path is or is not constructed by displacements consistent with the constraints. Virtual displacements, in particular, may or may not satisfy the constraints.

It appears that a reasonably straightforward treatment of nonholonomic systems by a variational principle is possible only when the equations of constraint can be put in the form

$$\sum_k a_{lk}\, dq_k + a_{lt}\, dt = 0, \qquad (2\text{--}20)$$

that is, a linear relation connecting the *differentials* of the q's. The index l indicates there may be more than one such equation; it will be assumed there are m equations in all, that is, $l = 1, 2, \ldots, m$. It should be noted that the coefficients a_{lk}, a_{lt} may be functions of the q's and time.

It would be expected that the varied paths, or equivalently, the displacements constructing the varied path, should satisfy the constraints of Eq. (2–20). However, it has been proven that no such varied path can be constructed* unless Eqs. (2–20) are integrable, in which case the constraints are actually holonomic. A variational principle leading to the correct equations of motion can nonetheless be obtained when the varied paths are constructed from the actual motion by virtual displacements. The constraint equations valid for the virtual displacements are then

$$\sum_k a_{lk}\, \delta q_k = 0, \qquad (2\text{--}21)$$

and the varied path will then in general not satisfy Eqs. (2–20).

We can now use Eqs. (2–21) to reduce the number of virtual displacements to independent ones. The procedure for eliminating these extra virtual displacements is the method of *Lagrange undetermined multipliers*. If Eqs. (2–21) hold, then it is also true that

$$\lambda_l \sum_k a_{lk}\, \delta q_k = 0, \qquad (2\text{--}22)$$

where the λ_l, $l = 1, 2, \ldots, m$, are some undetermined quantities, functions in general of the coordinates and of the time t. In addition, Hamilton's principle,

$$\delta \int_{t_1}^{t_2} L\, dt = 0, \qquad (2\text{--}2)$$

is assumed to hold for the nonholonomic system. Following the development of Section (2–3), Hamilton's principle then implies that

$$\int_1^2 dt \sum_k \left(\frac{\partial L}{\partial q_k} - \frac{d}{dt}\frac{\partial L}{\partial \dot q_k} \right) \delta q_k = 0. \qquad (2\text{--}23)$$

---

* See, for example, H. Rund, *The Hamilton–Jacobi Theory in the Calculus of Variations* (New York: Van Nostrand, 1966), Chapter 5.

We can combine Eq. (2–23) with the m equations of constraint on the virtual displacements δq_k by summing Eqs. (2–22) over l and integrating the result with respect to time from point 1 and point 2:

$$\int_1^2 \sum_{k,l} \lambda_l a_{lk} \, \delta q_k \, dt = 0. \tag{2–24}$$

The sum of Eqs. (2–23) and (2–24) is then the relation

$$\int_1^2 dt \sum_{k=1}^n \left(\frac{\partial L}{\partial q_k} - \frac{d}{dt}\frac{\partial L}{\partial \dot{q}_k} + \sum_l \lambda_l a_{lk} \right) \delta q_k = 0. \tag{2–25}$$

The δq_k's are still not independent, of course; they are connected by the m relations (2–21). That is, while the first $n - m$ of these may be chosen independently, the last m are then fixed by the Eqs. (2–21). However, the values of the λ_l's remain at our disposal. Suppose we now choose the λ_l's to be such that

$$\frac{\partial L}{\partial q_k} - \frac{d}{dt}\frac{\partial L}{\partial \dot{q}_k} + \sum_l \lambda_l a_{lk} = 0, \qquad k = n - m + 1, \ldots, n. \tag{2–26}$$

which are in the nature of equations of motion for the last m of the q_k variables. With the λ_l determined by (2–26), we can write (2–25) as

$$\int_1^2 dt \sum_{k=1}^{n-m} \left(\frac{\partial L}{\partial q_k} - \frac{d}{dt}\frac{\partial L}{\partial \dot{q}_k} + \sum_l \lambda_l a_{lk} \right) \delta q_k = 0. \tag{2–27}$$

Here the only δq_k's involved are the independent ones. Hence it follows that

$$\frac{\partial L}{\partial q_k} - \frac{d}{dt}\frac{\partial L}{\partial \dot{q}_k} + \sum_l \lambda_l a_{lk} = 0, \qquad k = 1, 2, \ldots, n - m. \tag{2–28}$$

Combining (2–26) and (2–28) we have finally the complete set of Lagrange's equations for nonholonomic systems:

$$\frac{d}{dt}\frac{\partial L}{\partial \dot{q}_k} - \frac{\partial L}{\partial q_k} = \sum_l \lambda_l a_{lk}, \qquad k = 1, 2, \ldots, n. \tag{2–29}$$

But this is not the whole story, for now we have $n + m$ unknowns, namely the n coordinates q_k and the m Lagrange multipliers λ_l, while (2–29) gives us a total of only n equations. The additional equations needed, of course, are exactly the equations of constraint linking up the q_k's, Eqs. (2–20), except that they are now to be considered as first-order differential equations:

$$\sum_k a_{lk} \dot{q}_k + a_{lt} = 0. \tag{2–30}$$

Equations (2–30) and (2–29) together constitute $n + m$ equations for $n + m$ unknowns.

In this process we have obtained more information than was originally sought. Not only do we get the q_k's we set out to find, but we also get m λ_l's. What is the physical significance of the λ_l's? Suppose one removed the constraints on the system, but instead applied external forces Q'_k in such a manner as to keep the motion of the system unchanged. The equations of motion would likewise remain the same. Clearly these extra applied forces must be equal to the forces of constraint, for they are the forces applied to the system so as to satisfy the condition of constraint. Under the influence of these forces Q'_k, the equations of motion are

$$\frac{d}{dt}\frac{\partial L}{\partial \dot{q}_k} - \frac{\partial L}{\partial q_k} = Q'_k. \qquad (2\text{–}31)$$

But these must be identical with Eqs. (2–29). Hence we can identify $\Sigma \lambda_l a_{lk}$ with Q'_k, the generalized forces of constraint. In this type of problem we really do not eliminate the forces of constraint from the formulation, and they are supplied as part of the answer.

Although it is not obvious, the version of Hamilton's principle adopted here for nonholonomic systems also requires that the constraints do no work in virtual displacements. This can be most easily seen by rewriting Hamilton's principle in the form

$$\delta \int_{t_1}^{t_2} L\, dt = \delta \int_{t_1}^{t_2} T\, dt - \delta \int_{t_1}^{t_2} U\, dt = 0.$$

If the variation of the integral over the generalized potential is carried out by the procedures of Section (2–3), the principle takes the form

$$\delta \int_{t_1}^{t_2} T\, dt = \int_{t_1}^{t_2} \sum_k \left(\frac{\partial U}{\partial q_k} - \frac{d}{dt}\left(\frac{\partial U}{\partial \dot{q}_k}\right) \right) \delta q_k\, dt;$$

or, by Eq. (1–58),

$$\delta \int_{t_1}^{t_2} T\, dt = - \int_{t_1}^{t_2} \sum_k Q_k \delta q_k\, dt. \qquad (2\text{–}32)$$

In this dress, Hamilton's principle says that the difference in the time integral of the kinetic energy between two neighboring paths is equal to the negative of the time integral of the work done in the virtual displacements between the paths. The work involved is that done only by the forces derivable from the generalized potential. If we take the same Hamilton's principle to hold for both holonomic and nonholonomic systems, it must be required that the additional forces of nonholonomic constraints do no work in the displacements δq_k. This restriction parallels the earlier condition that the virtual work of the forces of holonomic constraint also be zero (cf. Section 1–4). In practice, the restriction presents little handicap to the applications, as most problems in which the nonholonomic formalism is used relate to rolling without slipping, where the constraints are obviously workless.

Indeed, if the assumption of workless constraints is made from the start, then the physical arguments leading to Eqs. (2–31) can be directly extended to derive the full form of the nonholonomic Lagrange's equations, Eqs. (2–29). The condition that the virtual work of the constraint forces is zero can be written as

$$\sum_k Q'_k \, \delta q_k = 0. \tag{2–33}$$

At the same time the equations of constraint imply that

$$\sum_k a_{lk} \, \delta q_k = 0, \qquad l = 1, 2, \ldots, m. \tag{2–21}$$

Hence, Eq. (2–33) will be satisfied if the constant forces are given by

$$Q'_k = \sum_l \lambda_l a_{lk},$$

where the λ_l's are (as yet) undetermined multipliers. The rest of the treatment then follows the lines as given above from Eqs. (2–28) on.*

Notice that Eq. (2–20) is not the most general type of nonholonomic constraint, e.g., it does not include equations of constraint in the form of inequalities. On the other hand, it does include holonomic constraints. A holonomic equation of constraint,

$$f(q_1, q_2, q_3, \ldots, q_n, t) = 0 \tag{2–34}$$

is equivalent to a differential equation,

$$\sum_k \frac{\partial f}{\partial q_k} \, dq_k + \frac{\partial f}{\partial t} \, dt = 0, \tag{2–35}$$

which is identical in form with (2–20), with the coefficients

$$a_{lk} = \frac{\partial f}{\partial q_k}, \qquad a_{lt} = \frac{\partial f}{\partial t}. \tag{2–36}$$

Thus the Lagrange multiplier method can be used also for holonomic constraints when (1) it is inconvenient to reduce all the q's to independent coordinates or (2) we might wish to obtain the forces of constraint.

As an example of the method, consider the following somewhat trivial illustration—a hoop rolling, without slipping, down an inclined plane. In this example the constraint of "rolling" is actually holonomic, but this fact will be immaterial to the discussion. On the other hand, the holonomic constraint that the hoop be on the inclined plane will be contained implicitly in the choice of generalized coordinates.

---

* In view of the difficulties in formulating a variational principle for nonholonomic systems, and the relative ease with which the equations of motion can be obtained directly, it is natural to question the usefulness of the variational approach in this case. It is for this reason that discussions of variational principles and their consequences will be confined from here on to holonomic systems in which generalized coordinates are independent.

The two generalized coordinates are x, θ, as in Fig. 2–5, and the equation of rolling constraint is

$$r \, d\theta = dx.$$

The kinetic energy can be resolved into kinetic energy of motion of the center of mass plus the kinetic energy of motion about the center of mass:

$$T = \tfrac{1}{2} M \dot{x}^2 + \tfrac{1}{2} M r^2 \dot{\theta}^2.$$

The potential energy is

$$V = Mg(l - x) \sin \phi,$$

where l is the length of the inclined plane and the Lagrangian is

$$L = T - V.$$
$$= \frac{M \dot{x}^2}{2} + \frac{M r^2 \dot{\theta}^2}{2} - Mg(l - x) \sin \phi.$$

Since there is one equation of constraint, only one Lagrange multiplier λ is needed. The coefficients appearing in the constraint equation are

$$a_\theta = r,$$
$$a_x = -1.$$

The two Lagrange equations therefore are

$$M \ddot{x} - Mg \sin \phi + \lambda = 0, \tag{2–37}$$
$$M r^2 \ddot{\theta} - \lambda r = 0, \tag{2–38}$$

which along with the equation of constraint,

$$r \dot{\theta} = \dot{x}, \tag{2–39}$$

constitutes three equations for three unknowns, θ, x, λ.

Differentiating (2–39) with respect to time, we have

$$r \ddot{\theta} = \ddot{x}.$$

Hence from (2–38)

$$M \ddot{x} = \lambda$$

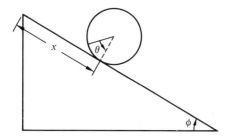

FIGURE **2–5**
A hoop rolling down an inclined plane.

and (2–37) becomes

$$\ddot{x} = \frac{g \sin \phi}{2},$$

along with

$$\lambda = \frac{Mg \sin \phi}{2}$$

and

$$\ddot{\theta} = \frac{g \sin \phi}{2r}.$$

Thus the hoop rolls down the incline with only one half the acceleration it would have slipping down a frictionless plane, and the friction force of constraint is $\lambda = Mg \sin \phi/2$.

From

$$\ddot{x} = v\frac{dv}{ds},$$

one gets $v = \sqrt{gl \sin \phi}$ at the bottom, which can, of course, be obtained by elementary means too.

2–5 ADVANTAGES OF A VARIATIONAL PRINCIPLE FORMULATION

Although it is thus possible to extend the original formulation of Hamilton's principle (2–2) to include some nonholonomic constraints, practically, this formulation of mechanics is most useful when a Lagrangian of independent coordinates can be set up for the system. The variational principle formulation has been justly described as "elegant," for in the compact Hamilton's principle is contained all of the mechanics of holonomic systems with forces derivable from potentials. The principle has the further merit that it involves only physical quantities that can be defined without reference to a particular set of generalized coordinates, namely, the kinetic and potential energies. The formulation is therefore automatically invariant with respect to the choice of coordinates for the system.

From the variational Hamilton's principle it is also obvious why the Lagrangian is always uncertain to a total time derivative of any function of the coordinates and time, as mentioned at the end of Section 1–4. The time integral of such a total derivative between points 1 and 2 depends only on the values of the arbitrary function at the end points. As the variation at the end points is zero, the addition of the arbitrary time derivative to the Lagrangian does not affect the variational behavior of the integral.

Another advantage is that the Lagrangian formulation can be extended easily to describe systems that are not normally considered in dynamics—such as the elastic field, the electromagnetic field, field properties of elementary particles. Some of these generalizations will be considered later, but as a simple example of its application outside the usual framework of mechanics, let us consider the following case.

Suppose we have a system for which there is a Lagrangian

$$L = \frac{1}{2}\sum_j L_j \dot{q}_j^2 + \frac{1}{2}\sum_{\substack{jk \\ j \neq k}} M_{jk}\dot{q}_j\dot{q}_k - \sum_j \frac{q_j^2}{2C_j} + \sum_j E_j(t)q_j \qquad (2\text{-}40)$$

and a dissipation function

$$\mathscr{F} = \frac{1}{2}\sum_j R_j \dot{q}_j^2. \qquad (2\text{-}41)$$

The Lagrange equations are

$$L_j\frac{d^2 q_j}{dt^2} + \sum_{\substack{k \\ j \neq k}} M_{jk}\frac{d^2 q_k}{dt^2} + R_j\frac{dq_j}{dt} + \frac{q_j}{C_j} = E_j(t). \qquad (2\text{-}42)$$

These equations of motion can be interpreted in at least two ways: One can say the q's are charges, the L_j's self-inductances, the M_{jk}'s mutual inductances, the R_j's resistances, the C_j's capacities, and the E_j's external emf's. Then Eqs. (2–42) are a set of equations describing a system of mutually inductively coupled networks, e.g., for $j = 1, 2, 3$ we would have three networks somewhat as in Fig. 2–6. On the other hand, it is seen that the first two terms in L together constitute an arbitrary homogeneous quadratic function of the generalized velocities. Whenever the system constraints (holonomic) are independent of time, the kinetic energy, T, can be put into such a form (cf. Section 1–6). The coefficients L_j, M_{jk} then partake of the character of masses—they are *inertial terms*. The next term in the Lagrangian exactly corresponds to the potential energy of a set of springs—harmonic oscillators—where the forces obey Hooke's law,

$$F = -kx,$$

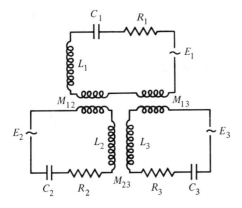

FIGURE 2–6
A system of coupled circuits to which the Lagrangian formulation can be applied.

with the resulting potential

$$V = \frac{kx^2}{2},$$

so that the $1/C_j$'s represent spring constants. The last term corresponds to the potential due to driving forces $E_j = Q_j$ that are independent of coordinates (as for example, gravitational forces) except that the E_j's may be time-varying forces. Finally, the dissipation function corresponds to the existence of dissipative or viscous forces, proportional to the generalized velocities. We are thus led to an alternative interpretation of Eqs. (2–40) and (2–41), or (2–42), which brings to mind a picture of a complicated system of masses on springs moving in some viscous fluid and driven by external forces.

This description of two different physical systems by Lagrangians of the same form means that all the results and techniques devised for investigating one of the systems can be taken over immediately and applied to the other. In this particular case the study of the behavior of electrical circuits has been pursued intensely and some special techniques have been developed; these can be directly applied to the corresponding mechanical systems. Much work has been done in formulating equivalent electrical problems for mechanical or acoustical systems, and vice versa. Terms normally reserved for electrical circuits (reactance, susceptance, etc.) are the accepted modes of expression in much of the theory of vibrations of mechanical systems.*

But, in addition, there is a type of generalization of mechanics that is due to a subtler form of equivalence. We have seen that the Lagrangian and Hamilton's principle together form a compact invariant way of implying the mechanical equations of motion. This possibility is not reserved for mechanics only; in almost every field of physics variational principles can be used to express the "equations of motion," whether they be Newton's equations, Maxwell's equations, or the Schrödinger equation. Consequently, when a variational principle is used as the basis of the formulation, all such fields will exhibit, at least to some degree, a *structural analogy*. When the results of experiments show the need for alteration of the physical content in the theory of one field, this degree of analogy has often indicated how similar alterations may be carried out in other fields. Thus, the experiments performed early in this century showed the need for quantization of both electromagnetic radiation and elementary particles. The methods of quantization, however, were first developed for particle mechanics, starting essentially from the Lagrangian formulation of classical mechanics. By describing the electromagnetic field by a Lagrangian and corresponding Hamilton's variational principle, it is possible to carry over the methods of particle quantization to construct a quantum electrodynamics (cf. Sections 12–5 and 12–6).

---

* For a detailed exposition, see H. F. Olson, *Solutions of Engineering Problems by Dynamic Analogues* (New York: Van Nostrand, 1966).

2–6 CONSERVATION THEOREMS AND SYMMETRY PROPERTIES

Thus far we have been concerned primarily with obtaining the equations of motion, and little has been said about how to solve them for a particular problem once they have been obtained. In general, this is a question of mathematics. A system of n degrees of freedom will have n differential equations that are second order in time. The solution of each equation will require two integrations resulting, all told, in $2n$ constants of integration. In a specific problem these constants will be determined by the initial conditions, i.e., the initial values of the $n q_j$'s and the $n \dot{q}_j$'s. Sometimes the equations of motion will be integrable in terms of known functions, but not always. In fact, the majority of problems are not completely integrable. However, even when complete solutions cannot be obtained, it is often possible to extract a large amount of information about the physical nature of the system motion. Indeed, such information may be of greater interest to the physicist than the complete solution for the generalized coordinates as a function of time. It is important, therefore, to see how much can be stated about the motion of a given system without requiring a complete integration of the problem.*

In many problems a number of first integrals of the equations of motion can be obtained immediately; by this we mean relations of the type

$$f(q_1, q_2, \ldots, \dot{q}_1, \dot{q}_2, \ldots, t) = \text{constant}, \qquad (2\text{–}43)$$

which are first-order differential equations. These first integrals are of interest because they tell us something physically about the system. They include, in fact, the conservation laws obtained in Chapter 1.

Consider as an example a system of mass points under the influence of forces derived from potentials dependent on position only. Then

$$\frac{\partial L}{\partial \dot{x}_i} \equiv \frac{\partial T}{\partial \dot{x}_i} - \frac{\partial V}{\partial \dot{x}_i} = \frac{\partial T}{\partial \dot{x}_i} = \frac{\partial}{\partial \dot{x}_i} \sum \frac{1}{2} m_i (\dot{x}_i^2 + \dot{y}_i^2 + \dot{z}_i^2)$$

$$= m_i \dot{x}_i = p_{ix},$$

which is the x component of the linear momentum associated with the ith particle. This result suggests an obvious extension to the concept of momentum. The generalized momentum associated with the coordinate q_j shall be defined as

$$p_j = \frac{\partial L}{\partial \dot{q}_j}. \qquad (2\text{–}44)$$

The terms *canonical momentum* or *conjugate momentum* are often also used for p_j. Notice that if q_j is not a Cartesian coordinate, p_j does not necessarily have the dimensions of a linear momentum. Further, if there is a velocity-dependent

---

* In this and succeeding sections it will be assumed, unless otherwise specified, the system is such that its motion is completely described by a Hamilton's principle of the form (2–2).

potential, then even with a Cartesian coordinate q_j the associated *generalized* momentum will not be identical with the usual *mechanical* momentum. Thus in the case of a group of particles in the electromagnetic field the Lagrangian is (cf. 1–66)

$$L = \sum_i \frac{1}{2} m_i \dot{r}_i^2 - \sum_i q_i \phi(x_i) + \sum_i \frac{q_i}{c} \mathbf{A}(x_i) \cdot \dot{\mathbf{r}}_i$$

(q_i here denotes charge) and the generalized momentum conjugate to x_i is

$$p_{ix} = \frac{\partial L}{\partial \dot{x}_i} = m_i \dot{x}_i + \frac{q_i A_x}{c}, \tag{2-45}$$

that is, mechanical momentum plus an additional term.

If the Lagrangian of a system does not contain a given coordinate q_j (although it may contain the corresponding velocity $\dot{q}_j$), then the coordinate is said to be *cyclic* or *ignorable*. This definition is not universal,* but it is the customary one and will be used here. The Lagrange equation of motion,

$$\frac{d}{dt}\frac{\partial L}{\partial \dot{q}_j} - \frac{\partial L}{\partial q_j} = 0,$$

reduces, for a cyclic coordinate, to

$$\dot{p}_j = \frac{\partial L}{\partial q_j}$$

$$\frac{d}{dt}\frac{\partial L}{\partial \dot{q}_j} = 0$$

or

$$\frac{dp_j}{dt} = 0,$$

which means that

$$p_j = \text{constant}. \tag{2-46}$$

Hence, we can state as a general conservation theorem that *the generalized momentum conjugate to a cyclic coordinate is conserved.*

Equation (2–46) constitutes a first integral of the form (2–43) for the equations of motion. It can be used formally to eliminate the cyclic coordinate from the problem, which can then be solved entirely in terms of the remaining generalized coordinates. Briefly, the procedure, originated by Routh, consists in

---

* The two terms are usually taken as interchangeable and as having the meaning assigned above. However, a few authors distinguish between them, defining a cyclic coordinate as one not in the kinetic energy, T, and an ignorable coordinate as one not in the Lagrangian (cf. Webster, *The Dynamics of Particles*, and Byerly, *Generalized Coordinates*). Ames and Murnaghan (*Theoretical Mechanics*) use the two terms interchangeably but apparently confine them to meaning a coordinate not in T. Lanczos (*Variational Principles of Mechanics*) has revived an older term, "kinosthenic," as equivalent to cyclic or ignorable. In addition, "cyclic" is sometimes used in a different sense in connection with periodic variables (cf. Section 9–5 and Synge, *Encyclopedia of Physics*, Vol. 3/1, p. 102).

modifying the Lagrangian so that it is no longer a function of the generalized velocity corresponding to the cyclic coordinate, but instead involves only its conjugate momentum. The advantage in so doing is that p_j can then be considered one of the constants of integration, and the remaining integrations involve only the noncyclic coordinates. We shall defer a detailed discussion of Routh's method until the Hamiltonian formulation (to which it is quite similar) is treated.

Note that the conditions for the conservation of generalized momenta are more general than the two momentum conservation theorems previously derived. For example, they furnish a conservation theorem for a case in which the law of action and reaction is violated, namely, when electromagnetic forces are present. Suppose we have a single particle in a field in which neither ϕ nor $\mathbf{A}$ depends on x. Then x nowhere appears in L and is therefore cyclic. The corresponding canonical momentum p_x must therefore be conserved. From (1–66) this momentum now has the form

$$p_x = m\dot{x} + \frac{qA_x}{c} = \text{constant}. \tag{2–47}$$

In this case it is not the mechanical linear momentum $m\dot{x}$ that is conserved but rather its sum with qA_x/c.* Nevertheless, it should still be true that the conservation theorems of Chapter 1 are contained within the general rule for cyclic coordinates; with proper restrictions (2–46) should reduce to the theorems of Section 1–2.

Consider first a generalized coordinate q_j, for which a change dq_j represents a translation of the system as a whole in some given direction. An example would be one of the Cartesian coordinates of the center of mass of the system. Then clearly q_j cannot appear in T, for velocities are not affected by a shift in the origin, and therefore the partial derivative of T with respect to q_j must be zero. Further we will assume conservative systems for which V is not a function of the velocities, so as to eliminate such anomalies as electromagnetic forces. The Lagrange equation of motion for a coordinate so defined then reduces to

$$\frac{d}{dt}\frac{\partial T}{\partial \dot{q}_j} \equiv \dot{p}_j = -\frac{\partial V}{\partial q_j} \equiv Q_j. \tag{2–48}$$

We will now show that (2–48) is the equation of motion for the total linear momentum, i.e., that Q_j represents the component of the total force along the direction of translation of q_j, and p_j is the component of the total linear momentum along this direction. In general, the generalized force Q_j is given by Eq. (1–49):

$$Q_j = \sum_i \mathbf{F}_i \cdot \frac{\partial \mathbf{r}_i}{\partial q_j}.$$

---

* It can be shown from classical electrodynamics that under these conditions, i.e., neither $\mathbf{A}$ nor ϕ depending on x, that qA_x/c is exactly the x component of the electromagnetic linear momentum of the field associated with the charge q.

Since dq_j corresponds to a translation of the system along some axis, the vectors $\mathbf{r}_i(q_j)$ and $\mathbf{r}_i(q_j + dq_j)$ are related as shown in Fig. 2–7. By definition of derivative we have

$$\frac{\partial \mathbf{r}_i}{\partial q_j} = \mathop{L}_{dq_j \to 0} \frac{\mathbf{r}_i(q_j + dq_j) - \mathbf{r}_i(q_j)}{dq_j} = \frac{dq_j}{dq_j}\, \mathbf{n} = \mathbf{n}, \qquad (2\text{–}49)$$

where $\mathbf{n}$ is the unit vector along the direction of translation. Hence

$$Q_j = \sum \mathbf{F}_i \cdot \mathbf{n} = \mathbf{n} \cdot \mathbf{F},$$

which, as was stated, is the component of the total force in the direction of $\mathbf{n}$. To prove the other half of the statement note that with the kinetic energy in the form

$$T = \frac{1}{2} \sum m_i \dot{\mathbf{r}}_i^2$$

the conjugate momentum is

$$p_j = \frac{\partial T}{\partial \dot{q}_j} = \sum_i m_i \dot{\mathbf{r}}_i \cdot \frac{\partial \dot{\mathbf{r}}_i}{\partial \dot{q}_j},$$

$$= \sum_i m_i \mathbf{v}_i \cdot \frac{\partial \mathbf{r}_i}{\partial q_j},$$

using Eq. (1–51). Then from Eq. (2–49)

$$p_j = \mathbf{n} \cdot \sum_i m_i \mathbf{v}_i,$$

which again, as predicted, is the component of the total system linear momentum along $\mathbf{n}$.

Suppose now that the translation coordinate q_j that we have been discussing is cyclic. Then q_j cannot appear in V and therefore

$$-\frac{\partial V}{\partial q_j} \equiv Q_j = 0.$$

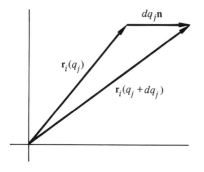

FIGURE 2–7
Change in a position vector under translation of the system.

But this is just the familiar conservation theorem for linear momentum—that if a given component of the total applied force vanishes, the corresponding component of the linear momentum is conserved.

In a similar fashion it can be shown that if a cyclic coordinate q_j is such that dq_j corresponds to a rotation of the system of particles around some axis, then the conservation of its conjugate momentum corresponds to conservation of an angular momentum. By the same argument as used above, T cannot contain q_j, for a rotation of the coordinate system cannot affect the magnitude of the velocities. Hence the partial derivative of T with respect to q_j must again be zero, and since V is independent of $\dot{q}_j$ we once more get Eq. (2–48). But now we wish to show that with q_j a rotation coordinate the generalized force is the component of the total applied torque about the axis of rotation, and p_j is the component of the total angular momentum along the same axis.

The generalized force Q_j is again given by

$$Q_j = \sum_i \mathbf{F}_i \cdot \frac{\partial \mathbf{r}_i}{\partial q_j},$$

only the derivative now has a different meaning. Here the change in q_j must correspond to an infinitesimal rotation of the vector $\mathbf{r}_i$, keeping the magnitude of the vector constant. From Fig. 2–8 the magnitude of the derivative can easily be obtained:

$$|d\mathbf{r}_i| = r_i \sin \theta \, dq_j$$

and

$$\left| \frac{\partial \mathbf{r}_i}{\partial q_j} \right| = r_i \sin \theta;$$

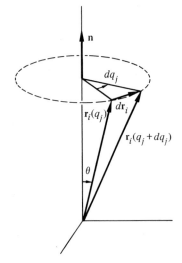

FIGURE 2–8
Change of a position vector under rotation of the system.

and its direction is perpendicular to both $\mathbf{r}_i$ and $\mathbf{n}$. Clearly the derivative can be written in vector form as

$$\frac{\partial \mathbf{r}_i}{\partial q_j} = \mathbf{n} \times \mathbf{r}_i. \tag{2–50}$$

With this result the generalized force becomes

$$Q_j = \sum_i \mathbf{F}_i \cdot \mathbf{n} \times \mathbf{r}_i$$

$$= \sum_i \mathbf{n} \cdot \mathbf{r}_i \times \mathbf{F}_i$$

reducing to

$$Q_j = \mathbf{n} \cdot \sum_i \mathbf{N}_i = \mathbf{n} \cdot \mathbf{N},$$

which proves the first part. A similar manipulation of p_j provides proof of the second part of the statement:

$$p_j = \frac{\partial T}{\partial \dot{q}_j} = \sum_i m_i \mathbf{v}_i \cdot \frac{\partial \mathbf{r}_i}{\partial q_j} = \sum_i \mathbf{n} \cdot \mathbf{r}_i \times m_i \mathbf{v}_i$$

$$= \mathbf{n} \cdot \sum_i \mathbf{L}_i = \mathbf{n} \cdot \mathbf{L}.$$

Summarizing these results we see that if the rotation coordinate q_j is cyclic, then Q_j, which is the component of the applied torque along $\mathbf{n}$, vanishes, and the component of $\mathbf{L}$ along $\mathbf{n}$ is constant. Here we have recovered the angular momentum conservation theorem out of the general conservation theorem relating to cyclic coordinates.

The significance of cyclic translation or rotation coordinates in relation to the properties of the system deserves some notice at this point. If a coordinate corresponding to a displacement is cyclic, it means that a translation of the system, as if rigid, has no effect on the problem. In other words, if the system is *invariant* under translation along a given direction, the corresponding linear momentum is conserved. Similarly, the fact that a rotation coordinate is cyclic (and therefore the conjugate angular momentum conserved) indicates that the system is invariant under rotation about the given axis. Thus the momentum conservation theorems are closely connected with the *symmetry properties* of the system. If the system is spherically symmetric we can say without further ado that all components of angular momentum are conserved. Or, if the system is symmetric only about the z axis, then only L_z will be conserved, and so on for the other axes. These symmetry considerations can often be used with relatively complicated problems to determine by inspection whether certain constants of the motion exist. Suppose, for example, the system consisted of a set of mass points moving in a potential field generated by fixed sources uniformly distributed on an infinite plane, say, the $z = 0$ plane. (The sources might be a mass distribution if the forces were gravitational, or a charge distribution for electrostatic forces.) Then the symmetry of the problem is such that the Lagrangian is invariant under a translation of the system of particles in the x or y directions (but not in the z direction) and also under a rotation about the z axis. It immediately follows that the x and y components of the total linear momentum, P_x and P_y, are constants of the motion along with L_z, the z component of the total angular momentum. However, if the sources were restricted only to the half plane, $x \geqslant 0$, then the symmetry for translation along the x axis and for rotation about the z axis would be destroyed. In that case, P_x, and L_z could not be conserved, but P_y would remain a constant of the motion. We will encounter the connections between the constants of motion and the symmetry properties of the system several times in the following chapters.

Another conservation theorem we should expect to obtain in the Lagrangian formulation is the conservation of total energy for systems where the forces are derivable from potentials dependent only on position. Indeed, it is possible to demonstrate a conservation theorem for which conservation of total energy represents only a special case. Consider a general Lagrangian, which will be a function of the coordinates q_j and the velocities $\dot{q}_j$ and may also depend explicitly on the time. (The explicit time dependence may arise from the time variation of external potentials, or from time-dependent constraints.) Then the total time derivative of L is

$$\frac{dL}{dt} = \sum_j \frac{\partial L}{\partial q_j}\frac{dq_j}{dt} + \sum_j \frac{\partial L}{\partial \dot{q}_j}\frac{d\dot{q}_j}{dt} + \frac{\partial L}{\partial t}. \qquad (2\text{--}51)$$

From Lagrange's equations,

$$\frac{\partial L}{\partial q_j} = \frac{d}{dt}\left(\frac{\partial L}{\partial \dot{q}_j}\right),$$

and (2–51) can be rewritten as

$$\frac{dL}{dt} = \sum_j \frac{d}{dt}\left(\frac{\partial L}{\partial \dot{q}_j}\right)\dot{q}_j + \sum_j \frac{\partial L}{\partial \dot{q}_j}\frac{d\dot{q}_j}{dt} + \frac{\partial L}{\partial t}$$

or

$$\frac{dL}{dt} = \sum_j \frac{d}{dt}\left(\dot{q}_j\frac{\partial L}{\partial \dot{q}_j}\right) + \frac{\partial L}{\partial t}.$$

It therefore follows that

$$\frac{d}{dt}\left(\sum_j \dot{q}_j\frac{\partial L}{\partial \dot{q}_j} - L\right) + \frac{\partial L}{\partial t} = 0. \qquad (2\text{--}52)$$

$$\sum_j \dot{q}_j P_j - L$$

or Hamiltonian

The quantity in the brackets is oftentimes called the *energy function** and will be denoted by h:

$$h(q, \dot{q}, t) = \sum_j \dot{q}_j \frac{\partial L}{\partial \dot{q}_j} - L, \qquad (2\text{–}53)$$

and Eq. (2–52) can be looked on as giving the total time derivative of h:

$L \neq L(t)$
H conserved

$$\frac{dh}{dt} = -\frac{\partial L}{\partial t}.$$

$\left(h = H \neq T + V \right)$ (2–54)

If the Lagrangian is not an explicit function of time, that is, if t does not appear in L explicitly but only implicitly through the time variation of q and $\dot{q}$, then Eq. (2–54) says that h is conserved; it is one of the first integrals of the motion and is sometimes referred to as Jacobi's integral.†

Under certain circumstances, the function h is the total energy of the system. To determine what these circumstances are, we recall that the total kinetic energy of a system can always be written as

$$T = T_0 + T_1 + T_2, \qquad (1\text{–}73)$$

where T_0 is a function of the generalized coordinates only, $T_1 (q, \dot{q})$ is linear in the generalized velocities, and $T_2 (q, \dot{q})$ is a quadratic function of the $\dot{q}$'s. For a very wide range of systems and sets of generalized coordinates, the Lagrangian can be similarly decomposed as regards its functional behavior in the $\dot{q}$ variables:

$$L(q, \dot{q}, t) = L_0(q, t) + L_1(q, \dot{q}, t) + L_2(q, \dot{q}, t). \qquad (2\text{–}55)$$

Here L_2 is a homogeneous function of the second degree (not merely quadratic) in $\dot{q}$ while L_1 is homogeneous of the first degree in $\dot{q}$. There is no reason intrinsic to mechanics that requires the Lagrangian to conform to Eq. (2–55), but in fact it does for almost all problems of interest. The Lagrangian clearly has this form when the forces are derivable from a potential not involving the velocities. Even with the velocity-dependent potentials we note that the Lagrangian for a charged particle in an electromagnetic field, Eq. (1–66), satisfies Eq (2–55). Now, it will be

---

* The energy function h is identical in value with the Hamiltonian H (see Chapter 8). It is here given a different name and symbol to emphasize that h is considered a function of n independent variables q_j and their time derivatives $\dot{q}_j$, (along with the time), whereas the Hamiltonian will be treated as a function of $2n$ independent variables, q_j, p_j (and possibly the time).

† This designation is most often confined to a first integral in the restricted three-body problem. However, the integral there is merely a special case of the energy function h, and there is some historical precedent to apply the name Jacobi integral to the more general situation.

remembered that Euler's theorem states that if f is a homogeneous function of degree n in the variables x_i, then*

$$\sum_i x_i \frac{\partial f}{\partial x_i} = nf. \tag{2-56}$$

Applied to the function h, Eq. (2–53), for Lagrangians of the form (2–55) this theorem implies that

$$h = 2L_2 + L_1 - L = L_2 - L_0. \tag{2-57}$$

If the transformation equations defining the generalized coordinates, Eqs. (1–38), do not involve the time explicitly, then by Eqs. (1–68) $T = T_2$. If, further, the potential does not depend on the generalized velocities, then $L_2 = T$ and $L_0 = -V$, so that

$$h = T + V = E, \tag{2-58}$$

and the energy function is indeed the total energy. Under these circumstances, if V does not involve the time explicitly neither will L, and by Eq. (2–54) h, which is here the total energy, will be conserved.

It should be emphasized that the conditions for conservation of h are in principle quite distinct from those that identify h as the total energy. One can have a set of generalized coordinates such that in a particular problem h is conserved but is not the total energy. On the other hand, it can occur that h is the total energy, in the form $T + V$, but is not conserved. It may also be noted that whereas the Lagrangian is uniquely fixed for each system by the prescription

$$L = T - U$$

independent of the choice of generalized coordinates, the energy function h depends in magnitude and functional form on the specific set of generalized coordinates. For one and the same system, various energy functions h of different physical content can be generated depending on how the generalized coordinates are chosen.

Finally it may be noted that where the system is not conservative, but there are frictional forces derivable from a dissipation function $\mathscr{F}$, it can be easily shown that $\mathscr{F}$ is related to the decay rate of h. When the equations of motion are given by Eq. (1–70), including dissipation, then Eq. (2–52) has the form

$$\frac{dh}{dt} + \frac{\partial L}{\partial t} = \sum_j \frac{\partial \mathscr{F}}{\partial \dot{q}_j} \dot{q}_j.$$

---

* See most textbooks on advanced calculus; e.g., W. Kaplan, *Advanced Calculus*, 2d ed., p. 139.

Define: $P_i = \frac{\partial L}{\partial \dot{q}_i}$; $\frac{d}{dt}\left(\frac{\partial L}{\partial \dot{q}_i}\right) - \frac{\partial L}{\partial q_i} = 0 \Rightarrow \dot{P}_i = \frac{\partial L}{\partial q_i}$

By the definition of $\mathscr{F}$, Eq. (1–67), it is a homogeneous function of the $\dot{q}$'s of degree 2. Hence, applying Euler's theorem again, we have

$$\frac{dh}{dt} = -2\mathscr{F} - \frac{\partial L}{\partial t}. \tag{2–59}$$

If L is not an explicit function of time, *and* the system is such that h is the same as the energy, then Eq. (2–59) says that $2\mathscr{F}$ is the rate of energy dissipation,

$$\frac{dE}{dt} = -2\mathscr{F}, \tag{2–60}$$

a statement proved above (cf. p.24) in less general circumstances.

SUGGESTED REFERENCES

R. COURANT AND D. HILBERT, *Methods of Mathematical Physics*, vol. 1. The literature on the calculus of variations is daunting in its volume and usually covers far more than is needed for the purposes of this chapter. Most treatises on "mathematics for the physicist" contain some brief discussion of the calculus of variations, and the classic work of Courant and Hilbert gives one of the clearest in their Chapter IV.

R. WEINSTOCK, *Calculus of Variations*. This is a book a physicist can feel comfortable with. Indeed over half the book is concerned with applications to problems of physics. The fundamental lemma is treated at the start of Chapter 3 and some of the difficulties of the continuous solution to the problem of the minimum surface of revolution are described in Section 3–7.

G. A. BLISS, *Calculus of Variations*. Of the older literature on the calculus of variations this little book is notable for its detailed discussion of the nature of the solutions to some of the standard problems, such as the brachistochrone. Chapter 4 has all that one would possibly want to know about surfaces of revolution of minimum area.

L. A. PARS, *An Introduction to the Calculus of Variations*. A careful and painstaking survey of the simpler mathematical aspects of the calculus of variations paying some attention (less than the author thinks) to physical applications. The fundamental lemma (and some cognate theorems) are presented in Chapter II. There is a brief note on nonholonomic systems on page 253.

C. LANCZOS, *The Variational Principles of Mechanics*. The first five chapters of this book are a leisurely survey, flavored by the author's original viewpoint, of the content of the present chapter. Chapter 10, containing historical notes, is particularly interesting. This is probably the best single reference for the entire chapter, although the approach differs considerably from that given here.

E. WHITTAKER, *Analytical Dynamics*. The treatment given for the topics in this chapter are still of interest, especially for many esoteric side notes not to be found elsewhere. Conservation theorems are discussed in Chapter III, while Hamilton's principle and its derivation from Lagrange's equations (the converse of the route taken in the present chapter) will be found in Chapter IX. The presence of a set of differential equations in the

Lagrangian form thus always implies the existence of an associated variational principle. It is therefore of some interest to know when a set of differential equations of second degree are, or can be put, in the Lagrangian form. This problem was first tackled by Helmholtz in 1887. The conditions he found, and some associated consequences, are described in an admirable and detailed review paper by P. Havas in *Nuovo Cimento Supplement*, vol. 5, p. 363 (1957).

L. D. LANDAU AND E. M. LIFSHITZ, *Mechanics*. This is the first volume of the *Course of Theoretical Physics*, that monument to the genius of Lev Landau. An incredible amount of material is contained within the brief compass of the 166 pages (in the English translation), and careful reading will be repaid in relation to almost any topic in the present book. The style might be described as that of "hand-waving arguments" written down on paper, and some holes are often left in the reasoning, but the physical insights are invaluable. Chapters I and II are particularly relevant to the present chapter.

W. E. BYERLY, *Generalized Coordinates*. Happily still available in reprint form, this little book is of value particularly for the many detailed examples of the Lagrangian technique for setting up and solving mechanical problems. The lack of an index is a deplorable defect that makes use of the book somewhat difficult.

D. W. WELLS, *Lagrangian Dynamics*. Although sometimes mislabeled as unsophisticated, this "outline" contains a wealth of detail and practical problems on a wide variety of aspects of Lagrangian mechanics. Chapter 6 has an unusually detailed treatment of frictional and dissipative sources within the Lagrangian framework (the author's "power function" is our dissipation function). Chapter 15 is devoted to Lagrange's equations for electrical systems and their interaction with mechanical systems. Chapters 12 (on constraint forces) and 17 (on Hamilton's principle) are also useful.

H. F. OLSON, *Solution of Engineering Problems by Dynamical Analogies*. This book discusses in great detail the electrical circuit problems equivalent to given mechanical and acoustical systems and illustrates the application of circuit theory to the solution of purely mechanical or acoustical problems. Lagrangians per se are introduced and used only briefly. A more pervasive Lagrangian viewpoint characterizes the following reference.

B. R. GOSSICK, *Hamilton's Principle and Physical Systems*. Although other nonmechanical systems are discussed, the emphasis is on applications from electrical engineering. The author's particular interest is in nonconservative (but linear) systems, and he enlarges the concept of a dissipation function to include energy loss by electromagnetic radiation.

H. RUND, *Hamilton–Jacobi Theory in the Calculus of Variations*. A good deal has been written about Hamilton's principle for nonholonomic systems, and most of it is wrong (including some things that were said in the first edition). Rund's book, though highly mathematical, has a number of interesting discussions on "pathological" problems encountered in the actual physical world and will be referred to here on several occasions. What is particularly relevant here is Section 5.5 on nonholonomic dynamical systems, which arrives at the flat conclusion that Hamilton's principle, in the form of Eq. (2–2), is applicable only to holonomic constraints. The Lagrange multiplier procedure used here, based on varied paths constructed from virtual displacements, is gone into much greater detail in a pair of papers by H. Jeffreys and L. A. Pars, respectively, published in 1954 in the *Quarterly Journal of Mechanics and Applied Mathematics*, vol. 7, p. 335 and p. 338.

EXERCISES

1. Prove that the shortest distance between two points in space is a straight line.

2. Show that the geodesics of a spherical surface are great circles, i.e., circles whose centers lie at the center of the sphere.

3. Complete the solution of the brachistochrone problem begun in Section 2–2 and show that the desired curve is a cycloid with a cusp at the initial point at which the particle is released. Show also that if the particle is projected with an initial kinetic energy $\frac{1}{2}mv_0^2$ that the brachistochrone is still a cycloid passing through the two points with a cusp at a height z above the initial point given by $v_0^2 = 2gz$.

4. Find the Euler–Lagrange equation describing the brachistochrone curve for a particle moving *inside* a spherical Earth of uniform mass density. Obtain a first integral for this differential equation by analogy to the Jacobi integral h. With the help of this integral show that the desired curve is a hypocycloid (the curve described by a point on a circle rolling on the inside of a larger circle). Obtain an expression for the time of travel along the brachistochrone between two points on the surface of the Earth. How long would it take to go from New York to Los Angeles (assumed to be 3,000 miles apart) along a brachistochrone tunnel (assuming no friction) and how far below the surface would the deepest point of the tunnel be?

5. In the problem of the minimum surface of revolution examine the symmetric case $y_2 = y_1, x_2 = -x_1 > 0$, and express the condition for the parameter a as a transcendental equation in terms of the dimensionless quantities $k = x_2/a$, and $\alpha = y_2/x_2$. Show that for α greater than a certain value α_0 two values of k are possible, for $\alpha = \alpha_0$ only one value of k is possible, while if $\alpha < \alpha_0$ no real value of k (or a) can be found, so that no catenary solution exists in this region. Find the value of α_0, numerically if necessary.

6. The broken-segment solution described in the text (cf. p. 42), in which the area of revolution is only that of the end circles of radius y_1 and y_2, respectively, is known as the *Goldschmidt solution*. For the symmetric situation discussed in Exercise 5 above, obtain an expression for the ratio of the area generated by the catenary solutions to that given by the Goldschmidt solution. Your result should be a function only of the parameters k and α. Show that for sufficiently large values of α at least one of the catenaries gives an area below that of the Goldschmidt solution. On the other hand show that if $\alpha = \alpha_0$, the Goldschmidt solution gives a lower area than the catenary.

7. A chain or rope of indefinite length passes freely over pulleys at heights y_1 and y_2 above the plane surface of the earth, with a horizontal distance $x_2 - x_1$ between them. If the chain or rope has a uniform linear mass density show that the problem of finding the curve assumed between the pulleys is identical with that of the problem of minimum surface of revolution. (The transition to the Goldschmidt solution as the heights y_1 and y_2 are changed makes for a striking lecture demonstration.)

8. Suppose that is was known experimentally that a particle fell a given distance y_0 in a time $t_0 = \sqrt{2y_0/g}$, but that the time of fall for distances other than y_0 were not known. Suppose further that the Lagrangian for the problem is known, but that instead of solving the equation of motion for y as a function of t, it is guessed that the functional form is

$$y = at + bt^2.$$

If the constants a and b are adjusted always so that the time to fall y_0 is correctly given by t_0, show directly that the integral

$$\int_0^{t_0} L \, dt$$

is an extremum for real values of the coefficients only when $a = 0$ and $b = g/2$.

9. When two billiard balls collide the instantaneous forces between them are very large but act only in an infinitesimal time Δt, in such a manner that the quantity

$$\int_{\Delta t} F \, dt$$

remains finite. Such forces are described as *impulsive* forces, and the integral over Δt is known as the *impulse* of the force. Show that if impulsive forces are present Lagrange's equations may be transformed into

$$\left(\frac{\partial L}{\partial \dot{q}_j}\right)_f - \left(\frac{\partial L}{\partial \dot{q}_j}\right)_i = S_j,$$

where the subscripts i and f refer to the state of the system before and after the impulse, S_j is the impulse of the generalized impulsive force corresponding to q_j, and L is the Lagrangian including all the nonimpulsive forces.

10. The term *generalized mechanics* has come to designate a variety of classical mechanics in which the Lagrangian contains time derivatives of q_i higher than the first. By applying the methods of the calculus of variations, show that if there is a Lagrangian of the form $L(q_i, \dot{q}_i, \ddot{q}_i, t)$, and Hamilton's principle holds with the zero variation of both q_i and $\dot{q}_i$ at the end points, then the corresponding Euler–Lagrange equations are

$$\frac{d^2}{dt^2}\left(\frac{\partial L}{\partial \ddot{q}_i}\right) - \frac{d}{dt}\left(\frac{\partial L}{\partial \dot{q}_i}\right) + \frac{\partial L}{\partial q_i} = 0, \qquad i = 1, 2, \ldots, n.$$

Apply this result to the Lagrangian

$$L = -\frac{m}{2}q\ddot{q} - \frac{k}{2}q^2.$$

Do you recognize the equations of motion?

11. A heavy particle is placed at the top of a vertical hoop. Calculate the reaction of the hoop on the particle by means of the Lagrange's undetermined multipliers and Lagrange's equations. Find the height at which the particle falls off.

12. A uniform hoop of mass m and radius r rolls without slipping on a fixed cylinder of radius R as shown in the figure. The only external force is that of gravity. If the smaller cylinder starts rolling from rest on top of the bigger cylinder, find by the method of Lagrange multipliers the point at which the hoop falls off the cylinder.

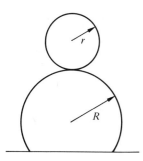

13. A form of the Wheatstone impedance bridge has, in addition to the usual four resistances, an inductance in one arm and a capacitance in the opposite arm. Set up L and $\mathscr{F}$ for the unbalanced bridge, with the charges in the elements as coordinates. Using the Kirchhoff junction conditions as constraints on the currents, obtain the Lagrange equations of motion, and show that eliminating the λ's reduces these to the usual network equations.

14. In certain situations, particularly one-dimensional systems, it is possible to incorporate frictional effects without introducing the dissipation function. As an example, find the equations of motion for the Lagrangian

$$L = e^{\gamma t}\left[\frac{m\dot{q}^2}{2} - \frac{kq^2}{2}\right].$$

How would you describe the system? Are there any constants of motion? Suppose a point transformation is made of the form

$$s = e^{\gamma t}q.$$

What is the effective Lagrangian in terms of s? Find the equation of motion for s. What do these results say about the conserved quantities for the system?

15. Show that if the potential in the Lagrangian contains velocity-dependent terms, the canonical momentum corresponding to a coordinate of rotation θ of the entire system is no longer the mechanical angular momentum L_θ but is given by

$$p_\theta = L_\theta - \sum_i \mathbf{n}\cdot\mathbf{r}_i \times \nabla_{v_i}U,$$

where ∇_v is the gradient operator in which the derivatives are with respect to the velocity components and $\mathbf{n}$ is a unit vector in the direction of rotation. If the forces are electromagnetic in character the canonical momentum is therefore

$$p_\theta = L_\theta + \sum_i \mathbf{n}\cdot\mathbf{r}_i \times \frac{q_i}{c}\mathbf{A}_i.$$

16. It sometimes occurs that the generalized coordinates appear separately in the kinetic energy and the potential energy in such a manner that T and V may be written in the form

$$T = \sum_i f_i(q_i)\dot{q}_i^2 \quad \text{and} \quad V = \sum_i V_i(q_i).$$

Show that Lagrange's equations then separate, and that the problem can always be reduced to quadratures.

17. A point mass is constrained to move on a massless hoop of radius a fixed in a vertical plane that is rotating about the vertical with constant angular speed ω. Obtain the Lagrange equations of motion assuming the only external forces arise from gravity. What are the constants of motion? Show that if ω is greater than a critical value ω_0, there can be a solution in which the particle remains stationary on the hoop at a point other than at the bottom, but that if $\omega < \omega_0$, the only stationary point for the particle is at the bottom of the hoop. What is the value of ω_0?

18. A particle moves in a conservative field of force produced by various mass distributions. In each instance the force generated by a volume element of the distribution is derived from a potential that is proportional to the mass of the volume element and is a function only of the scalar distance from the volume element. For the following fixed, homogeneous mass distributions, state the conserved quantities in the motion of the particle:

a) The mass is uniformly distributed in the plane $z = 0$.
b) The mass is uniformly distributed in the half-plane $z = 0, y > 0$.
c) The mass is uniformly distributed in a circular cylinder of infinite length, with axis along the z axis.
d) The mass is uniformly distributed in a circular cyclinder of finite length, with axis along the z axis.
e) The mass is uniformly distributed in a right cylinder of elliptical cross section and infinite length, with axis along the z axis.
f) The mass is uniformly distributed in a dumbbell whose axis is oriented along the z axis.
g) The mass is in the form of a uniform wire wound in the geometry of an infinite helical solenoid, with axis along the z axis.

19. A particle of mass m slides without friction on a wedge of angle α and mass M that can move without friction on a smooth horizontal surface (cf. figure). Treating the constraint of the particle on the wedge by the method of Lagrange multipliers, find the equations of motion for the particle and wedge. Also obtain an expression for the forces of constraint. Calculate the work done in time t by the forces of constraint acting on the particle and on the wedge. What are the constants of motion for the system? Contrast the results you have found with the situation when the wedge is fixed. [*Suggestion*: For the particle you may either use a Cartesian coordinate system with y vertical, or one with y normal to the wedge or, even more instructively, do it in both systems.]

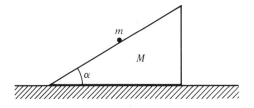

20. A carriage runs along rails on a rigid beam. The carriage is attached to one end of a spring, equilibrium length r_0 and force constant k, whose other end is fixed on the beam.

On the carriage there is another set of rails perpendicular to the first along which a particle of mass m moves, held by a spring fixed on the beam, of force constant k and zero equilibrium length. Beam, rails, springs and carriage are assumed to have zero mass. The whole system is forced to move in a plane about the point of attachment of the first spring, with a constant angular speed ω. The length of the second spring is at all times considered small compared to r_0.

a) What is the energy of the system? Is it conserved?
b) Using generalized coordinates in the laboratory system what is the Jacobi integral for the system? Is it conserved?
c) In terms of generalized coordinates relative to a system rotating with the angular speed ω, what is the Lagrangian? What is the Jacobi integral? Is it conserved? Discuss the relationship between the two Jacobi integrals.

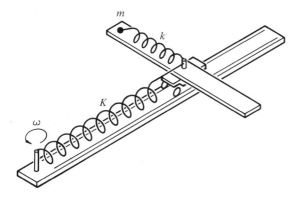

21. Suppose that a particle moves in space subject to a conservative potential $V(\mathbf{r})$ but is constrained always to move on a surface whose equation is $\sigma(\mathbf{r}, t) = 0$. (The explicit dependence on t indicates that the surface may be moving.) The instantaneous force of constraint is taken as always perpendicular to the surface. Show analytically that the energy of the particle is not conserved if the surface moves in time. What physically is the reason for nonconservation of the energy under this circumstance?

22. The one-dimensional harmonic oscillator has the Lagrangian $L = m\dot{x}^2/2 - kx^2/2$. Suppose you did not know the solution to the motion, but realized that the motion must be periodic and therefore could be described by a Fourier series of the form

$$x(t) = \sum_{j=0} a_j \cos j\omega t$$

(taking $t = 0$ at a turning point) where ω is the (unknown) angular frequency of the motion. This representation for $x(t)$ defines a many-parameter path for the system point in configuration space. Consider the action integral I for two points t_1 and t_2 separated by the period $T = 2\pi/\omega$. Show that with this form for the system path, I is an extremum for nonvanishing x only if $a_j = 0$, for $j \neq 1$, and only if $\omega^2 = k/m$.

CHAPTER 3
The Two-Body
Central Force Problem

In this chapter we shall discuss the problem of two bodies moving under the influence of a mutual central force as an application of the Lagrangian formulation. Not all the problems of central force motion are integrable in terms of well-known functions. However, we shall attempt to explore the problem as thoroughly as is possible with the tools already developed.

3-1 REDUCTION TO THE EQUIVALENT ONE-BODY PROBLEM

Consider a monogenic system of two mass points, m_1 and m_2, where the only forces are those due to an interaction potential U. It will be assumed at first that U is any function of the vector between the two particles, $\mathbf{r}_2 - \mathbf{r}_1$, or of their relative velocity, $\dot{\mathbf{r}}_2 - \dot{\mathbf{r}}_1$, or of any higher derivatives of $\mathbf{r}_2 - \mathbf{r}_1$. Such a system has six degrees of freedom and hence six independent generalized coordinates. Let us choose these to be the three components of the radius vector to the center of mass, $\mathbf{R}$, plus the three components of the difference vector $\mathbf{r} = \mathbf{r}_2 - \mathbf{r}_1$. The Lagrangian will then have the form

$$L = T(\dot{\mathbf{R}}, \dot{\mathbf{r}}) - U(\mathbf{r}, \dot{\mathbf{r}}, \ldots). \tag{3-1}$$

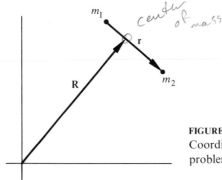

FIGURE 3-1
Coordinates for the two-body problem.

70

The kinetic energy T can be written as the sum of the kinetic energy of the motion of the center of mass, plus the kinetic energy of motion about the center of mass, T':

$$T = \tfrac{1}{2}(m_1 + m_2)\dot{\mathbf{R}}^2 + T'$$

with

$$T' = \tfrac{1}{2}m_1\dot{\mathbf{r}}_1'^2 + \tfrac{1}{2}m_2\dot{\mathbf{r}}_2'^2.$$

Here $\mathbf{r}_1'$ and $\mathbf{r}_2'$ are the radii vectors of the two particles relative to the center of mass and are related to $\mathbf{r}$ by

$$\mathbf{r}_1' = -\frac{m_2}{m_1 + m_2}\mathbf{r},$$

$$r = r_2' - r_1'$$

(3-2)

$$\mathbf{r}_2' = \frac{m_1}{m_1 + m_2}\mathbf{r}.$$

Expressed in terms of $\mathbf{r}$ by means of Eq. (3-2), T' takes on the form

Reduced mass

$$T' = \frac{1}{2}\frac{m_1 m_2}{m_1 + m_2}\dot{\mathbf{r}}^2$$

and the total Lagrangian (3-1) is

$$L = \frac{m_1 + m_2}{2}\dot{\mathbf{R}}^2 + \frac{1}{2}\frac{m_1 m_2}{m_1 + m_2}\dot{\mathbf{r}}^2 - U(\mathbf{r}, \dot{\mathbf{r}}, \dots). \qquad (3-3)$$

cm   RM

It is seen that the three coordinates $\mathbf{R}$ are cyclic, so that the center of mass is either at rest or moving uniformly. None of the equations of motion for $\mathbf{r}$ will contain terms involving $\mathbf{R}$ or $\dot{\mathbf{R}}$. Consequently the process of ignoration is particularly simple here. We merely drop the first term from the Lagrangian in all subsequent discussion.　Drop R

The rest of the Lagrangian is exactly what would be expected if we had a fixed center of force with a single particle at a distance $\mathbf{r}$ from it, having a mass

$$\mu = \frac{m_1 m_2}{m_1 + m_2}, \qquad (3-4)$$

where μ is known as the *reduced mass*. Frequently Eq. (3-4) is written in the form

$$\frac{1}{\mu} = \frac{1}{m_1} + \frac{1}{m_2}. \qquad (3-5)$$

Thus the central force motion of two bodies about their center of mass can always be reduced to an equivalent one-body problem.

3-2 THE EQUATIONS OF MOTION AND FIRST INTEGRALS

We now restrict ourselves to conservative central forces, where the potential is $V(r)$, a function of r only, so that the force is always along $\mathbf{r}$. By the results of the preceding section we need consider only the problem of a single particle of mass m moving about a fixed center of force, which will be taken as the origin of the coordinate system. Since potential energy involves only the radial distance, the problem has spherical symmetry, i.e., any rotation, about any fixed axis, can have no effect on the solution. Hence an angle coordinate representing rotation about a fixed axis must be cyclic. These symmetry properties result in a considerable simplification in the problem. Since the system is spherically symmetric, the total angular momentum vector,

$$\mathbf{L} = \mathbf{r} \times \mathbf{p},$$

is conserved. It therefore follows that $\mathbf{r}$ is always perpendicular to the fixed direction of $\mathbf{L}$ in space. This can be true only if $\mathbf{r}$ always lies in a plane whose normal is parallel to $\mathbf{L}$. While the reasoning breaks down if $\mathbf{L}$ is zero, the motion in that case must be along a straight line going through the center of force, for $\mathbf{L} = 0$ requires $\mathbf{r}$ to be parallel to $\dot{\mathbf{r}}$, which can be satisfied only in straight line motion.* Thus, central force motion is always motion in a plane. Now, the motion of a single particle in space is described by three coordinates; in spherical polar coordinates these are the azimuth angle θ, the zenith angle (or colatitude) ψ, and the radial distance r. By choosing the polar axis to be in the direction of $\mathbf{L}$, the motion is always in the plane perpendicular to the polar axis. The coordinate ψ then has only the constant value $\pi/2$ and can be dropped from the subsequent discussion. The conservation of the angular momentum vector furnishes three independent constants of motion (corresponding to the three Cartesian components). In effect, two of these, expressing the constant *direction* of the angular momentum, have been used to reduce the problem from three to two degrees of freedom. The third of these constants, corresponding to the conservation of the magnitude of $\mathbf{L}$, remains still at our disposal in completing the solution.

Expressed now in plane polar coordinates the Lagrangian is

$$L = T - V$$
$$= \tfrac{1}{2}m(\dot{r}^2 + r^2\dot{\theta}^2) - V(r). \tag{3-6}$$

As was foreseen θ is a cyclic coordinate, whose corresponding canonical momentum is the angular momentum of the system:

$$p_\theta = \frac{\partial L}{\partial \dot{\theta}} = mr^2\dot{\theta}.$$

One of the two equations of motion is then simply

$$\dot{p}_\theta = \frac{d}{dt}(mr^2\dot{\theta}) = 0 \tag{3-7}$$

---

* Formally: $\dot{\mathbf{r}} = \dot{r}\mathbf{n}_r + r\dot{\theta}\mathbf{n}_\theta$, hence $\mathbf{r} \times \dot{\mathbf{r}} = 0$ requires $\dot{\theta} = 0$.

cons. Ang. mom.

with the immediate integral

$$mr^2\dot\theta = l, \qquad (3\text{–}8)$$

where l is the constant magnitude of the angular momentum. From (3–7) it also follows that

$$\frac{d}{dt}\left(\frac{1}{2}r^2\dot\theta\right) = 0. \qquad (3\text{–}9)$$

The factor $\frac{1}{2}$ is inserted because $\frac{1}{2}r^2\dot\theta$ is just the *areal velocity*—the area swept out by the radius vector per unit time. This interpretation follows from the diagram, Fig. 3–2, the differential area swept out in time dt being

$$\boxed{dA = \tfrac{1}{2}r(rd\theta),}$$ *Areal velocity*

and hence

$$\frac{dA}{dt} = \frac{1}{2}r^2\frac{d\theta}{dt}. \qquad cons. \text{ of } L = \text{constant}$$
Areal velocity

The conservation of angular momentum is thus equivalent to saying the areal velocity is constant. Here we have the proof of the well-known Kepler's second law of planetary motion: the radius vector sweeps out equal areas in equal times. It should be emphasized, however, that the conservation of the areal velocity is a general property of central force motion and is not restricted to an inverse square law of force.

The remaining Lagrange equation, for the coordinate r, is

$$\frac{d}{dt}(m\dot r) - mr\dot\theta^2 + \frac{\partial V}{\partial r} = 0. \qquad (3\text{–}10)$$

Designating the value of the force along $\mathbf{r}$, $-\dfrac{\partial V}{\partial r}$, by $f(r)$ the equation can be rewritten as

$$m\ddot r - mr\dot\theta^2 = f(r). \qquad (3\text{–}11)$$

$$f(r) = -\frac{\partial}{\partial r}V(r)$$

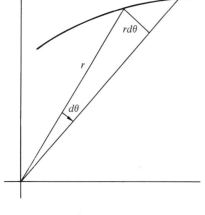

$rd\theta$

r

$d\theta$

FIGURE 3–2
The area swept out by the radius vector in a time dt.

By making use of the first integral, Eq. (3–8), $\dot{\theta}$ can be eliminated from the equation of motion, yielding a second order differential equation involving r only:

$$m\ddot{r} - \frac{l^2}{mr^3} = f(r). \tag{3–12}$$

There is another first integral of motion available, namely the total energy, since the forces are conservative. On the basis of the general energy conservation theorem we can immediately state that a constant of the motion is

$$E = \frac{1}{2}m(\dot{r}^2 + r^2\dot{\theta}^2) + V(r), \; = \; constant \tag{3–13}$$

where E is the energy of the system. Alternatively, this first integral could be derived again directly from the equations of motion (3–7) and (3–12). The latter can be written as

$$m\ddot{r} = -\frac{d}{dr}\left(V + \frac{1}{2}\frac{l^2}{mr^2}\right). = f(r) + \frac{l^2}{mr^3} \tag{3–14}$$

If both sides of Eq. (3–14) are multiplied by $\dot{r}$ the left-hand side becomes

$$m\dot{r}\ddot{r} = \frac{d}{dt}\left(\frac{1}{2}m\dot{r}^2\right).$$

The right-hand side similarly can be written as a total time derivative, for if $g(r)$ is any function of r, then the total time derivative of g has the form

$$\frac{d}{dt}g(r) = \frac{dg}{dr}\frac{dr}{dt}.$$

Hence Eq. (3–14) is equivalent to

$$\frac{d}{dt}\left(\frac{1}{2}m\dot{r}^2\right) = -\frac{d}{dt}\left(V + \frac{1}{2}\frac{l^2}{mr^2}\right)$$

or

$$\frac{d}{dt}\left(\frac{1}{2}m\dot{r}^2 + \frac{1}{2}\frac{l^2}{mr^2} + V\right) = 0,$$

and therefore

$$E = \frac{1}{2}m\dot{r}^2 + \frac{1}{2}\frac{l^2}{mr^2} + V = \text{constant}. \tag{3–15}$$

Equation (3–15) is the statement of the conservation of total energy, for by using (3–8) for l the middle term can be written

$$\frac{1}{2}\frac{l^2}{mr^2} = \frac{1}{2mr^2}m^2r^4\dot{\theta}^2 = \frac{mr^2\dot{\theta}^2}{2},$$

and (3–15) reduces to (3–13).

These two first integrals give us in effect two of the quadratures necessary to complete the problem. As there are two variables, r and θ, a total of four integrations are needed to solve the equations of motion. The first two integrations have left the Lagrange equations as two first order equations (3–8) and (3–15); the two remaining integrations can be accomplished (formally) in a variety of ways. Perhaps the simplest procedure starts from Eq. (3–15). Solving for $\dot{r}$ we have

$$\dot{r} = \sqrt{\frac{2}{m}\left(E - V - \frac{l^2}{2mr^2}\right)}, \tag{3–16}$$

or

$$dt = \frac{dr}{\sqrt{\frac{2}{m}\left(E - V - \frac{l^2}{2mr^2}\right)}}. \tag{3–17}$$

At time $t = 0$ let r have the initial value r_0. Then the integral of both sides of the equation from the initial state to the state at time t takes the form

Quadrature

$$t = \int_{r_0}^{r} \frac{dr}{\sqrt{\frac{2}{m}\left(E - V - \frac{l^2}{2mr^2}\right)}}. \tag{3–18}$$

As it stands Eq. (3–18) gives t as a function of r and the constants of integration E, l, and r_0. However, it may be inverted, at least formally, to give r as a function of t and the constants. Once the solution for r is thus found, the solution θ follows immediately from Eq. (3–8), which can be written as

$$d\theta = \frac{l\,dt}{mr^2}. \tag{3–19}$$

If the initial value of θ is θ_0, then the integral of (3–19) is simply

Integration constants.

$$\theta = l \int_0^t \frac{dt}{mr^2(t)} + \theta_0. \tag{3–20}$$

Equations (3–18) and (3–20) are the two remaining integrations, and formally the problem has been reduced to quadratures, with four constants of integration E, l, r_0, θ_0. These constants are not the only ones that can be considered. We might equally as well have taken $r_0, \theta_0, \dot{r}_0, \dot{\theta}_0$, but of course E and l can always be determined in terms of this set. For many applications, however, the set containing the energy and angular momentum is the natural one. In quantum mechanics such constants as the initial values of r and θ, or of $\dot{r}$ and $\dot{\theta}$, become meaningless, but we can still talk in terms of the system energy or of the system angular momentum. Indeed, the salient differences between classical and quantum mechanics appear in the properties of E and l in the two theories. In order to discuss the transition to quantum theories it is important therefore that the classical description of the system be in terms of its energy and angular momentum.

3–3 THE EQUIVALENT ONE-DIMENSIONAL PROBLEM, AND CLASSIFICATION OF ORBITS

While the problem has thus been solved formally, practically speaking the integrals (3–18) and (3–20) are usually quite unmanageable, and in any specific case it is often more convenient to perform the integration in some other fashion. But before obtaining the solution for specific force laws, let us see what can be learned about the motion in the general case, using only the equations of motion and the conservation theorems, without requiring explicit solutions.

For example, with a system of known energy and angular momentum, the magnitude and direction of the velocity of the particle can be immediately determined in terms of the distance r. The magnitude v follows at once from the conservation of energy in the form

$$E = \frac{1}{2}mv^2 + V(r)$$

or

$$v = \sqrt{\frac{2}{m}(E - V(r))}. \tag{3–21}$$

The radial velocity—the component of $\dot{r}$ along the radius vector—has already been given in Eq. (3–16). Combined with the magnitude v this is sufficient information to furnish the direction of the velocity.* These results, and much more, can also be obtained from consideration of an equivalent one-dimensional problem.

The equation of motion in r, with $\dot{\theta}$ expressed in terms of l, Eq. (3–12), involves only r and its derivatives. It is the same equation as would be obtained for a fictitious one-dimensional problem in which a particle of mass m is subject to a force

$$m\ddot{r} = f(r) + \frac{l^2}{mr^3} \qquad\qquad f' = f + \frac{l^2}{mr^3}. \tag{3–22}$$

The significance of the additional term is clear if it is written as $mr\dot{\theta}^2 = mv_\theta^2/r$, which is the familiar centrifugal force. An equivalent statement can be obtained from the conservation theorem for energy. By Eq. (3–15) the motion of the particle in r is that of a one-dimensional problem with a fictitious potential energy:

$$V' = V + \frac{1}{2}\frac{l^2}{mr^2}. \tag{3–22'}$$

As a check we note that

$$f' = -\frac{\partial V'}{\partial r} = f(r) + \frac{l^2}{mr^3},$$

---

* Alternatively, the conservation of angular momentum furnishes $\dot{\theta}$, the angular velocity, and this together with $\dot{r}$ gives both the magnitude and direction of $\dot{r}$.

which agrees with Eq. (3–22). The energy conservation theorem (3–15) can thus also be written as

$$E = V' + \frac{1}{2}m\dot{r}^2. \tag{3–15'}$$

As an illustration of this method of examining the motion, consider a plot of V' against r for the specific case of an attractive inverse square law of force:

$$f = -\frac{k}{r^2}.$$

((For positive k the minus sign ensures that the force is *toward* the center of force.) The potential energy for this force is

$$V = -\frac{k}{r}$$

and the corresponding fictitious potential is

$$V' = -\frac{k}{r} + \frac{l^2}{2mr^2}.$$

Such a plot is shown in Fig. 3–3; the two dotted lines represent the separate components

$$-\frac{k}{r} \quad \text{and} \quad \frac{l^2}{2mr^2},$$

and the solid line is the sum V'.

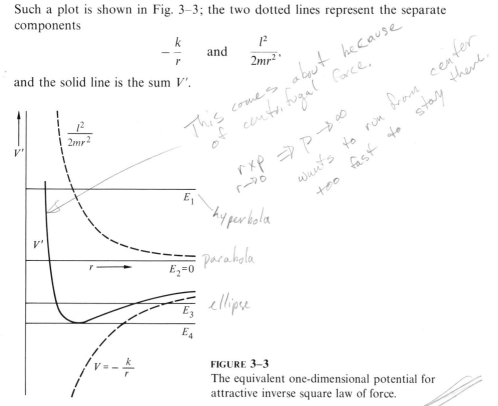

This comes about because of centrifugal force.

$r \times p \Rightarrow p \rightarrow \infty$
$r \rightarrow 0$
wants to run from center
too fast to stay there.

hyperbola

parabola

ellipse

FIGURE 3–3
The equivalent one-dimensional potential for attractive inverse square law of force.

No barrier
pene tration(s)
trunneling

Let us consider now the motion of a particle having the energy E_1, as shown in Figs. 3–3 and 3–4. Clearly this particle can never come closer than r_1 (cf. Fig. 3–4). Otherwise with $r < r_1$, V' exceeds E_1 and by Eq. (3–15') the kinetic energy would have to be negative, corresponding to an imaginary velocity! On the other hand, there is no upper limit to the possible value of r, so that the orbit is not bounded. A particle will come in from infinity, strike the "repulsive centrifugal barrier," be repelled, and travel back out to infinity (cf. Fig. 3–5). The distance between E and V' is $\frac{1}{2}m\dot{r}^2$, that is, proportional to the square of the radial velocity, and becomes zero, naturally, at the *turning point* r_1. At the same time the distance between E and V on the plot is the kinetic energy $\frac{1}{2}mv^2$ at the given value of r. Hence the distance between the V and V' curves is $\frac{1}{2}mr^2\dot{\theta}^2$. These curves therefore supply the magnitude of the particle velocity and its components for any distance r, at the given energy and angular momentum. This information is sufficient to provide an approximate picture of the form of the orbit.

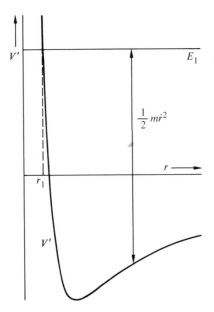

FIGURE 3–4
Unbounded motion at positive energies for inverse square law of force.

For the energy $E_2 = 0$ (cf. Fig. 3–3) a roughly similar picture of the orbit behavior is obtained. But for any lower energy, such as E_3 indicated in Fig. 3–6, we have a different story. In addition to a lower bound r_1, there is also a maximum value r_2 that cannot be exceeded by r with positive kinetic energy. The motion is then "bounded," and there are two turning points, r_1 and r_2, also known as *apsidal distances*. This does not necessarily mean that the orbits are closed. All that can be said is that they are bounded, contained between two circles of radius r_1 and r_2 with turning points always lying on the circles (cf. Fig. 3–7).

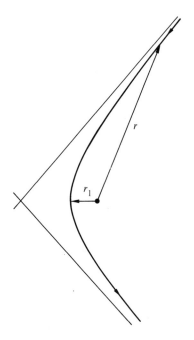

FIGURE 3–5
Schematic picture of the orbit for E_1
corresponding to unbounded motion.

If the energy is E_4 just at the minimum of the fictitious potential as shown in Fig. 3–8, then the two bounds coincide. In such case motion is possible at only one radius; $\dot{r} = 0$, and the orbit is a circle. Remembering that the effective "force" is the negative of the slope of the V' curve, the requirement for circular orbits is simply that f' be zero, or

$$f(r) = -\frac{l^2}{mr^3} = -mr\dot{\theta}^2.$$

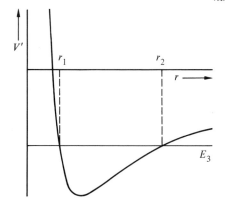

FIGURE 3–6
The equivalent one-dimensional potential
for inverse square law of force, illustrating
bounded motion at negative energies.

We have here the familiar elementary condition for a circular orbit, that the applied force be equal and opposite to the "reversed effective force" of centripetal

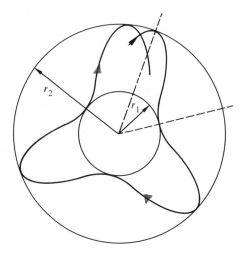

FIGURE 3–7
Schematic illustration of the nature of
the orbits for bounded motion.

acceleration.* The properties of circular orbits and the conditions for them will be studied in greater detail below in Section 3–6.

It is to be emphasized that all of this discussion of the orbits for various energies has been at one value of the angular momentum. Changing l will change the quantitative details of the V' curve but it will not affect the general classification of the types of orbits.

For the attractive inverse square law of force discussed above, we shall see that the orbit for E_1 is a hyperbola, for E_2 a parabola, and for E_3 an ellipse. With other forces the orbits may not have such simple forms. However, the same general qualitative division into open, bounded, and circular orbits will be true for any

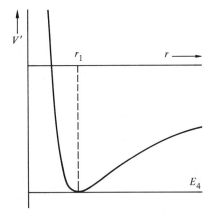

FIGURE 3–8
The equivalent one-dimensional potential
of inverse square law of force, illustrating
the condition for circular orbits.

---

* The case $E < E_4$ does not correspond to physically possible motion, for then $\dot{r}^2$ would have to be negative, or $\dot{r}$ imaginary.

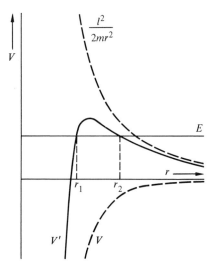

FIGURE 3-9
The equivalent one-dimensional potential for an attractive inverse fourth law of force.

attractive potential that (1) falls off slower than $1/r^2$ as $r \to \infty$; (2) becomes infinite slower than $1/r^2$ as $r \to 0$. The first condition ensures that the potential predominates over the centrifugal term for large r, while the second condition is such that for small r it is the centrifugal term that is the important one.

The qualitative nature of the motion will be altered if the potential does not satisfy these requirements, but we may still use the method of the equivalent potential to examine features of the orbits. As an example, consider the attractive potential

$$V(r) = -\frac{a}{r^3}, \quad \text{with} \quad f = -\frac{3a}{r^4}.$$

The energy diagram then is as shown in Fig. 3-9. For an energy E there are two possible types of motion, depending upon the initial value of r. If r_0 is less than r_1 the motion will be bounded, r will always remain less than r_1, and the particle will eventually pass through the center of force. If r is initially greater than r_2, then it will always remain so; the motion is unbounded, and the particle can never get inside the "potential" hole. The initial condition $r_1 < r_0 < r_2$ is again not physically possible.

Another interesting example of the method occurs for a linear restoring force (isotropic harmonic oscillator):

$$f = -kr, \quad V = \frac{1}{2}kr^2.$$

For zero angular momentum, corresponding to motion along a straight line, $V' = V$ and the situation is as shown in Fig. 3-10. For any positive energy the motion is bounded and, as we know, simple harmonic. If $l \neq 0$ we have the state of affairs shown in Fig. 3-11. The motion then is always bounded for all physically

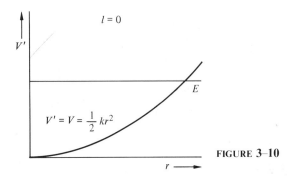

FIGURE 3–10

possible energies and does not pass through the center of force. In this particular case it is easily seen that the orbit is elliptic, for if $\mathbf{f} = -k\mathbf{r}$, the x and y components of the force are

$$f_x = -kx, \qquad f_y = -ky.$$

The total motion is thus the resultant of two simple harmonic oscillations at right angles, and of the same frequency, which in general leads to an elliptic orbit. A well-known example is the spherical pendulum for small amplitudes. The familiar Lissajous figures are obtained as the composition of two sinusoidal oscillations at right angles where the ratio of the frequencies is a rational number.* Central force motion under a linear restoring force therefore constitutes the simplest of the Lissajous figures.

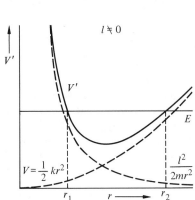

FIGURE 3–11

3–4 THE VIRIAL THEOREM

Another property of central force motion can be derived as a special case of a general theorem valid for a large variety of systems—the *virial theorem*. It differs

---

* See, for example, K. R. Symon, *Mechanics*, 3rd ed., (Reading, Massachusetts: Addison-Wesley, 1971), Section 3–10.

in character from the theorems previously discussed in being *statistical* in nature, i.e., it is concerned with the time averages of various mechanical quantities.

Consider a general system of mass points with position vectors $\mathbf{r}_i$ and applied forces $\mathbf{F}_i$ (including any forces of constraint). The fundamental equations of motion are then

$$\dot{\mathbf{p}}_i = \mathbf{F}_i. \tag{1-1}$$

We shall be interested in the quantity

[handwritten: $\dfrac{G}{|r|}$ = radial part of momentum]

$$G = \sum_i \mathbf{p}_i \cdot \mathbf{r}_i,$$

[handwritten: Not L. $L = \sum_i \mathbf{r}_i \times \mathbf{p}_i$]

where the summation is over all particles in the system. The total time derivative of this quantity is

$$\frac{dG}{dt} = \sum_i \dot{\mathbf{r}}_i \cdot \mathbf{p}_i + \sum_i \dot{\mathbf{p}}_i \cdot \mathbf{r}_i. \tag{3-23}$$

[handwritten: 2T]

The first term can be transformed to

$$\sum_i \dot{\mathbf{r}}_i \cdot \mathbf{p}_i = \sum_i m_i \dot{\mathbf{r}}_i \cdot \dot{\mathbf{r}}_i = \sum_i m_i v_i^2 = 2T,$$

while the second by Eq. (1–1) is

$$\sum_i \dot{\mathbf{p}}_i \cdot \mathbf{r}_i = \sum_i \mathbf{F}_i \cdot \mathbf{r}_i.$$

[handwritten: or $\dfrac{\dot{\mathbf{p}}_i \cdot \mathbf{r}}{|r|} = \dfrac{\mathbf{F}_i \cdot \mathbf{r}}{|r|}$ true]

Equations (3–23) therefore reduces to

$$\frac{d}{dt} \sum_i \mathbf{p}_i \cdot \mathbf{r}_i = 2T + \sum_i \mathbf{F}_i \cdot \mathbf{r}_i. \tag{3-24}$$

The time average of Eq. (3–24) over a time interval τ is obtained by integrating both sides with respect to t from 0 to τ, and dividing by τ:

$$\frac{1}{\tau} \int_0^\tau \frac{dG}{dt}\, dt \equiv \overline{\frac{dG}{dt}} = \overline{2T} + \overline{\sum_i \mathbf{F}_i \cdot \mathbf{r}_i}$$

or

[handwritten: for periodic]

$$\boxed{\overline{2T} + \overline{\sum_i \mathbf{F}_i \cdot \mathbf{r}_i} = \frac{1}{\tau}[G(\tau) - G(0)].} \quad \text{[handwritten: = 0]} \tag{3-25}$$

If the motion is periodic, i.e., all coordinates repeat after a certain time, and if τ is chosen to be the period, then the right-hand side of (3–25) vanishes. A similar conclusion can be reached even if the motion is not periodic, provided that the coordinates and velocities for all particles remain finite so that there is an upper bound to G. By choosing τ sufficiently long, the right-hand side of Eq. (3–25) can be made as small as desired. In both cases it then follows that

[handwritten: If $v = v(r)$]

$$\boxed{\overline{T} = -\frac{1}{2}\overline{\sum_i \mathbf{F}_i \cdot \mathbf{r}_i}.} \tag{3-26}$$

[handwritten: $T = \frac{1}{2} r \left| \vec{\nabla} v \right|$]

[handwritten: Virial thm]

Equation (3–26) is known as the *virial theorem,* and the right-hand side is called the *virial of Clausius.* In this form the theorem is very useful in the kinetic theory of gases. Thus the virial theorem can be used to derive Boyle's Law for perfect gases by means of the following brief argument.

Consider a gas consisting of N atoms confined within a container of volume V. The gas is further assumed to be at a temperature T (not to be confused with the symbol for kinetic energy). Then by the equipartition theorem of kinetic theory, the average kinetic energy of each atom is given by $\frac{3}{2}kT$, k being the Boltzmann constant, a relation that in effect is the definition of temperature. The left-hand side of Eq. (3–26) is therefore

$$\frac{3}{2}NkT.$$

On the right-hand side of Eq. (3–26), the forces $\mathbf{F}_i$ include both the forces of interaction between atoms and the forces of constraint on the system. A perfect gas is defined as one for which the forces of interaction contribute negligibly to the virial. This occurs, e.g., if the gas is so tenuous that collisions between atoms occur rarely, compared to collisions with the walls of the container. It is these walls that constitute the constraint on the system, and the forces of constraint, $\mathbf{F}_c$, are localized at the wall and come into existence whenever a gas atom collides with the wall. The sum on the right-hand side of Eq. (3–26) can therefore be replaced in the average by an integral over the surface of the container. The force of constraint represents the reaction of the wall to the collision forces exerted by the atoms on the wall, i.e., to the pressure P. With the usual outward convention for the unit vector $\mathbf{n}$ in the direction of the normal to the surface, we can write therefore

$$d\mathbf{F}_i = -P\mathbf{n}\,dA,$$

or

$$\frac{1}{2}\sum_i \mathbf{F}_i \cdot \mathbf{r}_i = -\frac{P}{2}\int \mathbf{n}\cdot\mathbf{r}\,dA.$$

But, by Gauss' theorem,

$$\int \mathbf{n}\cdot\mathbf{r}\,dA = \int \boldsymbol{\nabla}\cdot\mathbf{r}\,dV = 3V.$$

The virial theorem, Eq. (3–26), for the system representing a perfect gas therefore can be written

$$\frac{3}{2}NkT = \frac{3}{2}PV,$$

which, cancelling the common factor of $\frac{3}{2}$ on both sides, is the familiar Boyle's Law. Where the interparticle forces contribute to the virial, the perfect gas law of course no longer holds. The virial theorem is then the principal tool, in classical kinetic theory, for calculating the equation-of-state corresponding to such imperfect gases.

One can further show that if the forces $\mathbf{F}_i$ are the sum of nonfrictional forces $\mathbf{F}'_i$ and frictional forces $\mathbf{f}_i$ proportional to the velocity, then the virial depends only on the $\mathbf{F}'_i$; there is no contribution from the $\mathbf{f}_i$. Of course, the motion of the system must not be allowed to die down as a result of the frictional forces. Energy must constantly be pumped into the system to maintain the motion; otherwise *all* time averages would vanish as τ increases indefinitely. (See Exercise 4.)

If the forces are derivable from a potential, then the theorem becomes

$$\bar{T} = \frac{1}{2}\overline{\sum_i \nabla V \cdot \mathbf{r}_i},$$ (3–27)

and for a single particle moving under a central force it reduces to

$$\bar{T} = \frac{1}{2}\overline{\frac{\partial V}{\partial r}r}.$$ (3–28)

If V is a power-law function of r,

$$V = ar^{n+1},$$

where the exponent is chosen so that the force law goes as r^n, then

$$\frac{\partial V}{\partial r}r = (n+1)V,$$

and Eq. (3–28) becomes

$$\bar{T} = \frac{n+1}{2}\bar{V}.$$ (3–29)

By an application of Euler's theorem for homogeneous functions (cf. p. 61) it is clear that Eq. (3–29) holds also whenever V is a homogeneous function in r of degree $n+1$. For the further special case of inverse square law forces n is -2 and the virial theorem takes on a well-known form:

$$\boxed{\bar{T} = -\frac{1}{2}\bar{V}.}$$ *Inverse square laws*   (3–30)

3–5 THE DIFFERENTIAL EQUATION FOR THE ORBIT, AND INTEGRABLE POWER-LAW POTENTIALS

In treating specific details of actual central force problems a change in the orientation of our discussion is desirable. Hitherto solving a problem has meant finding r and θ as functions of time with E, l, and so on, as constants of integration. But most often what we really seek is the equation of the orbit, i.e., the dependence of r upon θ, eliminating the parameter t. For central force problems the elimination is particularly simple, since t occurs in the equations of motion only as a variable of differentiation. Indeed one equation of motion, (3–8), simply provides a definite relation between a differential change dt and the corresponding change $d\theta$:

$$l\,dt = mr^2\,d\theta. \tag{3-31}$$

The corresponding relation between derivatives with respect to t and θ is

$$\frac{d}{dt} = \frac{l}{mr^2}\frac{d}{d\theta}. \tag{3-32}$$

These relations may be used to convert the equation of motion (3–12) into a different equation for the orbit. Alternatively they can be applied to the formal solution of the equations of motion, given in Eq. (3–17), to furnish the equation of the orbit directly. For the moment we shall follow the former of these possibilities.

From (3–32) a second derivative with respect to t can be written

$$\frac{d^2}{dt^2} = \frac{l}{mr^2}\frac{d}{d\theta}\left(\frac{l}{mr^2}\frac{d}{d\theta}\right),$$

and the Lagrange equation for r, (3–12), becomes

$$\frac{l}{r^2}\frac{d}{d\theta}\left(\frac{l}{mr^2}\frac{dr}{d\theta}\right) - \frac{l^2}{mr^3} = f(r). \tag{3-33}$$

Now, to simplify (3–33) we notice that

$$\frac{1}{r^2}\frac{dr}{d\theta} = -\frac{d(1/r)}{d\theta};$$

hence if the variable is changed to $u = 1/r$, we have

$$\frac{l^2 u^2}{m}\left(\frac{d^2 u}{d\theta^2} + u\right) = -f\left(\frac{1}{u}\right). \tag{3-34a}$$

Since

$$\frac{d}{du} = \frac{dr}{du}\frac{d}{dr} = -\frac{1}{u^2}\frac{d}{dr},$$

Eq. (3–34a) can be written alternatively in terms of the potential as

$$\frac{d^2 u}{d\theta^2} + u = -\frac{m}{l^2}\frac{d}{du}V\left(\frac{1}{u}\right). \tag{3-34b}$$

Either form of Eq. (3–34) is thus a differential equation for the orbit if the force law f, or the potential V, is known. Conversely if the equation of the orbit is known, that is, r is given as a function of θ, then one can work back and obtain the force law $f(r)$.

Here, however, we want to obtain some rather general results. For example, it can be shown from (3–34) that the orbit is symmetrical about the turning points. To prove this statement it will be noted that if the orbit is symmetrical it should be possible to reflect it about the direction of the turning angle without producing any change. If the coordinates are so chosen that the turning point occurs for $\theta = 0$, then the reflection can be effected mathematically by the substitution of $-\theta$ for θ. The differential equation for the orbit, (3–34), is obviously invariant under such a substitution. Further the initial conditions, here

$$u = u(0), \qquad \left(\frac{du}{d\theta}\right)_0 = 0, \qquad \text{for } \theta = 0,$$

will likewise be unaffected. Hence the orbit equation must be the same whether expressed in terms of θ or $-\theta$, which is the desired conclusion. *The orbit is therefore invariant under reflection about the apsidal vectors.* In effect this means that the complete orbit can be traced if the portion of the orbit between any two turning points is known. Reflection of the given portion about one of the apsidal vectors produces a neighboring stretch of the orbit, and this process can be repeated indefinitely until the rest of the orbit is completed, as illustrated in Fig. 3–12.

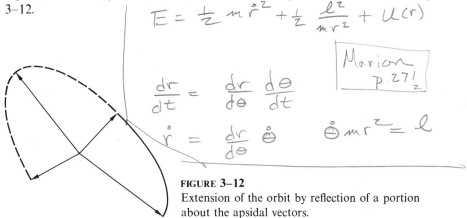

$$E = \tfrac{1}{2} m \dot{r}^2 + \tfrac{1}{2} \frac{l^2}{m r^2} + U(r)$$

$$\frac{dr}{dt} = \frac{dr}{d\theta}\frac{d\theta}{dt}$$

$$\dot{r} = \frac{dr}{d\theta}\dot{\theta}$$

Marion
p. 271
2

$$\dot{\theta} m r^2 = l$$

FIGURE 3–12
Extension of the orbit by reflection of a portion about the apsidal vectors.

For any particular force law the actual equation of the orbit must be obtained by integrating the differential equation Eq. (3–34), in either of its forms. However it is not necessary to go through all the details of the integration, as most of the work has already been done in solving the equation of motion (3–12). All that remains is to eliminate t from the solution (3–17) by means of (3–31), resulting in

$$d\theta = \frac{l\,dr}{mr^2 \sqrt{\dfrac{2}{m}\left(E - V(r) - \dfrac{l^2}{2mr^2}\right)}}. \tag{3–35}$$

With slight rearrangements the integral of (3–35) is

$$\theta = \int_{r_0}^{r} \frac{dr}{r^2 \sqrt{\dfrac{2mE}{l^2} - \dfrac{2mV}{l^2} - \dfrac{1}{r^2}}} + \theta_0 \tag{3–36}$$

or, if the variable of integration is changed to $u = 1/r$,

$$\theta = \theta_0 - \int_{u_0}^{u} \frac{du}{\sqrt{\dfrac{2mE}{l^2} - \dfrac{2mV}{l^2} - u^2}}. \tag{3-37}$$

As in the case of the equation of motion, Eq. (3–37), while solving the problem formally, is not always a practicable solution, because the integral often cannot be expressed in terms of well-known functions. In fact, only certain types of force laws have been investigated. The most important are the power-law functions of r,

$$V = ar^{n+1} \tag{3-38}$$

so that the force varies as the nth power of r.* With this potential (3–37) becomes

$$\theta = \theta_0 - \int_{u_0}^{u} \frac{du}{\sqrt{\dfrac{2mE}{l^2} - \dfrac{2ma}{l^2} u^{-n-1} - u^2}}. \tag{3-39}$$

This again will be integrable in terms of simple functions only in certain cases. If the quantity in the radical is of no higher power in u than quadratic, the denominator has the form $\sqrt{\alpha u^2 + \beta u + \gamma}$ and the integration can be directly effected in terms of circular functions. This restriction is equivalent to requiring that

$$-n - 1 = 0, 1, 2,$$

or, excluding the $n = -1$ case, for

$$n = -2, -3,$$

corresponding to inverse square or inverse cube force laws. One further easily integrable case is for $n = 1$, i.e., the linear force; for then Eq. (3–39) can be written as

$$\theta = \theta_0 - \int_{u_0}^{u} \frac{du}{\sqrt{\dfrac{2mE}{l^2} - \dfrac{2ma}{l^2} \dfrac{1}{u^2} - u^2}}. \tag{3-39'}$$

If now we make the substitution

$$u^2 = x, \qquad du = \frac{dx}{2\sqrt{x}},$$

---

* The case $n = -1$ is to be excluded from the following discussion. In the potential (3–38) it corresponds to a constant potential, i.e., no force at all. It is an equally anomalous case if the exponent is used in the force law directly, since a force varying as r^{-1} corresponds to a logarithmic potential, which is not a power law at all. A logarithmic potential is unusual for motion about a point; it is more characteristic of a *line* source.

Eq. (3–39′) becomes

$$\theta = \theta_0 - \frac{1}{2} \int_{x_0}^{x} \frac{dx}{\sqrt{\frac{2mE}{l^2}x - \frac{2ma}{l^2} - x^2}},\tag{3–40}$$

which again is in the desired form. Thus, a solution in terms of simple functions is obtained for the exponents.

$$n = 1, -2, -3.$$

This does not mean other powers are not integrable, merely that they lead to functions not as well known. For example, there is a range of exponents for which Eq. (3–39) involves *elliptic integrals*, with the solution expressed in terms of *elliptic functions*.

By definition an elliptic integral is

$$\int R(x, \omega)\, dx,$$

where R is any rational function of x and ω, and ω is defined as

$$\omega = \sqrt{\alpha x^4 + \beta x^3 + \gamma x^2 + \delta x + \eta}.$$

Of course α and β cannot simultaneously be zero, for then the integral could be evaluated in terms of circular functions. It can be shown (Whittaker and Watson, *Modern Analysis*, 4th ed., p. 512) that any such integral can be reduced to forms involving circular functions and the Legendre elliptic integrals of the first, second, and third kind. There exist complete and detailed tables of these standard elliptic integrals, and their properties and connections with elliptic functions have been discussed exhaustively in the literature. Intrinsically they do not require any higher logical concept for their use than do circular functions; they are just not as familiar. From the definition it is seen that the integral in (3–39) can be evaluated in terms of elliptic functions if

$$n = -4, -5.$$

We can attempt to put the integral in another form also leading to elliptic integrals by multiplying numerator and denominator by u^ρ where ρ is some undetermined exponent. The integral then becomes

$$\int \frac{u^\rho\, du}{\sqrt{\frac{2mE}{l^2}u^{2\rho} - \frac{2ma}{l^2}u^{-n-1+2\rho} - u^{2(\rho+1)}}},$$

where the expression in the radical will be a polynomial of higher order than a quartic except if $\rho = 1$. The integral will therefore be no worse than elliptic only if

$$-n - 1 + 2 = 0, 1, 2, 3, 4$$

or

$$n = +1, 0, -1, -2, -3.$$

For $n = +1, -2, -3$ the solutions reduce to circular functions, the case $n = -1$ has already been eliminated, so that this procedure leads to elliptic functions only for $n = 0$.

We again can obtain integrals of the elliptic type in certain cases by changing the variable to $u^2 = x$. The integral in question then appears as

$$\frac{1}{2} \int \frac{dx}{\sqrt{\frac{2mE}{l^2}x - \frac{2ma}{l^2}x^{(1-n)/2} - x^2}},$$

which reduces to the elliptic for for

$$\frac{1-n}{2} = 3, 4$$

leading to the exponents

$$n = -5, -7.$$

Finally we again can perform the trick of multiplying numerator and denominator by x, and the condition for obtaining elliptic integrals or simpler is then

$$\frac{1-n}{2} + 2 = 0, 1, 2, 3, 4$$

or

$$n = +5, +3, +1, -1, -3,$$

which leads to new possibilities only for $n = +5, +3$. The total number of integral exponents resulting in elliptic functions is thus

$$n = +5, +3, 0, -4, -5, -7.$$

Although this exhausts the possibilities for integral exponents, with suitable transformations some fractional exponents can also be shown to lead to elliptic integrals.

3-6 CONDITIONS FOR CLOSED ORBITS (BERTRAND'S THEOREM)

We have not yet extracted all the information that can be obtained from the equivalent one-dimensional problem or from the orbit equation without explicitly solving for the motion. In particular, it is possible to derive a powerful and thought-provoking theorem on the types of attractive central forces that lead to *closed orbits*, i.e., orbits in which the particle eventually retraces its own footsteps.

Conditions have already been described for one kind of closed orbit, namely a circle about the center of force. For any given l, this will occur if the equivalent potential $V'(r)$ has a minimum or maximum at some distance r_0 and if the energy E is just equal to $V'(r_0)$. The requirement that V' have an extremum is equivalent to the vanishing of f' at r_0, leading to the condition derived previously (cf. p. 79),

$$f(r_0) = -\frac{l^2}{mr_0^3}, \qquad (3\text{--}41)$$

which says the force must be attractive for circular orbits to be possible. In addition, the energy of the particle must be given by

$$E = V(r_0) + \frac{l^2}{2mr_0^2}, \qquad (3\text{--}42)$$

which, by Eq. (3–15), corresponds to the requirement that for a circular orbit $\dot{r}$ is zero. Equations (3–41) and (3–42) are both elementary and familiar. Between them they imply that for any attractive central force it is possible to have a circular orbit at some arbitrary radius r_0, provided the angular momentum l is given by Eq. (3–41) and the particle energy by Eq. (3–42).

The character of the circular orbit depends on whether the extremum of V' is a minimum, as in Fig. 3–8, or a maximum, as in Fig. 3–9. If the energy is slightly above that required for a circular orbit at the given value of l, then for a minimum in V' the motion, though no longer circular, will still be bounded. However if V' exhibits a maximum, then the slightest raising of E above the circular value, Eq. (3–34), results in motion that is unbounded, with the particle moving both through the center of force and out to infinity for the potential shown in Fig. 3–9. Borrowing the terminology from the case of static equilibrium the circular orbit arising in Fig. 3–8 is said to be *stable*; that in Fig. 3–9 is *unstable*. The stability of the circular orbit is thus determined by the sign of the second derivative of V' at the radius of the circle, being stable for positive second derivative (V' concave up) and unstable for V' concave down. A stable orbit therefore occurs if

$$\left.\frac{\partial^2 V'}{\partial r^2}\right|_{r=r_0} = -\left.\frac{\partial f}{\partial r}\right|_{r=r_0} + \frac{3l^2}{mr_0^4} > 0.$$

Using Eq. (3–41) this condition can be written

$$\left.\frac{\partial f}{\partial r}\right|_{r=r_0} < -\frac{3f(r_0)}{r_0}, \qquad (3\text{--}43)$$

or

$$\left.\frac{d \ln f}{d \ln r}\right|_{r=r_0} < -3. \qquad (3\text{--}43')$$

If the force behaves like a power law of r in the vicinity of the circular radius r_0,

$$f = -\frac{k}{r^{n+1}},$$

then the stability condition, Eq. (3–43), becomes

$$-\frac{(n+1)k}{r_0^{n+2}} < -\frac{3k}{r_0^{n+2}}$$

or

$$n < 2. \tag{3–44}$$

A power-law attractive potential varying more slowly than $1/r^2$ is thus capable of stable circular orbits for all values of r_0.

If the circular orbit is stable, then a small increase in the particle energy above the value for a circular orbit results in only a slight variation of r about r_0. It can be shown (cf. Appendix A) that for such small deviations from the circularity conditions, the particle executes a simple harmonic motion in $u(\equiv 1/r)$ about u_0:

$$u = u_0 + a \cos \beta\theta. \tag{3–45}$$

Here a is an amplitude that depends on the deviation of the energy from the value for circular orbits, and β is a quantity arising from a Taylor series expansion of the force law $f(r)$ about the circular orbit radius r_0. It is shown in Appendix A that β is given by

$$\beta^2 = 3 + \frac{r}{f}\frac{df}{dr}\bigg|_{r=r_0}. \tag{3–46}$$

As the radius vector of the particle sweeps completely around the plane, u goes through β cycles of its oscillation (cf. Fig. 3–13). If β is a rational number, the ratio of two integers, p/q, then after q revolutions of the radius vector the orbit would begin to retrace itself, i.e., the orbit would be *closed*.

At each r_0 such that the inequality in Eq. (3–43) is satisfied, it is possible to establish a stable circular orbit by giving the particle an initial energy and angular momentum prescribed by Eqs. (3–41) and (3–42). The question naturally arises as to what form must the force law take in order that the slightly perturbed orbit

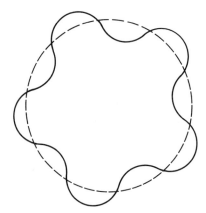

FIGURE **3–13**
Orbit for motion in a central force deviating slightly from a circular orbit.

about any of these circular orbits should be closed. It is clear that under these conditions β must not only be a rational number, but it must be the *same* rational number at all distances that a circular orbit is possible. Otherwise, since β can take on only discrete values, the number of oscillatory periods would change discontinuously with r_0, and indeed the orbits could not be closed at the discontinuity. With β^2 everywhere constant, the defining equation for β^2, Eq. (3–46), becomes in effect a differential equation for the force law f in terms of the independent variable r_0. We can indeed consider Eq. (3–46) to be written in terms of r if we keep in mind that the equation is valid only over the ranges in r for which stable circular orbits are possible. A slight rearrangement of Eq. (3–46) leads to the equation

$$\frac{d \ln f}{d \ln r} = \beta^2 - 3, \tag{3–47}$$

which can be immediately integrated to give a force law:

$$f(r) = -\frac{k}{r^{3 - \beta^2}}. \tag{3–48}$$

All force laws of this form, with β a rational number, lead to closed stable orbits for initial conditions that differ only *slightly* from conditions defining a circular orbit. Included within the possibilities allowed by Eq. (3–48) are some familiar forces such as the inverse square law ($\beta \equiv 1$), but of course many other behaviors, such as $f = -kr^{-2/9}$, ($\beta = \frac{5}{3}$) are also permitted.

Suppose the initial conditions deviate more than slightly from the requirements for circular orbits; will these same force laws still give circular orbits? The question can be answered directly by keeping an additional term in the Taylor series expansion of the force law and solving the resultant orbit equation. While the calculations involved are elementary they are somewhat lengthy. Details are given in Appendix A. What is found is that for more than first-order deviations from circularity, the orbits are closed only for $\beta^2 = 1$ and $\beta^2 = 4$. The first of these values of β^2, by Eq. (3–48), leads to the familiar attractive inverse square law; the second is an attractive force proportional to the radial distance—Hooke's law! These force laws, and only these, could possibly produce closed orbits for any arbitrary combination of l and $E(E < 0)$, and in fact we know from direct solution of the orbit equation that they do. Hence, *the only central forces that result in closed orbits for all bound particles are the inverse square law and Hooke's law.**

This is a remarkable result, well worth the tedious algebra required. It is a commonplace of astronomical observation that celestial objects that are bound move in orbits that are in first approximation closed. For the most part, the small

---

* This conclusion was apparently first derived by J. Bertrand, *Comptes Rendus* 77, 849–853 (1873), and is frequently referred to as Bertrand's theorem. See other pertinent literature referenced at the end of this chapter.

deviations from a closed orbit are traceable to perturbations such as the presence of other bodies. The prevalence of closed orbits holds true whether we consider only the solar system, or look out to the many examples of true binary stars that have been examined. Now, Hooke's law is a most unrealistic force law to hold at all distances, for it implies a force increasing indefinitely to infinity. Thus, the existence of closed orbits for a wide range of initial conditions by itself leads to the conclusion that the gravitational force varies as the inverse square of the distance. It is not necessary, for example, to use the elliptic character of the orbits to arrive at the gravitational force law.

We can phrase this conclusion in a slightly different manner, one that is of somewhat more significance in modern physics. The orbital motion in a plane can be looked on as compounded of two oscillatory motions, one in r and one in θ. That the orbit is closed is equivalent to saying that the periods of the two oscillations are commensurate—that they are *degenerate*. Hence, *the degenerate character of orbits in a gravitational field fixes the form of the force law.* Later on we shall encounter other formulations of the relation between degeneracy and the nature of the potential.

3–7 THE KEPLER PROBLEM: INVERSE SQUARE LAW OF FORCE

The inverse square law is the most important of all the central force laws and it deserves detailed treatment. For this case the force and potential can be written as

$$f = -\frac{k}{r^2}, \qquad V = -\frac{k}{r}. \tag{3–49}$$

There are several ways to integrate the equation for the orbit, the simplest being to substitute (3–49) in the differential equation for the orbit (3–34):

$$\frac{d^2u}{d\theta^2} + u = \frac{-mf\left(\dfrac{1}{u}\right)}{l^2u^2} = \frac{mk}{l^2}. \tag{3–50}$$

Changing the variable to $y = u - \dfrac{mk}{l^2}$, the differential equation becomes

$$\frac{d^2y}{d\theta^2} + y = 0,$$

which has the immediate solution

$$y = B\cos(\theta - \theta'),$$

B and θ' being the two constants of integration. In terms of r the solution is

$$\frac{1}{r} = \frac{mk}{l^2}(1 + e\cos(\theta - \theta')), \tag{3–51}$$

where

$$g = B\frac{l^2}{mk}.$$ (3-52)

It is instructive to obtain the orbit equation also from the formal solution (3-39). While this procedure is longer than the simple integration of the differential equation (3-50), it has the advantage that the significant constant of integration e is automatically evaluated in terms of the energy E and the angular momentum l of the system. We write (3-39) in the form

$$\theta = \theta' - \int \frac{du}{\sqrt{\frac{2mE}{l^2} + \frac{2mku}{l^2} - u^2}},$$ (3-53)

where the integral is now taken as indefinite. The quantity θ' appearing in (3-53) is a constant of integration determined by the initial conditions and will not necessarily be the same as the initial angle θ_0 at time $t = 0$. The indefinite integral is of the standard form,*

$$\int \frac{dx}{\sqrt{\alpha + \beta x + \gamma x^2}} = \frac{1}{\sqrt{-\gamma}} \arccos - \frac{\beta + 2\gamma x}{\sqrt{q}},$$ (3-54)

where

$$q = \beta^2 - 4\alpha\gamma.$$

To apply this to (3-53) we must set

$$\alpha = \frac{2mE}{l^2}, \qquad \beta = \frac{2mk}{l^2}, \qquad \gamma = -1,$$

and the discriminant q is therefore

$$q = \left(\frac{2mk}{l^2}\right)^2 \left(1 + \frac{2El^2}{mk^2}\right).$$

With these substitutions, Eq. (3-53) becomes

$$\theta = \theta' - \arccos \frac{\frac{l^2 u}{mk} - 1}{\sqrt{1 + \frac{2El^2}{mk^2}}}.$$

---

* Cf., for example, B. O. Pierce, *A Short Table of Integrals*, 3d ed. no 161; 4th ed. no. 166. See also I.S. Gradshtein and I. W. Ryzhik, *Table of Integrals*, no. 2.261, or M. R. Spiegel, *Mathematical Handbook* no. 14.280. (For full description of these books, see the section on 'Works of Reference' in the Bibliography, p. 6 0). A constant, $-\pi/2$, is to be added to the result as given in all of these tables of integrals in order to obtain (3-45), a procedure that is permissible since the integral is indefinite.

Finally, by solving for u, $\equiv \dfrac{1}{r}$, the equation of the orbit is found to be

$$\frac{1}{r} = \frac{mk}{l^2}\left(1 + \sqrt{1 + \frac{2El^2}{mk^2}}\cos(\theta - \theta')\right),$$ (3–55)

which agrees with (3–51), except that here e is evaluated in terms of E and l. The constant of integration θ' can now be identified from Eq. (3–55) as one of the turning angles of the orbit. It will be noted that only three of the four constants of integration appear in the orbit equation, and this is always a characteristic property of the orbit. In effect, the fourth constant locates the initial position of the particle on the orbit. If we are interested solely in the orbit equation this information is clearly irrelevant and hence does not appear in the answer. Of course, the missing constant has to be supplied if it is desired to complete the solution by finding r and θ as functions of time. Thus, if one chooses to integrate the conservation theorem for angular momentum,

$$mr^2\, d\theta = l\, dt,$$

by means of (3–55), one must specify in addition the initial angle θ_0.

Now, the general equation of a conic with one focus at the origin is

$$\frac{1}{r} = C(1 + e\cos(\theta - \theta')),$$ (3–56)

where e is the eccentricity of the conic section. By comparison with Eq. (3–55) it follows that the orbit is always a conic section, with the eccentricity

$$e = \sqrt{1 + \frac{2El^2}{mk^2}}.$$ (3–57)

The nature of the orbit depends on the magnitude of e according to the following scheme:

$$e > 1, \quad E > 0: \qquad \text{hyperbola,}$$

$$e = 1, \quad E = 0: \qquad \text{parabola,}$$

$$e < 1, \quad E < 0: \qquad \text{ellipse,}$$

$$e = 0, \quad E = -\frac{mk^2}{2l^2}: \ \text{circle.}$$

This classification agrees with the qualitative discussion of the orbits based on the energy diagram of the equivalent one-dimensional potential V'. The condition for circular motion appears here in a somewhat different form, but it can easily be derived as a consequence of the previous conditions for circularity. For a circular orbit, T and V are constant in time, and from the virial theorem

$$E \equiv T + V = -\frac{V}{2} + V = \frac{V}{2}.$$

Hence

$$E = -\frac{k}{2r_0}. \tag{3-58}$$

But from Eq. (3–41), the statement of equilibrium between the central force and the "effective force," we can write

$$-\frac{k}{r_0^2} = -\frac{l^2}{mr_0^3},$$

or

$$r_0 = \frac{l^2}{mk}. \tag{3-59}$$

With this formula for the orbital radius, Eq. (3–58) becomes

$$E = -\frac{mk^2}{2l^2},$$

the above condition for circular motion.

In the case of elliptic orbits it can be shown the major axis depends solely on the energy, a theorem of considerable importance in the Bohr theory of the atom. The semimajor axis is one half the sum of the two apsidal distances r_1 and r_2 (cf. Fig. 3–6). By definition the radial velocity is zero at these points, and the conservation of energy implies that the apsidal distances are therefore the roots of the equation

$$E - \frac{l^2}{2mr^2} + \frac{k}{r} = 0,$$

or

$$r^2 + \frac{k}{E}r - \frac{l^2}{2mE} = 0. \tag{3-60}$$

Now, the coefficient of the linear term in a quadratic equation is the negative of the sum of the roots. Hence the semimajor axis is given by

$$a = \frac{r_1 + r_2}{2} = -\frac{k}{2E}. \tag{3-61}$$

Note that in the circular limit, Eq. (3–61) agrees with Eq. (3–58). In terms of the semimajor axis, the eccentricity of the ellipse can be written

$$e = \sqrt{1 - \frac{l^2}{mka}}, \tag{3-62}$$

(a relation we will have use for in a later chapter). Further, from Eq. (3–62) we have the expression

$$\frac{l^2}{mk} = a(1 - e^2), \tag{3-63}$$

in terms of which the elliptical orbit equation (3–51) can be written

$$r = \frac{a(1 - e^2)}{1 + e\cos(\theta - \theta')}. \tag{3-64}$$

From Eq. (3–64) it follows that the two apsidal distances (which occur when $\theta - \theta'$ is 0 and π, respectively) are equal to $a(1 - e)$ and $a(1 + e)$, as is to be expected from the properties of an ellipse.

3–8 THE MOTION IN TIME IN THE KEPLER PROBLEM

The orbital equation for motion in a central inverse-square force law can thus be solved in a fairly straightforward manner with results that can be stated in simple closed expressions. To describe the motion of the particle in time as it traverses the orbit is, however, a much more involved matter. In principle the relation between the radial distance of the particle r and the time (relative to some starting point) is given by Eq. (3–18), which here takes on the form

$$t = \sqrt{\frac{m}{2}} \int_{r_0}^{r} \frac{dr}{\sqrt{\dfrac{k}{r} - \dfrac{l^2}{2mr^2} + E}}. \tag{3-65}$$

Similarly, the polar angle θ and the time are connected through the conservation of angular momentum,

$$dt = \frac{mr^2}{l} d\theta,$$

which combined with the orbit equation (3–51) leads to

$$t = \frac{l^3}{mk^2} \int_{\theta_0}^{\theta} \frac{d\theta}{[1 + e\cos(\theta - \theta')]^2}. \tag{3-66}$$

Either of these integrals can be carried out in terms of elementary functions. (For Eq. (3–66) see, for example, formula 14.391 in *Mathematical Handbook of Formulas and Tables*, ed. by M. R. Spiegel). But the relations are very complex, and their inversion to give r or θ as functions of t pose formidable problems, especially when one wants the high precision needed for astronomical observations.

    To illustrate some of these involvements let us consider the situation for parabolic motion ($e = 1$), where the integrations can be most simply carried out. It is customary to measure the plane polar angle from the radius vector at the point of closest approach—a point most usually designated as the *perihelion*.* This convention corresponds to setting θ' in the orbit equation (3–51) equal to

---

* Literally, the term should be restricted to orbits around the sun, while the more general term should be *periapsis*. However, it has become customary to use perihelion no matter where the center of force is. Even for space craft orbiting the moon, official descriptions of the orbital parameters refer to perihelion where pericynthion would be the pedantic term.

zero. Correspondingly, time is measured from the moment, T, of perihelion passage. Using the trigonometric identity

$$1 + \cos\theta = 2\cos^2\frac{\theta}{2},$$

Eq. (3–66) then reduces for parabolic motion to the form

$$t = \frac{l^3}{4mk^2}\int_0^\theta \sec^4\frac{\theta}{2}\,d\theta.$$

The integration is easily performed by a change of variable to $x = \tan(\theta/2)$, leading to the integral

$$t = \frac{l^3}{2mk^2}\int_0^{\tan(\theta/2)} (1 + x^2)\,dx,$$

or

$$t = \frac{l^3}{2mk^2}\left(\tan\frac{\theta}{2} + \frac{1}{3}\tan^3\frac{\theta}{2}\right). \tag{3–67}$$

This is a straightforward relation for t as a function of θ; inversion to obtain θ at a given time requires solving a cubic equation for $\tan(\theta/2)$, and then finding the corresponding arctan. The radial distance at the given time is then given through the orbital equation.

For elliptic motion, Eq. (3–65) is most conveniently integrated through an auxiliary variable ψ, denoted as the *eccentric anomaly*,* and defined by the relation

$$r = a(1 - e\cos\psi). \tag{3–68}$$

By comparison with the orbit equation, (3–64), it is clear that ψ also covers the interval 0 to 2π as θ goes through a complete revolution, and that the perihelion occurs at $\psi = 0$ (where $\theta = 0$ by convention) and the aphelion at $\psi = \pi = \theta$. A geometrical interpretation can be given to ψ, but it is of historical interest only (see, e.g., McCuskey, *Introduction to Celestial Mechanics*, p. 45). Expressing E and l in terms of a, e, and k, Eq. (3–65) can be rewritten for elliptic motion as

$$t = \sqrt{\frac{m}{2k}}\int_{r_0}^r \frac{r\,dr}{\sqrt{r - \dfrac{r^2}{2a} - \dfrac{a(1 - e^2)}{2}}}, \tag{3–69}$$

where, by the convention on the starting time, r_0 is the perihelion distance. Substitution of r in terms of ψ from Eq. (3–68) reduces this integral, after some

---

* The name connects with the terminology of medieval astronomy in which θ was called the *true anomaly*.

algebra, to the simple form

$$t = \sqrt{\frac{ma^3}{n}} \int_0^\psi (1 - e \cos \psi) \, d\psi. \tag{3-70}$$

First, we may note that Eq. (3–70) provides an expression for the period, τ, of elliptical motion, if the integral is carried over the full range in ψ of 2π:

$$\tau = 2\pi a^{3/2} \sqrt{\frac{m}{k}}. \tag{3-71}$$

This important result can also be obtained directly from the properties of an ellipse. From the conservation of angular momentum the areal velocity is constant and is given by

$$\frac{dA}{dt} = \frac{1}{2}r^2\theta = \frac{l}{2m}. \tag{3-72}$$

The area of the orbit, A, is to be found by integrating (3–72) over a complete period τ:

$$\int_0^\tau \frac{dA}{dt} \, dt = A = \frac{l\tau}{2m}.$$

Now, the area of an ellipse is

$$A = \pi ab,$$

where, by the definition of eccentricity, the semiminor axis b is related to a according to the formula

$$b = a\sqrt{1 - e^2}.$$

By (3–62) it is seen that the semiminor axis can also be written as

$$b = a^{1/2}\sqrt{\frac{l^2}{mk}},$$

and the period is therefore

$$\tau = \frac{2m}{l}\pi a^{3/2}\sqrt{\frac{l^2}{mk}} = 2\pi a^{3/2}\sqrt{\frac{m}{k}},$$

as was found previously. Equation (3–71) states that, other things being equal, the square of the period is proportional to the cube of the major axis, and this conclusion is often referred to as the third of Kepler's laws.* Actually, Kepler was

---

* Kepler's three laws of planetary motion, published around 1610, were the result of his pioneering analysis of planetary observations and laid the groundwork for Newton's great advances. The second law, the conservation of areal velocity, is a general theorem for central force motion, as has been noted previously. However, the first—that the planets move in elliptical orbits about the sun at one focus—and the third are restricted specifically to the inverse square law of force.

concerned with the specific problem of planetary motion in the gravitational field of the sun. A more precise statement of his law would therefore be: The square of the periods of the various planets are proportional to the cube of their major axes. In this form the law is only approximately true. It must be remembered that the motion of a planet about the sun is a two-body problem and m in (3–71) must be replaced by the reduced mass:

$$\mu = \frac{m_1 m_2}{m_1 + m_2},$$

where m_1 may be taken as referring to the planet and m_2 to the sun. Further, the gravitational law of attraction is

$$f = -G\frac{m_1 m_2}{r^2},$$

so that the constant k is

$$k = Gm_1 m_2. \tag{3–73}$$

Under these conditions (3–71) becomes

$$\tau = \frac{2\pi a^{3/2}}{\sqrt{G(m_1 + m_2)}} \approx \frac{2\pi a^{3/2}}{\sqrt{Gm_2}}, \tag{3–74}$$

if we neglect the mass of the planet compared to the sun. It is the approximate version of Eq. (3–74) that is Kepler's third law, for it states that τ is proportional to $a^{3/2}$, with the same constant of proportionality for all planets. However, the planetary mass m_1 is not always completely negligible compared to the sun's; for example, Jupiter has a mass about 0.1 % of the mass of the sun. On the other hand Kepler's third law is rigorously true for the electron orbits in the Bohr atom, since μ and k are then the same for all orbits in a given atom.

To return to the general problem of the position in time for an elliptic orbit, we may rewrite Eq. (3–70) slightly by introducing the frequency of revolution ω as

$$\omega = \frac{2\pi}{\tau} = \sqrt{\frac{k}{ma^3}}. \tag{3–75}$$

The integration in Eq. (3–70) is of course easily performed, resulting in the relation

$$\omega t = \psi - e \sin \psi, \tag{3–76}$$

known as *Kepler's equation*. The quantity ωt goes through the range 0 to 2π, along with ψ and θ, in the course of a complete orbital revolution and is therefore also denoted as an anomaly, specifically the *mean anomaly*.

To find the position in orbit at a given time t, Kepler's equation, (3–76), would first be inverted to obtain the corresponding eccentric anomaly ψ. Equation (3–68) then yields the radial distance, while the polar angle θ can be expressed in terms of ψ by comparing the defining equation (3–68) with the orbit equation (3–64):

$$1 + e\cos\theta = \frac{1 - e^2}{1 - e\cos\psi}.$$

With a little algebraic manipulation this can be simplified to

$$\cos\theta = \frac{\cos\psi - e}{1 - e\cos\psi}. \tag{3–77}$$

By successively adding and subtracting both sides of Eq. (3–77) from unity and taking the ratio of the resulting two equations, one is led to the alternate form

$$\tan\frac{\theta}{2} = \sqrt{\frac{1 + e}{1 - e}}\,\tan\frac{\psi}{2}. \tag{3–78}$$

Either Eq. (3–77) or (3–78) thus provides θ, once ψ is known. The solution of the transcendental Kepler's equation (3–76) to give the value of ψ corresponding to a given time is a problem that has attracted the attention of many famous mathematicians ever since Kepler posed the question early in the seventeenth century. Newton, for example, contributed what today would be called an analog solution. Indeed, it can be claimed that the practical need to solve Kepler's equation to accuracies of a second of arc over the whole range of eccentricity fathered many of the developments in numerical mathematics in the eighteenth and nineteenth centuries. A few of the more than 100 methods of solution developed in the pre-computer era are considered in the exercises to this chapter.

3–9 THE LAPLACE–RUNGE–LENZ VECTOR

The Kepler problem is also distinguished by the existence of an additional conserved vector besides the angular momentum. For a general central force, Newton's second law of motion can be written vectorially as

$$\dot{\mathbf{p}} = f(r)\frac{\mathbf{r}}{r}. \tag{3–79}$$

The cross product of $\dot{\mathbf{p}}$ with the constant angular momentum vector $\mathbf{L}$ therefore can be expanded as

$$\dot{\mathbf{p}} \times \mathbf{L} = \frac{mf(r)}{r}[\mathbf{r} \times (\mathbf{r} \times \dot{\mathbf{r}})]$$

$$= \frac{mf(r)}{r}[\mathbf{r}(\mathbf{r} \cdot \dot{\mathbf{r}}) - r^2\dot{\mathbf{r}}]. \tag{3–80}$$

Equation (3–80) can be further simplified by noting that

$$\mathbf{r} \cdot \dot{\mathbf{r}} = \frac{1}{2}\frac{d}{dt}(\mathbf{r} \cdot \mathbf{r}) = r\dot{r}$$

(or, in less forml terms, the component of the velocity in the radial direction is $\dot{r}$). As **L** is constant, Eq. (3–80) can then be rewritten, after a little manipulation, as

$$\frac{d}{dt}(\mathbf{p} \times \mathbf{L}) = -mf(r)r^2\left[\frac{\dot{\mathbf{r}}}{r} - \frac{\mathbf{r}\dot{r}}{r^2}\right],$$

or

$$\frac{d}{dt}(\mathbf{p} \times \mathbf{L}) = -mf(r)r^2\frac{d}{dt}\left(\frac{\mathbf{r}}{r}\right). \tag{3–81}$$

Without specifying the form of $f(r)$ we can go no further. But Eq. (3–81) can be immediately integrated if $f(r)$ is inversely proportional to r^2 —the Kepler problem. Writing then $f(r)$ in the form prescribed by Eq. (3–49), Eq. (3–81) becomes

$$\frac{d}{dt}(\mathbf{p} \times \mathbf{L}) = \frac{d}{dt}\left(\frac{mk\mathbf{r}}{r}\right),$$

which says that for the Kepler problem there exists a *conserved vector* **A** defined by

$$\mathbf{A} = \mathbf{p} \times \mathbf{L} - mk\frac{\mathbf{r}}{r}. \tag{3–82}$$

In recent times, the vector **A** has become known amongst physicists as the Runge–Lenz vector, but priority belongs to Laplace.*

From the definition of **A**, one can easily see that

$$\mathbf{A} \cdot \mathbf{L} = 0, \tag{3–83}$$

since **L** is perpendicular to $\mathbf{p} \times \mathbf{L}$ and **r** is perpendicular to $\mathbf{L} = \mathbf{r} \times \mathbf{p}$. It follows from this orthogonality of **A** to **L** that **A** must be some fixed vector in the plane of the orbit. If θ is used to denote the angle between **r** and the fixed direction of **A**, then the dot product of **r** and **A** is given by

$$\mathbf{A} \cdot \mathbf{r} = Ar \cos \theta = \mathbf{r} \cdot (\mathbf{p} \times \mathbf{L}) - mkr. \tag{3–84}$$

Now, by permutation of the terms in the triple dot product, we have

$$\mathbf{r} \cdot (\mathbf{p} \times \mathbf{L}) = \mathbf{L} \cdot (\mathbf{r} \times \mathbf{p}) = l^2,$$

---

* Laplace explicitly exhibited the components of **A** in the first part of his "*Traite de Mecanique Celeste*," which appeared in 1799. The designation as the Laplace vector, common in a number of treatises on celestial mechanics, is therefore probably the proper eponym. W. R. Hamilton apparently discovered **A** as a conserved quantity independently in 1845. The first derivation in vector language, substantially as given here, was that of J. W. Gibbs about 1900. C. Runge repeated the derivation in a popular German text on vector analysis (1919) and was quoted as a reference by W. Lenz in a 1924 paper on quantum mechanical treatment of the perturbed hydrogen atom. Since then the literature on the Laplace–Runge–Lenz vector and its uses has become enormous. For further historical details see H. Goldstein, *American Journal of Physics*, **43**, 735 (1975) and **44**, 1123 (1976).

$$[L_i, P_j(r)] = i\hbar\, \epsilon_{ijk}\, P_k(r_h)$$

so that Eq. (3–84) becomes

$$Ar \cos \theta = l^2 - mkr,$$

or

$$\frac{1}{r} = \frac{mk}{l^2}\left(1 + \frac{A}{mk}\cos \theta\right). \tag{3–85}$$

The Laplace–Runge–Lenz vector thus provides still another way of deriving the orbit equation for the Kepler problem! Comparison of Eq. (3–85) with the orbit equation in the form of Eq. (3–51) shows that **A** is in the direction of the radius vector to the perihelion point on the orbit, and has a magnitude

$$\mathbf{A} = mke. \tag{3–86}$$

For the Kepler problem we thus have identified two vector constants of the motion **L** and **A**, and a scalar E. Since a vector must have three independent components, this corresponds to seven conserved quantities in all. Now, a system such as this with three degrees of freedom has six independent constants of the motion, corresponding, say, to the three components each of the initial position and velocity of the particle. Further, the constants of the motion we have found are all algebraic functions of **r** and **p** that describe the orbit as a whole (orientation in space, eccentricity, etc.); none of these seven conserved quantities relate to where the particle is located in the orbit at the initial time. Since one of the constants of the motion must relate to this information, say in the form of T, the time of the perihelion passage, there can be only five independent constants of the motion describing the size, shape, and orientation of the orbit. It can therefore be concluded that not all of the quantities making up **L**, **A**, and E can be independent; there must in fact be two relations connecting these quantities. One such relation has already been obtained as the orthogonality of **A** and **L**, Eq. (3–83). The other follows from Eq. (3–86) when the eccentricity is expressed in terms of E and l from Eq. (3–57), leading to

$$A^2 = m^2k^2 + 2mEl^2, \tag{3–87}$$

thus confirming that there are only five *independent* constants out of the seven.*

The angular momentum vector and the energy alone contain only four independent constants of the motion: the Laplace–Runge–Lenz vector thus adds one more. It is natural to ask why there should not exist for any general central force law some conserved quantity that together with **L** and E serves to define the orbit in a manner similar to the Laplace–Runge–Lenz vector for the special case of the Kepler problem. The answer seems to be that such conserved quantities can in fact be constructed,† but that they are in general rather peculiar functions of the

---

* The arguments in the above paragraph were apparently first presented by Laplace in 1799. He also then explicitly demonstrated the relation between the magnitude of **A** and the eccentricity, Eq. (3–86).

† See, for example, D. M. Fradkin, *Progress of Theoretical Physics* **37**, 798, May 1967.

[handwritten annotations in top margin: "2a", "b semi major", "$\frac{1}{r_{min}} = (1+\varepsilon)A$", "$\frac{1}{r_{max}} = A(1-\varepsilon)$"]

motion. The constants of the motion relating to the orbit between them define the orbit, i.e., lead to the orbit equation giving r as a function of θ. We have seen that in general orbits for central force motion are not closed; the arguments of Section 3–6 showed that closed orbits implied rather stringent conditions on the form of the force law. It is a property of nonclosed orbits that the curve will eventually pass through any arbitrary (r, θ) point that lies between the bounds of the turning points of r. Intuitively this can be seen from the nonclosed nature of the orbit; as θ goes around a full cycle the particle must never retrace its footsteps on any previous orbit. Thus the orbit equation is such that r is a multivalued function of θ (modulo 2π), in fact it is an *infinite-valued function* of θ. The corresponding conserved quantity additional to L and E defining the orbit must similarly involve an infinite-valued function of the particle motion. Only where the orbits are closed, or more generally where the motion is *degenerate*, as in the Kepler problem, can we expect the additional conserved quantity to be a simple algebraic function of $\mathbf{r}$ and $\mathbf{p}$ such as the Laplace–Runge–Lenz vector. From these arguments we would expect a simple analog of such a vector to exist for the case of a Hooke's Law force, where, as we have seen, the orbits are also degenerate. This is indeed the case, except that the natural way to formulate the constant of the motion leads not to a vector but to a tensor of the second rank (see Section 9–7). Thus, the existence of an additional constant or integral of the motion, beyond $\mathbf{E}$ and $\mathbf{L}$, that is a simple algebraic function of the motion is sufficient to indicate that the motion is degenerate and the bounded orbits are closed.

3–10 SCATTERING IN A CENTRAL FORCE FIELD

Historically, the interest in central forces arose out of the astronomical problems of planetary motion. There is no reason, however, why central force motion must be thought of only in terms of such problems; mention has already been made of the orbits in the Bohr atom. Another field that can be investigated in terms of classical mechanics is the *scattering* of particles by central force fields. Of course, if the particles are on the atomic scale it must be expected that the specific results of a classical treatment will often be incorrect physically, for quantum effects are usually large in such regions. Nevertheless there are many classical predictions that remain valid to a good approximation. More important, the procedures for *describing* scattering phenomena are the same whether the mechanics is classical or quantum; one can learn to speak the language equally as well on the basis of classical physics.

In its one-body formulation the scattering problem is concerned with the scattering of particles by a *center of force*. We consider a uniform beam of particles—whether electrons, or α-particles, or planets is irrelevant—all of the same mass and energy incident upon a center of force. It will be assumed that the force falls off to zero for very large distances. The incident beam is characterized by specifying its *intensity* I (also called flux density), which gives the number of particles crossing unit area normal to the beam in unit time. As a particle

[handwritten annotation in right margin: "planets don't scatter"]

approaches the center of force it will be either attracted or repelled, and its orbit will deviate from the incident straight line trajectory. After passing the center of force, the force acting on the particle will eventually diminish so that the orbit once again approaches a straight line. In general the final direction of motion is not the same as the incident direction, and the particle is said to be scattered. The *cross section for scattering in a given direction*, $\sigma(\Omega)$, is defined by

$$\sigma(\Omega)\,d\Omega = \frac{\text{number of particles scattered into solid angle } d\Omega \text{ per unit time}}{\text{incident intensity}},$$

$$\tag{3-88}$$

where $d\Omega$ is an element of solid angle in the direction Ω. Often $\sigma(\Omega)$ is also designated as the *differential scattering cross section*. With central forces there must be complete symmetry around the axis of the incident beam, hence the element of solid angle can be written

$$d\Omega = 2\pi \sin \Theta\, d\Theta, \tag{3-89}$$

where Θ is the angle between the scattered and incident directions, known as the *scattering angle* (cf. Fig. 3–14, where repulsive scattering is illustrated). It will be noted that the name "cross section" is deserved in that $\sigma(\Omega)$ has the dimensions of an area.

For any given particle the constants of the orbit, and hence the amount of scattering, are determined by its energy and angular momentum. It is convenient to express the angular momentum in terms of the energy and a quantity known as the *impact parameter, s*, defined as the perpendicular distance between the center of force and the incident velocity. If v_0 is the incident speed of the particle, then

$$l = mv_0 s = s\sqrt{2mE}. \tag{3-90}$$

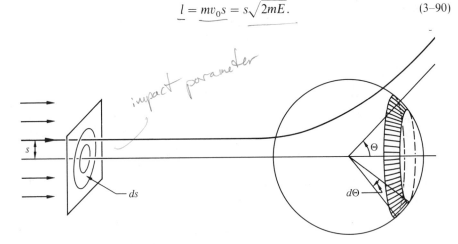

FIGURE 3–14
Scattering of an incident beam of particles by a center of force.

Once E and s are fixed, the angle of scattering Θ is then determined uniquely.*
For the moment it will be assumed that different values of s cannot lead to the
same scattering angle. Therefore, the number of particles scattered into a solid
angle $d\Omega$ lying between Θ and $\Theta + d\Theta$ must be equal to the number of the
incident particles with impact parameter lying between the corresponding s and
$s + ds$:

$$2\pi I s \, |ds| = 2\pi\sigma(\Theta) I \sin\Theta \, |d\Theta|. \tag{3-91}$$

Absolute signs are introduced in Eq. (3–91) because numbers of particles must of
course always be positive while s and Θ often vary in opposite directions. If s is
considered as a function of the energy and the corresponding scattering angle,

$$s = s(\Theta, E), \tag{3-92}$$

then the dependence of the differential cross section on Θ is given by

$$\sigma(\Theta) = \frac{s}{\sin\Theta}\left|\frac{ds}{d\Theta}\right|. \tag{3-93}$$

A formal expression for the scattering angle Θ as a function of s can be
directly obtained from the orbit equation, Eq. (3–36). Again, for simplicity, we
will consider the case of purely repulsive scattering (cf. Fig. 3–15). As the orbit
must be symmetric about the direction of the periapsis, the scattering angle is
given by

$$\Theta = \pi - 2\Psi, \tag{3-94}$$

where Ψ is the angle between the direction of the incoming asymptote and the
periapsis direction. In turn, Ψ can be obtained from Eq. (3–36) by setting $r_0 = \infty$

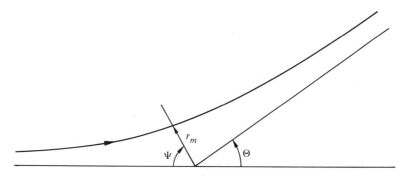

FIGURE **3–15**
Relation of orbit parameters and scattering angle in an example of repulsive
scattering.

---

* It is at this point in the formulation that classical and quantum mechanics part company.
Indeed it is fundamentally characteristic of quantum mechanics that one cannot
unequivocally predict the trajectory of any particular particle. One can only give
probabilities for scattering in various directions.

when $\theta_0 = \pi$ (the incoming direction), whence $\theta = \pi - \Psi$ when $r = r_m$, the distance of closest approach. A trivial rearrangement then leads to

$$\Psi = \int_{r_m}^{\infty} \frac{dr}{r^2 \sqrt{\dfrac{2mE}{l^2} - \dfrac{2mV}{l^2} - \dfrac{1}{r^2}}}. \tag{3–95}$$

Expressing l in terms of the impact parameter s (Eq. 3–90), the resultant expression for $\Theta(s)$ is

$$\Theta = \pi - 2 \int_{r_m}^{\infty} \frac{s\, dr}{r \sqrt{r^2 \left[1 - \dfrac{V(r)}{E}\right] - s^2}}, \tag{3–96}$$

or

$$\Theta = \pi - 2 \int_{0}^{u_m} \frac{s\, du}{\sqrt{1 - \dfrac{V(u)}{E} - s^2 u^2}}. \tag{3–97}$$

Equations (3–96) and (3–97) are rarely of use except for direct numerical computation of the scattering angle. However, when an analytic expression is available for the orbits, the relation between Θ and s can often be obtained almost by inspection. An historically important illustration of such a procedure is the repulsive scattering of charged particles by a coulomb field. The scattering force field is that produced by a fixed charge $-Ze$ acting on the incident particles having a charge $-Z'e$; so that the force can be written as

$$f = \frac{ZZ'e^2}{r^2},$$

that is, a repulsive inverse square law. The results of Section 3–7 can be taken over here with no more change than writing the force constant as

$$k = -ZZ'e^2. \tag{3–98}$$

The energy E is greater than zero, and the orbit is a hyperbola with the eccentricity given by*

$$\epsilon = \sqrt{1 + \frac{2El^2}{m(ZZ'e^2)^2}} = \sqrt{1 + \left(\frac{2Es}{ZZ'e^2}\right)^2}, \tag{3–99}$$

where use has been made of Eq. (3–90). If θ' in Eq. (3–51) is chosen to be π, periapsis corresponds to $\theta = 0$ and the orbit equation becomes

$$\frac{1}{r} = \frac{mZZ'e^2}{l^2}(\epsilon \cos \theta - 1). \tag{3–100}$$

---

* To avoid confusion with the electron charge e, the eccentricity will temporarily be denoted by ϵ.

The direction of the incoming asymptote, Ψ, is then determined by the condition $r \to \infty$:

$$\cos \Psi = \frac{1}{\epsilon}$$

or, by Eq. (3–94),

$$\sin \frac{\Theta}{2} = \frac{1}{\epsilon}.$$

Hence

$$\cot^2 \frac{\Theta}{2} = \epsilon^2 - 1,$$

and using Eq. (3–99)

$$\cot \frac{\Theta}{2} = \frac{2Es}{ZZ'e^2}.$$

The desired functional relationship between the impact parameter and the scattering angle is therefore

$$s = \frac{ZZ'e^2}{2E} \cot \frac{\Theta}{2}, \qquad (3\text{–}101)$$

so that on carrying through the manipulation required by Eq. (3–93), we find that $\sigma(\Theta)$ is given by

$$\sigma(\Theta) = \frac{1}{4} \left(\frac{ZZ'e^2}{2E} \right)^2 \csc^4 \frac{\Theta}{2}. \qquad (3\text{–}102)$$

Equation (3–102) gives the famous Rutherford scattering cross section, originally derived by Rutherford for the scattering of α particles by atomic nuclei. Quantum mechanics in the nonrelativistic limit yields a cross section identical with this classical result.

In atomic physics the concept of a *total scattering cross section* σ_T, defined as

$$\sigma_T = \int_{4\pi} \sigma(\Omega) \, d\Omega = 2\pi \int_0^\pi \sigma(\Theta) \sin \Theta \, d\Theta,$$

is of considerable importance. However, if we attempt to calculate the total cross section for coulomb scattering by substituting Eq. (3–102) in this definition we obtain an infinite result! The physical reason behind this behavior is not far to seek. From its definition the total cross section is the number of particles scattered in all directions per unit time for unit incident intensity. Now, the coulomb field is an example of a "long range" force; its effects extend to infinity. The very small deflections occur only for particles with very large impact parameters. Hence all particles in an incident beam of infinite lateral extent will be scattered to some extent and must be included in the total scattering cross section. It is clear therefore, that the infinite value for σ_T is not peculiar to the coulomb field; it occurs in classical mechanics whenever the scattering field is different from zero at all distances no

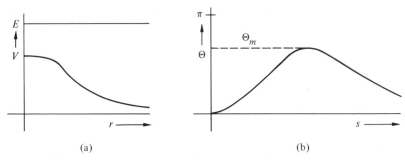

(a)   (b)

FIGURE 3-16

Repulsive nonsingular scattering potential and double-valued curve of
scattering angle Θ versus impact parameter s_0 for sufficiently high energy.

matter how large.* Only if the force field "cuts off," i.e., is zero beyond a certain
distance, will the scattering cross section be finite. Physically, such a cut-off occurs
for the coulomb field of a nucleus as a result of the presence of the atomic electrons,
which "screen" the nucleus and effectively cancel its charge at large distances.

In Rutherford scattering the scattering angle Θ is a smooth monotonic
function of the impact parameter s. From Eq. (3-101) we see that as s decreases
from infinity Θ increases monotonically from zero, reaching the value π as s goes
to zero. However, other types of behavior are possible in classical systems,
requiring some modification in the prescription, Eq. (3-93), for the classical cross
section. For example, with a repulsive potential and particle energy qualitatively
of the nature shown in Fig. 3-16(a) it is easy to see physically that the curve of Θ
versus s may behave as indicated in Fig. 3-16(b). Thus, with very large values of
the impact parameter, as has been noted above, the particle always remains at
large radial distances from the center of force and suffers only minor deflection. At
the other extreme, for $s = 0$, the particle travels in a straight line into the center of
force, and if the energy is greater than the maximum of the potential, it will
continue on through the center without being scattered at all. Hence, for both
limits in s the scattering angle goes to zero. For some intermediate value of s the
scattering angle must pass through a maximum Θ_m. When $\Theta < \Theta_m$ there will be
two values of s that can give rise to the same scattering angle. Each will contribute
to the scattering cross section at that angle, and Eq. (3-93) should accordingly be
modified to the form

$$\sigma(\Theta) = \sum_i \frac{s_i}{\sin \Theta} \left| \frac{ds}{d\Theta} \right|_i. \tag{3-103}$$

---

* σ_T is also infinite for the coulomb field in quantum mechanics, since it has been stated
that Eq. (3-98) remains valid there. However, not all "long range" forces give rise to infinite
total cross sections in quantum mechanics. It turns out that all potentials that fall off faster
at larger distances than $1/r^2$ produce a finite quantum-mechanical total scattering cross
section. Cf. Landau and Lifschitz, *Course of Theoretical Physics*, Vol. 3, *Quantum Mechanics*
(Reading, Massachusetts, Addison-Wesley, 1965), p. 473.

Here the subscript i distinguishes the various values of s giving rise to the same value of Θ.

Of particular interest is the cross section at the maximum angle of scattering Θ_m. As the derivative of Θ with respect to s vanishes at this angle, it follows from Eq. (3–93) or (3–103) that the cross section must become infinite at $\Theta \to \Theta_m$. But for all larger angles the cross section is zero, since the scattering angle cannot exceed Θ_m. The phenomenon of the infinite rise of the cross section followed by abrupt disappearance is very similar to what occurs in the geometrical optics of the scattering of sunlight by raindrops. On the basis of this similarity the phenomenon is called *rainbow scattering.*

So far the examples used have been for purely repulsive scattering. If the scattering involves attractive forces further complications may arise. The effect of attraction will be to pull the particle in toward the center instead of the repulsive deflection outward shown in Fig. 3–15. In consequence the angle Ψ between the incoming direction and the periapsis direction may be greater than $\pi/2$, and the scattering angle as given by Eq. (3–94) is then negative. This in itself is no great difficulty as clearly it is the magnitude of Θ that is involved in finding the cross section. But, under certain circumstances Θ as calculated by Eq. (3–96) may be greater than 2π. That is, the particle undergoing scattering may circle the center of force for one or more revolutions before going off finally in the scattered direction.

To see physically how this may occur, consider a scattering potential shown schematically as the $s = 0$ curve in Fig. 3–17. It is typical of the intermolecular potentials assumed in many kinetic theory problems—an attractive potential at

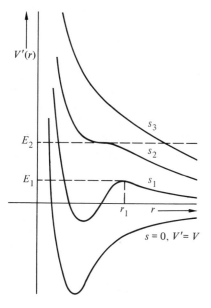

FIGURE 3–17
A combined attractive and repulsive scattering potential, and the corresponding equivalent one-dimensional potential at several values of the impact parameter s.

large distances falling off more rapidly than $1/r^2$, with a rapidly rising repulsive potential at small distances. The other curves in Fig. 3–17 show the effective one-dimensional potential $V'(r)$, Eq. (3–22'), for various values of the impact parameter s (equivalently: various values of l). Since the repulsive centrifugal barrier dominates at large r for all values of $s > 0$, the equivalent potential for small s will exhibit a hump. Consider now an incoming particle with impact parameter s_1 and at the energy E_1 corresponding to the maximum of the hump. As noted in Section 3–3, the difference between E_1 and $V'(r)$ is proportional to the square of the radial velocity at that distance. When the incoming particle reaches r_1, the location of the maximum in V', the radial velocity is zero. Indeed it will be remembered from the discussion in Section 3–6 that we have here the conditions for an unstable circular orbit at the distance r_1. In the absence of any perturbation the incoming particle with parameters E_1 and s_1, once having reached r_1, would circle around the center of force indefinitely at that distance without ever emerging! For the same impact parameter but at an energy E slightly higher than E_1, no true circular orbit would be established. However when the particle is in the immediate vicinity of r_1 the radial speed would be very small, and the particle would spend a disproportionately large time in the neighbourhood of the hump. The angular velocity, $\dot{\theta}$, meanwhile would not be affected by the maximum, being given at r, by

$$\dot{\theta} = \frac{l}{mr_1^2} = \frac{s_1}{r_1^2}\sqrt{\frac{2E}{m}}.$$

Thus in the time it takes the particle to get through the region of the hump the angular velocity may have carried the particle through angles larger than 2π or even multiples thereof. In such instances the classical scattering is said to exhibit *orbiting* or *spiraling*.

As the impact parameter is increased, the well and hump in the equivalent potential V' tend to flatten out, until at some parameter s_2 there is only a point of inflection in V' at an energy E_2 (cf. Fig. 3–17). For particle energies above E_2 there will no longer be orbiting. But the combined effects of the attractive and repulsive components of the effective potential can lead even in such cases to zero deflection for some finite value of the impact parameter. At large energies and small impact parameters the major scattering effects are caused by the strongly repulsive potentials at small distances, and the scattering qualitatively resembles the behavior of Rutherford scattering.

We have seen that the scattered particle may be deflected by more than π when orbiting takes place. On the other hand, the observed scattering angle in the laboratory lies between 0 and π. It is therfore helpful in such ambiguous cases to distinguish between a *deflection angle* Φ, as calculated by the right-hand sides of Eqs. (3–96) or (3–97), and the observed scattering angle Θ. For given Φ, the angle Θ is to be determined from the relation

$$\Theta = \pm\Phi - 2m\pi, \qquad m \text{ a positive integer.}$$

The sign and the value of m are to be chosen so that Θ lies between 0 and π. The sum in Eq. (3–103) then covers all values of Φ leading to the same Θ. Figure 3–18 shows schematic curves of Φ versus s for the potential of Fig. 3–17 at two different energies. The orbiting that takes place for $E = E_1$ shows up as a singularity in the curve at $s = s_1$. When $E > E_2$, orbiting no longer takes place but there is a rainbow effect at $\Theta = -\Phi'$ (although there is a nonvanishing cross section at higher scattering angles). It will be noticed that Θ vanishes at $s = s_3$, which means from Eq. 3–93 that the cross section becomes infinite in the forward direction through the vanishing of $\sin \Theta$. The cross section can similarly become infinite in the backward direction providing

$$s \left| \frac{ds}{d\Theta} \right|$$

remains finite at $\Theta = \pi$. These infinities in the forward or backward scattering angles are referred to as *glory scattering*, again in analogy to the corresponding phenomenon in meteorological optics.*

In recent years there has been a renewed interest in classical scattering and in calculations of the classical cross section. In some instances quantum effects are small, as in the scattering of low energy ions in crystal lattices, and the classical calculations are directly useful. Even when quantum mechanical corrections are important, it often suffices to use an approximation method (the "semiclassical" approximation) for which a knowledge of the classical trajectory is required. For almost all potentials of practical interest it is impossible to find an analytic form for

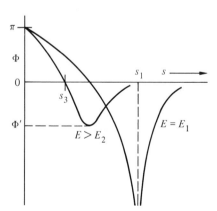

FIGURE 3–18
Schematic curves of deflection angle Φ versus s, for the potential of Fig. 3–17 at two different energies.

---

* The backward glory is familiar to airplane travelers as the ring of light observed to encircle the shadow of the plane projected on clouds underneath. Incidentally, the description of classical scattering phenomena by the terminology of geometrical optics is no coincidence. As will be shown in Section 10–8 classical mechanics is in a very real way the geometrical optics (ray-picture) limit of quantum mechanics, and for each scattering situation in classical mechanics a corresponding geometrical optics problem can be found.

the orbit, and Eq. (3–96) (or variant forms) is either approximated for particular regions of s or integrated numerically.

3–11 TRANSFORMATION OF THE SCATTERING PROBLEM TO LABORATORY COORDINATES

The previous section has been concerned with the one-body problem of the scattering of a particle by a fixed center of force. In practice the scattering always involves two bodies, e.g., in Rutherford scattering we have the α particle and the atomic nucleus. The second particle is not fixed but recoils from its initial position as a result of the scattering. Since it has been shown that any two-body central force problem can be reduced to a one-body problem it might be thought that the only change is to replace m by the reduced mass μ. However, the matter is not quite that simple. The scattering angle actually measured in the laboratory, which we shall denote by ϑ, is the angle between the final and incident directions of the scattered particle.* On the other hand, the angle Θ calculated from the equivalent one-body problem is the angle between the final and initial directions of the relative vector between the two particles. These two angles would be the same only if the second particle remains stationary through the scattering process. In general, however, the second particle, though initially at rest, is itself set in motion by the mutual force between the two particles, and, as is indicated in Fig. 3–19, the two angles then have different values. The equivalent one-body problem thus does not directly furnish the scattering angle as measured in the laboratory coordinate system.

The relationship between the scattering angles Θ and ϑ can be determined by examining how the scattering takes place in a coordinate system moving with the center of mass of both particles. In such a system the total linear momentum is

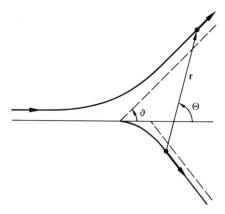

FIGURE 3–19
Scattering of two particles as viewed in the laboratory system.

---

* The scattering angle ϑ must not be confused with the angle coordinate θ of the relative vector, $\mathbf{r}$, between the two particles.

zero, of course, and the two particles always move with equal and opposite momenta. Figure 3–20 illustrates the appearance of the scattering process to an observer in the center-of-mass system. Before the scattering the particles are moving directly toward each other; after, they are moving directly away from each other. The angle between the initial and final directions of the relative vector, Θ, must therefore be the same as the scattering angle of either particle in the center-of-mass system. The connection between the two scattering angles Θ and ϑ can thus be obtained by considering the transformation between the center-of-mass system and the laboratory system. It is convenient here to use the terminology of Section 3–1, with slight modifications:

$\mathbf{r}_1$ and $\mathbf{v}_1$ are the position and velocity, after scattering, of the incident particle 1 in the laboratory system,

$\mathbf{r}_1'$ and $\mathbf{v}_1'$ are the position and velocity, after scattering, of particle 1 in the center-of-mass system, and

$\mathbf{R}$ and $\mathbf{V}$ are the position and (constant) velocity of the center-of-mass in the laboratory system.

At any instant, by definition

$$\mathbf{r}_1 = \mathbf{R} + \mathbf{r}_1',$$

and consequently

$$\mathbf{v}_1 = \mathbf{V} + \mathbf{v}_1'. \tag{3–104}$$

Figure 3–21 graphically portrays this vector relation evaluated *after* the scattering has taken place; at which time $\mathbf{v}_1$ and $\mathbf{v}_1'$ make the angles ϑ and Θ, respectively, with the vector $\mathbf{V}$ lying along the initial direction. Since the target is initially stationary in the laboratory system, the incident velocity of particle 1 in that system, $\mathbf{v}_0$, is the same as the initial relative velocity of the particles. By conservation of total linear momentum, the constant velocity of the center of mass is therefore given by

$$(m_1 + m_2)\mathbf{V} = m_1\mathbf{v}_0,$$

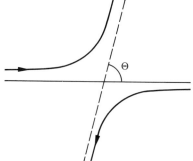

FIGURE 3–20
Scattering of two particles as viewed in the center-of-mass system.

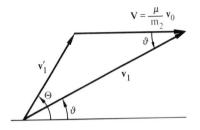

FIGURE 3–21
The relations between the velocities in the center-of-mass and laboratory coordinates.

or

$$V = \frac{\mu}{m_2} v_0. \tag{3–105}$$

From Fig. 3–21 it is readily seen that

$$v_1 \sin \vartheta = v_1' \sin \Theta$$

and

$$v_1 \cos \vartheta = v_1' \cos \Theta + V. \tag{3–106}$$

The ratio of these two equations gives a relation between ϑ and Θ:

$$\tan \vartheta = \frac{\sin \Theta}{\cos \Theta + \rho}, \tag{3–107}$$

where ρ is defined as

$$\rho \equiv \frac{\mu}{m_2} \frac{v_0}{v_1'}. \tag{3–108}$$

An alternative relation can be obtained by expressing v_1 in terms of the other speeds through the cosine law as applied to the triangle of Fig. 3–21:

$$v_1^2 = v_1'^2 + V^2 + 2v_1' V \cos \Theta. \tag{3–109}$$

When this is used to eliminate v_1 from Eq. (3–106) and V is expressed in terms of v_0 by Eq. (3–105), then we find

$$\cos \vartheta = \frac{\cos \Theta + \rho}{\sqrt{1 + 2\rho \cos \Theta + \rho^2}}. \tag{3–110}$$

Both these relations still involve a ratio of speeds through ρ. By the definition of center of mass the speed of particle 1 in the center-of-mass system, v_1', is connected with the relative speed v, by the equation (cf. Eq. 3–2):

$$v_1' = \frac{\mu}{m_1} v.$$

Hence ρ can also be written as

$$\rho = \frac{m_1}{m_2} \frac{v_0}{v}, \tag{3–108'}$$

where v, it should be emphasized, is the relative speed *after* the collision. When the collision is *elastic*, the total kinetic energy of the two particles remains unaltered and v must equal v_0, so that ρ is simply

$$\rho = \frac{m_1}{m_2}, \qquad \text{(elastic collision)} \qquad (3\text{–}111)$$

independent of energies or speeds. If the collision is *inelastic*, the total kinetic energy of the two particles is altered (e.g., some of the kinetic energy goes into the form of internal excitation energy of the target). Since the kinetic energy of the center-of-mass motion must remain constant, by conservation of linear momentum, the energy change resulting from the collision can be expressed as

$$\frac{\mu v^2}{2} = \frac{\mu v_0^2}{2} + Q. \qquad (3\text{–}112)$$

The so-called Q value of the inelastic collision is clearly negative in magnitude, but the sign convention is chosen to conform to that used in general for atomic and nuclear reactions. From Eq. (3–112) the ratio of relative speeds before and after collision can be written

$$\frac{v}{v_0} = \sqrt{1 + \frac{m_1 + m_2}{m_2}\frac{Q}{E}}, \qquad (3\text{–}113)$$

where E is the energy of the incoming particle (in the laboratory system). Thus, for inelastic scattering ρ becomes

$$\rho = \frac{m_1}{m_2\sqrt{1 + \dfrac{m_1 + m_2}{m_2}\dfrac{Q}{E}}}. \qquad \text{(inelastic scattering)} \qquad (3\text{–}114)$$

Not only are the scattering angles ϑ and Θ in general different in magnitude, but the values of the differential scattering cross section depend on which of the two angles is used as the argument of σ. The connection between the two functional forms is obtained from the observation that in a particular experiment the number of particles scattered into a given element of solid angle must be the same whether we measure the event in terms of ϑ or Θ. As an equation this statement can be written

$$2\pi I \sigma(\Theta) \sin \Theta \, |d\Theta| = 2\pi I \sigma'(\vartheta) \sin \vartheta \, |d\vartheta|,$$

or

$$\sigma'(\vartheta) = \sigma(\Theta)\frac{\sin \Theta}{\sin \vartheta}\left|\frac{d\Theta}{d\vartheta}\right| = \sigma(\Theta)\left|\frac{d(\cos \Theta)}{d(\cos \vartheta)}\right|, \qquad (3\text{–}115)$$

where $\sigma'(\vartheta)$ is the differential scattering cross section expressed in terms of the scattering angle in the laboratory system. The derivative can easily be evaluated from Eq. (3–110), leading to the result

$$\sigma'(\vartheta) = \sigma(\Theta)\frac{(1 + 2\rho \cos \Theta + \rho^2)^{3/2}}{1 + \rho \cos \Theta}. \qquad (3\text{–}116)$$

It should perhaps be emphasized that $\sigma(\Theta)$ is *not* the cross section an observer would measure in the center-of-mass system. Both $\sigma(\Theta)$ and $\sigma'(\vartheta)$ are cross sections measured in the laboratory system; they are merely expressed in terms of different coordinates. An observer fixed in the center-of-mass system would see a different flux density of incident particles from that measured in the laboratory system, and this transformation of flux density would have to be included if (for some reason) we wanted to relate the cross sections as measured in the two different systems.

The two scattering angles have a particularly simple relation for elastic scattering when the two masses of particles are equal. It then follows that $\rho = 1$, and from Eq. (3–110) we have

$$\cos \vartheta = \sqrt{\frac{1 + \cos \Theta}{2}} = \cos \frac{\Theta}{2},$$

or

$$\vartheta = \frac{\Theta}{2}, \qquad (\rho = 1).$$

Thus, with equal masses, scattering angles greater than $90°$ cannot occur in the laboratory system; all the scattering is in the forward hemisphere. Correspondingly, the scattering cross section is given in terms of Θ from Eq. (3–116) as

$$\sigma'(\vartheta) = 4 \cos \vartheta \cdot \sigma(\Theta), \qquad \vartheta \leqslant \frac{\pi}{2}, \quad (\rho = 1).$$

Even when the scattering is isotropic in terms of Θ, that is, $\sigma(\Theta)$ is constant, independent of Θ, then the cross section in terms of ϑ varies as the cosine of the angle!

We have seen that even in elastic collisions, where the total kinetic energy remains constant, a collision with an initially stationary target results in a transfer of kinetic energy to the target with a corresponding decrease in the kinetic energy of the incident particle. In other words, the collision *slows down* the incident particle. The degree of slowing down can be obtained from Eq. (3–109) if v_1' and V are expressed in terms of v_0 by Eqs. (3–108) and (3–105), respectively:

$$\frac{v_1^2}{v_0^2} = \left(\frac{\mu}{m_2 \rho}\right)^2 [1 + 2\rho \cos \Theta + \rho^2]. \tag{3–117}$$

For elastic collisions $\rho = m_1/m_2$, and Eq. (3–120) can be simplified to

$$\frac{E_1}{E_0} = \frac{1 + 2\rho \cos \Theta + \rho^2}{(1 + \rho)^2}, \qquad \text{(elastic collision)} \tag{3–117'}$$

where E_0 is the initial kinetic energy of the incident particle in the laboratory system and E_1 the corresponding energy after scattering. When the particles are of equal mass this relation becomes

$$\frac{E_1}{E_0} = \frac{1 + \cos \Theta}{2} = \cos \vartheta.$$

Thus at the maximum scattering angle ($\Theta = \pi, \vartheta = \pi/2$), the incident particle loses all its energy and is completely stopped in the laboratory system.

This transfer of kinetic energy by scattering is, of course, the principle behind the "moderator" in a thermal neutron reactor. Fast neutrons produced by fission make successive elastic scatterings until their kinetic energy is reduced to thermal energies, where they are more liable to cause fission than to be captured. Clearly the best moderators will be the light elements, ideally hydrogen ($\rho = 1$). For a nuclear reactor hydrogen is practical only when contained as part of a mixture or compound, such as water. Other light elements useful for their moderating properties include deuterium, of mass 2, and carbon, of mass 12. Hydrogen, as present in paraffin, water, or plastics, is frequently used in the laboratory to slow down neutrons.

Despite their very up-to-date applications these calculations of the transformation from laboratory to center of mass coordinates, and of the transfer of kinetic energy, are not particularly "modern" or "quantum" in nature. Nor is the classical mechanics involved particularly advanced or difficult. All that has been used, essentially, is the conservation of momentum and energy. Indeed, similar calculations may be found in freshman textbooks, usually in terms of elastic collisions between, say, billiard balls. But it is their very elementary nature that results in the widespread validity of these calculations. So long as momentum is conserved (and this will be true in quantum mechanics) and the Q value is known, the details of the scattering process are irrelevant. In effect the vicinity of the scattering particle is a "black box," and we are concerned only with what goes in and what comes out. It matters not at all whether the phenomena occurring inside the box are "classical" or "quantum." Consequently the formulae of this section may be used in the experimental analysis of phenomena essentially quantum in nature, as for example, neutron-proton scattering, so long as the energies are low enough that relativistic effects may be neglected. (See Section 7–7 for a discussion of the relativistic treatment of the kinematics of collisions.)

SUGGESTED REFERENCES

E. T. WHITTAKER, *Analytical Dynamics*. Almost every text on mechanics devotes considerable time to central force motion and only a few of the many references can be listed here. Sections 47–49 of Whittaker's treatise form a concise discussion of the subject that, despite its brevity, manages to consider many out-of-the-way aspects. It is practically the only source for the analysis of which force laws are soluble in elliptic functions.

W. D. MACMILLAN, *Statics and the Dynamics of a Particle*. Chapter XII of this reference provides a most elaborate discussion of central force motion, including detailed consideration of the orbits for some force laws other than the customary inverse square law. The treatment is elementary and does not use the Lagrangian formulation. Kepler's equation is derived, along with variant forms for various types of orbits, and several methods of solution are described. The Laplace–Runge–Lenz constant of motion is mentioned briefly, in a "throw-away" line as a "constant of integration" in Section 301.

L. D. LANDAU AND E. M. LIFSHITZ, *Mechanics.* As might be expected, in its coverage of the subject of central force motion and scattering, this reference includes many original and unusual insights. The constancy of the Laplace–Runge–Lenz vector is explicitly demonstrated and used to derive the orbit equation. Kinematics of collisions are described in some detail.

J. B. MARION, *Classical Dynamics of Particles and Systems.* This fine intermediate-level text has an unusually detailed chapter on central force motion, including discussions on Kepler's equation, stability of circular orbits, and some elementary results on the famous three-body problem. As befits the author's involvement in experimental nuclear physics the chapter on kinematics of collisions and cross section calculations is exceptionally complete in coverage.

J. O. HIRSCHFELDER; D. F. CURTISS; AND B. B. BIRD, *Molecular Theory of Gases and Liquids.* A veritable encyclopedia on the chemical physics of gases and liquids, this reference is probably the best source for applications of the classical virial theorem. There is also much material on interatomic potentials and scattering cross sections, although the substantial advances in the field since 1953 are naturally not included.

S. W. McCUSKEY, *Introduction to Celestial Mechanics.* Two-body motion in a mutual $1/r$ potential forms the first approximation to the motion of the planets, satellites, and space craft. All books on celestial mechanics therefore devote considerable attention to the "Kepler problem." Of the vast literature on celestial mechanics only a few can be cited here. McCuskey's book covers a wide range of information compactly, including a discussion on motion in time in various orbits, and approaches the subject on a relatively elementary level.

J. M. A. DANBY, *Fundamentals of Celestial Mechanics.* Written by a well-recognized and highly respected master of celestial mechanics, this text covers much of the classical area of celestial mechanics on an intermediate level, with a leavening of more recent material inspired by the advent of the computer and space age. Exercises and references are plentiful. The motion in time and Kepler's equation are discussed at length, but neither the Laplace–Runge–Lenz vector nor Bertrand's theorem is mentioned.

H. C. PLUMMER, *An Introductory Treatise on Dynamical Astronomy.* Although relatively old (1918, but with a 1960 reprint) this remains the most available reference on Bertrand's theorem, and the presentation in this Chapter is based on Plummer's approach. A different method of proof is briefly described in the classic treatise by F. Tisserand, *Traité de mécanique céleste,* Tome 1, Chapter I, Section 6. Plummer's book also presents some special tricks for approximate solution of Kepler's equation.

H. V. McINTOSH, "Symmetry and Degeneracy," in *Group Theory and its Applications,* Vol. II, E. M. Loebl, ed. This article contains an enthusiastic overview of the internal symmetries of simple physical systems, developed in a historical fashion. Though the discussion has a high words-to-formula ratio, it presupposes that the reader has some acquaintance with group theory and with quantum mechanics. But it is by far the best survey of what in 1970 was thought to be the connection between symmetries and degeneracies. Both the Kepler problem and the harmonic oscillator symmetries are described, as well as the implications of Bertrand's theorem.

R. G. NEWTON, *Scattering Theory of Waves and Particles.* Although this book is mainly involved with the quantum theory of scattering, Chapter 5 is concerned with scattering of classical particles and gives a concise discussion of "orbiting" and "glories," based for the most part on the fundamental paper of K. W. Ford and J. A. Wheeler, *Annals of Physics (N.Y.)* **7**, 259 (1959).

EXERCISES

1. A particle of mass m is constrained to move under gravity without friction on the inside of a paraboloid of revolution whose axis is vertical. Find the one-dimensional problem equivalent to its motion. What is the condition on the particle's initial velocity to produce circular motion? Find the period of small oscillations about this circular motion.

2. A particle moves in a central force field given by the potential

$$V = -k\frac{e^{-ar}}{r},$$

where k and a are positive constants. Using the method of the equivalent one-dimensional potential discuss the nature of the motion, stating the ranges of l and E appropriate to each type of motion. When are circular orbits possible? Find the period of small radial oscillations about the circular motion.

3. Two particles move about each other in circular orbits under the influence of gravitational forces, with a period τ. Their motion is suddenly stopped at a given instant of time, and they are then released and allowed to fall into each other. Prove that they collide after a time $\tau/4\sqrt{2}$.

4. Consider a system in which the total forces acting on the particles consist of conservative forces $\mathbf{F}_i'$ and frictional forces $\mathbf{f}_i$ proportional to the velocity. Show that for such a system the virial theorem holds in the form

$$\bar{T} = -\frac{1}{2}\overline{\sum_i \mathbf{F}_i' \cdot \mathbf{r}_i},$$

providing the motion reaches a steady state and is not allowed to die down as a result of the frictional forces.

5. Suppose that there are long-range interactions between atoms in a gas in the form of central forces derivable from a potential

$$U(r) = \frac{k}{r^m},$$

where r is the distance between any pair of atoms and m is a positive integer. Assume further that relative to any given atom the other atoms are distributed in space such that their volume density is given by the Boltzmann factor:

$$\rho(r) = \frac{N}{V}e^{-U(r)/kT},$$

where N is the total number of atoms in a volume V. Find the addition to the virial of Clausius resulting from these forces between pairs of atoms, and compute the resulting correction to Boyle's law. Take N so large that sums may be replaced by integrals. While closed results can be found for any positive m, if desired the mathematics can be simplified by taking $m \Rightarrow +1$.

6. a) Show that if a particle describes a circular orbit under the influence of an attractive central force directed toward a point on the circle, then the force varies as the inverse fifth power of the distance.

b) Show that for the orbit described the total energy of the particle is zero.

c) Find the period of the motion.

d) Find $\dot{x}$, $\dot{y}$, and v as a function of angle around the circle and show that all three quantities are infinite as the particle goes through the center of force.

7. Show that the central force problem is soluble in terms of elliptic functions when the force is a power law function of the distance with the following fractional exponents:

$$n = -\frac{3}{2}, \quad -\frac{5}{2}, \quad -\frac{1}{3}, \quad -\frac{5}{3}, \quad -\frac{7}{3}.$$

8. a) For circular and parabolic orbits in an attractive $1/r$ potential having the same angular momentum, show that the perihelion distance of the parabola is one half the radius of the circle.

b) Prove that in the same central force as in part (a) the speed of a particle at any point in a parabolic orbit is $\sqrt{2}$ times the speed in a circular orbit passing through the same point.

9. A meteor is observed to strike Earth with a speed v, making an angle ϕ with the zenith. Suppose that far from Earth the meteor's speed were v' and it was proceeding in a direction making a zenith angle ϕ'; the effect of Earth's gravity being to pull it into a hyperbolic orbit intersecting Earth's surface. Show how v' and ϕ' can be determined from v and ϕ in terms of known constants.

10. Prove that in a Kepler elliptic orbit with small eccentricity e the angular motion of a particle as viewed from the *empty* focus of the ellipse is uniform (the empty focus is the focus that is *not* the center of attraction) to first order in e. It is this theorem that enables the Ptolematic picture of planetary motion to be a reasonably accurate approximation. On this picture the Sun is assumed to move uniformly on a circle whose center is shifted from the Earth by a distance called the *equant*. If the equant is taken as the distance between the two foci of the correct elliptical orbit, then the angular motion is thus described by the Ptolemaic picture accurately to first order in e.

11. One of the classic themes of science fiction is a twin planet ("Planet X") to Earth that is identical in mass, energy, and momentum but is located on the orbit $90°$ out of phase with Earth so that it would be hidden by the Sun. However because of the elliptical nature of the orbit it would not always be completely hidden. Assume there is such a planet in the same Keplerian orbit as Earth in such a manner that it is in aphelion when Earth is in perihelion. Calculate to first order in the eccentricity e the maximum angular separation of the twin and the Sun as viewed from Earth. Could such a twin be visible from Earth? Suppose the twin planet were in an elliptical orbit having the same size and shape as that of Earth, but rotated $180°$ from the orbit of Earth, so that Earth and the twin would be in perihelion at the same time. Repeat your calculation and compare the visibility in the two situations.

12. At perigee of an elliptic gravitational orbit a particle experiences an impulse S (cf. Exercise 9, Chapter 2) in the radial direction, sending the particle into another elliptic orbit. Determine the new semimajor axis, eccentricity, and orientation of major axis in terms of the old.

13. A uniform distribution of dust in the solar system adds to the gravitational attraction of the sun on a planet an additional force

$$\mathbf{F} = -mC\mathbf{r},$$

where m is the mass of the planet, C is a constant proportional to the gravitational constant and the density of the dust, and $\mathbf{r}$ is the radius vector from the sun to the planet (both considered as points). This additional force is very small compared to the direct sun–planet gravitational force.

a) Calculate the period for a circular orbit of radius r_0 of the planet in this combined field.

b) Calculate the period of radial oscillations for slight disturbances from this circular orbit.

c) Show that nearly circular orbits can be approximated by a precessing ellipse and find the precession frequency. Is the precession in the same or opposite direction to the orbital angular velocity?

14. Show that the motion of a particle in the potential field

$$V(r) = -\frac{k}{r} + \frac{h}{r^2}$$

is the same as that of the motion under the Kepler potential alone when expressed in terms of a coordinate system rotating or precessing around the center of force.

For negative total energy show that if the additional potential term is very small compared to the Kepler potential, then the angular speed of precession of the elliptical orbit is

$$\dot{\Omega} = \frac{2\pi mh}{l^2 \tau}.$$

The perihelion of Mercury is observed to precess (after correction for known planetary perturbations) at the rate of about 40″ of arc per century. Show that this precession could be accounted for classically if the dimensionless quantity

$$\eta = \frac{h}{ka}$$

(which is a measure of the perturbing inverse square potential relative to the gravitational potential) were as small as 7×10^{-8}. (The eccentricity of Mercury's orbit is 0.206, and its period is 0.24 year.)

15. The additional term in the potential behaving as r^{-2} in the previous problem looks very much like the centrifugal barrier term in the equivalent one-dimensional potential. Why is it then that the additional force term causes a precession of the orbit, while an addition to the barrier, through a change in l, does not?

16. Evaluate approximately the ratio of the mass of the sun to that of the earth, using only the lengths of the year and of the lunar month (27.3 days), and the mean radii of the earth's orbit (1.49×10^8 km) and of the moon's orbit (3.8×10^5 km).

17. Show that for elliptical motion in a gravitational field the radial speed can be written as

$$\dot{r} = \frac{\omega a}{r}\sqrt{a^2 e^2 - (r - a)^2}.$$

Introduce the eccentric anomaly variable ψ in place of r and show that the resulting differential equation in ψ can be integrated immediately to give Kepler's equation.

18. If the eccentricity e is small, Kepler's equation for the eccentric anomaly ψ as a function of ωt, Eq. (3–76), is easily solved on a computer by an iterative technique that treats the $e \sin \psi$ term as of lower order than ψ. Denoting ψ_n by the nth iterative solution, the obvious iteration relation is

$$\psi_n = \omega t + e \sin \psi_{n-1}.$$

Using this iteration procedure find the analytic form for an expansion of ψ in powers of e at least through terms in e^2.

19. By expanding $e \sin \psi$ in a Fourier series in ωt, show that Kepler's equation has the formal solution

$$\psi = \omega t + \sum_{n=1}^{\infty} \frac{2}{n} J_n(ne) \sin \omega t,$$

where J_n is the Bessel function of order n. For small argument the Bessel function can be approximated in a power series of the argument. Accordingly, derive from this result the first few terms in the expansion of ψ in powers of e. A good source for information on the properties of Bessel functions is the "Handbook of Mathematical Functions" by Abramowitz and Stegun, especially page 360.

20. If the difference $\psi - \omega t$ is represented by ρ, Kepler's equation can be written

$$\rho = e \sin (\omega t + \rho).$$

Successive approximations to ρ can be obtained by expanding $\sin \rho$ in a Taylor series in ρ, and then replacing ρ by its expression given by Kepler's equation. Show that the first approximation to ρ is ρ_1 given by

$$\tan \rho_1 = \frac{e \sin \omega t}{1 - e \cos \omega t},$$

and that the next approximation is found from

$$\sin (\rho_2 - \rho_1) = -e^3 \sin (\omega t + \rho_1)(1 + e \cos \omega t),$$

an expression that is accurate through terms of order e^4.

21. Earth's period between successive perihelion transits (the "anomalistic year") is 365.2596 mean solar days, and the eccentricity of the orbit is 0.0167504. Assuming motion in a Keplerian elliptical orbit, how far does Earth move in angle in the orbit, starting from perihelion, in a time equal to one quarter of the anomalistic year? Give your result in degrees to an accuracy of one second of arc or better. Any method may be used, including numerical computation with a calculator or computer.

22. In hyperbolic motion in a $1/r$ potential the analogue of the eccentric anomaly is F defined by

$$r = a(e \cosh F - 1),$$

where $a(e - 1)$ is the distance of closest approach. Find the analogue to Kepler's equation giving t from the time of closest approach as a function of F.

23. A *magnetic monopole* is defined (if one exists) by a magnetic field singularity of the form $\mathbf{B} = b\mathbf{r}/r^3$, where b is a constant (a measure of the magnetic charge, as it were). Suppose a particle of mass m moves in the field of a magnetic monopole and a central force field derived from the potential $V(r) = -k/r$.

a) Find the form of Newton's equation of motion, using the Lorentz force given by Eq. (1–61). By looking at the product $\mathbf{r} \times \dot{\mathbf{p}}$ show that while the mechanical angular momentum is not conserved (the field of force is noncentral) there is a conserved vector

$$\mathbf{D} = \mathbf{L} - \frac{qb}{c} \frac{\mathbf{r}}{r}.$$

(b) By paralleling the steps leading from Eq. (3–79) to Eq. (3–82) show that for some $f(r)$ there is a conserved vector analogous to the Laplace–Runge–Lenz vector in which $\mathbf{D}$ plays the same role as $\mathbf{L}$ in the pure Kepler force problem.

24. If the velocity or momentum vector of a particle is translated so as to start from the center of force, then the head of the vectors traces out the particle's *hodograph*, a locus curve of considerable antiquity in the history of mechanics but one that has had something of a revival in connection with the dynamics of space vehicles. By taking the cross product of $\mathbf{L}$ with the Laplace–Runge–Lenz vector $\mathbf{A}$, show that the hodograph for elliptical Kepler motion is, in terms of the momentum, a circle of radius mk/l with origin on the y axis a distance A/l displaced from the center of force.

25. What changes, if any, would there be in Rutherford scattering if the coulomb force were attractive, instead of repulsive?

26. Examine the scattering produced by a repulsive central force $f = kr^{-3}$. Show that the differential cross section is given by

$$\sigma(\Theta)\,d\Theta = \frac{k}{2E}\frac{(1-x)\,dx}{x^2(2-x)^2\sin \pi x},$$

where x is the ratio Θ/π and E is the energy.

27. A central force potential frequently encountered in nuclear physics is the *rectangular well*, defined by the potential

$$V = 0, \qquad\qquad r > a.$$
$$= -V_0, \qquad\quad r \leqslant a.$$

Show that the scattering produced by such a potential in classical mechanics is identical with the refraction of light rays by a sphere of radius a and relative index of refraction

$$n = \sqrt{\frac{E + V_0}{E}}.$$

(This equivalence demonstrates why it was possible to explain refraction phenomena both by Huygens' waves and by Newton's mechanical corpuscles.) Show also that the differential cross section is

$$\sigma(\Theta) = \frac{n^2 a^2}{4\cos\dfrac{\Theta}{2}} \frac{\left(n\cos\dfrac{\Theta}{2} - 1\right)\left(n - \cos\dfrac{\Theta}{2}\right)}{\left(1 + n^2 - 2n\cos\dfrac{\Theta}{2}\right)^2}.$$

What is the total cross section?

28. Consider a truncated repulsive Coulomb potential defined as

$$V = \frac{k}{r}, \qquad r > a,$$
$$= \frac{k}{a}, \qquad r \leqslant a.$$

For a particle of total energy $E > k/a$, obtain expressions for the scattering angle Θ as a function of s/s_0, where s_0 is the impact parameter for which the periapsis occurs at the point $r = a$. (The formulas can be given in closed form but they are not simple!) Make a numerical plot of Θ versus s/s_0 for the special case $E = 2k/a$. What can you deduce about the angular scattering cross section from the dependence of Θ on s/s_0 for this particular case?

29. Another version of the truncated Coulomb potential has the form

$$V = \frac{k}{r} - \frac{k}{a}, \qquad r > a$$

$$= 0, \qquad r < a.$$

Obtain closed-form expressions for the scattering angle and the differential scattering cross section. These are most conveniently expressed in terms of a parameter measuring the distance of closest approach in units of a. What is the total cross section?

30. Show that for repulsive scattering, Eq. (3–96) for the angle of scattering as a function of the impact parameter, s, can be rewritten as

$$\Theta = \pi - 4s \int_0^1 \frac{\rho\, d\rho}{\sqrt{r_m^2 \left[1 - \dfrac{V}{E}\right] - s^2(1 - \rho^2)}},$$

or

$$\Theta = \pi - 4s \int_0^1 \frac{d\rho}{\sqrt{\dfrac{r_m^2}{\rho^2 E}[V(r_m) - V(r)] + s^2(2 - \rho^2)}},$$

by changing the variable of integration to some function $\rho(r)$. Show that for a repulsive potential the integrand is never singular in the limit $r \to r_m$. Because of the definite limits of integration these formulations have advantages for numerical calculations of $\Theta(s)$ and allow naturally for the use of Gauss–Legendre quadrature schemes.

31. Apply the formulation of the preceding exercise to compute numerically $\Theta(s)$ and the differential cross section $\sigma(\Theta)$ for the repulsive potential

$$V = \frac{V_0}{1 + r}$$

and for a total energy $E = 1.2V_0$. It is suggested that 16-point Gauss–Legendre quadrature will give adequate accuracy. Does the scattering exhibit a rainbow?

32. If a repulsive potential drops off monotonically with r, then for energies high compared to $V(r_m)$ the angle of scattering will be small. Under these conditions show that Eq. (3–97) can be manipulated so that the deflection angle is given approximately by

$$\Theta = \frac{1}{E} \int_0^1 \frac{[V(u_m) - V(u)]\, dy}{(1 - y^2)^{3/2}},$$

where y, obviously, is u/u_m.

Show further, that if $V(u)$ is of the form Cu^n, where n is a positive integer, then in the high energy limit the cross section is proportional to $\Theta^{-2(1 + 1/n)}$.

33. a) Show that the angle of recoil of the target particle relative to the incident direction of the scattered particle is simply $\Phi = \frac{1}{2}(\pi - \Theta)$.

b) It is observed that in elastic scattering the scattering cross section is isotropic in terms of Θ. What are the corresponding probability distributions for the scattered energy of the incident particle, E_1, and for the recoil energy of the target particle, E_2?

34. Show that the angle of scattering in the laboratory system, ϑ, is related to the energy before scattering, E_0, and the energy after scattering E_1, according to the equation

$$\cos \vartheta = \frac{m_2 + m_1}{2m_1} \sqrt{\frac{E_1}{E_0}} - \frac{m_2 - m_1}{2m_1} \sqrt{\frac{E_0}{E_1}} + \frac{m_2 Q}{2m_1 \sqrt{E_0 E_1}}.$$

CHAPTER 4
The Kinematics of
Rigid Body Motion

A rigid body was defined previously as a system of mass points subject to the holonomic constraints that the distances between all pairs of points remain constant throughout the motion. Although something of an idealization, the concept is quite useful, and the mechanics of rigid body motion deserves a full exposition. In this chapter we shall discuss principally the *kinematics* of rigid bodies, i.e., the nature and characteristics of their motions. We shall devote some time to developing the mathematical techniques involved, which are of considerable interest in themselves, and have many important applications to other fields of physics. Having learned how to describe the motion of rigid bodies, the next chapter will then discuss, within the framework of the Lagrangian formulation, how such motion is generated by applied forces and torques.

4–1 THE INDEPENDENT COORDINATES OF A RIGID BODY

Before discussing the motion of a rigid body we must first establish how many independent coordinates are necessary to specify its configuration. A rigid body with N particles can at most have $3N$ degrees of freedom, but these are greatly reduced by the constraints, which can be expressed as equations of the form

$$r_{ij} = c_{ij}. \tag{4-1}$$

Here r_{ij} is the distance between the ith and jth particles and the c's are constants. The actual number of degrees of freedom cannot be obtained simply by subtracting the number of constraint equations from $3N$, for there are $\frac{1}{2}N(N-1)$ possible equations of the form of Eq. (4–1), which exceeds $3N$ for large N. In truth, the Eqs. (4–1) are not all independent. To fix a point in the rigid body it is not necessary to specify its distances to *all* other points in the body; one need only state the distances to any three other noncollinear points, cf. Fig. 4–1. Thus, once the positions of three of the particles of the rigid body are determined the constraints fix the positions of all remaining particles. The number of degrees of freedom therefore cannot be more than nine. But the three reference points are

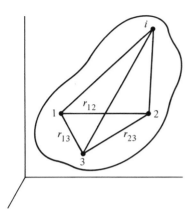

FIGURE 4-1
Diagram illustrating the location of a point in a rigid body by its distances from three reference points.

themselves not independent; there are in fact three equations of rigid constraint imposed on them,

$$r_{12} = c_{12}, \qquad r_{23} = c_{23}, \qquad r_{13} = c_{13},$$

that reduce the number of degrees of freedom to *six*. That only six coordinates are needed can also be seen from the following considerations. To establish the position of one of the reference points three coordinates must be supplied. But once point 1 is fixed, point 2 can be specified by only two coordinates, since it is constrained to move on the surface of a sphere centered at point 1. With these two points determined point 3 has only one degree of freedom, for it can only rotate about the axis joining the other two points. Hence a total of six coordinates is sufficient.

A rigid body in space thus needs six independent generalized coordinates to specify its configuration, no matter how many particles it may contain—even in the limit of a continuous body. Of course, there may be additional constraints on the body besides the constraint of rigidity. For example, the body may be constrained to move on a surface, or with one point fixed. In such case the additional constraints will further reduce the number of degrees of freedom, and hence the number of independent coordinates.

How shall these coordinates be assigned? It will be noticed that the configuration of a rigid body is completely specified by locating a Cartesian set of coordinates fixed in the rigid body (the primed axes shown in Fig. 4–2) relative to the coordinate axes of the external space. Clearly three of the coordinates are needed to specify the coordinates of the origin of this "body" set of axes. The remaining three coordinates must then specify the orientation of the primed axes relative to a coordinate system parallel to the external axes, but with the same origin as the primed axes.

There are many ways of specifying the orientation of a Cartesian set of axes relative to another set with common origin. A most fruitful procedure is to state the direction cosines of the primed axes relative to the unprimed. Thus the x' axis

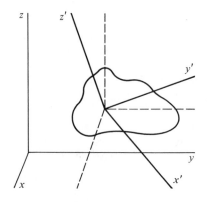

FIGURE 4–2
Unprimed axes represent an external reference set of axes; the primed axes are fixed in the rigid body.

could be specified by its three direction cosines $\alpha_1, \alpha_2, \alpha_3$ with respect to the x, y, z axes. If, as customary, $\mathbf{i}, \mathbf{j}, \mathbf{k}$ are three unit vectors along x, y, z, and $\mathbf{i}', \mathbf{j}', \mathbf{k}'$ perform the same function in the primed system (cf. Fig. 4–3), then these direction cosines are defined as

$$\alpha_1 = \cos(\mathbf{i}', \mathbf{i}) = \mathbf{i}' \cdot \mathbf{i}$$
$$\alpha_2 = \cos(\mathbf{i}', \mathbf{j}) = \mathbf{i}' \cdot \mathbf{j} \qquad (4\text{–}2)$$
$$\alpha_3 = \cos(\mathbf{i}', \mathbf{k}) = \mathbf{i}' \cdot \mathbf{k}.$$

The vector $\mathbf{i}'$ can be expressed in terms of $\mathbf{i}, \mathbf{j}, \mathbf{k}$ by the relation

$$\mathbf{i}' = (\mathbf{i}' \cdot \mathbf{i})\mathbf{i} + (\mathbf{i}' \cdot \mathbf{j})\mathbf{j} + (\mathbf{i}' \cdot \mathbf{k})\mathbf{k}$$

or

$$\mathbf{i}' = \alpha_1 \mathbf{i} + \alpha_2 \mathbf{j} + \alpha_3 \mathbf{k}. \qquad (4\text{–}3)$$

Similarly the direction cosines of the y' axis with x, y, z may be designated by $\beta_1, \beta_2,$ and β_3, and these will be the components of $\mathbf{j}'$ in the unprimed reference frame:

$$\mathbf{j}' = \beta_1 \mathbf{i} + \beta_2 \mathbf{j} + \beta_3 \mathbf{k}. \qquad (4\text{–}4)$$

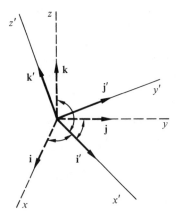

FIGURE 4–3
Direction cosines of the body set of axes relative to an external set of axes.

An equation analogous to (4–4) can be written for $\mathbf{k}'$, with the direction cosines of the z' axis designated by γ's. These sets of nine direction cosines then completely specify the orientation of the x', y', z' axes relative to the x, y, z set. One can equally well invert the process, and use the direction cosines to express the $\mathbf{i}$, $\mathbf{j}$, $\mathbf{k}$ unit vectors in terms of their components along the primed axes. Thus we can write

$$\mathbf{i} = (\mathbf{i} \cdot \mathbf{i}')\mathbf{i}' + (\mathbf{i} \cdot \mathbf{j}')\mathbf{j}' + (\mathbf{i} \cdot \mathbf{k}')\mathbf{k}'$$

or

$$\mathbf{i} = \alpha_1 \mathbf{i}' + \beta_1 \mathbf{j}' + \gamma_1 \mathbf{k}', \tag{4–5}$$

with analogous equations for $\mathbf{j}$ and $\mathbf{k}$.

The direction cosines also furnish directly the relations between the coordinates of a given point in one system and the coordinates in the other system. Thus, the coordinates of a point in a given reference frame are the components of the position vector, $\mathbf{r}$, along the axes of the system. The x' coordinate is then given in terms of x, y, z by

$$x' = (\mathbf{r} \cdot \mathbf{i}') = \alpha_1 x + \alpha_2 y + \alpha_3 z,$$

while for the other coordinates we obtain

$$y' = \beta_1 x + \beta_2 y + \beta_3 z$$

$$z' = \gamma_1 x + \gamma_2 y + \gamma_3 z. \tag{4–6}$$

What has been done here for the components of the $\mathbf{r}$ vector can obviously be done for any arbitrary vector. If $\mathbf{G}$ is some vector, then the component of $\mathbf{G}$ along the x' axis will be related to its x, y, z components by

$$G_{x'} = (\mathbf{G} \cdot \mathbf{i}') = \alpha_1 G_x + \alpha_2 G_y + \alpha_3 G_z,$$

and so on. The set of nine direction cosines thus completely spells out the transformation between the two coordinate systems.

If the primed axes are taken as fixed in the body, then the nine direction cosines will be functions of time as the body changes its orientation in the course of the motion. In this sense the α's, β's and γ's can be considered as coordinates describing the instantaneous orientation of the body, relative to a coordinate system fixed in space but with origin in common with the body system. But, clearly, they are not independent coordinates, for there are nine of them and it has been shown that only three coordinates are needed to specify an orientation.

The connections between the direction cosines arise from the fact that the basis vectors in both coordinate systems are orthogonal to each other and have unit magnitude; in symbols,

$$\mathbf{i} \cdot \mathbf{j} = \mathbf{j} \cdot \mathbf{k} = \mathbf{k} \cdot \mathbf{i} = 0,$$

and

$$\mathbf{i} \cdot \mathbf{i} = \mathbf{j} \cdot \mathbf{j} = \mathbf{k} \cdot \mathbf{k} = 1, \tag{4–7}$$

with similar relations for $\mathbf{i}'$, $\mathbf{j}'$, and $\mathbf{k}'$. We can obtain the conditions satisfied by the nine coefficients by forming all possible dot products among the three equations for $\mathbf{i}$, $\mathbf{j}$, and $\mathbf{k}$ in terms of $\mathbf{i}'$, $\mathbf{j}'$, and $\mathbf{k}'$ (as in Eq. 4–5), making use of the Eqs. (4–7):

$$\alpha_l \alpha_m + \beta_l \beta_m + \gamma_l \gamma_m = 0, \qquad l, m = 1, 2, 3; \, l \neq m,$$

$$\alpha_l^2 + \beta_l^2 + \gamma_l^2 = 1, \qquad l = 1, 2, 3. \tag{4–8}$$

These two sets of three equations each are exactly sufficient to reduce the number of independent quantities from nine to three. Formally, the six equations can be combined into one by using the Kronecker δ-symbol δ_{lm}, defined by

$$\delta_{lm} = 1 \qquad l = m$$

$$= 0 \qquad l \neq m.$$

Equations (4–8) can then be written as

$$\alpha_l \alpha_m + \beta_l \beta_m + \gamma_l \gamma_m = \delta_{lm}. \tag{4–9}$$

It is not possible, therefore, to set up a Lagrangian and subsequent equations of motion with the nine direction cosines as generalized coordinates. For this purpose we must use some set of three independent functions of the direction cosines. A number of such sets of independent variables will be described later, the most important being the Euler angles. The use of direction cosines to describe the connections between two Cartesian coordinate systems nevertheless has a number of important advantages. With their aid many of the theorems about the motion of rigid bodies can be expressed with great elegance and generality, and in a form naturally leading to the procedures necessarily used in special relativity and quantum mechanics. Such a mode of description therefore merits an extended discussion here.

4–2 ORTHOGONAL TRANSFORMATIONS

To study the properties of the nine direction cosines with greater ease it is convenient to change the notation and denote all coordinates by x, distinguishing the axes by subscripts:

$$x \rightarrow x_1$$

$$y \rightarrow x_2 \tag{4–10}$$

$$z \rightarrow x_3.$$

Thus the Eqs. (4–6) become

$$x_1' = \alpha_1 x_1 + \alpha_2 x_2 + \alpha_3 x_3$$

$$x_2' = \beta_1 x_1 + \beta_2 x_2 + \beta_3 x_3 \tag{4–11}$$

$$x_3' = \gamma_1 x_1 + \gamma_2 x_2 + \gamma_3 x_3.$$

Equations (4–11) constitute a group of transformation equations from a set of coordinates x_1, x_2, x_3 to a new set x'_1, x'_2, x'_3. In particular they form an example of a *linear* or *vector* transformation, defined by transformation equations of the form

$$x'_1 = a_{11}x_1 + a_{12}x_2 + a_{13}x_3$$
$$x'_2 = a_{21}x_1 + a_{22}x_2 + a_{23}x_3 \qquad (4\text{–}12)$$
$$x'_3 = a_{31}x_1 + a_{32}x_2 + a_{33}x_3,$$

where the $a_{11}, a_{12}, \ldots$ are any set of constant (independent of x, x') coefficients.* To simplify the appearance of many of the expressions we will also make use of the summation convention first introduced by Einstein: Whenever an index occurs two or more times in a term, it is implied, without any further symbols, that the terms are to be summed over all possible values of the index. Thus Eqs. (4–12) can be written most compactly in accordance with this convention as

$$x'_i = a_{ij}x_j, \qquad i = 1, 2, 3. \qquad (4\text{–}12')$$

The repeated appearance of the index j indicates that the left-hand side of Eq. (4–12′) is a sum over the dummy index j for all possible values (here, $j = 1, 2, 3$). Some ambiguity is possible where powers of an indexed quantity occur, and for that reason an expression such as

$$\sum_i x_i^2$$

appears under the summation convention as

$$x_i x_i.$$

For the rest of the book the summation convention should be automatically assumed in reading the equations unless otherwise explicitly indicated. Where convenient, or to remove ambiguity, the summation sign may be occasionally displayed explicitly, e.g., when certain values of the index are to be excluded from the summation.

The transformation represented by Eqs. (4–11) is only a special case of the general linear transformation, Eqs. (4–12), since the direction cosines are not all independent. The connections between the coefficients, Eqs. (4–8) may be rederived here in terms of the newer notation. Since both coordinate systems are Cartesian, the magnitude of a vector is given in terms of the sum of squares of the components. Further, since the actual vector remains unchanged no matter which coordinate system is used, the magnitude of the vector must be the same in both systems. In symbols we can state the invariance of the magnitude as

$$x'_i x'_i = x_i x_i. \qquad (4\text{–}13)$$

---

* Equations (4–12), of course, are not the most general set of transformation equations, cf., for example, those from the **r**'s to the **q**'s (1–38).

The left-hand side of Eq. (4–13) is therefore

$$a_{ij}a_{ik}x_jx_k,$$

and it will reduce to the right-hand side of Eq. (4–13), if, and only if

$$a_{ij}a_{ik} = 1 \qquad j = k$$
$$\qquad\qquad = 0 \qquad j \neq k, \tag{4–14}$$

or, in a more compact form, if

$$a_{ij}a_{ik} = \delta_{jk}, \qquad j, k = 1, 2, 3. \tag{4–15}$$

When the a_{ij} coefficients are expressed in terms of the α, β, γ's the six equations contained in Eq. (4–15) become identical with the Eqs. (4–9).

Any linear transformation, (4–12), that has the properties required by Eq. (4–15) is called an *orthogonal* transformation, and Eq. (4–15) itself is known as the *orthogonality condition*. Thus, the transition from coordinates fixed in space to coordinates fixed in the rigid body (with common origin) is accomplished by means of an orthogonal transformation. The array of transformation quantities (the direction cosines), written as

$$\begin{pmatrix} a_{11} & a_{12} & a_{13} \\ a_{21} & a_{22} & a_{23} \\ a_{31} & a_{32} & a_{33} \end{pmatrix}, \tag{4–16}$$

is called the *matrix of transformation*, and will be denoted by a capital letter **A**. The quantities a_{ij} are correspondingly known as the *matrix elements* of the transformation.

To make these formal considerations more meaningful consider the simple example of motion in a plane, so that we are restricted to two dimensional coordinate systems. Then the indices in the above relations can take on only the values 1, 2, and the transformation matrix reduces to the form

$$\begin{pmatrix} a_{11} & a_{12} \\ a_{21} & a_{22} \end{pmatrix}.$$

The four matrix elements are connected by three orthogonality conditions:

$$a_{ij}a_{ik} = \delta_{jk}, \qquad j, k = 1, 2,$$

and therefore only one independent parameter is needed to specify the transformation. But this conclusion is not surprising. A two-dimensional transformation from one Cartesian coordinate system to another corresponds to a rotation of the axes in the plane, cf. Fig. 4–4, and such a rotation can be specified completely by only one quantity, the rotation angle ϕ. Expressed in terms of this single parameter, the transformation equations become

$$x_1' = x_1 \cos \phi + x_2 \sin \phi$$
$$x_2' = -x_1 \sin \phi + x_2 \cos \phi.$$

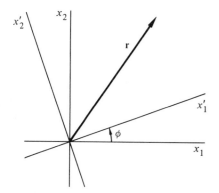

FIGURE 4-4

Rotation of the coordinate axes, as equivalent to two-dimensional orthogonal transformation.

The matrix elements are therefore

$$a_{11} = \cos \phi, \qquad a_{12} = \sin \phi,$$
$$a_{21} = -\sin \phi, \qquad a_{22} = \cos \phi,$$

$$(4\text{--}17)$$

so that the matrix **A** can be written

$$\mathbf{A} = \begin{pmatrix} \cos \phi & \sin \phi \\ -\sin \phi & \cos \phi \end{pmatrix}.$$

$$(4\text{--}17')$$

The three orthogonality conditions expand into the equations

$$a_{11}a_{11} + a_{21}a_{21} = 1$$
$$a_{12}a_{12} + a_{22}a_{22} = 1$$
$$a_{11}a_{12} + a_{21}a_{22} = 0.$$

These conditions are obviously satisfied by the matrix (4–17'), for in terms of the matrix elements (4–17) they reduce to the identities

$$\cos^2 \phi + \sin^2 \phi = 1$$
$$\sin^2 \phi + \cos^2 \phi = 1$$
$$\cos \phi \sin \phi - \sin \phi \cos \phi = 0.$$

The transformation matrix **A** can be thought of as an *operator* that, acting on the unprimed system, transforms it into the primed system. Symbolically the process might be written

$$(\mathbf{r})' = \mathbf{A}\mathbf{r},$$

$$(4\text{--}18)$$

which is to be read: The matrix **A** operating on the components of a vector in the unprimed system yields the components of the vector in the primed system. It is to be emphasized that in the development of the subject so far, **A** acts on the coordinate system only, the vector is unchanged, and we ask merely for its components in two different coordinate frames. A parenthesis has therefore been placed around **r** on the left in Eq. (4–18) to make clear that the same vector is involved on both sides of the equation. Only the components have changed. In two dimensions the transformation of coordinates, it has been seen, is simply a rotation, and **A** is then identical with the *rotation* operator in a plane.

Despite this, it must be pointed out that without changing the formal mathematics, **A** can also be thought of as an operator acting on the *vector* **r**, changing it to a different vector **r'**:

$$\mathbf{r'} = \mathbf{Ar}, \tag{4–19}$$

with both vectors expressed in the same coordinate system. Thus, in two dimensions, instead of rotating the coordinate system counterclockwise one can rotate the vector **r** *clockwise* by an angle ϕ to a new vector **r'**. The components of the new vector will then be related to the components of the old by the same Eqs. (4–12) that describe the transformation of coordinates. From a formal standpoint it is therefore not necessary to use the parenthesis in Eq. (4–18); rather, it can be written as in Eq. (4–19) and interpreted equally as an operation on the coordinate system or on the vector. The algebra remains the same no matter which of these two points of view is followed. The interpretation as an operator acting on the coordinates is the more pertinent one when using the orthogonal transformation to specify the orientation of a rigid body. On the other hand, the notion of an operator changing one vector into another has the more widespread application. In the mathematical discussion either interpretation will be freely used, as suits the convenience of the situation. Of course, it should be emphasized that the nature of the operation represented by **A** will change according to which interpretation is selected. Thus if **A** corresponds to a *counterclockwise* rotation by an angle ϕ when applied to the coordinate system, it will correspond to a *clockwise* rotation when applied to the vector.

The same duality of roles often occurs with other types of coordinate transformations that are more general than orthogonal transformations. They may at times be looked on as affecting only the coordinate system, expressing

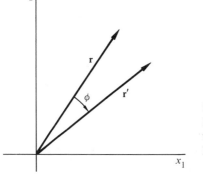

FIGURE 4–5
Illustrating the interpretation of an orthogonal transformation as a rotation of the vector, leaving the coordinate system unchanged.

some given quantity or function in terms of a new coordinate system. At other times they may be considered as operating on the quantity or functions themselves, changing them to new quantities in the same coordinate system. When the transformation is taken as acting only on the coordinate system we speak of the *passive* role of the transformation. In the *active* sense the transformation is looked on as changing the vector or other physical quantity. These alternate interpretations of a transformation will be encountered in various formulations of classical mechanics to be considered below (cf. Chap. 9) and indeed occur in many fields of physics.

To develop further the kinematics of rigid body motion about a fixed origin, we shall make much use of the algebra governing the manipulation of the transformation matrix. The following section is therefore by way of a brief summary of the elementary aspects of matrix algebra with specific application to orthogonal matrices. For those unacquainted with this branch of mathematics, the section should provide an introduction adequate for the immediate purpose. The material also serves to detail the particular terminology and notation we will employ. Those already thoroughly familiar with matrix algebra may however omit the section and proceed directly to Section 4–4.

4–3 FORMAL PROPERTIES OF THE TRANSFORMATION MATRIX

Consider what happens when two successive transformations are made—corresponding to two successive displacements of the rigid body. Let the first transformation from $\mathbf{r}$ to $\mathbf{r}'$ be denoted by $\mathbf{B}$:

$$x'_k = b_{kj}x_j, \tag{4–20}$$

and the succeeding transformation from $\mathbf{r}'$ to a third coordinate set $\mathbf{r}''$ by $\mathbf{A}$:

$$x''_i = a_{ik}x'_k. \tag{4–21}$$

The relation between x''_i and x_j can then be obtained by combining the two Eqs. (4–20) and (4–21):

$$x''_i = a_{ik}b_{kj}x_j.$$

This may also be written as

$$x''_i = c_{ij}x_j, \tag{4–22}$$

where

$$c_{ij} = a_{ik}b_{kj}. \tag{4–23}$$

The successive application of two orthogonal transformations $\mathbf{A}$, $\mathbf{B}$ is thus equivalent to a third linear transformation $\mathbf{C}$. It can be shown that $\mathbf{C}$ is also an orthogonal transformation in consequence of the orthogonality of $\mathbf{A}$ and $\mathbf{B}$. The detailed proof will be left for the exercises. Symbolically the resultant operator $\mathbf{C}$ can be considered as the product of the two operators $\mathbf{A}$ and $\mathbf{B}$:

$$\mathbf{C} = \mathbf{AB},$$

and the matrix elements c_{ij} are by definition the elements of the square matrix obtained by multiplying the two square matrices **A** and **B**.

Note that this "matrix" or operator multiplication is not commutative,

$$\mathbf{BA} \neq \mathbf{AB},$$

for, by definition, the elements of the transformation $\mathbf{D} = \mathbf{BA}$ are

$$d_{ij} = b_{ik}a_{kj}, \tag{4–24}$$

which generally do not agree with the matrix elements of **C**, Eq. (4–23). Thus, the final coordinate system depends on the order of application of the operators **A** and **B**, that is, whether first **A** then **B**, or first **B** and then **A**. However, matrix multiplication is associative; in a product of three or more matrices the order of the *multiplications* is unimportant:

$$(\mathbf{AB})\mathbf{C} = \mathbf{A}(\mathbf{BC}). \tag{4–25}$$

In Eq. (4–19) the juxtaposition of **A** and **r**, to indicate the operation of **A** on the coordinate system (or on the vector), was said to be merely symbolic. But, by extending our concept of matrices, it may also be taken as indicating an actual matrix multiplication. Up to the present the matrices used have been square, i.e., with equal number of rows and columns. However, we may also have one-column matrices, such as **x** and **x**' defined by

$$\mathbf{x} = \begin{pmatrix} x_1 \\ x_2 \\ x_3 \end{pmatrix}, \qquad \mathbf{x}' = \begin{pmatrix} x_1' \\ x_2' \\ x_3' \end{pmatrix}. \tag{4–26}$$

The product **Ax**, by definition, shall be taken as a one-column matrix, with the elements

$$(\mathbf{Ax})_i = a_{ij}x_j = x_i'.$$

Hence Eq. (4–19) can also be written as the matrix equation

$$\mathbf{x}' = \mathbf{Ax}.$$

The *addition* of two matrices, while not as important a concept as multiplication, is a frequently used operation. The sum $\mathbf{A} + \mathbf{B}$ is a matrix **C** whose elements are the sum of the corresponding elements of **A** and **B**:

$$c_{ij} = a_{ij} + b_{ij}.$$

Of greater importance is the transformation inverse to **A**, the operation that changes r' back to r. This transformation will be called $\mathbf{A}^{-1}$ and its matrix elements designated by a_{ij}'. We then have the set of equations

$$x_i = a_{ij}'x_j', \tag{4–27}$$

which must be consistent with

$$x_k' = a_{ki}x_i. \tag{4–28}$$

Substituting x_i from (4–27), Eq. (4–28) becomes

$$x'_k = a_{ki}a'_{ij}x'_j. \tag{4–29}$$

Since the components of $\mathbf{r}'$ are independent, Eq. (4–29) is correct only if the summation reduces identically to x'_k. The coefficient of x'_j must therefore be 1 for $j = k$ and zero for $j \neq k$; in symbols,

$$a_{ki}a'_{ij} = \delta_{kj}. \tag{4–30}$$

The left-hand side of Eq. (4–30) is easily recognized as the matrix element for the product $\mathbf{AA}^{-1}$, while the right-hand side is the element of the matrix known as the unit matrix $\mathbf{1}$:

$$\mathbf{1} = \begin{pmatrix} 1 & 0 & 0 \\ 0 & 1 & 0 \\ 0 & 0 & 1 \end{pmatrix}. \tag{4–31}$$

Equation (4–30) can therefore be written as

$$\mathbf{AA}^{-1} = \mathbf{1}, \tag{4–32}$$

which indicates the reason for the designation of the inverse matrix by $\mathbf{A}^{-1}$. The transformation corresponding to $\mathbf{1}$ is known as the *identity transformation*, producing no change in the coordinate system:

$$\mathbf{x} = \mathbf{1x}.$$

Similarly multiplying any matrix $\mathbf{A}$ by $\mathbf{1}$, in any order, leaves $\mathbf{A}$ unaffected:

$$\mathbf{1A} = \mathbf{A1} = \mathbf{A}.$$

By slightly changing the order of the proof of Eq. (4–32) it can be shown that $\mathbf{A}$ and $\mathbf{A}^{-1}$ commute. Instead of substituting x_i in Eq. (4–29) in terms of x', one could equally as well demand consistency by eliminating x' from the two equations, leading in analogous fashion to

$$a'_{ij}a_{jk} = \delta_{ik}.$$

In matrix notation this reads

$$\mathbf{A}^{-1}\mathbf{A} = \mathbf{1}, \tag{4–33}$$

which proves the statement.

Consider now the double sum

$$a_{kl}a_{ki}a'_{ij},$$

which can be written either as

$$c_{li}a'_{ij} \qquad \text{with } c_{li} = a_{kl}a_{ki}$$

or as

$$a_{kl}d_{kj} \qquad \text{with } d_{kj} = a_{ki}a'_{ij}.$$

Applying the orthogonality conditions, Eq. (4–15), the sum in the first form reduces to

$$\delta_{li}a'_{ij} = a'_{lj}.$$

On the other hand, the same sum from the second point of view, and with the help of Eq. (4–30), can be written

$$a_{kl}\,\delta_{kj} = a_{jl}.$$

Thus the elements of the direct matrix **A** and the reciprocal $\mathbf{A}^{-1}$ are related by

$$a'_{lj} = a_{jl}. \tag{4–34}$$

In general, the matrix obtained from **A** by interchanging rows and columns is known as the *transposed matrix*, indicated by the tilde thus: **Ã**. Equation (4–34) therefore states that for *orthogonal matrices* the reciprocal matrix is to be identified as the transposed matrix; symbolically:

$$\mathbf{A}^{-1} = \mathbf{\tilde{A}}. \tag{4–35}$$

If this result is substituted in Eq. (4–33), we obtain

$$\mathbf{\tilde{A}A} = \mathbf{1}, \tag{4–36}$$

which is identical with the set of orthogonality conditions, Eq. (4–15), written in abbreviated form, as can be verified by direct expansion.* Similarly, an alternative form of the orthogonality conditions can be obtained from Eq. (4–30) by substituting (4–34):

$$a_{ki}a_{ji} = \delta_{kj}. \tag{4–37}$$

In symbolic form (4–37) can be written

$$\mathbf{A\tilde{A}} = \mathbf{1}$$

and may be derived directly from (4–36) by multiplying it from the left by **A** and from the right by $\mathbf{A}^{-1}$.

A rectangular matrix is said to be of dimension $m \times n$ if it has m rows and n columns, i.e., if the matrix element is a_{ij}, then i runs from 1 to m, and j from 1 to n. Clearly, the transpose of such a matrix has the dimension $n \times m$. If a vector column matrix is considered as a rectangular matrix of dimension $m \times 1$, the transpose of a vector is of dimension $1 \times m$, that is, a one-row matrix. The product **AB** of two rectangular matrices exists only if the number of columns of **A**

---

* Indeed one may obtain (4–35) directly from the orthogonality conditions in the form (4–36) and the brevity of the proof is indicative of the power of the symbolic procedures. Multiply (4–36) by $\mathbf{A}^{-1}$ from the right:

$$\mathbf{\tilde{A}AA}^{-1} = \mathbf{A}^{-1},$$

and by (4–32) there results

$$\mathbf{\tilde{A}} = \mathbf{A}^{-1}.$$

is the same as the number of rows of **B**. This is an obvious consequence of the definition of the multiplication operation leading to a matrix element:

$$c_{ij} = a_{ik}b_{kj}.$$

From this viewpoint, the product of a vector column matrix with a square matrix does not exist. The only product between these quantities that can be formed is that of a square matrix with a single column matrix. But note that a single row matrix, that is, a vector transpose, can indeed pre-multiply a square matrix. For a vector, however, the distinction between the column matrix and its transpose is often of no consequence. The symbol **x** may therefore be used to denote either a column or a row matrix, as the situation warrants.* Thus in the expression **Ax**, where **A** is a square matrix, the symbol **x** stands for a column matrix, whereas in the expression **xA** it represents the same elements arranged in a single row. It should be noted that the ith component of **Ax** can be written as

$$A_{ij}x_j = x_j(\tilde{\mathbf{A}})_{ji}.$$

Hence we have a useful commutation property of the product of a vector and a square matrix that

$$\mathbf{Ax} = \mathbf{x\tilde{A}}.$$

A square matrix that is the same as its transpose,

$$A_{ij} = A_{ji},$$

is said (for obvious reasons) to be *symmetric*. When the transpose is the negative of the matrix,

$$A_{ij} = -A_{ji},$$

the matrix is *antisymmetric* or *skew symmetric*. It is clear that in an antisymmetric matrix, the diagonal elements are always zero. For any square matrix **A**, the matrix **A**$_s$ defined as

$$\mathbf{A_s} = \tfrac{1}{2}(\mathbf{A} + \mathbf{\tilde{A}})$$

is symmetric, and a corresponding antisymmetric matrix can be defined as

$$\mathbf{A_a} = \tfrac{1}{2}(\mathbf{A} - \mathbf{\tilde{A}}).$$

It obviously follows that

$$\mathbf{A} = \mathbf{A_s} + \mathbf{A_a},$$

and

$$\mathbf{\tilde{A}} = \mathbf{A_s} - \mathbf{A_a}.$$

---

* The transpose sign on vector matrices will occasionally be retained where it is useful to emphasize the distinction between column and row matrices.

Associated with the notion of the transposed matrix is its complex conjugate known to physicists as the *adjoint matrix*, and indicated by a dagger, †:

$$\mathbf{A}^\dagger = (\tilde{\mathbf{A}})^* \qquad (4\text{--}38)$$

where the * stands, as customary, for the complex conjugate. Analogous to the definition (4–36) for an orthogonal matrix, a *unitary matrix* **A** satisfies the condition

$$\mathbf{A}^\dagger \mathbf{A} = 1. \qquad (4\text{--}39)$$

In the problem of specifying the orientation of a rigid body the transformation matrix must be real, for both **x** and **x'** are real. There is then no distinction between the orthogonality and the unitary property or between transposed and adjoint matrices. In short, a real orthogonal matrix is unitary. But we shall soon have occasion in this chapter, and later in connection with relativity, to introduce complex matrices. There the difference is significant. Many of the properties of the transposed matrix have obvious analogs for the adjoint matrix. It should be noted, however, that the adjoint of a vector matrix is not equivalent to the vector because of the possible effect of complex-conjugation. A matrix that is identical with its adjoint is called *self-adjoint* or *hermitean*.*

The two interpretations of an operator as transforming the vector, or alternatively the coordinate system, are both involved if we seek to find the transformation of an operator under a change of coordinates. Let **A** be considered an operator acting upon a vector **F** (or a single-column matrix **F**) to produce a vector **G**:

$$\mathbf{G} = \mathbf{AF}.$$

If the coordinate system is transformed by a matrix **B** the components of the vector **G** in the new system will be given by

$$\mathbf{BG} = \mathbf{BAF},$$

which can also be written

$$\mathbf{BG} = \mathbf{BAB}^{-1}\mathbf{BF}. \qquad (4\text{--}40)$$

Equation (4–40) can be interpreted as stating that the operator $\mathbf{BAB}^{-1}$ acting upon the vector **F**, expressed in the new system, produces the vector **G**, likewise expressed in the new coordinates. We may therefore consider $\mathbf{BAB}^{-1}$ to be the form taken by the operator **A** when transformed to a new set of axes:

$$\mathbf{A'} = \mathbf{BAB}^{-1}. \qquad (4\text{--}41)$$

---

* The reader should beware that the term "adjoint matrix" is given an entirely different meaning in much of the mathematical literature, having to do rather with the inverse of a matrix. Mathematicians must often struggle therefore with clumsy designations such as "complex-conjugate transpose matrix." The meanings used here for adjoint and self-adjoint have become imbedded in physics through use in quantum mechanics.

Any transformation of a matrix having the form of Eq. (4–41) is known as a *similarity transformation.*

It is appropriate at this point to consider the properties of the determinant formed from the elements of a square matrix. As is customary, we shall denote such a determinant by vertical bars, thus: $|\mathbf{A}|$. It will be noticed that the definition of matrix multiplication is identical with that for the multiplication of determinants (cf. Bôcher, *Introduction to Higher Algebra*, p. 26). Hence

$$|\mathbf{AB}| = |\mathbf{A}| \cdot |\mathbf{B}|.$$

Since the determinant of the unit matrix is 1, the determinantal form of the orthogonality conditions, Eq. (4–36), can be written

$$|\tilde{\mathbf{A}}| \cdot |\mathbf{A}| = 1.$$

Further, as the value of a determinant is unaffected by interchanging rows and columns, we can write

$$|\mathbf{A}|^2 = 1, \tag{4-42}$$

which implies that the determinant of an orthogonal matrix can only be $+1$ or -1. (The geometrical significance of these two values will be considered in the next section.)

When the matrix is not orthogonal the determinant does not have these simple values, of course. It can be shown however that the value of the determinant is invariant under a similarity transformation. Multiplying the equation (4–41) for the transformed matrix from the right by $\mathbf{B}$, we obtain the relation

$$\mathbf{A'B} = \mathbf{BA},$$

or in determinantal form

$$|\mathbf{A'}| \cdot |\mathbf{B}| = |\mathbf{B}| \cdot |\mathbf{A}|.$$

Since the determinant of $\mathbf{B}$ is merely a number, and not zero,* we can divide by $|\mathbf{B}|$ on both sides to obtain the desired result:

$$|\mathbf{A'}| = |\mathbf{A}|.$$

In discussing rigid body motion later, all these properties of matrix transformations, especially of orthogonal matrices, will be employed. In addition, other properties are needed, and they will be derived as the occasion requires.

4-4 THE EULER ANGLES

It has already been noted (cf. p. 131) that the nine elements a_{ij} are not suitable as generalized coordinates because they are not independent quantities. The six

---

* If it were zero there could be no inverse operator $\mathbf{B}^{-1}$ (by Cramer's rule), which is required in order that Eq. (4–41) make sense.

relations that express the orthogonality conditions, Eqs. (4–9) or Eqs. (4–15), of course reduce the number of independent elements to three. But in order to characterize the motion of a rigid body there is an additional requirement the matrix elements must satisfy, beyond those implied by orthogonality. In the previous section it was pointed out that the determinant of a real orthogonal matrix could have the value $+1$ or -1. The following argument shows however that an orthogonal matrix whose determinant is -1 cannot represent a physical displacement of a rigid body.

Consider a simple matrix with the determinant -1:

$$\mathbf{S} = \begin{pmatrix} -1 & 0 & 0 \\ 0 & -1 & 0 \\ 0 & 0 & -1 \end{pmatrix} = -1.$$

The transformation $\mathbf{S}$ has the effect of changing the sign of each of the components or coordinate axes (cf. Fig. 4–6). Such an operation transforms a right-handed coordinate system into a left-handed one and is known as an *inversion* or *reflection* of the coordinate axes.

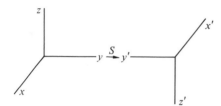

FIGURE 4–6
Inversion of the coordinate axes.

From the nature of this operation it is clear that an inversion of a right-handed system into a left-handed one cannot be accomplished by any *rigid* change in the orientation of the coordinate axes. An inversion therefore never corresponds to a physical displacement of a rigid body. What is true for $\mathbf{S}$ is equally valid for any matrix whose determinant is -1, for any such matrix can be written as the product of $\mathbf{S}$ with a matrix whose determinant is $+1$, and thus includes the inversion operation. Consequently it cannot describe a rigid change in orientation. Therefore, the transformations representing rigid body motion must be restricted to matrices having the determinant $+1$. Another method of reaching this conclusion starts from the fact that the matrix of transformation must evolve continuously from the unit matrix, which of course has the determinant $+1$. It would be incompatible with the continuity of the motion to have the matrix determinant change suddenly from its initial value $+1$ to -1 at some given time. Orthogonal transformations with determinant $+1$ are said to be *proper*, so naturally those with the determinant -1 are called *improper*.

In order to describe the motion of rigid bodies in the Lagrangian formulation of mechanics, it will therefore be necessary to seek three independent parameters specifying the orientation of a rigid body in such a manner that the corresponding orthogonal matrix of transformation has the determinant $+1$. Only when such generalized coordinates have been found can one write a Lagrangian for the system and obtain the Lagrangian equations of motion. A number of such sets of parameters have been described in the literature, but the most common and useful are the *Euler angles*.* We shall therefore define these angles at this point, and show how the elements of the orthogonal transformation matrix can be expressed in terms of them.

One can carry out the transformation from a given Cartesian coordinate system to another by means of three successive rotations performed in a specific sequence. The Euler angles are then defined as the three successive angles of rotation. Within limits, the choice of rotation angles is arbitrary. The main convention that will be followed here is used widely in celestial mechanics, applied mechanics, and frequently in molecular and solid state physics. Other conventions will be described below.

The sequence employed here is started by rotating the initial system of axes, xyz, by an angle ϕ counterclockwise about the z axis, and the resultant coordinate system is labeled the $\xi\eta\zeta$ axes. In the second stage the intermediate axes, $\xi\eta\zeta$, are rotated about the ξ axis counterclockwise by an angle θ to produce another intermediate set, the $\xi'\eta'\zeta'$ axes. The ξ' axis is at the intersection of the xy and $\xi'\eta'$ planes and is known as the *line of nodes*. Finally the $\xi'\eta'\zeta'$ axes are rotated counterclockwise by an angle ψ about the ζ' axis to produce the desired $x'y'z'$ system of axes. Figure 4–7 illustrates the various stages of the sequence. The Euler angles θ, ϕ, and ψ thus completely specify the orientation of the $x'y'z'$ system relative to the xyz and can therefore act as the three needed generalized coordinates.†

The elements of the complete transformation **A** can be obtained by writing the matrix as the triple product of the separate rotations, each of which has a relatively simple matrix form. Thus, the initial rotation about z can be described by a matrix **D**:

$$\xi = \mathbf{D}\mathbf{x},$$

where ξ and **x** stand for column matrices. Similarly the transformation from $\xi\eta\zeta$ to $\xi'\eta'\zeta'$ can be described by a matrix **C**,

$$\xi' = \mathbf{C}\xi,$$

and the last rotation to $x'y'z'$ by a matrix **B**

$$\mathbf{x}' = \mathbf{B}\xi'.$$

---

* Also denoted, interchangeably, as Euler's angles, or Eulerian angles.

† A number of minor variations will be found in the older literature even within this convention. The differences are not very great, but they are often sufficient to frustrate easy comparison of the end formulae, such as the matrix elements. Greatest confusion, perhaps, arises from the occasional use of left-handed coordinate systems (as by Osgood and by Margenau and Murphy). Some European authors agree with the practice given here except that the meanings of ϕ and ψ are interchanged.

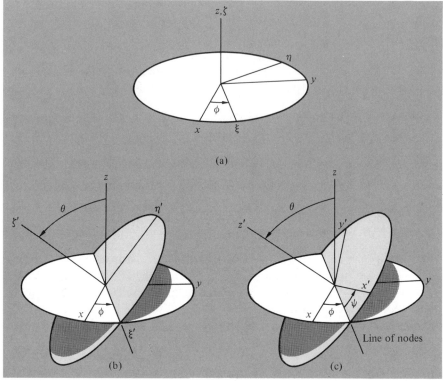

FIGURE 4–7
The rotations defining the Eulerian angles.

Hence the matrix of the complete transformation

$$\mathbf{x}' = \mathbf{A}\mathbf{x}$$

is the product of the successive matrices,

$$\mathbf{A} = \mathbf{BCD}.$$

Now the **D** transformation is a rotation about z, and hence has a matrix of the form (cf. Eq. (4–17))

$$\mathbf{D} = \begin{pmatrix} \cos\phi & \sin\phi & 0 \\ -\sin\phi & \cos\phi & 0 \\ 0 & 0 & 1 \end{pmatrix}. \qquad (4\text{–}43)$$

The **C** transformation corresponds to a rotation about ξ, with the matrix

$$\mathbf{C} = \begin{pmatrix} 1 & 0 & 0 \\ 0 & \cos\theta & \sin\theta \\ 0 & -\sin\theta & \cos\theta \end{pmatrix}, \qquad (4\text{–}44)$$

and finally **B** is a rotation about ζ' and therefore has the same form as **D**:

$$\mathbf{B} = \begin{pmatrix} \cos\psi & \sin\psi & 0 \\ -\sin\psi & \cos\psi & 0 \\ 0 & 0 & 1 \end{pmatrix}. \tag{4-45}$$

The product matrix $\mathbf{A} = \mathbf{BCD}$ then follows as

$$\mathbf{A} = \begin{pmatrix} \cos\psi\cos\phi - \cos\theta\sin\phi\sin\psi & \cos\psi\sin\phi + \cos\theta\cos\phi\sin\psi & \sin\psi\sin\theta \\ -\sin\psi\cos\phi - \cos\theta\sin\phi\cos\psi & -\sin\psi\sin\phi + \cos\theta\cos\phi\cos\psi & \cos\psi\sin\theta \\ \sin\theta\sin\phi & -\sin\theta\cos\phi & \cos\theta \end{pmatrix}. \tag{4-46}$$

The inverse transformation from body coordinates to space axes

$$\mathbf{x} = \mathbf{A}^{-1}\mathbf{x}'$$

is then given immediately by the transposed matrix $\tilde{\mathbf{A}}$:

$$\mathbf{A}^{-1} = \tilde{\mathbf{A}} = \begin{pmatrix} \cos\psi\cos\phi - \cos\theta\sin\phi\sin\psi & -\sin\psi\cos\phi - \cos\theta\sin\phi\cos\psi & \sin\theta\sin\phi \\ \cos\psi\sin\phi + \cos\theta\cos\phi\sin\psi & -\sin\psi\sin\phi + \cos\theta\cos\phi\cos\psi & -\sin\theta\cos\phi \\ \sin\theta\sin\psi & \sin\theta\cos\psi & \cos\theta \end{pmatrix}. \tag{4-47}$$

Verification of the multiplication, and demonstration that **A** represents a proper, orthogonal matrix will be left to the exercises.

It will be noted that the sequence of rotations used to define the final orientation of the coordinate system is to some extent arbitrary. The initial rotation could be taken about any of the three Cartesian axes. In the subsequent two rotations, the only limitation is that no two successive rotations can be about the same axis. A total of twelve conventions is therefore possible in defining the Euler angles (in a right-handed coordinate system). The two most frequently used conventions differ only in the choice of axis for the second rotation. In the Euler's angle definitions described above, and used throughout the book, the second rotation is about the intermediate x axis. We will refer to this choice as the *x-convention*. In quantum mechanics, nuclear physics, and particle physics, the custom has arisen to take the second defining rotation about the intermediate y axis,* and this form will be denoted as the *y-convention*.

A third convention is commonly used in engineering applications relating to the orientation of moving vehicles such as aircraft and satellites. Both the x- and y-conventions have the drawback that when the primed coordinate system is only slightly different from the unprimed system, the angles ϕ and ψ become indistinguishable, as their respective axes of rotation, z and z' are then nearly coincident. To get around this problem all three rotations are taken around different axes. The first rotation is about the vertical axis and gives the *heading* or *yaw* angle. The second is around a perpendicular axis fixed in the vehicle and

---

* The usage of Wigner in *Group Theory and Its Applications to the Quantum Mechanics of Atomic Spectra* and of Rose in *Elementary Theory of Angular Momentum* appears to have been decisive in this regards.

normal to the figure axis; it is measured by the *pitch* or *attitude* angle. Finally the third angle is one of rotation about the figure axis of the vehicle and is the *roll* or *bank* angle. Because all three axes are involved in the rotations it will be designated as the *xyz-convention* (although the order of axes chosen may actually be different). This last convention is sometimes referred to as the *Tait–Bryan* angles.

While only the *x*-convention will be used in the text, for reference purposes Appendix B lists all the formulas involving Euler's angles, such as rotation matrices, in both the *y*- and *xyz*-conventions.

4–5 THE CAYLEY–KLEIN PARAMETERS AND RELATED QUANTITIES

We have seen that only three independent quantities are needed to specify the orientation of a rigid body. Nonetheless, there are occasions when it is desirable to use sets of variables containing more than the minimum number of quantities to describe a rotation even though they are not suitable as generalized coordinates. Thus, Felix Klein introduced the set of four parameters bearing his name to facilitate the integration of complicated gyroscopic problems. The Euler angles are difficult to use in numerical computation because of the large number of trigonometric functions involved, and the four-parameter representations are much better adapted for use on computers. Further, the four-parameter sets are of great theoretical interest in branches of physics beyond the scope of this book, wherever rotations or rotational symmetry are involved. It therefore seems worthwhile to devote some space to describe these enlarged parameter sets. However, none of the results of this section will be directly used in the discussion of rigid body motion in the following chapter.

In the previous sections we employed on occasion a two-dimensional real space with axes x_1 and x_2 to illustrate the properties of orthogonal transformations. We shall now consider a different two-dimensional space, this time having complex axes denoted by u and v. A general linear transformation in such a space appears as

$$u' = \alpha u + \beta v,$$
$$v' = \gamma u + \delta v,$$

(4–48)

with the corresponding transformation matrix

$$Q = \begin{pmatrix} \alpha & \beta \\ \gamma & \delta \end{pmatrix}.$$

(4–49)

As it stands Q has eight quantities to be specified, since each of the four elements is complex. To reduce the transformation to three independent quantities, additional conditions must be imposed on Q. For much of the following discussion, it is sufficient to require that the transformation be such that Q is unitary:

$$Q^\dagger Q = 1 = QQ^\dagger.$$

(4–50)

The unitary condition also implies that the *magnitude* of the determinant of $\mathbf{Q}$ must be unity:

$$|\mathbf{Q}|^* |\mathbf{Q}| = 1. \tag{4–51}$$

In expanded form Eq. (4–50) can be written as the equations

$$\alpha\alpha^* + \gamma\gamma^* = 1,$$
$$\beta\beta^* + \delta\delta^* = 1, \tag{4–52}$$
$$\alpha^*\beta + \gamma^*\delta = 0.$$

The first two of Eqs. (4–52) are real, while the last is complex so that together they comprise four conditions. In order that the elements of $\mathbf{Q}$ involve only three independent quantities, an additional condition must be imposed, by requiring that the determinant be exactly $+1$; that is,

$$\alpha\delta - \beta\gamma = 1. \tag{4–53}$$

Equation (4–53) is complex, so it might be thought to involve two conditions. But the unitary property, as expressed by Eq. (4–50) or Eqs. (4–52), already fixes the magnitude of the determinant, and Eq. (4–53) only serves to fix the phase angle. Thus only one of the conditions implied by Eq. (4–53) is independent of the already imposed unitarity requirement. Matrices with determinant $+1$ will be called *unimodular*.* Transformations in a two-dimensional complex space with unitary unimodular transformation matrices therefore involve only three independent quantities, the same number required to specify the orientation of a rigid body.

Some of the reduction to independent parameters can be performed without much difficulty. Thus, from the last of Eqs. (4–52) one may write

$$\delta = -\alpha^* \frac{\beta}{\gamma^*}, \tag{4–54}$$

which when substituted in the determinant condition Eq. (4–53) yields

$$-\frac{\beta}{\gamma^*}(\alpha\alpha^* + \gamma\gamma^*) = 1.$$

The first of Eqs. (4–52) states that the quantity in parentheses is unity, and hence

$$\beta = -\gamma^*. \tag{4–55}$$

It follows then from (4–54) that

$$\delta = \alpha^*. \tag{4–56}$$

---

* The designation is not universal (nor even entirely consistent with the literal meaning) but appears to be widely followed.

As a result of these four conditions (Eqs. 4–55 and 4–56), the matrix $\mathbf{Q}$ could be written also as

$$\mathbf{Q} = \begin{pmatrix} \alpha & \beta \\ -\beta^* & \alpha^* \end{pmatrix} \tag{4-57}$$

with one remaining condition (derivable from either of the first two of Eqs. 4–52 or from Eq. 4–53):

$$\alpha\alpha^* + \beta\beta^* = 1. \tag{4-58}$$

However, we often prefer to leave the matrix in the form (4–49).

Consider now $\mathbf{Q}$ only as a unitary matrix in a two-dimensional complex space. Let $\mathbf{P}$ be a matrix operator in this space with the specific form

$$\mathbf{P} = \begin{pmatrix} z & x - iy \\ x + iy & -z \end{pmatrix}. \tag{4-59}$$

Mathematically x, y, z can be considered simply as any three real quantities; physically they will be interpreted as coordinates of a point in space. Suppose the $\mathbf{P}$ matrix is transformed by means of the $\mathbf{Q}$ matrix in the following manner:

$$\mathbf{P}' = \mathbf{QPQ}^\dagger. \tag{4-60}$$

From the unitary property of $\mathbf{Q}$, the adjoint $\mathbf{Q}^\dagger$ is the same as the inverse $\mathbf{Q}^{-1}$ and Eq. (4–60) merely represents the similarity transformation of $\mathbf{P}$ when the uv space is subjected to the unitary transformation $\mathbf{Q}$. It will be noted that $\mathbf{P}$ is a hermitean matrix. Further, the sum of the diagonal elements of $\mathbf{P}$, known as the *spur* or *trace* of the matrix, is here zero. Now, it can be shown that both the hermitean property and the trace of a matrix are invariant under similarity transformation (cf. the exercises at the end of the chapter). Hence $\mathbf{P}'$ must likewise be self-adjoint and have a vanishing trace, which can be true only if it has the form

$$\mathbf{P}' = \begin{pmatrix} z' & x' - iy' \\ x' + iy' & -z' \end{pmatrix}, \tag{4-61}$$

where x', y', and z' are real quantities. The determinant of $\mathbf{P}$ is also invariant under the similarity transformation (4–60), so that we can write the following equality:

$$|\mathbf{P}| = -(x^2 + y^2 + z^2) = -(x'^2 + y'^2 + z'^2) = |\mathbf{P}'|.$$

This statement will be recognized as the orthogonality condition; it requires that the length of the vector $\mathbf{r} = x\mathbf{i} + y\mathbf{j} + z\mathbf{k}$ shall be unchanged by the transformation. To each unitary matrix $\mathbf{Q}$ in the complex two-dimensional space there is therefore associated some real orthogonal transformation in ordinary three-dimensional space.

Some insight into the nature of this association is provided by the following considerations. Let the real orthogonal matrix transforming from coordinates $\mathbf{x}$ to $\mathbf{x}'$ be designated by $\mathbf{B}$,

$$\mathbf{x}' = \mathbf{Bx},$$

and denote the associated unitary matrix by $\mathbf{Q}_1$,

$$\mathbf{P}' = \mathbf{Q}_1 \mathbf{P} \mathbf{Q}_1{}^\dagger.$$

A second orthogonal transformation from $\mathbf{x}'$ to $\mathbf{x}''$ may be accomplished by the matrix $\mathbf{A}$,

$$\mathbf{x}'' = \mathbf{A}\mathbf{x}',$$

with the associated matrix $\mathbf{Q}_2$,

$$\mathbf{P}'' = \mathbf{Q}_2 \mathbf{P}' \mathbf{Q}_2{}^\dagger.$$

Now, the direct transformation from $\mathbf{x}$ to $\mathbf{x}''$ is produced by the matrix $\mathbf{C}$ defined by

$$\mathbf{C} = \mathbf{A}\mathbf{B}.$$

Correspondingly the direct transformation from $\mathbf{P}$ to $\mathbf{P}''$ will be effected by a similarity transformation with some matrix $\mathbf{Q}_3$, which must therefore be associated with $\mathbf{C}$. However we can also obtain the transformation from $\mathbf{P}$ to $\mathbf{P}''$ from the equation

$$\mathbf{P}'' = \mathbf{Q}_2 \mathbf{Q}_1 \mathbf{P} \mathbf{Q}_1{}^\dagger \mathbf{Q}_2{}^\dagger.$$

It can easily be shown that

$$\mathbf{Q}_1{}^\dagger \mathbf{Q}_2{}^\dagger = (\mathbf{Q}_2 \mathbf{Q}_1)^\dagger.$$

Since the product of two unitary matrices is also unitary it follows that $\mathbf{Q}_3$ must be identified with the product $\mathbf{Q}_2 \mathbf{Q}_1$:

$$\mathbf{Q}_3 = \mathbf{Q}_2 \mathbf{Q}_1.$$

Thus the correspondence between the 2×2 complex unitary matrices and the 3×3 real orthogonal matrices is such that any relation among the matrices of one set is satisfied also by the corresponding matrices of the other set. The two sets of matrices are said to be *homomorphic*. The arguments on the association of the $\mathbf{Q}$ matrices with orthogonal transformation have so far used only the unitary property of the $\mathbf{Q}$ matrices. Since a unitary 2×2 complex matrix has four independent quantities it is clear that there are many possible $\mathbf{Q}$ matrices corresponding to the same orthogonal transformation. The additional requirement that the determinant of $\mathbf{Q}$ be $+1$ reduces the multiplicity of equivalent $\mathbf{Q}$ matrices, as will be seen, down to a pair of matrices. Further, the unimodularity requirement will be found to restrict the association to orthogonal transformations with determinant $+1$.

The mathematical "jargon" of group theory has often been used, especially in recent years, to give an alternate description of the two types of matrices that are thus in correspondence. It is easy to show (cf. the exercises) that both the real, proper orthogonal matrices and the complex unimodular matrices have the "group property." The group of 3×3 real, proper orthogonal matrices is denoted as $O^+(3)$, the "plus" superscript standing for the sign of the determinant.* The group of 2×2 complex unitary unimodular matrices is correspondingly denoted by SU(2) (special unitary). Thus the correspondence we have derived above is between the $O^+(3)$ group and the SU(2) group.

One can write the elements of an orthogonal matrix in terms of the elements of the homomorphic **Q** matrix. From (4–55) and (4–56) the adjoint to **Q** is

$$\mathbf{Q}^\dagger = \begin{pmatrix} \alpha^* & \gamma^* \\ \beta^* & \delta^* \end{pmatrix} = \begin{pmatrix} \delta & -\beta \\ -\gamma & \alpha \end{pmatrix}.$$

To simplify the calculation we shall introduce the notational abbreviations x_+ and x_- defined as

$$x_+ = x + iy$$
$$x_- = x - iy$$

The transformed matrix **P'** is then written as

$$\mathbf{P'} = \begin{pmatrix} z' & x'_- \\ x'_+ & -z' \end{pmatrix} = \begin{pmatrix} \alpha & \beta \\ \gamma & \delta \end{pmatrix} \begin{pmatrix} z & x_- \\ x_+ & -z \end{pmatrix} \begin{pmatrix} \delta & -\beta \\ -\gamma & \alpha \end{pmatrix}$$

or, upon performing the indicated multiplications,

$$\mathbf{P'} = \begin{pmatrix} (\alpha\delta + \beta\gamma)z - \alpha\gamma x_- + \beta\delta x_+ & -2\alpha\beta z + \alpha^2 x_- - \beta^2 x_+ \\ 2\gamma\delta z - \gamma^2 x_- + \delta^2 x_+ & -(\alpha\delta + \beta\gamma)z + \alpha\gamma x_- - \beta\delta x_+ \end{pmatrix}. \tag{4–62}$$

By equating matrix elements the transformation equations between the primed and unprimed coordinate systems can be written in the form

$$\begin{aligned} x'_+ &= 2\gamma\delta z & -\gamma^2 x_- + \delta^2 x_+ \\ x'_- &= -2\alpha\beta z & +\alpha^2 x_- - \beta^2 x_+ \\ z' &= (\alpha\delta + \beta\gamma)z & -\alpha\gamma x_- + \beta\delta x_+. \end{aligned} \tag{4–63}$$

Finally, the matrix elements a_{ij} can be obtained in terms of α, β, γ, and δ by comparing Eqs. (4–63) with the customary transformation equations (4–14). Thus the last of Eqs. (4–63) may be written as

$$z' = (\beta\delta - \alpha\gamma)x + i(\alpha\gamma + \beta\delta)y + (\alpha\delta + \beta\gamma)z,$$

from which it follows immediately that

$$a_{31} = (\beta\delta - \alpha\gamma), \quad a_{32} = i(\alpha\gamma + \beta\delta), \quad a_{33} = \alpha\delta + \beta\gamma.$$

---

* The designation SO(3)—S for special—is frequently used for $O^+(3)$.

By this process the complete transformation matrix is easily found to be

$$
\mathbf{A} = \begin{pmatrix}
\dfrac{1}{2}(\alpha^2 - \gamma^2 + \delta^2 - \beta^2) & \dfrac{i}{2}(\gamma^2 - \alpha^2 + \delta^2 - \beta^2) & \gamma\delta - \alpha\beta \\[2mm]
\dfrac{i}{2}(\alpha^2 + \gamma^2 - \beta^2 - \delta^2) & \dfrac{1}{2}(\alpha^2 + \gamma^2 + \beta^2 + \delta^2) & -i(\alpha\beta + \gamma\delta) \\[2mm]
\beta\delta - \alpha\gamma & i(\alpha\gamma + \beta\delta) & \alpha\delta + \beta\gamma
\end{pmatrix}. \tag{4–64}
$$

Equation (4–64) provides a matrix that specifies the orientation of a rigid body, and that is expressed entirely in terms of the quantities α, β, γ, and δ. Like the Eulerian angles these four thus furnish a way of establishing the body's orientation; they are customarily known as the *Cayley–Klein parameters*.* Of course the four complex quantities are connected by the relations Eqs. (4–55) and (4–56) and the determinant condition Eq. (4–58). The reality of the matrix elements of **A** in Eq. (4–64) can be shown directly with the aid of Eqs. (4–55) and (4–56) by writing α and β explicitly in terms of their real and imaginary parts:

$$
\begin{aligned}
\alpha &= e_0 + ie_3, \\
\beta &= e_2 + ie_1.
\end{aligned} \tag{4–65}
$$

(Why the apparently odd choice of symbols will become clear later on, cf. p. 166). In terms of these four real quantities the determinant condition Eq. (4–58) becomes

$$
e_0^2 + e_1^2 + e_2^2 + e_3^2 = 1. \tag{4–66}
$$

A bit of algebraic manipulation then shows that the matrix **A** as given by Eq. (4–64) can be written in terms of the four real parameters in the form

$$
\mathbf{A} = \begin{pmatrix}
e_0^2 + e_1^2 - e_2^2 - e_3^2 & 2(e_1 e_2 + e_0 e_3) & 2(e_1 e_3 - e_0 e_2) \\
2(e_1 e_2 - e_0 e_3) & e_0^2 - e_1^2 + e_2^2 - e_3^2 & 2(e_2 e_3 + e_0 e_1) \\
2(e_1 e_3 + e_0 e_2) & 2(e_2 e_3 - e_0 e_1) & e_0^2 - e_1^2 - e_2^2 + e_3^2
\end{pmatrix}. \tag{4–67}
$$

The reality of the matrix elements is now manifest.† It can also be easily demonstrated that the matrix **A** in terms of these parameters cannot be put in the form of the inversion transformation **S**. An examination of the off-diagonal elements and their transposes shows that they all vanish only if at least three of the parameters are zero. One cannot then choose the remaining nonzero parameter such that all three of the diagonal elements (or only one of them) are -1. Hence the

---

* The matrix **A**, Eq. (4–64), does not agree with the corresponding form as given, say, in Whittaker, p. 12. Essentially this is because of a different initial choice of the matrix **P**. Clearly there are many ways of setting up a matrix whose determinant will be $-r^2$, and the specific choice is a matter of convention. The form used here, (4–59), was chosen to agree with customary usage in quantum mechanics.

† The four real parameters e_0, e_1, e_2, e_3 (or slight variants thereof) are occasionally also referred to in the literature as the Cayley–Klein parameters but the more correct practice historically appears to be to denote them as the *Euler parameters*.

representation of $\mathbf{A}$ by Eq. (4–64) or Eq. (4–67) cannot describe a coordinate inversion or indeed any improper orthogonal transformation.

One can express the Cayley–Klein parameters in terms of the corresponding Euler angles; if need be, by direct comparison of the elements of Eq. (4–64) with the elements expressed in terms of ϕ, θ, and ψ. However it is a simpler procedure, and more instructive, to first construct the $\mathbf{Q}$ matrices corresponding to the separate successive rotations that define the Euler angles and then combine them to form the complete matrix. Thus the angle ϕ has been defined in terms of a rotation about the z axis, where the transformation in terms of x_+, x_-, and z appears as

$$x'_+ = e^{-i\phi}x_+$$

$$x'_- = e^{i\phi}x_-$$

$$z' = z.$$

Comparing these equations with (4–63) it is clear that for this simple rotation the elements of the matrix $\mathbf{Q}$ must have the form

$$\gamma = \beta = 0, \qquad \alpha^2 = e^{i\phi}, \qquad \delta^2 = e^{-i\phi},$$

or

$$\mathbf{Q}_\phi = \begin{pmatrix} e^{i\phi/2} & 0 \\ 0 & e^{-i\phi/2} \end{pmatrix}. \tag{4–68}$$

It will be noted that these matrix elements automatically satisfy the conditions (4–55), (4–56), and (4–58).

The next rotation (in the x-convention) is about the new x axis *counterclockwise* by an angle θ, and the identification of the corresponding matrix elements proceeds in a similar fashion, but the calculations become rather tedious. It will simply be stated that the corresponding $\mathbf{Q}$ matrix is

$$\mathbf{Q}_\theta = \begin{pmatrix} \cos\dfrac{\theta}{2} & i\sin\dfrac{\theta}{2} \\ i\sin\dfrac{\theta}{2} & \cos\dfrac{\theta}{2} \end{pmatrix}. \tag{4–69}$$

To check, one can directly verify that

$$\begin{pmatrix} \cos\dfrac{\theta}{2} & i\sin\dfrac{\theta}{2} \\ i\sin\dfrac{\theta}{2} & \cos\dfrac{\theta}{2} \end{pmatrix} \begin{pmatrix} z & x_- \\ x_+ & -z \end{pmatrix} \begin{pmatrix} \cos\dfrac{\theta}{2} & -i\sin\dfrac{\theta}{2} \\ -i\sin\dfrac{\theta}{2} & \cos\dfrac{\theta}{2} \end{pmatrix}$$

$$= \begin{pmatrix} z\cos\theta - y\sin\theta & x - i(y\cos\theta + z\sin\theta) \\ x + i(y\cos\theta + z\sin\theta) & -z\cos\theta + y\sin\theta \end{pmatrix},$$

which leads to the desired transformation:

$$x' = x$$

$$y' = y\cos\theta + z\sin\theta$$

$$z' = -y\sin\theta + z\cos\theta.$$

The final rotation, defining ψ, is again about a z axis so that

$$\mathbf{Q}_\psi = \begin{pmatrix} e^{i\psi/2} & 0 \\ 0 & e^{-i\psi/2} \end{pmatrix}. \tag{4-70}$$

In Section 4-4 the orthogonal matrix for the complete transformation was obtained as the product of the separate matrices for each of the three rotations. It follows from the homomorphism of the 3×3 real orthogonal matrices with the $\mathbf{Q}$ matrices that $\mathbf{Q}$ for the complete transformation is likewise given by the product of the three rotation matrices $\mathbf{Q}_\psi$, $\mathbf{Q}_\theta$, $\mathbf{Q}_\phi$:

$$\mathbf{Q} = \mathbf{Q}_\psi \mathbf{Q}_\theta \mathbf{Q}_\phi = \begin{pmatrix} e^{i\psi/2} & 0 \\ 0 & e^{-i\psi/2} \end{pmatrix} \begin{pmatrix} \cos\dfrac{\theta}{2} & i\sin\dfrac{\theta}{2} \\ i\sin\dfrac{\theta}{2} & \cos\dfrac{\theta}{2} \end{pmatrix} \begin{pmatrix} e^{i\phi/2} & 0 \\ 0 & e^{-i\phi/2} \end{pmatrix},$$

or

$$\mathbf{Q} = \begin{pmatrix} e^{i(\psi + \phi)/2}\cos\dfrac{\theta}{2} & ie^{i(\psi - \phi)/2}\sin\dfrac{\theta}{2} \\ ie^{-i(\psi - \phi)/2}\sin\dfrac{\theta}{2} & e^{-i(\psi + \phi)/2}\cos\dfrac{\theta}{2} \end{pmatrix}. \tag{4-71}$$

The Cayley-Klein parameters in terms of the Euler angles are then

$$\alpha = e^{i(\psi + \phi)/2}\cos\frac{\theta}{2}, \qquad \beta = ie^{i(\psi - \phi)/2}\sin\frac{\theta}{2},$$

$$\gamma = ie^{-i(\psi - \phi)/2}\sin\frac{\theta}{2}, \qquad \delta = e^{-i(\psi + \phi)/2}\cos\frac{\theta}{2}, \tag{4-72}$$

completing the desired identification. Euler angle representations for the four real Euler parameters are immediately provided by Eqs. (4-72):

$$e_0 = \cos\frac{\phi + \psi}{2}\cos\frac{\theta}{2}, \qquad e_2 = \sin\frac{\phi - \psi}{2}\sin\frac{\theta}{2},$$

$$e_1 = \cos\frac{\phi - \psi}{2}\sin\frac{\theta}{2}, \qquad e_3 = \sin\frac{\phi + \psi}{2}\cos\frac{\theta}{2}. \tag{4-72'}$$

Appendix B lists the corresponding formulas for the Cayley-Klein and the Euler parameters in terms of the Euler angles as defined in the other conventions.

It will be noted that the **P** matrix can be written as the sum of three matrices having zero trace:

$$\mathbf{P} = x\boldsymbol{\sigma}_1 + y\boldsymbol{\sigma}_2 + z\boldsymbol{\sigma}_3, \tag{4-73}$$

where $\boldsymbol{\sigma}_1$, $\boldsymbol{\sigma}_2$, and $\boldsymbol{\sigma}_3$ are called the *Pauli spin matrices:*

$$\boldsymbol{\sigma}_1 = \begin{pmatrix} 0 & 1 \\ 1 & 0 \end{pmatrix}, \qquad \boldsymbol{\sigma}_2 = \begin{pmatrix} 0 & -i \\ i & 0 \end{pmatrix}, \qquad \boldsymbol{\sigma}_3 = \begin{pmatrix} 1 & 0 \\ 0 & -1 \end{pmatrix}. \tag{4-74}$$

These three matrices can conveniently be thought of as forming the components of a vector* so that Eq. (4–73) can be compactly written symbolically as

$$\mathbf{P} = \mathbf{r} \cdot \boldsymbol{\sigma}. \tag{4-73'}$$

Together with the unit matrix

$$\mathbf{1} = \begin{pmatrix} 1 & 0 \\ 0 & 1 \end{pmatrix},$$

the Pauli spin matrices form a set of four independent matrices. Consequently, any 2×2 matrix involving four independent quantities can be expressed as a linear function of them. Thus, expressed in terms of the Euler parameters the **Q** matrix can be written as

$$\mathbf{Q} = e_0 \mathbf{1} + i(e_1 \boldsymbol{\sigma}_1 + e_2 \boldsymbol{\sigma}_2 + e_3 \boldsymbol{\sigma}_3), \tag{4-74}$$

a form that begins to explain the choice of symbols for the parameters.† The **Q** matrices for rotation about a coordinate axis can be expressed in terms of the $\boldsymbol{\sigma}$'s in a particularly simple form. For example, $\mathbf{Q}_\theta$ for rotation about the x axis, Eq. (4–69), may be written as

$$\mathbf{Q}_\theta = \mathbf{1} \cos\frac{\theta}{2} + i\boldsymbol{\sigma}_1 \sin\frac{\theta}{2}. \tag{4-75}$$

Similarly the $\mathbf{Q}_\phi$ matrix for rotation about the z axis has the form

$$\mathbf{Q}_\phi = \begin{pmatrix} \cos\dfrac{\phi}{2} + i\sin\dfrac{\phi}{2} & 0 \\ 0 & \cos\dfrac{\phi}{2} - i\sin\dfrac{\phi}{2} \end{pmatrix} = \mathbf{1} \cos\frac{\phi}{2} + i\boldsymbol{\sigma}_3 \sin\frac{\phi}{2}, \tag{4-76}$$

---

* The notation is one of convenience, for of course the individual matrices do not transform as the component of a vector in configuration space. Indeed they must have the same representation for all Cartesian spatial coordinate systems, else Eq. (4–73) will not be valid.

† The connossieur of somewhat musty mathematics will recognize in Eq. (4–74) a representation of **Q** as a matrix *quaternion*, a quantity invented by Sir William R. Hamilton in 1843. Here e_0 is the (quaternion) scalar and the quantity in parentheses is the vector of the quaternion.

and it can be directly verified that a rotation about the y axis has the same matrix form as in (4–76) with $\boldsymbol{\sigma}_3$ replaced by $\boldsymbol{\sigma}_2$. All the elementary rotation matrices are thus given by similar expressions, involving only the unit matrix and the corresponding $\boldsymbol{\sigma}$ matrix. Each of the Pauli spin matrices is therefore associated with rotation about one particular axis and may be thought of as the *unit rotator* for that axis.

By making use of the multiplication properties of the Pauli spin matrices it can be shown (cf. Exercise 13) that the elementary rotation matrix $\mathbf{Q}_\theta$ as given by Eq. (4–75) can also be written symbolically in the form

$$\mathbf{Q}_\theta = e^{i\,\boldsymbol{\sigma}_1(\theta/2)}. \tag{4–77}$$

The exponential of a matrix is taken as a shorthand for the series representation of the exponential, where the first term is the unit matrix **1**. In this connection it may be noted that if **B** is a hermitean matrix, then

$$\mathbf{A} = e^{i\mathbf{B}}$$

is unitary (cf. Exercise 12). Since the Pauli spin matrices are manifestly hermitean the unitary condition on the **Q** matrices is clearly satisfied.

Characteristic of the Cayley–Klein parameters, and of the matrices containing them, is the ubiquitous presence of half angles, and this feature leads to some peculiar properties for the uv space. For example, a rotation in ordinary space about the z axis through the angle 2π merely reproduces the original coordinate system. Thus, if in the **D** matrix of the preceding section, ϕ is set equal to 2π, then $\cos\phi = 1$, $\sin\phi = 0$, and **D** properly reduces to the unit matrix **1** corresponding to the identity transformation. On the other hand if the same substitution is made in $\mathbf{Q}_\phi$, Eq. (4–68), we obtain

$$\mathbf{Q}_{2\pi} = \begin{pmatrix} e^{i\pi} & 0 \\ 0 & e^{-i\pi} \end{pmatrix} = \begin{pmatrix} -1 & 0 \\ 0 & -1 \end{pmatrix},$$

which is $-\mathbf{1}$ and not **1**. At the same time the 2×2 $-\mathbf{1}$ matrix must also correspond to the three-dimensional identity transformation. Hence there are two **Q** matrices, **1** and $-\mathbf{1}$, corresponding to the 3×3 unit matrix. In general, if a matrix **Q** corresponds to some real orthogonal matrix, then $-\mathbf{Q}$ also corresponds to the same matrix. The homomorphism between the two sets thus involves, in this case, a one-to-one correspondence, or *isomorphism*, between the single 3×3 matrix and the *pair* of matrices $(\mathbf{Q}, -\mathbf{Q})$, and not between the individual matrices. In this sense one may say that the **Q** matrix is a *double-valued* function of the corresponding three-dimensional orthogonal matrix.

Such a paradoxical situation plays no havoc with our common sense. As here presented the uv space is entirely a mathematical construct, devised solely to establish a correspondence between 3×3 and 2×2 matrices of a certain type. One would not require nor expect such a space to have the same properties as physical three-dimensional space. Mathematicians have paid considerable attention to the properties of the uv space and have designated the two-dimensional complex vector in the space by the term *spinor*. It turns out that in quantum mechanics the spinor space comes a bit closer to physical reality, for to include the effects of the "spin" of the electron, the wave function, or parts of it, must be made a spinor. Indeed, the half angles and resultant double-valued property are intimately connected with the fact that the spin is half integral.* To pursue the subject further would clearly take us outside the scope of classical mechanics.

4–6 EULER'S THEOREM ON THE MOTION OF A RIGID BODY

The discussions of the previous sections provide a complete mathematical technique for describing the motions of a rigid body. At any instant the orientation of the body can be specified by an orthogonal transformation, the elements of which may be expressed in terms of some suitable set of parameters. As time progresses the orientation will change and hence the matrix of transformation will be a function of time and may be written $\mathbf{A}(t)$. If the body axes are chosen coincident with the space axes at the time $t = 0$, then the trans-formation is initially simply the identity transformation:

$$\mathbf{A}(0) = \mathbf{1}.$$

At any later time $\mathbf{A}(t)$ will in general differ from the identity transformation, but since the physical motion must be continuous $\mathbf{A}(t)$ must be a continuous function of time. The transformation may thus be said to evolve *continuously from the identity transformation.*

With this method of describing the motion, and using only the mathematical apparatus already introduced, we are now in a position to obtain the important characteristics of rigid body motion. Of basic importance is

Euler's theorem: the general displacement of a rigid body with one point fixed is a rotation about some axis.

If the fixed point is taken as the origin of the body set of axes, then the displacement of the rigid body involves no translation of the body axes; the only change is in orientation. The theorem then states that the body set of axes at any time t can always be obtained by a single rotation of the initial set of axes (taken as coincident with the space set). In other words, the *operation* implied in the matrix $\mathbf{A}$ describing the physical motion of the rigid body is a *rotation*. Now it is

---

* Although the wave function may be double valued under rotation, all physically observable properties remain single valued, of course. A rotation through two whole turns, i.e., through 4π, would correspond to $\mathbf{Q}_{2\pi}^2$, and this is always the unit matrix $+\mathbf{1}$. Dirac has pointed out the corresponding topological curiosity that a braid of fibers made by twisting two whole turns can be disentangled, i.e., returned to its original state, without further rotations, but a braid formed by only one whole turn cannot! (See M. Gardner, *New Mathematical Diversions from Scientific American* (St. Louis, Missouri: Fireside, 1971) chap. 2; also *Scientific American*, Dec. 1959, p. 166.)

characteristic of a rotation that one direction, namely the axis of rotation, is left unaffected by the operation. Thus any vector lying along the axis of rotation must have the same components in both the initial and final axes. The other necessary condition for a rotation, that the magnitude of the vectors be unaffected, is automatically provided by the orthogonality conditions. Hence Euler's theorem will be proven if it can be shown that there exists a vector $\mathbf{R}$ having the same components in both systems. Using matrix notation for the vector,

$$\mathbf{R}' = \mathbf{AR} = \mathbf{R}. \tag{4-78}$$

Equation (4-78) constitutes a special case of the more general equation:

$$\mathbf{R}' = \mathbf{AR} = \lambda\mathbf{R}, \tag{4-79}$$

where λ is some constant, which may be complex. The values of λ for which Eq. (4-79) is soluble are known as the characteristic values, or *eigenvalues,** of the matrix. The problem of finding the vectors satisfying Eq. (4-79) is therefore called the *eigenvalue problem* for the given matrix, and Eq. (4-79) itself is referred to as the *eigenvalue equation*. Correspondingly, the vector solutions are the *eigenvectors* of $\mathbf{A}$. Euler's theorem can now be restated in the following language:

The real orthogonal matrix specifying the physical motion of a rigid body with one point fixed always has the eigenvalue $+1$.

The eigenvalue Eqs. (4-79) may be written

$$(\mathbf{A} - \lambda\mathbf{1})\mathbf{R} = 0, \tag{4-80}$$

or, in expanded form,

$$
\begin{aligned}
(a_{11} - \lambda)X + \quad a_{12}Y \quad + \quad a_{13}Z \quad &= 0, \\
a_{21}X + (a_{22} - \lambda)Y + \quad a_{23}Z \quad &= 0, \\
a_{31}X + \quad a_{32}Y \quad + (a_{33} - \lambda)Z &= 0.
\end{aligned}
\tag{4-81}
$$

Equations (4-81) comprise a set of three homogeneous simultaneous equations for the components X, Y, Z of the eigenvector $\mathbf{R}$. As such they can never furnish definite values for the three components, but only ratios of components. Physically, this corresponds to the circumstance that only the *direction* of the eigenvector can be fixed; the magnitude remains undetermined. The product of a constant with an eigenvector is also an eigenvector. In any case, being homogeneous, the Eqs. (4-81) can have a solution only when the determinant of the coefficients vanishes:

$$|\mathbf{A} - \lambda\mathbf{1}| = \begin{vmatrix} a_{11} - \lambda & a_{12} & a_{13} \\ a_{21} & a_{22} - \lambda & a_{23} \\ a_{31} & a_{32} & a_{33} - \lambda \end{vmatrix} = 0. \tag{4-82}$$

---

* This term is derived from the German *eigenwerte*, literally "proper values."

Equation (4–82) is known as the *characteristic* or *secular* equation of the matrix, and the values of λ for which the equation is satisfied are the desired eigenvalues. Euler's theorem reduces to the statement that, for the real orthogonal matrices under consideration, the secular equation must have the root $\lambda = +1$.

In general the secular equation will have three roots with three corresponding eigenvectors. For convenience in discussion, the notation X_1, X_2, X_3 will often be used instead of X, Y, Z. In such a notation the components of the eigenvectors might be labeled as X_{ik}; the first subscript indicating the particular component, the second denoting which of the three eigenvectors is involved. A typical member of the group of Equations (4–81) would then be written (with explicit summation) as

$$\sum_j a_{ij} X_{jk} = \lambda_k X_{ik}$$

or, alternatively, as

$$\sum_j a_{ij} X_{jk} = \sum_j X_{ij} \delta_{jk} \lambda_k. \tag{4–83}$$

Both sides of Eq. (4–83) then have the form of a matrix product element; the left side as the product of **A** with a matrix **X** having the elements X_{jk}, the right side as the product of **X** with a matrix whose jkth element is $\delta_{jk} \lambda_k$. The last matrix is diagonal, and its diagonal elements are the eigenvalues of **A**. We shall therefore designate the matrix by λ:

$$\lambda = \begin{pmatrix} \lambda_1 & 0 & 0 \\ 0 & \lambda_2 & 0 \\ 0 & 0 & \lambda_3 \end{pmatrix}. \tag{4–84}$$

Equation (4–83) thus implies the matrix equation

$$\mathbf{AX} = \mathbf{X}\lambda,$$

or, multiplying from the left by $\mathbf{X}^{-1}$,

$$\mathbf{X}^{-1}\mathbf{AX} = \lambda. \tag{4–85}$$

Now, the left side is in the form of a similarity transformation operating on **A**. (One has only to denote $\mathbf{X}^{-1}$ by the symbol **Y** to reduce it to the form Eq. (4–41).) Thus Eq. (4–85) provides the following alternate approach to the eigenvalue problem: We seek to diagonalize **A** by a similarity transformation. Each column of the matrix used to carry out the similarity transformation consists of the components of an eigenvector. The elements of the diagonalized form of **A** are the corresponding eigenvalues.

Euler's theorem can be directly proven by using the orthogonality property of $\tilde{\mathbf{A}}$. Consider the expression

$$(\mathbf{A} - 1)\tilde{\mathbf{A}} = 1 - \tilde{\mathbf{A}}.$$

If we take the determinant of the matrices forming both sides (cf. p. 143), we can write the equality

$$|\mathbf{A} - 1||\tilde{\mathbf{A}}| = |1 - \tilde{\mathbf{A}}|. \tag{4–86}$$

To describe the motion of a rigid body, the matrix $\mathbf{A}(t)$ must correspond to a proper rotation; therefore the determinant of $\mathbf{A}$, and of its transpose, must be $+1$. Further, since in general the determinant of the transpose of a matrix is the same as that of the matrix, the transpose signs in Eq. (4–86) can be removed:

$$|\mathbf{A} - 1| = |1 - \mathbf{A}|. \tag{4–87}$$

Equation (4–87) says that the determinant of a particular matrix is the same as the determinant of the negative of the matrix. Suppose $\mathbf{B}$ is some $n \times n$ matrix. Then it is a well-known property of determinants that

$$|-\mathbf{B}| = (-1)^n |\mathbf{B}|.$$

Since we are working in a three-dimensional space ($n = 3$), it is clear that Eq. (4–87) can hold for any arbitrary proper rotation only if

$$|\mathbf{A} - 1| = 0. \tag{4–88}$$

Comparing Eq. (4–88) with the secular equation (4–82), it is seen that one of the eigenvalues satisfying Eq. (4–82) must always be $\lambda = +1$, which is the desired result of Euler's theorem.

Note how the proof of Euler's theorem emphasizes the importance of the number of dimensions in the space considered. In spaces with an even number of dimensions Eq. (4–87) is an identity for all matrices and Euler's theorem doesn't hold. Thus, for two dimensions there is no vector *in the space* that is left unaltered by a rotation—the axis of rotation is perpendicular to the plane and therefore out of the space.

It is now a simple matter to determine the properties of the other eigenvalues in three dimensions. Designate the $+1$ eigenvalue as λ_3. The determinant of any matrix is unaffected by a similarity transformation (see p.143). Hence by Eqs. (4–84) and (4–85), and the properties of $\mathbf{A}$ as a proper rotation,

$$|\mathbf{A}| = \lambda_1 \lambda_2 \lambda_3 = \lambda_1 \lambda_2 = 1. \tag{4–89}$$

Further, since $\mathbf{A}$ is a real matrix, then if λ is a solution of the secular equation (4–82), the complex conjugate λ^* must also be a solution.

If a given eigenvalue λ_i is complex, then the corresponding eigenvector, $\mathbf{R}_i$, that satisfies Eq. (4–79) will in general also be complex. We have not previously dealt with the properties of complex vectors under (real) orthogonal transformations, and there are some modifications to previous definitions. The square of the length or magnitude of a complex vector $\mathbf{R}$ is $\mathbf{R} \cdot \mathbf{R}^*$, or in matrix notation $\tilde{\mathbf{R}}\mathbf{R}^*$, where the transpose sign on the left-hand vector indicates it is represented by a row matrix. Under a real orthogonal transformation the square of the magnitude is invariant:

$$\tilde{\mathbf{R}}'\mathbf{R}'^* = (\widetilde{\mathbf{A}\mathbf{R}})\mathbf{A}\mathbf{R}^* = \tilde{\mathbf{R}}\tilde{\mathbf{A}}\mathbf{A}\mathbf{R}^* = \tilde{\mathbf{R}}\mathbf{R}^*.$$

Suppose now that **R** is a complex eigenvector corresponding to a complex eigenvalue λ. Hence, by Eq. (4–79), we have

$$\tilde{\mathbf{R}}'\mathbf{R}'^* = \lambda\lambda^*\tilde{\mathbf{R}}\mathbf{R}^*,$$

which leads to the conclusion that *all eigenvalues have unit magnitude:*

$$\lambda\lambda^* = 1. \tag{4–90}$$

From these properties it may be concluded that there are three possible distributions of eigenvalues. If all of the eigenvalues are real, then only two situations are possible:

1. All eigenvalues are $+1$. The transformation matrix is then just **1**, a case we may justly call trivial.

2. One eigenvalue is $+1$ and the other two are both -1. Such a transformation may be characterized as an inversion in two coordinate axes with the third unchanged. Equally it is a rotation through the angle π about the direction of the unchanged axis.

If not all of the eigenvalues are real, there is only one additional possibility:

3. One eigenvalue is $+1$, and the other two are complex conjugates of each other of the form $e^{i\Phi}$ and $e^{-i\Phi}$.

A more complete statement of Euler's theorem thus is that any nontrivial real orthogonal matrix has one, *and only one*, eigenvalue $+1$.

The direction cosines of the axis of rotation can then be obtained by setting $\lambda = 1$ in the eigenvalue Eqs. (4–81) and solving for X, Y, and Z.* The angle of rotation can likewise be obtained without difficulty. By means of some similarity transformation it is always possible to transform the matrix **A** to a system of coordinates where the z axis lies along the axis of rotation. In such a system of coordinates **A**' represents a rotation about the z axis through an angle Φ, and therefore has the form

$$\mathbf{A}' = \begin{pmatrix} \cos\Phi & \sin\Phi & 0 \\ -\sin\Phi & \cos\Phi & 0 \\ 0 & 0 & 1 \end{pmatrix}.$$

---

* If there are multiple roots to the secular equation, then the corresponding eigenvectors cannot be found as simply, cf. Sections 5–4 and 6–2. Indeed, it is not always possible to completely diagonalize a general matrix if the eigenvalues are not all distinct. These exceptions are of no importance for the present considerations, as Euler's theorem shows that for all nontrivial orthogonal matrices $+1$ is a single root.

The trace of $\mathbf{A}'$ is simply

$$1 + 2\cos\Phi.$$

Since the trace is always invariant under a similarity transformation, the trace of $\mathbf{A}$ with respect to any initial coordinate system must have the same form,

$$\operatorname{Tr}\mathbf{A} = a_{ii} = 1 + 2\cos\Phi, \qquad (4\text{–}91)$$

which gives the value of Φ in terms of the matrix elements. The rotation angle Φ is to be identified also with the phase angle of the complex eigenvalues λ, as the sum of the eigenvalues is just the trace of $\mathbf{A}$ in its diagonal form, Eq. (4–84). By Euler's theorem and the properties of the eigenvalues this sum is

$$\operatorname{Tr}\mathbf{A} = \sum_i \lambda_i = 1 + e^{i\Phi} + e^{-i\Phi} = 1 + 2\cos\Phi.$$

We see that the situations in which the eigenvalues are all real are actually special cases of $\mathbf{A}$ having complex eigenvalues. All the $\lambda_i = +1$ corresponds to a rotation angle $\Phi = 0$ (the identity transformation), while the case with a double eigenvalue -1 corresponds to $\Phi = \pi$, as previously noted.

The prescriptions for the direction of the rotation axis and for the rotation angle are not unambiguous. Clearly if $\mathbf{R}$ is an eigenvector so is $-\mathbf{R}$, hence the sense of the direction of the rotation axis is not specified. Further, $-\Phi$ satisfies Eq. (4–91) if Φ does. Indeed, it is clear that the eigenvalue solution does not uniquely fix the orthogonal transformation matrix $\mathbf{A}$. From the determinantal secular equation (4–82), it follows that the inverse matrix $\mathbf{A}^{-1} = \mathbf{A}$ has the same eigenvalues and eigenvectors as $\mathbf{A}$. However the ambiguities can at least be ameliorated by assigning Φ to $\mathbf{A}$ and $-\Phi$ to $\mathbf{A}^{-1}$, and fixing the sense of the axes of rotation by the right-hand screw rule.

Finally, note should be made of an immediate corollary of Euler's theorem, sometimes called

Chasles' theorem: the most general displacement of a rigid body is a translation plus a rotation.

Detailed proof is hardly necessary. Simply stated, removing the constraint of motion with one point fixed introduces three translatory degrees of freedom for the origin of the body system of axes.*

---

* M. Chasles (1793–1881) also proved a stronger form of the theorem, namely that it is possible to choose the origin of the body-set of coordinates so that the translation is in the same direction as the axis of rotation. Such a combination of translation and rotation is called a *screw motion*. There seems to be little present use for this version of Chasles' theorem, nor for the elaborate mathematics of screw motions as developed in the nineteenth century. See Routh, *Elementary Dynamics of a Rigid Body*, 5th ed. (London: Macmillan, 1891), pp. 194–198.

4-7 FINITE ROTATIONS

The relative orientation of two Cartesian coordinate systems with common origin has been described by various representations, including the three successive Euler angles of rotation that transform one coordinate system to the other. In the previous section it was shown that the coordinate transformation can be carried through by a single rotation about a suitable direction. It is natural therefore to seek a representation of the coordinate transformation in terms of the parameters of the rotation–the angle of rotation and the direction cosines of the axis of rotation.

With the help of some simple vector algebra it is possible to derive such a representation. For the purpose it is convenient to treat the transformation in its *active* sense, i.e., as one which rotates the vector in a fixed coordinate system (see p. 136). It should be remembered that a counterclockwise rotation of the coordinate system then appears as a *clockwise* rotation of the vector. In Fig. 4–8(a) the initial position of the vector $\mathbf{r}$ is denoted by $\overrightarrow{OP}$ and the final position $\mathbf{r}'$ by $\overrightarrow{OQ}$, while the unit vector along the axis of rotation is denoted by $\mathbf{n}$. The distance between O and N has the magnitude $\mathbf{n} \cdot \mathbf{r}$, so that the vector $\overrightarrow{ON}$ can be written as $\mathbf{n}(\mathbf{n} \cdot \mathbf{r})$. Figure 4–8(b) is a plan of the vectors in the plane normal to the axis of rotation. The vector $\overrightarrow{NP}$ can be described also as $\mathbf{r} - \mathbf{n}(\mathbf{n} \cdot \mathbf{r})$, but its magnitude is the same as that of the vectors $\overrightarrow{NQ}$ and $\mathbf{r} \times \mathbf{n}$. To obtain the desired relation between $\mathbf{r}'$ and $\mathbf{r}$, we construct $\mathbf{r}'$ as the sum of three vectors:

$$\mathbf{r}' = \overrightarrow{ON} + \overrightarrow{NV} + \overrightarrow{VQ}$$

or

$$\mathbf{r}' = \mathbf{n}(\mathbf{n} \cdot \mathbf{r}) + [\mathbf{r} - \mathbf{n}(\mathbf{n} \cdot \mathbf{r})] \cos \Phi + (\mathbf{r} \times \mathbf{n}) \sin \Phi.$$

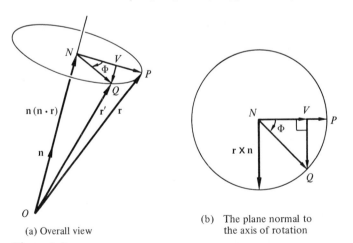

(a) Overall view

(b) The plane normal to the axis of rotation

Figure 4–8

Vector diagrams for derivation of the rotation formula.

A slight rearrangement of terms leads to the final result:

$$\mathbf{r}' = \mathbf{r}\cos\Phi + \mathbf{n}(\mathbf{n}\cdot\mathbf{r})[1 - \cos\Phi] + (\mathbf{r}\times\mathbf{n})\sin\Phi. \tag{4-92}$$

Equation (4–92) will be referred to as the *rotation formula*.* It is to be emphasized that Eq. (4–92) holds for any rotation, no matter what its magnitude, and thus is a finite-rotation version (in a clockwise sense) of the description given on page 58 for the change of a vector under infinitesimal rotation. (See also page 170.)

The rotation formula (4–92) can be put in a more useful form by introducing a scalar e_0 and a vector $\mathbf{e}$ with components e_1, e_2, and e_3 defined as

$$e_0 = \cos\frac{\Phi}{2},$$

$$\mathbf{e} = \mathbf{n}\sin\frac{\Phi}{2}. \tag{4-93}$$

These four quantities are obviously related by the condition

$$e_0^2 + |\mathbf{e}|^2 = e_0^2 + e_1^2 + e_2^2 + e_3^2 = 1. \tag{4-94}$$

It follows that the trigonometric functions of Φ can be expressed as

$$\cos\Phi = 2e_0^2 - 1 = e_0^2 - e^2 = e_0^2 - e_1^2 - e_2^2 - e_3^2,$$

and   (4–95)

$$\mathbf{n}\sin\Phi = 2e_0\mathbf{e}.$$

With the help of these results, and a little further manipulation, Eq. (4–92) can be rewritten as

$$\mathbf{r}' = \mathbf{r}(e_0^2 - e_1^2 - e_2^2 - e_3^2) + 2\mathbf{e}(\mathbf{e}\cdot\mathbf{r}) + 2(\mathbf{r}\times\mathbf{e})e_0. \tag{4-96}$$

Equation (4–96) states that the components of $\mathbf{r}'$ are linear functions of the components of $\mathbf{r}$. It is in fact a vector form of the linear transformation equations corresponding to the orthogonal matrix **A** of rotation. The matrix elements a_{ij} can be obtained in terms of the e's merely by expanding Eq. (4–96) for each x_i' and collecting coefficients of x_j. For example, from Eq. (4–96) x' is related to x, y, and z by

$$x' = x(e_0^2 - e_1^2 - e_2^2 - e_3^2) + 2e_1(e_1x + e_2y + e_3z) + 2(ye_3 - ze_2)e_0,$$

and hence

$$a_{11} = e_0^2 + e_1^2 - e_2^2 - e_3^2,$$

$$a_{12} = 2(e_1e_2 + e_0e_3),$$

$$a_{13} = 2(e_1e_3 - e_0e_2).$$

---

* Apparently, it does not have an eponymic designation. Hamel (*Theoretische Mechanik*, p. 103) ascribes it to the French mathematician O. Rodrigues (1794–1851), but that is probably an error. Presumably Gibbs was the first to put it in vector form (*Vector Analysis*, p. 338), but the underlying formula is much older.

Comparison with the matrix form in Eq. (4–67) shows that there is complete agreement with the matrix elements given there, and therefore the e's defined by Eq. (4–93) are identical with the Euler parameters introduced in Section (4–5) (hence the choice of notation). Equation (4–94) is thus the same as the determinantal condition, Eq. (4–66).

The identification of the Euler parameters now permits expressing many previous results in terms of the rotation angle and the direction of the axis of rotation. Thus, the first of Eqs. (4–72') for e_0, combined with Eq. (4–93), immediately gives the angle of rotation in terms of the Euler angles:

$$\cos\frac{\Phi}{2} = \cos\frac{\phi + \psi}{2}\cos\frac{\theta}{2}. \qquad (4\text{–}97)$$

Of course, this result can also be obtained, with a little bit of trigonometric manipulation from the trace of the matrix **A**, Eq. (4–46). The 2×2 unimodular matrix **Q**, homomorphic to the rotation matrix **A**, can also be rewritten in terms of the rotation parameters, starting from Eq. (4–74) in the form

$$\mathbf{Q} = e_0\mathbf{1} + i\mathbf{e}\cdot\boldsymbol{\sigma}.$$

By Eq. (4–93), **Q** then also appears as

$$\mathbf{Q} = \mathbf{1}\cos\frac{\Phi}{2} + i\mathbf{n}\cdot\boldsymbol{\sigma}\sin\frac{\Phi}{2}. \qquad (4\text{–}98)$$

Equation (4–98) is the straightforward generalization of Eq. (4–75) to an arbitrary axis of rotation and, indeed, could have been obtained from it by a suitable transformation of coordinates. Similarly, the corresponding analog to Eq. (4–77), the exponential representation of $\mathbf{Q}_\theta$, is

$$\mathbf{Q} = e^{i\mathbf{n}\cdot\boldsymbol{\sigma}(\Phi/2)}, \qquad (4\text{–}99)$$

with the same qualifications as to the meaning of the exponential of a matrix.

4–8 INFINITESIMAL ROTATIONS

In the previous sections various matrices have been associated with the description of the rigid body orientation. However, the number of matrix elements has always been larger than the number of independent variables, and various subsidiary conditions have had to be tagged on. Now that it has been established that any given orientation can be obtained by a single rotation about some axis, it is tempting to try to associate a vector, characterized by three independent quantities, with the finite displacement of a rigid body about a fixed point. Certainly a direction suggests itself obviously—that of the axis of rotation—and any function of the rotation angle would seem suitable as the magnitude. But it soon becomes evident that such a correspondence cannot be made successfully. Suppose **A** and **B** are two such "vectors" associated with transformations **A** and **B**. Then to qualify as vectors they must be commutative in addition:

$$\mathbf{A} + \mathbf{B} = \mathbf{B} + \mathbf{A}.$$

But the addition of two rotations, i.e., one rotation performed after another, it has been seen, corresponds to the product **AB** of the two matrices. However, matrix multiplication is not commutative, **AB** $\neq$ **BA**, and hence **A, B** are not commutative in addition and cannot be accepted as vectors. This conclusion, that the sum of finite rotations depends on the order of the rotations, is strikingly demonstrated by a simple experiment. Thus Fig. 4–9 illustrates the sequence of events in rotating a block first through 90° about the z' axis fixed in the block, and then 90° about the y' axis, while Fig. 4–10 presents the same rotations in reverse order. The final position is markedly different in the two sequences.

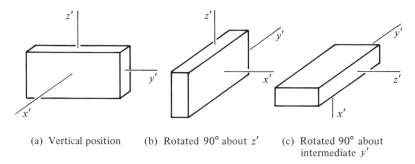

(a) Vertical position   (b) Rotated 90° about z' (c) Rotated 90° about intermediate y'

FIGURE 4–9
Illustrating the effect of two rotations performed in a given order.

While a finite rotation thus cannot be represented by a single vector, the same objections do not hold if only *infinitesimal rotations* are considered. An infinitesimal rotation is an orthogonal transformation of coordinate axes in which the components of a vector are almost the same in both sets of axes—the

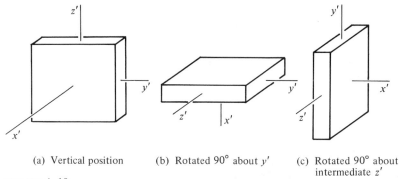

(a) Vertical position   (b) Rotated 90° about y' (c) Rotated 90° about intermediate z'

FIGURE 4–10
The two rotations shown in Fig. 4–9, but performed in reverse order.

change is infinitesimal. Thus the x_1' component of some vector $\mathbf{r}$ (on the passive interpretation of the transformation) would be practically the same as x_1, the difference being extremely small:

$$x_1' = x_1 + \epsilon_{11}x_1 + \epsilon_{12}x_2 + \epsilon_{13}x_3. \tag{4-100}$$

The matrix elements ϵ_{11}, ϵ_{12}, and so on are to be considered as infinitesimals, so that in subsequent calculations only the first nonvanishing order in ϵ_{ij} need be retained. For any general component x_i' the equations of infinitesimal transformation can be written as

$$x_i' = x_i + \epsilon_{ij}x_j$$

or

$$x_i' = (\delta_{ij} + \epsilon_{ij})x_j. \tag{4-101}$$

The quantity δ_{ij} will be recognized as the element of the unit matrix, and Eq. (4-101) appears in matrix notation as

$$\mathbf{x}' = (\mathbf{1} + \boldsymbol{\epsilon})\mathbf{x}. \tag{4-102}$$

Equation (4-102) states that the typical form for the matrix of an infinitesimal transformation is $\mathbf{1} + \boldsymbol{\epsilon}$, that is, it is almost the identity transformation, differing at most by an infinitesimal operator.

It can now be seen that the sequence of operations is unimportant for infinitesimal transformations; in other words, they *commute*. If $\mathbf{1} + \boldsymbol{\epsilon}_1$, and $\mathbf{1} + \boldsymbol{\epsilon}_2$ are two infinitesimal transformations, then one of the possible products is

$$(\mathbf{1} + \boldsymbol{\epsilon}_1)(\mathbf{1} + \boldsymbol{\epsilon}_2) = \mathbf{1}^2 + \boldsymbol{\epsilon}_1\mathbf{1} + \mathbf{1}\boldsymbol{\epsilon}_2 + \boldsymbol{\epsilon}_1\boldsymbol{\epsilon}_2$$

$$= \mathbf{1} + \boldsymbol{\epsilon}_1 + \boldsymbol{\epsilon}_2, \tag{4-103}$$

neglecting higher order infinitesimals. The product in reverse order merely interchanges $\boldsymbol{\epsilon}_1$ and $\boldsymbol{\epsilon}_2$; this has no effect on the result, as matrix addition is always commutative. The commutative property of infinitesimal transformations removes the objection to their representation by vectors.

The inverse matrix for an infinitesimal transformation is readily obtained. If $\mathbf{A} = \mathbf{1} + \boldsymbol{\epsilon}$ is the matrix of the transformation, then the inverse is

$$\mathbf{A}^{-1} = \mathbf{1} - \boldsymbol{\epsilon}. \tag{4-104}$$

As proof note that the product $\mathbf{A}\mathbf{A}^{-1}$ reduces to the unit matrix,

$$\mathbf{A}\mathbf{A}^{-1} = (\mathbf{1} + \boldsymbol{\epsilon})(\mathbf{1} - \boldsymbol{\epsilon}) = \mathbf{1},$$

in agreement with the definition for the inverse matrix, Eq. (4-32). Further, the orthogonality of $\mathbf{A}$ implies that $\tilde{\mathbf{A}} \equiv (\mathbf{1} + \tilde{\boldsymbol{\epsilon}})$ must be equal to $\mathbf{A}^{-1}$ as given by Eq.

(4–104). Hence the infinitesimal matrix is antisymmetric*:

$$\tilde{\boldsymbol\epsilon} = -\boldsymbol\epsilon.$$

Since the diagonal elements of an antisymmetric matrix are necessarily zero there can be only three distinct elements in any 3×3 antisymmetric matrix. Hence there is no loss of generality in writing $\boldsymbol\epsilon$ in the form

$$\boldsymbol\epsilon = \begin{pmatrix} 0 & d\Omega_3 & -d\Omega_2 \\ -d\Omega_3 & 0 & d\Omega_1 \\ d\Omega_2 & -d\Omega_1 & 0 \end{pmatrix}. \tag{4–105}$$

The three quantities $d\Omega_1,\ d\Omega_2,\ d\Omega_3$ are clearly to be identified with the three independent parameters specifying the rotation. It will now be shown that these three quantities also form the components of a particular kind of vector. By Eq. (4–102) the *change* in the components of a vector under the infinitesimal transformation of the coordinate system can be expressed by the matrix equation

$$\mathbf{x}' - \mathbf{x} \equiv d\mathbf{x} = \boldsymbol\epsilon \mathbf{x}, \tag{4–106}$$

which in expanded form, with $\boldsymbol\epsilon$ given by (4–105), becomes

$$dx_1 = x_2\, d\Omega_3 - x_3\, d\Omega_2$$
$$dx_2 = x_3\, d\Omega_1 - x_1\, d\Omega_3 \tag{4–107}$$
$$dx_3 = x_1\, d\Omega_2 - x_2\, d\Omega_1.$$

The right-hand side of each of Eqs. (4–107) is in the form of a component of the cross product of two vectors, namely, the cross product of $\mathbf{r}$ with a vector $d\boldsymbol\Omega$ having components† $d\Omega_1,\ d\Omega_2,\ d\Omega_3$. We can therefore write Eq. (4–107) equivalently as

$$d\mathbf{r} = \mathbf{r} \times d\boldsymbol\Omega. \tag{4–108}$$

The vector $\mathbf{r}$ transforms under an orthogonal matrix $\mathbf{B}$ according to the relations (cf. Eq. 4–20)

$$x_i' = b_{ij}x_j. \tag{4–109}$$

---

*It has been assumed implicitly in this section that an infinitesimal orthogonal transformation corresponds to a rotation. In a sense this assumption is obvious; an "infinitesimal inversion" is a contradiction in terms. Formally, the statement follows from the antisymmetry of $\boldsymbol\epsilon$. All the diagonal elements of $\mathbf{1} + \boldsymbol\epsilon$ are then unity and to first order in small quantities the determinant of the transformation is always $+1$, which is the mark of a rotation.

†It cannot be emphasized too strongly that $d\boldsymbol\Omega$ is *not* the differential of a vector. The combination $d\boldsymbol\Omega$ stands for a differential vector, i.e., a vector of differential magnitude. Unfortunately, notational convention results in having the vector characteristic applied only to Ω, but it should be clear to the reader there is no vector of which $d\boldsymbol\Omega$ represents a differential. As we have seen, a finite rotation *cannot* be represented by a single vector.

If $d\mathbf{\Omega}$ is to be a vector in the same sense as $\mathbf{r}$ it must transform under $\mathbf{B}$ in the same way. As we shall see, $d\mathbf{\Omega}$ passes most of this test for a vector, although in one respect it fails to make the grade. One way of examining the transformation properties of $d\mathbf{\Omega}$ is to find how the matrix $\boldsymbol{\epsilon}$ transforms under a coordinate transformation. As has been shown in Section 4–3 above, the transformed matrix $\boldsymbol{\epsilon}'$ is obtained by a similarity transformation:

$$\boldsymbol{\epsilon}' = \mathbf{B}\,\boldsymbol{\epsilon}\,\mathbf{B}^{-1}.$$

As the antisymmetry property of a matrix is preserved under an orthogonal similarity transformation (see Exercise 3), $\boldsymbol{\epsilon}'$ can also be put in the form of Eq. (4–105) with nonvanishing elements $d\Omega_i'$. A detailed study of these elements, given in Appendix C, shows that $\boldsymbol{\epsilon}$ transforms under the similarity transformation such that

$$d\Omega_i' = |\mathbf{B}|\, b_{ij}\, d\Omega_j. \tag{4–110}$$

The transformation of $d\mathbf{\Omega}$ is thus almost the same as for $\mathbf{r}$, but differs by the factor $|\mathbf{B}|$, the determinant of the transformation matrix.

There is, however, a simpler way to uncover the vector characteristics of $d\mathbf{\Omega}$, and indeed to verify its transformation properties as given by Eq. (4–110). In the previous section a vector formula was derived for the change in the components of $\mathbf{r}$ under a finite rotation Φ of the coordinate system. By letting Φ go to the limit of an infinitesimal angle $d\Phi$ the corresponding formula for an infinitesimal rotation can be obtained. In this limit $\cos\Phi$ in Eq. (4–92) approaches unity, and $\sin\Phi$ goes to Φ; the resultant expression for the infinitesimal change in $\mathbf{r}$ is then

$$\mathbf{r}' - \mathbf{r} \equiv d\mathbf{r} = \mathbf{r} \times \mathbf{n}\,d\Phi. \tag{4–111}$$

Comparison with Eq. (4–108) indicates that $d\mathbf{\Omega}$ is indeed a vector and is determined by

$$d\mathbf{\Omega} = \mathbf{n}\,d\Phi. \tag{4–112}$$

Equation (4–111) can of course be derived directly without recourse to the finite rotation formula. Considered in its active sense, the infinitesimal coordinate transformation corresponds to a rotation of a vector $\mathbf{r}$ *clockwise* through an angle $d\Phi$ about the axis of rotation, a situation that is depicted in Fig. 4–11.* The magnitude of $d\mathbf{r}$, to first order in $d\Phi$ is, from the figure

$$dr = r\sin\theta\,d\Phi,$$

and the direction $d\mathbf{r}$ is, in this limit, perpendicular to both $\mathbf{r}$ and $d\mathbf{\Omega} = \mathbf{n}\,d\Phi$. Finally, the sense of $d\mathbf{r}$ is in the direction a right-hand screw advances as $\mathbf{r}$ is turned into $d\mathbf{\Omega}$. Figure (4–11) thus shows that in magnitude, direction, and sense $d\mathbf{r}$ is the same as that predicted by Eq. 4–111.

---

* Figure 4–11 is the clockwise-rotation version of Fig. 2–8.

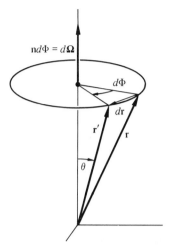

$nd\Phi = d\mathbf{\Omega}$

$d\Phi$

$d\mathbf{r}$

$\mathbf{r'}$ $\mathbf{r}$

θ

FIGURE 4–11
Change in a vector produced by an infinitesimal clockwise rotation of the vector.

The transformation properties of $d\mathbf{\Omega}$, as defined by Eq. (4–112), are still to be discussed. As is well known from elementary vector algebra,* there are two kinds of vectors in regard to transformation properties under an inversion. Vectors that transform according to Eq. (4–109) are known as *polar vectors*. Under a three-dimensional inversion, $S_{ij} = -\delta_{ij}$, all components of a polar vector change sign. On the other hand, the components of *axial vectors* or *pseudovectors* do not change sign under inversion. The simplest example of an axial vector is a cross product of two polar vectors,

$$\mathbf{C} = \mathbf{D} \times \mathbf{F},$$

where the components of the cross product are given, as customary, by the definitions:

$$C_i = D_j F_k - F_j D_k, \qquad i, j, k \text{ in cyclic order.} \tag{4–113}$$

The components of $\mathbf{D}$ and $\mathbf{F}$ change sign under inversion, hence those of $\mathbf{C}$ do not. Many familiar physical quantities are examples of axial vectors such as the angular momentum $\mathbf{L} = \mathbf{r} \times \mathbf{p}$, and the magnetic field intensity. The transformation law for an axial vector is of the form of Eq. (4–110). For proper orthogonal transformations axial and polar vectors are indistinguishable, but for improper transformations, i.e., involving inversion, the determinant $|\mathbf{B}|$ is -1, and the two types of vectors behave differently.†

On the passive interpretation of the transformation, it is easy to see why polar vectors behave as they do under inversion. The vector remains unaffected by the transformation, but the coordinate axes, and therefore the components, change sign. What then is different for an axial vector? It appears that an axial vector

---

* See, for example, J. B. Marion, *Principles of Vector Analysis*, pp. 42–49.

† The dot product of a pseudovector and a polar vector is called a *pseudoscalar*. Whereas a true scalar is completely invariant under an orthogonal transformation, a pseudoscalar changes sign under an improper rotation.

always carries with it a "handedness" convention, as implied, for example, by the definition, Eq. (4–113), of a cross product. Under inversion a right-handed coordinate system changes to a left-handed system, and the cyclic order requirement of Eq. (4–113) implies a similar change from the right-hand screw convention to a left-hand convention. Hence, even on the passive interpretation, there is an actual change in the direction of the cross product upon inversion.

It is clear now why $d\boldsymbol{\Omega}$ transforms as an axial vector according to Eq. (4–110). Algebraically, we see that since both $\mathbf{r}$ and $d\mathbf{r}$ in Eq. (4–111) are polar vectors, then $\mathbf{n}$, and therefore $d\boldsymbol{\Omega}$, must be axial vectors. Geometrically, the inversion of the coordinates corresponds to the switch from a right-hand screw law to a left-hand screw to define the sense of $\mathbf{n}$.

The discussion of the cross product provides an opportunity to introduce a notation that will be most useful on future occasions. The *permutation symbol* or *Levi–Civita density** ϵ_{ijk} is defined to be zero if any two of the indices ijk are equal, and otherwise either $+1$ or -1 according as ijk is an even or odd permutation of $1, 2, 3$. Thus, in terms of the permutation symbol, Eq. (4–113) for the components of a cross product can be written

$$C_i = \epsilon_{ijk} D_j F_k \qquad (4\text{–}113')$$

(where the usual summation convention has been employed). Further, Eq. (4–96), the rotation formula in terms of the Euler parameters, can be rewritten as

$$x'_i = x_i(e_0^2 - e_k e_k) + 2e_i e_j x_j + 2\epsilon_{ijk} e_0 x_j e_k, \qquad (4\text{–}114)$$

where the summation is over the values $1, 2, 3$ for the repeated indices. The elements of the orthogonal matrix of transformation $\mathbf{A}$ can now immediately be expressed in terms of the Euler parameters, by looking at the coefficient of x_j in Eq. (4–114):

$$a_{ij} = \delta_{ij}(e_0^2 - e_k e_k) + 2e_i e_j + 2\epsilon_{ijk} e_0 e_k. \qquad (4\text{–}115)$$

A little experimenting will show that this compact formula does indeed describe all the matrix elements in Eq. (4–67).

The descriptions of rotation presented so far in this chapter have been developed for the purpose of representing the orientation of a rigid body. It has been emphasized that the transformations primarily involve rotation of the *coordinate system*. The corresponding "active" interpretation of rotation of a vector in a fixed coordinate system therefore implies a rotation in the opposite direction, i.e., in a clockwise sense. But there are many areas of mechanics, or of physics in general for that matter, where we are concerned with the effects of rotating the *physical system* and associated vectors. The connection between invariance of the system under rotation and conservation of angular momentum has already been pointed out (Section 2–6). In such applications it is necessary to consider the consequences of rotation of vectors in the usual counterclockwise sense. For reference purposes, a number of rotation formulas given above will be

---

* Also known interchangeably as the *alternating tensor* or *isotropic tensor of rank 3.*

listed here, but for counterclockwise rotation of vectors. *All equations and statements from here to the end of this section apply only for such counterclockwise rotations.*

The rotation formula, Eq. (4–92), becomes

$$\mathbf{r'} = \mathbf{r}\cos\Phi + \mathbf{n}(\mathbf{n}\cdot\mathbf{r})[1 - \cos\Phi] + (\mathbf{n}\times\mathbf{r})\sin\Phi, \qquad (4\text{-}92')$$

and the corresponding infinitesimal rotation, Eq. (4–111), appears as

$$d\mathbf{r'} = d\boldsymbol{\Omega}\times\mathbf{r} = (\mathbf{n}\times\mathbf{r})\,d\Phi. \qquad (4\text{-}111')$$

Further, the associated 2×2 complex matrix $\mathbf{Q}$ takes the form

$$\mathbf{Q} = \mathbf{1}\cos\frac{\Phi}{2} - i\mathbf{n}\cdot\boldsymbol{\sigma}\sin\frac{\Phi}{2}, \qquad (4\text{-}98')$$

with the corresponding exponential representation as

$$\mathbf{Q} = e^{-i\mathbf{n}\cdot\boldsymbol{\sigma}(\Phi/2)}. \qquad (4\text{-}99')$$

The antisymmetric matrix of the infinitesimal rotation, Eq. (4–105), becomes

$$\boldsymbol{\epsilon} = \begin{pmatrix} 0 & -d\Omega_3 & d\Omega_2 \\ d\Omega_3 & 0 & -d\Omega_1 \\ -d\Omega_2 & d\Omega_1 & 0 \end{pmatrix} = \begin{pmatrix} 0 & -n_3 & n_2 \\ n_3 & 0 & -n_1 \\ -n_2 & n_1 & 0 \end{pmatrix} d\Phi, \quad (4\text{-}105')$$

where n_i are the components of the unit vector along the axis of rotation. Letting $d\mathbf{x}$ stand for the infinitesimal change $\mathbf{x'} - \mathbf{x}$, Eq. (4–102) can then take the form of a matrix differential equation with respect to the rotation angle:

$$\frac{d\mathbf{x}}{d\Phi} = -\mathbf{N}\mathbf{x}, \qquad (4\text{-}116)$$

where $\mathbf{N}$ is the matrix on the right in Eq. (4–105') with elements $N_{ij} = \epsilon_{ijk}n_k$. Another useful representation is to write $\boldsymbol{\epsilon}$ in Eq. (4–105') as

$$\boldsymbol{\epsilon} = n_i\mathbf{M}_i\,d\Phi$$

where $\mathbf{M}_i$ are the three matrices:

$$\mathbf{M}_1 = \begin{pmatrix} 0 & 0 & 0 \\ 0 & 0 & -1 \\ 0 & 1 & 0 \end{pmatrix}, \qquad \mathbf{M}_2 = \begin{pmatrix} 0 & 0 & 1 \\ 0 & 0 & 0 \\ -1 & 0 & 0 \end{pmatrix}, \qquad \mathbf{M}_3 = \begin{pmatrix} 0 & -1 & 0 \\ 1 & 0 & 0 \\ 0 & 0 & 0 \end{pmatrix}.$$

$$(4\text{-}117)$$

The matrices $\mathbf{M}_i$ are known as the *infinitesimal rotation generators* and have the interesting property that their products are such that

$$\mathbf{M}_i\mathbf{M}_j - \mathbf{M}_j\mathbf{M}_i \equiv [\mathbf{M}_i, \mathbf{M}_j] = \epsilon_{ijk}\mathbf{M}_k. \qquad (4\text{-}118)$$

The difference between the two matrix products, or *commutator*, is also called the *Lie bracket* of $\mathbf{M}_i$, and Eq. (4–118) defines the *Lie algebra* of the rotation group parametrized in terms of the rotation angle. To go further into the group theory of rotation would take us too far afield, but we shall have further occasion to refer to these properties of the rotation operation.

4–9 RATE OF CHANGE OF A VECTOR

The concept of an infinitesimal rotation provides a powerful tool for describing the motion of a rigid body in time. Consider some arbitrary vector **G** involved in the mechanical problem, such as the position vector of a point in the body, or the total angular momentum. Usually such a vector will vary in time as the body moves, but the change will often depend on the coordinate system to which the observations are referred. For example, if the vector happens to be the radius vector from the origin of the body set of axes to a point in the rigid body then, clearly, such a vector appears constant when measured by the body set of axes. However, to an observer fixed in the space set of axes the components of the vector, as measured on the space axes will vary in time, if the body is in motion.

The change in a time dt of the components of a general vector **G** as seen by an observer in the body system of axes will differ from the corresponding change as seen by an observer in the space system. A relation between the two differential changes in **G** can be derived on the basis of physical arguments. We can write that the only difference between the two is the effect of rotation of the body axes:

$$(d\mathbf{G})_{\text{space}} = (d\mathbf{G})_{\text{body}} + (d\mathbf{G})_{\text{rot}}.$$

Consider now a vector fixed in the rigid body. As the body rotates there is of course no change in the components of this vector as seen by the body observer, i.e., relative to body axes. The only contribution to $(d\mathbf{G})_{\text{space}}$ is then the effect of the rotation of the body. But since the vector is fixed in the body system, it rotates with it *counterclockwise*, and the change in the vector as observed in space is that given by Eq. (4–111'), and hence $(d\mathbf{G})_{\text{rot}}$ is given by

$$(d\mathbf{G})_{\text{rot}} = d\boldsymbol{\Omega} \times \mathbf{G}.$$

For an arbitrary vector the change relative to the space axes is the sum of the two effects:

$$(d\mathbf{G})_{\text{space}} = (d\mathbf{G})_{\text{body}} + d\boldsymbol{\Omega} \times \mathbf{G}. \tag{4–119}$$

The time *rate of change* of the vector **G** as seen by the two observers is then obtained by dividing the terms in Eq. (4–119) by the differential time element dt under consideration;

$$\left(\frac{d\mathbf{G}}{dt}\right)_{\text{space}} = \left(\frac{d\mathbf{G}}{dt}\right)_{\text{body}} + \boldsymbol{\omega} \times \mathbf{G}. \tag{4–120}$$

Here $\boldsymbol{\omega}$ is the instantaneous *angular velocity* of the body defined by the relation*

$$\boldsymbol{\omega} \, dt = d\boldsymbol{\Omega}. \tag{4-121}$$

The vector $\boldsymbol{\omega}$ lies along the axis of the infinitesimal rotation occurring between t and $t + dt$, a direction known as the *instantaneous axis of rotation*. In magnitude $\boldsymbol{\omega}$ measures the instantaneous rate of rotation of the body.

    A more formal derivation of the basic Eq. (4-120) can be given in terms of the orthogonal matrix of transformation between the space and body coordinates. The component of $\mathbf{G}$ along the ith space axis is related to the components along the body axes:

$$G_i = \tilde{a}_{ij} G_j' = a_{ji} G_j'.$$

As the body moves in time the components G_j' will change as will also the elements a_{ij} of the transformation matrix. Hence the change in G_i in a differential time element dt is

$$dG_i = a_{ji} \, dG_j' + da_{ji} G_j'. \tag{4-122}$$

It is no loss of generality to take the space and body axes as instantaneously coincident at the time t. Components in the two systems will then be the same instantaneously, but differentials will *not* be the same, since the two systems are moving relative to each other. Thus $G_j' = G_j$ but $a_{ji} \, dG_j' = dG_i'$, the prime emphasizing the differential is measured in the body axis system. The change in the matrix $\mathbf{A}$ in the time dt is thus a change from the unit matrix and therefore corresponds to the matrix $\boldsymbol{\epsilon}$ of the infinitesimal rotation. Hence

$$da_{ji} = (\tilde{\boldsymbol{\epsilon}})_{ij} = -\epsilon_{ij},$$

using the antisymmetry property of $\boldsymbol{\epsilon}$. In terms of the permutation symbol ϵ_{ijk} the elements of $\boldsymbol{\epsilon}$ are such that (cf. Eq. 4-105)

$$-\epsilon_{ij} = -\epsilon_{ijk} \, d\Omega_k = \epsilon_{ikj} \, d\Omega_k.$$

Equation (4-122) can now be written

$$dG_i = dG_i' + \epsilon_{ikj} \, d\Omega_k G_j.$$

The last term on the right will be recognized as the expression for the ith component of a cross product, so that the final expression for the relation between differentials in the two systems is

$$dG_i = dG_i' + (d\boldsymbol{\Omega} \times \mathbf{G})_i, \tag{4-123}$$

which is the same as the ith component of Eq. (4-119).

    Equation (4-119) is not so much an equation about a particular vector $\mathbf{G}$ as it is a statement of the transformation of the time derivative between the two coordinate systems. The arbitrary nature of the vector $\mathbf{G}$ made use of in the

---

* As $\boldsymbol{\omega}$ is *not* the derivative of any vector, it is sometimes described as a *nonholonomic* vector, in analogy to the nonintegrable differential constraints.

derivation can be emphasized by writing Eq. (4–120) as an operator equation acting on some given vector:

$$\left(\frac{d}{dt}\right)_s = \left(\frac{d}{dt}\right)_r + \boldsymbol{\omega} \times \tag{4-124}$$

Here the subscripts s and r indicate the time derivatives are to be those observed in the space and body (rotating) system of axes, respectively. The resultant vector equation can then of course be resolved along any desired set of axes, fixed or moving. But it must be emphasized again that the time rate of change is only relative to the specified coordinate system. When a time derivative of a vector is with respect to one coordinate system, components may be taken along another set of coordinate axes only *after* the differentiation has been carried out.

It is often convenient to express the angular velocity vector in terms of the Euler angles and their time derivatives. The general infinitesimal rotation associated with $\boldsymbol{\omega}$ can be considered as consisting of three successive infinitesimal rotations with angular velocities $\omega_\phi = \dot{\phi}$, $\omega_\theta = \dot{\theta}$, $\omega_\psi = \dot{\psi}$. In consequence of the vector property of infinitesimal rotations, the vector $\boldsymbol{\omega}$ can be obtained as the sum of the three separate angular velocity vectors. Unfortunately, the directions $\boldsymbol{\omega}_\phi$, $\boldsymbol{\omega}_\theta$, and $\boldsymbol{\omega}_\psi$ are not symmetrically placed: $\boldsymbol{\omega}_\phi$ is along the space z axis, $\boldsymbol{\omega}_\theta$ is along the line of nodes, while $\boldsymbol{\omega}_\psi$ alone is along the body z' axis. However, the orthogonal transformations **B, C, D** of Section 4–4 may be used to furnish the components of these vectors along any desired set of axes.

The body set of axes proves to be the most useful for discussing the equations of motion, and we shall therefore obtain the components of $\boldsymbol{\omega}$ for such a coordinate system. Since $\boldsymbol{\omega}_\phi$ is parallel to the space z axis, its components along the body axes are given by applying the complete orthogonal transformation **A = BCD** (4–46):

$$(\omega_\phi)_{x'} = \dot{\phi} \sin\theta \sin\psi, \qquad (\omega_\phi)_{y'} = \dot{\phi} \sin\theta \cos\psi, \qquad (\omega_\phi)_{z'} = \dot{\phi} \cos\theta.$$

The line of nodes, which is the direction of $\boldsymbol{\omega}_\theta$, coincides with the ξ' axis, so that the components of $\boldsymbol{\omega}_\theta$ with respect to the body axes are furnished by applying only the final orthogonal transformation **B** (4–45):

$$(\omega_\theta)_{x'} = \dot{\theta} \cos\psi, \qquad (\omega_\theta)_{y'} = -\dot{\theta} \sin\psi, \qquad (\omega_\theta)_{z'} = 0.$$

No transformation is necessary for the components of $\boldsymbol{\omega}_\psi$, which lies along the z' axis. Adding these components of the separate angular velocities, the components of $\boldsymbol{\omega}$ with respect to the body axes are

$$\omega_{x'} = \dot{\phi} \sin\theta \sin\psi + \dot{\theta} \cos\psi$$
$$\omega_{y'} = \dot{\phi} \sin\theta \cos\psi - \dot{\theta} \sin\psi \tag{4-125}$$
$$\omega_{z'} = \dot{\phi} \cos\theta + \dot{\psi}.$$

Similar techniques may be used to express the components of ω along the space set of axes in terms of the Euler angles.*

4–10 THE CORIOLIS FORCE

Equation (4–124) is the basic kinematical law upon which the dynamical equations of motion for a rigid body are founded. But its validity is not restricted solely to rigid body motion. It may be used whenever we wish to discuss the motion of a particle, or system of particles, relative to a rotating coordinate system. A most important problem in this latter category is the description of particle motion relative to coordinate axes rotating with the earth. It will be recalled that in Section 1–1 an inertial system was defined as one in which Newton's laws of motion were valid. For many purposes a system of coordinates fixed in the rotating earth is a sufficient approximation to an inertial system. However the system of coordinates in which the local stars are fixed comes still closer to the ideal inertial system. Detailed examination shows there are observable effects arising from the rotation of the earth relative to this nearly inertial system. Equation (4–124) provides the needed modifications of the equations of motion relative to the *noninertial* system fixed in the rotating earth.

The initial step is to apply Eq. (4–124) to the radius vector, $\mathbf{r}$, from the origin of the terrestrial system to the given particle:

$$\mathbf{v}_s = \mathbf{v}_r + \boldsymbol{\omega} \times \mathbf{r}, \tag{4–126}$$

where $\mathbf{v}_s$ and $\mathbf{v}_r$ are the velocities of the particle relative to the space and rotating set of axes, respectively, and ω is the (constant) angular velocity of the earth relative to the inertial system. In the second step Eq. (4–124) is used to obtain the time rate of change of $\mathbf{v}_s$:

$$\left(\frac{d\mathbf{v}_s}{dt}\right)_s = \mathbf{a}_s = \left(\frac{d\mathbf{v}_s}{dt}\right)_r + \boldsymbol{\omega} \times \mathbf{v}_s$$

$$= \mathbf{a}_r + 2(\boldsymbol{\omega} \times \mathbf{v}_r) + \boldsymbol{\omega} \times (\boldsymbol{\omega} \times \mathbf{r}), \tag{4–127}$$

where $\mathbf{v}_s$ has been substituted from Eq. (4–126), and where $\mathbf{a}_s$ and $\mathbf{a}_r$ are the accelerations of the particle in the two systems. Finally, the equation of motion, which in the inertial system is simply

$$\mathbf{F} = m\mathbf{a}_s,$$

expands, when expressed in the rotating coordinates, into the equation

$$\mathbf{F} - 2m(\boldsymbol{\omega} \times \mathbf{v}_r) - m\boldsymbol{\omega} \times (\boldsymbol{\omega} \times \mathbf{r}) = m\mathbf{a}_r. \tag{4–128}$$

---

* Equation (4–125) refers to the *x*-convention for the Euler angles. Corresponding formulas for the other Euler angle conventions are given in Appendix B.

To an observer in the rotating system it therefore appears as if the particle is moving under the influence of an effective force $\mathbf{F}_{eff}$:

$$\mathbf{F}_{eff} = \mathbf{F} - 2m(\boldsymbol{\omega} \times \mathbf{v}_r) - m\boldsymbol{\omega} \times (\boldsymbol{\omega} \times \mathbf{r}). \qquad (4\text{--}129)$$

Let us examine the nature of the terms occurring in Eq. (4–129). The last term is a vector normal to $\boldsymbol{\omega}$ and pointing outward. Further, its magnitude is $m\omega^2 r \sin\theta$. It will therefore be recognized that this term is simply the familiar centrifugal force. When the particle is stationary in the moving system the centrifugal force is the only added term in the effective force. However, when the particle is moving, the middle term known as the *Coriolis* force comes into play. The order of magnitude of both of these forces may easily be calculated for a particle on the earth's surface. The earth rotates counterclockwise about the North Pole with an angular velocity relative to the fixed stars:

$$\omega = \left(\frac{2\pi}{24 \times 3600}\right)\left(\frac{366.5}{365.5}\right) = 7.292 \times 10^{-5} \ \text{sec}^{-1}.$$

Here the first parenthesis gives the angular velocity relative to the radius vector to the sun. The quantity in the second parenthesis, the ratio of the number of sidereal days in a year to the corresponding number of solar days, is the correction factor to give the angular velocity relative to the fixed stars. With this value for ω, and with r equal to the equatorial radius, the maximum centripetal acceleration is

$$\omega^2 r = 3.38 \ \text{cm/sec}^2,$$

or about 0.3 % of the acceleration of gravity. While small, this acceleration is by no means negligible. However, the measured effects of gravity represent the combination of the gravitational field of the mass distribution of the earth and the effects of centripetal acceleration. It has become customary to speak of the sum of the two as the earth's *gravity* field, as distinguished from its *gravitational* field.

The situation is further complicated by the effect of the centripetal acceleration in flattening the rotating earth. If the earth were completely fluid, the effect of rotation would be to deform it into the shape of an ellipsoid whose surface would be an equipotential surface of the combined gravity field. The mean level of the earth's seas conforms very closely to this equilibrium ellipsoid (except for local variations of wind and tide) and defines what is called the *geoid* of the earth. Except for effects of local perturbations, the force of gravity will be perpendicular to the equipotential surface of the geoid. Accordingly, the local vertical is defined as the direction perpendicular to the geoid at the given point on the surface. For phenomena that occur in the vicinity of a particular spot on the earth, the centripetal acceleration terms in Eq. (4–129) can be considered as swallowed up in the gravitational acceleration $\mathbf{g}$, which will be oriented in the local vertical direction. The magnitude of $\mathbf{g}$ of course varies with the latitude on the earth. The effects of centripetal acceleration and the flattening of the earth combine to make $\mathbf{g}$ about 0.53 % less at the Equator than at the Poles.

Incidentally, the centrifugal force on a particle arising from the earth's revolution around the sun is appreciable compared to gravity, but it is almost exactly balanced by the gravitational attraction to the sun. It is, of course, just this balance between centrifugal force and gravitational attraction that keeps the earth (and all that are on it) in orbit around the sun.

The Coriolis force on a moving particle is perpendicular to both **ω** and **v**.* In the Northern Hemisphere, where **ω** points out of the ground, the Coriolis force, $2m(\mathbf{v} \times \boldsymbol{\omega})$ tends to deflect a projectile shot along the earth's surface, to the right of its direction of travel; cf. Fig. 4–12. The Coriolis deflection reverses direction in the Southern Hemisphere and is zero at the Equator, where **ω** is horizontal. The magnitude of the Coriolis acceleration is always less than

$$2\omega v \simeq 1.5 \times 10^{-4}v,$$

which for a velocity of 10^5 cm/sec (roughly 2000 mph) is 15 cm/sec², or about 0.015 g. Normally, such an acceleration is extremely small but there are instances where it becomes important. To take an artificial illustration, suppose a projectile were fired horizontally at the North Pole. The Coriolis acceleration would then have the magnitude $2\omega v$, so that the linear deflection after a time t is $\omega v t^2$; and the angular deflection would be the linear deflection divided by the distance of travel:

$$\theta = \frac{\omega v t^2}{vt} = \omega t, \tag{4–130}$$

which is the angle the earth rotates in the time t. Physically, this result means that a projectile shot off at the North Pole has no initial rotational motion and hence its trajectory in the inertial space is a straight line, the apparent deflection being due to the earth rotating beneath it. Some idea of the magnitude of the effect can

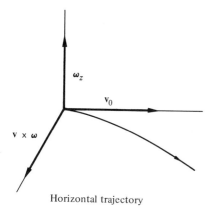

Horizontal trajectory

FIGURE 4–12
Direction of Coriolis deflection in the Northern Hemisphere.

---

* From here on, the subscript r will be dropped from v as all velocities will be taken with respect to the rotating coordinate axes only.

be obtained by substituting a time of flight of 100 seconds—not unusual for large projectiles—in Eq. (4–130). The angular deflection is then of the order of 7×10^{-3} radians, about $0.4°$, which is not inconsiderable. Clearly, the effect is even more important for long-range missiles that have a much longer time of flight.

The Coriolis force also plays a significant role in many oceanographic and meteorological phenomena involving displacements of masses of matter over long distances, such as the circulation pattern of the trade winds and the course of the Gulf stream. A full description of these phenomena requires the solution of complex hydrodynamic problems in which the Coriolis acceleration is only one among many terms involved. It is possible however to give some indication of the contribution of Coriolis forces by considering a highly simplified picture of one particular meteorological problem—the large scale horizontal wind circulation. Masses of air tend to move, other things be equal, from regions of high pressure to regions of low pressure—the so-called pressure gradient flow. In the vertical direction the pressure gradient is roughly balanced by gravitational forces so that it is only in the horizontal plane that there are persistent long-range motions of air masses—which we perceive as winds. The pressure gradient forces are quite modest, and comparable in magnitude to the Coriolis forces acting on air masses moving at usual speeds. In the absence of Coriolis forces the wind directions, ideally, would be perpendicular to the isobars, as shown in Fig. 4–13. However the Coriolis forces deflect the wind to the right of this direction in the sense indicated in the figure. The deflection to the right continues until the wind vector is parallel to the isobars and the Coriolis force is in the opposite direction to, and ideally just balances, the pressure-gradient force. The wind then continues parallel to the isobars, circulating in the Northern Hemisphere in a counterclockwise direction about a center of low pressure. In the Southern Hemisphere the Coriolis force acts in the opposite direction, and the cyclonic direction (i.e., the flow around a low pressure center) is clockwise. (Such a wind flow, which has been deflected parallel to the isobars, is known as a *geostrophic wind*.) In this simplified picture the effect of friction has been neglected. At

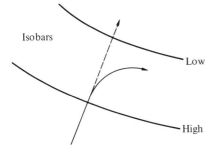

Isobars

Low

High

FIGURE 4–13
Deflection of wind from the direction of the pressure gradient by the Coriolis force (shown for the Northern Hemisphere).

atmospheric altitudes below several km the friction effects of eddy viscosity become important, and the equilibrium wind direction never becomes quite parallel to the isobars, as indicated in Fig. 4–14.

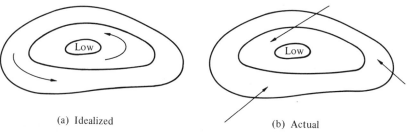

(a) Idealized

(b) Actual

FIGURE 4–14
Cyclone pattern in the Northern Hemisphere.

Another classical instance where Coriolis force produces a measurable effect is in the deflection from the vertical of a freely falling particle. Since the particle velocity is almost vertical and $\boldsymbol{\omega}$ lies in the north–south vertical plane, the deflecting force $2m(\mathbf{v} \times \boldsymbol{\omega})$ is in the east–west direction. Thus, in the Northern Hemisphere, a body falling freely will be deflected to the east. Calculation of the deflection is greatly simplified by choosing the z axis of the terrestrial coordinate system to be along the direction of the upward vertical as previously defined. If the y axis is taken as pointing north, then the equation of motion in the x (east) direction is

$$m\frac{d^2x}{dt^2} = -2m(\boldsymbol{\omega} \times \mathbf{v})_x$$

$$= -2m\omega v_z \sin\theta, \tag{4-131}$$

where θ is the colatitude. The effect of the Coriolis force on v_z would constitute a small correction to the deflection, which itself is very small. Hence the vertical velocity appearing in (4–131) may be computed as if Coriolis forces were absent:

$$v_z = -gt$$

and

$$t = \sqrt{\frac{2z}{g}}.$$

With these values, Eq. (4–131) may be easily integrated to give the deflection as

$$x = \frac{\omega g}{3}t^3 \sin\theta$$

or

$$x = \frac{\omega}{3}\sqrt{\frac{(2z)^3}{g}}\sin\theta.$$

An order of magnitude of the deflection can be obtained by assuming $\theta = \pi/2$ (corresponding to the Equator) and $z = 100$ m. The deflection is then, roughly,

$$x \simeq 2.2 \text{ cm}.$$

The actual experiment is difficult to perform, as the small deflection may often be masked by the effects of wind currents, viscosity, or other disturbing influences.*

More easily observable is the well-known experiment of the Foucault pendulum. If a pendulum is set swinging at the North Pole in a given plane in space, then its linear momentum perpendicular to the plane is zero, and it will continue to swing in this invariable plane while the earth rotates beneath it. To an observer on the earth the plane of oscillation appears to rotate once a day. At other latitudes the result is more complicated, but the phenomenon is qualitatively the same and detailed calculation will be left as an exercise.

Effects due to the Coriolis terms also appear in atomic physics. Thus, two types of motion may occur simultaneously in polyatomic molecules: The molecule *rotates* as a rigid whole, and the atoms *vibrate* about their equilibrium positions. As a result of the vibrations the atoms are in motion relative to the rotating coordinate system of the molecule. The Coriolis term will then be different from zero and will cause the atoms to move in a direction perpendicular to the original oscillations. Perturbations in molecular spectra due to Coriolis forces thus appear as interactions between the rotational and vibrational levels of the molecule.

SUGGESTED REFERENCES

R. R. STOLL, *Linear Algebra and Matrix Theory*. The textbook literature on linear algebra is voluminous, evanescent, and largely unusable for present purposes. For the reference given here a deliberate choice was made of one with a somewhat old-fashioned and leisurely approach, displaying a penchant for Euclidean spaces and frequent three-dimensional examples. The physics in the present chapter demands no higher level of mathematical development.

G. ARFKEN, *Mathematical Methods for Physicists*. Texts with titles such as this usually have some treatment of matrices and linear transformations, and Arfken's book is representative of the best of these. Chapter 4 on matrices and determinants is particularly relevant and has a collection of useful problems. Pauli matrices are discussed here, along with their four-dimensional generalizations, the Dirac matrices.

H. JEFFREYS AND B. S. JEFFREYS, *Methods of Mathematical Physics*. Though a rather aging text by now, this reference contains a wealth of physical applications born out of the authors' long research experience in theoretical physics. Many of the topics of the present chapter will be found discussed in Chapters 3 and 4, including a treatment of the Pauli spin matrices and their connection with the three-dimensional rotation matrices. The section on Euler angles is practically unintelligible, not solely because of a poor diagram. The apt and

---

* It is easy to show, using Eq. 4–131, that a particle projected upward will fall back to the ground *westward* of the original launching spot.

witty quotations heading each chapter are alone almost worth the price of admission. What's given for Chapter 7 is probably a misquotation but is all the more bitingly humorous for that.

J. B. MARION, *Principles of Vector Analysis*. Although this is but a pocket size paperback the author goes into extensive detail on the properties of the orthogonal transformation matrix and the transformation properties of axial and polar vectors. Here and in later sections, considerable use is made of the permutation symbol, e.g., in derivation of complicated vector identities.

J. L. SYNGE AND A. SCHILD, *Tensor Calculus*. This well-known monograph ranges widely over many aspects of tensors including a few of special interest here. Section 4.2 grudgingly considers properties of tensors in flat space (we will be considering only flat space in this book), especially those of the permutation symbol. Chapter 7 considers relative tensors, or pseudotensors, which are the tensor generalizations of pseudovectors. Two chapters consider physical applications, including dynamics, from a somewhat unusual viewpoint.

A. I. BORISENKO AND I. E. TARAPOV, *Vector and Tensor Analysis with Applications*. Although written throughout from the standpoint of physics' applications, the methods and viewpoints differ frequently from those used here. The authors have a great fondness for working in terms of the unit vectors $\mathbf{n}_i$ along orthogonal axes. They use, for example, the representation of the permutation symbol by $\mathbf{n}_i \cdot \mathbf{n}_j \times \mathbf{n}_k$ in a proof of the transformation properties of an antisymmetric tensor with the same characteristics as $\boldsymbol{\epsilon}$.

E. T. WHITTAKER, *Analytical Dynamics*. Chapter I contains the material pertinent to our purposes. The section on Eulerian angles is difficult to follow because of the lack of *any* diagram. Reference should be made to our footnote in Section 4–5 in comparing his results with our equations. Section 12 discusses the relation of the Cayley–Klein parameters to the so-called homographic transformation.

L. A. PARS, *A Treatise on Analytical Dynamics*. A monumental treatise indeed, this looks and feels like a modern day successor to Whittaker, in full scholarly Cambridge tradition. It is a book that, as it were, seems to wear cap and gown, and like Whittaker displays a wealth of erudition. Chapter VIII is devoted to the theory of rotations and contains, among many other items, three separate proofs of the rotation formula. The strong version of Chasles' theorem (see above, footnote p. 163) is given a simple geometric proof that would not be out of place in Newton's *Principia*.

G. HAMEL, *Theoretische Mechanik*. Those to whom the language is accessible will find that this reference contains an almost complete encyclopedia of information on the kinematics of rigid rotation, primarily in Sections 8 and 9 of Chapter 2. Examples are a thorough discussion of the relation of the stereographic projection to the Cayley–Klein and Euler parameters, and of the quaternion representation of rotation. The world of physics in which the treatment is imbedded is roughly that of 1925. The book is concluded by a 260 page section of exercises *and* their solutions.

T. C. BRADBURY, *Theoretical Mechanics*. Although labeled as an intermediate level text, many items of interest here are covered, some from rather unusual viewpoints. Chapters 1 and 3 cover much the same ground as here on matrices, but somewhat more intensively. The properties of the permutation symbol are discussed explicitly. For motion in a rotating system an effective Lagrangian is set up in which the Coriolis force is derived from a velocity-dependent potential, introduced in something of an adhoc manner. It does provide an efficient tool for the study of motion in such noninertial frames.

E. J. SALETAN AND A. H. CROMER, *Theoretical Mechanics*. Written from an advanced viewpoint this text makes full use of modern mathematics on a middling abstract level. Matrices and their applications in mechanics are presented in Chapter 4 as part of a discussion of linear vector spaces. Kinematics of rotation, including Euler angles in the *x*-convention, appear in the following chapter. A view of rotations from the standpoint of group theory and Lie algebras is presented briefly in Chapter 9.

F. D. STACEY, *Physics of the Earth*. The observed "gravity" field on the surface of the rotating earth is complicated by the deviation of the earth's figure from a sphere. Chapters 2 and 3 of this book contain, inter alia, a detailed yet compact discussion of these effects. Coriolis force phenomena are not mentioned.

G. HERZBERG, *Infrared and Raman Spectra*. The effect of Coriolis forces on the spectra of polyatomic molecules is exhaustively discussed in this book, especially in Chapter IV, Sections 1 and 2, although a grounding in small oscillation theory (see our Chapter 6) and quantum mechanics is necessary for a full understanding. See, however, pages 372–375 for a brief classical treatment.

S. L. HESS, *Introduction to Theoretical Meteorology*. This book presents, in Chapters 11 and 12, a careful and detailed presentation of the effects of Coriolis force on atmospheric winds. In Chapter 13 there is a discussion of Coriolis terms in the so-called circulation theorem of fluid dynamics. The derivations are painstaking and go back to first principles, albeit on an elementary level.

Orthogonal: $A^{-1} = \tilde{A}$

$\tilde{A}$ — transpose

EXERCISES

1. Prove that matrix multiplication is associative. Show that the product of two orthogonal matrices is also orthogonal.

2. Prove the following properties of the transposed and adjoint matrices:

transposed $A_{ij} \to A_{ji}$

$$\widetilde{AB} = \tilde{B}\tilde{A},$$

Adjoint $A^{\dagger} = \left(\tilde{A}\right)^{*}$

$$(AB)^{\dagger} = B^{\dagger}A^{\dagger}.$$

3. Show that the trace of a matrix is invariant under any similarity transformation. Show also that the antisymmetry property of a matrix is preserved under an orthogonal similarity transformation, while the hermitean property is invariant under any unitary similarity transformation.

similarity trans. $A' = BAB^{-1}$

4. a) By examining the eigenvalues of an antisymmetric 3×3 real matrix **A**, show that $1 \pm$ **A** is nonsingular.

  b) Show then that under the same conditions the matrix

$$B = (1 + A)(1 - A)^{-1}$$

is orthogonal.

5. Obtain the matrix elements of the general rotation matrix in terms of the Euler angles, Eq. (4–46), by performing the multiplications of the successive component rotation matrices. Verify directly that the matrix elements obey the orthogonality conditions.

6. a) Find the vector equation describing the reflection of **r** in a plane whose unit normal is **n**.

b) Show that if l_i, $i = 1, 2, 3$, are the direction cosines of **n**, then the matrix of transformation has the elements

$$A_{ij} = \delta_{ij} - 2l_i l_j,$$

and verify that **A** is an improper orthogonal matrix.

7. The body set of axes can be related to the space set in terms of Euler's angles by the following set of rotations:

1) Rotation about the x axis by an angle θ.
2) Rotation about the z' axis by an angle ψ.
3) Rotation about the *old* z axis by an angle ϕ.

Show that this sequence leads to the same elements of the matrix of transformation as the sequence of rotations given in the book. [*Hint:* It is not necessary to carry out the explicit multiplication of the rotation matrices.]

8. If **A** is the matrix of a rotation through $180°$ about any axis, show that if

$$P_{\pm} = \tfrac{1}{2}(1 \pm A),$$

then $P_{\pm}^2 = P_{\pm}$. Obtain the elements of $P_{\pm}$ in any suitable system, and find a geometric interpretation of the operation P_{+} and P_{-} on any vector **F**.

9. Express the "rolling" constraint of a sphere on a plane surface in terms of the Euler angles. Show that the conditions are nonintegrable and that the constraint is therefore nonholonomic.

10. a) Show that the rotation matrix in the form of Eq. (4–67) cannot be put in the form of the matrix of the inversion transformation **S**.

b) Verify by direct multiplication that the matrix in Eq. (4–67) is orthogonal.

11. Show that any rotation can be represented by successive reflection in two planes, both passing through the axis of rotation with the planar angle $\Phi/2$ between them.

12. If **B** is a square matrix and **A** is the exponential of **B**, defined by the infinite series expansion of the exponential,

$$A \equiv e^{B} = 1 + B + \tfrac{1}{2}B^2 + \cdots + \frac{B^n}{n!} + \cdots,$$

then prove the following properties:

a) $e^{B}e^{C} = e^{B+C}$, providing **B** and **C** commute.
b) $A^{-1} = e^{-B}$
c) $e^{CBC^{-1}} = CAC^{-1}$
d) **A** is orthogonal if **B** is antisymmetric.
e) e^{iB} is unitary if **B** is hermitean.

13. Show that Q_θ may be written symbolically in the form

$$Q_\theta = e^{i\sigma_1(\theta/2)}$$

where the exponential is taken as standing for its series expansion, whose first term is **1**.

14. a) Show that the three Pauli spin matrices anticommute with each other, i.e., that

$$\sigma_i \sigma_j = -\sigma_j \sigma_i, \qquad i \neq j$$

and that they obey the commutation relations

$$\sigma_i\sigma_j - \sigma_j\sigma_i \equiv [\sigma_i,\sigma_j] = 2i\epsilon_{ijk}\sigma_k.$$

b) Prove that $\sigma_i^2 = \mathbf{1}$, for all values of i.

c) If σ stands for the vector with components σ_i, show that if $\mathbf{A}$ and $\mathbf{B}$ are any two vectors, then

$$(\sigma\cdot\mathbf{A})(\sigma\cdot\mathbf{B}) = \mathbf{1}\mathbf{A}\cdot\mathbf{B} + i\sigma\cdot(\mathbf{A}\times\mathbf{B}),$$

and that therefore

$$[\sigma\cdot\mathbf{A},\sigma\cdot\mathbf{B}] = 2i\sigma\cdot(\mathbf{A}\times\mathbf{B}).$$

15. In a set of axes where the z axis is the axis of rotation of a finite rotation, the rotation matrix is given by Eq. (4–43) with θ replaced by the angle of finite rotation Φ. Derive the rotation formula, Eq. (4–92), by transforming to an arbitrary coordinate system, expressing the orthogonal matrix of transformation in terms of the direction cosines of the axis of the finite rotation.

16. a) On the basis of Eq. (4–73′) show that

$$\mathbf{r}'\cdot\sigma = \mathbf{r}\cdot\sigma + [\mathbf{Q},\sigma\cdot\mathbf{r}]\mathbf{Q}^\dagger.$$

b) With this result, and using the representation (4–98) for $\mathbf{Q}$ and Exercise 12 above, derive the rotation formula, Eq. (4–92).

17. a) Suppose two successive coordinate rotations through angles Φ_1 and Φ_2 are carried out, equivalent to a single rotation through an angle Φ. Show that Φ_1, Φ_2, and Φ can be considered as the sides of a spherical triangle with the angle opposite to Φ given by the angle between the two axes of rotation.

b) Show that a rotation about any given axis can be obtained as the product of two successive rotations, each through 180°.

18. a) Verify that the permutation symbol satisfies the following identity in terms of Kronecker delta symbols:

$$\epsilon_{ijp}\epsilon_{rmp} = \delta_{ir}\delta_{jm} - \delta_{im}\delta_{jr}.$$

b) Show that

$$\epsilon_{ijp}\epsilon_{ijk} = 2\delta_{pk}.$$

19. Show that the components of the angular velocity along the space set of axes are given in terms of the Euler angles by

$$\omega_x = \dot\theta\cos\phi + \dot\psi\sin\theta\sin\phi,$$

$$\omega_y = \dot\theta\sin\phi - \dot\psi\sin\theta\cos\phi,$$

$$\omega_z = \dot\psi\cos\theta + \dot\phi.$$

20. Show that the Euler parameter e_0 has the equation of motion

$$-2\dot e_0 = e_1\omega_{x'} + e_2\omega_{y'} + e_3\omega_{z'},$$

where the prime denotes the body set of axes. Find the corresponding equations for the other three Euler parameters and for the complex Cayley–Klein parameters α and β.

21. Verify directly that the matrix generators of infinitesimal rotation, $\mathbf{M}_i$, as given by Eq. (4–117) obey the commutation relations

$$[\mathbf{M}_i,\mathbf{M}_j] = \epsilon_{ijk}\mathbf{M}_k.$$

22. A particle is thrown up vertically with initial speed v_0, reaches a maximum height and falls back to ground. Show that the Coriolis deflection when it again reaches the ground is opposite in direction, and four times greater in magnitude, than the Coriolis deflection when it is dropped at rest from the same maximum height.

23. A projectile is fired horizontally along the earth's surface. Show that to a first approximation the angular deviation from the direction of fire resulting from the Coriolis force varies linearly with time at a rate

$$\omega \cos\theta,$$

where ω is the angular frequency of the earth's rotation and θ is the colatitude, the direction of deviation being to the right in the Northern Hemisphere.

24. The Foucault pendulum experiment consists in setting a long pendulum in motion at a point on the surface of the rotating earth with its momentum originally in the vertical plane containing the pendulum bob and the point of suspension. Show that its subsequent motion may be described by saying that the plane of oscillation rotates uniformly $2\pi \cos\theta$ radians per day, where θ is the colatitude. What is the direction of rotation? The approximation of small oscillations may be used, if desired.

23

$$2m\left(\vec{w} \times \vec{v}\right) =$$

$$F_c = 2m\, w v \sin\left(\frac{\pi}{2} - \theta\right)$$

24

spherical pendulum with
initial cond⁼s and small angles
w/ coriolis terms.

CHAPTER 5
The Rigid Body
Equations of Motion

Chapter 4 has presented all the kinematical tools needed in the discussion of rigid body motion. In the Euler angles we have a set of three coordinates, defined rather unsymmetrically it is true, yet suitable for use as the generalized coordinates describing the orientation of the rigid body. In addition, the method of orthogonal transformations, and the associated matrix algebra, furnish a powerful and elegant technique for investigating the characteristics of rigid body motion. We have already had one application of the technique in deriving Eq. (4–124), the relation between the rates of change of a vector as viewed in the space system and in the body system. These tools will now be applied to obtain the dynamical equations of motion of the rigid body in their most convenient form. With the help of the equations of motion, some simple but highly important problems of rigid body motion can be discussed.

5–1 ANGULAR MOMENTUM AND KINETIC ENERGY OF MOTION ABOUT A POINT

Chasles' theorem states that any general displacement of a rigid body can be represented by a translation plus a rotation. The theorem suggests that it ought to be possible to split the problem of rigid body motion into two separate phases, one concerned solely with the translational motion of the body, the other, with its rotational motion. Of course, if one point of the body is fixed the separation is obvious, for then there is only a rotational motion about the fixed point, without any translation. But even for a general type of motion such a separation is often possible. The six coordinates needed to describe the motion have already been formed into two sets in accordance with such a division: the three Cartesian coordinates of a point fixed in the rigid body to describe the translational motion and, say, the three Euler angles for the motion about the point. If, further, the origin of the body system is chosen to be the center of mass, then by Eq. (1–28) the total angular momentum divides naturally into contributions from the translation of the center of mass and from the rotation about the center of mass. The former term will involve only the Cartesian coordinates of the center of mass, the latter only the angle coordinates. By Eq. (1–31) a similar division holds for the total kinetic energy T, which can be written in the form

$$T = \tfrac{1}{2}Mv^2 + T'(\phi, \theta, \psi),$$

as the sum of the kinetic energy of the entire body as if concentrated at the center of mass, plus the kinetic energy of motion about the center of mass.

Often the potential energy can be similarly divided, each term involving only one of the coordinate sets, either the translational or rotational. Thus, the potential energy in a uniform gravitational field will depend only upon the Cartesian vertical coordinate of the center of gravity.* Or if the force on a body is due to a uniform magnetic field, $\mathbf{B}$, acting on its magnetic dipole moment, $\mathbf{M}$, then the potential is proportional to $\mathbf{M} \cdot \mathbf{B}$, which involves only the orientation of the body. Certainly, almost all problems soluble in practice will allow of such a separation. In such case the entire mechanical problem does indeed split into two, for the Lagrangian, $L = T - V$, divides into two parts, one involving only the translational coordinates, the other only the angle coordinates. These two groups of coordinates will then be completely separated, and the translational and rotational problems can be solved independently of each other. It is of obvious importance, therefore, to obtain expressions for the angular momentum and kinetic energy of the motion about some point fixed in the body. To do so we will make abundant use of Eq. (4–124) linking derivatives relative to a coordinate system fixed at some point in the rigid body. It is intuitively obvious that the rotation angle of a rigid body displacement, as also the instantaneous angular velocity vector, is independent of the choice of origin of the body system of axes. The essence of the rigid body constraint is that all particles of the body move and rotate together. However, a formal proof is easily constructed. Let $\mathbf{R}_1$ and $\mathbf{R}_2$ be the position vectors, relative to a fixed set of coordinates, of the origins of two sets of body coordinates (cf. Fig. 5–1). The

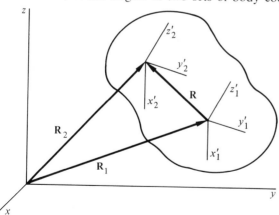

FIGURE 5–1

Vectorial relation between sets of rigid body coordinates with different origins.

---

* The center of gravity, of course, coincides with the center of mass in a uniform gravitational field.

difference vector is denoted by $\mathbf{R}$:

$$\mathbf{R}_2 = \mathbf{R}_1 + \mathbf{R}.$$

If the origin of the second set of axes is considered as a point defined relative to the first, then the time derivative of $\mathbf{R}_2$ relative to the space axes is given by

$$\left(\frac{d\mathbf{R}_2}{dt}\right)_s = \left(\frac{d\mathbf{R}_1}{dt}\right)_s + \left(\frac{d\mathbf{R}}{dt}\right)_s = \left(\frac{d\mathbf{R}_1}{dt}\right)_s + \boldsymbol{\omega}_1 \times \mathbf{R}.$$

The last step follows from Eq. (4–124), recalling that the derivatives of $\mathbf{R}$ relative to any rigid body axes must vanish, and with $\boldsymbol{\omega}_1$ as being the angular velocity vector appropriate to the first coordinate system. Alternatively, the origin of the first coordinate system can be considered as fixed in the second system with the position vector $-\mathbf{R}$. In the same manner, then, the derivative of the position vector $\mathbf{R}_1$ to this origin relative to the fixed space axes can be written as

$$\left(\frac{d\mathbf{R}_1}{dt}\right)_s = \left(\frac{d\mathbf{R}_2}{dt}\right)_s - \left(\frac{d\mathbf{R}}{dt}\right)_s = \left(\frac{d\mathbf{R}_2}{dt}\right)_s - \boldsymbol{\omega}_2 \times \mathbf{R}.$$

A comparison of these two expressions shows, as expected, that the two angular velocity vectors must be equal:

$$\boldsymbol{\omega}_1 = \boldsymbol{\omega}_2.$$

The angular velocity vector is the same for all coordinate systems in the rigid body.

When a rigid body moves with one point stationary, the total angular momentum about that point is

$$\mathbf{L} = m_i(\mathbf{r}_i \times \mathbf{v}_i), \tag{5–1}$$

(employing the summation convention) where $\mathbf{r}_i$ and $\mathbf{v}_i$ are the radius vector and velocity, respectively, of the ith particle relative to the given point. Since $\mathbf{r}_i$ is a fixed vector relative to the body, the velocity $\mathbf{v}_i$ with respect to the space set of axes arises solely from the rotational motion of the rigid body about the fixed point. From Eq. (4–124), $\mathbf{v}_i$ is then

$$\mathbf{v}_i = \boldsymbol{\omega} \times \mathbf{r}_i. \tag{5–2}$$

Hence Eq. (5–1) can be written as

$$\mathbf{L} = m_i(\mathbf{r}_i \times (\boldsymbol{\omega} \times \mathbf{r}_i)),$$

or, expanding the triple cross product,

$$\mathbf{L} = m_i(\boldsymbol{\omega} r_i^2 - \mathbf{r}_i(\mathbf{r}_i \cdot \boldsymbol{\omega})). \tag{5–3}$$

Again expanding, the x component of the angular momentum becomes

$$L_x = \omega_x m_i(r_i^2 - x_i^2) - \omega_y m_i x_i y_i - \omega_z m_i x_i z_i, \tag{5–4}$$

with similar equations for the other components of **L**. Thus each of the components of the angular momentum is a linear function of all the components of the angular velocity. *The angular momentum vector is related to the angular velocity by a linear transformation.* To emphasize the similarity of (5–4) with the equations of a linear transformation, (4–12), we may write L_x as

$$L_x = I_{xx}\omega_x + I_{xy}\omega_y + I_{xz}\omega_z. \tag{5–5}$$

Analogously, for L_y and L_z we have

$$L_y = I_{yx}\omega_x + I_{yy}\omega_y + I_{yz}\omega_z,$$
$$L_z = I_{zx}\omega_x + I_{zy}\omega_y + I_{zz}\omega_z. \tag{5–5}$$

The nine coefficients I_{xx}, I_{xy}, etc., are the nine elements of the transformation matrix. The diagonal elements are known as *moment of inertia coefficients*, and have the form illustrated by

$$I_{xx} = m_i(r_i^2 - x_i^2), \tag{5–6}$$

while the off-diagonal elements are designated as *products of inertia*, a typical one being

$$I_{xy} = -m_i x_i y_i. \tag{5–7}$$

In Eqs. (5–6) and (5–7) the matrix elements appear in the form suitable if the rigid body is composed of discrete particles. For continuous bodies the summation is replaced by a volume integration, with the particle mass becoming a mass density. Thus, the diagonal element I_{xx} appears as

$$I_{xx} = \int_V \rho(\mathbf{r})(r^2 - x^2)\,dV. \tag{5–6'}$$

With a slight change in notation an expression for all matrix elements can be stated for continuous bodies. If the coordinate axes are denoted by $x_j, j = 1, 2, 3$, then the matrix element I_{jk} can be written

$$I_{jk} = \int_V \rho(\mathbf{r})(r^2\delta_{jk} - x_j x_k)\,dV. \tag{5–8}$$

Up to the present, the coordinate system used in resolving the components of **L** has not been specified. From now on, it will be convenient to take it to be a system fixed in the body.* The various distances x_i, y_i, z_i are then constant in time, so that the matrix elements are likewise constants, peculiar to the body involved, and dependent on the origin and orientation of the particular body set of axes in which they are expressed.

---

* In Chapter 4, such a system was denoted by primes. As components along spatial axes will rarely be used here, this convention will be dropped from now on to simplify the notation. Unless otherwise specified all coordinates used for the rest of the chapter will refer to systems fixed in the rigid body.

The equations (5–5) relating the components of **L** and **ω** can be summarized by a single operator equation,

$$\mathbf{L} = \mathsf{I}\boldsymbol{\omega}, \tag{5-9}$$

where the symbol **I** stands for the operator whose matrix elements are the inertia coefficients appearing in (5–5). Of the two interpretations that have been given to the operator of a linear transformation (cf. Section 4–2), it is clear that here **I** must be thought of as acting upon the vector **ω**, and not upon the coordinate system. The vectors **L** and **ω** are two physically different vectors, having different dimensions, and are not merely the same vector expressed in two different coordinate systems. Unlike the operator of rotation, **I** will have dimensions— mass times length squared—and it is not restricted by any orthogonality conditions. Equation (5–9) is to be read as saying that the operator **I** acting upon the vector **ω** results in the physically new vector **L**. While full use will be made of the matrix algebra techniques developed in the discussion of the rotation operator, more attention must be paid here to the nature and physical character of the operator per se. However, a certain amount of preliminary mathematical matter needs first to be discussed. Those already familiar with tensors and dyadics can proceed immediately to Section 5–3.

5–2 TENSORS AND DYADICS

The quantity **I** may be considered as defining the quotient of **L** and **ω**:

$$\mathsf{I} = \frac{\mathbf{L}}{\boldsymbol{\omega}},$$

for the product of **I** and **ω** gives **L**. Now, the quotient of two quantities is often not a member of the same class as the dividing factors, but may belong to a more complicated class. Thus, the quotient of two integers is in general not an integer but rather a rational number. Similarly, the quotient of two vectors, as is well known, cannot be defined consistently within the class of vectors. It is not surprising, therefore to find that **I** is a new type of quantity, a *tensor of the second rank*.

In a Cartesian three-dimensional space, a tensor **T** of the Nth rank may be defined for our purposes as a quantity having 3^N components $T_{ijk\ldots}$ (with N indices) that transform under an orthogonal transformation of coordinates, **A**, according to the following scheme:*

$$T'_{ijk\ldots}(\mathbf{x}') = a_{il}a_{jm}a_{kn}\ldots T_{lmn\ldots}(\mathbf{x}). \tag{5-10}$$

---

* In a Cartesian space (i.e., with orthogonal straight line axes) there is no distinction between "covariant" and "contravariant" indices and the terminology will not be needed. Indeed, strictly speaking the tensors defined here should be denoted as "Cartesian tensors" (cf. J. L. Synge and A. Schild, *Tensor Calculus*, Toronto 1949, pp. 127–136). However, as this is the only type of tensor that will be used in this book (except in Section 7–3) the adjective will be omitted in subsequent discussions.

No need for contravariant - covariant.

By this definition, a tensor of the zero rank has one component, which is invariant under orthogonal transformation. Hence, *a scalar is a tensor of zero rank*. A tensor of the first rank has three components transforming as

$$T'_i = a_{ij}T_j.$$

Comparison with the transformation equations for a vector, (4–12′), shows that a *tensor of the first rank is completely equivalent to a vector.** Finally, the nine components of a tensor of the second rank transform as

$$T'_{ij} = a_{ik}a_{jl}T_{kl}. \tag{5–11}$$

Rigorously speaking, one must distinguish between a second rank tensor **T** and the square matrix formed from its components. A tensor is defined only in terms of its transformation properties under orthogonal coordinate transformations. On the other hand, a matrix is in no way restricted in the types of transformations it may undergo and indeed may be considered entirely independently of its properties under some particular class of transformation. Nevertheless, the distinction must not be stressed unduly. Within the restricted domain of orthogonal transformations there is a practical identity. The tensor components and the matrix elements are manipulated in the same fashion; and for every tensor equation there will be a corresponding matrix equation, and vice versa. By Eq. (4–41) the components of a square matrix **V** transform under a linear change of coordinates defined by the matrix **A** according to a similarity transformation:

$$\mathbf{V}' = \mathbf{AVA}^{-1}.$$

For an orthogonal transformation we therefore have

$$\mathbf{V}' = \mathbf{AV\tilde{A}}$$

orthogonal trans

$$A^{-1} = \tilde{A}$$

$$= A^T$$

or

$$V'_{ij} = a_{ik}v_{kl}a_{jl}. \tag{5–13}$$

Comparison with Eq. (5–11) above thus shows that the matrix components transform identically, under an orthogonal transformation, with the components of a tensor of the second rank. All the terminology and operations of matrix algebra, such as "transpose," "hermitean," "antisymmetrical," etc., can be applied to tensors without change. The equivalence between the tensor and the matrix is not restricted to tensors of the second rank. For example, we already know that the components of a vector, which is a tensor of the first rank, form a column or row

---

* A *pseudotensor* in three dimensions transforms as a tensor except under inversion. In general, the transformation equation for a pseudotensor of the Nth rank is (cf. Eq. 4–110)

$$T'_{ijk\ldots} = |\mathbf{A}|a_{il}a_{jm}a_{kn}\ldots T_{lmn\ldots}.$$

As rigid body motion involves only proper rotations no further use will be made here of the general pseudotensor.

matrix, and vector manipulation may be treated completely in terms of these associated matrices.

Still another useful representation of the operator **I** is as a dyadic. A *dyad* is simply a pair of vectors, written in a definite order **AB**, **A** being known as the *antecedent* and **B** the *consequent*. The scalar dot product of a dyad with a vector **C** can be performed in two ways, either as

$$\mathbf{AB} \cdot \mathbf{C} = \mathbf{A}(\mathbf{B} \cdot \mathbf{C})$$

or as

$$\mathbf{C} \cdot \mathbf{AB} = \mathbf{B}(\mathbf{A} \cdot \mathbf{C}).$$

In the first case **C** is called the *postfactor*, in the second the *prefactor*. The two products, in general, will not be equal—dyad scalar multiplication is not commutative. In both cases, it should be noted, the result of the dot product is to produce a vector having a direction and magnitude in general different from **C**. One can also define the double dot product of two dyads as the scalar given by

$$\mathbf{AB}:\mathbf{CD} = (\mathbf{A} \cdot \mathbf{C})(\mathbf{B} \cdot \mathbf{D})$$

A more convenient notation is to write the double dot product as

$$\mathbf{AB}:\mathbf{CD} \equiv \mathbf{C} \cdot \mathbf{AB} \cdot \mathbf{D}.$$

A *dyadic* is defined as a linear polynomial of dyads:

$$\mathbf{AB} + \mathbf{CD} + \cdots.$$

Actually, any dyad **AB** can be expressed as a dyadic by writing the vectors **A** and **B** in component form in terms of the unit vectors **i, j,** and **k**. When expanded in this manner, the dyad appears as

$$
\begin{aligned}
\mathbf{AB} = \;& A_x B_x \mathbf{ii} + A_x B_y \mathbf{ij} + A_x B_z \mathbf{ik} \\
& + A_y B_x \mathbf{ji} + A_y B_y \mathbf{jj} + A_y B_z \mathbf{jk} \\
& + A_z B_x \mathbf{ki} + A_z B_y \mathbf{kj} + A_z B_z \mathbf{kk}.
\end{aligned}
\tag{5-14}
$$

Equation (5–14) is the *nonion* form of the dyad, so called from the nine coefficients involved. Obviously dyadics in like fashion can always be reduced to a nonion form. It is easy to show that the coefficients of the nonion representation of a dyadic transform under an orthogonal transformation exactly as do the components of a tensor of the second rank (see Exercise 1 to this chapter). There is also an equivalence in their effect as operators acting on vectors for we have seen that the dot product of a dyad or a dyadic with a vector results in a new vector, just as the product of **I** and **ω** gives **L**. A dyadic is therefore in all ways equivalent to a tensor of the second rank.

A useful dyadic is the unit dyadic **1**, defined by the nonion representation

$$\mathbf{1} = \mathbf{ii} + \mathbf{jj} + \mathbf{kk}. \tag{5-15}$$

The designation is certainly well merited, for the matrix of **1** is exactly the unit matrix, and direct multiplication shows that

$$1 \cdot A = A \cdot 1 = A.$$

5–3 THE INERTIA TENSOR AND THE MOMENT OF INERTIA

Considered as a linear operator that transforms ω into **L**, the matrix **I** has elements that behave as the elements of a second-rank tensor. The quantity **I** is therefore identified as a second-rank tensor and is usually called the *moment of inertia tensor* or briefly as the *inertia tensor*. It can simply be written in dyadic notation as

$$\mathbf{I} = m_i(r_i^2 \mathbf{1} - \mathbf{r}_i \mathbf{r}_i), \qquad particle \; \# \tag{5–16}$$

for then

$$\mathbf{I} \cdot \omega = m_i(r_i^2 \omega - \mathbf{r}_i(\mathbf{r}_i \cdot \omega)) = \mathbf{L},$$

in agreement with Eq. (5–3). The advantage of using the dyadic form for **I** is that the familiar methods of vector manipulation can still be employed. Thus, one is led in a natural fashion to express the kinetic energy of rotation in terms of the dyadic **I**. The kinetic energy of motion about a point is

$$T = \tfrac{1}{2}m_i v_i^2,$$

where v_i is the velocity of the ith particle relative to the fixed point as measured in the space axes. By Eq. (5–2), T may also be written as

$$T = \tfrac{1}{2}m_i \mathbf{v}_i \cdot (\omega \times \mathbf{r}_i),$$

which, upon permuting the vectors in the triple dot product, becomes

$$T = \frac{\omega}{2} \cdot m_i(\mathbf{r}_i \times \mathbf{v}_i).$$

The quantity summed over i will be recognized as the angular momentum of the body about the origin, and in consequence the kinetic energy can be written in the form

$$T = \frac{\omega \cdot \mathbf{L}}{2} = \frac{\omega \cdot \mathbf{I} \cdot \omega}{2}. \tag{5–17}$$

Let **n** be a unit vector in the direction of ω so that $\omega = \omega\mathbf{n}$. Then an alternative form for the kinetic energy is

$$T = \frac{\omega^2}{2}\mathbf{n} \cdot \mathbf{I} \cdot \mathbf{n} = \tfrac{1}{2}I\omega^2, \tag{5–18}$$

where I is a scalar, defined by

$$I = \mathbf{n} \cdot \mathbf{I} \cdot \mathbf{n} = m_i(r_i^2 - (\mathbf{r}_i \cdot \mathbf{n})^2), \tag{5–19}$$

and known as the *moment of inertia about the axis of rotation.*

In the usual elementary discussions the moment of inertia about an axis is defined as the sum, over the particles of the body, of the product of the particle mass and the square of the perpendicular distance from the axis. It must be shown that this definition is in accord with the expression given in Eq. (5–19). The perpendicular distance is equal to the magnitude of the vector $\mathbf{r}_i \times \mathbf{n}$ (cf. Fig. 5–2). Therefore, the customary definition of I may be written as

$$I = m_i(\mathbf{r}_i \times \mathbf{n}) \cdot (\mathbf{r}_i \times \mathbf{n}).$$

Multiplying and dividing by ω^2, this definition of I may also be written as

$$I = \frac{m_i}{\omega^2}(\boldsymbol{\omega} \times \mathbf{r}_i) \cdot (\boldsymbol{\omega} \times \mathbf{r}_i).$$

But each of the vectors in the dot product is exactly the relative velocity $\mathbf{v}_i$ as measured in the space system of axes. Hence I so defined is related to the kinetic energy by

$$I = \frac{2T}{\omega^2},$$

which is the same as Eq. (5–18), and therefore I must be identical with the scalar defined by Eq. (5–19).

The value of the moment of inertia depends upon the direction of the axis of rotation. As $\boldsymbol{\omega}$ usually changes its direction with respect to the body in the course of time, the moment of inertia must also be considered a function of time. When the body is constrained so as to rotate only about a fixed axis then the moment of inertia is a constant. In such case the kinetic energy in (5–17) is almost in the form required to fashion the Lagrangian and the equations of motion. The one further step needed is to express ω as the time derivative of some angle, which can usually be done without difficulty.

Along with the inertia tensor, the moment of inertia depends also upon the choice of origin of the body set of axes. However, the moment of inertia about some given axis is related simply to the moment about a parallel axis through the

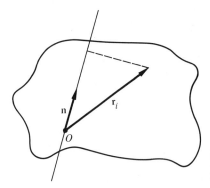

FIGURE 5–2
Illustrating the definition of the moment of inertia.

center of mass. Let the vector from the given origin O to the center of mass be $\mathbf{R}$, and let the radii vectors from O and the center of mass to the ith particle be $\mathbf{r}_i$ and $\mathbf{r}_i'$, respectively. The three vectors so defined are connected by the relation (cf. Fig. 5-3)

$$\mathbf{r}_i = \mathbf{R} + \mathbf{r}_i'. \tag{5-20}$$

The moment of inertia about the axis a is therefore

$$I_a = m_i(\mathbf{r}_i \times \mathbf{n})^2 = m_i[(\mathbf{r}_i' + \mathbf{R}) \times \mathbf{n}]^2$$

or

$$I_a = M(\mathbf{R} \times \mathbf{n})^2 + m_i(\mathbf{r}_i' \times \mathbf{n})^2 + 2m_i(\mathbf{R} \times \mathbf{n})\cdot(\mathbf{r}_i' \times \mathbf{n}),$$

where M is the total mass of the body. The last term in this expression can be rearranged as

$$-2(\mathbf{R} \times \mathbf{n})\cdot(\mathbf{n} \times m_i\mathbf{r}_i').$$

By the definition of center of mass, the summation $m_i\mathbf{r}_i'$ vanishes. Hence I_a can be expressed in terms of the moment about the parallel axis b as

$$I_a = I_b + M(\mathbf{R} \times \mathbf{n})^2. \tag{5-21}$$

The magnitude of $\mathbf{R} \times \mathbf{n}$ is the perpendicular distance of the center of mass from the axis passing through O. Consequently the moment of inertia about a given axis is equal to the moment of inertia about a parallel axis through the center of mass plus the moment of inertia of the body, as if concentrated at the center of mass, with respect to the original axis.

A similar decomposition can be obtained for the inertia tensor itself. From Eq. (5-20) it follows that

$$r_i^2 = R^2 + r_i'^2 + 2\mathbf{R}\cdot\mathbf{r}_i'.$$

The inertia tensor for the origin O, in the dyadic form of Eq. (5-14), can be written

$$\mathbf{I} = MR^2\mathbf{1} + m_i[(r_i'^2 + 2\mathbf{R}\cdot\mathbf{r}_i')\mathbf{1} - (\mathbf{r}_i'\mathbf{r}_i' + \mathbf{R}\mathbf{r}_i' + \mathbf{r}_i'\mathbf{R})] - M\mathbf{R}\mathbf{R}.$$

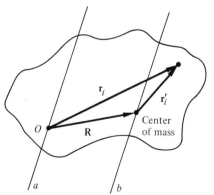

FIGURE 5-3

The vectors involved in the relation between moments of inertia about parallel axes.

All sums of the form $m_i \mathbf{r}_i'$ vanish and the expression reduces to

$$\mathbf{I} = m_i(r_i'^2 \mathbf{1} - \mathbf{r}_i' \mathbf{r}_i') + M(R^2 \mathbf{1} - \mathbf{R}\mathbf{R}). \tag{5-22}$$

The first term is the inertia tensor $\mathbf{I}'$ relative to the center of mass and the second is the inertia tensor, relative to O, of a single particle of mass M located at the center of mass. Thus, both the moment of inertia and the inertia tensor possess a type of resolution, relative to the center of mass, very similar to that found for the linear and angular momentum and the kinetic energy in Section (1–2).

5-4 THE EIGENVALUES OF THE INERTIA TENSOR AND THE PRINCIPAL AXIS TRANSFORMATION

The preceding discussion has served to emphasize the important role the inertia tensor plays in the discussion of the motion of rigid bodies. An examination, at this point, of the properties of this tensor and its associated matrix will therefore prove of considerable interest. From the defining equation, (5–7), it is seen that the components of the tensor are symmetrical, that is

$$I_{xy} = I_{yx}. \qquad xy \ \text{commute}$$

Since the components are real it follows that the tensor is equal to its adjoint (cf. Eq. 4–38), which is to say $\mathbf{I}$ is *self-adjoint* or *hermitean*. Thus, while the inertia tensor will in general have nine components, only six of them will be independent—the three along the diagonal plus three of the off-diagonal elements.

The inertia coefficients depend both upon the location of the origin of the body set of axes and upon the orientation of these axes with respect to the body. It would be very convenient, of course, if for a given origin one could find a particular orientation of the body axes for which the inertia tensor is diagonal, so that the dyadic could be written as

$$\mathbf{I}' = I_1 \mathbf{ii} + I_2 \mathbf{jj} + I_3 \mathbf{kk}. \qquad \text{find eigen values and vectors}$$

With respect to such a set of axes each of the components of $\mathbf{L}$ would involve only the corresponding component of $\boldsymbol{\omega}$, thus*

$$L_1 = I_1 \omega_1, \qquad L_2 = I_2 \omega_2, \qquad L_3 = I_3 \omega_3. \tag{5-23}$$

A similar simplification would also occur in the form of the kinetic energy:

$$T = \frac{1}{2}(\omega_1\ \omega_2\ \omega_3)\begin{pmatrix} I_1 & & 0 \\ & I_2 & \\ 0 & & I_3 \end{pmatrix}\begin{pmatrix} \omega_1 \\ \omega_2 \\ \omega_3 \end{pmatrix} = T = \frac{\boldsymbol{\omega} \cdot \mathbf{I} \cdot \boldsymbol{\omega}}{2} = \frac{1}{2}I_1\omega_1^2 + \frac{1}{2}I_2\omega_2^2 + \frac{1}{2}I_3\omega_3^2. \tag{5-24}$$

One can show that it is always possible to find such axes, and the proof is based essentially on the hermitean nature of the inertia tensor.

---

* With an eye to future applications, components relative to these axes will be denoted by subscripts 1, 2, 3.

It has been remarked, in Section 4–6, that the eigenvalue equation of a matrix can be solved by transforming the matrix to a diagonal form, the elements of which are then the desired eigenvalues. Hence the problem of finding a set of axes in which $\mathbf{I}$ is diagonal is equivalent to the eigenvalue problem for the matrix of $\mathbf{I}$, I_1, I_2, and I_3 being the three eigenvalues. It also follows that for the coordinate system in which $\mathbf{I}$ is diagonal the direction of the axes coincides with the direction of the eigenvectors. For example, suppose $\boldsymbol{\omega}$ is along one of the axes, say the x axis. Then, by (5–23) the angular momentum $\mathbf{L} = \mathbf{I} \cdot \boldsymbol{\omega}$ is also along the x axis. The effect of $\mathbf{I}$ on any vector parallel to the three coordinate axes is thus to produce another vector in the same direction. By definition such a vector must therefore be one of the eigenvectors of $\mathbf{I}$.

In Section 4–6 we outlined the scheme for diagonalizing any matrix and finding its eigenvalues. By itself, however, this procedure does not constitute a proof for the existence of a *real Cartesian* coordinate system in which $\mathbf{I}$ is diagonal. Thus, it will be recalled that an orthogonal matrix, except for trivial cases, has only one real eigenvalue and therefore only one real direction corresponding to an eigenvector (namely, the axis of rotation). In contrast, what we seek to prove here is that *all* the eigenvalues of $\mathbf{I}$ are real, and that the three real directions of the eigenvectors are mutually orthogonal.*

Let $\mathbf{R}_j$ be the jth eigenvector of $\mathbf{I}$, with eigenvalue I_j. The corresponding eigenvalue equation is then†

$$\mathbf{I} \cdot \mathbf{R}_j = I_j \mathbf{R}_j.\tag{5–25}$$

If $\mathbf{R}_l$ similarly stands for the lth eigenvector, the dot product of $\mathbf{R}_l^*$ with Eq. (5–25) from the left is

$$\mathbf{R}_l^* \cdot \mathbf{I} \cdot \mathbf{R}_j = I_j \mathbf{R}_l^* \cdot \mathbf{R}_j.\tag{5.26}$$

The complex conjugate of the eigenvalue equation for $\mathbf{R}_l$, following the pattern of Eq. (5–25) is

$$\mathbf{I}^* \cdot \mathbf{R}_l^* = I_l^* \mathbf{R}_l^*.\tag{5–27}$$

Now, the order of a dot product of a dyad and a vector can be interchanged by taking the transpose, just as with the product of a square matrix and a column matrix. Formally, if $\mathbf{A}$ is any dyad and $\mathbf{P}$ any vector, then the ith component of the dot product is

$$(\mathbf{A} \cdot \mathbf{P})_i = A_{ik} P_k = P_k (\tilde{\mathbf{A}})_{ki} = (\mathbf{P} \cdot \tilde{\mathbf{A}})_i.$$

It follows that Eq. (5–27) can be written as

$$\mathbf{R}_l^* \cdot \mathbf{I}^\dagger = I_l^* \mathbf{R}_l^*.\tag{5–27'}$$

---

* In terms of the matrix $\mathbf{X}$ that diagonalizes $\mathbf{I}$ by means of a similarity transformation, these conditions state that $\mathbf{X}$ must be a *real orthogonal* matrix, i.e., $\mathbf{X}$ transforms one real Cartesian coordinate system into another.

† In Eqs. (5–25) through (5–29) there is no summation over the indices j or l, as is obvious from the argument.

Taking the dot product of this result with $\mathbf{R}_j$ from the right we have

$$\mathbf{R}_l^* \cdot \mathbf{I}^\dagger \cdot \mathbf{R}_j = I_l^* \mathbf{R}_l^* \cdot \mathbf{R}_j. \tag{5–28}$$

Since $\mathbf{I}$ is hermitean the left-hand sides of Eqs. (5–26) and (5–28) are the same, so that the difference of the two equations reduces to

$$(I_j - I_l^*) \mathbf{R}_l^* \cdot \mathbf{R}_j = 0 \tag{5–29}$$

If l is equal to j, then $\mathbf{R}_l^* \cdot \mathbf{R}_j = |\mathbf{R}_j|^2$, which must be positive definite. Hence Eq. (5–29) can vanish in that case only if $I_j = I_j^*$; which proves one of the desired results. Notice that the proof has used only the hermitean property of $\mathbf{I}$—*the eigenvalues of any hermitean matrix are real.* Since $\mathbf{I}$ is also real, the direction cosines of the eigenvectors $\mathbf{R}_j$ must likewise be real, as can be seen from Eq. (5–25).

If l is chosen different from j, and the eigenvalues are distinct, then Eq. (5–29) can be satisfied only if $\mathbf{R}_l \cdot \mathbf{R}_j$ vanishes, which verifies the second requirement that the eigenvectors be orthogonal.* If the eigenvalues are not all distinct this orthogonality proof falls through, but it may be mended with little difficulty. Suppose, for example, two of the eigenvalues are equal, $I_2 = I_3$. It is always possible to find at least one eigenvector which satisfies Eq. (5–25) for this double eigenvalue. By Eq. (5–29) this eigenvector is orthogonal to the eigenvector for the distinct eigenvalue I_1. Thus one can find a right-handed Cartesian coordinate system for which the unit vectors i and j are along the directions of the two eigenvectors found so far. It follows from the eigenvector equations

$$\mathbf{I} \cdot \mathbf{i} = I_1 \mathbf{i}, \qquad \mathbf{I} \cdot \mathbf{j} = I_2 \mathbf{j}$$

that in this coordinate system $I_{12} = I_{13} = I_{23} = 0$. But the symmetry of $\mathbf{I}$ then implies that $I_{31} = I_{32} = 0$, i.e., the inertia tensor is diagonal in this system and $\mathbf{k}$ is also an eigenvector of $\mathbf{I}$ with eigenvalue I_2. It immediately follows that *any* vector in the plane defined by $\mathbf{j}$ and $\mathbf{k}$ is an eigenvector with I_2 as eigenvalue.† Similarly if all eigenvalues are the same then all directions in space are eigenvectors. But then $\mathbf{I}$ will be diagonal to start with, and no diagonalization is necessary.

---

* The proof given here has a close analog in the theory of the Sturm–Liouville problem in ordinary differential equations. There one shows that the eigenvalues of a hermitean differential operator are real and the corresponding eigenfunction solutions of the differential equation are orthogonal. The similarity is not accidental; one can always construct matrix quantities to corrrespond to any given problem involving a Sturm–Liouville differential equation, and this is exactly what happens in the correspondence between the matrix mechanics and wave mechanics formulations of quantum theory. See, e.g., G. F. Arfken, *Mathematical Methods for Physicists*, Section 9.2.

† An equivalent, if more general procedure, of constructing an orthogonal set of eigenvectors out of a possible nonorthogonal set is the Gram–Schmidt orthogonalization process in linear algebra. See, e.g., J. A. Eisele and R. M. Mason, *Applied Matrix and Tensor Analysis*, Wiley-Interscience, N.Y. 1970, pp. 16–21, or M. C. Pease, *Methods of Matrix Algebra*, Academic, N.Y. 1965, pp. 59–61. Note also the analogous discussion in the construction of normal modes, pp. 253 below.

The methods of matrix algebra thus enable us to show that for any point in a rigid body one can find a set of Cartesian axes for which the inertia tensor will be diagonal. The axes are called the *principal axes* and the corresponding diagonal elements I_1, I_2, I_3 are known as the *principal moments of inertia*. Given some initial body set of coordinates one can transform to the principal axes by a particular orthogonal transformation, which is therefore known as the *principal axis transformation*. In practice, of course, the principal moments of inertia, being the eigenvalues of **I**, are found as the roots of the secular equation. To recall the steps leading to the secular equation, it will be noted that the eigenvalue equation (cf. Eq. 5–25), in the form

$$(\mathbf{I} - I\mathbf{1}) \cdot \mathbf{R} = 0, \tag{5–30}$$

corresponds to a set of three homogeneous linear equations for the components of the eigenvector. These equations are consistent only if the determinant of the coefficients vanishes:

$$\begin{vmatrix} I_{xx} - I & I_{xy} & I_{zx} \\ I_{xy} & I_{yy} - I & I_{yz} \\ I_{zx} & I_{yz} & I_{zz} - I \end{vmatrix} = 0, \tag{5–31}$$

where the symmetry of **I** has been displayed explicitly. Equation (5–31) is the secular equation, in the form of a cubic in I, whose three roots are the desired principal moments. For each of these roots the Eqs. (5–25) can be solved to obtain the direction of the corresponding principal axis. In most of the easily soluble problems in rigid dynamics the principal axes can be determined by inspection. For example, one almost always has to deal with rigid bodies that are solids of revolution about some axis, with the origin of the body system on the symmetry axis. All directions perpendicular to the axis of symmetry are then alike, which is the mark of a double root to the secular equation. The principal axes are then the symmetry axis and any two perpendicular axes in the plane normal to the symmetry axis.

The principal moments of inertia cannot be negative, because as the diagonal elements in the principal axes system they have the form of sums of squares. Thus I_{xx} is given by (cf. Eq. 5–6)

$$I_{xx} = m_i(y_i^2 + z_i^2).$$

In order for one of the principal moments to vanish, all the points of the body must be such that two coordinates of each particle are zero. Clearly this can happen only if all of the points of the body are collinear with the principal axis corresponding to the zero principal moment. Any two axes perpendicular to the line of the body will then be the other principal axes. Indeed this is clearly a limiting case of a body with an axis of symmetry passing through the origin.

One may also be led to the concept of principal axes through some geometrical considerations that historically formed the first approach to the subject. The moment of inertia about a given axis has been defined as $I = \mathbf{n} \cdot \mathbf{I} \cdot \mathbf{n}$. Let the direction cosines of the axis be α, β, and γ so that

$$\mathbf{n} = \alpha \mathbf{i} + \beta \mathbf{j} + \gamma \mathbf{k};$$

I then can be written as

$$I = I_{xx}\alpha^2 + I_{yy}\beta^2 + I_{zz}\gamma^2 + 2I_{xy}\alpha\beta + 2I_{yz}\beta\gamma + 2I_{zx}\gamma\alpha, \qquad (5\text{-}32)$$

using explicitly the symmetry of $\mathbf{I}$. It is convenient to define a vector $\boldsymbol{\rho}$ by the equation

$$\boldsymbol{\rho} = \frac{\mathbf{n}}{\sqrt{I}}. \qquad (5\text{-}33)$$

The magnitude of $\boldsymbol{\rho}$ is thus related to the moment of inertia about the axis whose direction is given by $\mathbf{n}$. In terms of the components of this new vector Eq. (5–32) takes on the form

$$1 = I_{xx}\rho_1^2 + I_{yy}\rho_2^2 + I_{zz}\rho_3^2 + 2I_{xy}\rho_1\rho_2 + 2I_{yz}\rho_2\rho_3 + 2I_{zx}\rho_3\rho_1. \qquad (5\text{-}34)$$

Considered as a function of the three variables ρ_1, ρ_2, ρ_3, Eq. (5–34) is the equation of some surface in ρ space. In particular Eq. (5–34) is the equation of an ellipsoid designated as the *inertia ellipsoid*. It is well known that one can always transform to a set of Cartesian axes in which the equation of an ellipsoid takes on its normal form:

$$1 = I_1\rho_1'^2 + I_2\rho_2'^2 + I_3\rho_3'^2, \qquad (5\text{-}35)$$

with the principal axes of the ellipsoid along the new coordinate axes. But (5–34) is just the form Eq. (5–34) has in a system of coordinates in which the inertia tensor $\mathbf{I}$ is diagonal. Hence the coordinate transformation that puts the equation of ellipsoid into its normal form is exactly the principal axis transformation previously discussed. The principal moments of inertia determine the lengths of the axes of the inertia ellipsoid. If two of the roots of the secular equation are equal the inertia ellipsoid thus has two equal axes and is an ellipsoid of revolution. If all three principal moments are equal the inertia ellipsoid is a sphere.

A quantity closely related to the moment of inertia is the *radius of gyration*, R_0, defined by the equation

$$I = MR_0^2. \qquad (5\text{-}36)$$

In terms of the radius of gyration the vector $\boldsymbol{\rho}$ can be written as

$$\boldsymbol{\rho} = \frac{\mathbf{n}}{R_0\sqrt{M}}.$$

The radius vector to a point on the inertia ellipsoid is thus inversely proportional to the radius of gyration about the direction of the vector.

It is perhaps worth reemphasizing that the inertia tensor $\mathbf{I}$ and all the quantities associated with it—principal axes, principal moments, inertia ellipsoid, etc.—are only relative to some particular point fixed in the body. If the point is shifted elsewhere in the body all the quantities will in general be changed. Thus Eq. (5–22) gives the effect of moving the reference point from the center of mass to some other point. The principal axis transformation which diagonalizes $\mathbf{I}'$ at the center of mass will not necessarily diagonalize the difference term $M(R^2\mathbf{1} - \mathbf{RR})$, and hence is not in general the principal axis transformation for the shifted tensor $\mathbf{I}$. Only if the shift vector $\mathbf{R}$ is along one of the principal axes relative to the center of mass will the difference tensor be diagonal in that system. The new inertia tensor $\mathbf{I}$ will in that special case have the same principal axes as at the center of mass. However the principal moments of inertia are changed, except for that corresponding to the shift axis, where the diagonal element of the difference tensor is clearly zero. The "parallel axis" theorem for the diagonalized form of the inertia tensor thus has a rather specialized and restricted form.

5–5 METHODS OF SOLVING RIGID BODY PROBLEMS AND THE EULER EQUATIONS OF MOTION

Practically all the tools necessary for setting up and solving problems in rigid body dynamics have by now been assembled. If nonholonomic constraints are present then special means must be taken to include the effects of these constraints in the equations of motion. For example, if there are "rolling constraints" these must be introduced into the equations of motion by the method of Lagrange undetermined multipliers, as in Section 2–4. As has been discussed in Section 5–1, one usually seeks a particular reference point in the body such that the problem can be split into two separate parts, one purely translational and the other purely rotational about the reference point. Of course, if one point of the rigid body is fixed in an inertial system then that is the obvious reference point. All that has to be considered then is the rotational problem about the fixed point. For bodies without a fixed point the most useful reference point is almost always the center of mass. We have already seen that the total kinetic energy and angular momentum then split neatly into one term relating to the translational motion of the center of mass and another involving rotation *about* the center of mass. Thus, Eq. (1–31) can now be written

$$T = \tfrac{1}{2}Mv^2 + \tfrac{1}{2}I\omega^2.$$

For many problems (certainly all those that will be considered here) a similar sort of division can be made for the potential energy. One can then solve individually for the translational motion of the center of mass and for the rotational motion about the center of mass. For example, the Newtonian equations of motion can be used directly; Eq. (1–22) for the motion of the center of mass and Eq. (1–26) for the motion about that point. With holonomic conservative systems the Lagrangian formulation is available, with the Lagrangian taking the form

$$L(q, \dot{q}) = L_c(q_c, \dot{q}_c) + L_b(q_b, \dot{q}_b).$$

(1-22) $M\dfrac{d^2R}{dt^2} = \sum_i F_i^{(e)}$

(1-26) $\dfrac{dL}{dt} = N^{(e)}$

center of mass

orientation of body rotation, etc.

Here L_c is that part of the Lagrangian involving the generalized coordinates q_c (and velocities $\dot{q}_c$) of the center of mass, and L_b the part relating to the orientation of the body about the center of mass, as described by $q_b, \dot{q}_b$. In effect then there are two distinct problems, one with Lagrangian L_c and the other with Lagrangian L_b. In both the Newtonian and Lagrangian formulations, it is convenient to work in terms of the principal axes system of the point of reference, so that the kinetic energy of rotation takes the simple form given in Eq. (5–24). So far the only suitable generalized coordinates we have for the rotational motion of the rigid body are the Euler angles. Of course, the motion is often effectively confined to two dimensions, as in the motion of a rigid lamina in a plane. The axis of rotation is then fixed in the direction perpendicular to the plane; only one angle of rotation is necessary and one may dispense with the cumbersome machinery of the Euler angles.

For the rotational motion about a fixed point or the center of mass, the direct Newtonian approach leads to a set of equations known as Euler's equations of motion. We consider either an inertial frame whose origin is at the fixed point of the rigid body, or a system of space axes with origin at the center of mass. In these two situations either Eq. (1–26) or (1–26′) holds, which here appears simply as

$$\left(\frac{d\mathbf{L}}{dt}\right)_s = \mathbf{N}.$$

The subscript s is used because the time derivative is with respect to axes that do not share the rotation of the body. However, Eq. (4–124) can be used to obtain the derivatives with respect to axes fixed in the body:

$$\left(\frac{d\mathbf{L}}{dt}\right)_s = \left(\frac{d\mathbf{L}}{dt}\right)_b + \boldsymbol{\omega} \times \mathbf{L},$$

or, dropping the "body" subscript:

$$\frac{d\mathbf{L}}{dt} + \boldsymbol{\omega} \times \mathbf{L} = \mathbf{N}. \tag{5–37}$$

Equation (5–37) is thus the appropriate form of the Newtonian equation of motion relative to body axes. The ith component of Eq. (5–37) can be written

$$\frac{dL_i}{dt} + \epsilon_{ijk}\omega_j L_k = N_i. \tag{5–38}$$

If now the body axes are taken as the principal axes relative to the reference point then the angular momentum components are $L_i = I_i\omega_i$. Equation (5–38) then takes the form

$$I_i\frac{d\omega_i}{dt} + \epsilon_{ijk}\omega_j\omega_k I_k = N_i, \tag{5–39}$$

since the principal moments of inertia are of course time independent.* In expanded form the three equations making up Eq. (5–39) look like

$$I_1 \dot{\omega}_1 - \omega_2 \omega_3 (I_2 - I_3) = N_1,$$
$$I_2 \dot{\omega}_2 - \omega_3 \omega_1 (I_3 - I_1) = N_2, \qquad (5\text{–}39')$$
$$I_3 \dot{\omega}_3 - \omega_1 \omega_2 (I_1 - I_2) = N_3.$$

Equations (5–39) or Eqs. (5–39′) are the so-called Euler's equations of motion for a rigid body with one point fixed. They can also be derived from Lagrange's equations in the form of Eq. (1–53) where the generalized forces Q_j are the torques corresponding to the Euler angles of rotation. However only one of the Euler angles has its associated torque along one of the body axes, and the remaining two Euler's equations must be obtained by cyclic permutation (cf. Exercise 8).

5–6 TORQUE-FREE MOTION OF A RIGID BODY

One problem in rigid dynamics where Euler's equations are applicable is in the motion of a rigid body not subject to any net forces or torques. The center of mass is then either at rest or moving uniformly, and it does not decrease the generality of the solution to discuss the rotational motion in a reference frame in which the center of mass is stationary. In such case the angular momentum arises only from rotation about the center of mass and Euler's equations are the equations of motion for the complete system. In the absence of any net torques they reduce to

$$I_1 \dot{\omega}_1 = \omega_2 \omega_3 (I_2 - I_3),$$
$$I_2 \dot{\omega}_2 = \omega_3 \omega_1 (I_3 - I_1), \qquad (5\text{–}40)$$
$$I_3 \dot{\omega}_3 = \omega_1 \omega_2 (I_1 - I_2).$$

The same equations, of course, will also describe the motion of a rigid body when one point is fixed and there are no net applied torques. We know two immediate integrals of the motion, for both the kinetic energy and the total angular momentum vector must be constant in time. With these two integrals it is possible to integrate (5–40) completely in terms of elliptic functions, but such a treatment is not very illuminating. However, it is also possible to derive an elegant geometrical description of the motion, known as Poinsot's construction, without requiring a complete solution to the problem.

Consider a coordinate system oriented along the principal axes of the body but whose axes measure the components of a vector $\boldsymbol{\rho}$ along the instantaneous axis of rotation as defined by Eq. (5–33). For our purposes it is convenient to make use

---

* It is obvious that Eq. (5–39), as the ith component of a vector equation, does not involve a summation over i, although summation *is* implied over the repeated indices j and k.

of Eq. (5–18) for the kinetic energy (here constant) and write the definition of ρ in the form

$$\rho = \frac{\omega}{\omega\sqrt{I}} = \frac{\omega}{\sqrt{2T}}. \qquad (5\text{–}41)$$

In this ρ space we define a function

$$F(\rho) = \rho \cdot \mathbf{I} \cdot \rho = \tfrac{1}{2}\rho_i^2 I_i, \qquad (5\text{–}42)$$

where the surfaces of constant F are ellipsoids, the particular surface $F = 1$ being the inertia ellipsoid. As the direction of the axis of rotation changes in time the parallel vector ρ moves accordingly, its tip always defining a point on the inertia ellipsoid. The gradient of F, evaluated at this point, furnishes the direction of the corresponding normal to the inertia ellipsoid. From Eq. (5–42) for $F(\rho)$ the gradient of F with respect to ρ has the form

$$\nabla_\rho F = 2\mathbf{I} \cdot \rho = \frac{2\mathbf{I} \cdot \omega}{\sqrt{2T}},$$

or

$$\nabla_\rho F = \sqrt{\frac{2}{T}}\mathbf{L}. \qquad (5\text{–}43)$$

Thus the ω vector will always move such that the corresponding normal to the inertia ellipsoid is in the direction of the angular momentum. In the particular case under discussion the direction of $\mathbf{L}$ is fixed in space and it is the inertia ellipsoid (fixed with respect to the body) that must move in space in order to preserve this connection between ω and $\mathbf{L}$ (cf. Fig. 5–4).

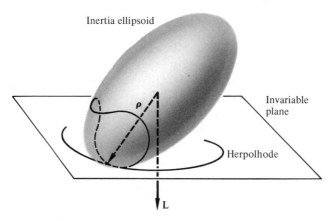

FIGURE 5–4
The motion of the inertia ellipsoid relative to the invariable plane.

It can also be shown that the distance between the origin of the ellipsoid and the plane tangent to it at the point ρ must similarly be constant in time. This distance is equal to the projection of ρ on L and is given by

$$\frac{\rho \cdot L}{L} = \frac{\omega \cdot L}{L\sqrt{2T}}$$

or

$$\frac{\rho \cdot L}{L} = \frac{\sqrt{2T}}{L}, \tag{5–44}$$

where use has been made of Eq. (5–17). Both T, the kinetic energy, and L, the angular momentum, are constants of the motion and the tangent plane is therefore always a fixed distance from the origin of the ellipsoid. Since the normal to the plane, being along L, also has a fixed direction, the tangent plane is known as the *invariable plane*. We can picture the force-free motion of the rigid body as being such that the inertia ellipsoid rolls, without slipping, on the invariable plane, with the center of the ellipsoid a constant height above the plane. The rolling occurs without slipping because the point of contact is defined by the position of ρ which, being along the instantaneous axis of rotation, is the one direction in the body momentarily at rest. The curve traced out by the point of contact on the inertia ellipsoid is known as the *polhode*, while the similar curve on the invariable plane is called the *herpolhode*.*

Poinsot's geometrical discussion is quite adequate to describe completely the force-free motion of the body. The direction of the invariable plane and the height of the inertia ellipsoid above it are determined by the values of T and L, which are among the initial conditions of the problem. It is then a matter of geometry to trace out the polhode and the herpolhode. The direction of the angular velocity in space is given by the direction of ρ, while the instantaneous orientation of the body is provided by the orientation of the inertia ellipsoid, which is fixed in the body. Elaborate descriptions of force-free motion obtained in this fashion are to be found frequently in the literature.† In the special case of a symmetrical body, the inertia ellipsoid is an ellipsoid of revolution, so that the polhode on the ellipsoid is clearly a circle about the symmetry axis. The herpolhode on the invariable plane is likewise a circle. An observer fixed in the body sees the angular velocity vector ω move on the surface of a cone—called the *body cone*—whose intersection with the inertia ellipsoid is the polhode. Correspondingly an observer fixed in the space axes sees ω move on the surface of a *space cone* whose

---

* Hence the jabberwockian sounding statement: the polhode rolls without slipping on the herpolhode lying in the invariable plane.

† See especially Webster, *Dynamics of Particles and Rigid Bodies*, Macmillan, *Theoretical Mechanics—Dynamics of a Rigid Body*, Routh, *Advanced Rigid Dynamics*, and Gray, *Treatise on Gyrostatics and Rotational Motion*. Among the many properties of the Poinsot construction given in these references is the curious fact that the herpolhode is always concave to the origin, belying its name, which means "snakelike."

intersection with the invariable plane is the herpolhode. Thus the free motion of the symmetrical rigid body is sometimes described as the rolling of the body cone on the space cone. If the moment of inertia about the symmetry axis is less than that about the other two principal axes then from Eq. (5–35) the inertia ellipsoid is prolate, i.e., football shaped—somewhat as is shown in Fig. 5–4. In that case the body cone is outside the space cone. When the moment of inertia about the symmetry axis is the greater, the ellipsoid is oblate and the body cone rolls around the inside of the space cone. In either case the physical description of the motion is that the direction of $\boldsymbol{\omega}$ *precesses* in time about the axis of symmetry of the body.

The Poinsot construction shows how $\boldsymbol{\omega}$ moves, but gives no information as to how the **L** vector appears to move in the body system of axes. Another geometrical description is available, however, to describe the path of the **L** vector as seen by an observer in the principal axes system. Equations (5–23) and (5–24) imply that in this system the kinetic energy is related to the components of the angular momentum by the equation

$$T = \frac{L_1^2}{2I_1} + \frac{L_2^2}{2I_2} + \frac{L_3^2}{2I_3}. \tag{5–45}$$

Since T is constant this relation defines an ellipsoid* also fixed in the body axes but *not* the same as the inertia ellipsoid. At the same time, the conservation of the magnitude of the angular momentum means that **L** must lie on a sphere defined by

$$L^2 = L_1^2 + L_2^2 + L_3^2. \tag{5–46}$$

For given initial conditions—kinetic energy and angular momentum—the path of L in the body system is given by the intersection of the ellipsoid (5–45) with the sphere (5–46). That such an intersection always exists can be seen by writing Eq. (5–45) in a form employing the summation convention:

$$1 = \frac{L_i^2}{a_i^2},$$

where

$$a_i = \sqrt{2TI_i}$$

are the semimajor axes of the ellipsoid. It follows that for any j

$$a_j^2 = L_i^2 \frac{a_j^2}{a_i^2}.$$

Comparison with Eq. (5–46) shows that $a_j^2 < L^2$ if a_j is the smallest semimajor axis, and $a_j^2 > L^2$ if a_j is the largest. If, as conventional, the axes are ordered such that

$$I_3 > I_2 > I_1,$$

---

* This ellipsoid is sometimes referred to as the *Binet* ellipsoid.

then the inequalities are summarized by the relations

$$\sqrt{2TI_1} < L < \sqrt{2TI_3}.$$

Hence the momentum sphere, of radius L, always intersects the kinetic energy ellipsoid. Figure 5-5 illustrates the appearance of the kinetic energy ellipsoid and some possible paths of the **L** vector on its surface.

     With the help of this geometrical construction something can be said about the possible motions of a free asymmetric body. It is easy to see that a steady rotation of such a body is possible only about one of the principal axes. From the Euler equations (5-40) all the components of ω can be constant only if

$$\omega_1\omega_2(I_1 - I_2) = \omega_2\omega_3(I_2 - I_3) = \omega_3\omega_1(I_3 - I_1) = 0,$$

which requires that at least two of the components ω_i be zero, i.e., ω is along only one of the principal axes. However, not all of these possible motions are stable— that is, not moving far from the principal axis under small perturbation. For example, steady motion about the x_3 axis will occur when $L^2 = 2TI_3$. When there are slight deviations from this condition the radius of the angular momentum sphere is just slightly smaller than this value, and the intersection with the kinetic energy ellipsoid is a small circle about the x_3 axis. The motion is thus stable; the **L** vector never being far from the axis. Similarly, at the other extreme, when the motion about the axis of smallest I is perturbed, the radius of the angular momentum sphere is just larger than the smallest semimajor axis. The intersection is again a small closed figure around the principal axis and the

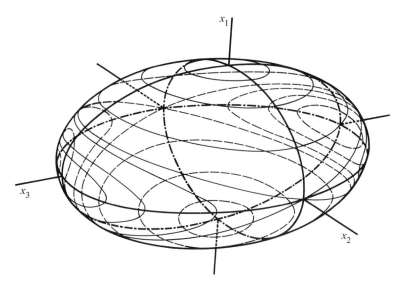

FIGURE 5-5
The kinetic energy, or Binet, ellipsoid fixed in the body axes, and some possible paths of the **L** vector in its surface.

motion is stable. However, the motion about the intermediate axis is *unstable*; for small deviations the intersections with the sphere are not small closed curves, but range over the ellipsoid. This behavior can best be understood by recognizing that at the intermediate axis the radius of curvature of the ellipsoid in one direction is greater than that of the contact sphere, and less in the perpendicular direction. At the other two extremes, the radii of curvature are either greater or smaller than the sphere radius in all directions. These conclusions on the stability of free body motion have been known for a long time, but recent applications, e.g., to the stability of spinning spacecraft, have brought them out of the obscurity of old monographs on rigid body dynamics.*

For a symmetrical rigid body the analytical solution for the force-free motion is not difficult to obtain, and one may directly confirm the precessing motion predicted by the Poinsot construction. Let the symmetry axis be taken as the z principal axis so that $I_1 = I_2$. Euler's equations (5–40) reduce then to

$$I_1 \dot{\omega}_1 = (I_1 - I_3)\omega_3 \omega_2,$$

$$I_1 \dot{\omega}_2 = -(I_1 - I_3)\omega_3 \omega_1, \tag{5–47}$$

$$I_3 \dot{\omega}_3 = 0.$$

The last of these equations states that ω_3 is a constant and it can therefore be treated as one of the known initial conditions of the problem. The remaining two equations can now be written

$$\dot{\omega}_1 = -\Omega \omega_2, \qquad \dot{\omega}_2 = \Omega \omega_1, \tag{5–48}$$

where Ω is an angular frequency

$$\Omega = \frac{I_3 - I_1}{I_1} \omega_3. \tag{5–49}$$

Elimination of ω_2 between the two equations (5–48) leads to the standard differential equation for simple harmonic motion

$$\ddot{\omega}_1 = -\Omega^2 \omega_1,$$

with the typical solution

$$\omega_1 = A \cos \Omega t.$$

The corresponding solution for ω_2 can be found by substituting this expression

---

* If there are dissipative mechanisms present, these stability arguments have to be modified. It is easy to see that for a body with constant L, but slowly decreasing T, the only stable rotation is about the principal axis with the largest moment of inertia. The kinetic energy of rotation about the ith principal axis for given L is $T = L^2/2I_i$, which is least for the axis with the largest I_i. If a body is set spinning about any other principal axis the effect of a slowly decreasing kinetic energy is to cause the angular velocity vector to shift until the spinning is about the axis requiring the least value of T for the given L. Such dissipative effects are present in spacecraft because of the flexing of various members in the course of the motion, especially of the long booms carried by many of them. These facts were learned the hard way by the early designers of spacecraft!

for ω_1, back in the first of Eqs. (5–48):

$$\omega_2 = A \sin \Omega t.$$

The solutions for ω_1 and ω_2 show that the vector $\omega_1 \mathbf{i} + \omega_2 \mathbf{j}$ has a constant magnitude and rotates uniformly about the z axis of the body with the angular frequency Ω (cf. Fig. 5–6). Hence the total angular velocity $\boldsymbol{\omega}$ is also constant in magnitude and *precesses* about the z axis with the same frequency, exactly as predicted by the Poinsot construction.* It should be remembered that the precession described here is relative to the body axes, which are themselves rotating in space with the larger frequency $\boldsymbol{\omega}$. From Eq. (5–49) it is seen that the closer I_1 is to I_3 the slower will be the precession frequency Ω compared to the rotation frequency ω. The constants A (the amplitude of the precession) and ω_3 can be evaluated in terms of the more usual constants of the motion, namely the kinetic energy and the magnitude of the angular momentum. Both T and L^2 can be written as functions of A and ω_3:

$$T = \frac{1}{2}I_1 A^2 + \frac{1}{2}I_3 \omega_3^2,$$

$$L^2 = I_1^2 A^2 + I_3^2 \omega_3^2,$$

and these relations in turn may be solved for A and ω_3 in terms of T and L.

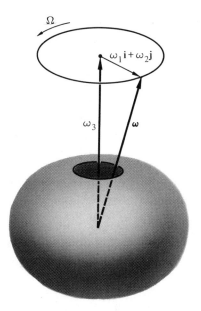

FIGURE 5–6
Precession of the angular velocity about the axis of symmetry in the force-free motion of a symmetrical rigid body.

---

* The precession can be demonstrated in another fashion by defining a vector $\boldsymbol{\Omega}$ lying along the z axis with magnitude given by (5–49). Equations (5–47) are then essentially equivalent to the vector equation

$$\dot{\boldsymbol{\omega}} = \boldsymbol{\omega} \times \boldsymbol{\Omega},$$

which immediately reveals the precession of $\boldsymbol{\omega}$ with the frequency Ω.

It would be expected that the axis of rotation of the earth should exhibit this precession, for the external torques acting on the earth are so weak that the *rotational* motion may be considered as that of a free body. The earth is symmetrical about the polar axis and slightly flattened at the poles so that I_1 is less than I_3. Numerically the ratio of the moments is such that

$$\frac{I_3 - I_1}{I_1} = 0.00327$$

and the magnitude of the precession angular frequency should therefore be

$$\Omega = \frac{\omega_3}{306}.$$

Since ω_3 is practically the same as the magnitude of ω this result predicts a period of precession of 306 days or about 10 months. If some circumstance disturbed the axis of rotation from the figure axis of the earth, we would therefore expect the axis of rotation to precess around the figure axis (i.e. around the North Pole) once every ten months. Practically such a motion should show up as a periodic change in the apparent latitude of points on the earth's surface. Careful measurements of latitude at a network of locations around the world, carried out now for about a century, have shown that the axis of rotation is indeed moving about the pole with an amplitude of the order of a few tenths of a second of latitude (about 10 m). But the situation is far more complicated (and interesting) than the above simple analysis would suggest. The deviations between the figure and rotation axes are very irregular so that it's more a "wobble" than a precession. Careful frequency analysis has shown the existence of an annual period in the motion, thought to arise from the annual cycle of seasons and the corresponding mean displacement of atmospheric masses about the globe. But in addition there is a strong frequency component centered about a period of 420 days, known as the *Chandler wobble*. The present belief is that this motion represents the free-body precession derived above. It is thought that the difference in period arises from the fact that the earth is not a rigid body but is to some degree elastic. In effect some part of the earth follows along with the shift in the rotation axis, which has the effect of reducing the difference in the principal moments of inertia and therefore increasing the period. (If, for example, the earth were completely fluid then the figure axis would instantaneously adjust to the rotation axis and there could be no precession.)

There are still other obscure features to the observed wobble. The frequency analysis indicates there are strong damping effects present, believed to arise from either tidal friction or dissipative effects in the coupling between the mantle and the core. The damping period ought to be the order of 10–20 years. But no such decay of the amplitude of the Chandler wobble has been observed; some sort of random excitation must be present to keep the wobble going. Various sources of the excitation have been suggested. Present speculation points to deep earthquakes, or the mantle phenomena underlying them, as possibly producing discontinuous changes in the inertia tensor large enough to keep exciting the free-body precession.*

5-7 THE HEAVY SYMMETRICAL TOP WITH ONE POINT FIXED

As a further and more complicated example of the application of the methods of rigid dynamics we shall consider the motion of a symmetrical body in a uniform gravitational field when one point on the symmetry axis is fixed in space. A wide variety of physical systems, ranging from a child's top to complicated gyroscopic navigational instruments, are approximated by such a *heavy symmetrical top.* Both for its practical applications and as an illustration of many of the techniques previously developed, the motion of the heavy symmetrical top deserves a detailed exposition.

The symmetry axis is of course one of the principal axes and will be chosen as the z axis of the coordinate system fixed in the body.† Since one point is stationary the configuration of the top is completely specified by the three Euler angles: θ gives the inclination of the z axis from the vertical, ϕ measures the azimuth of the top about the vertical, while ψ is the rotation angle of the top about its own z axis (cf. Fig. 5–7). The distance of the center of gravity (located on the symmetry axis) from the fixed point will be denoted by l.

The Lagrangian procedure, rather than Euler's equations, will be used to obtain a solution for the motion of the top. Since the body is symmetrical the kinetic energy can be written as

$$T = \frac{1}{2}I_1(\omega_1^2 + \omega_2^2) + \frac{1}{2}I_3\omega_3^2,$$

or, in terms of Euler's angles, and using Eqs. (4–125), as

$$T = \frac{I_1}{2}(\dot{\theta}^2 + \dot{\phi}^2\sin^2\theta) + \frac{I_3}{2}(\dot{\psi} + \dot{\phi}\cos\theta)^2, \tag{5-50}$$

where the cross terms in ω_1^2 and ω_2^2 cancel. It is a well-known elementary theorem that in a constant gravitational field the potential energy is the same as if the body were concentrated at the center of mass, but a brief formal proof may be given here. The potential energy of the body is the sum over all particles:

$$V = -m_i\mathbf{r}_i \cdot \mathbf{g}$$

---

* The free precession of the earth's axis is not to be confused with its slow precession about the normal to the ecliptic. This *astronomical* precession of the equinoxes is due to the gravitational torques of the sun and the moon, which were considered negligible in the above discussion. That the assumption is justified is shown by the long period of the precession of the equinoxes (26,000 years) compared to a period of roughly one year for the force-free precession. The astronomical precession is discussed further below.

† Only the body axes need specific identification here; it will therefore be convenient to designate them in this section as the *xyz* axes, without fear of confusing them with the space axes.

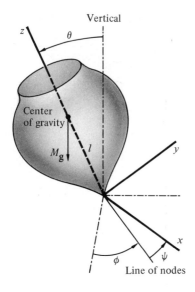

Vertical

FIGURE 5-7
Euler's angles specifying the orientation of a symmetrical top.

where **g** is the constant vector for the acceleration of gravity. By Eq. (1–21) defining the center of mass, this is equivalent to

$$V = -M\mathbf{R} \cdot \mathbf{g}, \tag{5-51}$$

which proves the theorem. In terms of the Euler angles,

$$V = Mgl \cos\theta, \tag{5-51'}$$

so that the Lagrangian is

$$L = \frac{I_1}{2}(\dot\theta^2 + \dot\phi^2 \sin^2\theta) + \frac{I_3}{2}(\dot\psi + \dot\phi\cos\theta)^2 - Mgl\cos\theta. \tag{5-52}$$

It will be noticed that ϕ and ψ do not appear explicitly in the Lagrangian; they are therefore cyclic coordinates, indicating that the corresponding generalized momenta are constant in time. Now, we have seen that the momentum conjugate to a rotation angle is the component of the total angular momentum along the axis of rotation, which for ϕ is the vertical axis, and for ψ, the z axis in the body. One can in fact show from elementary principles that these components of the angular momentum must be constant in time. Since the torque of gravity is along the line of nodes, there is no component of the torque along either the vertical or the body z axis, for by definition both of these axes are perpendicular to the line of nodes. Hence the components of the angular momentum along these two axes must be constant in time.

We therefore have two immediate first integrals of the motion:

$$p_\psi = \frac{\partial L}{\partial \dot\psi} = I_3(\dot\psi + \dot\phi\cos\theta) = I_3\omega_3 = I_1 a \tag{5-53}$$

and

$$p_\phi = \frac{\partial L}{\partial \dot\phi} = (I_1 \sin^2 \theta + I_3 \cos^2 \theta)\dot\phi + I_3\dot\psi \cos\theta = I_1 b. \tag{5–54}$$

Here the two constants of the motion are expressed in terms of new constants a and b. There is one further first integral available; since the system is conservative the total energy E is constant in time:

$$E = T + V = \frac{I_1}{2}(\dot\theta^2 + \dot\phi^2 \sin^2 \theta) + \frac{I_3}{2}\omega_3^2 + Mgl\cos\theta. \tag{5–55}$$

Only three additional quadratures are needed to solve the problem, and they can easily be obtained from these three first integrals without directly using the Lagrange equations. From Eq. (5–53) ψ is given in terms of ϕ by

$$I_3\dot\psi = I_1 a - I_3\dot\phi \cos\theta, \tag{5–56}$$

and this result can be substituted in (5–54) to eliminate ψ:

$$I_1\dot\phi \sin^2 \theta + I_1 a \cos\theta = I_1 b,$$

or

$$\dot\phi = \frac{b - a\cos\theta}{\sin^2 \theta}. \tag{5–57}$$

Thus if θ were known as a function of time Eq. (5–57) could be integrated to furnish the dependence of ϕ on time. Substituting Eq. (5–57) back in Eq. (5–56) results in a corresponding expression for $\dot\psi$:

$$\dot\psi = \frac{I_1 a}{I_3} - \cos\theta \frac{b - a\cos\theta}{\sin^2 \theta}, \tag{5–58}$$

which furnishes ψ if θ is known. Finally, Eqs. (5–57) and (5–58) can be used to eliminate $\dot\phi$ and $\dot\psi$ from the energy equation, resulting in a differential equation involving θ alone. First notice that Eq. (5–53) says ω_3 is constant in time and equal to $(I_1/I_3)a$. Therefore $E - I_3\omega_3^2/2$ is a constant of the motion, which we shall designate as E'. Making use of Eq. (5–57) the energy equation can thus be written as

$$E' = \frac{I_1\dot\theta^2}{2} + \frac{I_1}{2}\frac{(b - a\cos\theta)^2}{\sin^2 \theta} + Mgl\cos\theta. \tag{5–59}$$

Equation (5–59) has the form of an equivalent one-dimensional problem in the variable θ, with the effective potential $V'(\theta)$ given by

$$V'(\theta) = Mgl\cos\theta + \frac{I_1}{2}\left(\frac{b - a\cos\theta}{\sin\theta}\right)^2. \tag{5–60}$$

We will in effect use this one-dimensional problem to discuss the motion in θ, very similarly to what was done in Section 3–3 in describing the radial motion in the central force problem. It is somewhat more convenient, however, to change the variable to $u = \cos\theta$, and rewrite Eq. (5–59) as

$$E'(1 - u^2) = \frac{I_1}{2}\dot{u}^2 + \frac{I_1}{2}(b - au)^2 + Mgl\, u(1 - u^2).$$

Introducing two new constants

$$\alpha = \frac{2E'}{I_1}, \qquad \beta = \frac{2Mgl}{I_1} \tag{5–61}$$

and rearranging terms, this becomes

$$\dot{u}^2 = (1 - u^2)(\alpha - \beta u) - (b - au)^2, \tag{5–62}$$

which can be reduced immediately to a quadrature:

$$t = \int_{u(o)}^{u(t)} \frac{du}{\sqrt{(1 - u^2)(\alpha - \beta u) - (b - au)^2}}. \tag{5–63}$$

With this result, and Eqs. (5–57) and (5–58), ϕ and ψ can also be reduced to quadratures. However, the polynomial in the radical is a cubic so that we have to deal with elliptic integrals. Extensive discussions of these solutions involving elliptic functions are to be found in the literature,* but, as in the case of the force-free motion, the physics tends to be obscured in the profusion of mathematics. Fortunately, the general nature of the motion can be discovered without actually performing the integrations.

Let the right-hand side of Eq. (5–62) be denoted by $f(u)$. The roots of this cubic polynomial furnish the angles at which $\dot{\theta}$ changes sign, i.e., the "turning angles" in θ. For u large, the dominant term in $f(u)$ is βu^3. Since β (cf. Eq. (5–61)) is always greater than zero, $f(u)$ is positive for large positive u and negative for large negative u. At the points $u = \pm 1$, $f(u)$ becomes equal to $-(b \mp a)^2$ and is therefore always negative, except for the unusual case where $u = \pm 1$ is a root (corresponding to a vertical top). Hence at least one of the roots must lie in the region $u > 1$, a region that does not correspond to real angles. Indeed, physical motion of the top can occur only when $\dot{u}^2$ is positive somewhere in the interval between $u = -1$ and $u = +1$, i.e., θ between 0 and $+\pi$. We must conclude, therefore, that for any actual top $f(u)$ will have two roots, u_1 and u_2, between -1 and $+1$ (cf. Fig. 5–8), and that the top moves such that $\cos\theta$ always remains between these two roots. The location of these roots, and the behavior of ϕ and ψ for values of θ between them, provide much qualitative information about the motion of the top.

It is customary to depict the motion of the top by tracing the curve of the intersection of the figure axis on a sphere of unit radius about the fixed point. This curve will be known as the *locus* of the figure axis. The polar coordinates of a point

---

* See, for example, the treatise by F. Klein and A. Sommerfeld; Whittaker, *Analytical Dynamics*; or the very detailed treatment in Macmillan, *Dynamics of Rigid Bodies*.

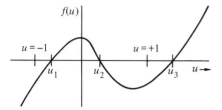

FIGURE 5–8

Illustrating the location of the turning angles of θ in the motion of a heavy symmetrical top.

on the locus are identical with the Euler angles θ, ϕ for the body system. From the discussion of the above paragraph it is seen that the locus lies between the two bounding circles of colatitude $\theta_1 = \arccos u_1$ and $\theta_2 = \arccos u_2$, with $\dot{\theta}$ vanishing at both circles. The shape of the locus curve is in large measure determined by the value of the root of $b - au$, which will be denoted by u':

$$u' = \frac{b}{a}. \tag{5–64}$$

Suppose, for example, the initial conditions are such that u' is larger than u_2. Then, by Eq. (5–57), $\dot{\phi}$ will always have the same sign for the allowed inclination angles between θ_1 and θ_2. Hence the locus of the figure axis must be tangent to the bounding circles in such a manner that $\dot{\phi}$ is in the same direction at both θ_1 and θ_2, as is shown in Fig. 5–9(a). Since ϕ therefore increases secularly in one direction or the other, the axis of the top may be said to *precess* about the vertical axis. But it is not the regular precession encountered in force-free motion, for as the figure axis goes around it nods up and down between the bounding angles θ_1 and θ_2—the top *nutates* during the precession.

Should b/a be such that u' lies between u_1 and u_2 the direction of the precession will be different at the two bounding circles and the locus of the figure axis exhibits loops, as shown in Fig. 5–9(b). The average of $\dot{\phi}$ will not vanish, however, so that there is always a net precession in one direction or the other. It can also happen that u' coincides with one of the roots of $f(u)$. At the corresponding

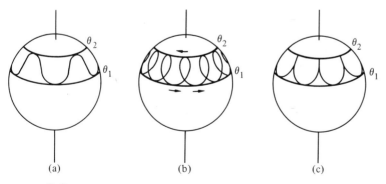

(a)   (b)   (c)

FIGURE 5–9

The possible shapes for the locus of the figure axis on the unit sphere.

bounding circles both $\dot{\theta}$ and $\dot{\phi}$ must then vanish, which requires that the locus have cusps touching the circle, as shown in Fig. (5–9(c)).

This last case is not as exceptional as it sounds; it corresponds in fact to the initial conditions usually stipulated in elementary discussions of tops: one assumes that initially the symmetrical top is spinning about its figure axis, which is fixed in some direction θ_0. At time $t = 0$ the figure axis is released and the problem is to describe the subsequent motion. Explicitly these initial conditions are that at $t = 0, \theta = \theta_0$ and $\dot{\theta} = \dot{\phi} = 0$. The angle θ_0 must therefore be one of the roots of $f(u)$, in fact it corresponds to the upper circle:

$$u_0 = u_2 = u' = \frac{b}{a}. \tag{5–65}$$

For proof note that with these initial conditions E' is equal to $Mgl \cos \theta_0$, and that the terms in E' derived from the top's kinetic energy can never be negative. Hence as $\dot{\theta}$ and $\dot{\phi}$ begin to differ from their initial zero values, energy can be conserved only by a decrease in $Mgl \cos \theta$, that is, by an increase in θ. The initial θ_0 is therefore the same as θ_2, the minimum value θ can have. When released in this manner, *the top always starts to fall*, and continues to fall until the other bounding angle θ_1 is reached, precessing the meanwhile. The figure axis then begins to rise again to θ_2, the complete motion being as shown in Fig. 5–9(c).

Some quantitative predictions can be made about the motion of the top under these initial conditions of vanishing $\dot{\theta}$ and $\dot{\phi}$, provided that the initial kinetic energy of rotation is assumed large compared to the maximum change in potential energy:

$$\frac{1}{2} I_3 \omega_3^2 \gg 2Mgl. \tag{5–66}$$

The effects of the gravitational torques, namely the precession and accompanying nutation, will then be only small perturbations on the dominant rotation of the top about its figure axis. In this situation, we speak of the top as being a "fast top." With this assumption we can obtain expressions for the extent of the nutation, the nutation frequency, and the average frequency of precession.

The extent of the nutation under these initial conditions is given by $u_1 - u_0$, where u_1 is the other physical root of $f(u)$. The initial condition $E' = Mgl \cos \theta_0$ is equivalent to the equality

$$\alpha = \beta u_0.$$

With this relation, and the conditions of Eq. (5–65), $f(u)$ may be rewritten more simply as

$$f(u) = (u_0 - u)\{\beta(1 - u^2) - a^2(u_0 - u)\}. \tag{5–67}$$

The roots of $f(u)$ other than u_0 are given by the roots of the quadratic expression in the curly brackets, and the desired root u_1 therefore satisfies the equation

$$(1 - u_1^2) - \frac{a^2}{\beta}(u_0 - u_1) = 0. \tag{5–68}$$

Denoting $u_0 - u$ by x and $u_0 - u_1$ by x_1, Eq. (5-68) can be rewritten as

$$x_1^2 + px_1 - q = 0, \tag{5-69}$$

where

$$p = \frac{a^2}{\beta} - 2\cos\theta_0, \quad q = \sin^2\theta_0.$$

The condition for a "fast" top, Eq. (5-66), implies that p is much larger than q. This can be seen by writing the ratio a^2/β as

$$\frac{a^2}{\beta} = \left(\frac{I_3}{I_1}\right)\frac{I_3\omega_3^2}{2Mgl}.$$

Except in the case that $I_3 \ll I_1$ (which would correspond to a top in the unusual shape of a cigar), the ratio is much greater than unity, and $p \gg q$. To first order in the small quantity q/p the only physically realizable root of Eq. (5-68) is then

$$x_1 = \frac{q}{p}.$$

Neglecting $2\cos\theta_0$ compared to a^2/β this result can be written

$$x_1 = \frac{\beta\sin^2\theta_0}{a^2} = \frac{I_1}{I_3}\frac{2Mgl}{I_3\omega_3^2}\sin^2\theta_0. \tag{5-70}$$

Thus the extent of the nutation, as measured by $x_1 = u_0 - u_1$, goes down as $1/\omega_3^2$; the faster the top is spun the less is the nutation.

The *frequency* of nutation likewise can easily be found for the "fast" top. Since the amount of nutation is small the term $(1 - u^2)$ in Eq. (5-67) can be replaced by its initial value, $\sin^2\theta_0$. Equation (5-67) then reads, with the help of Eq. (5-70);

$$f(u) = \dot{x}^2 = a^2 x(x_1 - x).$$

If we shift the origin of x to the midpoint of its range, by changing variable to

$$y = x - \frac{x_1}{2},$$

then the differential equation becomes

$$\dot{y}^2 = a^2\left(\frac{x_1^2}{4} - y^2\right),$$

which on differentiation again reduces to the familiar equation for simple harmonic motion

$$\ddot{y} = -a^2 y.$$

In view of the initial condition $x = 0$ at $t = 0$, the complete solution is

$$x = \frac{x_1}{2}(1 - \cos at), \tag{5-71}$$

where x_1 is given by (5–70). The angular frequency of nutation of the figure axis between θ_0 and θ_1 is therefore

$$a = \frac{I_3}{I_1}\omega_3, \tag{5–72}$$

which *increases* the faster the top is spun initially.

Finally, the angular velocity of precession, from (5–57), is given by

$$\dot{\phi} = \frac{a(u_0 - u)}{\sin^2\theta} \approx \frac{ax}{\sin^2\theta_0},$$

or, substituting Eqs. (5–72) and (5–70),

$$\dot{\phi} = \frac{\beta}{2a}(1 - \cos at). \tag{5–73}$$

The rate of precession is therefore not uniform but varies harmonically with time, with the same frequency as the nutation. The *average* precession frequency, however, is

$$\bar{\dot{\phi}} = \frac{\beta}{2a} = \frac{Mgl}{I_3\omega_3}, \tag{5–74}$$

which indicates that the rate of precession decreases as the initial rotational velocity of the top is increased.

We are now in a position to present a complete picture of the motion of the fast top when the figure axis initially has zero velocity. Immediately after the figure axis is released, the initial motion of the top is always to fall under the influence of gravity. But as it falls the top picks up a precession velocity, directly proportional to the extent of its fall, which starts the figure axis moving sideways about the vertical. The initial fall results in a periodic nutation of the figure axis in addition to the precession. As the top is spun faster and faster initially, the extent of the nutation decreases rapidly, although the frequency of nutation increases, while at the same time the precession about the vertical becomes slower. In practice, for a sufficiently fast top the nutation is damped out by the friction at the pivot and becomes unobservable. The top then *appears* to process uniformly about the vertical axis. Because the precession is regular only in appearance, Klein and Sommerfeld have dubbed it a *pseudoregular* precession. In most of the elementary discussions of precession the phenomenon of nutation is neglected. In consequence such derivations seem to lead to the paradoxical conclusion that upon release the top *immediately* begins to process uniformly; a motion that is *normal* to the forces of gravity that are the ultimate cause of the precession. Our discussion of pseudoregular precession serves to resolve the paradox; the precession builds up continuously from rest without any infinite accelerations, and the initial tendency of the top *is* to move in the direction of the forces of gravity.

It is of interest to determine exactly what initial conditions will result in a true regular precession. In such case the angle θ remains constant at its initial value θ_0, which means that $\theta_1 = \theta_2 = \theta_0$, or in other words, $f(u)$ must have a double root at u_0 (cf. Fig. 5–10), or

$$f(u) = 0, \qquad \frac{df}{du} = 0; \qquad u = u_0.$$

[handwritten: $u = \cos\theta$]

[handwritten: $f(u) = \alpha(1-u^2) - (b-au)^2$]

The first of these conditions, from Eq. (5–62), implies

$$(\alpha - \beta u_0) = \frac{(b - au_0)^2}{1 - u_0^2}; \tag{5–75}$$

[handwritten: $\alpha = \frac{(b-au_0)^2}{1-u_0^2}$]

the second corresponds to

$$\frac{\beta}{2} = \frac{a(b - au_0)}{1 - u_0^2} - u_0 \frac{(\alpha - \beta u_0)}{1 - u_0^2}. \tag{5–76}$$

[handwritten: $\alpha' = -2au + 2a(b-au)$]

[handwritten: $\alpha = \frac{a(b-au_0)}{u_0}$]

Substitution of Eq. (5–75) in Eq. (5–76) leads, in view of Eq. (5–57) for $\dot{\phi}$, to a quadratic equation for $\dot{\phi}$:

$$\frac{\beta}{2} = a\dot{\phi} - \dot{\phi}^2 \cos\theta_0. \tag{5–76'}$$

With the definitions of β, Eq. (5–61), and of a, Eq. (5–53), this can be written in two alternate forms, depending on whether a is expressed in terms of ω_3 or the (constant) $\dot{\psi}$ and $\dot{\phi}$:

$$Mgl = \dot{\phi}(I_3\omega_3 - I_1\dot{\phi}\cos\theta_0), \tag{5–77a}$$

or

$$Mgl = \dot{\phi}(I_3\dot{\psi} - (I_1 - I_3)\dot{\phi}\cos\theta_0). \tag{5–77b}$$

The initial conditions for the problem of the heavy top require the specification of $\theta, \phi, \psi, \dot{\theta}, \dot{\phi}$, and, say, either $\dot{\psi}$ or ω_3 at the time $t = 0$. Because they are cyclic the initial values of ϕ and ψ are largely irrelevant and in general we can choose any desired value for each of the four others. But if in addition we require that the motion of the figure axis be one of uniform precession without nutation then our choice of these four initial values is no longer completely unrestricted, instead they must satisfy either of Eqs. (5–77). For $\dot{\theta} = 0$, one may still choose initial values of θ and ω_3, almost arbitrarily, but the value of $\dot{\phi}$ is then determined. The

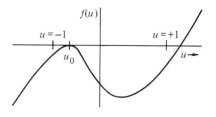

FIGURE 5–10

Appearance of $f(u)$ for a regular precession.

phrase "almost arbitrarily" is used because Eqs. (5–77) are quadratic, and in order that $\dot\phi$ be real the discriminant of Eq. (5–77a) must be positive:

$$I_3^2\omega_3^2 > 4MglI\cos\theta_0. \tag{5–78}$$

For $\theta_0 > \pi/2$ (a top mounted so its center of mass is below the fixed point) then any value of ω_3 can lead to uniform precession. But for $\theta_0 < \pi/2$, ω_3 must be chosen to be above a minimum value:

$$\omega_3 > \omega_3' = \frac{2}{I_3}\sqrt{Mgl\cos\theta_0}, \tag{5–79}$$

to achieve the same situation. Similar conditions can be obtained from Eq. (5–77b) for the allowable values of ψ. As a result of the quadratic nature of Eq. (5–77) there will in general be two solutions for $\dot\phi$, known as the "fast" and "slow" precession. It will also be noticed that (5–77) can never be satisfied by $\dot\phi = 0$ for finite $\dot\psi$ or ω_3; to obtain uniform precession we must always give the top a shove to start it on its way. Without this correct initial precessional velocity one can obtain at best only a pseudoregular precession.

If the precession is slow, so that $\dot\phi\cos\theta_0$ may be neglected compared to a, then an approximate solution for $\dot\phi$ is

$$\dot\phi \approx \frac{\beta}{2a} = \frac{Mgl}{I_3\omega_3} \qquad \text{(slow)},$$

which agrees with the average rate of pseudoregular precession for a fast top. This result is to be expected, of course; if the rate of precession is slow there is little difference between starting the gyroscope off with a little shove or with no shove at all. Note that with this value for $\dot\phi$, the neglect of $\dot\phi\cos\theta_0$ compared to a is equivalent to requiring that ω_3 be much greater than the minimum allowed value. For such large values of ω_3 the "fast" precession is obtained when $\dot\phi$ is so large that Mgl is small compared to the other terms in Eq. (5–77a):

$$\dot\phi = \frac{I_3\omega_3}{I_1\cos\theta_0} \qquad \text{(fast)}.$$

The fast precession is independent of the gravitational torques and can in fact be related to the precession of a free body (see Exercise 19a).

One further case deserves some attention, namely, when $u = 1$ corresponds to one of the roots of $f(u)$.* Suppose, for instance, a top is set spinning with its figure axis initially vertical. Clearly then $b = a$, for $I_1 b$ and $I_1 a$ are the constant components of the angular momentum about the vertical axis and the figure axis respectively, and these axes are initially coincident. Since the initial angular

---

* Note that this must be treated as a special case, since in the previous discussions factors of $\sin^2\theta$ were repeatedly divided out of the expressions.

velocity is only about the figure axis the energy equation (5–59) evaluated at time $t = 0$ states that

$$E' = E - \frac{1}{2}I_3\omega_3^2 = Mgl.$$

By the definitions of α and β (Eq. 5–61) it follows that $\alpha = \beta$.

The energy equation at any angle may therefore be written as

$$\dot{u}^2 = (1 - u^2)\beta(1 - u) - a^2(1 - u)^2$$

or

$$\dot{u}^2 = (1 - u)^2\{\beta(1 + u) - a^2\}.$$

The form of the equation indicates that $u = 1$ is always a double root, with the third root given by

$$u_3 = \frac{a^2}{\beta} - 1.$$

If $a^2/\beta > 2$ (which corresponds to the condition for a "fast" top), u_3 is larger than 1 and the only possible motion is for $u = 1$; the top merely continues to spin about the vertical. For this state of affairs the plot of $f(u)$ appears as shown in Fig. 5–11(a). On the other hand if $a^2/\beta < 2$, u_3 is then less than 1, $f(u)$ takes on the form shown in Fig. 5–11(b), and the top will nutate between $\theta = 0$ and $\theta = \theta_3$. There is thus a critical angular velocity, ω', above which only vertical motion is possible, whose value is given by

$$\frac{a^2}{\beta} = \left(\frac{I_3}{I_1}\right)\frac{I_3\omega'^2}{2Mgl} = 2$$

or

$$\omega'^2 = 4\frac{MglI_1}{I_3^2}, \tag{5–80}$$

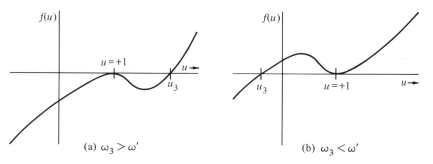

(a) $\omega_3 > \omega'$ (b) $\omega_3 < \omega'$

FIGURE 5–11
Plot of $f(u)$ when the figure axis is initially vertical.

which is identical with Eq. (5–79) for the minimum frequency for uniform precession with $\theta_0 = 0$. In practice, if a top is started spinning with its axis vertical and with ω_3 greater than the critical angular velocity, it will continue to spin quietly for a while about the vertical (hence the designation as a "sleeping" top*). However, friction gradually reduces the frequency of rotation below the critical value and the top then begins to wobble in ever larger amounts as it slows down. The effects of friction (which of course cannot be directly included in the Lagrangian framework) can give rise to unexpected phenomena in the behavior of tops. A notable example is the "tippie-top," which consists basically of somewhat more than half a sphere with a stem added on the flat surface. When set rotating with the spherical surface downwards on a hard surface, it proceeds to skid and nutate until it eventually turns upside down, pivoting on the stem, where it then behaves as a normal "sleeping" top. The complete reversal of the angular momentum vector is the result of the frictional torque occurring as the top skids on its spherical surface.†

A large and influential technology is based on the applications of rapidly spinning rigid bodies, particularly through the use of what are called "gyroscopes." Basically, a gyroscope is a symmetrical top rotated very rapidly by external means about the figure axis and mounted in gimbals so that the motion of the figure axis is unrestricted about one or more spatial axes while the center of gravity remains stationary. If external torques are suitably exerted on the gyroscope it will undergo the precession and nutation motions described above for the heavy top. However, the condition for the "fast" top is abundantly satisfied, so that the extent of the nutation is always very small, and moreover is deliberately damped out by the method of mounting. The only gyroscopic phenomenon then observed is precession and the mathematical treatment required to describe this precession can be greatly simplified. One can see how to do this by generalization from the case of the heavy symmetrical top. If $\mathbf{R}$ is the radius vector along the figure axis from the fixed point to the center of gravity, then the gravitational torque exerted on the top is

$$\mathbf{N} = \mathbf{R} \times M\mathbf{g}, \tag{5–81}$$

where $\mathbf{g}$ is the *downward* vector of the acceleration of gravity. If $\mathbf{L}_3$ is a vector along the figure axis, describing the angular momentum of rotation about the figure axis, and $\boldsymbol{\omega}_p$, known as the precession vector, is aligned along the vertical

---

* As in   "... we have come, last and best,
        From the wide zone in dizzying circles hurled
        To that still centre where the spinning world
        Sleeps on its axis, to the heart of rest."

                                  From *Gaudy Night* by D. L. Sayers
                                Reprinted by permission of Harper & Row.

† A fad of the 1950s, the "tippie-top" engendered a voluminous literature. See, for example, A. R. Del Campo, *American Journal of Physics* **23**, 544 (1955), and also Barger and Olsson, *Classical Mechanics*, pp. 254–257.

with magnitude equal to the mean precession angular velocity $\bar{\dot{\phi}}$, Eq. (5–74), then the sense and magnitude of the (pseudoregular) precession is given by

$$\boldsymbol{\omega}_p \times \mathbf{L}_3 = \mathbf{N}. \tag{5–82}$$

Since any torque about the fixed point or center of mass can be put in the form $\mathbf{R} \times \mathbf{F}$, similar to Eq. (5–81), the resulting average precession rate for a "fast" top can always be derived from Eq. (5–82), with the direction of the force $\mathbf{F}$ defining the precession axis. For almost all of the engineering applications of gyroscopes, the equilibrium behavior (i.e., neglecting transients) can be derived from Eq. (5–82).

Free from any torques a gyroscope spin axis will always preserve its original direction relative to an inertial system. Gyros can therefore be used to indicate or maintain specific directions, e.g., provide stabilized platforms. As indicated by Eq. (5–82), through the precession phenomena they can sense and measure angular rotation rates and applied torques. Note from Eq. (5–82) that the precession rate is proportional to the torque, whereas in a nonspinning body it is the angular acceleration that is given by the torque. Once the torque is removed, a nonspinning body will continue to move; under similar conditions a gyro simply stops precessing.

The gyrocompass involves more complicated considerations because here we are dealing with the behavior of a gyroscope fixed in the noninertial system, while the earth rotates underneath it. In a gyrocompass an additional precession is automatically applied by an external torque at a rate just enough to balance the rotation rate of the earth. Once set in the direction of the earth's rotation, i.e., the North direction, the gyrocompass then preserves this direction, at least in slowly moving vehicles. What has been presented here is admittedly an oversimplified, highly compressed view of the fascinating technological uses of fast spinning bodies. To continue further in this direction would regrettably lead us too far afield.

There are, however, two examples of precession phenomena in nature for which a somewhat fuller discussion would be valuable, both for the great interest in the phenomena themselves and as examples of the techniques derived in this chapter. The first concerns the types of precession that arise from torques induced by the earth's equatorial "bulge," and the second is the precession of moving charges in a magnetic field. The next two sections are concerned with these examples.

5–8 PRECESSION OF THE EQUINOXES AND OF SATELLITE ORBITS

It has been mentioned previously that the earth is a top whose figure axis is precessing about the normal to the ecliptic, a motion known astronomically as the precession of the equinoxes. Were the earth completely spherical, none of the other members of the solar system could exert a gravitational torque on it. But, as has been pointed out, the earth deviates slightly from a sphere, being closely approximated by an oblate spheroid of revolution. It is just the net torque on the resultant equatorial "bulge" arising from gravitational attraction, chiefly of the sun and the moon, that sets the earth's axis precessing in space.

To calculate the rate of this precession a slight excursion into potential theory is needed to find the mutual gravitational potential of a mass point (representing the sun or the moon) and a nonspherical distribution of matter. We will find the properties of the inertia tensor as obtained above very useful in the derivation of this potential.

Consider a distribution of mass points forming one body, and a single mass point, mass M, representing the other (cf. Fig. 5–12). If r_i is the distance between the ith point in the distribution and the mass point M, then the mutual gravitational potential between the two bodies is*

$$V = -\frac{GMm_i}{r_i},$$

$$= -\frac{GMm_i}{r\sqrt{1 + \left(\dfrac{r_i'}{r}\right)^2 - 2\dfrac{r_i'}{r}\cos\psi_i}}. \tag{5–83}$$

In this last expression the terminology of Fig. 5–12 is used: r_i' is the radius vector to the ith particle from a particular point, which will later be taken to be the center of mass of the first body, r is the corresponding radius vector to the mass point M, and ψ_i is the angle between the two vectors. It is well known that a simple expansion in terms of Legendre polynomials can be given for Eq. (5–83); in fact the reciprocal of the square root in Eq. (5–83) is known as the *generating function* for Legendre polynomials,† so that

$$V = -\frac{GM}{r} \sum_{n=0} m_i \left(\frac{r_i'}{r}\right)^n P_n(\cos\psi_i), \tag{5–84}$$

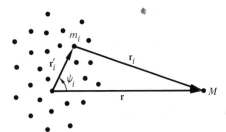

FIGURE 5–12
Geometry involved in gravitational potential between an extended body and a mass point.

---

* It may be worth a reminder that summation is implied over repeated subscripts.

† See for example, J. Mathews and R. L. Walker, *Mathematical Methods of Physics*, first edition, p. 164, and G. Arfken, *Mathematical Methods for Physicists*, p. 418.

providing r, the distance from the origin to M, is greater than any r'_i. We shall make use of only the first three Legendre polynomials that, for reference, are

$$P_0(x) = 1, \qquad P_1(x) = x, \qquad P_2(x) = \tfrac{1}{2}(3x^2 - 1). \tag{5–85}$$

For a continuous spherical body, with only a radial variation of density, all terms except the first in Eq. (5–84) can easily be shown to vanish. Thus the nth term inside the summation, for a body with spherical symmetry and mass density $\rho(r')$, can be written

$$\int \int \int dV' \rho(r') \left(\frac{r'}{r}\right)^n P_n(\cos \psi).$$

Using spherical polar coordinates, with the polar axis along $\mathbf{r}$, this becomes

$$\int r'^2 \, dr' \rho(r') \left(\frac{r'}{r}\right)^n \int_{-1}^{+1} d(\cos \psi) P_n(\cos \psi).$$

From the orthonormal properties of P_n with respect to P_0, the integral over $\cos \psi$ vanishes except for $n = 0$, which proves the statement.

If the body deviates only slightly from spherical symmetry, as is the case with the earth, one would expect the terms in Eq. (5–84) beyond $n = 0$ to decrease rapidly with increasing n. It will therefore be sufficient to retain only the first nonvanishing correction term in Eq. (5–48) to the potential for a sphere. Now, the choice of the center of mass as origin causes the $n = 1$ term to vanish identically, since it can be written

$$-\frac{GM}{r^2} m_i r'_i \cos \psi_i = -\frac{GM}{r^3} \mathbf{r} \cdot m_i \mathbf{r}'_i,$$

which is zero, by definition of the center of mass. The next term, for $n = 2$, can be written

$$\frac{GM}{2r^3} m_i r'^2_i (1 - 3 \cos^2 \psi_i).$$

It is useful to write the $r'^2_i \cos^2 \psi_i$ expression in dyadic form:

$$r'^2_i \cos^2 \psi_i = \frac{\mathbf{r} \cdot \mathbf{r}'_i \mathbf{r}'_i \cdot \mathbf{r}}{r^2}$$

so that, with a little judicious addition and subtraction, the $n = 2$ term in the potential takes the form

$$\frac{3}{2} \frac{GM}{r^5} m_i \mathbf{r} \cdot [r'^2_i \mathbf{1} - \mathbf{r}'_i \mathbf{r}'_i] \cdot \mathbf{r} - \frac{GM}{r^3} m_i r'^2_i.$$

In the first part of this expression we can recognize the dyadic form for the inertia tensor, Eq. (5–15), while from Eq. (5–6) the second part is seen to involve the trace of the inertia tensor. We can therefore write the $n = 2$ term as

$$\frac{3}{2}\frac{GM}{r^5}\mathbf{r}\cdot\mathbf{I}\cdot\mathbf{r} - \frac{GM}{2r^3}\mathrm{Tr}\,\mathbf{I},$$

and the complete approximation to the nonspherical potential as

$$V = -\frac{GMm}{r} + \frac{GM}{2r^3}[3I_r - \mathrm{Tr}\,\mathbf{I}], \qquad (5\text{--}86)$$

where m is the mass of the first body (earth) and I_r is the moment of inertia about the direction of $\mathbf{r}$. From the diagonal representation of the inertia tensor in the principal axis system, its trace is just the sum of the principal moments of inertia, so that V can be written as

$$V = \frac{GMm}{r} + \frac{GM}{2r^3}[3I_r - (I_1 + I_2 + I_3)]. \qquad (5\text{--}87)$$

Equation (5–87) is sometimes known as *MacCullagh's formula*. So far no assumption of rotational symmetry has been made. Let us now take the axis of symmetry to be along the third principal axis, so that $I_1 = I_2$. If α, β, γ are the direction cosines of $\mathbf{r}$ relative to the principal axes, then the moment of inertia I_r can be expressed as

$$I_r = I_1(\alpha^2 + \beta^2) + I_3\gamma^2 = I_1 + (I_3 - I_1)\gamma^2.$$

With this form for I_r, the potential, Eq. (5–87), becomes

$$V = -\frac{GMm}{r} + \frac{GM(I_3 - I_1)}{2r^3}[3\gamma^2 - 1],$$

or

$$V = -\frac{GMm}{r} + \frac{GM(I_3 - I_1)}{2r^3}P_2(\gamma). \qquad (5\text{--}88)$$

The general form of Eq. (5–88) could have been foretold from the start, for the potential from a mass distribution obeys Poisson's equation. The solution appropriate to the symmetry of the body, as is well known, is an expansion of terms of the form $P_n(\gamma)/r^{n+1}$, of which Eq. (5–88) shows the first two nonvanishing terms. However this approach does not give the coefficients of the terms any more simply than the derivation employed here. It should also be remarked that the expansion of V is the gravitational analog of the multipole expansion of, say, the electrostatic potential of an arbitrary charged body.* The $n = 1$ term is absent here because there is only one sign of gravitational "charge" and there can be no gravitational dipole moment. Further, the inertia tensor is defined analogously to

---

* See, for example, J. D. Jackson, *Classical Electrodynamics*, Section 4.1.

the quadrupole moment tensor. Therefore the mechanical effects we are seeking can be said to arise from the gravitational quadrupole moment of the oblate earth.*

Of the terms in Eq. (5–88) for the potential, the only one that depends on the orientation of the body, and thus could give rise to torques, is

$$V_2 = \frac{GM(I_3 - I_1)}{2r^3} P_2(\gamma).$$ (5–89)

For the example of the earth's precession, it should be remembered that γ is the direction cosine between the figure axis of the earth and the radius vector from the earth's center to the sun or moon. As these bodies go around their apparent orbits γ will change. The relation of γ to the more customary astronomical angles can be seen from Fig. 5–13 where the orbit of the sun or moon is taken as being in the xy plane, and the figure axis of the body in the xz plane. The angle θ between the figure axis and the z direction is the obliquity of the figure axis. The dot product of a unit vector along the figure axis with the radius vector to the celestial body involves only the products of their x components, so that

$$\gamma = \sin \theta \cos \eta.$$

Hence V_2 can be written

$$V_2 = \frac{GM(I_3 - I_1)}{2r^3} [3 \sin^2 \theta \cos^2 \eta - 1].$$

As we shall see, the orbital motion is very rapid compared to the precessional motion and for the purpose of obtaining the mean precession rate it will be adequate to average V_2 over a complete orbital period of the celestial body considered. Since the apparent orbits of the sun and moon have low eccentricities, r can be assumed constant and the only variation is in $\cos \eta$. The average of $\cos^2 \eta$ over a complete period is 1/2, and the averaged potential is then

$$\overline{V}_2 = \frac{GM(I_3 - I_1)}{2r^3} \left[\frac{3}{2} \sin^2 \theta - 1 \right] = \frac{GM(I_3 - I_1)}{2r^3} \left[\frac{1}{2} - \frac{3}{2} \cos^2 \theta \right],$$

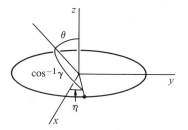

FIGURE 5–13
Diagram of figure axis of the earth relative to orbit of mass point.

---

*Note that so far nothing in the argument restricts the potential of Eq. (5–88) to *rigid* bodies. The constraint of rigidity enters only when we require from here on that the principle axes be fixed in the body and the associated moments of inertia be constant in time.

or, finally,

$$\overline{V}_2 = -\frac{GM(I_3 - I_1)}{2r^3}P_2(\cos\theta). \qquad (5\text{–}90)$$

The torque derived from Eq. (5–90) is perpendicular to both the figure axis and the normal to the orbit (which plays the same role as the vertical axis for the heavy top). Hence the precession is about the direction of the orbit normal vector. The magnitude of the precession rate can be obtained from Eq. (5–82), but because the potential differs in form from that for the heavy top, it may be more satisfying to obtain a more formal derivation. For any symmetric body in which the potential is a function of $\cos\theta$ only, the Lagrangian can be written, following Eq. (5–52), as

$$L = \frac{I_1}{2}(\dot{\theta}^2 + \dot{\phi}^2\sin^2\theta) + \frac{I_3}{2}(\dot{\psi} + \dot{\phi}\cos\theta)^2 - V(\cos\theta). \qquad (5\text{–}91)$$

If we are to assume only uniform precession and are not concerned about the necessary initial conditions, we can simply take $\dot{\theta}$ and $\ddot{\theta}$ to be zero in the equations of motion. The Lagrange equation corresponding to θ is then

$$\frac{\partial L}{\partial\theta} = I_1\dot{\phi}^2\sin\theta\cos\theta - I_3\dot{\phi}\sin\theta(\dot{\psi} + \dot{\phi}\cos\theta) - \frac{\partial V}{\partial\theta} = 0$$

or

$$I_3\omega_3\dot{\phi} - I_1\dot{\phi}^2\cos\theta = \frac{\partial V}{\partial(\cos\theta)}, \qquad (5\text{–}92)$$

which is the analog of Eq. (5–76') for a more general potential. For slow precession, which means basically that $\dot{\phi} \ll \omega_3$, the $\dot{\phi}^2$ terms in Eq. (5–92) can be neglected, and the rate of uniform precession is given by

$$\dot{\phi} = \frac{1}{I_3\omega_3}\frac{\partial V}{\partial(\cos\theta)}. \qquad (5\text{–}93)$$

It may easily be checked that for the heavy top Eq. (5–93) agrees with the result of Eq. (5–74). With the potential of Eq. (5–90) the precession rate is

$$\dot{\phi} = -\frac{3GM}{2\omega_3 r^3}\frac{I_3 - I_1}{I_3}\cos\theta. \qquad (5\text{–}94)$$

For the case of the precession due to the sun, this formula can be put in a simpler form, by taking r as the semimajor axis of the earth's orbit and using Kepler's law, Eq. (3–71), in the form

$$\omega_0^2 = \left(\frac{2\pi}{\tau}\right)^2 = \frac{GM}{r^3}.$$

The precession rate, relative to the orbital angular velocity, ω_0, is then

$$\frac{\dot{\phi}}{\omega_0} = -\frac{3}{2}\frac{\omega_0}{\omega_3}\frac{I_3 - I_1}{I_3}\cos\theta. \tag{5-95}$$

With the value of $(I_3 - I_1)/I_3$ as given in Section 5–6, and $\theta = 23°27'$, Eq. (5–95) says that the solar-induced precession would be such as to cause a complete rotation of the figure axis about normal to the ecliptic in about 81,000 years. The moon is far less massive than the sun but it is also much closer; the net result is that the lunar-induced precession rate is over twice that caused by the sun. Since the lunar orbit is close to the ecliptic and has the same sense as the apparent solar orbit, the two precessions nearly add together arithmetically, and the combined lunisolar precession rate is 50.25″/year, or one complete rotation in about 26,000 years. Note that this rate of precession is so slow that the approximation of neglecting $\dot{\phi}$ compared to ω_3 is abundantly satisfied. Because the sun, moon, and earth are in constant relative motion, and the moon's orbit is inclined about 5° to the ecliptic, the precession exhibits irregularities designated as *astronomical nutation*. The extent of these periodic irregularities is not large—about 9″ of arc in θ and about 18″ in ϕ. Even so, they are far larger than the true nutation that, as Klein and Sommerfeld have shown, is manifested by the Chandler wobble whose amplitude is never more than a few tenths of an arc second.

One further application can be made of the potential, Eq. (5–88), and associated uniform precession rate, Eq. (5–93). It has been stressed that the potential represents a *mutual* gravitational interaction; if it results in torques acting on the spinning earth, it also gives rise to (noncentral) forces acting on the mass point M. The effect of these small forces appears as a precession of the plane of the orbit of the mass point, relative to an inertial space. It is possible to obtain an approximate formula for this precession by an argument again based on the behavior of spinning rigid bodies. Since the precession rates are small compared to the orbital angular velocity, we can again average over the orbit. The averaging in effect replaces the particle by a rigid ring of mass M with the same radius as the (assumed circular) orbit, spinning about the figure axis of the ring with the orbital frequency. Equation (5–90) gives the potential field in which this ring is located, with θ the angle between the figure axes of the ring and the earth. The average precession rate is still given by Eq. (5–93), *but now I_3 and ω_3 refer to the spinning ring and not the earth*. It would therefore be better to rewrite Eq. (5–93) for this application as

$$\dot{\phi} = \frac{\tau}{2\pi M r^2}\frac{\partial V}{\partial(\cos\theta)}, \tag{5-93'}$$

and Eq. (5–94) appears as

$$\dot{\phi} = -\frac{\tau}{2\pi}\frac{3}{2}\frac{G(I_3 - I_1)}{r^5}\cos\theta. \tag{5-94'}$$

Equation (5–94′) could be used, for example, to find the precession of the orbit of the moon due to the earth's oblateness. A more current application would be to the precession of nearly circular orbits of artificial satellites revolving about the earth. The fraction of a complete precession rotation in one period of the satellite is

$$\frac{\dot{\phi}\tau}{2\pi} = -\left(\frac{\tau}{2\pi}\right)^2 \frac{3}{2} \frac{G(I_3 - I_1)}{r^5} \cos\theta.$$

An application of Kepler's law, this time for the period of the satellite, reduces this result to

$$\frac{\dot{\phi}\tau}{2\pi} = -\frac{3}{2} \frac{I_3 - I_1}{mr^2} \cos\theta, \tag{5-96}$$

where m is the mass of the earth. If the earth were a uniform sphere then the principal moments of inertia would be

$$I_3 \sim I_1 = \frac{2}{5} mR^2,$$

with R the radius of the earth. Because the core is much more dense than the outer layers the moment of inertia is smaller, such that in fact*

$$I_3 = 0.331 mR^2 \approx \frac{1}{3} mR^2.$$

The approximate precession is thus given by

$$\frac{\dot{\phi}\tau}{2\pi} = -\frac{1}{2} \frac{I_3 - I_1}{I_3} \left(\frac{R}{r}\right)^2 \cos\theta. \tag{5-97}$$

For a "close" satellite where r is very close to R, and the inclination of the satellite orbit to the equator is, say, 30°, Eq. (5–97) says that the plane of the orbit precesses completely around 2π in about 700 orbits of the satellite. Since the period of a close satellite is about $1\frac{1}{2}$ hours, complete rotation of the orbital plane occurs in a little over six weeks time. Clearly, the effect is quite significant. We shall rederive the precession of the satellite orbit later on, when we discuss the subject of perturbation theory.

5–9 PRECESSION OF SYSTEMS OF CHARGES IN A MAGNETIC FIELD

The motion of systems of charged particles in magnetic fields does not normally involve rigid body motion. In a number of particular instances, the motion is however most elegantly discussed using the techniques developed here for rigid body motion. For this reason, and because of their importance in atomic and nuclear physics, brief consideration of a few examples will be given here.

---

* The best values of I_3 are now obtained from observation of just such effects on satellite orbits. See F. D. Stacey, *Physics of the Earth*, p. 26.

The *magnetic moment* of a system of moving charges (relative to a particular origin) is defined as*

$$\mathbf{M} = \frac{1}{2c} q_i(\mathbf{r}_i \times \mathbf{v}_i) \rightarrow \frac{1}{2c} \int dV \rho_e(\mathbf{r})(\mathbf{r} \times \mathbf{v}). \tag{5-98}$$

Here the first expression is a sum over discrete particles with charge q_i, while the second is the corresponding generalization to a continuous distribution of charge density $\rho_e(\mathbf{r})$. The angular momentum of the system under corresponding conventions is

$$\mathbf{L} = m_i(\mathbf{r}_i \times \mathbf{v}_i) \rightarrow \int dV \rho_m(\mathbf{r})(\mathbf{r} \times \mathbf{v}).$$

Both the magnetic moment and the angular momentum thus have similar form and are related to each other through a dyadic. Here, however, we shall restrict the discussion to situations in which $\mathbf{M}$ is directly proportional to $\mathbf{L}$:

$$\mathbf{M} = \Gamma \mathbf{L}, \tag{5-99}$$

most naturally by having a uniform q/m ratio for all particles or at all points in the continuous system. In such cases the *gyromagnetic ratio* Γ is given by

$$\Gamma = \frac{q}{2mc}, \tag{5-100}$$

but, with an eye to models of particle and atomic spin, Γ will often be left unspecified. The forces and torques on a magnetic dipole may be considered as derived from a potential†

$$V = -(\mathbf{M} \cdot \mathbf{B}). \tag{5-101}$$

It is implied along with Eq. (5-101) that the magnetic field is substantially constant over the system. Indeed, the picture applies best to a pointlike magnetic moment whose magnitude is not affected by the motion it undergoes—a picture appropriate to permanent magnets or systems on an atomic or smaller scale. With uniform $\mathbf{B}$, the potential depends only on the orientation of $\mathbf{M}$ relative to $\mathbf{B}$; no forces are exerted on the magnetic moment, but there is a torque

$$\mathbf{N} = \mathbf{M} \times \mathbf{B}, \tag{5-102}$$

(Compare with Eqs. (5-51) and (5-81).) The time rate of change of the total angular momentum is equal to this torque, so that in view of Eq. (5-99) we can write

$$\frac{d\mathbf{L}}{dt} = \mathbf{L} \times \Gamma \mathbf{B}. \tag{5-103}$$

---

* See, for example, J. D. Jackson, *Classical Electrodynamics*, 2d ed., pp. 180–184, or W. K. H. Panofsky and M. Phillips, *Classical Electricity and Magnetism*, 2d ed., pp. 130–133.

† Jackson, op. cit., pp. 185–187.

But this is exactly the equation of motion for a vector of constant magnitude rotating in space about the direction of **B** with an angular velocity $\boldsymbol{\omega} = -\Gamma\mathbf{B}$. The effect of a uniform magnetic field on a permanent magnetic dipole is to cause the angular momentum vector (and the magnetic moment) to *precess* uniformly. For the classical gyromagnetic ratio, Eq. (5–100), the precession angular velocity is

$$\omega_l = -\frac{q\mathbf{B}}{2mc}, \tag{5–104}$$

known as the *Larmor frequency*. For electrons q is negative, and the Larmor precession is counterclockwise around the direction of **B**.

As a second example, consider a collection of moving charged particles, without restrictions on the nature of their motion, but assumed to all have the same q/m ratio, and to be in a region of uniform constant magnetic field. It will also be assumed that any interaction potential between particles depends only on the scalar distance between the particles. The Lagrangian for the system can be written (cf. Eq. 1–66)

$$L = \frac{1}{2}m_i v_i^2 + \frac{q}{mc}m_i \mathbf{v}_i \cdot \mathbf{A}_i(\mathbf{r}_i) + V(|\mathbf{r}_i - \mathbf{r}_j|), \tag{5–105}$$

where the constant magnetic field **B** is generated by a vector potential **A**

$$\mathbf{A} = \frac{1}{2}\mathbf{B} \times \mathbf{r}. \tag{5–106}$$

In terms of **B** the Lagrangian has the form (permuting dot and cross products)

$$L = \frac{1}{2}m_i v_i^2 + \frac{q\mathbf{B}}{2mc} \cdot \mathbf{r}_i \times m_i \mathbf{v}_i + V(|\mathbf{r}_i - \mathbf{r}_j|). \tag{5–107}$$

The interaction term with the magnetic field can be variously written (cf. Eq. 5–101 and 5–104)

$$\frac{q\mathbf{B} \cdot \mathbf{L}}{2mc} = \mathbf{M} \cdot \mathbf{B} = -\boldsymbol{\omega}_l \cdot \mathbf{r}_i \times m_i \mathbf{v}_i. \tag{5–108}$$

Suppose now we express the Lagrangian in terms of coordinates relative to "primed" axes having a common origin with the original set, but rotating uniformly about the direction of **B** with angular velocity $\boldsymbol{\omega}_l$. Distance vectors from the origin are unchanged as, of course, are scalar distances such as $|\mathbf{r}_i - \mathbf{r}_j|$. However velocities relative to the new axes differ from the original velocities by the relation

$$\mathbf{v}_i = \mathbf{v}_i' + \boldsymbol{\omega}_l \times \mathbf{r}_i.$$

The two terms in the Lagrangian affected by the transformation are

$$\frac{m_i v_i^2}{2} = \frac{m_i v_i'^2}{2} + m_i \mathbf{v}_i' \cdot (\boldsymbol{\omega}_l \times \mathbf{r}_i) + \frac{m_i}{2}(\boldsymbol{\omega}_l \times \mathbf{r}_i) \cdot (\boldsymbol{\omega}_l \times \mathbf{r}_i),$$

$$-\boldsymbol{\omega}_l \cdot \mathbf{r}_i \times m_i \mathbf{v}_i = -\boldsymbol{\omega}_l \cdot (\mathbf{r}_i \times m_i \mathbf{v}_i') - \boldsymbol{\omega}_l \cdot (\mathbf{r}_i \times m_i(\boldsymbol{\omega}_l \times \mathbf{r}_i)).$$

By permuting dot and cross product it is seen that the terms linear in $\boldsymbol{\omega}_l$ and $\mathbf{v}'_i$ are just equal and opposite and therefore cancel in the Lagrangian. A similar permutation in the terms quadratic in ω_l show that they are of the same form and are related to the moment of inertia of the system about the axis defined by $\boldsymbol{\omega}_l$ (see p. 195). The quadratic term in the Lagrangian can in fact be written as

$$-\frac{m_i}{2}(\boldsymbol{\omega}_l \times \mathbf{r}_i) \cdot (\boldsymbol{\omega}_l \times \mathbf{r}_i) = -\frac{1}{2}\boldsymbol{\omega}_l \cdot \mathbf{I} \cdot \boldsymbol{\omega}_l = -\frac{1}{2}I_l\omega_l^2, \qquad (5\text{--}109)$$

where I_l denotes the moment of inertia about the axis of $\boldsymbol{\omega}_l$. In terms of coordinates in the rotating system the Lagrangian thus has the simple form

$$L = \frac{1}{2}m_i v_i'^2 + V(|\mathbf{r}_i - \mathbf{r}_j|) - \frac{1}{2}I_l\omega_l^2, \qquad (5\text{--}110)$$

from which all linear terms in the magnetic field have disappeared. We can get an idea of the relative magnitude of the quadratic term by considering a situation in which the motion of the system consists of a rotation with some frequency ω, e.g., an electron revolving around the atomic nucleus. Then for systems not too far from spherical symmetry the kinetic energy is approximately $\frac{1}{2}I\omega^2$ (without subscripts on the moment of inertia) and the linear term in ω_l is the order of $\boldsymbol{\omega}_l \cdot \mathbf{L} \approx I\omega_l\omega$. Hence the quadratic term in Eq. (5–110) is the order of $(\omega_l/\omega)^2$ compared to the kinetic energy, and of the order (ω_l/ω) relative to the linear term. In most systems on the atomic or smaller scale the natural frequencies are much larger than the Larmor frequency. Compare, for example, the frequency of a spectral line (which is a difference of natural frequencies) to the frequency shift in the simple Zeeman effect, which is proportional to the Larmor frequency. Thus for such systems the motion in the rotating system is the same as in the laboratory system when there is no magnetic field. What we have is *Larmor's theorem* that states that to first order in $\mathbf{B}$, the effect of a constant magnetic field on a classical system is to superimpose on its normal motion a uniform precession with angular frequency $\boldsymbol{\omega}_l$.

SUGGESTED REFERENCES

L. BRAND, *Vector and Tensor Analysis*. J. W. Gibbs introduced dyads and dyadics, and the fullest exposition of this subject is in the classic work *Vector Analysis* by Gibbs and Wilson (1901). An extensive treatment of dyadics is to be found in the somewhat later treatise by Wills, *Vector and Tensor Analysis*. The texts on tensor and matrix analysis referenced in Chapter 4 above may be used for most of the pertinent material of the present chapter. Brand's book is added here because in addition to the standard material on tensors it has a good deal on dyadics in his Chapter 4 on linear vector functions. The last chapter provides an easily accessible brief introduction to quaternions.

K. SYMON, *Mechanics*. The pragmatic usefulness of dyadics is exemplified in Symon's treatment in his Chapter 10 of the properties of the inertia tensor, where a number of interesting examples are given. The relevant algebra of dyadics and tensors is developed explicitly.

E. J. ROUTH, *Dynamics of Rigid Bodies, Elementary and Advanced*. In the nineteenth century the dynamics of rigid bodies formed one of the main topics at the research frontier of mechanics. Routh's two-volume treatise, the latest edition of which appeared in the 1890s, forms a convenient and elaborate presentation of the achievements made in this field up to that time. For many topics it remains an almost unique source in English. Routh, a colleague of Maxwell, was a pioneer in the study of the stability of small oscillations, and his work there is still relevant today.

W. D. MACMILLAN, *Dynamics of Rigid Bodies*. While not recommended for a systematic study of rigid body dynamics, this work contains much material not readily available elsewhere. Chapter VII, in particular, has long and elaborate discussions of Poinsot motion, and of the motion of the heavy symmetrical top, including the explicit solutions in terms of elliptic functions. The chapter on the complex problems of rolling rigid bodies is also worthy of note.

A. GRAY, *A Treatise on Gyrostatics and Rotational Motion*. The product of World War I interest in gyroscopic devices, Gray's treatise represents the culmination of the British tradition in the field of rigid body dynamics. In a somewhat dense and mostly nonvectorial treatment it covers a wide range of topics, from wandering of the earth's rotational poles to the theory of the boomerang and the operation of the diabolo. A more systematic discussion of many of the same areas will be found in the works of Klein and Sommerfeld.

F. KLEIN AND A. SOMMERFELD, *Theorie des Kreisels*. This monumental work on the theory of the top, in four volumes, has all the external appearances of the typical stolid and turgid German "Handbuch." Appearances are deceiving, however, for it is remarkably readable, despite the handicap of being written in the German language. The graceful, informal style has the fluency and attention to pedagogic details characteristic of all of Sommerfeld's later writings. Although the treatment becomes highly mathematical at times, the physical world is never lost sight of, and one does not founder in a maze of formula. Although limited by the title to tops and gyroscopes, the treatise actually provides a liberal education in all of rigid body mechanics, with excursions into other branches of physics and mathematics. Thus, Chapter I discusses, among other items, Euler angles, infinitesimal rotations, and the Cayley–Klein parameters and their connections with the homographic transformation and with the theory of quaternions. The later notes to this chapter (in Vol. IV) discuss also the connections with electrodynamics and special relativity (quantum mechanics was still far in the future). By and large, Vol. I lays the necessary foundations in rigid dynamics and gives a physical description of top motion with little mathematics.

Volume II is devoted to the detailed exposition of the heavy symmetrical top, although there is also much on Poinsot motion, and it contains a summary of what was then known about the asymmetric top. The distinction between regular and pseudoregular precession was first introduced here and the authors spend much time in examining the two motions, and the approach to regular precession. Many pages are given to a thorough demolishing of the popular or elementary "derivations" of gyroscopic precession. (The authors remark that it was the unsatisfactory nature of these derivations that led them to write the treatise!) There is a long discussion on questions of the stability of motion. Most of the treatment is based on the solution in terms of elliptic integrals and not merely on the approximate small nutation, as was done here.

Volume III is mainly on perturbing forces (chiefly friction) and astronomical applications (nutation of the earth, precession of the equinoxes, etc.). The discussion of the wandering of the earth's poles is especially complete for the time it was written, including an estimation of the effects of the earth's elasticity and the transport of atmospheric masses by the wind circulation. Volume IV is on technical applications and is rather out of date by now.

F. KLEIN, *The Mathematical Theory of the Top.*. In 1896 Felix Klein gave a series of lectures at Princeton, the notes for which constitute this slim volume, recently reprinted along with some unconnected mathematical articles. Most of the book is concerned with highly abstract mathematical details of the theory, but the first Lecture provides a readable account of Cayley–Klein parameters. It is interesting to note that both in this work and in the larger treatise with Sommerfeld use was made of a four-dimensional non-Euclidean space in which time is the fourth dimension—anticipating the use in special relativity by many years (see next chapter). However, the space was solely for mathematical convenience and no physical significance was intended.

A. SOMMERFELD, *Mechanics*. Sommerfeld's work with Klein about the top was one of his first publications, while this text, part of a famous series of *Lectures on Theoretical Physics*, was published more than forty years later, as one of the last of his writings. His interest in the top had apparently not diminished in that time and he devoted considerable space to qualitative discussions of a wide range of gyroscopic and top phenomena—even to a page or two on the asymmetrical top. The thirty or so pages on the entire subject occupy almost all of the chapter on rigid bodies and practically forms an abstract of the larger work! The treatment is extensive, rather than intensive, and there is little detailed discussion.

V. D. BARGER AND M. G. OLSSON, *Classical Mechanics, A Modern Perspective*. This intermediate level text is referenced because it is about the only book that contains even a brief description of what the "tippie-top" is and how it works (p. 254). For pictorial evidence of how the tippie-top has fascinated royalty and the great of physics alike, see the photographs opposite p. 208 in *Niels Bohr*, edited by S. Rozental. To the references given on p. 224 above may be added a paper by T. R. Kane and D. A. Levinson, *Journ. Applied Mech.* **45**, 903 (Dec. 1978), which presents a modern computer solution of the tippie-top accompanied by an extensive bibliography.

J. AHARONI, *Lectures on Mechanics*. A discursive set of essays on relatively isolated topics in mechanics, this reference is noteworthy for its extensive collections of diagrams. The carefully thought-out figures in Chapter 15–17 may be found to illuminate some sticky points about the inertia tensor, Poinsot motion, and gyrocompasses, among other items.

L. MEIROVITCH, *Methods of Analytical Dynamics*, and S. W. GROESBERG, *Advanced Mechanics*. In modern technological applications of mechanics, the dynamics of a rigid body plays a central role, not only for such devices as gyroscopes, but also as a first approximation to systems that are not entirely rigid—such as a space ship. The need to solve actual problems—and not merely derive formulations—imposes a perspective on the methods of mechanics that often contrasts with the viewpoint of the physicist. These two texts give an introduction to the modern methods needed for tackling engineering problems.

E. LEIMANIS, *Motion of Coupled Rigid-Bodies*. Russian applied mathematicians have given much attention to the general problem of the motion of one or more rigid bodies about a fixed point. Their efforts stretch from the days of Sonya Kovalevskaya in the 1880s to the present time. Much of their work is not accessible in English. The reference of Leimanis is a modern treatment of the motion of rigid bodies, as viewed by an applied mathematician, with a full understanding of the Russian literature. Up to date mathematical techniques, such as use of Lie series, are incorporated.

W. H. MUNK AND G. J. F. MACDONALD, *The Rotation of the Earth*. One of the most fascinating applications of the dynamics of rigid bodies is to the phenomena of the rotating earth—although the first step is to realize to what an extent the earth and its appurtenances deviate from a rigid body. The treatise of Munk and Macdonald provides a well-written, lucid introduction to these geophysical applications, which range from continental drift through ancient historical records of eclipses to questions of earthquake excitation of Chandler wobble. Unfortunately—or is it happily?—the treatment cannot be considered as definitive because the field is undergoing intense development and significant advances have been made since the 1960 publication date. But any study of the features of the earth's rotation would do well to start with Munk and Macdonald.

F. D. STACEY, *Physics of the Earth*. Chapter 2 provides a compact discussion of such topics as precession of the equinoxes and the Chandler wobble, with references to the modern literature.

W. WRIGLEY, W. M. HOLLISTER, AND W. G. DENHARD, *Gyroscopic Theory, Design and Instrumentation*. Gyroscopic devices are at the heart of modern advances in inertial navigation. The published literature is considerable, and the unpublished report literature is even more voluminous (not to mention the substantial oral tradition amongst the practitioners of the art). This treatise, a product of the famous M.I.T. Draper Instrumentation Laboratory, gives a reasonably modern overview of the field starting from a discussion of rotation matrices and ranging to detailed blueprints of actual devices. A short review paper by Wrigley and Hollister, "The Gyroscope: Theory and Application," *Science* **149**, 713 (Aug. 13, 1965), may prove illuminating.

EXERCISES

1. If R_i is an antisymmetric matrix associated with the coordinates of the ith mass point of a system, with elements $R_{mn} = \epsilon_{mnl}x_l$ show that the matrix of the inertia tensor can be written as

$$I = -m_i(R_i)^2.$$

2. Show directly by vector manipulation that the definition of the moment of inertia as

$$I = m_i(r_i \times n) \cdot (r_i \times n)$$

reduces to Eq. (5–19).

3. What is the height-to-diameter ratio of a right cylinder such that the inertia ellipsoid at the center of the cylinder is a sphere?

4. Find the principal moments of inertia about the center of mass of a flat rigid body in the shape of a 45° right triangle with uniform mass density. What are the principal axes?

5. Three equal mass points are located at $(a, 0, 0)$, $(0, a, 2a)$, and $(0, 2a, a)$. Find the principal moments of inertia about the origin and a set of principal axes.

6. A uniform right circular cone of height h, half-angle α, and density ρ rolls on its side without slipping on a uniform horizontal plane in such a manner that it returns to its original position in a time τ. Find expressions for the kinetic energy and the components of the angular momentum of the cone.

7. Prove that for a general rigid body motion about a fixed point the time variation of the kinetic energy T is given by

$$\frac{dT}{dt} = \boldsymbol{\omega} \cdot \mathbf{N}.$$

8. Derive Euler's equations of motion, Eq. (5–39′), from the Lagrange equation of motion, in the form of Eq. (1–53), for the generalized coordinate ψ.

9. a) A bar of negligible weight and length l has equal mass points m at the two ends. The bar is made to rotate uniformly about an axis passing through the center of the bar and making an angle θ with the bar. From Euler's equations find the components along the principal axes of the bar of the torque driving the bar.

b) From the fundamental torque equation (1–26) find the components of the torque along axes fixed in space. Show that these components are consistent with those found in part (a).

10. Equation (5–38) holds for the motions of systems that are not rigid, relative to a chosen rotating set of coordinates. For general nonrigid motion, if the rotating axes are chosen to coincide with the (instantaneous) principal axes of the continuous system, show that Eqs. (5–39) are to be replaced by

$$\frac{d(I_i \omega_i)}{dt} + \epsilon_{ijk} \omega_j \omega_k I_k + \frac{dl_i}{dt} + \epsilon_{ijk} l_j \omega_k = N_i, \qquad i = 1, 2, 3,$$

where

$$l_i = \int dV \rho(\mathbf{r}) \epsilon_{ijk} x_j v'_k$$

with $\rho(\mathbf{r})$ the mass density at point $\mathbf{r}$, and $\mathbf{v}'$ the velocity of the system point at $\mathbf{r}$ relative to the rotating axes. These equations are sometimes known as the *Liouville equations* and have applications for discussing almost-rigid motion, such as that of the earth including the atmosphere and oceans.

11. A plane pendulum consists of a uniform rod of length l and negligible thickness with mass m, suspended in a vertical plane by one end. At the other end a uniform disc of radius a and mass M is attached so it can rotate freely in its own plane, which is the vertical plane. Set up the equations of motion in the Lagrangian formulation.

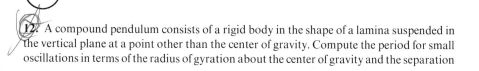

12. A compound pendulum consists of a rigid body in the shape of a lamina suspended in the vertical plane at a point other than the center of gravity. Compute the period for small oscillations in terms of the radius of gyration about the center of gravity and the separation of the point of suspension from the center of gravity. Show that if the pendulum has the same period for two points of suspension at unequal distances from the center of gravity, then the sum of these distances is equal to the length of the equivalent simple pendulum.

13. A uniform bar of mass M and length $2l$ is suspended from one end by a spring of force constant k. The bar can swing freely only in one vertical plane, and the spring is constrained to move only in the vertical direction. Set up the equations of motion in the Lagrangian formulation.

14. A uniform rod slides with its ends on a smooth vertical circle. If the rod subtends an angle of 120° at the center of the circle, show that the equivalent simple pendulum has a length equal to the radius of the circle.

15. An automobile is started from rest with one of its doors initially at right angles. If the hinges of the door are toward the front of the car, the door will slam shut as the automobile picks up speed. Obtain a formula for the time needed for the door to close if the acceleration f is constant, the radius of gyration of the door about the axis of rotation is r_0, and the center-of-mass is at a distance a from the hinges. Show that if f is 1 ft/sec^2 and the door is a uniform rectangle 4 ft wide, the time will be approximately 3.04 seconds.

16. A wheel rolls down a flat inclined surface that makes an angle α with the horizontal. The wheel is constrained so that its plane is always perpendicular to the inclined plane, but it may rotate about the axis normal to the surface. Obtain the solution for the two-dimensional motion of the wheel, using Lagrange's equations and the method of undetermined multipliers.

17. a) Express in terms of Euler's angles the constraint conditions for a uniform sphere rolling without slipping on a flat horizontal surface. Show that they are nonholonomic.

 b) Set up the Lagrangian equations for this problem by the method of Lagrange multipliers. Show that the translational and rotational parts of the kinetic energy are separately conserved. Are there any other constants of motion?

18. For the symmetrical body precessing uniformly in the absence of torques, find analytical solutions for the Euler angles as a function of time.

19. a) Show that the angular momentum of the torque-free symmetrical top rotates in the body coordinates about the symmetry axis with an angular frequency Ω. Show also that the symmetry axis rotates in space about the fixed direction of the angular momentum with the angular frequency

$$\dot{\phi} = \frac{I_3 \omega_3}{I_1 \cos \theta},$$

where ϕ is the Euler angle of the line of nodes with respect to the angular momentum as the space z axis.

 b) Using the results of Exercise 19, Chapter 4, show that $\boldsymbol{\omega}$ rotates in space about the angular momentum with the same frequency $\dot{\phi}$, but that the angle θ' between $\boldsymbol{\omega}$ and $\mathbf{L}$ is given by

$$\sin \theta' = \frac{\Omega}{\dot{\phi}} \sin \theta'',$$

where θ'' is the inclination of $\boldsymbol{\omega}$ to the symmetry axis. Using the figures given in Section 5–6, show therefore that the earth's rotation axis and the axis of angular momentum are never more than 0.6 inch apart on the surface of the earth.

c) Show from parts (a) and (b) that the motion of the force-free symmetrical top can be described in terms of the rotation of a cone fixed in the body whose axis is the symmetry axis, rolling on a fixed cone in space whose axis is along the angular momentum. The angular velocity vector is along the line of contact of the two cones. Show that the same description follows immediately from the Poinsot construction in terms of the inertia ellipsoid.

20. For the general asymmetric rigid body verify analytically the stability theorem shown geometrically above on p. 209 by examining the solution of Euler's equations for small deviations from rotation about each of the principal axes. The direction of $\boldsymbol{\omega}$ is assumed to differ so slightly from a principal axis that the component of $\boldsymbol{\omega}$ along the axis can be taken as constant, while the product of components perpendicular to the axis can be neglected. Discuss the boundedness of the resultant motion for each of the three principal axes.

21. When the rigid body is not symmetrical, an analytic solution to Euler's equation for the torque-free motion cannot be given in terms of elementary functions. Show, however, that the conservation of energy and angular momentum can be used to obtain expressions for the body components of $\boldsymbol{\omega}$ in terms of elliptic integrals.

22. Apply Euler's equations to the problem of the heavy symmetrical top, expressing ω_i in terms of the Euler angles. Show that the two integrals of motion, Eqs. (5–53) and (5–54), can be obtained directly from Euler's equations in this form.

23. Obtain from Euler's equations of motion the condition (5–77) for the uniform precession of a symmetrical top in a gravitational field, by imposing the requirement that the motion be a uniform precession without nutation.

24. Show that the magnitude of the angular momentum for a heavy symmetrical top can be expressed as a function of θ and the constants of the motion only. Prove that as a result the angular momentum vector precesses uniformly only when there is uniform precession of the symmetry axis. ~~optional~~

25. In Section 5–6 the precession of the earth's axis of rotation about the pole was calculated on the basis that there were no torques acting on the earth. Section 5–8, on the other hand, showed that the earth is undergoing a forced precession due to the torques of the sun and moon. Actually, both results are valid: the motion of the axis of rotation about the symmetry axis appears as the nutation of the earth in the course of its forced precession. To prove this statement, calculate θ and ϕ as a function of time for a heavy symmetrical top that is given an initial velocity $\dot{\phi}_0$, which is large compared with the net precession velocity $\beta/2a$, but which is small compared with ω_3. Under these conditions, the bounding circles for the figure axis still lie close together, but the orbit of the figure axis appears as in Fig. 5–9 (b), i.e., shows large loops that move only slowly around the vertical. Show for this case that (5–71) remains valid but now

$$x_1 = \left(\frac{\beta}{a^2} - \frac{2\dot{\phi}_0}{a}\right) \sin^2 \theta_0.$$

From these values of θ and $\dot{\phi}$ obtain ω_1 and ω_2, and show that for $\beta/2a$ small compared with $\dot{\phi}_0$, the vector $\boldsymbol{\omega}$ precesses around the figure axis with an angular velocity

$$\Omega = \frac{I_1 - I_3}{I_1}\omega_3$$

in agreement with Eq. (5–49). Verify from the numbers given in Section 5–6 that $\dot{\phi}_0$ corresponds to a period of about 1600 years, so that $\dot{\phi}_0$ is certainly small compared with the daily rotation and is sufficiently large compared with $\beta/2a$, which corresponds to the precession period of 26,000 years.

26. a) Consider a primed set of axes coincident in origin with an inertial set of axes but rotating with respect to the inertial frame with fixed angular velocity $\boldsymbol{\omega}_0$. If a system of mass points is subject to forces derived from a conservative potential V depending only on the distance to the origin, show that the Lagrangian for the system in terms of coordinates relative to the primed set can be written as

$$L = T' + \boldsymbol{\omega}_0 \cdot \mathbf{L}' + \frac{1}{2}\boldsymbol{\omega}_0 \cdot \mathbf{I}' \cdot \boldsymbol{\omega}_0 - V,$$

where primes indicate the quantities are evaluated relative to the primed set of axes. What is the physical significance of each of the two additional terms?

b) Suppose that $\boldsymbol{\omega}_0$ is in the $x_2' x_3'$ plane, and that a symmetric top is constrained to move with its figure axis in the $x_3' x_1'$ plane, so that only two Euler angles are needed to describe its orientation. If the body is mounted so that the center of mass is fixed at the origin and $V = 0$, show that the figure axis of the body oscillates about the x_3' axis according the the plane-pendulum equation of motion and find the frequency of small oscillations. This illustrates the principle of the gyro compass.

27. Suppose that in a symmetric top each element of mass has a proportionate charge associated with it, so that the e/m ratio is constant—the so-called charged symmetric top. If such a body rotates in a uniform magnetic field the Lagrangian, from (5–108), is

$$L = T - \boldsymbol{\omega}_l \cdot \mathbf{L}.$$

Show that T is a constant (which is a manifestation of the property of the Lorentz force that a magnetic field does no work on a moving charge) and find the other constants of motion. Under the assumption that ω_l is much smaller than the initial rotational velocity about the figure axis obtain expressions for the frequencies and amplitudes of nutation and precession. Where do the kinetic energies of nutation and precession come from?

$\omega_x = \dot{\phi}\sin\theta\sin\psi + \dot{\theta}\cos\psi$

$\omega_y = \dot{\phi}\sin\theta\cos\psi - \dot{\theta}\sin\psi$

$\omega_z = \dot{\phi}\cos\theta + \dot{\psi}$

body rotations

space

$\omega_x = \dot{\theta}\cos\phi + \dot{\psi}\sin\theta\sin\phi$

$\omega_y = \dot{\theta}\sin\phi - \dot{\psi}\sin\theta\cos\phi$

$\omega_z = \dot{\psi}\cos\theta + \dot{\phi}$

CHAPTER 6
Small Oscillations

A class of mechanical motions that can best be treated in the Lagrangian formulation is that of the small oscillations of a system about positions of equilibrium. The theory of such small oscillations finds widespread physical applications in acoustics, molecular spectra, vibrations of mechanisms, and coupled electrical circuits. If the deviations of the system from stable equilibrium conditions are small enough, the motion can generally be described as that of a system of coupled linear harmonic oscillators. It will be assumed the reader is familiar with the properties of a simple harmonic oscillator of one degree of freedom, both in free and forced oscillation, with and without damping. Here the emphasis will be on methods appropriate to discrete systems with more than one degree of freedom. As will be seen, the mathematical techniques required turn out to be very similar to those employed in studying rigid body motion, although the mechanical systems considered need not involve rigid bodies at all. Analogous treatments of oscillations about stable motions can also be developed, but these are most easily done in the Hamiltonian formulation presented in Chapter 8.

6-1 FORMULATION OF THE PROBLEM

We consider conservative systems in which the potential energy is a function of position only. It will be assumed that the transformation equations defining the generalized coordinates of the system, $q_1, \ldots, q_n$, do not involve the time explicitly. Thus, time-dependent constraints are to be excluded. The system is said to be in *equilibrium* when the generalized forces acting on the system vanish:

$$Q_i = \left(\frac{\partial V}{\partial q_i} \right)_0 = 0. \tag{6-1}$$

The potential energy therefore has an extremum at the equilibrium configuration of the system, $q_{01}, q_{02}, \ldots, q_{0n}$. If the configuration is initially at the equilibrium position, with zero initial velocities $\dot{q}_i$, then the system will continue in equilibrium indefinitely. Examples of the equilibrium of mechanical systems are legion—a pendulum at rest, a suspension galvanometer at its zero position, an egg standing on end.

An equilibrium position is classified as *stable* if a small disturbance of the system from equilibrium results only in small bounded motion about the rest position. The equilibrium is *unstable* if an infinitesimal disturbance eventually produces unbounded motion. A pendulum at rest is in stable equilibrium, but the egg standing on end is an obvious illustration of unstable equilibrium. It can be readily seen that when the extremum of V is a minimum the equilibrium must be stable. Suppose the system is disturbed from the equilibrium by an increase in energy dE above the equilibrium energy. If V is a minimum at equilibrium, any deviation from this position will produce an increase in V. By the conservation of energy the velocities must then decrease and eventually come to zero, indicating bound motion. On the other hand, if V decreases as the result of some departure from equilibrium, the kinetic energy and the velocities increase indefinitely, corresponding to unstable motion. The same conclusion may be arrived at graphically by examining the shape of the potential energy curve, as shown symbolically in Fig. 6–1. A more rigorous mathematical proof that stable equilibrium requires a minimum in V will be given in the course of the discussion.

We shall be interested in the motion of the system within the immediate neighborhood of a configuration of stable equilibrium. Since the departures from equilibrium are to be small, all functions may be expanded in a Taylor series about the equilibrium, retaining only the lowest order terms. The deviations of the generalized coordinates from equilibrium will be denoted by η_i:

$$q_i = q_{0i} + \eta_i, \qquad (6\text{--}2)$$

and these may be taken as the new generalized coordinates of the motion. Expanding the potential energy about q_{0i}, we obtain

$$V(q_1,\ldots,q_n) = V(q_{01},\ldots,q_{0n}) + \left(\frac{\partial V}{\partial q_i}\right)_0 \eta_i + \frac{1}{2}\left(\frac{\partial^2 V}{\partial q_i \partial q_j}\right)_0 \eta_i \eta_j + \cdots, \qquad (6\text{--}3)$$

where the summation convention has been invoked, as usual. The terms linear in η_i vanish automatically in consequence of the equilibrium conditions (6–1). The

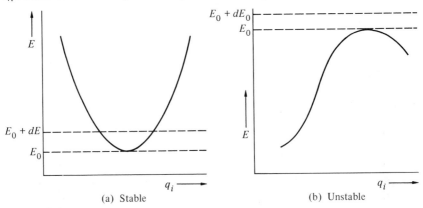

(a) Stable   (b) Unstable

FIGURE 6–1
Shape of the potential energy curve at equilibrium.

first term in the series is the potential energy of the equilibrium position, and by shifting the arbitrary zero of potential to coincide with the equilibrium potential, this term may also be made to vanish. We are therefore left with the quadratic terms as the first approximation to V:

$$V = \frac{1}{2}\left(\frac{\partial^2 V}{\partial q_i \, \partial q_j}\right)_0 \eta_i \eta_j = \frac{1}{2} V_{ij} \eta_i \eta_j, \tag{6–4}$$

where the second derivatives of V have been designated by the constants V_{ij} depending only on the equilibrium values of the q_i's. It is obvious from their definition that the V_{ij}'s are symmetrical, i.e., that $V_{ij} = V_{ji}$. The V_{ij} coefficients can vanish under a variety of circumstances. Thus, the potential can simply be independent of a particular coordinate, so that equilibrium occurs at any arbitrary value of that coordinate. We speak of such cases as *labile, neutral* or *indifferent* equilibrium. It may also happen, e.g., that the potential behaves like a quartic at that point, again causing one or more of the V_{ij}'s to vanish. Either situation calls for special treatment in the mathematical discussion that follows.

A similar series expansion can be obtained for the kinetic energy. Since the generalized coordinates do not involve the time explicitly, the kinetic energy is a homogeneous quadratic function of the velocities (cf. Eq. 1–71):

$$T = \frac{1}{2} m_{ij} \dot{q}_i \dot{q}_j = \frac{1}{2} m_{ij} \dot{\eta}_i \dot{\eta}_j. \tag{6–5}$$

The coefficients m_{ij} are in general functions of the coordinates q_k, but they may be expanded in a Taylor series about the equilibrium configuration:

$$m_{ij}(q_1,\ldots,q_n) = m_{ij}(q_{01},\ldots,q_{0n}) + \left(\frac{\partial m_{ij}}{\partial q_k}\right)_0 \eta_k + \cdots,$$

As Eq. (6–5) is already quadratic in the $\dot{\eta}_i$'s, the lowest nonvanishing approximation to T is obtained by dropping all but the first term in the expansions of m_{ij}. Denoting the constant values of the m_{ij} functions at equilibrium by T_{ij}, we can therefore write the kinetic energy as

$$T = \frac{1}{2} T_{ij} \dot{\eta}_i \dot{\eta}_j. \tag{6–6}$$

It is again obvious that the constants T_{ij} must be symmetric, since the individual terms in Eq. (6–6) are unaffected by an interchange of indices. From Eqs. (6–4) and (6–6) the Lagrangian is given by

$$L = \frac{1}{2}(T_{ij} \dot{\eta}_i \dot{\eta}_j - V_{ij} \eta_i \eta_j). \tag{6–7}$$

Taking the η's as the generalized coordinates, the Lagrangian of Eq. (6–7) leads to the following n equations of motion:

$$T_{ij} \ddot{\eta}_j + V_{ij} \eta_j = 0, \tag{6–8}$$

where explicit use has been made of the symmetry property of the V_{ij} and T_{ij} coefficients. Each of the equations (6–8) will involve, in general, all of the coordinates η_i, and it is this set of simultaneous differential equations that must be solved to obtain the motion near the equilibrium.

6–2 THE EIGENVALUE EQUATION AND THE PRINCIPAL AXIS TRANSFORMATION

The equations of motion (6–8) are linear differential equations with constant coefficients, of a form familiar from electrical circuit theory. We are therefore led to try an oscillatory solution of the form

$$\eta_i = Ca_i e^{-i\omega t}. \tag{6–9}$$

Here Ca_i gives the complex amplitude of the oscillation for each coordinate η_i, the factor C being introduced for convenience as a scale factor, the same for all coordinates. It is understood, of course, that it is the real part of Eq. (6–9) that is to correspond to the actual motion. Substitution of the trial solution (6–9) into the equations of motion leads to the following equations for the amplitude factors:

$$(V_{ij}a_j - \omega^2 T_{ij}a_j) = 0. \tag{6–10}$$

Equations (6–10) constitute n linear homogeneous equations for the a_i's, and consequently can have a solution only if the determinant of the coefficients vanishes:

$$\begin{vmatrix} V_{11} - \omega^2 T_{11} & V_{12} - \omega^2 T_{12} & \cdots \\ V_{21} - \omega^2 T_{21} & V_{22} - \omega^2 T_{22} & \\ V_{31} - \omega^2 T_{31} & & \\ \vdots & & \end{vmatrix} = 0. \tag{6–11}$$

This determinantal condition is in effect an algebraic equation of the nth degree for ω^2, and the roots of the determinant provide the frequencies for which Eq. (6–9) represents a correct solution to the equations of the motion. For each of these values of ω^2 the equations (6–10) may be solved for the amplitudes of a_i, or more precisely, for $n - 1$ of the amplitudes in terms of the remaining a_i.

All this has a familiar ring, and we may obtain the proper mathematical perspective by briefly considering a simple variant of the general problem. Suppose the appropriate generalized coordinates were the Cartesian coordinates of the system particles. The kinetic energy then contains only the squares of the velocity components. By introducing generalized coordinates that are the Cartesian components multiplied by the square root of the particle mass,* the kinetic energy can be put in the form

$$T = \frac{\dot{\eta}_i \dot{\eta}_i}{2}, \tag{6–12}$$

---

* Sometimes referred to as *mass-weighted coordinates*.

so that in this case $T_{ij} = \delta_{ij}$. If ω^2 is denoted by λ, the homogeneous equations (6–10) simplify to

$$V_{ij}a_j = \lambda a_i. \qquad (6\text{–}13)$$

But this is precisely the formulation of the eigenvalue problem familiar to us from Chapters 4 and 5; the only difference is that the vector space has n dimensions rather than three. Considering V_{ij} as an element of an $n \times n$ matrix V and a_i as the component of an n-dimensional vector a, Eq. (6–13) can be put in the form

$$Va = \lambda a, \qquad (6\text{–}14)$$

which resembles the eigenvalue equation (4–79). Under these conditions the determinantal equation (6–11) similarly reduces to the secular equation for the eigenvalues λ.

Since V is symmetrical and real, the corresponding eigenvalues are real (cf. Section 5–4). If the n sets of the a_i's corresponding to the n eigenvalues are formed into a matrix A, then, as in Section 4–6, A must diagonalize V by means of a similarity transformation. Further, the n eigenvectors a are orthogonal to each other (Section 5–4) and the diagonalizing matrix A must therefore be orthogonal.

These conclusions are valid beyond the special case in which T_{ij} is diagonal; similar results can be proved for the general problem. Equations (6–10) do represent a type of eigenvalue equation, for writing T_{ij} as an element of the matrix T, the equations may be written

$$Va = \lambda Ta. \qquad (6\text{–}15)$$

Here the effect of V on the eigenvector a is not merely to reproduce the vector times the factor λ, as in the ordinary eigenvalue problem. Instead, the eigenvector is such that V acting on a produces a multiple of the result of T acting on a. We shall show that the eigenvalues λ for which Eq. (6–15) can be satisfied are all real in consequence of the hermitean property of T and V, and, in fact, must be positive. It will also be shown that the eigenvectors a are orthogonal—in a sense. In addition, the matrix of the eigenvectors, A, diagonalizes *both* T and V, the former to the unit matrix 1 and the latter to a matrix whose diagonal elements are the eigenvalues λ.

Proceeding as in Section (5–4), let a_k be a column matrix representing the kth eigenvector, satisfying the eigenvalue equation*

$$Va_k = \lambda_k Ta_k. \qquad (6\text{–}16)$$

The adjoint equation, i.e., the transposed complex conjugate equation, for λ_l has the form

$$a_l^\dagger V = \lambda_l^* a_l^\dagger T. \qquad (6\text{–}17)$$

---

* It hardly need be added that there is *no* summation over k in Eq. (6–16). Indeed, in this chapter the summation convention will apply only to the components of matrices or tensors (of any rank) and *not* to the matrices and tensors themselves.

Here $\mathbf{a}_l^\dagger$ stands for the adjoint vector—the complex conjugate row matrix—and explicit use has been made of the fact that the $\mathbf{V}$ and $\mathbf{T}$ matrices are real and symmetric, in other words hermitean. Multiply Eq. (6–17) from the right by $\mathbf{a}_k$ and subtract the result from the similar product of Eq. (6–16) from the left with $\mathbf{a}_l$. The left-hand side of the difference equation vanishes, leaving only

$$0 = (\lambda_k - \lambda_l^*)\mathbf{a}_l^\dagger \mathbf{T}\mathbf{a}_k. \tag{6–18}$$

When $l = k$, Eq. (6–18) takes on the special form

$$(\lambda_k - \lambda_k^*)\mathbf{a}_k^\dagger \mathbf{T}\mathbf{a}_k = 0. \tag{6–19}$$

That the matrix product in Eq. (6–19) is real can be shown immediately by taking its complex conjugate and using the hermitean property of $\mathbf{T}$. However, we want to prove that the matrix product is not only real but is positive definite. For this purpose separate $\mathbf{a}_k$ into its real and imaginary components:

$$\mathbf{a}_k = \mathbf{\alpha}_k + i\mathbf{\beta}_k.$$

The matrix product can then be written as

$$\mathbf{a}_k^\dagger \mathbf{T}\mathbf{a}_k = \tilde{\mathbf{\alpha}}_k \mathbf{T}\mathbf{\alpha}_k + \tilde{\mathbf{\beta}}_k \mathbf{T}\mathbf{\beta}_k + i(\tilde{\mathbf{\alpha}}_k \mathbf{T}\mathbf{\beta}_k - \tilde{\mathbf{\beta}}_k \mathbf{T}\mathbf{\alpha}_k). \tag{6–20}$$

The imaginary term vanishes by virtue of the symmetry of $\mathbf{T}$ and therefore, as has been said, the matrix product is real. Further, it may be noted that the kinetic energy in Eq. (6–6) can be rewritten in terms of a column matrix $\dot{\mathbf{\eta}}$ as

$$T = \frac{1}{2}\tilde{\dot{\mathbf{\eta}}}\mathbf{T}\dot{\mathbf{\eta}}. \tag{6–6'}$$

Hence the first two terms in Eq. (6–20) are twice the kinetic energies when the velocity matrix $\dot{\mathbf{\eta}}_k$ has the values $\mathbf{\alpha}_k$ and $\mathbf{\beta}_k$ respectively. Now, a kinetic energy by its physical nature must be positive definite for real velocities, and therefore the matrix product in Eq. (6–19) cannot be zero. It follows that the eigenvalues λ_k must be real.

Since the eigenvalues are real, the ratios of the eigenvector components a_{jk} determined by Eqs. (6–16) must all be real. There is still some indeterminateness, of course, since the value of a particular one of the a_{jk}'s can still be chosen at will without violating Eqs. (6–16). We can require, however, that this component shall be real, and the reality of λ_k then ensures the reality of all the other components. (Any complex phase factor in the amplitude of the oscillation will be thrown into the factor C, Eq. (6–9).) Multiply now Eq. (6–16) by $\tilde{\mathbf{a}}_k$ from the left and solve for λ_k:

$$\lambda_k = \frac{\tilde{\mathbf{a}}_k \mathbf{V}\mathbf{a}_k}{\tilde{\mathbf{a}}_k \mathbf{T}\mathbf{a}_k}. \tag{6–21}$$

The denominator of this expression is equal to twice the kinetic energy for velocities a_{ik} and since the eigenvectors are real the sum must be positive definite. Similarly, the numerator is the potential energy for coordinates a_{ik}, and the condition that V be a minimum at equilibrium requires that the sum must be positive or zero. Neither numerator nor denominator can be negative, and the denominator cannot be zero, hence λ is always finite and positive. (It may, however, be zero.) It will be remembered that λ stands for ω^2, so that positive λ corresponds to real frequencies of oscillation. Were the potential not a local minimum, the numerator in Eq. (6–21) might be negative, giving rise to imaginary frequencies that would produce an unbounded exponential increase of the η_i with time. Such motion would obviously be unstable, and we have here the promised mathematical proof that a minimum of the potential is required for stable motion.

Let us return for the moment to Eq. (6–18) which, in view of the reality of the eigenvalues and eigenvectors, can be written

$$(\lambda_k - \lambda_l)\tilde{\mathbf{a}}_l \mathbf{T} \mathbf{a}_k = 0. \tag{6–18'}$$

If all the roots of the secular equation are distinct, then Eq. (6–18') can hold only if the matrix product vanishes for l not equal to k:

$$\tilde{\mathbf{a}}_l \mathbf{T} \mathbf{a}_k = 0, \qquad l \neq k. \tag{6–22a}$$

It has been remarked several times that the values of the a_{jk}'s are not completely fixed by the eigenvalue equations (6–10). We can remove this indeterminacy by requiring further that

$$\tilde{\mathbf{a}}_k \mathbf{T} \mathbf{a}_k = 1. \tag{6–22b}$$

There are n such equations (6–22) and they uniquely fix the one arbitrary component of each of the n eigenvectors $\mathbf{a}_k$.* If we form all the eigenvectors $\mathbf{a}_k$ into a square matrix $\mathbf{A}$ with components a_{jk} (cf. Section 4–6) then the two equations (6–22a and b) can be combined into one matrix equation:

$$\tilde{\mathbf{A}} \mathbf{T} \mathbf{A} = \mathbf{1}. \tag{6–23}$$

When two or more of the roots are repeated, the argument leading to Eq. (6–22a) falls through for $\lambda_l = \lambda_k$. We shall reserve a discussion of this exceptional case of *degeneracy* for a later time.† Suffice it for the present to state that a set of a_{jk} coefficients can always be found that satisfies both the eigenvalue conditions Eqs. (6–10), and Eq. (6–22a), so that Eq. (6–23) always holds.

---

* Equation (6–22b) may be put in a form that explicitly shows that it suffices to remove the indeterminacy in the a_{jk}'s. Suppose it is the magnitude of a_{1k} that is to be evaluated; the ratio of all the other a_{jk}'s to a_{1k} is obtained from Eqs. (6–10). Then Eq. (6–22b) can be written as

$$\sum_{i,j} T_{ij} \frac{a_{ik}}{a_{1k}} \frac{a_{jk}}{a_{1k}} = \frac{1}{a_{1k}^2}.$$

The left-hand side is completely determined from the eigenvalue equations and may be evaluated directly to provide a_{1k}.

† The usage here for the word "degeneracy" (to denote multiple roots of the secular equation) differs from meaning given in later chapters, particularly Chapter 10.

Equation (6–23) is reminiscent of the condition that a matrix **B** be orthogonal (cf. Eq. 4–36):

$$\tilde{B}B = 1.$$

To see the relation between the condition on **A** and the orthogonality condition, let us note that the difference equation (6–19) is analogous to the corresponding equation for the eigenvectors of the inertia tensor (cf. Eq. 5–29):

$$(I_j - I_l)R_l \cdot R_j = 0.$$

From this relation we deduced that distinct eigenvectors are orthogonal:

$$R_l \cdot R_j = 0.$$

If we had decided to normalize the length of the eigenvectors;

$$R_j \cdot R_j = 1$$

(analogous to Eq. 6–22b), the orthogonality and normalization conditions could be summarized by stating that the matrix **X** containing all the eigenvectors (cf. footnote, p. 199) had to be orthogonal:

$$\tilde{X}X = 1.$$

In like manner Eq. (6–23) is equivalent to the condition that the eigenvectors a_j be orthogonal and of unit magnitude—but in a particular Riemannian space that is not necessarily Cartesian.

A *Riemannian space* is defined such that the element of path length ds is given by

$$ds^2 = g_{ik}dx_i\,dx_k, \tag{6–24}$$

where g_{ik} is the element of the *metric tensor* **G** of the space. Correspondingly, the dot product of two vectors **x**, **y** in such a space is

$$x \cdot y = x_i g_{ik} y_k = \tilde{x}Gy, \tag{6–25}$$

so that the square of the magnitude of a vector **x** is

$$x \cdot x = \tilde{x}Gx. \tag{6–26}$$

Clearly for Cartesian coordinates the metric tensor is the unit matrix **1**. In general, **G** will be diagonal for curvilinear orthogonal coordinates (cf. the example of three-dimensional spherical coordinates).

The kinetic energy, Eq. (6–6'), can now be described as such that $2T$ is the square of the magnitude of the velocity vector *in a configuration space for which* **T** *is the metric tensor*. Similarly Eq. (6–23), or its component parts Eqs. (6–22a, b), say that the eigenvectors a_k are orthogonal in such a space and that they have been normalized to have unit magnitude in the space.

In Chapter 4 the *similarity* transformation of a matrix **C** by a matrix **B** was defined by the equation (cf. Eq. 4–41):

$$\mathbf{C}' = \mathbf{BCB}^{-1}.$$

We now introduce the related concept of the *congruence* transformation of **C** by **A** according to the relation

$$\mathbf{C}' = \tilde{\mathbf{A}}\mathbf{C}\mathbf{A}. \tag{6–27}$$

If **A** is orthogonal, so that $\tilde{\mathbf{A}} = \mathbf{A}^{-1}$, there is no essential difference between the two types of transformation (as may be seen by denoting $\mathbf{A}^{-1}$ by the matrix **E**). Equation (6–23) can therefore be read as the statement that **A** transforms **T** by a congruence transformation into a diagonal matrix, in particular into the unit matrix.

If a diagonal matrix λ with elements $\lambda_{lk} = \lambda_k \delta_{lk}$ be introduced, the eigenvalue equations (6–16) may be written

$$V_{ij}a_{jk} = T_{ij}a_{jl}\lambda_{lk},$$

which becomes in matrix notation

$$\mathbf{VA} = \mathbf{TA}\lambda. \tag{6–28}$$

Multiplying by $\tilde{\mathbf{A}}$ from the left, Eq. (6–28) takes the form

$$\tilde{\mathbf{A}}\mathbf{VA} = \tilde{\mathbf{A}}\mathbf{TA}\lambda,$$

which by Eq. (6–23) reduces to

$$\tilde{\mathbf{A}}\mathbf{VA} = \lambda. \tag{6–29}$$

Our final equation (6–29) states that a congruence transformation of **V** by **A** changes it into a diagonal matrix whose elements are the eigenvalues λ_k.

The matrix **A** thus simultaneously diagonalizes both **T** and **V**. Remembering the interpretation of **T** as a metric tensor in configuration space, we can give the following meaning to the diagonalization process. **A** is the matrix of a linear transformation from a system of *inclined* axes to *Cartesian orthogonal* axes, as evidenced by the fact that the transformed metric tensor is **1**. At the same time, the new axes are the orthogonal *principal axes* of **V**, so that the matrix **V** is diagonal in the transformed coordinate system. The entire process of obtaining the fundamental frequencies of small oscillation is thus a particular type of *principal axis transformation*, such as was discussed in Chapter 5.*

It remains only to consider the case of multiple roots to the secular equation, a situation that is more annoying in the mathematical theory than it is in practice. If one or more of the roots is repeated, it is found that the number of independent equations among the eigenvalues is insufficient to determine even the ratio of the

---

* An alternative geometric interpretation of the principal axis transformation will be given in connection with normal coordinates in Section 6–3.

eigenvector components. Thus, if the eigenvalue λ is a double root, any two of the components a_j may be chosen arbitrarily, the rest being fixed by the eigenvalue equations. To illustrate, let us consider a two-dimensional system, in which the secular equation appears as

$$\begin{vmatrix} V_{11} - \lambda T_{11} & V_{12} - \lambda T_{12} \\ V_{12} - \lambda T_{12} & V_{22} - \lambda T_{22} \end{vmatrix} = 0,$$

or

$$(V_{12} - \lambda T_{12})^2 - (V_{11} - \lambda T_{11})(V_{22} - \lambda T_{22}) = 0.$$

Suppose now that the matrix elements are such that

$$\frac{V_{12}}{T_{12}} = \frac{V_{11}}{T_{11}} = \frac{V_{22}}{T_{22}} = \lambda_0. \tag{6-30}$$

Then the secular equation can be written

$$(T_{12}^2 - T_{11}T_{22})(\lambda_0 - \lambda)^2 = 0,$$

indicating that λ_0 is a double root of the secular equation. But the eigenvalue equations (6–10) for this root are

$$(V_{11} - \lambda_0 T_{11})a_1 + (V_{12} - \lambda_0 T_{12})a_2 = 0,$$

$$(V_{12} - \lambda_0 T_{12})a_1 + (V_{22} - \lambda_0 T_{22})a_2 = 0,$$

and by virtue of the conditions (6–30) all of the coefficients of the a's vanish identically. Any set of values for the two a's will then satisfy the eigenvalue equations. Even with the normalization requirement (6–22b) there will thus be a single infinity of eigenvectors corresponding to a double root of the secular equation, a double infinity for a triple root, and so on.

In general, any pair of eigenvectors randomly chosen out of the infinite set of allowed vectors will not be orthogonal. Nevertheless, it is always possible to construct a pair of allowed vectors that are orthogonal, and these can be used to form the orthogonal matrix $\mathbf{A}$. Consider for simplicity the procedure to be followed for a double root. Let $\mathbf{a}_k'$ and $\mathbf{a}_l'$ be any two allowable eigenvectors for a given double root λ, which have been normalized so as to satisfy Eq. (6–22b). Any linear combination of $\mathbf{a}_k'$ and $\mathbf{a}_l'$ will also be an eigenvector for the root λ. We therefore seek to construct a vector $\mathbf{a}_l$,

$$\mathbf{a}_l = c_1 \mathbf{a}_k' + c_2 \mathbf{a}_l', \tag{6-31}$$

where c_1 and c_2 are constants such that $\mathbf{a}_l$ is orthogonal to $\mathbf{a}_k'$. The orthogonality condition, Eq. (6–22a), then requires that

$$\tilde{\mathbf{a}}_l \mathbf{T} \mathbf{a}_k' = c_1 + c_2 \tilde{\mathbf{a}}_l' \mathbf{T} \mathbf{a}_k' = 0,$$

where use has been made of the normalization of $\mathbf{a}_k'$. It therefore follows that the ratio of c_1 to c_2 must be given by

$$\frac{c_1}{c_2} = -\tilde{\mathbf{a}}_l' \mathbf{T} \mathbf{a}_k' \equiv -\tau_l. \tag{6-32}$$

In addition, the requirement that $\mathbf{a}_l$ be normalized provides another condition on the two coefficients, which in terms of τ_l defined by Eq. (6–32) takes the form

$$\tilde{\mathbf{a}}_l \mathbf{T} \mathbf{a}_l = 1 = c_1^2 + c_2^2 + 2c_1 c_2 \tau_l. \tag{6–33}$$

Together the two equations fix the coefficients c_1 and c_2, and therefore the vector $\mathbf{a}_l$. Both $\mathbf{a}_l$ and $\mathbf{a}_k \equiv \mathbf{a}_k'$ are automatically orthogonal to the eigenvectors of the other distinct eigenvalues, for then the argument based on Eq. (6–18') remains valid. Hence we have a set of n eigenvectors $\mathbf{a}_j$ whose components form the matrix $\mathbf{A}$ satisfying Eq. (6–23).

A similar procedure is followed for a root of higher multiplicity. If λ is an m-fold root, then orthogonal normalized eigenvectors are formed out of linear combinations of any of the m corresponding eigenvectors $\mathbf{a}_1', \ldots, \mathbf{a}_m'$. The first of the "ortho-normal" eigenvectors $\mathbf{a}_1$ is then chosen as a multiple of $\mathbf{a}_1'$; $\mathbf{a}_2$ is taken as a linear combination of $\mathbf{a}_1'$ and $\mathbf{a}_2'$; and so on. In this manner the number of constants to be determined is equal to the sum of the first m integers, or $\frac{1}{2}m(m + 1)$. The normalization requirements provide m conditions, while there are $\frac{1}{2}m(m - 1)$ orthogonality conditions, and together these are just enough to fix the constants uniquely.

This process of constructing orthogonalized eigenvectors in the case of multiple roots is completely analogous to the Gram–Schmidt method of constructing a sequence of orthogonal functions out of any arbitrary set of functions. Phrased in geometrical language, it is also seen to be identical with the procedure followed in Chapter 5 for multiple eigenvalues of the inertia tensor. For example, the added indeterminacy in the eigenvector components for a double root means that all of the vectors in a *plane* are eigenvectors. We merely choose any two perpendicular directions in the plane as being the new principal axes, with the eigenvectors in $\mathbf{A}$ as unit vectors along these axes.*

6–3 FREQUENCIES OF FREE VIBRATION, AND NORMAL COORDINATES

The somewhat lengthy arguments of the preceding section demonstrate that the equations of motion will be satisfied by an oscillatory solution of the form (6–9) not merely for one frequency but in general for a set of n frequencies ω_k. A complete solution of the equations of motion therefore involves a superposition of oscillations with all the allowed frequencies. Thus, if the system is displaced slightly from equilibrium and then released, the system performs small oscillations about the equilibrium with the frequencies $\omega_1, \ldots, \omega_n$. The solutions of the secular equation are therefore often designated as the frequencies of *free vibration* or as the *resonant frequencies* of the system.

---

* See also footnote on p. 200.

The general solution of the equations of motion may now be written as a summation over an index k:

$$\eta_i = C_k a_{ik} e^{-i\omega_k t}, \qquad (6\text{-}34)$$

there being a complex scale factor C_k for each resonant frequency. It might be objected that for each solution λ_k of the secular equation there are two resonant frequencies $+\omega_k$ and $-\omega_k$. The eigenvector $\mathbf{a}_k$ would be the same for the two frequencies, but the scale factors C_k^+ and C_k^- could conceivably be different. On this basis the general solution should appear as

$$\eta_i = a_{ik}(C_k^+ e^{+i\omega_k t} + C_k^- e^{-i\omega_k t}). \qquad (6\text{-}35)$$

It is to be remembered, however, that the actual motion is the real part of the complex solution, and the real part of either (6–34) or (6–35) can be written in the form

$$\eta_i = f_k a_{ik} \cos(\omega_k t + \delta_k), \qquad (6\text{-}36)$$

where the amplitude f_k and the phase δ_k are determined from the initial conditions. Either of the solutions (6–34 and 6–35) will therefore represent the actual motion, and the former, of course, is the more convenient.

The orthogonality properties of $\mathbf{A}$ greatly facilitate the determination of the scale factors C_k in terms of the initial conditions. At $t = 0$ the real part of Eq. (6–34) reduces to

$$\eta_i(0) = \operatorname{Re} C_k a_{ik}, \qquad (6\text{-}37)$$

where Re stands for "real part of." Similarly, the initial value of the velocities is obtained as

$$\dot{\eta}_i(0) = \operatorname{Im} C_k a_{ik} \omega_k, \qquad (6\text{-}38)$$

where Im C_k denotes the imaginary part of C_k. From these $2n$ equations the real and imaginary parts of the n constants C_k may be evaluated. To solve Eq. (6–37), for example, let us first write it in terms of column matrices $\mathbf{\eta}(0)$ and $\mathbf{C}$:

$$\mathbf{\eta}(0) = \mathbf{A} \operatorname{Re} \mathbf{C}. \qquad (6\text{-}37')$$

If we multiply by $\mathbf{\tilde{A}T}$ from the left and use Eq. (6–23), we immediately obtain a solution for Re $\mathbf{C}$:

$$\operatorname{Re} \mathbf{C} = \mathbf{\tilde{A}T}\mathbf{\eta}(0),$$

or, taking the lth component,

$$\operatorname{Re} C_l = a_{jl} T_{jk} \eta_k(0), \qquad (6\text{-}39)$$

A similar procedure leads to the imaginary part of the scale factors as*

$$\operatorname{Im} C_l = \frac{1}{\omega_l} \sum_{j,k} a_{jl} T_{jk} \dot{\eta}_k(0). \qquad (6\text{-}40)$$

---

*The summation over j and k is shown explicitly because there is no summation over the repeated subscript l.

Equations (6–39) and (6–40) thus permit the direct computation of the complex factors C_l (and therefore the amplitudes and phases) in terms of the initial conditions and the matrices **T** and **A**.

The solution for each coordinate, Eq. (6–34), is in general a sum of simple harmonic oscillations in all of the frequencies ω_k satisfying the secular equation. Unless it happens that all of the frequencies are commensurable, i.e., rational fractions of each other, η_i never repeats its initial value and is therefore not itself a periodic function of time. However it is possible to transform from the η_i to a new set of generalized coordinates that are all simple periodic functions of time—a set of variables known as the *normal coordinates*.

We define a new set of coordinates ζ_j related to the original coordinates η_i by the equations

$$\eta_i = a_{ij}\zeta_j, \tag{6–41}$$

or, in terms of the single-column matrices **η** and **ζ**,

$$\boldsymbol{\eta} = \mathbf{A}\boldsymbol{\zeta}. \tag{6–41'}$$

The potential energy, Eq. (6–4), is written in matrix notation as

$$V = \frac{1}{2}\tilde{\boldsymbol{\eta}}\mathbf{V}\boldsymbol{\eta}. \tag{6–42}$$

Now, the single-row transpose matrix $\tilde{\boldsymbol{\eta}}$ is related to $\boldsymbol{\zeta}$ by the equation

$$\tilde{\boldsymbol{\eta}} = \widetilde{\mathbf{A}\boldsymbol{\zeta}} = \boldsymbol{\zeta}\tilde{\mathbf{A}},$$

so that the potential energy can be written also as

$$V = \frac{1}{2}\boldsymbol{\zeta}\tilde{\mathbf{A}}\mathbf{V}\mathbf{A}\boldsymbol{\zeta}.$$

But **A** diagonalizes **V** by a congruence transformation (cf. Eq. 6–29), and the potential therefore reduces simply to

$$V = \frac{1}{2}\boldsymbol{\zeta}\boldsymbol{\lambda}\boldsymbol{\zeta} = \frac{1}{2}\omega_k^2\zeta_k^2. \tag{6–43}$$

The kinetic energy has an even simpler form in the new coordinates. Since the velocities transform as the coordinates, T as given in Eq. (6–6') transforms to

$$T = \frac{1}{2}\dot{\tilde{\boldsymbol{\zeta}}}\tilde{\mathbf{A}}\mathbf{T}\mathbf{A}\dot{\boldsymbol{\zeta}},$$

which by virtue of Eq. (6–23) reduces to

$$T = \frac{1}{2}\dot{\tilde{\boldsymbol{\zeta}}}\dot{\boldsymbol{\zeta}} = \frac{1}{2}\dot{\zeta}_i\dot{\zeta}_i. \tag{6–44}$$

Equations (6–43) and (6–44) state that in the new coordinates both the potential and kinetic energies are sums of squares only, without any cross terms. Of course, this result is simply another way of saying that **A** produces a principal axis transformation. It will be remembered that the principal axis transformation of the inertia tensor was specifically designed to reduce the moment of inertia to a sum of squares; the new axes being the principal axes of the inertia ellipsoid. Here the kinetic and potential energies are also quadratic forms (as was the moment of inertia) and both are diagonalized by **A**. For this reason the principal axis transformation employed here is a particular example of the well-known algebraic process of the *simultaneous diagonalization of two quadratic forms.*

There is another way of looking at the principal axis transformation of T and V, which is closer in language to the process of diagonalizing the inertia tensor as described in Chapter 5. It does not provide any simplification of the computational process, but it does help to explain why it is possible to diagonalize two quadratic forms simultaneously—and why one cannot in general do it with three quadratic forms. The matrix **T** is real and symmetric, just as is the inertia tensor **I**. If we now consider the η space to be a Cartesian space of n dimensions, it is therefore possible to find a real orthogonal transformation **B** to a new system of Cartesian coordinates,

$$\mathsf{y} = \mathsf{B}\boldsymbol{\eta},$$

in which T is diagonal. The matrix **B** must transform **T** by a similarity transformation to a diagonal matrix **C**:

$$\mathsf{BTB}^{-1} = \mathsf{BT}\tilde{\mathsf{B}} = \mathsf{C}.$$

Since the inverse transformation is

$$\boldsymbol{\eta} = \tilde{\mathsf{B}}\mathsf{y}, \qquad \tilde{\boldsymbol{\eta}} = \tilde{\mathsf{y}}\mathsf{B},$$

the kinetic energy transforms as

$$2T = \tilde{\boldsymbol{\eta}}\mathsf{T}\dot{\boldsymbol{\eta}} = \tilde{\dot{\mathsf{y}}}\mathsf{BT}\tilde{\mathsf{B}}\dot{\mathsf{y}} = \tilde{\dot{\mathsf{y}}}\mathsf{C}\dot{\mathsf{y}} = C_i\dot{y}_i^2.$$

As expected the new axes are the principal axes of the kinetic energy ellipsoid. Since the kinetic energy can never be zero for any finite velocities, i.e., is positive definite, the principal values C_i are always greater than zero. It is therefore always possible to introduce new coordinates z_i defined by the relations

$$z_i = y_i\sqrt{C_i}, \qquad \text{(no summation)}. \tag{6–45}$$

In terms of these new coordinates the kinetic energy becomes

$$2T = \dot{z}_i\dot{z}_i = \tilde{\dot{\mathsf{z}}}\mathsf{1}\dot{\mathsf{z}}.$$

The coordinate transformation, Eq. (6–45), does not involve any rotation of the axes; it is solely a change of scale along each of the y_i axes. What has been done in effect is to stretch or compress each of the principal axes until the kinetic energy ellipsoid becomes a sphere!

The two successive coordinate transformations do not, in general, diagonalize **V**, but the potential energy will have the form

$$2V = \tilde{z}\mathbf{D}z,$$

where **D** is a symmetric real matrix. It is therefore possible to find a third, final, coordinate transformation by a real orthogonal matrix **F**:

$$\zeta = \mathbf{F}z,$$

which diagonalizes **D** by a similarity transformation to a diagonal matrix λ so that

$$2V = \tilde{\zeta}\lambda\zeta = \lambda_i\zeta_i^2.$$

This final rotation of coordinates does not effect the form of the kinetic energy, because a sphere is always diagonal in a rotated system. The trick, thus, to achieve simultaneous diagonalization of two quadratic forms, one positive definite, is to find a coordinate system in which the positive definite form defines a sphere, so that all directions are principal axes for it. It's clear now why in general three quadratic forms cannot be simultaneously diagonalized. The final rotation can be used to diagonalize one of them, but it will normally leave the remaining quadratic form still in a nondiagonal form. This regrettable circumstance will be of importance when we consider the effects of dissipation.

To return to the properties of the ζ_i system of coordinates, it will be noted that the equations of motion share in the simplification resulting from their use. The new Lagrangian is

$$L = \frac{1}{2}(\dot{\zeta}_k\dot{\zeta}_k - \omega_k^2\zeta_k^2) \tag{6–46}$$

so that the Lagrange equations for ζ_k are

$$\ddot{\zeta}_k + \omega_k^2\zeta_k = 0. \tag{6–47}$$

Equations (6–47) have the immediate solutions

$$\zeta_k = C_k e^{-i\omega_k t}, \tag{6–48}$$

which could have been seen, of course, directly from Eqs. (6–34) and (6–41). Each of the new coordinates is thus a simply periodic function involving only *one* of the resonant frequencies. As has been mentioned, it is therefore customary to call the ζ's the *normal coordinates* of the system.

Each normal coordinate corresponds to a vibration of the system with only one frequency, and these component oscillations are spoken of as the *normal modes of vibration*. All of the particles in each mode vibrate with the same frequency and with the same phase;* the relative amplitudes being determined by the matrix elements a_{ik}. The complete motion is then built up out of the sum of the

---

* Particles may be exactly out of phase if the a's have opposite sign.

normal modes weighted with appropriate amplitude and phase factors contained in the C_k's.

Harmonics of the fundamental frequencies are absent in the complete motion essentially because of the stipulation that the amplitude of oscillation be small. We are then allowed to represent the potential as a quadratic form, which is characteristic of simple harmonic motion. The normal coordinate transformation emphasizes this point, for the Lagrangian in the normal coordinates (6–46) is seen to be the sum of Lagrangians for harmonic oscillators of frequencies ω_k. We can thus consider the complete motion for small oscillations as being obtained by exciting the various harmonic oscillators with different intensities and phases.*

6–4 FREE VIBRATIONS OF A LINEAR TRIATOMIC MOLECULE

To illustrate the technique for obtaining the resonant frequencies and normal modes, we shall consider in detail a model based on a linear symmetrical triatomic molecule. In the equilibrium configuration of the molecule two atoms of mass m are symmetrically located on each side of an atom of mass M (cf. Fig. 6–2). All three atoms are on one straight line, the equilibrium distances apart being denoted by b. For simplicity we shall first consider only vibrations along the line of the molecule, and the actual complicated interatomic potential will be approximated by two springs of force constant k joining the three atoms. There are three obvious coordinates marking the position of the three atoms on the line. In these coordinates the potential energy is

$$V = \frac{k}{2}(x_2 - x_1 - b)^2 + \frac{k}{2}(x_3 - x_2 - b)^2.$$

We now introduce coordinates relative to the equilibrium positions:

$$\eta_i = x_i - x_{0i},$$

where

$$x_{02} - x_{01} = b = x_{03} - x_{02}.$$

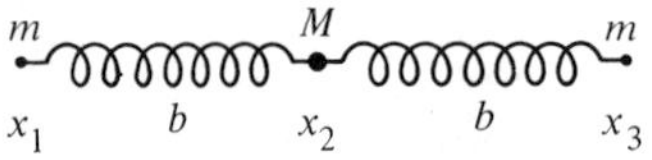

$m \quad\quad M \quad\quad\quad m$
$x_1 \quad b \quad x_2 \quad b \quad x_3$

FIGURE 6–2
Model of a linear symmetrical triatomic molecule.

---

* It might be mentioned for future reference that the same sort of picture appears in the quantization of the electromagnetic field. The frequencies of the harmonic oscillators are identified with the photon frequencies, and the amplitudes of excitation become the discrete quantized "occupation numbers"—the number of photons of each frequency.

The potential energy then reduces to

$$V = \frac{k}{2}(\eta_2 - \eta_1)^2 + \frac{k}{2}(\eta_3 - \eta_2)^2,$$

expand.

or

$$V = \frac{k}{2}(\eta_1^2 + 2\eta_2^2 + \eta_3^2 - 2\eta_1\eta_2 - 2\eta_2\eta_3). \tag{6-49}$$

Hence the **V** matrix has the form

$$\mathbf{V} = \begin{pmatrix} k & -k & 0 \\ -k & 2k & -k \\ 0 & -k & k \end{pmatrix}. \tag{6-50}$$

The kinetic energy has an even simpler form:

$$T = \frac{m}{2}(\dot{\eta}_1^2 + \dot{\eta}_3^2) + \frac{M}{2}\dot{\eta}_2^2, \tag{6-51}$$

so that the **T** matrix is diagonal:

$$\mathbf{T} = \begin{pmatrix} m & 0 & 0 \\ 0 & M & 0 \\ 0 & 0 & m \end{pmatrix}. \tag{6-52}$$

$$(\mathbf{V} - \omega^2\mathbf{T})\,a = 0$$

Combining these two matrices, the secular equation appears as

not diag.

$$|\mathbf{V} - \omega^2\mathbf{T}| = \begin{pmatrix} k - \omega^2 m & -k & 0 \\ -k & 2k - \omega^2 M & -k \\ 0 & -k & k - \omega^2 m \end{pmatrix} = 0. \tag{6-53}$$

Direct evaluation of the determinant leads to the cubic equation in ω^2:

$$\omega^2(k - \omega^2 m)(k(M + 2m) - \omega^2 Mm) = 0, \tag{6-54}$$

with the obvious solutions

$$\omega_1 = 0, \qquad \omega_2 = \sqrt{\frac{k}{m}}, \qquad \omega_3 = \sqrt{\frac{k}{m}\left(1 + \frac{2m}{M}\right)}. \tag{6-55}$$

The first eigenvalue, $\omega_1 = 0$, may appear somewhat surprising and even alarming at first sight. Such a solution does not correspond to an oscillatory motion at all, for the equation of motion for the corresponding normal coordinate is

$$\ddot{\zeta}_1 = 0,$$

which produces a uniform translational motion. But this is precisely the key to the difficulty. The vanishing frequency arises from the fact that the molecule may be translated rigidly along its axis without any change in the potential energy, an example of neutral equilibrium mentioned previously. Since the restoring force against such motion is zero, the effective "frequency" must also vanish. We have made the assumption that the molecule has three degrees of freedom for vibrational motion, whereas in reality one of them is a rigid body degree of freedom.

A number of interesting points can be discussed in connection with a vanishing resonant frequency. It is seen from Eq. (6–21) that a zero value of ω can occur only when the potential energy is positive but is not positive definite, i.e., it can vanish even when not all the η_i's are zero. An examination of V, Eq. (6–49), shows that it is not positive definite and that, in fact, V does vanish when all the η's are equal (uniform translation).

Since the zero frequency found here is of no consequence for the vibration frequencies of interest, it is often desirable to phrase the problem so that the root is eliminated from the outset. We can do this here most simply by imposing the condition or constraint that the center of mass remain stationary at the origin:

$$m(x_1 + x_3) + Mx_2 = 0. \qquad (6\text{–}56)$$

Equation (6–56) can then be used to eliminate one of the coordinates from V and T, reducing the problem to one of two degrees of freedom (cf. Exercise 3, this chapter).

The restriction of the motion to be along the molecular axis allows only one possible type of uniform rigid body motion. However, if the more general problem of vibrations in all three directions is considered, the number of rigid body degrees of freedom will be increased, in general, to six. The molecule may then translate uniformly along the three axes or perform uniform rotations about the axes. Hence in any general system of n degrees of freedom there will be six vanishing frequencies and only $n - 6$ true vibration frequencies. Again, the reduction in the number of degrees of freedom can be performed beforehand by imposing the conservation of linear and angular momentum upon the coordinates.

In addition to rigid body motion, it has been pointed out that zero resonant frequencies may also arise when the potential is such that both the first *and* second derivatives of V vanish at equilibrium. Small oscillations may still be possible in this case if the fourth derivatives do not also vanish (the third derivatives must vanish for a stable equilibrium), but the vibrations will not be simple harmonic. Such a situation therefore constitutes a breakdown of the customary method of small oscillations, but fortunately it is not of frequent occurrence.

Returning now to the examination of the resonant frequencies, ω_2 will be recognized as the well-known frequency of oscillation for a mass m suspended by a spring of force constant k. We are led to expect, therefore, that only the end atoms partake in this vibration; the center molecule remains stationary. It is only in the third mode of vibration, ω_3, that the mass M can participate in the

$(\bar{V} - \lambda_j T) A$

oscillatory motion. These predictions are verified by examining the eigenvectors for the three normal modes.

The components a_{ij} are determined for each frequency by the equations

$$(k - \omega_j^2 m)a_{1j} \qquad\qquad -ka_{2j} \qquad\qquad = 0,$$
$$-ka_{1j} + (2k - \omega_j^2 M)a_{2j} \qquad\qquad -ka_{3j} = 0, \qquad (6\text{-}57)$$
$$-ka_{2j} + (k - \omega_j^2 m)a_{3j} = 0,$$

$(\bar{V} - \omega^2 T)A = 0$

along with the normalization condition:

$$m(a_{1j}^2 + a_{3j}^2) + Ma_{2j}^2 = 1. \qquad (6\text{-}58)$$

For $\omega_1 = 0$, it follows immediately from the first and third of Eqs. (6-57) that all three coefficients are equal: $a_{11} = a_{21} = a_{31}$. This, of course, is exactly what was expected from the translational nature of the motion (cf. Fig. 6-3a). The normalization condition then fixes the value of a_{1j} so that

$$a_{11} = \frac{1}{\sqrt{2m + M}}, \qquad a_{12} = \frac{1}{\sqrt{2m + M}}, \qquad a_{13} = \frac{1}{\sqrt{2m + M}}. \qquad (6\text{-}59a)$$

The factors $(k - \omega_2^2 m)$ vanish for the second mode, and Eqs. (6-57) show immediately that $a_{22} = 0$ (as predicted) and $a_{12} = -a_{32}$. The numerical value of these quantities is then determined by Eq. (6-58):

$$a_{12} = \frac{1}{\sqrt{2m}}, \qquad a_{22} = 0, \qquad a_{32} = -\frac{1}{\sqrt{2m}}. \qquad (6\text{-}59b)$$

In this mode the center atom is at rest, while the two outer ones vibrate exactly out of phase (as they must in order to conserve linear momentum), cf. Fig. 6-3b. Finally, when $\omega = \omega_3$ it can be seen from the first and third of Eqs. (6-57) that a_{13} and a_{33} must be equal. The rest of the calculation for this mode is not quite as simple as for the others, and it will be sufficient to state the final result:

$$a_{13} = \frac{1}{\sqrt{2m\left(1 + \dfrac{2m}{M}\right)}}, \qquad a_{23} = \frac{-2}{\sqrt{2M\left(2 + \dfrac{M}{m}\right)}}, \qquad a_{33} = \frac{1}{\sqrt{2m\left(1 + \dfrac{2m}{M}\right)}}. \qquad (6\text{-}59c)$$

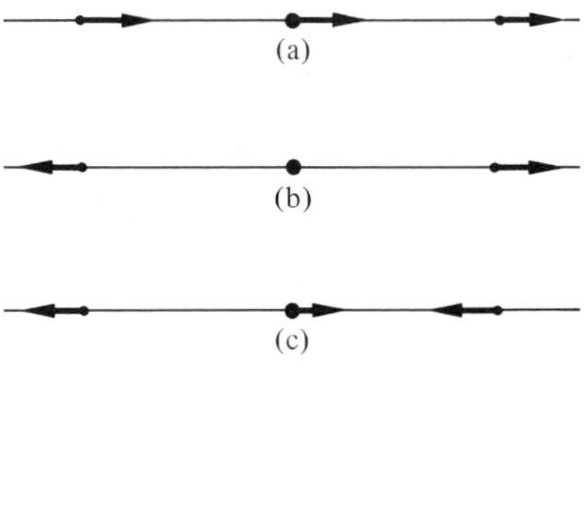

So: A) Determine $V_{ij}^{i} T$
A) obtained ω's
B) Found A
c) find C by 6.39, 40
D) Have sol's 6.34
Everything.

(a)

(b)

(c)

FIGURE 6-3
Longitudinal normal modes of the linear symmetric triatomic molecule.

Here the two outer atoms vibrate with the same amplitude, while the inner one oscillates out of phase with them and has a different amplitude, cf. Fig. 6–3c. Any general longitudinal vibration of the molecule that does not involve a rigid translation will be some linear combination of the normal modes ω_2 and ω_3. The amplitudes of the normal modes, and their phases relative to each other, will of course be determined by the initial conditions (cf. Exercise 5).

We have spoken so far only of vibrations along the axis; in the actual molecule there will also be normal modes of vibration perpendicular to the axis. The complete set of normal modes is naturally more difficult to determine than merely the longitudinal modes, for the general motion in all directions corresponds to nine degrees of freedom. While the procedure is straightforward, the algebra rapidly becomes quite complicated, and it is not feasible to present the detailed calculation here. However, it is possible to give a qualitative discussion on the basis of general principles, and most of the conclusions of the complete solution can be predicted beforehand.

The general problem will have a number of zero resonant frequencies corresponding to the possibility of rigid body motion. For the linear molecule there will be three degrees of freedom for rigid translation, but rigid rotation can account for only *two* degrees of freedom. Rotation about the axis of the molecule is obviously meaningless and will not appear as a mode of rigid body motion. We are therefore left with four true modes of vibration. Two of these are the longitudinal modes, which have already been examined, so that there can only be two modes of vibration perpendicular to the axis. However, the symmetry of the molecule about its axis shows that these two modes of perpendicular vibration must be degenerate. There is nothing to distinguish a vibration in the y direction from a vibration in the z direction, and the two frequencies must be equal. The additional indeterminacy of the eigenvectors of a degenerate mode appears here in that all directions perpendicular to the molecular axis are alike. Any two orthogonal axes in the plane normal to the molecule may be chosen as the directions of the degenerate modes of vibration. The complete motion of the atoms normal to the molecular axis will depend on the amplitudes and relative phases of the two degenerate modes. If both are excited, and they are exactly in phase, then the atoms will move on a straight line passing through the equilibrium configuration. But if they are out of phase, the composite motion is an elliptical Lissajous figure, exactly as in a two-dimensional isotropic oscillator. The two modes then represent a rotation, rather than a vibration.

It is obvious from the symmetry of the molecules that the amplitudes of the end atoms must be identical in magnitude. The complete calculation shows that the end atoms also travel in the same direction along the Lissajous figure. Hence the center atom must revolve in the opposite direction, in order to conserve angular momentum. Figure 6–4 illustrates the motion when the degenerate modes are ninety degrees out of phase.

As the complexity of the molecule increases, the size of the secular determinant becomes very large, and finding the normal frequencies and

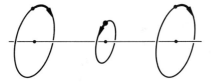

FIGURE 6–4
Degenerate modes of the symmetrical triatomic molecule.

amplitudes becomes a problem of considerable magnitude. We have seen, however, that even in a situation as simple as the linear triatomic molecule, a study of the symmetries to be expected in the vibrations greatly simplifies the calculations. Considerable mathematical ingenuity has been devoted to exploiting the symmetries inherent in complex molecules to reduce the labor involved in finding their vibration frequencies. The theory of symmetry groups has been applied with great success in factoring the large secular determinant into smaller blocks that may be diagonalized separately. It has been pointed out, however, that such elaborate mathematical manipulation was more appropriate in a time when numerical computations were difficult and tedious. Considering the speed and memory capacity of present-day computers, a straightforward approach may be easier and more accurate in the long run. Fast and accurate routines for solving the eigenvalue problems of large matrices are the stock-in-trade today of scientific computers of even moderate size. There has therefore been a trend toward a more brute-force approach in which mass-weighted Cartesian coordinates (see p. 246) are used to formulate the problem. The kinetic energy ellipsoid for the molecular vibrations is then already a sphere, and finding the normal modes reduces to diagonalizing the potential energy. It seems likely that this is the direction that will be taken in future calculations of vibrational frequencies for problems in chemical physics.

6–5 FORCED VIBRATIONS AND THE EFFECT OF DISSIPATIVE FORCES

Free vibrations occur when the system is displaced initially from the equilibrium configuration and is then allowed to oscillate by itself. Very often, however, the system is set into oscillation by an external driving force that continues to act on the system after $t = 0$. The frequency of such a *forced oscillation* is then determined by the frequency of the driving force and not by the resonant frequencies. Nevertheless, the normal modes are of great importance in obtaining the *amplitudes* of the forced vibration, and the problem is greatly simplified by use of the normal coordinates obtained from the free modes.

If F_j is the generalized force corresponding to the coordinate η_j, then by Eq. (1–49) the generalized force Q_i for the normal coordinate ζ_i is

$$Q_i = a_{ji}F_j. \tag{6–60}$$

The equations of motion when expressed in normal coordinates now become

$$\ddot{\zeta}_i + \omega_i^2\zeta_i = Q_i. \tag{6–61}$$

Equations (6–61) are a set of n inhomogeneous differential equations that can be solved only when we know the dependence of Q_i on time. While the solution will not be as simple as in the free case, note that the normal coordinates preserve their advantage of separating the variables, and each equation involves only a single coordinate.

Frequently the driving force varies sinusoidally with time. In an acoustic problem, for example, the driving force might arise from the pressure of a sound wave impinging on the system, and Q_i then has the same frequency as the sound wave. Or, if the system is a polyatomic molecule, a sinusoidal driving force is present if the molecule is illuminated by a monochromatic light beam. Each atom in the molecule is then subject to an electromagnetic force whose frequency is that of the incident light. Even where the driving force is not sinusoidal with a single frequency, it can often be considered as built up as a superposition of such sinusoidal terms. Thus, if the driving force is periodic, it can be represented by a Fourier series; other times a Fourier integral representation is suitable. Since Eqs. (6–61) are linear equations, its solutions for particular frequencies can be superposed to find the complete solution for given Q_i. It is therefore of general importance to study the nature of the oscillations when the force Q_i can be written as

$$Q_i = Q_{0i} \cos(\omega t + \delta_i), \tag{6-62}$$

where ω is the angular frequency of an external force. The equations of motion now appear as

$$\ddot{\zeta}_i + \omega_i^2 \zeta_i = Q_{0i} \cos(\omega t + \delta_i). \tag{6-63}$$

A complete solution of Eq. (6–63) consists of the general solution to the homogeneous equation (i.e., the free modes of vibrations) plus a particular solution to the inhomogeneous equation. By a proper choice of initial conditions, the superimposed free vibrations can be made to vanish,* centering our interest on the particular solution of Eqs. (6–63) that will obviously have the form

$$\zeta_i = B_i \cos(\omega t + \delta_i). \tag{6-64}$$

Here the amplitudes B_i are determined by substituting the solution in Eqs. (6–63):

$$B_i = \frac{Q_{0i}}{\omega_i^2 - \omega^2} \tag{6-65}$$

The complete motion is then

$$\eta_j = a_{ji}\zeta_i = \frac{a_{ji}Q_{0i}\cos(\omega t + \delta_i)}{\omega_i^2 - \omega^2}. \tag{6-66}$$

---

* The free vibrations are essentially the transients generated by the application of the driving forces. If we consider the system to be initially in an equilibrium configuration, and then slowly build up the driving forces from zero, these transients will not appear. Alternatively, dissipative forces can be assumed present (see pages following) that will damp out the free vibrations.

Thus the vibration of each particle is again composed of linear combinations of the normal modes, but now each normal oscillation occurs at the frequency of the driving force.

Two factors determine the extent to which each normal mode is excited. One is the amplitude of the generalized driving force, Q_{0i}. If the force on each particle has no component in the direction of vibration of some particular normal mode, then obviously the generalized force corresponding to the mode will vanish and Q_{0i} will be zero. *An external force can excite a normal mode only if it tends to move the particles in the same direction as in the given mode.* The second factor is the closeness of the driving frequency to the free frequency of the mode. In consequence of the denominators in Eq. (6–66), the closer ω approaches to any ω_i, the stronger will that mode be excited relative to the other modes. Indeed, Eq. (6–66) apparently predicts infinite amplitude when the driving frequency agrees exactly with one of the ω_i's—the familiar phenomenon of resonance. Actually, of course, the theory behind Eq. (6–66) presumes only small oscillations about equilibrium positions; when the amplitude predicted by the formula becomes large this assumption breaks down and Eq. (6–66) is then no longer valid. Note that the oscillations are in phase with the driving force when the frequency is less than the resonant frequency, but that there is a phase change of π in going through the resonance.

Our discussion has been unrealistic in that the absence of dissipative or frictional forces has been assumed. In many physical systems these forces, when present, are proportional to the particle velocities and can therefore be derived from a dissipation function $\mathscr{F}$ (cf. Section 1–5). Let us first consider the effects of frictional forces on the free modes of vibration.

From its definition, $\mathscr{F}$ must be a homogeneous quadratic function of the velocities:

$$\mathscr{F} = \frac{1}{2}\mathscr{F}_{ij}\dot{\eta}_i\dot{\eta}_j. \tag{6-67}$$

The coefficients $\mathscr{F}_{ij}$ are clearly symmetric, $\mathscr{F}_{ij} = \mathscr{F}_{ji}$, and in general will be functions of the coordinates. Since we are concerned with only small vibrations about equilibrium, it is sufficient to expand the coefficients about equilibrium and retain only the first, constant term, exactly as was done for the kinetic energy. In future applications of Eq. (6–67) we shall take $\mathscr{F}_{ij}$ as denoting these constant factors. It will be remembered that $2\mathscr{F}$ is the rate of energy dissipation due to the frictional forces (cf. Eq. 2–60). The dissipation function $\mathscr{F}$ therefore can never be negative. The complete set of Lagrange equations of motion now become (cf. Section 1–5)

$$T_{ij}\ddot{\eta}_j + \mathscr{F}_{ij}\dot{\eta}_j + V_{ij}\eta_j = 0. \tag{6-68}$$

Clearly, in order to find normal coordinates for which the equations of motion would be decoupled, it would be necessary to find a principal axis transformation that simultaneously diagonalizes the three quadratic forms T, V, and $\mathscr{F}$. As was shown above this is not in general possible; normal modes cannot usually be found for any arbitrary dissipation function. There are, however, some exceptional cases when simultaneous diagonalization is possible. For example, if the frictional force is proportional both to the particle's velocity *and* its mass, then $\mathscr{F}$ will be diagonal whenever T is. When such simultaneous diagonalization is feasible, then the equations of motions are decoupled in the normal coordinates with the form

$$\ddot{\zeta}_i + \mathscr{F}_i \dot{\zeta}_i + \omega_i^2 \zeta_i = 0 \qquad \text{(no summation)}. \qquad (6\text{–}69)$$

Here the $\mathscr{F}_i$'s are the nonnegative coefficients in the diagonalized form of $\mathscr{F}$ when expressed in terms of ζ_i. Being a set of linear differential equations with constant coefficients, Eqs. (6–69) may be solved by functions of the form

$$\zeta_i = C_i e^{-i\omega_i' t},$$

where ω_i' satisfies the quadratic equation

$$\omega_i'^2 + i\omega_i' \mathscr{F}_i - \omega_i^2 = 0 \qquad \text{(no summation)}. \qquad (6\text{–}70)$$

Equation (6–70) has the two solutions

$$\omega_i' = \pm \sqrt{\omega_i^2 - \frac{\mathscr{F}_i^2}{4}} - i\frac{\mathscr{F}_i}{2}. \qquad (6\text{–}71)$$

The motion is therefore not a pure oscillation, for ω' is complex. It is seen from Eq. (6–71) that the imaginary part of ω_i' results in a factor $\exp[-\mathscr{F}_i t/2]$, and by reason of the nonnegative nature of the $\mathscr{F}_i$'s this is always an exponentially decreasing function of time.* The presence of a damping factor due to the friction is hardly unexpected. As the particles vibrate, they do work against the frictional forces, and the energy of the system (and hence the vibration amplitudes) must decrease with time. The real part of Eq. (6–71) corresponds to the oscillatory factor in the motion, and it will be noted that the presence of friction also affects the frequency of the vibration. However, if the dissipation is small, the squared term in $\mathscr{F}_i$ may be neglected, and the frequency of oscillation reduces to the friction-free value. The complete motion is then simply an exponential damping of the free modes of vibration:

$$\zeta_i = C_i e^{-\mathscr{F}_i t/2} e^{-i\omega_i t}. \qquad (6\text{–}72)$$

If the dissipation function cannot be diagonalized along with T and V, the solution is much more difficult to obtain. The general nature of the solution remains pretty much the same, however: an exponential damping factor times an oscillatory exponential function. Suppose we seek a solution to Eqs. (6–68) of the form

$$\eta_j = C a_j e^{-i\omega t} = C a_j e^{-\kappa t} e^{-2\pi i \nu t}. \qquad (6\text{–}73)$$

---

* Some (but not all) $\mathscr{F}_i$'s may be zero, which simply means there are no frictional effects in the corresponding normal modes. The important point is that the $\mathscr{F}_i$'s cannot be negative.

With this solution Eqs. (6–68) become a set of simultaneous linear equations

$$V_{ij}a_j - i\omega \mathscr{F}_{ij}a_j - \omega^2 T_{ij}a_j = 0. \tag{6–74}$$

It is convenient to write ω as $i\gamma$, so that

$$\gamma = -i\omega = -\kappa - 2\pi i\nu, \tag{6–75}$$

and thus $-\kappa$ is the real part of γ. In terms of square matrices of V, T, and $\mathscr{F}$, the set of equations (6–74) becomes a column matrix equation involving γ:

$$\mathbf{Va} + \gamma \mathbf{Fa} + \gamma^2 \mathbf{Ta} = 0. \tag{6–76}$$

The set of homogeneous equations (6–74) or (6–76) can be solved for the a_j only for certain values of ω or γ. Without actually evaluating the corresponding secular equation we can show that κ must always be nonnegative. Convert the matrix equation (6–76) into a scalar equation for γ by multiplying from the left with $\mathbf{a}^\dagger$:

$$\mathbf{a}^\dagger \mathbf{Va} + \gamma \mathbf{a}^\dagger \mathbf{Fa} + \gamma^2 \mathbf{a}^\dagger \mathbf{Ta} = 0. \tag{6–77}$$

Equation (6–77) is a quadratic equation for γ with coefficients that are matrix products of the same general type as those encountered in Eq. (6–19). By virtue of the symmetry of $\mathbf{V}$, $\mathbf{F}$, and $\mathbf{T}$ the matrix products are all real, as can be seen by expanding $\mathbf{a}$ as $\boldsymbol{\alpha} + i\boldsymbol{\beta}$ (cf. Eq. 6–20). Hence if γ is a solution of the quadratic equation its complex conjugate γ^* must also be a solution. Now, the sum of the two roots of a quadratic equation is the negative of the coefficient of the linear term divided by the coefficient of the square term

$$\gamma + \gamma^* = -2\kappa = -\frac{\mathbf{a}^\dagger \mathbf{Fa}}{\mathbf{a}^\dagger \mathbf{Ta}}. \tag{6–78}$$

Hence κ can be expressed in terms of the real and imaginary parts of a_j as

$$\kappa = \frac{1}{2} \frac{\mathscr{F}_{ij}(\alpha_i\alpha_j + \beta_i\beta_j)}{T_{ij}(\alpha_i\alpha_j + \beta_i\beta_j)}. \tag{6–79}$$

The dissipation function $\mathscr{F}$ must always be positive, and T is positive definite; hence κ cannot be negative. The oscillations of the system may decrease exponentially with time, but they can never increase with time. Note that if $\mathscr{F}$ is positive definite, κ *must* be different from zero (and positive), and all modes will have an exponential damping factor. The frequencies of oscillation, given by the real part of ω, will of course be affected by the dissipative forces, but the change will be small if the damping is not very large during a period of oscillation.

Finally, we may consider forced sinusoidal oscillations in the presence of dissipative forces. Representing the variation of the driving force with time by

$$F_j = F_{0j}e^{-i\omega t},$$

where F_{0j} may be complex, the equations of motion are

$$V_{ij}\eta_j + \mathscr{F}_{ij}\dot{\eta}_j + T_{ij}\ddot{\eta}_j = F_{0i}e^{-i\omega t}. \tag{6–80}$$

If we seek a particular solution to these equations of the form

$$\eta_j = A_j e^{-i\omega t},$$

we obtain the following set of inhomogeneous linear equations for the amplitudes A_j:

$$(V_{ij} - i\omega \mathscr{F}_{ij} - \omega^2 T_{ij})A_j - F_{0i} = 0. \tag{6-81}$$

The solution to these equations* may easily be obtained from Cramer's rule:

$$A_j = \frac{D_j(\omega)}{D(\omega)}, \tag{6-82}$$

where $D(\omega)$ is the determinant of the coefficients of A_j in Eq. (6–81) and $D_j(\omega)$ is the modification in $D(\omega)$ resulting when the jth column is replaced by $F_{01} \ldots F_{0n}$. It is the denominator $D(\omega)$ that is of principal interest to us here, for the resonances arise essentially out of the algebraic form of the denominator. Now, D is the determinant appearing in the secular equation corresponding to the homogeneous equations (6–74); its roots are the complex frequencies of the free modes of vibration. The requirement that both γ and γ^* are roots of Eq. (6–77) means, on the basis of Eq. (6–75), that if ω_i is a root of $D(\omega)$ then $-\omega_i^*$ is a root. For a system of n degrees of freedom it is therefore possible to represent $D(\omega)$ as

$$D(\omega) = G(\omega - \omega_1)(\omega - \omega_2)\ldots(\omega - \omega_n)(\omega + \omega_1^*)(\omega + \omega_2^*)\ldots(\omega + \omega_n^*),$$

where G is some constant. Using product notation, and denoting ω by $2\pi\nu$, this representation can be written as

$$D(\omega) = G \prod_{i=1}^{n} (2\pi(\nu - \nu_i) + i\kappa_i)(2\pi(\nu + \nu_i) + i\kappa_i). \tag{6-83}$$

When we rationalize Eq. (6–83) to separate A_j into its real and imaginary parts, the denominator will be

$$D^*(\omega)D(\omega) = GG^* \prod_{i=1}^{n} (4\pi^2(\nu - \nu_i)^2 + \kappa_i^2)(4\pi^2(\nu + \nu_i)^2 + \kappa_i^2). \tag{6-84}$$

The amplitudes of the forced oscillation thus exhibit typical resonance behavior in the neighborhood of the frequencies of free oscillations $\pm \nu_i$. As a result of the presence of the damping constants κ_i, the resonance denominators no longer vanish at the free mode frequencies, and the amplitudes remain finite. The driving frequency at which the amplitude peaks is no longer exactly at the free frequencies because of frequency dependence of terms in A_j other than the particular resonance denominator. However, so long as the damping is small enough to preserve a recognizable resonant peak, the shift in the resonance frequencies is usually small.

---

* They are of course merely the inhomogeneous version of Eqs. (6–74).

We have discussed the properties of small oscillations solely in terms of mechanical systems. The reader, however, has undoubtedly noticed the similarity with the theory of the oscillations of electrical networks. The equations of motion (6–68) become the circuit equations for n coupled circuits if we read the V_{ij} coefficients as reciprocal capacitances, the $\mathscr{F}_{ij}$'s as resistances, and the T_{ij}'s as inductances. Driving forces are replaced by generators of frequency ω applied to one or more of the circuits, and the equations of forced vibration (6–80) reduce to the electrical circuit equations (2–39) mentioned in Chapter 2. We have presented here only a fraction of the techniques that have been devised for handling small oscillations, and of the general theorems about the motion. For example, space does not permit a discussion of the powerful Laplace transform techniques to study the response of a linearly oscillating system to driving forces with arbitrary time dependences. Nor is it appropriate here to consider the still expanding subject of nonlinear oscillations, where the potential energy contains terms beyond the quadratic and the motion is no longer simple harmonic. (Some relevant portions of this field will be introduced later when we treat perturbation theory). As has been mentioned, a formal development of the theory of small oscillations about steady motion will be given later in connection with the Hamiltonian version of mechanics. Another generalization that will deserve our attention relates to the oscillation of systems with continuously infinite numbers of degrees of freedom. The question is how one can construct a way of handling continuous systems that is analogous to the classical mechanics of discrete systems. We shall postpone such considerations of continuous systems to Chapter 12—after we have developed the canonical formulation of discrete mechanics, and after we have seen how the structure of Newtonian mechanics must be modified in the special theory of relativity.

SUGGESTED REFERENCES

H. JEFFREYS AND B. S. JEFFREYS, *Methods of Mathematical Physics*. Diagonalization of a matrix or of a pair of positive-definite quadratic forms is a common topic in texts on mathematical methods for physicists. This particular reference is chosen because the pertinent discussion in Chapter 4 has obviously been written with the problem of small oscillations particularly in mind. A number of theorems are given about the properties of roots of the secular determinant that are helpful in actually finding the eigenfrequencies.

E. T. WHITTAKER, *Analytical Dynamics*. Chapter VII is on the theory of vibrations and gives the explicit proof that T and V can be diagonalized together, but the treatment of this point is not very clear. More valuable are the later sections of the chapter on the effects of constraints, and on vibrations about steady motion, which is discussed in considerable detail. Section 94 of Chapter VIII on vibrations in the presence of dissipative forces is fragmentary and restricted to two degrees of freedom.

D. TER HAAR, *Elements of Hamiltonian Mechanics*. Despite its title this book is by no means confined to Hamiltonian mechanics, and Chapter 3 on small vibrations assumes no more than Lagrangian mechanics. The treatment is distinguished by a matrix approach—

with all the elements of the matrices written out in detail. There is an abundance of examples, with useful diagrams of the normal modes of vibration. Some of the conclusions on the symmetry properties of vibrations of the linear triatomic molecule are questionable.

J. L. SYNGE AND A. SCHILD, *Tensor Calculus.* Chapter 2 of this treatise provides an introduction to Riemannian spaces and associated metric tensors that is more than adequate for our purposes. In particular, they introduce immediately the differentiation into covariant and contravariant quantities, a distinction not needed here, though it is raised in Chapter 7 below.

L. D. LANDAU AND E. M. LIFSHITZ, *Mechanics.* As would be expected, the subject of small vibrations is presented compactly but covers a lot of ground. Oscillations of the triatomic molecule, both linear and "bent," are considered in some detail. In the present chapter only linear oscillations are considered, and the parametric elements of the potential and kinetic energy matrices are assumed constant in time. Study of parametric excitation of oscillators, in which these elements vary in time, and of nonlinear oscillations has grown rapidly in the last few decades and has become an area calling for separate and massive treatment. Landau and Lifshitz in their Sections 27–30 provide a concise introduction to the subject, one in which the Russian contributions have been particularly notable. Only the tip of the iceberg, however, is revealed in the discussion.

L. MEIROVITCH, *Methods of Analytical Dynamics.* Modern approaches to oscillatory systems are presented here in a manner that doesn't lose touch with practical reality. The subject of the present chapter appears here under the guise of linear autonomous systems, but the emphasis is on whether the motion is stable, something we have presupposed from the start. Both the classical and modern (e.g., Liapunov method) criteria are discussed. Parametric oscillations (nonautonomous systems) get a separate chapter.

Y. CHEN, *Vibrations: Theoretical Methods.* Much of this engineering-oriented text is concerned with introducing the basic ideas of mechanics, on the one hand, and in treating vibrations of continuous systems on the other. The rest covers the field of the present chapter with many examples worked out and the elements of the pertinent matrices explicitly (and sometimes clumsily) presented. Of particular interest is the application of Laplace and Fourier transform methods for handling problems of forced or driven oscillation.

H. C. CORBEN AND P. STEHLE, *Classical Mechanics.* The brief chapter on small oscillations in this well-known text concentrates on a number of examples rather than general theory. Special mention should be made of Appendix 3, which gives a brief introduction to the use of group-theoretical methods to reduce the complexity of the eigenfrequency problem for molecules, based on the intrinsic symmetries displayed by the systems. In the precomputer age when problems had to be solved by hand, any method for reducing the amount of computation was naturally the subject of intense concern and study. The advent of the high-speed computer has lessened the calculational importance of these approaches, although the use of symmetry properties to identify zero or degenerate frequencies is of obvious interest. See in this connection the paper by W. D. Gwinn: "Normal Coordinates: General Theory, Redundant Coordinates, and General Analysis Using Electronic Computers," *Jour. Chem. Phys.* **55**, 477 (July 15, 1971).

G. HERZBERG, *Infrared and Raman Spectra of Polyatomic Molecules.* This treatise provides many illustrations of the application of classical small vibration theory to molecular structure. The techniques of using constants of the motion and symmetry properties to reduce the complexity of the calculation are applied here to find explicit solutions for many molecular models. The various normal modes are shown diagrammatically for many of the molecules.

E. B. WILSON, JR.; J. C. DECIUS; AND P. C. CROSS, *Molecular Vibrations*. This 1955 treatise apparently remains the standard treatment of the molecular vibration problems for chemists. It is entirely precomputer in spirit—the long chapter on the benzene molecule ends with the reader adjured to gain facility in the matrix manipulation by checking the calculations of the chapter on a "standard desk-type computing machine." The use of symmetry groups is gone into in considerable detail. Some knowledge of quantum mechanics is needed.

LORD RAYLEIGH, *Theory of Sound*. One of the classics of physics literature, this treatise contains a wealth of theorems and physical illustrations on all of the aspects of vibration theory. Rayleigh himself was responsible for developing much of the theory, especially the introduction of the dissipation function. His treatment is smooth-flowing and clear and contains rarely discussed topics, as on the effects of constraints and the stationary properties of the eigenfrequencies. Rayleigh leans heavily on the work of Routh, who in his Adams Prize Essay of 1877 and in his text *Rigid Dynamics* was one of the first to give a systematic discussion of small vibrations.

E. A. GUILLEMIN, *The Mathematics of Circuit Analysis*. This reference is included to indicate the importance of small vibration theory in modern electrical engineering. Considerable attention is paid to quadratic forms and their principal axis transformations. The treatment, which makes abundant use of matrix algebra, is advanced and elegant.

EXERCISES

1. A mass particle moves in a constant vertical gravitational field along the curve defined by $y = ax^4$, where y is the vertical direction. Find the equation of motion for small oscillations about the position of equilibrium.

2. Obtain the normal modes of vibration for the double pendulum shown in Fig. 1–4, assuming equal lengths, but not equal masses. Show that when the lower mass is small compared to the upper one the two resonant frequencies are almost equal. If the pendula are set in motion by pulling the upper mass slightly away from the vertical and then releasing it, show that subsequent motion is such that at regular intervals one pendulum is at rest while the other has its maximum amplitude. This is the familiar phenomenon of "beats."

3. The problem of the linear triatomic molecule can be reduced to one of two degrees of freedom by introducing coordinates $y_1 = x_2 - x_1$, $y_2 = x_3 - x_2$, and eliminating x_2 by requiring that the center of mass remain at rest. Obtain the frequencies of the normal modes in these coordinates and show that they agree with the results of Section 6–4. The distances between the atoms, y_1 and y_2, are known as the *internal coordinates* of the molecule.

4. Obtain the frequencies of longitudinal vibration of the molecule discussed in Section 6–4, except that now the center atom is to be considered bound to the origin by a spring of force constant k. Show that the translational mode disappears under these conditions.

5. a) In the linear triatomic molecule suppose the initial condition is that the center atom is at rest but displaced by an amount a_0 from equilibrium, the other two being at their equilibrium points. Find the amplitudes of the longitudinal small oscillations about the center of mass. Give the amplitudes of the normal modes.

b) Repeat part (a) but with the center atom initially at equilibrium position but with an initial speed v_0.

6. A 5-atom linear molecule is simulated by a configuration of masses and ideal springs that looks like the following diagram:

| m | | M | | m | | M | | m |
| 1 | b | 2 | b | 3 | b | 4 | b | 5 |

All force constants are equal. Find the eigenfrequencies and normal modes for longitudinal vibration. [*Hint:* transform the coordinates η_i to ζ_i defined by

$$\eta_3 = \zeta_3, \qquad \eta_1 = \frac{\zeta_1 + \zeta_5}{\sqrt{2}}, \qquad \eta_5 = \frac{\zeta_1 - \zeta_5}{\sqrt{2}}$$

with symmetrical expressions for η_2 and η_4. The secular determinant will then factor into determinants of lower rank.]

7. In the linear triatomic molecule suppose that motion in the y and z directions is governed by the potentials

$$V_y = \frac{k}{2}(y_2 - y_1)^2 + \frac{k}{2}(y_3 - y_2)^2,$$

$$V_z = \frac{k}{2}(z_2 - z_1)^2 + \frac{k}{2}(z_3 - z_2)^2.$$

Find the eigenfrequencies for small vibrations in three dimensions and describe the normal modes. What symmetries do the zero frequencies represent? You may want to use the kind of intermediate coordinates suggested in Exercise 6.

8. The equilibrium configuration of a molecule is represented by three atoms of equal mass at the vertices of a 45° right triangle connected by springs of equal force constant. Obtain the secular determinant for the modes of vibration in the plane and show by rearrangement of the columns that the secular equation has a triple root $\omega = 0$. Reduce the determinant to one of the third rank and obtain the nonvanishing frequencies of free vibration.

9. Show directly that the equations of motion of the preceding problem are satisfied by (a) a uniform translation of all atoms along the x axis, (b) a uniform translation along the y axis, and (c) a uniform rotation about the z axis.

10. a) Three equal mass points have equilibrium positions at the vertices of an equilateral triangle. They are connected by equal springs that lie along the arcs of the circle circumscribing the triangle. Mass points and springs are constrained to move only on the circle, so that, e.g., the potential energy of a spring is determined by the arc length covered. Determine the eigenfrequencies and normal modes of small oscillations in the plane. Identify physically any zero frequencies.

b) Suppose one of the springs has a change in force constant δk, the others remaining unchanged. To first order in δk what are the changes in the eigenfrequencies and normal modes?

c) Suppose what is changed is the mass of one of the particles by an amount δm. Now how do the normal eigenfrequencies and normal modes change?

11. A uniform bar of length l and mass m is suspended by two equal springs of equilibrium length b and force constant k, as shown in the diagram.

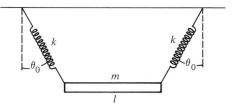

Find the normal modes of small oscillation in the plane.

12. Two particles move in one dimension at the junction of three springs, as shown in the figure. The springs all have unstretched lengths equal to a, and the force constants and masses are shown.

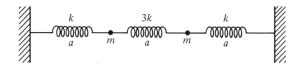

Find the eigenfrequencies and normal modes of the system.

13. Two mass points of equal mass m are connected to each other and to fixed points by three equal springs of force constant k, as shown in the diagram:

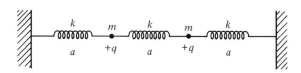

The equilibrium length of each spring is a. Each mass point has a positive charge $+q$, and they repel each other according to the Coulomb law. Set up the secular equation for the eigenfrequencies.

14. Find expressions for the eigenfrequencies of the following coupled circuit:

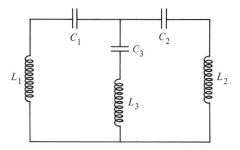

15. If the generalized driving forces Q_i are not sinusoidal, show that the forced vibrations of the normal coordinates in the absence of damping are given by

$$\zeta_i = \frac{1}{\sqrt{2\pi}} \int_{-\infty}^{+\infty} \frac{G_i(\omega)}{\omega_i^2 - \omega^2} e^{-i\omega t} \, d\omega,$$

where $G_i(\omega)$ is the Fourier transform of Q_i defined by

$$Q_i(t) = \frac{1}{\sqrt{2\pi}} \int_{-\infty}^{+\infty} G_i(\omega) e^{-i\omega t} \, d\omega.$$

If the dissipation function is simultaneously diagonalized along with T and V, show that the forced vibrations are given by

$$\zeta_i = \frac{1}{\sqrt{2\pi}} \int_{-\infty}^{+\infty} \frac{G_i(\omega)(\omega_i^2 - \omega^2 + i\omega \mathcal{F}_i)}{(\omega_i^2 - \omega^2)^2 + \omega^2 \mathcal{F}_i^2} e^{-i\omega t} \, dt,$$

which has the typical resonance denominator form. These results are simple illustrations of the powerful technique of the *operational calculus* for handling transient vibrations.

CHAPTER 7
Special Relativity in Classical Mechanics

Our development of classical mechanics has been based on a number of definitions and postulates that were presented in Chapter 1. However, when the velocities involved approach the speed of light these postulates, as is well known, no longer represent the experimental facts, and they must then be altered to conform to what is called the *special theory of relativity*. This is a modification of the structure of mechanics that must not be confused with the far more violent recasting required by quantum theory. There are many physical instances where quantum effects are important but relativistic corrections are negligible. And conversely, phenomena frequently occur involving relativistic velocities where the refinements of quantum mechanics do not affect the discussion. There is no inherent connection between special relativity and quantum mechanics, and the effects of one may be discussed without the other. It is therefore of considerable practical importance to examine the changes in the formulation of classical mechanics required by special relativity.

It is not intended, however, to present a comprehensive discussion of the theory of special relativity and its consequences. We shall not be greatly concerned with the events and experiments that led to the construction of the theory, far less with its philosophical implications, its apparent paradoxes "which gaily mock at common sense." It will be assumed that the reader has had at least an elementary introduction to the relativistic transformations between uniformly moving systems and to the more striking physical phenomena arising therefrom. The emphasis here will be on how special relativity may be fitted into the framework of classical mechanics and only as much of the theory as is needed for this task will be presented.

7–1 THE BASIC PROGRAM OF SPECIAL RELATIVITY

In the discussions of the previous chapters, frequent use has been made of such phrases as "space system" or "system fixed in space." By these phrases we have meant no more than an inertial system, one in which Newton's law of motion,

$$\mathbf{F} = m\mathbf{a}, \tag{7–1}$$

is valid. A system fixed in a body rotating with respect to an inertial system does not satisfy this qualification; one must add to (7–1) terms describing the effect of the rotation. On the other hand, it would appear that a system moving uniformly with respect to a "space system" should itself be an inertial system. If $\mathbf{r}'$ represents a radius vector from the origin of the second system to a given point, and $\mathbf{r}$ the corresponding vector in the first system, cf. Fig. 7–1, then if the two systems are coincident at $t = 0$, it seems obvious that these two vectors are connected by the relation:

$$\mathbf{r}' = \mathbf{r} - \mathbf{v}t. \tag{7–2}$$

Since the relative velocity is constant, the first time derivative of Eq. (7–2) is

$$\dot{\mathbf{r}}' = \dot{\mathbf{r}} - \mathbf{v}, \tag{7–3}$$

and another differentiation gives

$$\mathbf{a}' = \mathbf{a} \tag{7–4}$$

so that the acceleration is the same in both systems. If Newton's law, Eq. (7–1), holds in one system it should hold in the other, if the force has the same form in both systems. In many of the usual problems of mechanics the force is obviously unchanged between the two systems, as for example, with a constant force field such as $\mathbf{F} = m\mathbf{g}$. Whenever the forces in a system are derived from a potential that is a function only of the scalar distances between particles in the system, then the forces clearly have the same form expressed in the primed or unprimed systems, and the corresponding equations of motion are unaltered.*

On the other hand, the transformation represented by Eqs. (6–2) and (6–4), known as the *Galilean transformation*, predicts that the velocity of light should be different in the two systems. Thus, suppose there is a source of light at the origin of the unprimed system emitting spherical waves traveling with the speed c. Let the

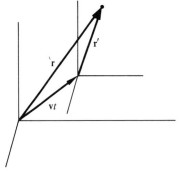

FIGURE 7–1
Sketch illustrating the Galilean transformation.

---

* It is interesting to note that with such a potential the Lagrangian does *not* have the same form in the two systems. But the difference can be shown to be a total derivative of a function of position only (cf. Exercise 1) and the equations of motion look alike in the two systems.

radius vector **r** be the position vector of a point on some given wave surface. Then in the unprimed system the velocity of the point on the wave surface is $\dot{\mathbf{r}} = c\mathbf{n}$, where **n** is a unit vector along **r**. According to (7–2), however, the corresponding wave velocity in the primed system is $\dot{\mathbf{r}}' = c\mathbf{n} - \mathbf{v}$. In the system moving with respect to the source of light the magnitude of the wave velocity will in general no longer be c; indeed, since it depends on direction, the waves will no longer be spherical.

A long series of investigations, especially the famous experiments of Michelson and Morley, have indicated that the velocity of light is always the same in all directions and is independent of the relative uniform motions of the observer, the transmitting medium, and the source.* Since the propagation of light in a vacuum with the speed c is a consequence of Maxwell's equations, it must be concluded that the Galilean transformation does not preserve the form of Maxwell's equations. Now, it is a postulate of physics, implicit since the time of Galileo and Newton, that *all* phenomena of physics should appear the same in all systems moving uniformly relatively to each other. Measurements made entirely *within* a given system must be incapable of distinguishing that system from all others moving uniformly with respect to it. This *postulate of equivalence* requires that physical laws must be phrased in an identical manner for all uniformly moving systems, i.e., be *covariant* when subjected to a Galilean transformation. The paradox confronting physics at the turn of the twentieth century was that experimentally both Newton's laws and Maxwell's equations seemed to satisfy the equivalence postulate but that theoretically, i.e., according to the Galilean transformation, Maxwell's equations did not. Einstein, affirming explicitly the postulate of equivalence, concluded that it is the form of Maxwell's equations that must be kept invariant and therefore the Galilean transformation could not be correct. A new relationship between uniformly moving systems, the *Lorentz transformation*, must be found that preserves the speed of light in all uniformly moving systems. Einstein showed that such a transformation requires revision of the usual concepts of time and simultaneity.

It is highly probable, a priori, that Newton's equations of motion also require revision, since they satisfy the equivalence postulate only under the Galilean transformation, now seen to be incorrect. Possibly other commonly accepted laws of physics might not preserve their form under a Lorentz transformation and must therefore also be generalized to have the proper transformation properties. Of course the generalizations must be such as to reduce to the more customary forms for velocities much smaller than that of light, when the Galilean transformation is approximately correct.

---

*One of the more striking of these experimental proofs involves measurements of the speed of photons emitted by π^0 mesons decaying in flight. Although the source was itself moving almost as fast as light, the velocity of the emitted photons as measured in the laboratory system agreed with c to within the experimental uncertainty, less than 0.01 %. See T. Alväger *et al.*, *Physics Letters* **12**, 260 (October 1, 1964).

The program of the theory of special relativity is therefore twofold. First, there must be obtained a transformation between two uniformly moving systems that will preserve the velocity of light. Second, the laws of physics must be examined as to their transformation properties under this Lorentz transformation. Those laws that do not keep their form invariant are to be generalized so as to obey the equivalence postulate. Abundant experimental verification has by now been obtained for the physical picture resulting from this program, and in the last analysis this is the only justification needed for Einstein's fundamental assumptions.

7–2 THE LORENTZ TRANSFORMATION

It will be assumed the reader has met up with the equations of the Lorentz transformation before.* Consider two systems of coordinates whose origins coincide at zero time, as seen by observers in both system. Let one system, call it the primed system, move unformly with velocity v relative to the other, unprimed system, in a direction parallel to the z axis. Then the Lorentz transformation between the coordinates, and times, measured in the two systems is given by the equations

$$x' = x,$$

$$y' = y,$$

$$z' = \frac{z - vt}{\sqrt{1 - \beta^2}}, \tag{7-5}$$

$$t' = \frac{t - \frac{vz}{c^2}}{\sqrt{1 - \beta^2}}$$

Here β is v/c, the ratio of v to the speed of light. It is simple to see that these equations of transformation satisfy all the properties that might, a priori, have been demanded of them. Above all, they preserve the speed of light in the two systems. Suppose that a point source located at the origin of the unprimed system emits a light wave at time $t = 0$. The equation of the wave front as seen in the unprimed system is that of a sphere:

$$x^2 + y^2 + z^2 = c^2t^2. \tag{7-6}$$

The transformation equations (7–5) then show, with a little algebraic manipulation, that the transformed wave front observed in the primed system is also a sphere expanding with the speed c:

$$x'^2 + y'^2 + z'^2 = c^2t'^2. \tag{7-7}$$

---

* Derivations of the Lorentz transformation abound at all levels of sophistication and degrees of rigor. See the annotated list of references at the end of the chapter.

Further, keeping only first order terms in v/c, Eqs. (7–5) reduce to

$$x' = x,$$
$$y' = y,$$
$$z' = z - vt, \tag{7-8}$$
$$t' = t.$$

In the limit of small relative speed, the Lorentz transformation equations thus do reduce to the Galilean transformation, Eq. (7–2), with the explicit statement that the times are the same in both systems.

Equations (7–5) say that the x and y coordinates are unaltered by the transformation, and this is also as we would have expected. The directions perpendicular to the direction of relative motion should not be affected by the motion. More formally, the isotropy of space presupposes that a relative motion in one direction should not induce a twist of the other directions. The transformation equations are also linear. While it might not be obvious that they have to be linear, the property helps in assuring that the transformation reduces to the linear Galilean transformation in the limit of small velocities. It can in fact be shown* that the transformation *has* to be linear in order to preserve the inertial quality of a system under Lorentz transformation. Since the validity of Newton's equations of motion is suspect, Einstein gave a new definition of an inertial system as one in which an isolated particle, free from interactions, remains at rest or moves uniformly. The Lorentz transformation equations must be linear if uniform motion in one system is to transform into uniform motion in the other system.

Finally it would be expected that the Lorentz transformation equations are truly relative, i.e., that the inverse transformation has the same form with the sign of v changed. That Eqs. (7–5) satisfy this condition can easily be checked by direct solution of x, y, z, and t in terms of x', y', z', and t'.

The equations of the Lorentz transformation can easily be put into a vector form that does not single out any particular direction for the relative velocity. With $\mathbf{v}$ along the z axis we can reformulate the spatial part of the equations of transformation, (7–5) as

$$\mathbf{r}' = \mathbf{r} + \left(\frac{z - vt}{\sqrt{1 - \beta^2}} - z \right) \frac{\mathbf{v}}{v}.$$

Explicit appearance of z can be eliminated by using the relationship

$$\mathbf{v} \cdot \mathbf{r} = vz,$$

---

* See for example R. D. Sard, *Relativistic Mechanics*, p. 596, or V. Fock, *The Theory of Space, Time and Gravitation*, 2d ed., pp. 20–24.

leading to

$$\mathbf{r}' = \mathbf{r} + (\mathbf{v} \cdot \mathbf{r})\frac{\mathbf{v}}{v^2}\left(\frac{1}{\sqrt{i - \beta^2}} - 1\right) - \frac{\mathbf{v}t}{\sqrt{1 - \beta^2}}. \tag{7-9}$$

Equation (7–9) can be put in a more compact form by using the vector $\boldsymbol{\beta} = \mathbf{v}/c$ and defining γ as

$$\gamma = \frac{1}{\sqrt{1 - \beta^2}}, \tag{7-10}$$

resulting in the equation

$$\mathbf{r}' = \mathbf{r} + \frac{(\boldsymbol{\beta} \cdot \mathbf{r})\boldsymbol{\beta}}{\beta^2}(\gamma - 1) - \boldsymbol{\beta}\gamma ct. \tag{7-11}$$

The corresponding form of the transformation equation for time is

$$t' = \gamma t - (\boldsymbol{\beta} \cdot \mathbf{r})\frac{\gamma}{c}. \tag{7-12}$$

Equations (7–11) and (7–12) constitute the Lorentz equations of transformation between two coordinate systems with parallel axes moving uniformly relative to each other with velocity $\mathbf{v}$ in any arbitrary direction.

   The general Lorentz transformation, Eqs. (7–11) and (7–12), is in the form of a linear transformation from one set of four coordinates to another set of four. Hence all of the mathematical apparatus developed in the earlier chapters for handling linear transformations can be applied to the Lorentz transformation. Minkowski pointed out that if the fourth coordinate is taken to be ict, in a four-dimensional Cartesian space, then the Lorentz transformation takes on a particularly simple and familiar form. The square of the magnitude of the position vector in such a 4-space (known as *Minkowski space*) has the form

$$x_1^2 + x_2^2 + x_3^2 + x_4^2 = x^2 + y^2 + z^2 - c^2t^2. \tag{7-13}$$

Now, the Lorentz transformation is designed to preserve the speed of light in the two systems, but by Eqs. (7–7) and (7–13), that condition is equivalent to requiring that the magnitude of the position vector in 4-space be held invariant under the transformation. We have seen in Chapter 4 that linear transformations that do not affect the magnitudes of vectors are orthogonal transformations. *Thus, the Lorentz transformation is an orthogonal transformation in Minkowski 4-space.*

   The Lorentz transformation can be described in terms of other four-dimensional coordinate systems (cf. Section 7–3). Any four-dimensional space involving time in one of the axes is referred to as *world-space*. For the most part our discussion will refer only to Minkowski space.

   Since the fourth coordinate is imaginary, Minkowski space is complex, and the elements of the Lorentz transformation matrix in this space may be imaginary. Explicit forms for the matrix elements may be obtained directly from the vector forms of the transformation, Eqs. (7–11) and (7–12). We shall denote the matrix of the Lorentz transformation in Minkowski space by **L**:

$$\mathbf{x}' = \mathbf{L}\mathbf{x}, \tag{7-14}$$

with $L_{\mu\nu}$ as the general element.* It is conventional to use Greek letters for a subscript that runs over all four coordinates, and Roman letters for subscripts only over the space components. As previously, the summation convention will be employed unless explicitly disowned.

In Minkowski space, Eqs. (7–11) and (7–12) take on the form

$$x'_j = x_j + \frac{\beta_j\beta_k x_k}{\beta^2}(\gamma - 1) + i\beta_j\gamma x_4, \tag{7-15}$$

$$x'_4 = -i\beta_k x_k\gamma + \gamma x_4. \tag{7-16}$$

Hence the elements of **L** for arbitrary direction of **β** are

$$L_{jk} = \delta_{jk} + \frac{\beta_j\beta_k}{\beta^2}(\gamma - 1),$$

$$L_{j4} = i\beta_j\gamma,$$

$$L_{4k} = -i\beta_k\gamma, \tag{7-17}$$

$$L_{44} = \gamma.$$

For example, the particular Lorentz transformation with which we started this section has the relative velocity along the z axis, and the matrix **L** then takes the simple form

$$\mathbf{L} = \begin{pmatrix} 1 & 0 & 0 & 0 \\ 0 & 1 & 0 & 0 \\ 0 & 0 & \gamma & i\beta\gamma \\ 0 & 0 & -i\beta\gamma & \gamma \end{pmatrix}. \tag{7-18}$$

It was seen in Chapter 4 that the basic property of an orthogonal matrix is that the transpose is equal to the inverse. For the Lorentz transformation with relative velocity vector **β**, the inverse must be the same matrix, but for velocity $-\mathbf{\beta}$. Inspection of Eq. (7–17) immediately shows the elements satisfy this condition; changing the sign of β_j is equivalent to transposing the subscripts.

The analogy of **L** with three-dimensional orthogonal transformations can be exploited further. Looking at Eq. (7–18) we see that the 2×2 submatrix of the

---

* It is regrettable that L is already used for Lagrangian and for angular momentum, but in practice the context almost always removes any chance of confusion.

third and fourth coordinates resembles the corresponding submatrix in the rotation of a three-dimensional coordinate system about one axis:

$$\begin{pmatrix} \cos\phi & \sin\phi \\ -\sin\phi & \cos\phi \end{pmatrix}.$$

Correspondingly, Eq. (7–18) can be said to be a rotation in the $x_3 x_4$ plane of Minkowski space but through an angle ϕ that is imaginary:

$$\cos\phi = \frac{1}{\sqrt{1-\beta^2}}, \qquad \sin\phi = \frac{i\beta}{\sqrt{1-\beta^2}}. \tag{7–19}$$

If the use of an imaginary angle leaves one uncomfortable, then a real angle ψ, defined by $\phi = i\psi$, can be introduced for which

$$\cosh\psi = \frac{1}{\sqrt{1-\beta^2}}, \qquad \sinh\psi = \frac{\beta}{\sqrt{1-\beta^2}}, \tag{7–20}$$

so that the (x_3, x_4) submatrix in Eq. (7–18) can be written

$$\begin{pmatrix} \cosh\psi & i\sinh\psi \\ -i\sinh\psi & \cosh\psi \end{pmatrix}.$$

With the help of the equivalent rotation angle, manipulation of the Lorentz transformation matrices can sometimes be simplified. Suppose, for example, we have a Lorentz transformation **L** between two systems with relative velocity β along the x_3 axis, and a second transformation **L′** with relative velocity β' also along the x_3 axis. What is the Lorentz transformation equivalent to the successive applications of **L** and **L′**, that is, **L″ = L′L**? Since the two Lorentz transformations are rotations in the same plane, the rotation angles simply add (as may be verified by multiplying the two matrices), and ϕ'' for **L″** is $\phi + \phi'$. By eqs. (7–19)

$$\tan\phi = i\beta, \tag{7–21}$$

and since

$$\tan\phi'' = \tan(\phi + \phi') = \frac{\tan\phi + \tan\phi'}{1 - \tan\phi\tan\phi'},$$

we have immediately that the two successive Lorentz transformations correspond to a single Lorentz transformation with relative speed:

$$\beta'' = \frac{\beta + \beta'}{1 + \beta\beta'}. \tag{7–22}$$

Equation (7–22) is the well-known Einstein addition law for parallel velocities.* Two points may be made about this result. (1) In the corresponding non-

---

* The derivation can equally well be carried out in terms of the real angle ψ and the formulas for hyperbolic functions of the sum of two angles.

relativistic problem the equivalent velocity would merely be the sum of the velocities of the two transformations; relativistically there is a correction through the term in the denominator. (2) The effect of the correction is that even if β and β' are arbitrarily close to unity (speeds near c), β'' never reaches, much less exceeds, unity. One cannot go faster than the speed of light by means of successive Lorentz transformations. Equivalently, starting from a system in which the observer is at rest there is no way to transform to a system going faster than the speed of light. In that sense the speed of light is the "ultimate speed."

It is quite easy to see that one can find orthogonal matrices in Minkowski space other than those of the form given by Eqs. (7–17). Suppose, for example, $L_{44} = 1$, $L_{4i} = L_{i4} = 0$, and the remaining nine elements of **L** formed a 3×3 matrix **R** that is orthogonal in ordinary 3-space. The matrix **L** so defined is clearly orthogonal in Minkowski space. But the transformation involves no relative motion of the two systems; all it effects is a rotation of the spatial coordinates. *All spatial rotations thus belong to a subclass of the Lorentz transformation.* Equally well, Eq. (7–14) does not define the most general coordinate transformation under which the equations of physics should be invariant in form. A shift in the origin of the coordinates in Minkowski space—spatial translation plus a change in the zero of time—should also not affect the phrasing of the physical laws. Nor have the possibilities of generalization of the Lorentz transformation beyond Eqs. (7–17) yet been exhausted. Some systematization of these possibilities therefore seems in order at this point.

The most general transformation in Minkowski space that would preserve the velocity of light thus has the form

$$\mathbf{x}' = \mathbf{Lx} + \mathbf{a}, \tag{7–23}$$

where **a** is an arbitrary translation vector of the origin and **L** is an orthogonal matrix. A transformation such as represented by Eq. (7–23) is spoken of as a *Poincaré transformation* or *inhomogeneous Lorentz transformation.* The orthogonality condition

$$\tilde{\mathbf{L}}\mathbf{L} = \mathbf{L}\tilde{\mathbf{L}} = \mathbf{1}, \tag{7–24}$$

corresponds in 4-space to ten constraints on the elements of **L** (four diagonal and six off-diagonal conditions) leaving only six independent quantities in **L**. Another four independent elements are required for **a**, so the Poincaré transformation in general requires the specification of ten independent quantities. For present purposes we shall need to be concerned only with the *homogeneous Lorentz transformation,* that is,

$$\mathbf{x}' = \mathbf{Lx}, \tag{7–14}$$

involving six independent elements. As with the ordinary spatial orthogonal transformation, the orthogonality condition requires that the square of the determinant of **L** be unity:

$$|\mathbf{L}|^2 = 1. \tag{7–25}$$

Of the possible determinants satisfying Eq. (7–25), only matrices for which $|\mathbf{L}| = 1$ can go continuously into the identity transformation. Such forms of $\mathbf{L}$ are, as in the three-dimensional case, known as *proper* Lorentz transformations.* However a transformation involving simultaneous inversion of all spatial coordinates *and* the time coordinate would also have a determinant $+1$. It is easy to see by the following argument that in general L_{44}^2 must be greater than or equal to unity. The 44 element of Eqs. (7–24) can be written

$$L_{44}^2 + L_{4j}L_{4j} = 1. \tag{7–26}$$

Now, the elements L_{4j} connect an imaginary fourth coordinate with real spatial coordinates. Therefore L_{4j} (and similarly L_{j4}) must be pure imaginary so that the sum $L_{4j}L_{4j}$ must be negative. Hence

$$L_{44}^2 \geq 1. \tag{7–27}$$

By the same type of argument L_{44} must be real. Of the two possibilities Eq. (7–27) presents, $L_{44} \leq -1$ involves a time inversion, and only $L_{44} \geq +1$ allows continuity with the identity transformation. Lorentz transformations for which $L_{44} \geq +1$ are called *orthochronous*, and if $L_{44} \leq -1$, they are designated *nonorthochronous*. Of the four possible choices of the signs of $|\mathbf{L}|$ and L_{44}, only the orthochronous proper Lorentz transformations can be continuously derived from the identity transformation. Such transformations are known as *restricted Lorentz transformations†*—only they can involve proper rotations and reduce to the Galilean transformation in the limit of small relative velocities. It is obvious that the signs of the square roots in the matrix elements in Eq. (7–17) are such that they represent a restricted Lorentz transformation. Like Moliere's hero who was delighted to learn he had been talking prose all his life without knowing it, we have really been talking about the restricted Lorentz transformation from the beginning of this section without realizing it. We shall continue to confine our discussions to restricted Lorentz transformations, and the adjective will be implied unless otherwise explicitly stated.

A restricted Lorentz transformation that describes the relation between two systems with parallel axes moving uniformly relatively to each other, i.e., without any spatial rotation, is called a *pure* Lorentz transformation or, in jargon, a "boost." Clearly the transformation matrix given by Eqs. (7–17) is that of a pure Lorentz transformation. It is intuitively obvious that any restricted Lorentz transformation can be decomposed into a pure Lorentz transformation and a

---

* The determinant $|\mathbf{L}|$ is real, as can be seen from the matrix elements in Eq. (7–17)—the determinant expansion involves at most only products of two imaginary matrix elements at a time. Hence the only other possibility is $|\mathbf{L}| = -1$, corresponding to *improper* transformations involving coordinate inversion.

† It can be shown (cf. Exercise 9) that the products of restricted transformations are also restricted. Products of the other three classes do not remain within their class. Thus only the restricted Lorentz transformations can have the group property.

spatial rotation without relative motion (in either order). The procedure for constructing the component transformations can be sketched simply. Suppose a Lorentz transformation is to be represented as the product of a pure transformation followed by a spatial rotation:

$$\mathbf{L} = \mathbf{RP}. \tag{7-28}$$

The coordinates of a point in the transformed space, x'_ν, are related to the coordinates in the original space by

$$x_\mu = L_{\nu\mu} x'_\nu.$$

How fast does an observer in the unprimed system see the origin of the primed system $(x'_j = 0)$ move by? The unprimed coordinates of that origin are

$$x_j = L_{4j} x'_4, \qquad x_4 = L_{44} x'_4.$$

The relative velocity vector of the primed origin therefore has components

$$\beta_j = \frac{x_j}{ct} = \frac{ix_j}{x_4} = \frac{iL_{4j}}{L_{44}}. \tag{7-29}$$

From Eq. (7–26), it then follows that

$$L_{44}^2 (1 - \beta_j \beta_j) = 1.$$

The reality of L_{44} immediately implies that β defined by Eq. (7–29) lies between 0 and 1, and that in terms of this value of β, L_{44} must have the value

$$L_{44} = (1 - \beta^2)^{1/2} \equiv \gamma. \tag{7-30}$$

Construct now a pure Lorentz transformation $\mathbf{P}(\boldsymbol{\beta})$ corresponding to the relative velocity vector, Eq. (7–29). The inverse transformation has the matrix $\mathbf{P}(-\boldsymbol{\beta})$. It therefore follows that the matrix $\mathbf{R}$ can be constructed as the product

$$\mathbf{R} = \mathbf{LP}(-\boldsymbol{\beta}). \tag{7-31}$$

A formal proof that the 4×4 matrix $\mathbf{R}$ has the form of a spatial rotation, as deduced from the matrix elements of $\mathbf{P}(-\boldsymbol{\beta})$ and the orthogonality of $\mathbf{L}$, will be left to the exercises. It should however be clear that $\mathbf{R}$ can only be a spatial rotation. The intermediate coordinate system defined by $\mathbf{P}(\boldsymbol{\beta})$ is at rest relative to the final set of axes and all $\mathbf{R}$ can do is rotate the coordinates. Note, that by this decomposition, of the six independent elements of $\mathbf{L}$, three are given by the components of the relative velocity vector $\boldsymbol{\beta}$ and the other three by the parameters defining the spatial rotation represented by $\mathbf{R}$, e.g., three Euler angles.

Successive application of two pure Lorentz transformations is of course equivalent to a single Lorentz transformation between the initial and final coordinate systems. But unless the relative velocities of the successive transformations are parallel, the final equivalent transformation will not be a pure Lorentz transformation. The decomposition process described above can be carried through on the product of two pure Lorentz transformations to obtain explicitly the rotation of the coordinate axes resulting from the two successive "boosts." In general, the algebra involved is quite forbidding, more than enough, usually, to discourage any actual demonstration of the rotation matrix. There is, however, one specific situation where allowable approximations reduce the calculational complexity, while the result obtained has important applications in many areas of modern physics. What is involved is a phenomenon known as the *Thomas precession.*

Consider three coordinate systems: O_1, the laboratory system; O_2, a system moving with velocity $\boldsymbol{\beta}$ relative to O_1; and O_3, a system moving with velocity $\Delta\boldsymbol{\beta} \equiv \boldsymbol{\beta}'$ relative to O_2. It is no loss of generality to take $\boldsymbol{\beta}$ along the z axis of O_1 and $\boldsymbol{\beta}'$ to be in the y-z plane of O_2, i.e., $\boldsymbol{\beta}, \boldsymbol{\beta}'$ define the y-z plane of O_2. The components of $\boldsymbol{\beta}'$ will be assumed to be infinitesimals, which need be retained only through the lowest nonvanishing order. Thus, γ' for the transformation between O_2 and O_3 can be replaced by unity. On this basis the matrix for **L**, the pure Lorentz transformation between O_1 and O_2, has the form of Eq. (7–18), while the corresponding matrix of transformation from O_2 to O_3 is

$$
\cdot\mathbf{L}' = \begin{pmatrix} 1 & 0 & 0 & 0 \\ 0 & 1 & 0 & i\beta_2' \\ 0 & 0 & 1 & i\beta_3' \\ 0 & -i\beta_2' & -i\beta_3' & 1 \end{pmatrix},
\tag{7–32}
$$

where β_2' and β_3' are the components of $\boldsymbol{\beta}'$. The product matrix, to the same approximation, is

$$
\mathbf{L}'' = \mathbf{L}'\mathbf{L} = \begin{pmatrix} 1 & 0 & 0 & 0 \\ 0 & 1 & \beta\beta_2'\gamma & i\beta_2'\gamma \\ 0 & 0 & \gamma & i\beta\gamma \\ 0 & -i\beta_2' & -i\beta\gamma & \gamma \end{pmatrix}
\tag{7–33}
$$

It is clear that $\mathbf{L}''$ cannot represent a pure Lorentz transformation, for which, as Eq. (7–17) shows, the elements L''_{jk} must be symmetric. By Eq. (7–29) the nonvanishing components of the effective relative velocity between O_1 and O_3 are*

$$
\beta_2'' = \frac{\beta_2'}{\gamma}, \qquad \beta_3'' = \beta.
\tag{7–34}
$$

---

* The effect of treating β' as small is that β_3' has been dropped in comparison to β.

Since $(\beta'')^2 = (\beta_2'')^2 + (\beta_3'')^2 \simeq \beta^2$ and therefore $\gamma'' \simeq \gamma$, the pure Lorentz transformation for the relative velocity $-\boldsymbol{\beta}''$ is to be approximated as

$$
\mathbf{P}(-\boldsymbol{\beta}'') = \begin{pmatrix}
1 & 0 & 0 & 0 \\
0 & 1 & \dfrac{\beta_2'}{\beta\gamma}(\gamma-1) & -i\beta_2' \\
0 & \dfrac{\beta_2'}{\beta\gamma}(\gamma-1) & \gamma & -i\beta\gamma \\
0 & i\beta_2' & i\beta\gamma & \gamma
\end{pmatrix}. \tag{7–35}
$$

Finally, the rotation matrix characterizing the rotation of the O_3 axes as seen from O_1 is found, after some algebraic simplification and the dropping of higher order terms in β', to be

$$
\mathbf{R} = \mathbf{L}''\mathbf{P}(-\boldsymbol{\beta}'') = \begin{pmatrix}
1 & 0 & 0 & 0 \\
0 & 1 & \dfrac{\beta_2'}{\beta\gamma}(\gamma-1) & 0 \\
0 & \dfrac{-\beta_2'}{\beta\gamma}(\gamma-1) & 1 & 0 \\
0 & 0 & 0 & 1
\end{pmatrix}. \tag{7–36}
$$

Comparison with Eq. (4–105) shows that $\mathbf{R}$ implies O_3 is rotated with respect to O_1 about the x axis through an infinitesimal angle:

$$
\overrightarrow{\Delta x} = \frac{\gamma-1}{\beta^2}\, \vec{\beta} \times \overrightarrow{\delta\beta}
$$

$$
\Delta\Omega_1 = \frac{\beta_2'}{\beta\gamma}(\gamma-1) = \beta_2''\beta\frac{(\gamma-1)}{\beta^2}. \tag{7–37}
$$

The spatial rotation resulting from the successive application of two parallel axes Lorentz transformations has been declared every bit as paradoxical as the more frequently discussed apparent violations of common sense, such as the "twin paradox." But this particular paradox has important applications, especially in atomic physics, and therefore has been abundantly verified experimentally. Consider a particle moving in the laboratory system with a velocity $\mathbf{v}$ that is not constant. Since the system in which the particle is at rest is accelerated with respect to the laboratory, the two systems should not be connected by a Lorentz transformation. We can circumvent this difficulty by a frequently used strategem (elevated by some to the status of an additional postulate of relativity). We imagine an infinity of inertial systems moving uniformly relative to the laboratory system, one of which instantaneously matches the velocity of the particle. The particle is thus instantaneously at rest in an inertial system that can be connected to the laboratory system by a Lorentz transformation. It is assumed that this Lorentz transformation will also describe the properties of the particle and its true rest system as seen from the laboratory system.

Suppose now, that O_1 is the laboratory system with O_2 and O_3 two of the instantaneous rest systems a time Δt apart in the particle's motion. By Eq. (7–34)

$$
\vec{W} = \frac{\overrightarrow{\Delta\Omega}}{\Delta t} = \frac{\gamma-1}{\beta^2}\,\vec{\beta}\times\frac{d\vec{\beta}}{dt} = \frac{\gamma^2}{\gamma+1}\frac{\vec{a}\times\vec{v}}{c^2}
$$

unprimed accelerating.

$A(\beta) A(-\beta) = ?$

$c\beta = v(t)$

$q(\beta + \delta\beta) = v(t+\delta t)$

the laboratory observer will see a change in the particle's velocity in this time, Δv, which has only a y component $\beta_2'' c$. Since the initial z axis has been chosen along the direction of v, the vector of the infinitesimal rotation in this time can be written

$X' = A(\beta)\, X(t)$

$$\Delta\Omega = -(\gamma - 1)\frac{v \times \Delta v}{v^2}. \tag{7-38}$$

$X'' = A(\beta + \delta\beta)\, X(t+\delta t)$

Hence, if the particle has some specific direction attached to it (such as a spin vector) it will be observed from the laboratory system that this direction *precesses* with an angular velocity

$X'' = A X'$

$A = ?$

$$\omega = -(\gamma - 1)\frac{v \times a}{v^2}, \tag{7-39}$$

$A(\beta + \delta\beta) X$

where a is the particle's acceleration as seen from O_1. Equation (7–39) is frequently encountered in the form it takes when v is small enough that γ can be approximated:

$= A X' = A(\beta + \delta\beta)\, A^{-1}(\beta)\, X'$

$$\omega = \frac{1}{2c^2}(a \times v). \tag{7-39'}$$

In either form, ω is known as the *Thomas precession frequency.*

$A = A(\beta + \delta\beta)(A(\beta)) = A(\beta + \delta\beta) A(-\beta) \implies 7-33 ?$

7–3 LORENTZ TRANSFORMATIONS IN REAL FOUR DIMENSIONAL SPACES*

While retaining its property of preserving the velocity of light, the Lorentz transformation can also be considered as a linear transformation in a *real* four-dimensional space. The fourth coordinate in such a space is usually denoted by $x_0 = ct$. In order that light should propagate in all systems isotropically with the speed c, the square of the magnitude of a vector should still be given by Eq. (7–13). Hence the real space cannot be Euclidean, but must be Riemannian with a diagonal metric tensor G of the form

$$G = \begin{pmatrix} 1 & 0 & 0 & 0 \\ 0 & 1 & 0 & 0 \\ 0 & 0 & 1 & 0 \\ 0 & 0 & 0 & -1 \end{pmatrix}, \tag{7-40}$$

---

*Almost all of the kinematical features of the Lorentz transformation needed for the rest of the book have been presented in the previous section. While this section, and the next, are of importance for the use of Lorentz transformation in various parts of modern physics, they may be omitted without affecting the discussion of classical mechanics of relativistic particles.

where the index on the coordinates runs 1230. It will be remembered from Eq. (6–26) that the square of the magnitude of a vector in such a space is given by

$$\tilde{x}Gx = x_1^2 + x_2^2 + x_3^2 - x_0^2 = x^2 + y^2 + z^2 - c^2 t^2, \tag{7-41}$$

exactly as in Eq. (7–13).* A homogeneous Lorentz transformation is a linear transformation in this real space that preserves the magnitude of vectors. Because the elements of the transformation are real and thus not identical with **L**, the real Lorentz transformation will be designated by **Λ**. The condition that vectors have the same magnitude before and after transformation is that

$$\tilde{x}'Gx' = \tilde{x}Gx. \tag{7-42}$$

But, we have

$$\tilde{x}'Gx' = \widetilde{\Lambda x}G\Lambda x = \tilde{x}\tilde{\Lambda}G\Lambda x.$$

Thus in order for Eq. (7–42) to hold, **Λ** must be such that

$$\tilde{\Lambda}G\Lambda = G. \tag{7-43}$$

Equation (7–43) can be read as saying the congruence transformation **Λ** leaves the metric tensor unchanged. A more significant view of Eq. (7–43) is to see its analogy to the condition for an orthogonal matrix, e.g., Eq. (7–24), in a Euclidean space (where the metric tensor is **1**). Indeed, Eq. (7–43) can be considered as the orthogonality condition on **Λ** in the real Riemannian space whose metric tensor is **G**.

Translations between formulas expressed in Minkowski space and in the real four-space are easily accomplished since

$$x_4 = i x_0. \tag{7-44}$$

It follows that

$$\Lambda_{j0} = i L_{j4}, \quad \Lambda_{0k} = -i L_{4k}, \tag{7-45}$$

while all other elements remain unchanged. Thus the pure Lorentz transformation with relative velocity along the z axis, corresponding to Eq. (7–18), has the real symmetric matrix

$$\Lambda = \begin{pmatrix} 1 & 0 & 0 & 0 \\ 0 & 1 & 0 & 0 \\ 0 & 0 & \gamma & -\beta\gamma \\ 0 & 0 & -\beta\gamma & \gamma \end{pmatrix}. \tag{7-46}$$

---

* Because the square may be either positive or negative, metric tensors such as Eq. (7–40) are said to define an indefinite metric.

All of the kinematic formulas of the preceding section can similarly be rewritten without difficulty. It should be remembered that the dot product of two vectors represented by **x** and **y** must now be written (cf. Eq. 6–25) as

$$\tilde{\mathbf{x}}\mathbf{G}\mathbf{y} = \tilde{\mathbf{y}}\mathbf{G}\mathbf{x} = x_\mu g_{\mu\nu} y_\nu.$$

(The orthogonality condition Eq. (7–43) guarantees that such a dot product is invariant under the Λ transformation.)

It is possible to get rid of the ubiquitous presence of the metric tensor and make the formulas resemble more those for Cartesian spaces by a change in notation, which requires a brief excursion into Riemannian geometry. Suppose we form a vector out of the coordinate elements dx_μ and consider its behavior under a general coordinate transformation of the kind

$$y_\nu = f_\nu(x_1, x_2, \ldots). \tag{7–47}$$

Then the transformation properties of dx_μ are

$$dy_\nu = \frac{\partial f_\nu}{\partial x_\mu} dx_\mu = \frac{\partial y_\nu}{\partial x_\mu} d\dot{x}_\mu. \tag{7–48}$$

Here the derivatives are the elements of the Jacobian matrix of transformation between (x) and (y). For a linear transformation **A** they would simply be the matrix elements $A_{\nu\mu}$. On the other hand the elements of a gradient vector would transform according to the equation

$$\frac{\partial}{\partial y_\nu} = \frac{\partial x_\mu}{\partial y_\nu} \frac{\partial}{\partial x_\mu}. \tag{7–49}$$

In Eq. (7–49) the coefficients are now the Jacobian matrix elements of the inverse transformation from y to x. Vectors that transform as in Eq. (7–48) are called *contravariant* vectors and by convention are given superscripts:

$$D'^\nu = \frac{\partial y_\nu}{\partial x_\mu} D^\mu, \tag{7–48'}$$

while *covariant* vectors transform as in Eq. (7–49) and are indicated by subscripts:

$$F'_\nu = \frac{\partial x_\mu}{\partial y_\nu} F_\mu, \tag{7–49'}$$

If one goes from (x) to (y) and then back to (x), the final result is of course an identity transformation. Hence the product of the Jacobian matrices for a transformation and its inverse must be the unit matrix. It therefore follows that the inner product (dot product) of a contravariant and a covariant vector is invariant under transformation, for

$$D'^\nu F'_\nu = \frac{\partial y_\nu}{\partial x_\mu} \frac{\partial x_\rho}{\partial y_\nu} D^\mu F_\rho = \delta_{\mu\rho} D^\mu F_\rho = D^\mu F_\mu.$$

For Cartesian spaces there is no difference between covariant and contravariant vectors under orthogonal linear transformations. Thus, if the matrix **A** describes such a transformation, a contravariant vector transforms as

$$D'^v = A_{v\mu}D^\mu,$$

and a covariant vector as

$$F'_v = (\mathbf{A}^{-1})_{\mu v}F_\mu = (\tilde{\mathbf{A}})_{\mu v}F_\mu = A_{v\mu}F_\mu.$$

We have therefore had little need to distinguish between the two types of transformation behavior till now.

The qualities of covariance and contravariance can easily be generalized to tensors of higher rank. Thus, a covariant tensor **G** of the second rank transforms as

$$G'_{\mu v} = G_{\rho\lambda}\frac{\partial x_\rho}{\partial y_\mu}\frac{\partial x_\lambda}{\partial y_v}, \tag{7-50}$$

and the contraction of a covariant tensor of the second rank with two contravariant vectors, e.g., $G_{\mu v}D^\mu F^v$, can similarly be shown to transform invariantly in form. In a Riemannian space, where the element of path length (cf. Eq. 6-24)

$$(ds)^2 = g_{\mu v}\,dx^\mu\,dx^v \tag{7-51}$$

is invariant under the transformations of interest, it follows that the metric tensor is a covariant tensor.* By a similar argument the contraction of a covariant tensor of the second rank and a contravariant vector transforms as a covariant vector:

$$F'_\mu = G'_{\mu v}D'^v = G_{\rho\lambda}\frac{\partial x_\rho}{\partial y_\mu}\frac{\partial x_\lambda}{\partial y_v}\frac{\partial y_v}{\partial x_\tau}D^\tau$$

$$= G_{\rho\lambda}\frac{\partial x_\rho}{\partial y_\mu}\delta_{\lambda\tau}D^\tau = G_{\rho\lambda}D^\lambda\frac{\partial x_\rho}{\partial y_\mu}$$

$$= F_\rho\frac{\partial x_\rho}{\partial y_\mu}.$$

Thus (and finally) the dot product of two contravariant vectors A^μ, B^v in our real four-space,

$$g_{\mu v}A^\mu B^v,$$

---

* For the Lorentz transformation this can also be seen directly from the orthogonality condition, Eq. (7-43), written as

$$\mathbf{G} = \tilde{\Lambda}^{-1}\mathbf{G}\Lambda^{-1}$$

and viewed as a congruence transformation of **G**.

can be rewritten as

$$A_v B^v$$

where A_v is the covariant vector:

$$A = g_{\mu\nu} A^\mu. \tag{7-52}$$

In this manner the square of the magnitude of the position vector in real four-space can be written as

$$x_\mu x^\mu$$

without having to introduce the metric tensor specifically. One just has to remember that one member of the dot product is replaced by the covariant vector formed by contraction with the metric tensor.*

A variant on the use of a real space is to introduce one with the metric tensor

$$\mathbf{G'} = \begin{pmatrix} -1 & 0 & 0 & 0 \\ 0 & -1 & 0 & 0 \\ 0 & 0 & -1 & 0 \\ 0 & 0 & 0 & +1 \end{pmatrix}. \tag{7-53}$$

The square of the magnitude of a position vector is now

$$\mathbf{x}\,\mathbf{G'x} \equiv x_\mu G'_{\mu\nu} x_\nu = -x_1^2 - x_2^2 - x_3^2 + c^2 t^2. \tag{7-41'}$$

Clearly a transformation preserving this magnitude will also keep the speed of light invariant, and the corresponding Lorentz transformation must be identical with Λ above. All the previous formalism is unaltered, only the value of dot products changes sign. Advocates of the metric tensor (7-53) point out that it will eliminate some minus signs in later formulas arising in particle dynamics. On the other hand, this metric represents a discontinuous change from what is considered the magnitude of a vector in ordinary three-space. The metric tensor $\mathbf{G'}$ is sometimes said to have the "signature" $(---+)$ whereas $\mathbf{G}$, Eq. (7-40), has the signature $(+++-)$. The tensors can also be identified by their traces, $+2$ for $\mathbf{G}$ and -2 for $\mathbf{G'}$.

In their massive treatise on gravity,† Misner, Thorne, and Wheeler call for the extinction of the complex Minkowski space: "One sometime participant in special relativity will have to be put to the sword: '$x^4 = ict$'." It is claimed that the use of a Cartesian space hides the basic indefinite nature of the metric—that the square of the magnitude of vectors can be positive, zero, or negative. Further, the artifice of a complex Cartesian space is feasible only in special relativity; in the

---

* If we are concerned with the dot product of two covariant vectors, the index can be "raised" by contraction with the inverse of the metric tensor, which can be shown to be contravariant. The metric tensors employed here, which are diagonal with elements ± 1, are their own inverses and there is no difference between covariant and contravariant metric tensors.

† C. W. Misner, K. S. Thorne, and J. A. Wheeler, "Gravitation," 1973, p. 51.

theory of general relativity the space is curved and the use of a non-Cartesian metric tensor is inescapable. It is also pointed out that in quantum mechanics where wave functions or state vectors are complex, the use of an imaginary coordinate will complicate the operation of complex, conjugation. These arguments appear to be persuasive; the real four-space with indefinite metric (often with trace -2) is widely used in quantum mechanics and particle physics.

It is difficult to oppose such distinguished authority. Nonetheless, we shall here make a stand against the euthanasia of Minkowski space. Neither general relativity nor quantum mechanics will be discussed in this book; we need not bow to their special requirements. The indefinite character of the line element is as manifest or as hidden in the complex space as in the real space; we shall have it thrust upon us only too often. On the other hand, the formulas in complex Minkowski space are usually particularly simple and neat, without the encumbrances of metric tensors or the (here) artificial distinction between covariant and contravariant quantities. It also permits natural extensions from our experience with ordinary three-dimensional space. For these reasons we shall use the formalism of the Minkowski space almost exclusively from here on. Most of the equations look the same in either space; in any case, conversion from one space to the other is a simple matter. We will make use of real space in only a few instances, where it is particularly convenient (as in the next section), and then almost always with the metric having the trace $+2$.

7-4 FURTHER DESCRIPTIONS OF THE LORENTZ TRANSFORMATION

One of the instances where the use of a real time coordinate leads to simpler derivations is in showing the homomorphism between the Lorentz transformations and a class of 2×2 complex matrices. It will be remembered from Section 4–5 that a homomorphism exists between the proper spatial rotations and unitary complex matrices with determinant $+1$ ("unimodular"). A similar homomorphism can be demonstrated for the Lorentz transformation, using much the same methods as were employed for spatial rotations.

Consider a 2×2 matrix **S** defined as

$$\mathbf{S} \equiv \begin{pmatrix} z + ct & x - iy \\ x + iy & -z + ct \end{pmatrix} \equiv \begin{pmatrix} x_0 + x_3 & x_1 - ix_2 \\ x_1 + ix_2 & x_0 - x_3 \end{pmatrix}. \tag{7-54}$$

Here **S**, a generalization of **P** given in Eq. (4–59), is a hermitean matrix. Indeed, it is in the form of the most general hermitean 2×2 matrix possible for real x_μ. We saw in Section (4–5) that the Pauli spin matrices, Eq. (4–74), together with the unit matrix form a complete basis set of 2×2 hermitean matrices. One can express **S** simply in terms of this basis set if the unit matrix is denoted as $\boldsymbol{\sigma}_0$; **S** can then be written as

$$\mathbf{S} = x_\mu \boldsymbol{\sigma}_\mu, \tag{7-55}$$

where the summation is from 0 through 3. The determinant of **S** is easily found to be

$$|\mathbf{S}| = -x^2 - y^2 - z^2 + c^2 t^2, \tag{7-56}$$

and thus proportional to the square of the position vector in four-space. Let **Q** be the same general complex 2×2 matrix as in Section 4–5 (Eq. 4–49) with only the restriction that the determinant of **Q** shall be $+1$, that is, **Q** is unimodular. Since the elements of **Q** are in general complex, the condition on the determinant provides two equations of constraint, reducing the number of the independent elements of **Q** to 6—as many as are needed in the homogeneous Lorentz transformation. With the aid of **Q** let us transform **S** according to the scheme (cf. Eq. 4–60)

$$\mathbf{S}' = \mathbf{Q}\mathbf{S}\mathbf{Q}^\dagger. \tag{7-57}$$

Because **Q** is not unitary this is not a similarity transformation, but it is simple to show that it also preserves the hermitean character of **S**. Hence the transformed matrix **S**′ must have a form similar to Eq. (7–55),

$$\mathbf{S}' = x'_\nu \sigma_\nu,$$

where the x'_ν must be real. Further, because **Q** is unimodular, the determinant of **S**′ must be the same as the determinant of **S**. We can therefore say that **Q** generates a transformation from coordinates x_μ to coordinates x'_ν in such a manner as to preserve the velocity of light—**Q** *is homomorphic to at least some subclass of the Lorentz transformation.*

It is possible to show that the Lorentz transformation equivalent to **Q** must be both proper and orthochronous, but the formal proof is involved.* However a plausible argument can be sketched briefly. Following the procedure used in Section 4–5 it is easy to show that if $\mathbf{L}_1$ and $\mathbf{L}_2$ are two Lorentz transformations corresponding to $\mathbf{Q}_1$ and $\mathbf{Q}_2$, then $\mathbf{L}_1\mathbf{L}_2$ corresponds to $\mathbf{Q}_3 = \mathbf{Q}_1\mathbf{Q}_2$, which is also unimodular. Hence the subclass of the Lorentz transformation homomorphic to the Q matrices is such that the product of two Lorentz transformations is still a member of the same subclass—one of the most important of the "group properties." Now, we have seen that only two classes of Lorentz transformation have this property: the general homogeneous Lorentz transformation and the restricted Lorentz transformation. To decide between these two possibilities it may be noted that brute force evaluation of the matrix products in Eq. (7–57) shows that L_{44} is given in terms of the elements of **Q** (Eq. 4–49) by the expression

$$L_{44} = \Lambda_{44} = \frac{1}{2}(\alpha\alpha^* + \beta\beta^* + \gamma\gamma^* + \delta\delta^*),$$

which is positive definite. For nonorthochronous transformations, L_{44} is negative, in fact $\leqslant -1$. Of the two classes, the **Q** matrices can therefore be

---

* See, for example, A. O. Barut, *Electrodynamics and Classical Theory of Fields*, p. 23, or A. J. Macfarlane, *Jour. Math. Phys.* **3**, 1116 (1962).

homomorphic only to the restricted Lorentz transformations, just as the unitary unimodular matrices are homomorphic only to proper spatial rotations.

Again, as in the three-dimensional case, there is a double-valued association between $\mathbf{L}$ and $\mathbf{Q}$, because both $+\mathbf{Q}$ and $-\mathbf{Q}$ lead to the same Lorentz transformation. Hence the relationship is spoken of as a homomorphism rather than isomorphism. The actual expressions for the elements $L_{\mu\nu}$ in terms of the elements of $\mathbf{Q}$ can be obtained either by brute force evaluation of Eq. (7–57) or, more elegantly, by methods described in the references of Barut and MacFarlane cited above. For *pure* Lorentz transformations, however, there is a simple description of the $\mathbf{Q}$ analogous to Eq. (4–98) for proper spatial rotations. Direct evaluation of Eq. (7–57) shows that the unimodular matrix

$$\mathbf{Q} = \begin{pmatrix} e^{-\psi/2} & 0 \\ 0 & e^{\psi/2} \end{pmatrix}, \tag{7–58}$$

where ψ is real, corresponds to the Lorentz transformation

$$
\begin{aligned}
x_1' &= x_1 \\
x_2' &= x_2 \\
x_3' &= x_3 \cosh\psi - x_0 \sinh\psi \\
x_0' &= -x_3 \sinh\psi + x_0 \cosh\psi.
\end{aligned}
\tag{7–59}
$$

Equation (7–59) will be recognized as the pure Lorentz transformation, Eq. (7–46), for relative motion along the x_3 axis, with ψ defined by Eq. (7–20). The matrix in Eq. (7–58) can also be written as

$$\mathbf{Q} = \mathbf{1} \cosh\frac{\psi}{2} - \sigma_3 \sinh\frac{\psi}{2}, \tag{7–60}$$

analogously to Eq. (4–76) for spatial rotation about the z axis. We can easily generalize this result to an arbitrary pure Lorentz transformation. Just as a proper rotation could be parameterized in terms of a direction of the axis of rotation and a finite angle of rotation, so a pure Lorentz transformation can be described in terms of a unit vector $\boldsymbol{\kappa}$ in the direction of relative motion and an "angle" ψ defined by Eq. (7–20). The vector form of the Lorentz transformation, Eqs. (7–11) and (7–12), can be rewritten in these parameters as

$$\mathbf{r}' = \mathbf{r} + (\boldsymbol{\kappa}\cdot\mathbf{r})\boldsymbol{\kappa}(\cosh\psi - 1) - \boldsymbol{\kappa}x_0\sinh\psi, \tag{7–61}$$

$$x_0' = -(\boldsymbol{\kappa}\cdot\mathbf{r})\sinh\psi + x_0\cosh\psi. \tag{7–62}$$

Equation (7–61) is quite reminiscent of the form of the finite rotation formula, Eq. (4–92). The obvious extension of Eq. (7–60) to relative motion in a given direction $\boldsymbol{\kappa}$ is

$$\mathbf{Q}(\boldsymbol{\kappa},\psi) = \mathbf{1}\cosh\frac{\psi}{2} - \boldsymbol{\kappa}\cdot\boldsymbol{\sigma}\sinh\frac{\psi}{2}, \tag{7–63}$$

which is to be compared with Eq. (4–98) for spatial rotations.* Note that the form of Eq. (7–63) shows that the $\mathbf{Q}$ matrices for pure Lorentz transformations are hermitean, whereas for proper rotations they are unitary. As with pure rotations, Eq. (7–63) implies an exponential representation for the $\mathbf{Q}$ matrix of a pure Lorentz transformation:

$$\mathbf{Q}(\boldsymbol{\kappa}, \psi) = \exp\left[-\boldsymbol{\kappa} \cdot \boldsymbol{\sigma}\,(\psi/2)\right]. \tag{7–64}$$

Since a restricted Lorentz transformation is the product of a boost and a spatial rotation, we can combine Eq. (7–64) with Eq. (4–99) to obtain the corresponding form of the Q matrix:

$$\mathbf{Q}(\mathbf{n}\Phi; \boldsymbol{\kappa}, \psi) = \exp\left[i\mathbf{n} \cdot \boldsymbol{\sigma}(\Phi/2)\right] \exp\left[-\boldsymbol{\kappa} \cdot \boldsymbol{\sigma}\,(\psi/2)\right]. \tag{7–65}$$

It should be remembered, of course, that the exponentials simply stand for the series representations.

The advantages of the unimodular representations of the Lorentz transformation become apparent when we try to resolve the product of two pure Lorentz transformations into a spatial rotation and a pure Lorentz transformation. As was pointed out, to disentangle the effective velocity and rotation vectors out of the straightforward multiplication is a formidable algebraic exercise. With the $\mathbf{Q}$ matrix form, it becomes much simpler although considerable manipulation is still required (see Exercise 13).

Infinitesimal transformations play so great a role in spatial rotations it is natural to seek their counterparts for the Lorentz transformation. We have already made use of an infinitesimal Lorentz transformation in deriving the Thomas precession (cf. Eq. 7–32). While not as immediately useful as in spatial rotations, the infinitesimal Lorentz transformations are valuable in applications where (as we shall see in Section 9–7) they provide the identifying mark for the mathematical group structure of the Lorentz transformation. In terms of κ and ψ, the real pure Lorentz transformation takes the form (either directly from Eqs. (7–61) and (7–62) or via Eqs. (7–17))

$$\Lambda_{ij} = \delta_{ij} + \kappa_i \kappa_j (\cosh \psi - 1),$$

$$\Lambda_{i0} = \Lambda_{0i} = -\kappa_i \sinh \psi, \tag{7–66}$$

$$\Lambda_{00} = \cosh \psi.$$

An infinitesimal pure Lorentz transformation is characterized by an infinitesimal relative velocity between the two systems or equivalently by an infinitesimal "angle" $\Delta\psi$. From Eq. (7–66) an infinitesimal pure Lorentz transformation can be written as

$$\Lambda = \mathbf{1} + \boldsymbol{\delta}, \tag{7–67}$$

---

* We cannot introduce any quantities similar to "Euler angles" because unlike proper rotations, the product of two noncollinear pure Lorentz transformations is not a pure Lorentz transformation.

where

$$\delta = \begin{pmatrix} 0 & 0 & 0 & -\kappa_1\,\Delta\psi \\ 0 & 0 & 0 & -\kappa_2\,\Delta\psi \\ 0 & 0 & 0 & -\kappa_3\,\Delta\psi \\ -\kappa_1\,\Delta\psi & -\kappa_2\,\Delta\psi & -\kappa_3\,\Delta\psi & 0 \end{pmatrix}.$$

In analogy to Eq. (4–117) for spatial rotations, we can introduce the notion of generator matrices for infinitesimal pure Lorentz transformations as

$$\mathbf{K}_1 = \begin{pmatrix} 0 & 0 & 0 & 1 \\ 0 & 0 & 0 & 0 \\ 0 & 0 & 0 & 0 \\ 1 & 0 & 0 & 0 \end{pmatrix}, \qquad \mathbf{K}_2 = \begin{pmatrix} 0 & 0 & 0 & 0 \\ 0 & 0 & 0 & 1 \\ 0 & 0 & 0 & 0 \\ 0 & 1 & 0 & 0 \end{pmatrix},$$

$$\mathbf{K}_3 = \begin{pmatrix} 0 & 0 & 0 & 0 \\ 0 & 0 & 0 & 0 \\ 0 & 0 & 0 & 1 \\ 0 & 0 & 1 & 0 \end{pmatrix}. \tag{7–68}$$

In terms of $\mathbf{K}_i$ the infinitesimal matrix δ can be written very simply as

$$\delta = -\kappa_i\mathbf{K}_i\,\Delta\psi,$$

and the change in a vector in real four-space, written as a matrix $\mathbf{x}$, under a pure infinitesimal Lorentz transformation is given completely as

$$\Delta\mathbf{x} = -\kappa_i\mathbf{K}_i\mathbf{x}\,\Delta\psi. \tag{7–69}$$

Equation (7–69) constitutes in effect a differential equation for generating a finite pure Lorentz transformation.

Because the spatial rotations are a subgroup of the Lorentz transformation, we should add to these the generators of the infinitesimal spatial rotation in the Λ formalism, which are identical with the $\mathbf{M}_i$ defined in Eq. (4–117) except for an added fourth row and column of zeros. The commutators or Lie brackets of $\mathbf{M}_i$ remain unchanged in this guise as

$$[\mathbf{M}_i, \mathbf{M}_j] = \epsilon_{ijk}\mathbf{M}_k. \tag{4–118}$$

Direct multiplication will verify that the $\mathbf{K}_i$ satisfy the commutation relations

$$[\mathbf{K}_i, \mathbf{K}_j] = -\epsilon_{ijk}\mathbf{M}_k,$$

and

$$[\mathbf{K}_i, \mathbf{M}_j] = \epsilon_{ijk}\mathbf{K}_k. \tag{7–70}$$

Equations (4–118) and (7–70) together define the Lie algebra of the restricted Lorentz group and act as a hallmark to identify that group. We shall find they appear, surprisingly, in association with the *nonrelativistic* unbound motion in the Kepler problem (see Section 9–7).

7–5 COVARIANT FOUR-DIMENSIONAL FORMULATIONS

Having obtained the Lorentz transformation to replace the incorrect Galilean transformation, we can now proceed to the second stage and require that the laws of mechanics, in common with all of physics, shall have the same form in all uniformly moving systems. The task of examining the laws of physics for invariance in form under Lorentz transformation is greatly facilitated by writing them in terms of the four-dimensional world introduced in the previous sections. Indeed, as will be shown, it is then possible to verify the Lorentz invariance of a given equation merely by inspection.

Invariance of form under Lorentz transformation is not the only invariant property demanded of physical laws. Clearly the physical content of any given relation cannot be affected by the particular orientation chosen for the spatial axes; the laws of physics must also be invariant in form under rigid rotations, i.e., proper spatial orthogonal transformations. An examination of this more familiar invariance requirement will clarify the procedure to be followed in establishing invariance under Lorentz transformation.

Normally we do not worry about the invariance of our theories under spatial rotations. In constructing any equation it is always required that the terms of the equation be *all* scalars, or *all* vectors; in general all terms must be tensors of the same rank, and this requirement automatically ensures the desired invariance under rotation. Thus a scalar relation will have the general form

$$a = b,$$

and since both sides of the equation, being scalars, are invariant under spatial rotations the relation obviously holds in all coordinate systems. A vector relation, of the form

$$\mathbf{F} = \mathbf{G},$$

really stands for three separate relations between the components of the vector:

$$F_i = G_i.$$

The values of these components, of course, are not invariant under the spatial rotation; rather they are transformed to new values F_i', G_i' that are the components of the transformed vectors, $\mathbf{F}', \mathbf{G}'$. But because both sides of the component relations transform in identical fashion, the same relation must hold between the transformed components:

$$F_i' = G_i'.$$

The relationship between the two vectors is thus undisturbed by the spatial rotation; in the new system we still have

$$\mathbf{F}' = \mathbf{G}'.$$

Note that the invariance in the form of the relationship is entirely in consequence of the fact that both sides transform as vectors. We speak of the terms of the equation as being *covariant*.* Similarly, an equality between two tensors of the second rank

$$\mathbf{C} = \mathbf{D}$$

necessarily implies the same equality between the two transformed tensors

$$\mathbf{C}' = \mathbf{D}'$$

because the two tensors transform covariantly under a spatial rotation. On the other hand, an equation involving separately a component of a vector and, say, a component of a tensor obviously cannot remain invariant in form under a three-dimensional orthogonal transformation. *Invariance of a physical law under rotation of the spatial coordinate system requires covariance of the terms of the equation under three-dimensional orthogonal transformation.*

Now, the restricted Lorentz transformation has been identified as an orthogonal transformation in Minkowski, or world, space. We have already handled scalars and vectors in this four-dimensional space. Similarly one can set up tensors of other ranks in this space, with transformation properties that are obvious generalizations of the three-dimensional transformations. These tensors of various ranks will be spoken of as *world tensors*, starting with *world scalars*, *world vectors* (or *four-vectors*), etc. The invariance of the form of any physical law under Lorentz transformation will then be immediately evident once it is expressed in a *covariant four-dimensional form*, all terms being world tensors of the same rank. A law failing to meet the requirements of the equivalence principle cannot be put into a covariant form. The four-dimensional transformation properties of the terms of a physical law thus act as the touchstone for examining its relativistic validity.

The simplest example of a four-vector is the position vector of a "point" in Minkowski space, with components (x_1, x_2, x_3, x_4). Since the four coordinates of a world point tell *where* in ordinary space something happened and *when* it happened, it is perhaps more descriptive to speak of a point in four-dimensional space as an *event*. Though we shall often use three-dimensional terminology in speaking of world space, it is well to remember the important physical differences in meanings between the two spaces.

As a particle moves in ordinary space its corresponding point in four-dimensional space will describe a path known as the *world line*. The four-vector dx_μ represents the change in the position four-vector for a differential motion along the world line. From the dot product of dx_μ with itself we can form a world scalar (and hence a Lorentz invariant), denoted by $d\tau$ and defined by the equation

$$(d\tau)^2 = -\frac{1}{c^2} dx_\mu \, dx_\mu. \tag{7-71}$$

---

* As used here, the term "covariant" *has nothing whatever to do with "covariant transformation"* as used in Section 7–3. Unfortunately we have here another instance of the mathematicians and physicists using the same term for two entirely separate concepts. Both conventions are too entrenched to be changed now.

The significance of $d\tau$ can be made clear by evaluating Eq. (7–71) in the Lorentz system in which the particle is momentarily at rest. In such a system the components of the transformed vector dx'_μ are $(0, 0, 0, icdt')$ and the invariant $d\tau$ is given by

$$(d\tau)^2 = -\frac{1}{c^2} dx'_\mu dx'_\mu = (dt')^2.$$

Thus $d\tau$ is the time interval as measured on a clock traveling with the particle,* hence it is referred to as an interval of the particle's *proper time* or *world time*.

The relation between $d\tau$ and an interval of time as measured in a given Lorentz system can be derived directly by expanding the defining equation (7–71):

$$(d\tau)^2 = -\frac{1}{c^2}((dx)^2 + (dy)^2 + (dz)^2 - c^2(dt)^2),$$

or

$$d\tau = dt \sqrt{1 - \frac{1}{c^2}\left[\left(\frac{dx}{dt}\right)^2 + \left(\frac{dy}{dt}\right)^2 + \left(\frac{dz}{dt}\right)^2\right]},$$

which is equivalent to the relation

$$\frac{d\tau}{\sqrt{1 - \beta^2}} = dt. \tag{7–72}$$

It may seem the use of the symbol β in Eq. (7–72) would lead to confusion. Hitherto β has referred solely to the relative velocity (in units of c) of two inertial systems connected by a Lorentz transformation. Here it is being used to represent the velocity of a particle as observed in one inertial system (the laboratory system). Actually the usage is made consistent by thinking in terms of Lorentz systems instantaneously at rest with respect to the particle, as mentioned in connection with the Thomas precession. Then β is both the relative velocity between the observer's system and the instantaneous rest system, and the observed velocity of the particle. Equation (7–72) says that a time interval measured in the rest system is always longer than the corresponding time interval observed in a system in which the particle is not at rest. This is an example of the by-now familiar "time-dilatation" which has been abundantly verified experimentally, most notably by observations of the lifetimes of unstable particles decaying in flight.†

---

* By definition $d\tau$ is taken as the positive square root of the expression given in Eq. (7–71). The minus sign inside the square root would have been eliminated if we had used the metric of trace -2, Eq. (7–53).

† Cf. the discussion in R. D. Sard, *Relativistic Mechanics*, 1970, pp. 96–100.

As has been mentioned, the square of the magnitude of a four-vector is not necessarily positive definite. Four-vectors for which the square of the magnitude is greater than or equal to zero are called *space-like*; when the squares of the magnitudes are negative they are known as *time-like* vectors. Since these characteristics arise from the dot products of the vectors with themselves, which are world scalars, the designations are obviously invariant under Lorentz transformation. The names stem from the fact that the square of a spatial vector is always positive definite, and a space-like four-vector can always be transformed so that its fourth component vanishes, as we shall see. On the other hand, a time-like four-vector must always have a fourth component, but it can be transformed so that the first three components vanish. As an example of these concepts it may be noted that the difference vector between two world points can be either space-like or time-like. Let X_μ be the difference vector, defined as

$$X_\mu = x_{1\mu} - x_{2\mu},$$

the subscripts 1 and 2 denoting the two events. The magnitude of X_μ is given by

$$X_\mu X_\mu = |\mathbf{r}_1 - \mathbf{r}_2|^2 - c^2(t_1 - t_2)^2.$$

Thus, X_μ is space-like if the two world points are separated such that

$$|\mathbf{r}_1 - \mathbf{r}_2|^2 \geq c^2(t_1 - t_2)^2,$$

while it is time-like if

$$|\mathbf{r}_1 - \mathbf{r}_2|^2 < c^2(t_1 - t_2)^2.$$

The condition for a time-like difference vector is equivalent to stating that it is possible to bridge the distance between the two events by a light signal, while if the points are separated by a space-like difference vector they cannot be connected by any wave traveling with the speed c.

The spatial axes can always be so oriented that the spatial difference vector $\mathbf{r}_1 - \mathbf{r}_2$ is along the z axis, such that $|\mathbf{r}_1 - \mathbf{r}_2|$ is equal to $z_1 - z_2$. Under a Lorentz transformation with velocity v parallel to the z axis, the fourth component of X_μ then transforms according to Eq. (7–5):

$$c(t_1' - t_2') = \frac{c(t_1 - t_2) - \dfrac{v}{c}(z_1 - z_2)}{\sqrt{1 - \beta^2}}.$$

If X_μ is space-like and the events are designated such that $t_2 > t_1$, then

$$c(t_1 - t_2) < z_1 - z_2,$$

and it is therefore possible to find a velocity $v < c$ such that $ic(t_1' - t_2') \equiv X_4'$ vanishes, as was stated above. Physically the vanishing of X_4' means that if the distance between two events is space-like, then one can always find a Lorentz system in which the two events are simultaneous. On the other hand, for time-like separations between events one cannot find a Lorentz transformation that will make them simultaneous or, a forteriori, change the order of the time sequence of the two events. Thus, "before" and "after" are concepts invariant under a Lorentz transformation, and causality is preserved. That the sequence of events with space-like separations can be reversed does not violate causality, because by definition there is no way one event can influence the other. For example, nothing we can do on Earth now will affect what happens within the next ten years on a planet around a star ten light-years away.

Examples of four-vectors may be easily multiplied. Thus, the four-velocity u_v is defined as the rate of change of the position vector of a particle with respect to its proper time:

$$u_v = \frac{dx_v}{d\tau}, \tag{7-73}$$

with space and time components

$$u_i = \frac{v_i}{\sqrt{1-\beta^2}} \quad \text{and} \quad u_4 = \frac{ic}{\sqrt{1-\beta^2}}. \tag{7-74}$$

The world velocity has a constant magnitude, for the sum $u_v u_v$ is given by

$$u_v u_v = \frac{v^2}{1-\beta^2} - \frac{c^2}{1-\beta^2} = -c^2, \tag{7-75}$$

and it is thus also time-like.

It can be shown that the electric current density $\mathbf{j}$ and $ic\rho$, where ρ is the electric charge density, also form a four-vector, j_μ.* The equation of continuity for charge and current,

$$\nabla \cdot \mathbf{j} = -\frac{\partial \rho}{\partial t},$$

can then be written in the language of Minkowski space as

$$\frac{\partial j_\mu}{\partial x_\mu} = 0. \tag{7-76}$$

We have already seen (p. 290) that the four-gradient operator transforms in Minkowski space as a four-vector,† since

$$\frac{\partial}{\partial x_v'} = \frac{\partial x_\mu}{\partial x_v'} \frac{\partial}{\partial x_\mu} = L_{\mu v}' \frac{\partial}{\partial x_\mu} = L_{v\mu} \frac{\partial}{\partial x_\mu}, \tag{7-77}$$

by the orthogonality condition on **L**. Hence the left-hand side of Eq. (7–76) is a four-divergence of a four-vector, i.e., a world scalar, and therefore invariant under a

---

* j_μ is simply the four-vector $\rho_0 u_\mu$, where ρ_0 is the charge density in the system in which the charges are at rest, i.e., the "proper charge density." See Panofsky and Phillips, *Classical Electricity and Magnetism*, or Jackson, *Classical Electrodynamics*.

† In non-Cartesian spaces it transforms covariantly (in the mathematician's sense) and Eq. (7–76) is thus the invariant dot product of a covariant vector and a contravariant vector.

Lorentz transformation. Here we have an example of how restating a law of physics in the language of Minkowski space can manifestly show its Lorentz covariance.

Another illustration involves the vector and scalar electromagnetic potentials, which together form a four-vector $A_\mu \rightarrow (\mathbf{A}, i\phi)$. If the potentials satisfy the Lorentz gauge condition

$$\nabla \cdot \mathbf{A} + \frac{1}{c}\frac{\partial \phi}{\partial t} = 0, \tag{7-78}$$

then they separately satisfy wave equations of the form

$$\nabla^2 \mathbf{A} - \frac{1}{c^2}\frac{\partial^2 \mathbf{A}}{\partial t^2} = -\frac{4\pi}{c}\mathbf{j},$$

$$\nabla^2 \phi - \frac{1}{c^2}\frac{\partial^2 \phi}{\partial t^2} = -4\pi\rho. \tag{7-79}$$

(See the above cited reference to Panofsky and Phillips or Jackson.) In the language of Minkowski space the Lorentz condition can be written in a manifestly covariant form as

$$\frac{\partial A_\mu}{\partial x_\mu} = 0. \tag{7-80}$$

In obvious generalization of the three-dimensional del operator ∇, the four-dimensional gradient operator may be denoted by the symbol $\square$. The dot product of $\square$ with itself, $\square^2$ (known as the *D'Alembertian*), is therefore a world scalar differential operator:

$$\square^2 = \frac{\partial^2}{\partial x_\mu \partial x_\mu} = \nabla^2 - \frac{1}{c^2}\frac{\partial^2}{\partial t^2}.$$

Hence the set of wave equations, Eqs. (7-78), can be written as a clearly covariant world vector equation:

$$\square^2 A_\mu = -\frac{4\pi j_\mu}{c}. \tag{7-81}$$

Just being able to write the gauge and wave equations in the form of Eqs. (7-80) and (7-81) is enough to show that Maxwell's electromagnetic theory is in accord with special relativity, and not with Galilean relativity.

7-6 THE FORCE AND ENERGY EQUATIONS IN RELATIVISTIC MECHANICS

We have seen that Newton's equations of motion, being invariant under a Galilean transformation, cannot be invariant under a Lorentz transformation; they must be suitably generalized to provide a law of force satisfying the covariance requirements of special relativity. Of course, the generalizations we seek must be such that for velocities small compared to c the new equations for a single particle reduce to the familiar form

$$\frac{d}{dt}(mv_i) = F_i. \tag{7–82}$$

Now, the space components of a four-vector by themselves constitute a spatial vector, for a Lorentz transformation for which $L_{4i} = L_{i4} = 0, L_{44} = 1$, is merely an ordinary spatial rotation and affects only the space components of a four-vector. The converse is not true, however; the components of a spatial vector do not necessarily transform as the spatial components of a four-vector. One may multiply the components of an ordinary vector by any function of β without changing their spatial rotation properties. But such a multiplication vitally affects the way in which they change under a Lorentz transformation.* Thus the spatial components of the world velocity u_v form a vector $\mathbf{v}/\sqrt{1 - \beta^2}$; but note that $\mathbf{v}$ itself is not part of a four-vector, it must first be divided by $\sqrt{1 - \beta^2}$.

While Eq. (7–82) is itself not Lorentz-invariant, we can expect, therefore, that its relativistic generalization will be a four-vector equation, whose spatial component reduces to (7–82) in the limit as $\beta \to 0$. It is not difficult to find a four-vector generalization of the left-hand side of the equation. The only four-vector whose space part reduces to $\mathbf{v}$ for small velocities is the world velocity u_v. Further, while m can be taken as an invariant property of the particle, we know that the time t is not a Lorentz-invariant but it can obviously be replaced by the scalar proper time τ, which approaches t as $\beta \to 0$. The desired generalization of Newton's equations of motion for a single particle must therefore have the form

$$\frac{d}{d\tau}(mu_v) = K_v, \tag{7–83}$$

where K_v is some four-vector, known as the *Minkowski force*.

It must not be thought that the spatial components of K_v are to be identified with the components of the force. All that is required by Eq. (7–82) is that K_i reduce to F_i in the limit of small velocities. Thus K_i may be equal to the product of F_i with any function of β that reduces to unity as $\beta \to 0$; the exact relation clearly depends on the Lorentz transformation properties of the force components. Two types of approach have been used in the past to determine the behavior of $\mathbf{F}$ under Lorentz transformation.

One procedure begins by pointing out that, fundamentally, forces arise from only a few physical sources—forces are either gravitational, electromagnetic, or possibly nuclear. It is the duty of a correct theory of these physical phenomena to provide expressions for the forces involved, and these expressions, if stated in covariant form, automatically tell us the transformation properties of the force

---

* As before β (and $\mathbf{v}$) are used interchangeably for the velocity of a particle as seen in the laboratory system, or as the velocity of a Lorentz system in which the particle is instantaneously at rest.

components. Unfortunately we don't have covariant theories of all the possible sources of force, indeed for nuclear forces we have hardly any theory worth speaking about. Only classical electromagnetic theory can be expected to provide a covariant force equation, since the Lorentz transformation was expressly constructed so as to preserve the invariance of the theory. But this is sufficient for our purposes; the transformation properties must be the same for all forces no matter what their origin. The statement "a particle is in equilibrium under the influence of two forces" must hold true in all Lorentz systems, which can only be the case if all forces transform in the same manner.

In Section 1–5 it was pointed out that the electromagnetic force on a particle is given by (cf. Eq. (1–64′))

$$F_i = -q\left(\frac{\partial}{\partial x_i}\left(\phi - \frac{1}{c}\mathbf{v}\cdot\mathbf{A}\right) + \frac{1}{c}\frac{dA_i}{dt}\right). \tag{7–84}$$

In terms of the four-vector potential A_μ, the expression $\phi - \frac{1}{c}\mathbf{v}\cdot\mathbf{A}$ can be written covariantly as

$$\phi - \frac{1}{c}\mathbf{v}\cdot\mathbf{A} = -\frac{1}{c}\sqrt{1 - \beta^2}\,u_\nu A_\nu, \tag{7–85}$$

and the force components F_i become

$$F_i = -\frac{q}{c}\sqrt{1 - \beta^2}\left(-\frac{\partial}{\partial x_i}(u_\nu A_\nu) + \frac{dA_i}{d\tau}\right). \tag{7–86}$$

The expression in the parentheses transforms as the spatial component of a four-vector, so that F_i is equal to the product of $\sqrt{1 - \beta^2}$ and the spatial component of a four-vector, which is to be identified as the Minkowski force K_μ. Hence the connection between the ordinary and Minkowski forces must be

$$F_i = K_i\sqrt{1 - \beta^2}, \tag{7–87}$$

irrespective of the origin of the forces. A by-product to this derivation is the particular form of the Minkowski force on charged particles:

$$K_\mu = \frac{q}{c}\left(\frac{\partial}{\partial x_\mu}(u_\nu A_\nu) - \frac{dA_\mu}{d\tau}\right). \tag{7–88}$$

The alternative procedure attempts to avoid the necessity of using a physical theory beyond mechanics itself; it simply *defines* force as being the time rate of change of momentum, in all Lorentz systems:

$$\frac{dp_i}{dt} = F_i. \tag{7–89}$$

The momentum indicated in (7–89), however, is not mv_i, but rather some relativistic generalization that reduces to it in the limit of small velocities. Lewis and Tolman* have obtained an expression for the relativistic momentum, independent of the form (7–83) for the force law, by noting that a Lorentz-invariant consequence of the definition (7–89) is the conservation of momentum in the absence of external forces. They examine an elastic collision between two particles and find the form of p_i such that it is conserved in such a collision for all Lorentz systems. But having accepted (7–83) as the form of the force law, we can find the relativistic momentum and the meaning of K_i at once, by putting (7–83) in a form resembling (7–89) as closely as possible. From the relation between τ and t, and the definition of world velocity, we can write the spatial components of Eq. (7–83) as

$$\frac{d}{dt}\left(\frac{mv_i}{\sqrt{1-\beta^2}}\right) = K_i\sqrt{1-\beta^2}.$$

Comparison with (7–89) shows that the conservation of momentum theorem is invariant providing the momentum is defined by

$$p_i = \frac{mv_i}{\sqrt{1-\beta^2}}, \tag{7–90}$$

and that F_i and K_i are related according to Eq. (7–87). It will be noted that Eq. (7–90) properly reduces to mv_i as $\beta \to 0$. The two approaches thus lead to the same conclusion. Comparison of Eq. (7–90) with Eq. (7–74) defining the four-velocity shows that p_i forms the spatial part of a four-vector that is called the four-momentum:

$$p_\nu = mu_\nu. \tag{7–91}$$

The generalized equation of motion for a single particle thus can also be written as

$$\frac{dp_\nu}{d\tau} = K_\nu. \tag{7–83'}$$

So far only the space part of the four-vector equation (7–83, 7–83') has been discussed; nothing has been said about the physical significance of the fourth equation. The time-like part of the four-vector K_μ can be obtained directly from the dot product of (7–83) with the world velocity:

$$u_\nu\frac{d}{d\tau}(mu_\nu) = \frac{d}{d\tau}\left(\frac{m}{2}u_\nu u_\nu\right) = K_\nu u_\nu. \tag{7–92}$$

Since the square of the magnitude of u_ν is a constant, $-c^2$ (cf. Eq. 7–75), and m is here likewise a constant, the left-hand side of (7–92) vanishes, leaving

$$K_\nu u_\nu \equiv \frac{F\cdot v}{1-\beta^2} + \frac{icK_4}{\sqrt{1-\beta^2}} = 0. \tag{7–93}$$

---

* See R. D. Sard, *Relativistic Mechanics*, pp. 146–152.

The fourth component of the Minkowski force is therefore

$$K_4 = \frac{i}{c} \frac{\mathbf{F} \cdot \mathbf{v}}{\sqrt{1 - \beta^2}}, \tag{7-94}$$

and the corresponding fourth component of Eq. (7–83) appears as

$$\frac{d}{dt} \frac{mc^2}{\sqrt{1 - \beta^2}} = \mathbf{F} \cdot \mathbf{v}. \tag{7-95}$$

Now, the kinetic energy T is defined in general to be such that $\mathbf{F} \cdot \mathbf{v}$, the rate at which the force does work on the particle, is the time rate of increase of T:

$$\frac{dT}{dt} = \mathbf{F} \cdot \mathbf{v}. \tag{7-95'}$$

This is a definition of kinetic energy that agrees with the customary nonrelativistic form $\frac{1}{2}mv^2$; the relativistic expression, by comparison, is furnished by Eq. (7–95), which is thus seen to be the energy equation with

$$T = \frac{mc^2}{\sqrt{1 - \beta^2}}. \tag{7-96}$$

In the limit as β^2 becomes much less than 1, Eq. (7–96) can be expanded as

$$T \rightarrow mc^2\left(1 + \frac{\beta^2}{2}\right) = mc^2 + \frac{mv^2}{2}. \tag{7-97}$$

This limiting value does not agree with the expected nonrelativistic form; there is here the added term mc^2. It would seem at first sight, however, that the term is of no importance, for clearly any constant of integration can be added to the right of Eq. (7–96) without affecting the validity of Eq. (7–95'). In particular the constant could be $-mc^2$, which would bring T in line with the nonrelativistic value.

However, it is preferable to retain the quantity T as given by Eq. (7–96), for comparison of Eq. (7–91) with Eq. (7–96) shows that iT/c is the fourth component of the four-momentum. In order to preserve the correct transition to nonrelativistic values, it is preferable to transfer the name "kinetic energy" to a separate quantity, K, defined by

$$K = T - mc^2 = mc^2(\gamma - 1). \tag{7-96'}$$

No uniform designation for T exists in the literature. It is sometimes referred to as the "total energy" although strictly this is appropriate only for a free particle, without interactions. It will often be referred to simply as the energy.

By whatever name, T possesses interesting and useful properties. For example, we can show that T as given by Eq. (7–96) has the advantageous property that any situation that conserves the spatial linear momentum must also conserve T. To verify this theorem we need only note that the *statement* "spatial linear momentum is conserved" must be invariant under a Lorentz transformation. (Indeed, the invariance is implicit in Einstein's definition of an inertial system, cf. p. 279). The transformed components of the linear momentum p'_j are given as linear functions of p_i and T. Hence, conservation of p'_j for all Lorentz systems requires joint conservation of all components of p_v, which is the conclusion we were seeking.

In nonrelativistic mechanics, conservation of linear momentum and conservation of kinetic energy were separate matters. Indeed, in Chapter 3 we saw that for inelastic collisions or reactions the linear momentum of the reactants was conserved, but not their kinetic energy. However the relativistic kinetic energy, because of its transformation properties as the fourth component of a world vector, must be conserved even in inelastic encounters, along with the spatial momentum. The term mc^2, known as the *rest energy*, therefore has an important physical significance.

A simple example will show that the conservation of T implies a change in the rest mass of the system as the result of reactions or collisions that are not elastic. Consider, for example, two particles or systems of equal mass moving, in the laboratory system, toward each other with equal and opposite velocities. The total spatial linear momentum is therefore zero in the laboratory system. We can form a total four-momentum for the system

$$P_\mu = p_\mu(1) + p_\mu(2),$$

which therefore has a zero spatial component, but a fourth component $2iT/c$, where T is the energy of each particle as defined by Eq. (7–96). Suppose now that the collision is completely inelastic—after the collision, the particles are brought completely to rest in the laboratory system, like two putty balls that collide and stick together. The total energy is then the rest energy of the final composite system which has a mass

$$M = 2m + \Delta M.$$

Conservation of P_4 in the collision implies

$$2T = Mc^2. \tag{7–98}$$

Now, to give the particles their initial velocity from rest in the laboratory system requires an energy

$$\Delta E = 2(T - mc^2) = \Delta Mc^2. \tag{7–99}$$

The energy of the initial motion in the laboratory system is thus all converted by the inelastic collision into the increased rest mass of the system, according to Einstein's famous relationship, Eq. (7–99).

In the case of the two putty balls colliding, we normally say the incident energy of motion lost in the collision is converted into heat. What special relativity tells us is that the rest mass, or inertia, of the system is increased in proportion to the heat produced. In principle the mass increase could be detected if the system were set in motion by a known force.* On an atomic or nuclear scale the incident energy of motion may, for example, go into the creation of a new third particle. Modern physics abounds in similar instances; something of the relativistic kinematics of such collisions will be found in the next section. Of course the process can go the other way—conversion of rest energy into energy of motion. The most striking example on a terrestrial scale is probably provided by man-made nuclear explosions. Again, the total energy T of a nuclear device or weapon remains constant through the explosion; a large amount of kinetic energy is released only by virtue of a decrease in the rest mass of the contents. Despite the awesome energies produced, the mass loss is only around 0.1 % of the original mass.

Formally the connection between the energy T and the momentum is expressed in the statement that the magnitude of the momentum four-vector is constant:

$$p_\mu p_\mu = -m^2 c^2 = p^2 - \frac{T^2}{c^2}$$

or

$$T^2 = p^2 c^2 + m^2 c^4. \tag{7–100}$$

Equation (7–100) is the relativistic analog of the relation $T = p^2/2m$ in nonrelativistic mechanics, except that T here includes the rest energy. From the definition of T, Eq. (7–96), it is obvious that the energy of a particle with finite rest mass tends to infinity as the speed approaches that of light. In other words, it takes an infinite amount of energy to increase the speed of a mass particle (or a space ship) from rest to c, the speed of light. Here we have another proof that it is impossible to attain or exceed the speed of light starting from any finite speed less than c.†

7–7 RELATIVISTIC KINEMATICS OF COLLISIONS AND MANY-PARTICLE SYSTEMS

The formulations of the previous sections enable us now to generalize relativistically the discussion of Section 3–11 on the transformation of collision phenomena between various systems. The subject is of considerable interest in

---

* The mass changes involved in such macroscopic collisions are, of course, quite small since a joule of energy corresponds to a mass of only 1.1×10^{-14} gm.

† It has been pointed out that it would not violate this statement to have particles *born* with speeds greater than that of light. For such "tachyons" the energy, by Eq. (7–96), could be real only if the "mass" associated with the particle were imaginary. In effect this means that a tachyon is described by a real parameter m' such that the total energy is given by $T = m'c^2/\sqrt{\beta^2 - 1}$. Speculation on the nature of tachyons and their interactions has given rise to a flourishing debate as to how they could be fitted into our normal views of causality. To go into the matter further would be pointless inasmuch as there is absolutely no experimental evidence at the moment for the existence of such particles.

experimental high energy physics. While the forces between elementary particles are only imperfectly known, and are certainly far from classical, so long as the particles involved in a reaction are outside the region of mutual interaction their mean motion can be described by classical mechanics. Further, the main principle involved in the transformations—conservation of the four-vector of momentum—is valid in both classical and quantum mechanics. The actual collision or reaction is taken as occurring at a point—or inside a very small black box—and one looks only at the behavior of the particles before and after. Because of the importance to high energy physics, this aspect of relativistic kinematics has become an elaborately developed field.* It would be impossible to give a comprehensive discussion here. All that can be done is to provide some of the important tools, and to cite a few simple examples that may illustrate the flavor of the techniques employed.

The notion of a point designated as the center-of-mass obviously presents difficulties in a Lorentz-invariant theory. But the center-of-mass system can be suitably generalized as the Lorentz frame of reference in which the total spatial linear momentum of all particles is zero. That such a Lorentz frame can always be found follows from the theorem that the total momentum four-vector is time-like for a system of mass points. In demonstrating this result, it is convenient to introduce a notation in which four-vector components have two indices; one to designate the particle (usually r, s, t, etc.) and the other the customary Greek letter index indicating the component. The total four-momentum of a system is thus written as

$$P_\mu = \sum_r p_{r\mu}. \tag{7-101}$$

The square of the magnitude of this vector, in view of Eqs. (7–91), (7–74) and (7–75), is given by

$$P_\mu P_\mu = \sum_{r,s} p_{r\mu} p_{s\mu}$$

$$= -\sum_r m_r^2 c^2 + \sum_{r \neq s} p_{r\mu} p_{s\mu}$$

$$= -\sum_r m_r^2 c^2 - \sum_{r \neq s} m_r m_s \gamma_r \gamma_s (c^2 - \mathbf{v}_r \cdot \mathbf{v}_s).$$

As the velocities of material particles will always be less than c, the square of the magnitude of P_μ is always negative. Hence there exists some Lorentz system of coordinates in which the spatial components of the transformed vector P_μ are all zero. This coordinate frame will be spoken of as the *center-of-momentum system*, or more loosely as the center-of-mass system, and will be designated by the abbreviation "C-O-M system."

---

* The description as "kinematics" is something of a misnomer, because it is actually concerned with dynamical quantities such as momentum and energy, but the name is too firmly entrenched to be dislodged now.

The major dynamical principle that can be employed—indeed, almost all the physics of the situation permits us—is the conservation of the total momentum four-vector before and after the collision. As we have seen, this automatically implies both conservation of spatial linear momentum and conservation of total energy (including rest mass energy). Our major tools for making use of the conservation principle are Lorentz transformations to and from the center-of-momentum system, and the formation of Lorentz invariants (world scalars) having the same value in all Lorentz frames. Many of the procedures are analogous to those used for nonrelativistic collisions. But, in a way, results in relativistic kinematics are more easily obtained. One does not have to worry about separate conservation of momentum and energy. And the transformations between laboratory system and center-of-momentum system are merely special cases of the Lorentz transformation.

As an example of the use of Lorentz invariants, consider a reaction initiated by two particles that produces another set of particles with masses m_r, $r = 3, 4, 5 \ldots$. In the C-O-M system the transformed total momentum P'_μ has zero spatial components and a fourth component iT'/c. It is often convenient to look on the C-O-M system as the proper or rest system of a composite mass particle of mass $M = T'/c^2$.* The square of the magnitude of P_μ must be invariant in all Lorentz systems (and conserved in the reaction). Hence, we have

$$P_\mu P_\mu = P'_\mu P'_\mu = -\frac{T'^2}{c^2} = -M^2 c^2. \qquad (7\text{–}102)$$

But for the initial particles, $P_\mu P_\mu$ can be evaluated as

$$P_\mu P_\mu = -(m_1^2 + m_2)c^2 + 2p_{1\mu} p_{2\mu}. \qquad (7\text{–}103)$$

The energy in the C-O-M system, or equivalent mass M, is therefore given in terms of the incident particles as

$$T'^2 \equiv M^2 c^4 = (m_1^2 + m_2^2)c^4 + 2(T_1 T_2 - c^2 \mathbf{p}_1 \cdot \mathbf{p}_2). \qquad (7\text{–}104)$$

Suppose now that as is customary, one particle, say 2, was stationary in the laboratory system. Since then $\mathbf{p}_2 = 0$ and $T_2 = m_2 c^2$, the C-O-M energy becomes

$$T'^2 \equiv M^2 c^4 = (m_1^2 + m_2^2)c^4 + 2m_2 c^2 T_1. \qquad (7\text{–}105)$$

If the excess of T_1 over the rest mass energy be denoted by K_1, that is, the kinetic energy, this can be written

$$T'^2 \equiv M^2 c^4 = (m_1 + m_2)^2 c^4 + 2m_2 c^2 K_1. \qquad (7\text{–}105')$$

It is clear that the available energy in the C-O-M system increases only slowly with incident kinetic energy. Even in the "ultrarelativistic" region, where the

---

* Although it is customary in high energy physics to use units in which $c = 1$, it seems more helpful in an introductory exposition such as this to retain the powers of c throughout.

kinetic energy of motion is very large compared to the rest mass energy, T' increases only as the square root of K_1.

The effect of the proportionally small amount of incident energy available in the C-O-M system is shown dramatically in terms of the threshold energies. It is obvious that the lowest energy at which a reaction (other than elastic scattering) is possible is when the reaction products are at rest in the C-O-M system. Any finite kinetic energy requires a higher T' or equivalently higher incident energy. The total four-momentum in the C-O-M system after the reaction, denoted by P''_μ, has the magnitude at threshold given by

$$P''_\mu P''_\mu = -c^2 \left(\sum_r m_r \right)^2, \tag{7-106}$$

which, by conservation of momentum, must be the same as Eq. (7–102). For a stationary target, the incident energy of motion at threshold is then given in consequence of Eq. (7–105') by

$$\frac{K_1}{m_1 c^2} = \frac{\left(\sum_r m_r \right)^2 - (m_1 + m_2)^2}{2m_1 m_2}.$$

If the Q of the reaction is defined as*

$$Q = \sum_r m_r - (m_1 + m_2), \tag{7-107}$$

this threshold energy becomes

$$\frac{K_1}{m_1 c^2} = \frac{Q^2 + 2Q(m_1 + m_2)}{2m_1 m_2}. \tag{7-108}$$

A common illustration of the application of Eq. (7–108) is in the historic production of an antiproton by the reaction

$$P + N \rightarrow P + N + P + \overline{P},$$

where N is a nucleon, either neutron or proton. The masses of all particles involved are nearly equal at 938 MeV equivalent rest mass energy and $Q = 2m$. Equation (7–108) then says that the incident particle kinetic energy at threshold must be

$$K_1 = 6mc^2 = 5.57 \, \text{GeV},$$

which is three times the energy represented by Q! If, however, the reaction was initiated by two nucleons incident on each other with equal and opposite velocity, then the laboratory system is the same as the C-O-M system. All of the kinetic energy is available in this case to go into production of the proton–antiproton pair and each of the incident particles at threshold need have a kinetic energy of

---

* Q here has the opposite sign to the convention adopted in Eq. (3–112).

motion equivalent to only the mass of one proton, 938 MeV. It is no wonder so much effort is put into constructing colliding beam machines!

Another instructive example of threshold calculation is for photomeson production, say, by the reaction

$$\gamma + P = \Sigma^0 + K^+, \tag{7-109}$$

where γ stands for a photon. For the purposes of classical mechanics, a photon is a zero-mass particle with spatial momentum $^0\mathbf{p}$ and energy ^0pc.* In calculating Q, the mass of m_1 is zero:

$$Q = (\Sigma^0 + K^+ - P) = 748 \text{ MeV}.$$

Equation (7–108) is to be rewritten for a reaction involving an incident photon as

$$K_1 = {}^0pc = c^2 \frac{Q^2 + 2Qm_2}{2m_2}.$$

From the value of Q and the rest mass energy m_2 of the proton, the threshold energy for the reaction Eq. (7–109) is then

$$K_1 = 1.05 \text{ GeV},$$

which is only slightly higher than Q.

We can also easily find the energy of the reaction products in the laboratory system at the threshold. The C-O-M system is the rest system for the mass M, with $P_4' = iMc$. In any other system the fourth component of the four-vector is $P_4 = iMc\gamma$. But in the laboratory system

$$P_4 = \frac{i}{c}(T_1 + T_2) = \frac{i}{c}(T_1 + m_2c^2),$$

where the last form holds only for a stationary target particle. Hence the C-O-M system moves relative to the laboratory system such that

$$\gamma = \frac{T_1 + m_2c^2}{Mc^2}. \tag{7-110}$$

But at threshold all the reaction products are at rest in the C-O-M so that $M = \sum_r m_r$, and therefore

$$\gamma = \frac{K_1 + (m_1 + m_2)c^2}{\sum_r m_r c^2} \qquad \text{(threshold)}. \tag{7-111}$$

---

* The square of the magnitude of the photon momentum four-vector is zero; the vector is sometimes described as "light-like." The center-of-momentum theorem (p. 310) is imperiled only if all of the particles are photons, and even then only if the photons are going in the same direction.

The kinetic energy of the sth reaction product in the laboratory system is then

$$K_s = m_s c^2 (\gamma - 1). \tag{7–112}$$

Thus the antiproton at threshold has a kinetic energy of motion $K_{\bar{p}} = mc^2 = 938$ MeV. In contrast the K^+ meson emerges at threshold with 88.4 MeV.

In Section 3–11 the kinematic transformations of a two-body nonrelativistic collision were investigated. Equation (3–117') gives the reduction in energy of an incident particle after elastic scattering from a stationary target, as function of the scattering angle in the C-O-M system. The derivation of the relativistic analog provides another interesting example of the methods of relativistic kinematics. Use of Lorentz invariants here is not particularly helpful; instead direct Lorentz transformations are made between the laboratory and C-O-M systems. Figure 7–2 illustrates the relations of the incident and scattered *spatial* momentum vectors in both systems. The incident and scattered momentum vectors define a plane, invariant in orientation under Lorentz transformation, here taken to be the xz plane with the incident direction along the z axis. As the collision is elastic the masses of the incident particle, m_1, and of the stationary target, m_2, remain unchanged, that is, $m_3 = m_1$, $m_4 = m_2$. Primes on the vectors denote C-O-M values, unprimed vectors are in the laboratory system. To distinguish clearly between before and after the scattering, the indexes 3 and 4 will be retained for the vectors *after* scattering. One only has to remember that 3 denotes the scattered incident particle, and 4 the recoiling target particle. Components of the separate particle four-vectors will always have two indices: the first for the particle, the second for the component.

The Lorentz transformation from the laboratory to the C-O-M system is defined by the γ of Eq. (7–110) with M given by Eq. (7–105):

$$\gamma = \frac{T_1 + m_2 c^2}{\sqrt{2m_2 c^2 T_1 + (m_1^2 + m_2^2)c^4}} = \frac{K_1 + (m_1 + m_2)c^2}{\sqrt{2m_2 c^2 K_1 + (m_1 + m_2)^2 c^4}}. \tag{7–113}$$

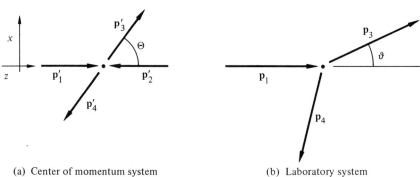

(a) Center of momentum system   (b) Laboratory system

FIGURE 7–2
Diagram of momentum vectors for relativistic elastic scattering in C-O-M and laboratory Lorentz frames.

The quantity β can be found from γ; or more directly by arguments similar to those used to obtain γ. In the C-O-M system the spatial part of the total momentum four-vector is zero; in any other system the spatial part is $Mc\beta\gamma$. However, in the laboratory system the spatial part is $\mathbf{p}_1$. Hence, by Eq. (7–110) β must be given as

$$\beta = \frac{\mathbf{p}_1 c}{T_1 + m_2 c^2} = \frac{\mathbf{p}_1 c}{K_1 + (m_1 + m_2)c^2}. \tag{7–114}$$

Because β is along the z axis, the Lorentz transformation takes the form of Eq. (7–18), and the components of $p'_{1\mu}$ in the C-O-M system are given by

$$p'_1 = p'_{13} = \gamma\left(p_1 - \frac{\beta T_1}{c}\right),$$

$$\frac{iT'_1}{c} = p'_{14} = i\gamma\left(\frac{T_1}{c} - \beta p_1\right). \tag{7–115}$$

After the collision $\mathbf{p}'_3$ is no longer along the z axis, but since the collision is elastic its magnitude is the same as that of $\mathbf{p}'_1$. If Θ is the angle between $\mathbf{p}'_3$ and the incident direction, as in Section 3–11, then the components of $p'_{3\mu}$ in the C-O-M system are

$$p'_{31} = p'_1 \sin\Theta, \qquad p'_{33} = p'_1 \cos\Theta, \qquad p'_{34} = p'_{14} = \frac{iT'_1}{c}. \tag{7–116}$$

The transformation back to the laboratory system is the same Lorentz transformation but with relative velocity $-\beta$. Hence the components of $\mathbf{p}_3$ are

$$p_{31} = p'_{31} = p'_1 \sin\Theta,$$

$$p_{33} = \gamma(p'_{33} - i\beta p'_{34}) = \gamma\left(p'_1 \cos\Theta + \frac{\beta T'_1}{c}\right), \tag{7–117}$$

$$p_{34} = \gamma(p'_{34} + i\beta p'_{33}) = i\gamma\left(\frac{T'_1}{c} + \beta p'_1 \cos\Theta\right).$$

If T'_1 and p'_1 are substituted in the last of Eqs. (7–117) from Eqs. (7–115) we obtain, after a little simplification, an expression for the energy of the scattered particle in terms of its incident properties:

$$T_3 = T_1 - \gamma^2\beta(1 - \cos\Theta)(p_1 c - \beta T_1). \tag{7–118}$$

In Eq. (7–118) γ and β must be expressed terms of the incident quantities through Eqs. (7–113) and (7–114), resulting in the relation

$$\gamma^2\beta(p_1 c - \beta T_1) = \frac{m_2 p_1^2 c^2}{2m_2 T_1 + (m_1^2 + m_2^2)c^2}.$$

With the help of the relation between p_1 and T_1, Eq. (7–100), this can be written

$$\gamma^2\beta(p_1 c - \beta T_1) = \frac{m_2 K_1(K_1 + 2m_1 c^2)}{2m_2 K_1 + (m_1 + m_2)^2 c^2}. \tag{7–119}$$

Some further algebraic manipulation then enables one to rewrite Eq. (7–118) as

$$\frac{K_3}{K_1} = 1 - \frac{2\rho(1 + \mathscr{E}_1/2)}{(1 + \rho)^2 + 2\rho\mathscr{E}_1}(1 - \cos\Theta), \tag{7–120}$$

where $\rho = m_2/m_1$, as in Section 3–11 for elastic scattering, and $\mathscr{E}_1$ is the kinetic energy of the incident particle in units of the rest mass energy,

$$\mathscr{E}_1 = \frac{K_1}{m_1 c^2}. \tag{7–121}$$

Equation (7–120) is the relativistic counterpart of Eq. (3–117′). It is easy to see that Eq. (7–120) reduces to the nonrelativistic case as $\mathscr{E}_1 \to 0$, and that if $\rho = 1$ (equal masses), the relativistic corrections cancel completely. Equation (7–120) implies that the minimum energy after scattering, in units of $m_1 c^2$, is given by

$$(\mathscr{E}_3)_{\min} = \mathscr{E}_1 \frac{(1 - \rho)^2}{(1 + \rho)^2 + 2\rho\mathscr{E}_1}. \tag{7–122}$$

In the nonrelativistic limit, the minimum fractional energy after scattering is

$$\frac{(\mathscr{E}_3)_{\min}}{\mathscr{E}_1} = \left(\frac{1 - \rho}{1 + \rho}\right)^2; \qquad \mathscr{E}_1 \ll 1, \tag{7–123}$$

which is a well-known result, easily obtained from Eq. (3–117′). Equation (7–123) says that in the nonrelativistic region a particle of mass m_1 cannot lose much kinetic energy through scattering from a much heavier particle, i.e., when $\rho \ll 1$, which clearly fits in with common sense. However, in the ultrarelativistic region, when $\rho\mathscr{E}_1 \gg 1$, the minimum energy after scattering is independent of $\mathscr{E}_1$:

$$(K_3)_{\min} = \frac{(m_2 - m_1)^2 c^2}{2m_2}; \qquad \rho\mathscr{E}_1 \gg 1 \tag{7–124}$$

Since the condition on $\mathscr{E}_1$ is equivalent to requiring $K_1 \gg m_2 c^2$, it follows from Eq. (7–124) that such a particle can lose a large fraction of its energy even when scattered by a much heavier particle. This behavior is unexpected, but it should be remembered that for particles at these energies, traveling very close to the speed of light, even a slight change in velocity corresponds to a large change in energy.

Finally, we may easily obtain the relation between the scattering angles in the C-O-M and laboratory systems by noting that

$$\tan\vartheta = \frac{p_{31}}{p_{33}} = \frac{\sin\Theta}{\gamma\left(\cos\Theta + \dfrac{\beta T_1'}{p_1' c}\right)}.$$

By Eqs. (7–90) and (7–96),

$$\frac{p_1' c}{T_1'} = \frac{v_1'}{c} \equiv \beta_1',$$

so that $\tan \vartheta$ can also be written

$$\tan \vartheta = \frac{\sin \Theta}{\gamma(\cos \Theta + \beta/\beta_1')}.$$

In terms of initial quantities, Eqs. (7–115) show that

$$\frac{\beta T_1'}{p_1' c} = \frac{\beta\left(\dfrac{T_1}{c} - \beta p_1\right)}{p_1 - \dfrac{\beta T_1}{c}}.$$

This can be further reduced by employing the relations (cf. Eq. 7–114)

$$\frac{\beta}{p_1 - \dfrac{\beta T_1}{c}} = \frac{1}{m_2 c},$$

$$T_1 - \beta p_1 = \frac{m_1(m_1 + m_2)c^4 + m_2 c^2 K_1}{(m_1 + m_2)c^2 + K_1}.$$

The final expression for $\tan \vartheta$ can then be written as

$$\tan \vartheta = \frac{\sin \Theta}{\gamma(\cos \Theta + \rho g(\rho, \mathscr{E}_1))}, \tag{7–125}$$

where $g(\rho, \mathscr{E}_1)$ is the function

$$g(\rho, \mathscr{E}_1) = \frac{1 + \rho(1 + \mathscr{E}_1)}{(1 + \mathscr{E}_1) + \rho}, \tag{7–126}$$

and γ, by Eq. (7–113), takes the form

$$\gamma(\rho, \mathscr{E}_1) = \frac{1 + \mathscr{E}_1 + \rho}{\sqrt{(1 + \rho)^2 + 2\rho\mathscr{E}_1}}. \tag{7–127}$$

Again in the nonrelativistic region, γ and g tend to unity, and Eq. (7–125) reduces to Eq. (3–107). The correction function $g(\rho, \mathscr{E}_1)$ never really amounts to much, approaching the constant limit ρ as $\mathscr{E}_1$ becomes very large. The important factor affecting the transformed angle is γ, which of course increases indefinitely as $\mathscr{E}_1$ increases. It does not affect the bounds of the angular distribution, when $\Theta = 0$ or π, but its presence means that at other angles ϑ is always smaller than it would be nonrelativistically. The Lorentz transformation from C-O-M to laboratory system, which does not affect the transverse component of the momentum, thus always tends to distort the scattered angular distribution into the forward direction.

One further topic will be mentioned here which deals only peripherally with collisions but which is closely connected with the notion of the center-of-momentum system—the relativistic generalization of angular momentum. In

Chapter 1 it was proven that the nonrelativistic angular momentum obeys an equation of motion much like that for the linear momentum, but with torques replacing forces. It was shown that for an isolated system obeying the law of action and reaction the total angular momentum is conserved, and that in the center-of-mass system it is independent of the point of reference. All of these statements have their relativistic counterparts, at times involving some additional restrictions.

For a single particle define an antisymmetric tensor of the second rank in Minkowski space with elements $m_{\mu\nu}$ given by*

$$m_{\mu\nu} = x_\mu p_\nu - x_\nu p_\mu. \tag{7-128}$$

The 3×3 subtensor m_{ij} clearly corresponds, as was seen in Section 4–7, with the spatial angular momentum of the particle. An equation of motion for $m_{\mu\nu}$ can be found by taking its derivative with respect to the particle's proper time and making use of Eq. (7–83′):

$$\frac{dm_{\mu\nu}}{d\tau} = (u_\mu p_\nu - u_\nu p_\mu) + (x_\mu K_\nu - x_\nu K_\mu),$$

where K_ν here is the component of the Minkowski force. The first term on the right obviously vanishes by the definition of the four-momentum as $p_\lambda = mu_\lambda$, while the remainder of the expression is equally obviously the best candidate for the relativistic generalization of torque:

$$N_{\mu\nu} = x_\mu K_\nu - x_\nu K_\mu, \tag{7-129}$$

Thus $m_{\mu\nu}$ obeys the equation of motion

$$\frac{dm_{\mu\nu}}{d\tau} = N_{\mu\nu}, \tag{7-130}$$

with Eq. (1–11) as the nonrelativistic limiting form.

For a system involving a collection of particles, a total angular momentum four-tensor can be defined (analogously to the total linear momentum four-vector) as

$$M_{\mu\nu} = \sum_s m_{s\mu\nu}, \tag{7-131}$$

where the index s denotes the sth particle. It is more difficult to form an equation of motion for $M_{\mu\nu}$ because each particle has its own proper time. (For the same reason, we did not attempt it even for P_μ.) But plausible arguments can be given for the conservation of $M_{\mu\nu}$ under certain circumstances. If the system is completely isolated and the particles do not interact with each other or with the outside world (including fields), then $m_{\mu\nu}$ for each particle is conserved by Eq.

---

* The double subscript will clearly distinguish $m_{\mu\nu}$ from the completely unrelated particle mass.

(7–130), and therefore $M_{\mu\nu}$ is also conserved. Even if the particles interact, but the interaction takes place only through binary collisions at a point, there still could be conservation as can be seen from the following argument. Instantaneously when the two particles collide they are traveling together and have the same proper time. (In other words, their world lines cross and they share the same *event*.) One can therefore write an equation of motion of the form of Eq. (7–130) for the sum of their angular momenta. If the impulsive forces of contact are equal and opposite—as we would expect from conservation of linear momentum in the collision—then the sum of the impulsive torques cancel. Hence relativistic angular momentum is also conserved through such collisions. Note that unlike the nonrelativistic case the interactions must be confined to instantaneous point collisions.

The relativistic angular momentum obeys the same kind of theorem regarding translation of the reference point as does its nonrelativistic counterpart. In the definition, Eq. (7–128) or Eq. (7–131), the reference point (really reference "event") is the arbitrary origin of the Lorentz system. With respect to some other reference event a_λ, the total angular momentum is

$$M_{\mu\nu}(a_\lambda) = \sum_s \left[(x_{s\mu} - a_\mu)p_{s\nu} - (x_{s\nu} - a_\nu)p_{s\mu}\right],$$

(7–132)

$$= M_{\mu\nu}(0) - (a_\mu P_\nu - a_\nu P_\mu).$$

As in the nonrelativistic case, the change in the angular momentum components is equal to the angular momentum, relative to the origin, that the whole system would have if it were located at a_λ.

In Chapter 1, one particular reference point played an important role—the center-of-mass. We can find something similar here, at least in one Lorentz frame, by examining the nature of the nonspatial three components of $M_{\mu\nu}$, namely $M_{4j} = -M_{j4}$. By definition, in some particular Lorentz frame, these components are given by

$$M_{4j} = \sum_s (x_{s4}p_{sj} - x_{sj}p_{s4})$$

$$= ic \sum_s \left(tp_{sj} - \frac{x_{sj}T_s}{c^2}\right).$$

In the center-of-momentum frame the total linear momentum P_j vanishes and M_{4j} in this frame has the form

$$M_{4j} = -ic \sum_s \frac{x_{sj}T_s}{c^2}.$$

(7–133)

If the system is such that the total angular momentum is conserved, as described above, then along with other components M_{4j} is conserved and hence

$$\sum_s x_{sj}T_s = \text{constant}.$$

Conservation of total linear momentum means that $T = \Sigma\, T_s$ is also conserved. It is therefore possible to define a *spatial* point R_j,

$$R_j = \frac{\sum\limits_s x_{sj} T_s}{\sum\limits_s T_s}, \tag{7-134}$$

associated with the system, which is stationary in the center-of-momentum coordinate frame. In the nonrelativistic limit, where to first approximation $T_s = m_s c^2$, Eq. (7–134) reduces to the usual definition, Eq. (1–21). Thus a meaningful center-of-mass (sometimes called *center-of-energy*) can be defined in special relativity only in terms of the angular-momentum tensor, and only for a particular frame of reference. Finally, it should be noted that by Eq. (7–132) the spatial part of the angular momentum tensor, M_{ij}, is independent of reference point in the center-of-momentum system, exactly as in the nonrelativistic case.

Except for the special case of point collisions, we have so far carefully skirted the problem of finding the motion of a relativistic particle given the Minkowski forces. To this more general problem we address ourselves in the next section, within the (nominal) framework of the Lagrangian formulation.

7–8 THE LAGRANGIAN FORMULATION OF RELATIVISTIC MECHANICS

Having established the appropriate generalization of Newton's equation of motion for special relativity, we can now seek to establish a Lagrangian formulation of the resulting relativistic mechanics. Generally speaking, there are two ways in which this has been attempted. One procedure makes no pretense at a manifestly covariant formulation and instead concentrates on reproducing, for some particular Lorentz frame, the spatial part of the equation of motion, Eq. (7–89). The forces F_i may or may not be suitably related to a covariant Minkowski force. The other method sets out to obtain a covariant Hamilton's principle and ensuing Lagrange's equations in which space and time are treated in common fashion as coordinates in a four-dimensional configuration space. The basis for the first method is at times quite shaky, especially when the forces are not relativistically well formulated. Most of the time, however, the equations of motions so obtained, while not manifestly covariant, are relativistically correct for some particular Lorentz frame. The second method, on the other hand, seems clearly to be the proper approach but quickly runs into difficulties that require skillful handling to circumvent, even for a single particle. For a system of more than one particle, it breaks down almost from the start. No satisfactory formulation for an interacting multiparticle system exists in classical relativistic mechanics except for some few special cases. The subject is still an active area of research.

This section will be concerned with the first approach, seeking to find a Lagrangian that leads to the relativistic equations of motion in terms of the coordinates of some particular inertial system. Within these limitations there is no great difficulty in constructing a suitable Lagrangian. It is true that the method of Section (1–4), deriving the Lagrangian from D'Alembert's principle, will not work here. While the principle itself remains valid (in any given Lorentz frame), the derivation there is based on $p_i = m_i v_i$, which is no longer valid relativistically. But one may also approach the Lagrangian formulation from the alternative route of Hamilton's principle (Section 2–1) and attempt simply to find a function L for which the Euler–Lagrange equations, as obtained from the variational principle

$$\delta I = \delta \int_{t_1}^{t_2} L \, dt = 0, \tag{7–135}$$

agree with the known relativistic equations of motion, Eq. (7–89).

A suitable relativistic Lagrangian for a single particle acted on by conservative forces independent of velocity would be

$$L = -mc^2 \sqrt{1 - \beta^2} - V, \tag{7–136}$$

where V is the potential, depending only upon position, and $\beta^2 = v^2/c^2$, with v the speed of the particle in the Lorentz frame under consideration. That this is the correct Lagrangian can be shown by demonstrating that the resultant Lagrange equations,

$$\frac{d}{dt}\left(\frac{\partial L}{\partial v_i}\right) - \frac{\partial L}{\partial x_i} = 0,$$

agree with Eq. (7–89). Since the potential is velocity independent v_i occurs only in the first term of (7–136) and therefore

$$\frac{\partial L}{\partial v_i} = \frac{mv_i}{\sqrt{1 - \beta^2}} = p_i. \tag{7–137}$$

The equations of motion derived from the Lagrangian (7–136) are then

$$\frac{d}{dt}\frac{mv_i}{\sqrt{1 - \beta^2}} = -\frac{\partial V}{\partial x_i} = F_i,$$

which agree with (7–84). Note that the Lagrangian is no longer $L = T - V$ but that the partial derivative of L with velocity is still the momentum. Indeed it is just this last fact that ensures the correctness of the Lagrange equations, and one could have worked backward from Eq. (7–137) to supply at least the velocity dependence of the Lagrangian.

One can readily extend the Lagrangian (7–136) to systems of many particles and change from Cartesian coordinates to any desired set of generalized coordinates q_j. The canonical momenta will still be defined by

$$p_j = \frac{\partial L}{\partial \dot{q}_j}, \tag{7-138}$$

so that the connection between cyclic coordinates and conservation of the corresponding momenta remains just as in the nonrelativistic theory. Further, just as in Section (2-6), if L does not contain the time explicitly there exists a constant of the motion

$$h = \dot{q}_j \, p_j - L. \tag{7-139}$$

However, the identification of h with the energy for, say, a Lagrangian of the form of Eq. (7-136) cannot proceed along the same route as in Section (2-6). It will be noted that L in Eq. (7-136) is not at all a homogeneous function of the velocity components. Nonetheless, direct evaluation of Eq. (7-139) from Eq. (7-136) shows that in this case h is indeed the total energy:

$$h = \frac{m v_i v_i}{\sqrt{1 - \beta^2}} + mc^2 \sqrt{1 - \beta^2} + V,$$

which, on collecting terms, reduces to

$$h = \frac{mc^2}{\sqrt{1 - \beta^2}} + V = T + V = E. \tag{7-140}$$

The quantity h is thus again seen to be the total energy E, which is therefore a constant of the motion under these conditions.

The introduction of velocity-dependent potentials produces no particular difficulty here and can be performed in exactly the same manner as in Section 1-5 for nonrelativistic mechanics. Thus the Lagrangian for a single particle in an electromagnetic field is

$$L = -mc^2 \sqrt{1 - \beta^2} - q\phi + \frac{q}{c} \mathbf{A} \cdot \mathbf{v}. \tag{7-141}$$

Note that the *canonical* momentum is no longer mu_i; there are now additional terms arising from the velocity-dependent part of the potential:

$$p_i = mu_i + \frac{q}{c} A_i. \tag{7-142}$$

This phenomenon is not a relativistic one, of course; exactly the same additional term was found in the earlier treatment (cf. Eq. 2-47). The formulation of Eq. (7-141) is not manifestly covariant. But we can confidently expect that the results will hold in all Lorentz frames as a consequence of the relativistic covariance of the Lorentz force derivable from the velocity dependent potential in Eq. (7-141).

Almost all of the procedures devised previously for the solution of specific mechanical problems thus can be carried over into relativistic mechanics. A few simple examples will be considered here by way of illustration.

1. *Motion under a constant force; hyperbolic motion.* It will be no loss of generality to take the x axis as the direction of the constant force. The Lagrangian is therefore

$$L = -mc^2\sqrt{1 - \beta^2} - max, \tag{7-143}$$

where β here is $\dot{x}/c$ and a is the constant magnitude of the force per unit mass. Either from Eq. (7-143) or directly on the basis of Eq. (7-89) the equation of motion is easily found to be

$$\frac{d}{dt}\left(\frac{\beta}{\sqrt{1 - \beta^2}}\right) = \frac{a}{c}.$$

The first integration leads to

$$\frac{\beta}{\sqrt{1 - \beta^2}} = \frac{at + \alpha}{c}$$

or

$$\beta = \frac{at + \alpha}{\sqrt{c^2 + (at + \alpha)^2}},$$

where α is a constant of integration. A second integration over t from 0 to t and x from x_0 to x,

$$x - x_0 = c\int_0^t \frac{(at' + \alpha)\,dt'}{\sqrt{c^2 + (at' + \alpha)^2}},$$

leads to the complete solution

$$x - x_0 = \frac{c}{a}(\sqrt{c^2 + (at + \alpha)^2} - \sqrt{c^2 + \alpha^2}). \tag{7-144}$$

If the particle starts at rest from the origin so that $x_0 = 0$ and $v_0 = 0 = \alpha$, then Eq. (7-144) can be written as

$$\left(x + \frac{c^2}{a}\right)^2 - c^2t^2 = \frac{c^4}{a^2},$$

which is the equation of a hyperbola in the x, t plane. (Under the same conditions the nonrelativistic motion is of course a parabola in the x, t plane). The nonrelativistic limit is obtained from Eq. (7-144) by considering $(at + \alpha)$ small compared to c; the usual freshman-physics formula for x as a function of t is then easily obtained, recognizing that in this limit $\alpha \to v_0$.

The motion described in this example arises in reasonably realistic situations. It corresponds, for example, to the acceleration of electrons to relativistic speeds in the laboratory system by means of a constant and uniform electric field. The illustration considered next is more academic but is of interest as an example of the techniques employed:

2. *The relativistic one-dimensional harmonic oscillator.* The Lagrangian in this case is of the form of Eq. (7–136) with

$$V(x) = \frac{1}{2}kx^2. \tag{7–145}$$

Since L is then not explicitly a function of time and not velocity dependent, the total energy E is constant. Equation (7–140) may now be solved for the velocity x as

$$\frac{1}{c^2}\left(\frac{dx}{dt}\right)^2 = 1 - \frac{m^2c^4}{(E-V)^2}. \tag{7–146}$$

For the moment we may postpone substituting in the particular form of $V(x)$ and generalize the problem slightly to include any potential sharing the qualitative characteristics of Eq. (7–145). Thus, suppose that $V(x)$ is any potential function symmetric about the origin and possessing a minimum at that point. Then providing E lies between $V(0)$ and the maximum of V, the motion will be oscillatory between limits $x = -b$ and $x = +b$, determined by

$$V(\pm b) = E.$$

The period of the oscillatory motion is, by Eq. (7–146), to be obtained from

$$\tau = \frac{4}{c}\int_0^b \frac{dx}{\sqrt{1 - \dfrac{m^2c^4}{(E-V(x))^2}}}. \tag{7–147}$$

Equation (7–147) when specialized to the particular Hooke's law form for $V(x)$ can be expressed in terms of elliptic integrals.* We shall instead examine the first order relativistic corrections when the potential energy is always small compared to the rest mass energy mc^2. A change of notation is helpful. The energy E can be written as

$$E = mc^2(1 + \mathscr{E})$$

so that here

$$\frac{E - V(x)}{mc^2} = 1 + \mathscr{E} - \kappa x^2 = 1 + \kappa(b^2 - x^2), \tag{7–148}$$

where

$$\kappa = \frac{k}{2mc^2}. \tag{7–149}$$

---

* See J. L. Synge, *Classical Dynamics*, p. 211, in Vol. III/1, *Encyclopedia of Physics*, 1960.

To the order $(\kappa b^2)^2$, the period, Eq. (7–147) then reduces to

$$\tau \simeq \frac{4}{c} \int_0^b \frac{dx}{\sqrt{2\kappa(b^2 - x^2)}} \left(1 - \frac{3\kappa}{4}(b^2 - x^2)\right). \tag{7–150}$$

The integral in Eq. (7–150) can be evaluated by elementary means, most simply by changing variable through $x = b \sin \phi$; the final result is

$$\tau \simeq \frac{2\pi}{c} \frac{1}{\sqrt{2\kappa}} \left[1 - \frac{3}{8}\kappa b^2\right] = 2\pi \sqrt{\frac{m}{k}} \left[1 - \frac{3kb^2}{16mc^2}\right].$$

It will be recognized that the expression in front of the bracket is τ_0, the nonrelativistic period of the harmonic oscillator. In special relativity the period of the harmonic oscillator is thus not independent of the amplitude; instead there is an amplitude dependent correction given approximately by

$$\frac{\Delta v}{v_0} = -\frac{\Delta \tau}{\tau_0} \simeq \frac{3}{16} \frac{kb^2}{mc^2} = \frac{3}{8}\mathscr{E}. \tag{7–151}$$

3. *Motion of a charged particle in a constant magnetic field.* In principle one should start from a Lagrangian of the form of Eq. (7–141) with $\phi = 0$ and $\mathbf{A}$ appropriate to a constant magnetic field (Eq. 5–106). But we know such a Lagrangian corresponds to the Lorentz force on the charged particle given by

$$\mathbf{F} = \frac{q}{c}(\mathbf{v} \times \mathbf{B}) \tag{7–152}$$

(cf. Eq. 1–61). Hence the equation of motion must be

$$\frac{d\mathbf{p}}{dt} = \frac{q}{c}(\mathbf{v} \times \mathbf{B}) = \frac{q}{mc\gamma}(\mathbf{p} \times \mathbf{B}). \tag{7–153}$$

The nature of the force, Eq. (7–152), is clearly such that the magnetic field does no work on the particle: $\mathbf{F} \cdot \mathbf{v} = 0$. Hence by Eq. (7–95) T must be a constant, as also p and γ by Eq. (7–100). Further, by Eq. (7–152), there is no component of the force parallel to $\mathbf{B}$, and the momentum component along that direction must remain constant. It is therefore no loss of generality to consider the motion only in the plane perpendicular to $\mathbf{B}$ and to let $\mathbf{p}$ represent the projection of the total linear momentum on to that plane. Equation (7–153) then says that the vector $\mathbf{p}$ (whose magnitude is constant) is precessing around the direction of the magnetic field with a frequency

$$\Omega = \frac{qB}{mc\gamma}. \tag{7–154}$$

Because γ is constant, the velocity vector in the plane is also of constant magnitude and rotating with the same frequency. The particle must therefore move in the plane uniformly in a circular orbit with angular speed Ω. It follows that the magnitude of the linear momentum in the plane must be given by

$$p = m\gamma r\Omega.$$

Combining this expression with Eq. (7–154) leads to the relation between the circle radius and the momentum:

$$r = \frac{p}{qB/c}. \qquad (7\text{–}155)$$

The radius of curvature into which the particle motion is bent depends only on the particle properties through the ratio pc/q $(= Br)$, which is sometimes called the *magnetic rigidity* of the particle. Note that while Ω (Eq. (7–154)) shows relativistic corrections through the presence of γ, the relation between r and p is the same both relativistically and nonrelativistically. It should be remembered that in both Eqs. (7–154) and (7–155) p is the magnitude of the momentum perpendicular to B, but in calculating γ one must use both the perpendicular and parallel components to find β.*

7–9 COVARIANT LAGRANGIAN FORMULATIONS

The Lagrangian procedure as given above certainly predicts the correct relativistic equations of motion. Yet it is a relativistic formulation only "in a certain sense." No effort has been made to keep to the ideal of a covariant four-dimensional form for all the laws of mechanics. Thus the time t has been treated as a parameter entirely distinct from the spatial coordinates, while a covariant formulation would require that space and time be considered as entirely similar coordinates in world space. Clearly some invariant parameter should be used, instead of t, to trace the progress of the system point in configuration space. Further, the examples of Lagrangian functions discussed in the previous section do not have any particular Lorentz transformation properties. Hamilton's principle must itself be manifestly covariant, which can only mean in this case that the action integral must be a world scalar. If the parameter of integration is a Lorentz invariant, then the Lagrangian function itself must be a world scalar in any covariant formulation. Finally, instead of being a function of x_i and $\dot{x}_i$, the Lagrangian should be a function of the coordinates in Minkowski space and of their derivatives with respect to the invariant parameter.

---

* The gyration frequency (7–154) must not be confused with the Larmor frequency given by Eq. (5–104). Indeed, the Larmor Theorem of Section (5–9) is inapplicable here because the kinetic energy of motion in the plane is of the same order of magnitude as the ω_l^2 term in Eq. (5–110) and both are comparable to the term linear in B. The conditions for the Larmor Theorem are therefore not met in this case.

We shall consider primarily a system of only one particle. The natural choice of the invariant parameter in such a system would seem to be the particle's proper time τ. But the various components of the generalized velocity, u_v, must then obey the relation

$$-u_v u_v = c^2, \tag{7-75}$$

which shows they are not independent. Therefore we shall instead assume the choice of some Lorentz-invariant quantity θ with no further specification than that it be a monotone function of the progress of the world point along the particle's world line. For the purposes of this discussion a superscript prime will be used to denote differentiation with respect to θ:

$$x_v' \equiv \frac{dx_v}{d\theta},$$

while a dot over the letter indicates differentiation with respect to t. A suitably covariant Hamilton's principle must therefore appear as

$$\delta I = \delta \int_{\theta_1}^{\theta_2} \Lambda(x_\mu, x_\mu') \, d\theta, \tag{7-156}$$

where the Lagrangian function Λ must be a world scalar. Note that this formulation includes what would have ordinarily been called "time-dependent Lagrangians," because Λ is considered a function of x_4. The Euler–Lagrange equations corresponding to Eq. (7–156) are

$$\frac{d}{d\theta}\left(\frac{\partial \Lambda}{\partial x_\mu'}\right) - \frac{\partial \Lambda}{\partial x_\mu} = 0. \tag{7-157}$$

The problem is to find the form of Λ such that Eqs. (7–157) are equivalent to the equations of motion, Eq. (7–83).

One way of seeking Λ is to transform the action integral from the usual integral over t to one over θ, and to treat the time t appearing explicitly in the Lagrangian not as a parameter but as an additional generalized coordinate. Since θ must be a monotone function of t as measured in some Lorentz frame, we have

$$\frac{dx_i}{dt} = \frac{dx_i}{d\theta}\frac{d\theta}{dt} = ic\frac{x_i'}{x_4'}. \tag{7-158}$$

Hence the action integral is transformed as

$$I = \int_{t_1}^{t_2} L(x_j, t, \dot{x}_j) \, dt = -\frac{i}{c} \int_{\theta_1}^{\theta_2} L\left(x_\mu, ic\frac{x_j'}{x_4'}\right) x_4' \, d\theta.$$

It would seem therefore that a recipe for a suitable Λ is given by the relation

$$\Lambda(x_\mu, x_\mu') = -\frac{ix_4'}{c} L\left(x_\mu, ic\frac{x_j'}{x_4'}\right). \tag{7-159}$$

The Lagrangian obtained this way is however a strange creature, unlike any Lagrangian we have so far met. Note that no matter what the functional form of L, the new Lagrangian Λ is a homogeneous function of the generalized velocities in the first degree:

$$\Lambda(x_\mu, ax'_\mu) = a\Lambda(x_\mu, x'_\mu). \tag{7-160}$$

This is not a phenomenon of relativistic physics per se; it is a mathematical consequence of enlarging configuration space to include t as a dynamical variable and using some other parameter to mark the system-point's travel through the space. A Lagrangian obeying Eq. (7–160) is often called (somewhat misleadingly) a homogeneous Lagrangian and the corresponding "homogeneous" problem of the calculus of variations requires special treatment.* The most serious of the resulting difficulties will arise in the Hamiltonian formulation, but we can glimpse some of them by noting that in consequence the energy function h, according to Eq. (2–56), is identically zero. It follows from Euler's Theorem on homogeneous functions that if Λ is homogeneous to first degree in x'_μ, then

$$\Lambda = x'_\mu \frac{\partial \Lambda}{\partial x'_\mu}.$$

One can then show (cf. Exercise 29 at the end of the chapter) that as a result the function Λ *identically* satisfies the relation

$$\left(\frac{d}{d\theta} \left(\frac{\partial \Lambda}{\partial x'_\mu} \right) - \frac{\partial \Lambda}{\partial x_\mu} \right) x'_\mu = 0. \tag{7-161}$$

Thus if any three of the Lagrangian Eqs. (7–157) are satisfied it will follow, solely as a consequence of the homogeneous property of Λ, that the fourth is satisfied identically.

Being thus forewarned to tread carefully, so to speak, let us carry out this transformation for a free particle. From Eq. (7–136), the "noncovariant" Lagrangian for the free particle is

$$L = -mc\sqrt{c^2 - \dot{x}_c \dot{x}_i}.$$

By the transformation of Eq. (7–159) a possible covariant Lagrangian is then†

$$\Lambda = -mc\sqrt{-x'_\mu x'_\mu}. \tag{7-162}$$

---

* For a full exposition, see H. Rund, *The Hamilton–Jacobi Theory in the Calculus of Variations* (New York: Van Nostrand, 1966), Chapter 3.

† In the algebraic manipulations leading to Eq. (7–162) there is an ambiguity of sign that must be decided so that $L\,dt$ has the same value as $\Lambda\,d\theta$. The final step in the derivation must be written as

$$\Lambda = imc\sqrt{x_4'^2 + x_j' x_j'} = -mc\sqrt{(-i)^2}\sqrt{x_4'^2 + x_j' x_j'} = -mc\sqrt{-x'_\mu x'_\mu}.$$

Note also that the choice of the special metric, Eq. (7–53), would eliminate the minus sign in the square root, but at the price of other complications in notation.

With this Lagrangian the Euler–Lagrange equations are equivalent to

$$\frac{d}{d\theta}\left(\frac{mcx_\nu'}{\sqrt{-x_\mu'x_\mu'}}\right) = 0.$$

The parameter θ must be a monotone function of the proper time τ so that derivatives with respect to θ are related to those in terms of τ according to

$$x_\nu' \equiv \frac{dx_\nu}{d\theta} = \frac{d\tau}{d\theta}u_\nu.$$

Hence the Lagrangian equations correspond to

$$\frac{d}{d\tau}\left(\frac{mcu_\nu}{\sqrt{-u_\nu u_\nu}}\right) = \frac{d(mu_\nu)}{d\tau} = 0,$$

which are Eqs. (7–83) for a free particle. As we have seen above, the fourth of these equations says that the kinetic energy T is conserved (cf. Eq. 7–95), which is indeed not new but can be derived from the other three equations.

We have thus been led to a covariant Lagrangian procedure that works, at least for a single free particle, but only in a tortuous fashion. The elaborate superstructure can be greatly simplified, however, by a few bold pragmatic steps. First of all we can avoid using θ and work in terms of the proper time τ directly by a procedure introduced in a slightly different context by Dirac. The constraint on the generalized velocities in terms of τ, Eq. (7–75), is not a true dynamical constraint on the motion; rather it is a geometric consequence of the way in which τ is defined. Equation (7–75) says in effect that we cannot roam over the full four-dimensional u_ν space; we are confined to a particular three-dimensional surface in the space. Dirac calls relations such as Eq. (7–75) *weak equations*. One can with impunity treat u_ν as unconstrained quantities and only *after* all differentiation operations have been carried out need the condition of Eq. (7–75) be imposed. Certainly the procedure would have worked above for the free particle Lagrangian. There would have been no difference if θ were set equal to τ from the start and Eq. (7–75) applied only in the last step. The covariant Lagrange equations can with this proviso therefore be written directly in terms of τ:

$$\frac{d}{d\tau}\left(\frac{\partial\Lambda}{\partial u_\nu}\right) - \frac{\partial\Lambda}{\partial x_\nu} = 0. \tag{7–163}$$

Secondly it is not a sacrosanct physical law that the action integral in Hamilton's principle must have the same value whether expressed in terms of t or in terms of θ (or τ). It *needn't* be given by the prescription of Eq. (7–159). All that is required is that Λ be a world scalar (or function of a world scalar) that leads to the correct equations of motion. It doesn't *have* to be homogeneous to first degree in the generalized velocities. For example, a suitable Λ for a free particle would clearly be the quadratic expression

$$\Lambda = \frac{1}{2} m u_\nu u_\nu. \tag{7-164}$$

Many other possibilities are available.* We shall use Eq. (7–162) for the "kinetic energy" part of the Lagrangian in all subsequent discussions; many present and future headaches will thereby be avoided.

If the particle is not free, but is acted on by external forces, then interaction terms have to be added to the Lagrangian of Eq. (7–164) that would lead to the corresponding Minkowski forces. Very little can be said at this time about the additional terms other than they must be Lorentz-invariant. For example, if G_μ were some (external) four-vector, then $G_\mu x_\mu$ would be a suitable interaction term. If in some particular Lorentz frame $G_1 = ma$ and all other components vanish, then we would have an example of a constant force such as discussed in the previous section. In general, these terms will represent the interaction of the particle with some external field. The specific form will depend on the covariant formulation of the field theory. We have only one example of a field already expressed in a covariant way—the electromagnetic field—and it is instructive therefore to examine the Lagrangian for a particle in an electromagnetic field.

A suitable Lagrangian can easily be seen to be

$$\Lambda(x_\mu, u_\mu) = \frac{1}{2} m u_\mu u_\mu + \frac{q}{c} u_\mu A_\mu(x_\lambda). \tag{7-165}$$

The corresponding Lagrange's equations are then

$$\frac{d}{d\tau}(m u_\nu) = -\frac{q}{c}\frac{dA_\nu}{d\tau} + \frac{\partial}{\partial x_\nu}\left(\frac{q}{c} u_\mu A_\mu\right),$$

which are exactly the generalized equations of motion Eq. (7–83), with the Minkowski force K_ν on a charged particle, Eq. (7–88). Note that again the "mechanical momentum" four-vector p_μ differs from the *canonical* momentum p_μ:

$$\mathsf{p}_\mu = \frac{\partial \Lambda}{\partial u_\mu} = m u_\mu + \frac{q}{c} A_\mu = p_\mu + \frac{q}{c} A_\mu \tag{7-166}$$

---

* In general Λ can have the form $mf(u_\nu u_\nu)$ where $f(y)$ is any function of y such that

$$\left.\frac{\partial f}{\partial y}\right|_{y=-c^2} = \frac{1}{2}.$$

In Eq. (7–164) we have used $f(u_\nu u_\nu) = \frac{1}{2} u_\nu u_\nu$. The choice

$$f(u_\nu u_\nu) = -c\sqrt{-u_\nu u_\nu}$$

corresponds to Eq. (7–162).

by a term linear in the electromagnetic potential. The canonical momentum conjugate to x_4 is now

$$p_4 = \frac{iT}{c} + \frac{iq\phi}{c} = \frac{i}{c}E,$$

where E is the total energy of the particle, $T + q\phi$. Thus the momentum conjugate to the *time* coordinate is proportional to the total *energy*. A similar conjugate connection between these two quantities will recur later in nonrelativistic theory. The connection between the magnitude of the spatial "mechanical" momentum and the energy T is still given by Eq. (7–100). From Eq. (7–166) it is seen that the canonical momenta conjugate to x_i form the components of a spatial Cartesian vector $\mathbf{p}$ related to $\mathbf{p}$ by

$$\mathbf{p} = \mathbf{p} + \frac{q}{c}\mathbf{A}. \tag{7–167}$$

In terms of $\mathbf{p}$, Eq. (7–100) can be rewritten as

$$T^2 = \left(\mathbf{p} - \frac{q}{c}\mathbf{A}\right)^2 + m^2c^4, \tag{7–168}$$

which is a useful relation between the energy T and the canonical momentum vector $\mathbf{p}$.

The interaction term in the Lagrangian of Eq. (7–165) is an example of a vector field interaction (as is also a term of the form $G_\mu x_\mu$). One could also have a simple scalar field interaction where the term added to the Lagrangian would be some world scalar $\psi(x_\mu)$. Or more complicated invariant interaction terms can be created involving an external tensor field. The nature of such Lagrangians properly stems from the physical field theory involved and cannot concern us further here.

So far we have spoken only of systems comprising a single mass particle. Multiparticle systems introduce new complications. One obvious problem is finding an invariant parameter to describe the evolution of the system—each particle in the system has its own proper time. With a little thought, however, one could imagine ways of solving this difficulty. For example, the proper time associated with the center-of-momentum system involves a symmetric treatment of all the particles and might prove suitable. One could also include in the picture interactions of the particles with external fields very much as was done for a single particle. The great stumbling block, however, is the treatment of the type of interaction that is so natural and common in nonrelativistic mechanics—direct interaction between particles.

At first sight, it would seem indeed that such interactions are impossible in relativistic mechanics. To say that the force on a particle depends on the positions or velocities of other particles at the same time implies propagation of effects with infinite velocity from one particle to another—"action at a distance." In special relativity, where signals cannot travel faster than the speed of light, action-at-a-distance seems outlawed. And in a certain sense this seems to be the correct picture. It has been formally proven, first in 1963, that if we require certain properties of the system to behave in the normal way (such as conservation of total linear momentum), then there can be no covariant direct interaction between particles except through contact forces.

There have been many attempts in recent years to get around this "no-interaction" theorem. After all, we have seen that electromagnetic forces can be expressed covariantly, and a static electric field gives rise to the Coulomb law of attraction, which has the same form as the supposedly banned Newtonian gravitational attraction. Some of these attempts have led to approximately covariant Lagrangians, correct through orders of v^2/c^2.* Others involve formulations of mechanics at variance with our normal structures; most for example cannot be stated in terms of a simple Hamilton's principle. Active research in the field is still going on and it is too early to say what picture will finally emerge. Nor is it clear what consequences, if any, these developments will have on other branches of physics, such as particle physics. The status of the field as of 1973 is described in some of the references given below.

SUGGESTED REFERENCES

A. P. FRENCH, *Special Relativity*. The literature on relativity has been one of the world's growth industries, especially during the last few decades. It would overstrain all the constraints of space limitation to list even a goodly fraction of the worthwhile references, and only a highly individualistic selection can be given here. French's book has been justly praised as one of the best introductory treatments. The mathematics is at the level of freshman or sophomore physics; there is a heavy emphasis on the experimental phenomena that led up to, and verified, the theory of special relativity. Two chapters relate to mechanics, mainly on conservation theorems and relativistic forces.

ALBERT EINSTEIN, *The Meaning of Relativity*. This is *not* a treatment designed for popular audiences. Little more than a third of this brief book is concerned with special relativity, but it contains a great deal of information. A considerable background in electrodynamics is assumed.

R. D. SARD, *Relativistic Mechanics: Special Relativity and Classical Particle Dynamics*. This is relativistic mechanics at the level of an intermediate mechanics course; Lagrangian mechanics is not discussed. There is otherwise an incredible amount of material here. In deriving the Lorentz transformation special care is taken to reduce the necessary presuppositions to the minimum. Considerable emphasis is given to particle kinematics. Minkowski space is used almost throughout.

K. R. SYMON, *Mechanics*. The last two chapters are an unusually elaborate treatment of relativity for an intermediate level text. Four-space is introduced early with a metric tensor of trace $+2$ and thus provides a gentle introduction to manipulations in a non-Euclidean space.

---

* A number of Lagrangians covariant to v^2/c^2 antedate the "no-interaction" theorem by many years, e.g., the Breit–Darwin Lagrangian for the interaction between two moving charged particles, published in 1920.

J. L. SYNGE AND A. SCHILD, *Tensor Calculus*. Chapter 2 of this book is one of the best compact references for the manipulation of tensors in Riemannian spaces. The senior author's books on relativity, both special and general, are so voluminous as to discourage the incidental reader, but that should not divert one from this text.

J. D. JACKSON, *Classical Electrodynamics*. The first edition of this renowned text covered almost all topics in special relativity, from the Michelson–Morley experiment to relativistic motion in particle accelerators, with an unusually extensive section on reaction kinematics. It used a complex Minkowski space with $x_4 = ict$. The second edition has changed to a space with trace -2 and has dropped most of the early experiments and all of the section on kinematics. Discussions on Lagrangian formulations have been extended, including Lagrangians for more than one particle that are only approximately relativistic, and Lagrangians for fields, to be treated in Chapter 12, below. Between the two editions one has almost all that would be wished for on special relativity. For present purposes the first edition version is in fact more useful.

H. M. SCHWARTZ, *Introduction to Special Relativity*. This reference is representative of the full-scale treatises on special relativity, in particular one with an approach roughly corresponding to that taken here. Notable are the treatments of group properties of the Lorentz transformations and of Thomas precession. Initially, the discussion is based on Minkowski space. Although the use of tensors in flat spaces is gone into in great detail, it is not clear what trace or signature is finally decided on.

V. FOCK, *Theory of Space, Time and Gravitation*, Most of this treatise by a distinguished Russian physicist is devoted to what we would call general relativity. The first 100 pages, however, are on special relativity, with a number of exceptional features, such as detailed analysis of the "paradoxical" experiments and the decomposition of a Lorentz transformation into a rotation and a pure Lorentz transformation. Where tensors are used the space mostly has a trace or signature -2.

C. W. MISNER; K. S. THORNE; AND J. A. WHEELER, *Gravitation*. This massive treatise (1279 pages! (the pun is irresistible)) is to be praised for the great efforts made to help the reader through the maze. The pedagogic apparatus includes separately marked tracks, boxes of various kinds, marginal comments, and cleverly designed diagrams. An angel blowing a trumpet marks the end of the book, celebrated, among other items, with a couple of French songs and a diagram of the phrenology of a devoted "relativist." It makes the reading great fun, if not completely painless for all that. The physics of flat space time (i.e., special relativity) covers only 193 pages, and there is a refreshing new viewpoint on every one of them. The death of *ict* is proclaimed on p. 51; the metric with signature $+2$ is used instead.

A. O. BARUT, *Electrodynamics and Classical Theory of Fields and Particles*. As a preliminary to a treatment of Lorentz covariant field theories, Barut gives a brief introduction to the covariant dynamics of a particle. Noteworthy topics include the group structure of the Lorentz transformatin, the rotation matrix that lurks in the Lorentz transformation, and the variety of covariant Lagrangians. The metric used has a trace -2.

R. HAGEDORN, *Relativistic Kinematics*. Naturally, this book covers much more ground than our discussion of kinematics, but it does include all the topics treated here. The equations of the Lorentz transformation are used directly, for the most part, along with the invariants, but there is a brief chapter on tensor notation, where formulas are given for *both* signatures -2 and $+2$. As is customary in particle physics, β is used throughout instead of

v (though it's written as v) and masses are relative to the proton mass. The typography is directly from typescript and is abominable.

H. Rund, *Hamilton–Jacobi Theory in the Calculus of Variations*. This reference is one of the few books that openly confronts the homogeneous problem and considers it at length (in Chapter 3). The particular solutions proposed are not adopted here, but the discussion of the mathematical aspects provides a useful orientation. In talking about relativity Minkowski space is used.

R. A. Mann, *The Classical Dynamics of Particles: Galilean and Lorentz Relativity*. Special relativity occupies only a fraction of this relatively brief book, which is primarily a general text on mechanics, so the treatment is sketchier than might be expected. It is one of the few books that says anything about tachyons and includes a proof of the "no-interaction" theorem. Special relativity is described in terms of a space with signature -2. The author is infatuated with group theory.

E. H. Kerner, ed. *Theory of Action-at-a-Distance in Relativistic Particle Dynamics*. Primarily a collection of reprints of basic papers on the question, there is a brief introductory essay by the editor surveying the state of affairs as of 1972. Since then the field has developed, and is continuing to develop. One must keep up with the journal literature.

EXERCISES

1. Consider a mechanical system of n particles, with a conservative potential consisting of terms dependent only on the scalar distance between pairs of particles. Show explicitly that the Lagrangian for the system when expressed in coordinates derived by a Galilean transformation differs in form from the original Lagrangian only by a term that is a total time derivative of a function of the position vectors. This is a special case of invariance under a point transformation (c.f. Exercise 15, Chapter 1).

2. Obtain the Lorentz transformation in which the velocity is at an infinitesimal angle $d\theta$ counterclockwise from the z axis, by means of a similarity transformation applied to Eq. (7–18). Show directly that the resulting matrix is orthogonal and that the inverse matrix is obtained by substituting $-v$ for v.

3. Show that if $\mathbf{K}(\boldsymbol{\beta})$ is the dyadic

$$\mathbf{K}(\boldsymbol{\beta}) = 1 + \frac{\boldsymbol{\beta}\boldsymbol{\beta}(\gamma - 1)}{\beta^2},$$

then the dyadic form of the space part of the pure Lorentz transformation is

$$\mathbf{r}' = \mathbf{K}(\boldsymbol{\beta}) \cdot (\mathbf{r} - \boldsymbol{\beta}ct).$$

4. A rocket of length l_0 in its rest system is moving with constant speed along the z axis of an inertial system. An observer at the origin of this system observes the apparent length of the rocket at any time by noting the z coordinates that can be seen for the head and tail of the rocket. How does this apparent length vary as the rocket moves from the extreme left of the observer to the extreme right?

5. In special relativity it is not necessarily obvious that the velocity of system B as observed in system A is the negative of the velocity vector of system A observed in system B. From the orthogonality properties of $\mathbf{L}$ prove that the two vectors have the same magnitude and are in fact the negative of each other. For simplicity a pure Lorentz transformation may be assumed, although this condition is not necessary for the proof.

6. The Einstein addition law can also be obtained by remembering that the second velocity is related directly to the space components of a four-velocity, which may then be transformed back to the initial system by a Lorentz transformation. If the second system is moving with a speed v' relative to the first in the direction of their z axes, while a third system is moving relative to the second with an arbitrarily oriented velocity $\mathbf{v}''$, show by this procedure that the magnitude of the velocity $\mathbf{v}$ between the first and third system is given by

$$\sqrt{1 - \beta^2} = \frac{\sqrt{1 - \beta'^2}\sqrt{1 - \beta''^2}}{1 + \beta'\beta''_{\mathbf{z}}},$$

and that the components of $\mathbf{v}$ are

$$\beta_x = \frac{\beta''_x\sqrt{1 - \beta'^2}}{1 + \beta'\beta''_z}, \qquad \beta_y = \frac{\beta''_y\sqrt{1 - \beta'^2}}{1 + \beta'\beta''_z}, \qquad \beta_z = \frac{\beta' + \beta''_z}{1 + \beta'\beta''_z}.$$

Here $\beta''_x = v''_x/c$, and so forth. Note that the equation for β_z correctly reduces to Eq. (7–22) (with a change in notation) when $\mathbf{v}''$ is along the z axis.

7. Show that the magnitude of the velocity of the preceding exercise between the first and the third systems can be given in general by

$$\beta^2 = \frac{(\boldsymbol{\beta}' + \boldsymbol{\beta}'')^2 - (\boldsymbol{\beta}' \times \boldsymbol{\beta}'')^2}{(1 + \boldsymbol{\beta}' \cdot \boldsymbol{\beta}'')^2}.$$

8. A beam of particles moving with uniform velocity collides with a collection of target particles that are at rest in a particular system. Let σ_0 be the collision cross section observed in this system. In another system the incident particles have a normalized velocity $\boldsymbol{\beta}_1$ and the target particles a normalized velocity $\boldsymbol{\beta}_2$. If σ is the observed cross section in this system, show that

$$\sigma = \sigma_0\sqrt{1 - \frac{(\boldsymbol{\beta}_1 \times \boldsymbol{\beta}_2)^2}{(\boldsymbol{\beta}_1 - \boldsymbol{\beta}_2)^2}}.$$

Remember that collision rate must be invariant under a Lorentz transformation.

9. A set of transformations are said to have the group property if they possess the following four characteristics:

1) The transformation equivalent to two successive transformations ("product" of transformations) is a member of the set.
2) The product operation obeys the associative law.
3) The identity transformation is a member of the set.
4) The inverse of each transformation in the set is also a member of the set.

Prove that the sets of full Lorentz transformations and of restricted Lorentz transformation have (separately) the group property, but that the other kinds of Lorentz transformations do not.

10. Show that the matrix **R** defined by Eq. (7–31) has the form of a spatial rotation by forming explicitly the elements R_{4i}, R_{i4}, and R_{44}, and by examining the properties of the

3×3 matrix with elements R_{ij}. Prove that there cannot be two rotation matrices such that Eq. (7–28) is satisfied, that is, **R** is unique. Finally, show that **L** can similarly be uniquely factored into a rotation and a pure Lorentz transformation in the form

$$\mathbf{L} = \mathbf{P'R'}.$$

11. For a "close" satellite of Earth (semimajor axis approximately the radius of Earth) calculate numerically the value of the Thomas precession rate. Compare the result with the precession rate induced in the orbit because of the oblate figure of Earth. Assume the satellite orbital plane is inclined at $30°$ to the equator.

12. Show by direct multiplication of the vector form of the Lorentz transformation, Eqs. (7–61) and (7–62), that

$$r'^2 - c^2 t'^2 = r^2 - c^2 t^2.$$

13. Examine the transformation corresponding to the product of two pure Lorentz transformations by multiplying the appropriate **Q** matrices, Eq. (7–63). Show that the product involves a spatial rotation along the direction given by $\kappa_1 \times \kappa_2$, through an angle θ that occurs in the equation.

$$\cos\frac{\theta}{2} = \frac{1 + \cosh\psi_1 + \cosh\psi_2 + \cosh\psi_3}{4\cosh\dfrac{\psi_1}{2}\cosh\dfrac{\psi_2}{2}\cosh\dfrac{\psi_3}{2}}$$

where ψ_3 gives the effective relative speed of the two successive transformations in the form

$$\cosh\psi_3 = \cosh\psi_1 \cosh\psi_2 + \kappa_1 \cdot \kappa_2 \sinh\psi_1 \sinh\psi_2.$$

14. Show that to each plane wave there is associated a covariant four-vector involving the frequency and the wave number. From the consequent transformation equations of the components of the four-vector, derive the Doppler effect equations.

15. From the transformation properties of the world acceleration show that the components of the acceleration **a** are given in terms of the transformed acceleration **a′** in a system momentarily at rest with respect to the particle by the formulas

$$a'_x = \frac{a_x}{1 - \beta^2}, \qquad a'_y = \frac{a_y}{1 - \beta^2}, \qquad a'_z = \frac{a_z}{(1 - \beta^2)^{3/2}},$$

the z axis being chosen in the direction of the relative velocity.

16. By expanding the equation of motion, Eq. (7–89), with Eq. (7–90) for the momentum show that the force is parallel to the acceleration only when the velocity is either parallel or perpendicular to the acceleration. Obtain expressions for the coefficients of the acceleration in these two cases. In the older literature these coefficients were known as the longitudinal and transverse masses, respectively.

17. Two particles with rest masses m_1 and m_2 are observed to move along the observer's z axis toward each other with speeds v_1 and v_2, respectively. Upon collision they are observed to coalesce into one particle of rest mass m_3 moving with speed v_3 relative to the observer. Find m_3 and v_3 in terms of m_1, m_2, v_1, and v_2. Would it be possible for the resultant particle to be a photon, that is, $m_3 = 0$, if neither m_1 nor $\dot{m}_2$ are zero?

18. In the β disintegration considered in Exercise 1, Chapter 1, the electron has a mass equivalent to a rest energy of 0.511 MeV, while the neutrino has no mass. What are the total energies carried away by the electron and neutrino? What fraction of the nuclear mass is converted into kinetic energy (including the electron rest energy)?

19. A meson of mass π comes to rest and disintegrates into a meson of mass μ and a neutrino of zero mass. Show that the kinetic energy of motion of the μ meson (i.e., without the rest mass energy) is

$$T = \frac{(\pi - \mu)^2}{2\pi} c^2.$$

20. A π^+ meson of rest mass 139.6 MeV collides with a neutron (rest mass 939.6 MeV) stationary in the laboratory system to produce a K^+ meson (rest mass 494 MeV) and a Λ hyperon (rest mass 1115 MeV). What is the threshold energy for this reaction in the laboratory system?

21. A photon may be described classically as a particle of zero mass possessing nevertheless a momentum $h/\lambda = h\nu/c$, and therefore a kinetic energy $h\nu$. If the photon collides with an electron of mass m at rest it will be scattered at some angle θ with a new energy $h\nu'$. Show that the change in energy is related to the scattering angle by the formula

$$\lambda' - \lambda = 2\lambda_c \sin^2\frac{\theta}{2},$$

where $\lambda_c = h/mc$, known as the Compton wave length. Show also that the kinetic energy of the recoil motion of the electron is

$$T = h\nu \frac{2\left(\frac{\lambda_c}{\lambda}\right)\sin^2\theta/2}{1 + 2\left(\frac{\lambda_c}{\lambda}\right)\sin^2\theta/2}.$$

22. A photon of energy $\mathscr{E}$ collides at angle θ with another photon of energy E. Prove that the minimum value of $\mathscr{E}$ permitting formation of a pair of particles of mass m is

$$\mathscr{E}_{th} = \frac{2m^2c^4}{E(1 - \cos\theta)}.$$

23. The theory of rocket motion developed in Exercise 3, Chapter 1, no longer applies in the relativistic region, in part because there is no longer conservation of mass. Instead, all the conservation laws are combined into the conservation of the world momentum; the change in each component of the rocket's world momentum in an infinitesimal time dt must be matched by the value of the same component of p_v for the gases ejected by the rocket in that time interval. Show that if there are no external forces acting on the rocket the differential equation for its velocity as a function of the mass is

$$m\frac{dv}{dm} + a\left(1 - \frac{v^2}{c^2}\right) = 0,$$

where a is the constant velocity of the exhaust gases *relative to the rocket*. Verify that the solution can be put in the form

$$\beta = \frac{1 - \left(\frac{m}{m_0}\right)^{\frac{2a}{c}}}{1 + \left(\frac{m}{m_0}\right)^{\frac{2a}{c}}},$$

m_0 being the initial mass of the rocket. Since mass is not conserved, what happens to the mass that is lost?

24. In hyperbolic motion starting from the origin at rest, find the time t_0 such that if a photon is emitted from the origin after t_0 it will never catch up with the particle.

25. A particle of rest mass m, charge q, and initial velocity $\mathbf{v_0}$ enters a uniform electric field $\mathbf{E}$ perpendicular to $\mathbf{v_0}$. Find the subsequent trajectory of the particle and show that it reduces to a parabola as the limit c becomes infinite.

26. Show that the relativistic motion of a particle in an attractive inverse square law of force is a precessing ellipse. Compute the precession of the perihelion of Mercury resulting from this effect. (The answer, about $7''$ per century, is much smaller than the actual precession of $40''$ per century which can be accounted for correctly only by general relativity.)

27. Starting from the equation of motion (7–89), derive the relativistic analog of the virial theorem, which states that for motions bounded in space and such that the velocities involved do not approach c indefinitely close, then

$$\overline{L_0 + T} = -\overline{\mathbf{F} \cdot \mathbf{r}},$$

where L_0 is the form the Lagrangian takes in the absence of external forces. Note that although neither L_0 nor T corresponds exactly to the kinetic energy in nonrelativistic mechanics, their sum, $L + T$, plays the same role as twice the kinetic energy in the nonrelativistic virial theorem, Eq. (3–26).

28. A generalized potential suitable for use in a covariant Lagrangian for a single particle is

$$\mathcal{U} = -A_{\lambda v}(x_\mu) u_\lambda u_v,$$

where $A_{\lambda v}$ stands for a symmetric world tensor of the second rank and u_v are the components of the world velocity. If the Lagrangian is made up of Eq. (7–164) minus $\mathcal{U}$, obtain the Lagrange equations of motion. What is the Minkowski force? Give the components of the force as observed in some Lorentz frame.

29. Show that if Λ satisfies the Lagrange equations, it identically satisfies Eq. (7–161) on the basis of the homogeneity of Λ, by explicitly forming the total derivative with respect to θ that occurs in the equation.

30. Covariant Lagrange equations for a single particle in terms of the proper time have been constructed incorporating the constraint of Eq. (7–75) by a method of Lagrange multipliers. The Lagrangian Λ (assumed not to depend explicitly on τ) is replaced in the variational principle by

$$\Lambda' = \Lambda + \frac{\lambda(\tau)}{2}(c^2 + u_\mu u_\mu).$$

Show that the Euler–Lagrange equation for λ gives Eq. (7–75). The Euler–Lagrange equations for x_μ involve the derivative of λ with respect to τ. Show that these can be integrated to give an expression for λ leading to the Lagrange equations:

$$\frac{d}{d\tau}\left[\frac{\partial \Lambda}{\partial u_v}\left(\delta_{\mu v} + \frac{u_\mu u_v}{c^2}\right) - \frac{\Lambda u_\mu}{c^2}\right] - \frac{\partial \Lambda}{\partial x_\mu} = 0.$$

CHAPTER 8

The Hamilton Equations
of Motion

The Lagrangian formulation of mechanics was developed largely in the first two chapters, and most of the subsequent discussion has been in the nature of application, but still within the framework of the Lagrangian procedure. In this chapter we resume the formal development of mechanics, turning our attention to an alternative statement of the structure of the theory known as the Hamiltonian formulation. Nothing new is added to the physics involved; we simply gain another (and more powerful) method of working with the physical principles already established. The Hamiltonian methods are not particularly superior to Lagrangian techniques for the direct solution of mechanical problems. Rather, the usefulness of the Hamiltonian viewpoint lies in providing a framework for theoretical extensions in many areas of physics. Within classical mechanics it forms the basis for further developments, such as Hamilton–Jacobi Theory and perturbation approaches. Outside classical mechanics, the Hamiltonian formulation provides much of the language with which present day statistical mechanics and quantum mechanics is constructed. We shall assume in the following chapters that the mechanical systems are holonomic and that the forces are monogenic, that is, derived either from a potential dependent on position only, or from velocity-dependent generalized potentials of the type discussed in Section 1–5.

8–1 LEGENDRE TRANSFORMATIONS AND THE
HAMILTON EQUATIONS OF MOTION

In the Lagrangian formulation (nonrelativistic) a system with n degrees of freedom possesses n equations of motion of the form

$$\frac{d}{dt}\left(\frac{\partial L}{\partial \dot{q}_i}\right) - \frac{\partial L}{\partial q_i} = 0. \tag{8–1}$$

As the equations are of second order, the motion of the system is determined for all time only when $2n$ initial values are specified, e.g., the n q_i's and n $\dot{q}_i$'s at a particular time t_1, or the n q_i's at *two* times, t_1 and t_2. We represent the state of the system by a point in an n-dimensional *configuration space* whose coordinates are the n generalized coordinates q_i and follow the motion of the system point in time as it traverses its path in configuration space. Physically, in the Lagrangian viewpoint a system with n independent degrees of freedom is a problem in n independent variables $q_i(t)$, and $\dot{q}_i$ appears only as a shorthand for the time derivative of q_i.

The Hamiltonian formulation is based on a fundamentally different picture. We seek to describe the motion in terms of *first-order* equations of motion. Since the number of initial conditions determining the motion must of course still be $2n$, there must be $2n$ independent first order equations expressed in terms of $2n$ *independent variables*. Hence the $2n$ equations of the motion describe the behavior of the system point in a *phase space* whose coordinates are the $2n$ independent variables. In thus doubling our set of independent quantities, it is natural (though not inevitable) to choose half of them to be the n generalized coordinates q_i. As we shall see, the formulation is nearly symmetric if we choose the other half of the set to be the generalized or *conjugate momenta* p_i already introduced by the definition (cf. Eq. 2–44):

$$p_i = \frac{\partial L(q_j, \dot{q}_j, t)}{\partial \dot{q}_i}. \tag{8-2}$$

The quantities (q, p) are known as the *canonical variables.**

From the mathematical viewpoint it can, however, be claimed that the q's and $\dot{q}$'s have been treated as distinct variables. In Lagrange's equations, Eq. (8–1), the partial derivative of L with respect to q_i means a derivative taken with all other q's and all $\dot{q}$'s constant. Similarly, in the partial derivatives with respect to $\dot{q}$, the q's are kept constant. Treated strictly as a mathematical problem, the transition from Lagrangian to Hamiltonian formulation corresponds to changing the variables in our mechanical functions from $(q, \dot{q}, t)$ to (q, p, t), where p is related to q and $\dot{q}$ by Eqs. (8–2). The procedure for switching variables in this manner is provided by the *Legendre transformation,*† which is tailored for just this type of change of variable.

Consider a function of only two variables $f(x, y)$, so that a differential of f has the form

$$df = u\,dx + v\,dy, \tag{8-3}$$

where

$$u = \frac{\partial f}{\partial x}, \qquad v = \frac{\partial f}{\partial y}. \tag{8-4}$$

---

* Unless otherwise specified, in this and subsequent chapters the symbol p will be used only for the conjugate or canonical momentum. When the forces are velocity dependent the canonical momentum will differ from the corresponding mechanical momentum (cf. Eq. 2–47).

† For a geometrical interpretation of the Legendre transformation and the role it plays in the theory of differential equations see R. Courant and D. Hilbert, *Methods of Mathematical Physics*, Vol. II, pp. 32–39, 1962.

We wish now to change the basis of description from x, y to a new distinct set of variables u, y, so that differential quantities are expressed in terms of the differentials du and dy. Let g be a function of u and y defined by the equation

$$g = f - ux. \qquad (8-5)$$

A differential of g is then given as

$$dg = df - u \, dx - x \, du,$$

or, by (8–3), as

$$dg = v \, dy - x \, du,$$

which is exactly in the form desired. The quantities x and v are now functions of the variables u and y given by the relations

$$x = -\frac{\partial g}{\partial u}, \qquad v = \frac{\partial g}{\partial y}, \qquad (8-6)$$

which are in effect the converse of Eqs. (8–4).

The Legendre transformation so defined is used frequently in thermodynamics. For example, the enthalpy X is a function of the entropy S and the pressure P with the properties that

$$\frac{\partial X}{\partial S} = T, \qquad \frac{\partial X}{\partial P} = V,$$

so that

$$dX = T \, dS + V \, dP,$$

where T and V are temperature and volume, respectively. The enthalpy is useful in considering isentropic and isobaric processes, but often one has to deal rather with isothermal and isobaric processes. In such case one wants a thermodynamic function of T and P alone. The Legendre transformation shows that the desired function may be defined as

$$G = X - TS$$

with

$$dG = -S \, dT + V \, dP, \qquad (8-7)$$

where G is the well-known Gibbs function, or free energy, whose properties are correctly given by Eq. (8–7).

The transformation from $(q, \dot{q}, t)$ to (q, p, t) differs from the type considered in Eqs. (8–3) to (8–5) only in that more than one variable is to be transformed. In place of the Lagrangian one deals with a function defined in analogy to Eq. (8–5), except for a minus sign:

$$H(q, p, t) = \dot{q}_i p_i - L(q, \dot{q}, t) \qquad (8-8)$$

(where, of course, the summation convention has been employed). Here H is known as the *Hamiltonian*. Considered as a function of q, p, and t only, the differential of H is given by

$$dH = \frac{\partial H}{\partial q_i} dq_i + \frac{\partial H}{\partial p_i} dp_i + \frac{\partial H}{\partial t} dt, \tag{8-9}$$

but from the defining equation (8–8) we can also write

$$dH = \dot{q}_i dp_i + p_i d\dot{q}_i - \frac{\partial L}{\partial \dot{q}_i} d\dot{q}_i - \frac{\partial L}{\partial q_i} dq_i - \frac{\partial L}{\partial t} dt. \tag{8-10}$$

The terms in $d\dot{q}_i$ in Eq. (8–10) cancel in consequence of the definition of generalized momentum, and from Lagrange's equation it follows that

$$\frac{\partial L}{\partial q_i} = \dot{p}_i.$$

Equation (8–10) therefore reduces to the simple form

$$dH = \dot{q}_i dp_i - \dot{p}_i dq_i - \frac{\partial L}{\partial t} dt. \tag{8-11}$$

Comparison with (8–9) furnishes the following set of $2n + 1$ relations, in analogy with Eqs. (8–6):

$$\dot{q}_i = \frac{\partial H}{\partial p_i},$$

$$\tag{8-12}$$

$$-\dot{p}_i = \frac{\partial H}{\partial q_i},$$

$$-\frac{\partial L}{\partial t} = \frac{\partial H}{\partial t}. \tag{8-13}$$

Equations (8–12) are known as the *canonical equations of Hamilton*; they constitute the desired set of $2n$ first order equations of motion replacing the Lagrange equations.*

The first half of Hamilton's equations give the $\dot{q}_i$'s as functions of (q, p, t). They form therefore the inverse of the constitutive equations (8–2), which define the momenta p_i as functions of $(q, \dot{q}, t)$. It may therefore be said that they provide no new information. In terms of solving mechanical problems by means of the

---

*Canonical is used here presumably in the sense of designating a simple, general set of standard equations. It appears that the term was first introduced by C. G. J. Jacobi in 1837 (*Comptes rendus de l'Academie des Sciences de Paris*, 5, p. 61) but in a slightly different context referring to an application of Hamilton's equations of motion to perturbation theory. Although the term rapidly gained common usage, the reason for its introduction apparently remained obscure even to contemporaries. By 1879, only 45 years after Hamilton explicitly introduced his equations, Thomson (Lord Kelvin) and Tait were moved by the adjective 'canonical' to exclaim: "Why it has been so called would be hard to say." (*Treatise on Natural Philosophy*, 1879, Vol. 1, p. 307.)

canonical equations, the statement is correct. But within the framework of the Hamiltonian picture, where $H(q, p, t)$ is some given function obtained no matter how, the two halves of the set of Hamiltonian equations are equally independent and meaningful. The first half says how $\dot{q}$ depends on q, p, and t; the second says the same thing for $\dot{p}$.

Of course the Hamiltonian H is constructed in the same manner, and has identically the same value, as h, the energy function defined in Eq. (2–53). But they are functions of different variables: like the Lagrangian, h is a function of q, $\dot{q}$ (and possibly t), while H must always be expressed as a function of q, p (and possibly t). It is to emphasize this difference in functional behavior that different symbols have been given to the quantities even though they have the same numerical values.

Nominally, the Hamiltonian for each problem must be constructed via the Lagrangian formulation. The formal procedure calls for a lengthy sequence of steps:

1. With a chosen set of generalized coordinates, q_i, the Lagrangian $L(q_i, \dot{q}_i, t)$ is constructed.

2. The conjugate momenta are defined as functions of q_i, $\dot{q}_i$, and t by Eqs. (8–2).

3. Equation (8–8) is used to form the Hamiltonian. At this stage one has h instead of H, or rather some mixed function of q_i, $\dot{q}_i$, p_i, and t.

4. Equations (8–2) are then inverted to obtain $\dot{q}_i$ as functions of (q, p, t). Possible difficulties in the inversion will be discussed below.

5. The results of the previous step are then applied to eliminate $\dot{q}$ from H so as to express it solely as a function of (q, p, t).

Now we are ready to use the Hamiltonian in the canonical equations of motion.

For many physical systems it is possible to shorten this drawn-out sequence quite appreciably. As has been described in Section 2–6, in many problems the Lagrangian is the sum of functions each homogeneous in the generalized velocities of degree 0, 1, and 2, respectively. In that case H by the prescription of Eq. (8–8) is given by (cf. Eq. 2–57)

$$H = L_2 - L_0, \tag{8–14}$$

where L_0 is the part of the Lagrangian independent of the generalized velocities and L_2 is the part that is homogeneous in $\dot{q}_i$ in the second degree. Further, if the equations defining the generalized coordinates don't depend on time explicitly then $L_2 = T$, and if the forces are derivable from a conservative potential V then $L_0 = -V$. When both these conditions are satisfied, the Hamiltonian is then *automatically* the total energy:

$$H = T + V = E. \tag{8–15}$$

If either Eqs. (8–14) or (8–15) holds, then much of the algebra in step 3 above is eliminated.

One can at times go further. In large classes of problems it happens that L_2 is a quadratic function of the generalized velocities and L_1 is a linear function of the same variables. The algebraic manipulations required in steps 2 through 5 can then be carried out, at least formally, once and for all. To show this let us form the $\dot{q}_i$'s into a single column matrix $\dot{\mathbf{q}}$. Under the given assumptions the Lagrangian can be written as

$$L(q, \dot{q}, t) = L_0(q, t) + \tilde{\dot{\mathbf{q}}}\mathbf{a} + \frac{1}{2}\tilde{\dot{\mathbf{q}}}\mathbf{T}\dot{\mathbf{q}}, \tag{8-16}$$

where the single row matrix has been written explicitly as the transpose of a single column matrix in view of operations to be performed subsequently. Here $\mathbf{a}$ is a column matrix and $\mathbf{T}$ is a square $n \times n$ matrix (much like the corresponding matrix introduced in Chapter 6). The elements of both are in general functions of q and t. The conjugate momenta, considered as a row matrix $\mathbf{p}$, is then, by Eq. (8–2), given as

$$\mathbf{p} = \mathbf{T}\dot{\mathbf{q}} + \mathbf{a}, \tag{8-17}$$

which can be inverted (step 4) as

$$\dot{\mathbf{q}} = \mathbf{T}^{-1}(\mathbf{p} - \mathbf{a}). \tag{8-18}$$

This step presupposes $\mathbf{T}^{-1}$ exists, which it normally does by virtue of the positive definite property of the kinetic energy. By the prescription of Eq. (8–14) the Hamiltonian, identical with the energy function h, is given by

$$h = \frac{1}{2}\tilde{\dot{\mathbf{q}}}\mathbf{T}\dot{\mathbf{q}} - L_0.$$

To obtain the right functional form for H, Eq. (8–18) must be substituted for $\dot{\mathbf{q}}$ in the quadratic part of h (step 5). Now $\mathbf{T}$ is obviously a symmetric matrix, and its inverse must also be symmetric. It therefore follows that

$$\tilde{\dot{\mathbf{q}}}\mathbf{T}\dot{\mathbf{q}} = (\tilde{\mathbf{p}} - \tilde{\mathbf{a}})\mathbf{T}^{-1}\mathbf{T}\mathbf{T}^{-1}(\mathbf{p} - \mathbf{a}).$$

Hence final form for the Hamiltonian is

$$H(q, p, t) = \frac{1}{2}(\tilde{\mathbf{p}} - \tilde{\mathbf{a}})\mathbf{T}^{-1}(\mathbf{p} - \mathbf{a}) - L_0(q, t). \tag{8-19}$$

If the Lagrangian can be written in the form of Eq. (8–16), then one can immediately skip the intervening steps and write the Hamiltonian as Eq. (8–19). The inverse matrix $\mathbf{T}^{-1}$ can usually most easily be obtained straightforwardly as*

$$\mathbf{T}^{-1} = \frac{\tilde{\mathbf{T}}_c}{|\mathbf{T}|}. \tag{8-20}$$

---

* See almost any book on mathematical methods in physics or on matrices, e.g., Margenau and Murphy, *The Mathematics of Physics and Chemistry*, 1943 (p. 295); Hildebrand, *Methods of Applied Mathematics*, 2d ed. 1965 (p. 16); or Nering, *Linear Algebra and Matrix Theory*, 1963 (p. 83). Incidentally, $\mathbf{T}_c$ is what the mathematicians call the adjoint matrix to $\mathbf{T}$ (cf. p. 142 above).

Here $\mathbf{T}_c$ is the cofactor matrix whose elements $(\mathbf{T}_c)_{jk}$ are $(-1)^{j+k}$ times the determinant of the matrix obtained by striking out the jth row and kth column of $\mathbf{T}$. It is easy to see that if $\mathbf{T}$ is diagonal, then $\mathbf{T}^{-1}$ is also diagonal with elements that are just the reciprocals of the corresponding elements of $\mathbf{T}$.

A number of exercises in applying this formalism to various mechanical systems will be found in the problems at the end of the chapter. Two very simple examples may be considered here, particularly because they illustrate some important aspects of the technique. First consider the spatial motion of a particle in a central force field, using spherical polar coordinates (r, θ, ϕ) for the generalized coordinates. The potential energy is some $V(r)$ and the kinetic energy is

$$T = \frac{mv^2}{2} = \frac{m}{2}(\dot{r}^2 + r^2 \sin^2\theta \dot{\phi}^2 + r^2 \dot{\theta}^2).$$

Clearly the Hamiltonian is the same as the total energy $T + V$, and since $\mathbf{T}$ is diagonal the form of H, by inspection, is

$$H(r, \theta, p_r, p_\theta, p_\phi) = \frac{1}{2m}\left(p_r^2 + \frac{p_\theta^2}{r^2} + \frac{p_\phi^2}{r^2 \sin^2\theta}\right) + V(r). \tag{8-21}$$

Note that the Hamiltonian would have a different functional form if the generalized coordinates were chosen to be the Cartesian coordinates x_i of the particle. The kinetic energy then has the form

$$T = \frac{mv^2}{2} = \frac{m\dot{x}_i\dot{x}_i}{2}$$

so that the Hamiltonian is now

$$H(x_i, p_i) = \frac{p_i p_i}{2m} + V(\sqrt{x_i x_i}). \tag{8-22}$$

It is sometimes convenient to form the canonical momenta p_i conjugate to x_i into a vector $\mathbf{p}$ such that the Hamiltonian can be written as

$$H(x_i, p_i) = \frac{\mathbf{p} \cdot \mathbf{p}}{2m} + V(\sqrt{|r|}). \tag{8-23}$$

We can of course take the components of $\mathbf{p}$ relative to any coordinate system we desire, curvilinear spherical coordinates, for example. But it is important not to confuse, say, p_θ with the θ component of $\mathbf{p}$, designated as $(\mathbf{p})_\theta$. The former is the canonical momentum conjugate to the coordinate θ; the latter is the θ component of the momentum vector conjugate to the Cartesian coordinates. Dimensionally it is clear they are quite separate quantities; p_θ is an angular momentum, $(\mathbf{p})_\theta$ is a linear momentum. *Whenever a vector is used from here on to represent canonical momenta it will refer to the momenta conjugate to Cartesian position coordinates.*

For a second example consider a single (nonrelativistic) particle moving in an electromagnetic field. By Eq. (1–66), the Lagrangian for this system is

$$L = T - V = \frac{1}{2}mv^2 - q\phi + \frac{q}{c}\mathbf{A}\cdot\mathbf{v}.$$

Using Cartesian position coordinates as generalized coordinates the Lagrangian can also be written as

$$L = \frac{m\dot{x}_i\dot{x}_i}{2} + \frac{q}{c}A_i\dot{x}_i - q\phi, \tag{8–24}$$

where the potentials ϕ and $\mathbf{A}$ are in general functions of x_i and the time (q here of course is the particle's charge, not a generalized coordinate). There is now a linear term in the generalized velocities such that the matrix $\mathbf{a}$ has the elements qA_i/c. Because of this linear term in U, the Hamiltonian is *not* $T + U$. However, it is still in this case the total energy since the "potential" energy in an electromagnetic field is determined by ϕ alone. The canonical momenta, either by Eq. (8–2) or Eq. (8–17), are

$$p_i = m\dot{x}_i + \frac{q}{c}A_i, \tag{8–25}$$

and the Hamiltonian (cf. Eq. 8–19) is

$$H = \frac{\left(p_i - \frac{q}{c}A_i\right)\left(p_i - \frac{q}{c}A_i\right)}{2m} + q\phi, \tag{8–26}$$

which is the total energy of the particle. Again the momenta p_i can be formed into a vector $\mathbf{p}$ and H written as

$$H = \frac{1}{2m}\left(\mathbf{p} - \frac{q}{c}\mathbf{A}\right)^2 + q\phi, \tag{8–27}$$

and it must be remembered again that $\mathbf{p}$ refers only to momenta conjugate to x_i.

It will have been noticed that Hamilton's equations of motion do not treat the coordinates and momenta in a completely symmetric fashion. The equation for $\dot{p}$ has a minus sign that is absent in the equation for $\dot{q}$. Considerable ingenuity has been exercised in devising nomenclature schemes that result in entirely symmetric equations, or combine the two sets into one. Most of these schemes have only oddity value, but one has proved to be an elegant and powerful tool for manipulating the canonical equations and allied expressions.

For a system of n degrees of freedom we construct a column matrix $\boldsymbol{\eta}$ with $2n$ elements such that

$$\eta_i = q_i, \qquad \eta_{i+n} = p_i; \qquad i \le n. \tag{8–28}$$

Similarly the column matrix $\dfrac{\partial H}{\partial \boldsymbol{\eta}}$ has the elements

$$\left(\frac{\partial H}{\partial \boldsymbol{\eta}}\right)_i = \frac{\partial H}{\partial q_i}, \qquad \left(\frac{\partial H}{\partial \boldsymbol{\eta}}\right)_{i+n} = \frac{\partial H}{\partial p_i}; \qquad i \le n. \tag{8-29}$$

Finally, let **J** be the $2n \times 2n$ square matrix composed of the $n \times n$ zero and unit matrices according to the scheme

$$\mathbf{J} = \begin{pmatrix} \mathbf{0} & \mathbf{1} \\ -\mathbf{1} & \mathbf{0} \end{pmatrix}. \tag{8-30}$$

Here **0** is the $n \times n$ matrix all of whose elements are zero, and **1** is the standard $n \times n$ unit matrix. Hamilton's equations of motion can then be written in compact form as

$$\dot{\boldsymbol{\eta}} = \mathbf{J}\frac{\partial H}{\partial \boldsymbol{\eta}}. \tag{8-31}$$

This method of displaying the canonical equations of motion will be referred to as Hamilton's equations in matrix or *symplectic** notation. In subsequent chapters we shall frequently employ this matrix form of the equations. For later use, some easily verified properties of **J** may be noted. The matrix (it has no standard name) is a sort of $2n \times 2n$ version of i times the Pauli matrix $\boldsymbol{\sigma}_2$ (cf. Eq. 4–74), and its square is therefore the negative of the $2n \times 2n$ unit matrix:

$$\mathbf{J}^2 = -\mathbf{1}. \tag{8-32}$$

It is also orthogonal:

$$\tilde{\mathbf{J}}\mathbf{J} = \mathbf{1} \tag{8-33}$$

so that

$$\tilde{\mathbf{J}} = -\mathbf{J} = \mathbf{J}^{-1}. \tag{8-34}$$

From the orthogonality property it follows the square of the determinant is 1, but in fact one can prove (cf. Exercise 25) the stronger statement that

$$|\mathbf{J}| = +1. \tag{8-34'}$$

8-2 CYCLIC COORDINATES AND CONSERVATION THEOREMS

According to the definition given in Section 2–6, a cyclic coordinate q_j is one that does not appear explicitly in the Lagrangian; by virtue of Lagrange's equations

---

* The term *symplectic* comes from the Greek for "intertwined," particularly appropriate for Hamilton's equations where $\dot{q}$ is matched with a derivative with respect to p and $\dot{p}$ similarly with the negative of a q derivative. H. Weyl first introduced the term in 1939 in his book *The Classical Groups* (p. 165 in both the first edition, 1939, and second edition, 1946).

its conjugate momentum p_j is then a constant. But comparison of Eq. (8–9) with Eq. (8–10) has already told us that

$$\dot{p}_j = \frac{\partial L}{\partial q_j} = -\frac{\partial H}{\partial q_j}.$$

A coordinate that is cyclic will thus also be absent from the Hamiltonian.* Conversely if a generalized coordinate does not occur in H, the conjugate momentum is conserved. The momentum conservation theorems of Section 2–6 can thus be transferred to the Hamiltonian formulation with no more than a substitution of H for L. In particular the connection between the invariance or symmetry properties of the physical system and the constants of the motion can also be derived in terms of the Hamiltonian. For example, if a system is completely self-contained, with only internal forces between the particles, then the system can be moved as a rigid ensemble without affecting the forces or subsequent motion. The system is said to be invariant under a rigid displacement. Hence a generalized coordinate describing such a rigid motion will not appear explicitly in the Hamiltonian and the corresponding conjugate momentum will be conserved. If the rigid motion is a translation along some particular direction, then the conserved momentum is the corresponding Cartesian component of the total linear (canonical) momentum of the system. Since the direction is arbitrary, the total vector linear momentum is conserved. The rigid displacement may be a rotation, from whence it follows that the total angular momentum vector is conserved. Even if the system interacts with external forces, there may be a symmetry in the situation that leads to a conserved canonical momentum. Suppose the system is symmetrical about a given axis so that H is invariant under rotation about that axis. Then H obviously cannot involve the rotation angle about the axis and the particular angle variable must be a cyclic coordinate. It follows, as in Section 2–6, that the component of the angular momentum about that axis is conserved.

The considerations concerning h in Section 2–6 have already shown that if L (and in consequence of Eq. (8–13), also H) is not an explicit function of t, then H is a constant of motion. This can also be seen directly from the equations of motion (8–12) by writing the total time derivative of the Hamiltonian as

$$\frac{dH}{dt} = \frac{\partial H}{\partial q_i}\dot{q}_i + \frac{\partial H}{\partial p_i}\dot{p}_i + \frac{\partial H}{\partial t}.$$

In consequence of the equations of motion (8–12) the first two sums on the right cancel each other and it therefore follows that

$$\frac{dH}{dt} = \frac{\partial H}{\partial t} = -\frac{\partial L}{\partial t}. \tag{8–35}$$

---

* This conclusion also follows from the definition of Eq. (8–8), for H differs from $-L$ only by $p_i\dot{q}_i$, which does not involve q_i explicitly.

Thus if t doesn't appear explicitly in L, it will also not be present in H, and H will be constant in time.

Further, it was proved in Section 2–6 that if the equations of transformation that define the generalized coordinates (1–38),

$$\mathbf{r}_m = \mathbf{r}_m(q_1,\ldots,q_n;t),$$

do not depend explicitly on the time, and if the potential is velocity-independent then H is the total energy, $T + V$. The identification of H as a constant of the motion and as the total energy are two separate matters, and the conditions sufficient for the one are not enough for the other. It can happen that the Eqs. (1–38) do involve time explicitly but that H does not. In this case H is a constant of the motion but it is *not* the total energy. As was also emphasized in Section (2–6), the Hamiltonian is dependent both in magnitude and in functional form on the initial choice of generalized coordinates. For the Lagrangian we have a specific prescription, $L = T - V$, and a change of generalized coordinates within that prescription may change the functional appearance of L but cannot alter its magnitude. On the other hand, use of a different set of generalized coordinates in the definition for the Hamiltonian, Eq. (8–8), may lead to an entirely different quantity for the Hamiltonian. It may be that for one set of generalized coordinates H is conserved, but that for another it varies in time.

To illustrate some of these points in a simple example we may consider a somewhat artificial one-dimensional system. Suppose a point mass m is attached to a spring, of force constant k, the other end of which is fixed on a massless cart that is being moved uniformly by an external device with speed v_0 (cf. Fig. 8–1). If we take as generalized coordinate the position x of the mass particle in the stationary system, then the Lagrangian of the system is obviously

$$L(x, \dot{x}, t) = T - V$$

$$= \frac{m\dot{x}^2}{2} - \frac{k}{2}(x - v_0 t)^2. \tag{8–36}$$

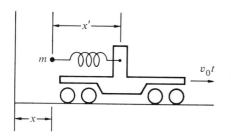

FIGURE 8–1
Example of a harmonic oscillator fixed to a uniformly moving cart.

(For simplicity the origin has been chosen so that the cart passed through it at $t = 0$.) The corresponding equation of motion is clearly

$$m\ddot{x} = -k(x - v_0 t).$$

An obvious way of solving this equation is to change the unknown to $x'(t)$ defined as

$$x' = x - v_0 t, \tag{8-37}$$

so that the equation of motion becomes

$$m\ddot{x}' = -kx'. \tag{8-38}$$

From Eq. (8–37) x' is the displacement of the particle relative to the cart; Eq. (8–38) says that to an observer on the cart the particle exhibits simple harmonic motion, as would be expected on the principle of equivalence in Galilean relativity.

Having looked at the nature of the motion, let us consider the Hamiltonian formulation. Since x is the Cartesian coordinate of the particle, and the potential does not involve generalized velocities, the Hamiltonian relative to x is the sum of the kinetic and potential energies, i.e., the total energy. In functional form the Hamiltonian is given by

$$H(x, p, t) = T + V = \frac{p^2}{2m} + \frac{k}{2}(x - v_0 t)^2. \tag{8-39}$$

The Hamiltonian *is* the total energy of the system, but since it is explicitly a function of t, it is *not* conserved. Physically this is understandable; energy must flow into and out of the "external physical device" to keep the cart moving uniformly against the reaction of the oscillating particle.*

Suppose now we formulated the Lagrangian from the start in terms of the relative coordinate x'. The same prescription gives the Lagrangian as

$$L(x', \dot{x}') = \frac{m\dot{x}'^2}{2} + m\dot{x}'v_0 + \frac{mv_0^2}{2} - \frac{kx'^2}{2}. \tag{8-40}$$

In setting up the corresponding Hamiltonian we note there is now a term linear in x', with the single component of **a** being mv_0. The new Hamiltonian is now

$$H'(x', p') = \frac{(p' - mv_0)^2}{2m} + \frac{kx'^2}{2} - \frac{mv_0^2}{2}. \tag{8-41}$$

Note that the last term is a constant involving neither x' nor p'; it could, if one wished, be dropped from H' without affecting the resultant equations of motion. Now H' is *not* the total energy of the system, but it *is* conserved. Except for the last

---

* Put another way, the moving cart constitutes a time-dependent constraint on the particle and the force of the constraint does do work in actual (*not* virtual) displacement of the system.

term it can be easily identified as the total energy of motion of the particle relative to the moving cart. The two Hamiltonian's are different in magnitude, time-dependence, and functional behavior. But the reader can easily verify both lead to the same motion for the particle.

8-3 ROUTH'S PROCEDURE AND OSCILLATIONS ABOUT STEADY MOTION

It has been remarked that the Hamiltonian formulation is not particularly helpful in direct solution of mechanical problems. Often one can solve the $2n$ first order equations only by eliminating some of the variables, e.g., the p variables, which speedily leads back to the second order Lagrangian equations of motion. But an important exception should be noted. The Hamiltonian procedure is especially adapted to the treatment of problems involving cyclic coordinates. Consider the situation in Lagrangian formulation when some coordinate, say q_n, is cyclic. The Lagrangian as a function of q and $\dot q$ can then be written

$$L = L(q_1, \ldots, q_{n-1}; \dot q_1, \ldots, \dot q_n; t).$$

All the generalized velocities still occur in the Lagrangian and in general will be functions of the time. We still have to solve a problem of n degrees of freedom even though one degree of freedom corresponds to a cyclic coordinate. A cyclic coordinate in the Hamiltonian formulation, on the other hand, truly deserves its alternative description as "ignorable," for in the same situation p_n is some constant α, and H has the form

$$H = H(q_1, \ldots, q_{n-1}; p_1, \ldots, p_{n-1}; \alpha; t).$$

In effect the Hamiltonian now describes a problem involving only $n - 1$ coordinates, which may be solved completely ignoring the cyclic coordinate except as it is manifested in the constant of integration α, to be determined from the initial conditions. The behavior of the cyclic coordinate itself with time is then found by integrating the equation of motion

$$\dot q_n = \frac{\partial H}{\partial \alpha}.$$

The advantages of the Hamiltonian formulation in handling cyclic coordinates may be combined with the Lagrangian procedure by a method devised by Routh. Essentially, one carries out a mathematical transformation from the $q, \dot q$ basis to the q, p basis only for those coordinates that are cyclic, obtaining their equations of motion in the Hamiltonian form, while the remaining coordinates are governed by Lagrange equations. If the cyclic coordinates are labeled $q_{s+1}, \ldots, q_n$, then a new function R, known as the Routhian,* may be introduced, defined as

$$R(q_1, \ldots, q_n; \dot{q}_1, \ldots, \dot{q}_s; p_{s+1}, \ldots, p_n; t) = \sum_{i=s+1}^{n} p_i \dot{q}_i - L. \qquad (8\text{-}42)$$

A differential of R is therefore given by

$$dR = \sum_{i=s+1}^{n} \dot{q}_i dp_i - \sum_{i=1}^{s} \frac{\partial L}{\partial \dot{q}_i} d\dot{q}_i - \sum_{i=1}^{n} \frac{\partial L}{\partial q_i} dq_i - \frac{\partial L}{\partial t} dt,$$

from which it follows that

$$\frac{\partial R}{\partial q_i} = -\frac{\partial L}{\partial q_i}, \qquad \frac{\partial R}{\partial \dot{q}_i} = -\frac{\partial L}{\partial \dot{q}_i}, \qquad i = 1, \ldots, s; \qquad (8\text{-}43)$$

$$\frac{\partial R}{\partial q_i} = -\dot{p}_i, \qquad \frac{\partial R}{\partial p_i} = \dot{q}_i, \qquad i = s + 1, \ldots, n. \qquad (8\text{-}44)$$

Equations (8–44) for the $n - s$ ignorable coordinates $q_{s+1}, \ldots, q_n$ are in the form of Hamilton's equations of motion with R as the Hamiltonian, while Eqs. (8–43) show that the s nonignorable coordinates obey the Lagrange equations

$$\frac{d}{dt}\left(\frac{\partial R}{\partial \dot{q}_i}\right) - \frac{\partial R}{\partial q_i} = 0, \qquad i = 1, \ldots, s, \qquad (8\text{-}45)$$

with R as the Lagrangian! Up to this point no explicit use has been made of the cyclic nature of the coordinates q_{s+1} to q_n. A coordinate absent from L will likewise not appear in the Routhian. The $n - s$ momenta p_{s+1} to p_n conjugate to the cyclic coordinates are constants and may be replaced in the Routhian by a set of constants $\alpha_1, \ldots, \alpha_r$, where $r = n - s$, to be determined by the initial conditions. With these modifications the only variables in the Routhian are the s noncyclic coordinates and their generalized velocities:

$$R = R(q_1, \ldots, q_s; \dot{q}_1, \ldots, \dot{q}_s; \alpha_1, \ldots, \alpha_r; t). \qquad (8\text{-}46)$$

The Lagrange equations for the noncyclic coordinates, (8–45), can now be solved without any regard for the behavior of the cyclic coordinates, exactly as in the Hamiltonian formulation. In effect the problem has been reduced to a Lagrangian problem for a system of s degrees of freedom, and except for the r constant parameters, α_j, we can "ignore" the other degrees of freedom.

The prime example where Routh's procedure may be usefully applied is in the examination of deviations from steady motion along with the question of the stability of such deviations. We have already considered several instances of steady motion. In Section 3–6 we investigated the stability of the steady motion of a particle in a circular orbit about a center of force for various central force laws.

---

* The function R as defined here is the negative of the form usually quoted, in order to bring it in line with the definition of H, Eq. (8–8).

The uniform precession of a heavy top, discussed in Section 5–7, constituted a steady motion and the phenomenon of nutation appeared as one of the deviations from this steady motion. In these and other examples steady motion is characterized by *all of the noncyclic coordinates being constant.* Thus in the central force problem, with two coordinates r and θ, a circular orbit means that the only noncyclic coordinate is constant. Again in the heavy top problem, of the three Euler angle coordinates θ, ϕ, ψ in the Lagrangian Eq. (5–50) the only noncyclic coordinate is θ, which is constant in uniform precession. If we restrict ourselves to situations where the Lagrangian is not an explicit function of time, then it follows that in steady motion *the cyclic coordinates are linear functions of time.* This can be seen from the equations of motion, (8–44), for the cyclic or ignorable coordinates. With a Routhian of the form of Eq. (8–46), these equations imply that a generalized velocity of an ignorable coordinate q_i is given in terms of the noncyclic variables by some relation of the type

$$\dot{q}_i = \dot{q}_i(q_1,\ldots,q_s;\dot{q}_1,\ldots,\dot{q}_s;\alpha_1,\ldots,\alpha_r), \qquad i = s+1,\ldots,n. \qquad (8\text{–}47)$$

For steady motion $q_1,\ldots,q_s$ are constant, $\dot{q}_1,\ldots,\dot{q}_s$ are zero, and therefore $\dot{q}_i$ for a cyclic variable is constant and q_i varies linearly with time. Thus in the steady motion in a circular orbit the cyclic coordinate θ increases uniformly with time. Again when the heavy top precesses uniformly the angular velocities of the cyclic coordinates, namely ψ and ϕ, are constant.

    As far as the "modified" system of s degrees of freedom is concerned, the problem has been reduced to that of oscillations about equilibrium positions q_{0i}, already studied in Chapter 6. As there, the generalized force corresponding to q_i must vanish at equilibrium, indicating that

$$\frac{\partial R}{\partial q_{0i}} \equiv \left(\frac{\partial R}{\partial q_i}\right)_0 = 0, \qquad i = 1,\ldots,s. \qquad (8\text{–}48)$$

(Cf. Eq. 6–1.) Indeed much of the procedure followed in Chapter 6 can be taken over here without significant change. The major difference is that we must start from specific generalized variables so chosen that the coordinates in steady motion are ignorable. These may be such that R is not simply a quadratic function of the generalized velocities but may have more complicated forms.

    To see both the similarities and the differences with small oscillations about an equilibrium we will carry the procedure through at least part of the way. For the remainder of this section we will resume the summation convention for sums from 1 through s, and as is suggested by Eq. (8–48), the subscript 0 will indicate values and derivatives evaluated at the "equilibrium" position corresponding to steady motion. As in Chapter 6, new variables ρ_i will denote the deviation from this equilibrium:

$$\begin{aligned} q_i &= q_{0i} + \rho_i, \\ \dot{q}_i &= \dot{\rho}_i, \end{aligned} \qquad i = 1,\ldots,s. \qquad (8\text{–}49)$$

The quantities ρ_i and $\dot{\rho}_i$ are treated as small in the sense that R can be expanded in a Taylor series about the steady motion configuration retaining only terms only through the second order. In light of Eq. (8–48) the expansion has the form

$$R = R_0 + \frac{\partial R}{\partial \dot{q}_{0i}}\dot{\rho}_i + \frac{1}{2}\frac{\partial^2 R}{\partial q_{0i}\partial q_{0j}}\rho_i\rho_j + \frac{\partial^2 R}{\partial q_{0i}\partial \dot{q}_{0j}}\rho_i\dot{\rho}_j + \frac{1}{2}\frac{\partial^2 R}{\partial \dot{q}_{0i}\partial \dot{q}_{0j}}\dot{\rho}_i\dot{\rho}_j. \qquad (8\text{–}50)$$

R and all its derivatives are functions only of the noncyclic q_i's, the corresponding $\dot{q}_i$'s, and constants. In steady motion the q_i's are constant and the $\dot{q}_i$'s zero. Hence R_0 and all the coefficients in Eq. (8–50) are constant in time, and R_0 as an additive constant can be dropped from R. The linear term in $\dot{\rho}_i$ automatically satisfies the Lagrangian equations of motion since the coefficient of $\dot{\rho}_i$ is constant in time and independent of ρ_i. This term too can therefore be dropped, and R may be rewritten as

$$R = \frac{1}{2}a_{ij}\dot{\rho}_i\dot{\rho}_j + g_{ij}\rho_i\dot{\rho}_j - \frac{1}{2}c_{ij}\rho_i\rho_j. \qquad (8\text{–}51)$$

Note that the matrix elements a_{ij} and c_{ij} are symmetric, but that is not necessarily so for g_{ij}. The terms in R bilinear in ρ_i and $\dot{\rho}_j$ are sometimes referred to as "gyroscopic terms" because they frequently arise in problems of gyroscopic motion. The Lagrange equations of motion for the variables ρ_i then take the form

$$a_{ij}\ddot{\rho}_j - b_{ij}\dot{\rho}_j + c_{ij}\rho_j = 0, \qquad (8\text{–}52)$$

where

$$b_{ij} = g_{ij} - g_{ji}.$$

Comparison with the equations of motion in Chapter 6, Eq. (6–8), shows that a_{ij} plays the role of T_{ij} and c_{ij} corresponds to V_{ij} (hence the choice of sign) but the term in $\dot{\rho}_j$ has no counterpart. (It does not correspond to the dissipative force terms that are not in R; typically the gyroscopic forces do no work and do not destroy the conservation of H.) The bilinear term in R makes it impossible to simultaneously diagonalize all terms in R by a transformation in configuration space as was done in Chapter 6, so normal modes cannot be obtained that way.* However, we can again seek oscillatory exponential solutions of the form of Eq. (6–9), with angular frequencies ω that will be the roots of the secular determinant:

$$|\mathbf{c} + i\omega\mathbf{b} - \omega^2\mathbf{a}| = 0. \qquad (8\text{–}53)$$

(cf. Eq. 6–11.) Despite the appearance of terms linear in ω it can be shown† from the antisymmetry of $\mathbf{b}$ and the symmetry of the other matrices that the secular

---

* The corresponding Hamiltonian *can* be diagonalized by a transformation in phase space, but that requires the theory of canonical transformations, to be discussed in the next chapter. See the article by J. L. Synge in Vol. III/1 of the Encyclopedia of Physics, p. 192, and Exercise 20 in Chapter 9 below.

† See K. R. Symon, *Mechanics*, 3d ed. (Reading, Massachusetts: Addison–Wesley, 1971). p. 483.

equation is a function of ω^2 only (as would be expected on physical grounds from symmetry under time reversal). If the roots of Eq. (8–53) correspond to real ω, the deviations from steady motion can be bounded and stability is possible. Imaginary values of ω can lead to unbounded increase of ρ with time, indicating instability of the steady motion.

It must be acknowledged that these statements give a highly oversimplified picture of the general problem of stability of motion. There are many complications, the least of which are the questions about the nature of the motion when there are multiple or zero roots to Eq. (8–53). More serious is the uncertainty as to what extent the linearized equations of motion (8–52) can represent the oscillations even for small amplitudes, and as to the nature of the nonlinear solutions. The general subject of stability of motion is today a very active field of investigation, often using sophisticated mathematical tools that are far beyond the scope of this book.

A simple, almost trivial, example may clarify Routh's procedure and the physical significance of the quantities involved. Consider the situation investigated in Section 3–6, that of a single particle moving in a plane under the influence of a central force $f(r)$ derived from a potential $V(r)$. The Lagrangian is then

$$L = \frac{m}{2}(\dot{r}^2 + r^2\dot{\theta}^2) - V(r).$$

As noted before the ignorable coordinate is θ, and if the constant conjugate momentum is denoted by l, the corresponding Routhian is

$$R = \frac{l^2}{2mr^2} + V(r) - \frac{m}{2}\dot{r}^2.$$

Physically we see that the Routhian is the equivalent one-dimensional potential $V'(r)$ *minus* the kinetic energy of radial motion. (The minus sign appears because the Routhian as defined by Eq. (8–42) is the negative of the Lagrangian for the effective one-dimensional problem.) In steady motion the noncyclic coordinate r is constant (at r_0) and the cyclic coordinate, θ, increases uniformly with time, i.e., the particle moves in a circular orbit with constant angular velocity. The steady state condition is given by Eq. (8–48), which is equivalent here to saying $V'(r)$ is an extremum at $r = r_0$:

$$\frac{\partial R}{\partial r_0} \equiv \frac{\partial V'(r)}{\partial r}\bigg|_{r=r_0} = -\frac{l^2}{mr_0^3} + \frac{\partial V}{\partial r}\bigg|_{r=r_0} = 0,$$

or

$$f(r_0) = -\frac{l^2}{mr_0^3},$$

which is Eq. (3–41). Expanding R in a Taylor series about the steady state leads to a linearized Routhian here of the form

$$R = \frac{a_{11}\dot{\rho}^2}{2} - \frac{c_{11}\rho^2}{2},$$

where

$$a_{11} = \frac{\partial^2 R}{\partial \dot{r}_0^2} = -m, \qquad c_{11} = -\frac{\partial^2 R}{\partial r_0^2} = -\frac{3l^2}{mr_0^4} + \frac{\partial f}{\partial r_0}.$$

The secular determinantal equation, Eq. (8–53), reduces to a single term

$$c_{11} - \omega^2 a_{11} = 0.$$

Hence, the condition that the motion be stable, that is, ω real, is equivalent to requiring $c_{11}/a_{11} > 0$, or, using the condition for steady motion,

$$\frac{\partial f}{\partial r_0} + \frac{3f(r_0)}{r_0} < 0,$$

which is Eq. (3–43), the criterion for stability of the steady motion.

Typically, Routh's procedure has not added to the physics of the analysis presented earlier in Section (3–6), but it has made the analysis automatic. In complicated problems with many degrees of freedom this feature can be a considerable advantage. It is not surprising therefore that Routh's procedure finds its greatest usefulness in the direct solution of problems relating to engineering applications. But basically, the Routhian is a sterile hybrid, combining some of the features of both the Lagrangian and the Hamiltonian pictures. For the development of various formalisms of classical mechanics, the complete Hamiltonian formulation is more fruitful.

8–4 THE HAMILTONIAN FORMULATION OF RELATIVISTIC MECHANICS

As with the Lagrangian picture in special relativity, two attitudes can be taken to the Hamiltonian formulation of relativistic mechanics. The first makes no pretense at a covariant description but instead works in some specific Lorentz or inertial frame. Time as measured in the particular Lorentz frame is then not treated on a common basis with other coordinates but serves, as in nonrelativistic mechanics, as a parameter describing the evolution of the system. Nonetheless, if the Lagrangian that leads to the Hamiltonian is itself based on a relativistically invariant physical theory, e.g., Maxwell's equations and the Lorentz force, then the resultant Hamiltonian picture will be relativistically correct. The second approach, of course, attempts a fully covariant description of the Hamiltonian picture, but the difficulties that plagued the corresponding Lagrangian approach (cf. Section 7–9) are even fiercer here. We shall consider the noncovariant method first.

For a single-particle Lagrangian of the form of Eq. (7–136),

$$L = -mc^2\sqrt{1 - \beta^2} - V,$$

we have already shown that the Hamiltonian (in the guise of the energy function h) is the total energy of the system:

$$H = T + V.$$

The energy T can be expressed in terms of the canonical momenta p_i (Eq. 7–137) through Eq. (7–100):

$$T^2 = p^2c^2 + m^2c^4,$$

so that a suitable form for the Hamiltonian is

$$H = \sqrt{p^2c^2 + m^2c^4} + V. \tag{8–54}$$

When the system consists of a single particle moving in an electromagnetic field, the Lagrangian has been given as (cf. Eq. 7–141)

$$L = -mc^2\sqrt{1 - \beta^2} + q\mathbf{A}\cdot\boldsymbol{\beta} - q\phi.$$

The term in L linear in the velocities drops out of the Hamiltonian, as we have seen, whereas the first term leads to the appearance of T in the Hamiltonian. Thus the Hamiltonian is again the total particle energy:

$$H = T + q\phi. \tag{8–55}$$

For this system the canonical momenta conjugate to the Cartesian coordinates of the particle are defined by (cf. Eq. 7–142)

$$p_i = mu_i + \frac{q}{c}A_i,$$

so that the relation between T and p_i is given by Eq. (7–168), and the Hamiltonian has the final form

$$H = \sqrt{\left(\mathbf{p} - \frac{q}{c}\mathbf{A}\right)^2 c^2 + m^2c^4} + q\phi. \tag{8–56}$$

It should be emphasized again that $\mathbf{p}$ here is the vector of the canonical momenta conjugate to the *Cartesian* position coordinates of the particle. We may also note that iH/c is the fourth component of the world vector

$$mu_v + \frac{q}{c}A_v$$

(Cf. Eqs. 7–74, 7–96, and 7–166.) While the Hamiltonian (8–56) is not expressed in covariant fashion, it does have a definite transformation behavior under a Lorentz transformation as being, in some Lorentz frame, the fourth component of a world vector.

In a covariant approach to the Hamiltonian formulation, time must be treated in the same fashion as the space coordinates, i.e., time must be taken as one of the canonical coordinates having an associated conjugate momentum. The foundations of such an extension of the dimensionality of phase space can in fact be constructed even in nonrelativistic mechanics. Following the pattern of Section 7–9, the progress of the system point along its trajectory in phase space can be marked by some parameter θ, and t "released," so to speak, to serve as an additional coordinate. If derivatives with respect to θ are denoted by a superscript prime, the Lagrangian in the $(q_1, \ldots, q_n; t)$ configuration space is (cf. Eq. 7–159)

$$\Lambda(q, q', t, t') = t' L\left(q, \frac{q'}{t'}, t\right). \qquad (8\text{–}57)$$

The momentum conjugate to t is then

$$p_t = \frac{\partial \Lambda}{\partial t'} = L + t' \frac{\partial L}{\partial t'}.$$

If we make explicit use of the connection $\dot{q} = q'/t'$, then this relation becomes

$$p_t = L - \frac{q_i'}{t'} \frac{\partial L}{\partial \dot{q}_i} = L - \dot{q}_i \frac{\partial L}{\partial \dot{q}_i} = -H. \qquad (8\text{–}58)$$

The momentum conjugate to the time "coordinate" is therefore the negative of the ordinary Hamiltonian.* While the framework of this derivation is completely nonrelativistic, the result is consistent with the identification of the time component of the four-vector momentum with iE/c. As can be seen from the definition, Eq. (8–2), if q is multiplied by a constant α, then the conjugate momentum is divided by α. Hence, the canonical momentum conjugate to ict is iH/c.

Thus, there seems to be a natural route available for constructing a relativistically covariant Hamiltonian. But the route turns out to be mined with booby traps. It will be recalled that the covariant Lagrangian used to start the process, Eq. (7–159) or Eq. (8–57), is homogeneous in first degree in the generalized velocities q', and for such a Lagrangian the recipe described above (cf. p. 343) for constructing the Hamiltonian formulation breaks down irreparably. If L is of type L_1, the corresponding Hamiltonian, call it $|H(q, t, p, p_t)$, is identically zero! Even before we reach that stage, however, there is a breakdown in step 4 that calls for inverting the set of Eqs. (8–2) to obtain the generalized velocities as

---

* The remaining momenta are unchanged by the shift from t to θ, as can be seen by evaluating the corresponding derivative:

$$\frac{\partial \Lambda}{\partial q_i'} = t' \frac{\partial L}{\partial q_i'} = t' \left(\frac{\partial L_1}{\partial \dot{q}}\frac{1}{t'}\right) = p_i.$$

functions of (q_i, p_i). For L_1 Lagrangians this inversion cannot be carried out. This is most directly seen if L is homogeneous in the first degree through being simply linear in the $\dot{q}_i$'s. In that case p_i is independent of all $\dot{q}_j$'s, and there can be no way of finding the $\dot{q}_j$'s in terms of p_i. When the Lagrangian is of type L_1 in a more general fashion, the proof of the impossibility of inversion is more involved. It will be recalled from the general implicit function theorem, that if n variables y_i are given as functions of n variables x_j,

$$y_i = y_i(x_1, \ldots, x_n),$$

then the functions can be inverted to find x_j as functions of y_i only if the determinant of the Jacobian matrix $\mathbf{M}$, with elements

$$M_{ij} = \frac{\partial y_i}{\partial x_j} \tag{8-59}$$

does not vanish.* Applied to Eqs. (8-2) we see that they can be inverted only for a nonvanishing determinant

$$|\mathbf{W}| = \left| \frac{\partial^2 L}{\partial \dot{q}_i \partial \dot{q}_j} \right|.$$

Now, if L is of type L_1 we have

$$\dot{q}_i \frac{\partial L}{\partial \dot{q}_i} = L.$$

Differentiation with respect to $\dot{q}_j$ then leads to the condition

$$\dot{q}_i \frac{\partial^2 L}{\partial \dot{q}_i \partial \dot{q}_j} = 0, \tag{8-60}$$

or in matrix notation

$$\dot{\mathbf{q}} \mathbf{W} = 0. \tag{8-60'}$$

Equation (8-60) can be interpreted as saying that one column of the determinant can be expressed in terms of the other columns, which means that the determinant vanishes. Alternatively Eq. (8-60') can be read as stating that $\mathbf{W}$ has only null eigenvalues, which again implies vanishing of the determinant. Of course, the vanishing of the Jacobian determinant means that not all the momenta are independent. For the general formulation in which phase space is extended to include t and p_t, Eq. (8-58) written in the form

$$p_t = -H(q_1, \ldots, q_n; p_1, \ldots, p_n; t)$$

explicitly says p_t is not independent but is given in terms of all the other coordinates and momenta. In the relativistic case, say, of a free particle, we have

---

* Many books on advanced calculus discuss the implicit function theorem; cf. S. Lang, *A Second Course in Calculus*, 2d. ed., 1968, p. 529. A particularly careful proof is to be found in E. Hille, *Analysis*, Vol. 2, 1966, p. 367.

seen that this constraint on the momenta corresponds physically to the fact that the four-vector p_μ has a constant magnitude.

It must be concluded therefore that for Lagrangians homogeneous in the $\dot{q}_i$'s in first degree the usual Hamiltonian procedure does not work. A number of alternate ways of formulating the Hamiltonian in the "homogeneous problem" have been devised.* All take into account explicitly the constraint implied by the vanishing of the Jacobian determinant, but they are all complicated, and no one has been accepted as the standard technique. It appears that if at all possible the homogeneous "noninvertible" Lagrangian should be avoided like the plague.

Fortunately there does not seem to be any compelling reason why the covariant Lagrangian has to be homogeneous in the first degree, at least for classical relativistic mechanics. It has already been seen that for a single free particle the covariant Lagrangian

$$\Lambda(x_\mu, u_\mu) = \frac{1}{2} m u_\mu u_\mu$$

leads to the correct equations of motion. Of course the four-velocity components, u_μ, are still not all independent, but the constraint can be treated as a "weak condition" to be imposed only *after* all the differentiations have been carried through.† There is now no difficulty in obtaining a Hamiltonian from this Lagrangian, by the same route as in nonrelativistic mechanics; the result is clearly

$$\mathbb{H} = \frac{p_\mu p_\mu}{2m}. \tag{8-61}$$

For a single particle in an electromagnetic field, a covariant Lagrangian has been found previously:

$$\Lambda(x_\mu, u_\mu) = \frac{1}{2} m u_\mu u_\mu + \frac{q}{c} u_\mu A_\mu(x_\lambda), \tag{7-165}$$

with the canonical momenta,

$$p_\mu = m u_\mu + \frac{q}{c} A_\mu. \tag{7-166}$$

---

* The literature dealing with the "homogeneous problem" more or less from the physicist's viewpoint is by now quite extensive, and only a few references can be cited. Lanczos, *The Variational Principles of Mechanics*, 4th ed., 1970, p. 187 gives about the simplest version of all. Dirac, in a number of papers starting in 1933, has developed a distinctive approach that is summarized in a brief book, *Lectures on Quantum Mechanics*, 1964. For still another way of looking at the problem, see H. Rund, *The Hamilton–Jacobi Theory in the Calculus of Variations*, 1966 Chapter 3.

It may be noted that the Legendre transformation process is reversible: given a Hamiltonian one can obtain the corresponding Lagrangian (cf. Exercise 1). But the difficulties also arise in either direction. If a given Hamiltonian is postulated to be homogeneous in first degree in the momenta, then it is not possible to find an equivalent Lagrangian.

† See Dirac, *Lectures on Quantum Mechanics*, for this classification of constraints.

In the corresponding Hamiltonian the term linear in u_μ disappears as usual, and the remaining L_2 part in terms of the canonical momenta is

$$\mathsf{H} = \frac{\left(p_\mu - \dfrac{q}{c}A_\mu\right)\left(p_\mu - \dfrac{q}{c}A_\mu\right)}{2m}. \tag{8-62}$$

Both Hamiltonians, Eqs. (8–61) and 8–62), are constant, with the same value, $-mc^2/2$, but to obtain the equations of motion it is the *functional* dependence on the four-vectors of position and momenta that is important. With a system of one particle, the covariant Hamiltonian leads to eight first order equations of motion

$$\frac{dx_\nu}{d\tau} = \frac{\partial \mathsf{H}}{\partial p_\nu}, \qquad \frac{dp_\nu}{d\tau} = -\frac{\partial \mathsf{H}}{\partial x_\nu}. \tag{8-63}$$

We know that these equations cannot be all independent. The space parts of Eqs. (8–63) obviously lead to the spatial equations of motion. We should expect therefore that the remaining two equations tell us nothing new, exactly as in the Lagrangian case. This can be verified by examining the $\nu = 4$ equations in some particular Lorentz frame. One of them is the constitutive equation for p_4:

$$u_4 = \frac{\partial \mathsf{H}}{\partial p_4} = \frac{1}{m}\left(p_4 - \frac{q}{c}A_4\right)$$

or

$$p_4 = \frac{i}{c}(T + q\phi) = \frac{iH}{c},$$

a general conclusion that has been noted before. The other can be written as

$$\frac{1}{\sqrt{1 - \beta^2}}\frac{dp_4}{dt} = -\frac{1}{ic}\frac{\partial \mathsf{H}}{\partial t}$$

or

$$\frac{dH}{dt} = \sqrt{1 - \beta^2}\frac{\partial \mathsf{H}}{\partial t}.$$

Comparison of the form of (8–60) with that of H (8–56) shows that

$$\frac{\partial \mathsf{H}}{\partial t} = \frac{T}{mc^2}\frac{\partial H}{\partial t},$$

so that the equation of motion reduces to

$$\frac{dH}{dt} = \frac{\partial H}{\partial t},$$

which is the general result already expressed in Eq. (8–35).

As with the covariant Lagrangian formulation one has the problem of finding suitable covariant potential terms in the Lagrangian to describe forces other than electromagnetic. And in multiparticle systems we are confronted in full measure with the critical difficulties of including interactions other than with fields. In Hamiltonian language, the "no-interaction" theorem already referred to says that only in the absence of direct particle interaction can Lorentz invariant systems be described in terms of the usual position coordinates and corresponding canonical momenta. The scope of the relativistic Hamiltonian framework is therefore quite limited and so for the most part we shall confine ourselves to nonrelativistic mechanics.

8-5 DERIVATION OF HAMILTON'S EQUATIONS FROM A VARIATIONAL PRINCIPLE

Lagrange's equations have been shown to be the consequence of a variational principle, namely, the Hamilton's principle of Section 2–1. Indeed, the variational method has often proved to be the preferable method of deriving Lagrange's equations, for it is applicable to types of systems not usually comprised within the scope of mechanics. It would be similarly advantageous if a variational principle could be found that leads directly to the Hamilton's equations of motion. Hamilton's principle,

$$\delta I \equiv \delta \int_{t_1}^{t_2} L\,dt = 0, \qquad (8\text{-}64)$$

lends itself to this purpose, but as formulated originally it refers to paths in configuration space. The first modification therefore is that the integral must be evaluated over the trajectory of the system point in phase space, and the varied paths must be in the neighborhood of this phase space trajectory. In the spirit of the Hamiltonian formulation, both q and p must be treated as independent coordinates of phase space, to be varied independently. To this end the integrand in the action integral, Eq. (8–64) must be expressed as a function of both q and p, and their time derivatives, through Eq. (8–8). Equation (8–64) then appears as

$$\delta I = \delta \int_{t_1}^{t_2} \left(p_i \dot{q}_i - H(q, p, t) \right) dt = 0. \qquad (8\text{-}65)$$

As a variational principle in phase space, Eq. (8–65) is sometimes referred to as the *modified Hamilton's principle*. Although it will be used most frequently in connection with transformation theory (see Chapter 9) the main interest in it here is to show that the principle leads to Hamilton's canonical equations of motion.

The modified Hamilton's principle is exactly of the form of the variational problem in a space of $2n$ dimensions considered in Section 2–3 (cf. Eq. 2–14):

$$\delta I = \delta \int_{t_1}^{t_2} f(q, \dot{q}, p, \dot{p}, t)\,dt = 0, \qquad (8\text{-}66)$$

for which the $2n$ Euler–Lagrange equations are

$$\frac{d}{dt}\left(\frac{\partial f}{\partial \dot{q}_j}\right) - \frac{\partial f}{\partial q_j} = 0, \qquad j = 1,\ldots,n, \tag{8-67}$$

$$\frac{d}{dt}\left(\frac{\partial f}{\partial \dot{p}_j}\right) - \frac{\partial f}{\partial p_j} = 0, \qquad j = 1,\ldots,n. \tag{8-68}$$

The integrand f as given in Eq. (8–65) contains $\dot{q}_j$ only through the $p_i\dot{q}_i$ term, and q_j only in H. Hence Eqs. (8–67) lead to

$$\dot{p}_j + \frac{\partial H}{\partial q_j} = 0. \tag{8-69}$$

On the other hand there is no explicit dependence of the integrand in Eq. (8–65) on $\dot{p}_j$. Equations (8–68) therefore reduce simply to

$$\dot{q}_j - \frac{\partial H}{\partial p_j} = 0. \tag{8-70}$$

Equations (8–69) and (8–70) are exactly Hamilton's equations of motion, Eqs. (8–12). The Euler–Lagrange equations of the modified Hamilton's principle are thus the desired canonical equations of motion.

This derivation of Hamilton's equations from the variational principle is so brief as to give the appearance of a sleight-of-hand trick. One wonders whether something extra has been sneaked in while we were being misdirected by the magician's patter. Is the modified Hamilton's principle equivalent to Hamilton's principle, or does it contain some additional physics? The question is largely irrelevant; the primary justification for the modified Hamilton's principle is that it leads to the canonical equations of motion in phase space. After all, no further argument was given for the validity of Hamilton's principle than that it corresponded to the Lagrangian equations of motion. So long as a Hamiltonian can be constructed, the Legendre transformation procedure shows that the Lagrangian and Hamiltonian formulations, and therefore their respective variational principles, have the same physical content.

One question that can be raised, however, is whether the derivation puts limitations on the variation of the trajectory that are not present in Hamilton's principle. The variational principle leading to the Euler–Lagrange equations is formulated, as in Section 2–2, such that the variations of the independent variables vanish at the endpoints. In phase space that would require $\delta q_i = 0$ *and* $\delta p_i = 0$ at the endpoints, whereas Hamilton's principle requires only the vanishing of δq_i under the same circumstances. A look at the derivation as spelled out in Section 2–2 will show, however, that the variation is required to be zero at the endpoints only in order to get rid of the integrated terms arising from the variations in the time derivatives of the independent variables. While the f function in Eq. (8–66) that corresponds to the modified Hamilton's principle, Eq. (8–65), is indeed a function of $\dot{q}_j$, there is no explicit appearance of $\dot{p}_j$. Equations

(8–68) and therefore (8–70) follow from Eq. (8–65) without stipulating the variations of p_j at the endpoints. The modified Hamilton's principle, with the integrand L defined in terms of the Hamiltonian by Eq. (8–8), leads to Hamilton's equations under the same variation conditions as those in Hamilton's principle.*

Nonetheless, there are advantages to requiring that the varied paths in the modified Hamilton's principle return to the same endpoints in both q and p, for we then have a more generalized condition for Hamilton's equations of motion. As with Hamilton's principle, if there is no variation at the endpoints we can add a total time derivative of any arbitrary (twice-differentiable) function $F(q, p, t)$ to the integrand without affecting the validity of the variational principle. Suppose, for example, we subtract from the integrand of Eq. (8–65) the quantity

$$\frac{d}{dt}(q_i p_i).$$

The modified Hamilton's principle would then read

$$\delta \int_{t_1}^{t_2} \left(-\dot{p}_i q_i - H(q, p, t) \right) dt = 0. \tag{8–71}$$

Here the f integrand of Eq. (8–66) is a function of $\dot{p}$, and it is easily verified that the Euler–Lagrange equations (8–67) and (8–68) with this f again correspond to Hamilton's equations of motion, Eqs. (8–12). Yet the integrand in Eq. (8–71) is not the Lagrangian nor can it, in general, be simply related to the Lagrangian by a point transformation in configuration space. By restricting the variation of both q and p to be zero at the endpoints, the modified Hamilton's principle provides an independent and general way of setting up Hamilton's equations of motion without a prior Lagrangian formulation. If you will, it does away with the necessity of a linkage between the Hamiltonian canonical variables and a corresponding Lagrangian set of generalized coordinates and velocities. This will be very important to us in the next chapter where we examine transformations of phase space variables that preserve the Hamiltonian form of the equations of motion.

The requirement of independent variation of q and p, so essential for the above derivation, highlights the fundamental difference between the Lagrangian

---

* It may be objected that q and p cannot be varied independently, because the defining Eqs. (8–2) link p with q and $\dot{q}$. One could not then have a variation of q (and $\dot{q}$) without a corresponding variation of p. An attempt is sometimes made to accomodate this objection by substituting a new independent variable, say r_j, for $\dot{x}_j$ in Eqs. (8–2) so that the q and r variables are independent in the modified Hamilton's principle, with p_j a derived function. The equations of motion are then obtained subject to the constraints $r_j = \dot{x}_j$, and turn out to be Hamilton's equations of motion! (Cf. D. Ter Haar, *Elements of Hamiltonian Mechanics*, 1961, p. 100.) But this procedure is not necessary, and indeed the entire objection is completely at variance with the intent and the spirit of the Hamiltonian picture. Once the Hamiltonian formulation has been set up, Eqs. (8–2) *form no part of it*. The momenta have been elevated to the status of independent variables, on an equal basis with the coordinates and connected with them and the time *only through the medium of the equations of motion themselves* and not by any a priori defining relationship.

and Hamiltonian formulations. Neither the coordinates q_i nor the momenta p_i are to be considered there as the more fundamental set of variables; both are equally independent. Only by broadening the field of independent variables from n to $2n$ quantities are we enabled to obtain equations of motion that are of first order. In a sense the names "coordinates" and "momenta" are unfortunate, for they bring to mind pictures of spatial coordinates and linear, or at most, angular momenta. A wider meaning must now be given to the terms. The division into coordinates and momenta corresponds to no more than a separation of the independent variables describing the motion into two groups having an almost symmetrical relationship to each other through Hamilton's equations.

8–6 THE PRINCIPLE OF LEAST ACTION

Another variational principle associated with the Hamiltonian formulation is known as *the principle of least action.* It involves a new type of variation, which we shall call the Δ-variation, requiring detailed explanation. In the δ-variation process used in the discussion of Hamilton's principle in Chapter 2, the varied path in configuration space always terminated at endpoints representing the system configuration at the same time t_1 and t_2 as the correct path. To obtain Lagrange's equations of motion we also required that the varied path return to the same endpoints in configuration space, that is, $\delta q_i(t_1) = \delta q_i(t_2) = 0$. The Δ-variation is less constrained; in general the varied path over which an integral is evaluated may end at different times than the correct path, and there may be a variation in the coordinates at the endpoints. We can however use the same parameterization of the varied path as in the δ-variation. In the notation of Section (2–3), a family of possible varied paths is defined by functions

$$q_i(t, \alpha) = q_i(t, 0) + \alpha \eta_i(t), \qquad (8\text{–}72)$$

where α is an infinitesimal parameter that goes to zero for the correct path. Here the functions η_i do not necessarily have to vanish at the endpoints, either the original or the varied. All that is required is that they be continuous and differentiable. Figure 8–2 schematically illustrates the correct and varied path for a Δ-variation in configuration space.

Let us evaluate the Δ-variation of the action integral:

$$\Delta \int_{t_1}^{t_2} L \, dt \equiv \int_{t_1 + \Delta t_1}^{t_2 + \Delta t_2} L(\alpha) \, dt - \int_{t_1}^{t_2} L(0) \, dt, \qquad (8\text{–}73)$$

where $L(\alpha)$ means the integral is evaluated along the varied path and $L(0)$ correspondingly refers to the actual path of motion. The variation is clearly composed of two parts. One arises from the change in the limits of the integral; to first order infinitesimals this part is simply the integrand on the actual path times the difference in the limits in time. The second part is caused by the change in the

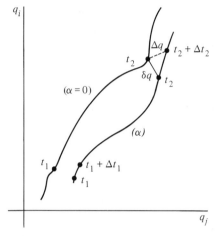

FIGURE 8–2
Schematic illustration of the Δ-variation in configuration space.

integrand on the varied path, but now between the same time limits as the original integral. One may therefore write the Δ-variation of the action integral as

$$\Delta \int_{t_1}^{t_2} L\,dt = L(t_2)\,\Delta t_2 - L(t_1)\,\Delta t_1 + \int_{t_1}^{t_2} \delta L\,dt. \tag{8–74}$$

Here the variation in the second integral can be carried out through a parameterization of the varied path, exactly as for Hamilton's principle except that the variation in q_i does not vanish at the endpoints. The endpoint terms arising in the integration by parts must be retained and the integral term on the right appears as

$$\int_{t_1}^{t_2} \delta L\,dt = \int_{t_1}^{t_2} \left[\frac{\partial L}{\partial q_i} - \frac{d}{dt}\left(\frac{\partial L}{\partial \dot{q}_i}\right) \right] \delta q_i\,dt + \frac{\partial L}{\partial \dot{q}_i} \delta q_i \bigg|_1^2.$$

By Lagrange's equations the quantities in the square brackets vanish, and the Δ-variation therefore takes the form

$$\Delta \int_{t_1}^{t_2} L\,dt = \left(L\,\Delta t + p_i\,\delta q_i\right)\bigg|_1^2. \tag{8–75}$$

In Eq. (8–75), δq_i refers to the variation in q_i at the original endpoint times t_1 and t_2. We would like to express the Δ-variation in terms of the change Δq_i between q_i at the endpoints of the actual path and q_i at the endpoints of the varied path, including the change in endpoint times. It is clear from Fig. (8–2) that these two variations are connected by the relation*

$$\Delta q_i = \delta q_i + \dot{q}_i \, \Delta t. \tag{8-76}$$

Hence Eq. (8–75) can be rewritten as

$$\Delta \int_{t_1}^{t_2} L \, dt = \left. (L \Delta t - p_i \dot{q}_i \, \Delta t + p_i \, \Delta q_i) \right|_1^2$$

or

$$\Delta \int_{t_1}^{t_2} L \, dt = \left. (p_i \, \Delta q_i - H \, \Delta t) \right|_1^2. \tag{8-77}$$

To obtain the principle of least action we restrict our further considerations by three important qualifications:

1. Only systems are considered for which L, and therefore H, are not explicit functions of time, and in consequence H is conserved.

2. The variation is such that H is conserved on the varied path as well as on the actual path.

3. The varied paths are further limited by requiring that Δq_i vanish at the endpoints (but not Δt).

The nature of the resultant variation may be illustrated by noting that the varied path satisfying these conditions might very well describe the same curve in configuration space as the actual path. The difference will be the speed with which the system point traverses this curve; i.e., the functions $q_i(t)$ will be altered in the varied path. In order then to preserve the same value of the Hamiltonian at all points on the varied path the times of the endpoints must be changed. With these three qualifications satisfied, the Δ-variation of the action integral, Eq. (8–77), reduces to

$$\Delta \int_{t_1}^{t_2} L \, dt = -H(\Delta t_2 - \Delta t_1). \tag{8-78}$$

But under the same conditions the action integral itself becomes

$$\int_{t_1}^{t_2} L \, dt = \int_{t_1}^{t_2} p_i \dot{q}_i \, dt - H(t_2 - t_1),$$

---

* Equation (8–76) may be derived formally from the parameter form, Eq. (8–72), of the varied path. Thus, at the upper endpoint we have

$$\Delta q_i(2) = q_i(t_2 + \Delta t_2, \alpha) - q_i(t_2, 0) = q_i(t_2 + \Delta t_2, 0) - q_i(t_2, 0) + \alpha \eta_i(t + \Delta t_2),$$

which to first order in small quantities α and Δt_2 is

$$\Delta q_i(2) = \dot{q}_i(2) \Delta t_2 + \delta q_i(2),$$

which is what Eq. (8–76) predicts.

the Δ-variation of which is

$$\Delta \int_{t_1}^{t_2} L \, dt = \Delta \int_{t_1}^{t_2} p_i \dot{q}_i \, dt - H(\Delta t_2 - \Delta t_1). \tag{8-79}$$

Comparison of Eqs. (8–78) and (8–79) finally gives the *principle of least action*:*

$$\Delta \int_{t_1}^{t_2} p_i \dot{q}_i \, dt = 0. \tag{8-80}$$

By way of caution, it may be noted that the modified Hamilton's principle can be written in a form with a superficial resemblance to Eq. (8–80). If the trajectory of the system point is described by a parameter θ, as in Sections (7–9) and (8–4), the modified Hamilton's principle appears as

$$\delta \int_{\theta_1}^{\theta_2} (p_i \dot{q}_i - H) t' \, d\theta = 0. \tag{8-81}$$

It will be recalled (cf. footnote on p.358) that the momenta p_i do not change under the shift from t to θ, and that $\dot{q}_i t' = q_i'$. Further, the momentum conjugate to t is $-H$. Hence Eq. (8–81) can be rewritten as

$$\delta \int_{\theta_1}^{\theta_2} \sum_{i=1}^{n+1} p_i q_i' \, d\theta = 0, \tag{8-82}$$

where t has been denoted by q_{n+1}. There should, however, be no confusion between Eq. (8–82) and the principle of least action. Equations (8–82) refers to a phase space of $(2n + 2)$ dimensions, as is indicated by the explicit summation to $i = n + 1$, whereas Eq. (8–80) is in the usual configuration space. But most important, the principle of least action is in terms of a Δ-variation for constant H, while Eq. (8–82) employs the δ-variation, and H in principle could be a function of time. Equation (8–82) is nothing more than the modified Hamilton's principle, and the absence of a Hamiltonian merely reflects the phenomenon that the Hamiltonian vanishes identically for the "homogeneous problem."

---

* The integral in Eq. (8–80) is usually referred to in the older literature as the action, or action integral, and the first edition of this book followed the same practice. In recent years it has become more customary to refer to the integral in Hamilton's principle as the action, and we have accepted this usage here. Following the example of Landau and Lifshitz (*Mechanics*, 1960, p. 141) the integral in Eq. (8–80) will be designated as the *abbreviated action*.

The principle of least action is commonly associated with the name of Maupertuis, and some authors call the integral involved the *Maupertuis action*. However the original statement of the principle by Maupertuis first in 1744 was vaguely teleological and could hardly pass muster today. The objective statement of the principle we owe to Euler and Lagrange. Of course, the name of the theorem should rather be *the principle of stationary action*, but the historical name is too firmly ingrained in the literature to be changed now. For the circumstances under which the abbreviated action is indeed a minimum, see E. T. Whittaker, *A Treatise on Analytical Dynamics*, 4th ed., 1936, p. 250.

The least action principle itself can be exhibited in a variety of forms. In nonrelativistic mechanics, if the defining equations for the generalized coordinates do not involve the time explicitly, then the kinetic energy is a quadratic function of the $\dot{q}_i$'s (cf. Eq. 1–71):

$$T = \frac{1}{2} M_{jk}(q)\dot{q}_j\dot{q}_k. \tag{8–83}$$

When, in addition, the potential is not velocity dependent the canonical momenta are derived from T only, and in consequence

$$p_i\dot{q}_i = 2T.$$

The principle of least action for such systems can therefore be written as

$$\Delta \int_{t_1}^{t_2} T\,dt = 0. \tag{8–84}$$

If, further, there are no external forces on the system, as, for example, a rigid body with no net applied forces, then T is conserved along with the total energy H. The least action principle then takes the special form

$$\Delta(t_2 - t_1) = 0. \tag{8–85}$$

Equation (8–85) states that of all paths possible between two points, consistent with conservation of energy, the system moves along that particular path for which the time of transit is the least (more strictly, an extremum). In this form the principle of least action recalls Fermat's principle in geometrical optics that a light ray travels between two points along such a path that the time taken is the least. We shall have occasion to return to these considerations in Chapter 10 when we discuss the connection between the Hamiltonian formulation and geometrical optics.

In Chapter 6 concerning small oscillations we introduced the notion (cf. p. 250) of a curvilinear configuration space for which the elements of the matrix **T** formed the metric tensor. Such a configuration space has the property that $2T$ is the square of the magnitude of the velocity vector of the system point. We can do something entirely similar here whenever T is of the form of Eq. (8–83).* A configuration space is therefore constructed for which the M_{jk} coefficients form the metric tensor. In general the space will be curvilinear and nonorthogonal. The element of path length in the space is then defined by (cf. Eq. 6–24)

$$(d\rho)^2 = M_{jk}\,dq_j\,dq_k \tag{8–86}$$

---

* The elements of **T** were independent of the displacements η_i from equilibrium positions. Here the elements M_{jk} are in general functions of the q's. But the difference in no way affects the formulation.

so that the kinetic energy has the form

$$T = \frac{1}{2}\left(\frac{d\rho}{dt}\right)^2, \tag{8-87}$$

or equivalently

$$dt = \frac{d\rho}{\sqrt{T}}. \tag{8-88}$$

Equation (8–88) enables us to change the variable in the abbreviated action integral from t to ρ, and the principle of least action becomes

$$\Delta \int_{t_1}^{t_2} T\,dt = 0 = \Delta \int_{\rho_1}^{\rho_2} \sqrt{T}\,d\rho,$$

or, finally

$$\Delta \int_{\rho_1}^{\rho_2} \sqrt{H - V(q)}\,d\rho = 0. \tag{8-89}$$

Equation (8–89) is often described as *Jacobi's form of the least action principle*. It now refers to the path of the system point in a special curvilinear configuration space characterized by the metric tensor with elements M_{jk}. The system point traverses the path in this configuration space with a speed given by $\sqrt{2T}$. If there are no forces acting on the body, T is constant, and Jacobi's principle says the system point travels along the shortest path length in the configuration space. Equivalently stated, the motion of the system is then such that the system point travels along the geodesics of the configuration space.

It should be emphasized that the Jacobi form of the principle of least action is concerned with the *path* of the system point rather than with its motion in *time*. Equation (8–89) is a statement about the element of path length $d\rho$; the time nowhere appears, since H is a constant and V depends on q_i only. Indeed, it is possible to use the Jacobi form of the principle to furnish the differential equations for the path, by a procedure somewhat akin to that leading to Lagrange's equations. In the form of Fermat's principle, the Jacobi version of the principle of least action finds many fruitful applications in geometrical optics and in electron optics. To go into any detail here would lead us too far afield.

A host of other similar, variational principles for classical mechanics can be derived in bewildering variety. To give one example out of many, the principle of least action leads immediately to *Hertz's principle of least curvature*, which states that a particle not under the influence of external forces travels along the path of least curvature. By Jacobi's principle such a path must be a geodesic, and the geometrical property of minimum curvature is one of the well-known characteristics of a geodesic. It has been pointed out that variational principles in themselves contain no new physical content, and they rarely simplify the practical solution of a given mechanical problem. Their value lies chiefly as starting points for new formulations of the theoretical structure of classical mechanics. For this purpose Hamilton's principle is especially fruitful, and to a lesser extent, so also is the principle of least action. The others have proved to be of little use, except as they have led to fruitless teleological speculations, and further discussion of them here seems pointless.

SUGGESTED REFERENCES

P. H. BADGER, *Equilibrium Thermodynamics.* Many textbooks on thermodynamics give some mention of the application of the Legendre transformation to the thermodynamic potentials. This reference has an exceptionally long and explicit treatment in Chapter 10.

R. COURANT AND D. HILBERT, *Methods of Mathematical Physics.* As has been mentioned the Legendre transformation has formal mathematical applications to the theory of partial differential equations. Courant, in Vol. II of this work, provides a readable discussion of this aspect and of the geometrical significance of the Legendre transformation (Chapter 1, Section 6).

E. T. WHITTAKER, *Analytical Dynamics.* The subject of the variational principles encountered in classical mechanics can become quite involved and possesses many far-flung roots in apparently unrelated fields. For example, there is a close connection between Hamilton's principle and the general theory of second order partial differential equations. Some of these topics we will discuss in the following chapters, but many are inappropriate for the treatment intended here. Similarly, such questions as whether the extremum in Hamilton's principle is a minimum or maximum cannot be taken up here, nor will it be feasible (or desirable) to consider the many subspecies of variational principles. The student interested in problems of this nature will find an abundant literature; indeed, there is an "embarras des richesses." Only a small fraction of the available references can be mentioned in this list, and of them Whittaker is one of the chief sources. Chapter IX and the first two sections of Chapter X are the portions pertinent to this chapter. The last four sections of Chapter VII are also a rich source of material and examples on oscillations about steady motion although, as often, he makes it seem more complicated than it is.

C. LANCZOS, *Variational Principles of Mechanics.* Noteworthy in Chapter 6 on the canonical equations is the emphasis that the shift to the Hamiltonian formulation doubles the dimensionality of the variable domain—from configuration to phase space. The principle of least action gets a somewhat different viewpoint from what's given here, in his Chapter 5.

J. L. SYNGE, *Classical Dynamics.* This is a respectable-size book (224 pages) tucked away as an article in Vol III/1 of the *Encyclopedia of Physics.* It covers much of the usual topics, often from a highly original viewpoint. Section E on general dynamical theory builds the formulation about the various kinds of variable spaces used, including phase space enlarged by time and energy. Often tough going, but worth the effort.

L. A. PARS, *Treatise on Analytical Dynamics.* Hamilton's equations get a mention only as a section in a chapter on "Further applications of Lagrange's equations." But the material on Routh's method applied to vibrations about steady motion (in Chapter 9) is voluminous and valuable, if somewhat disorganized.

K. R. SYMON, *Mechanics.* The discussion on small vibrations about steady motion (Section 12.6) is somewhat hampered by a refusal to mention the Routhian, but the author speaks on the subject with the voice of the expert. His research interests have included the theory of accelerators, where stability of small oscillations is an important question. Betatron oscillations are discussed as one example. Another unusual example is the stability of the so-called Lagrangian points in the three-body problem.

D. A. WELLS, *Lagrangian Dynamics*. There *is* a discussion of Hamilton's equations but it is brief and not too informative. However there is an extensive chapter on oscillations about steady motion from the Routhian point of view, replete with figures and examples, mostly with an engineering orientation.

P. A. M. DIRAC, *Lectures on Quantum Mechanics*. A slim book, reprinting some lectures he gave, reporting on his concern with the transition from classical to quantum mechanics. He outlines his method of dealing with the homogeneous problem, for which he devised a hierarchy of weak and strong constraints. Some acquaintance with canonical transformations (given in the next chapter) would be helpful in understanding the material.

H. RUND, *Hamilton–Jacobi Theory in the Calculus of Variations*. Mention was made in the previous chapter of the extensive treatment given here of the homogeneous problem, although the approach is not one that seems to have been generally accepted.

D. TER HAAR, *Elements of Hamiltonian Mechanics*. Reasons have been given above (p. 364) for the disagreement with the author's treatment of the modified Hamilton's principle. What is noteworthy, however, is the treatment of variational principles where time is varied, as in the principle of least action, and the formal extension of time and energy as canonical variables—although the dangers inherent in the homogeneous problem get only scant warning.

P. BRUNET, *Étude Historique sur le Principe de la Moindre Action*. Those interested in the early history of the principle of least action will find it here in a smooth urbane treatment, from the teleological beginnings of Maupertuis to the time of Lagrange when it had been transformed into a solid tool for developing mechanics.

R. L. LINDSAY AND H. MARGENAU, *Foundations of Physics*. The statements of Hamilton's principle and the principle of least action appear to endow the mechanical system with conscious knowledge of the final state toward which the motion is directed. Such an appearance is of course illusory; the motion of the system is determined only by the initial conditions. But the view has given rise in the past to much philosophical speculation. Chapter 3 of this text presents an adequate discussion of this and similar points, and furnishes references for further reading for those inclined.

EXERCISES

1. a) Reverse the Legendre transformation to derive the properties of $L(q_i, \dot{q}_i, t)$ from $H(q_i, p_i, t)$, treating the $\dot{q}_i$ as independent quantities, and show that it leads to the Lagrangian equations of motion.

b) By the same procedure find the equations of motion in terms of the function

$$L'(p, \dot{p}, t) = -\dot{p}_i q_i - H(q, p, t).$$

2. Write the problem of central force motion of two mass points in Hamiltonian formulation, eliminating the cyclic variables, and reducing the problem to quadratures.

3. Formulate the double-pendulum problem illustrated by Fig. 1–4, in terms of the Hamiltonian and Hamilton's equations of motion. It is suggested you find the Hamiltonian both directly from L and by Eq. (8–19).

4. The Lagrangian for a system can be written as

$$L = a\dot{x}^2 + b\frac{\dot{y}}{x} + c\dot{x}\dot{y} + f y^2 \dot{x}\dot{z} + g\dot{y} - k\sqrt{x^2 + y^2},$$

where a, b, c, f, g, and k are constants. What is the Hamiltonian? What quantities are conserved?

5. A dynamical system has the Lagrangian

$$L = \dot{q}_1^2 + \frac{\dot{q}_2^2}{a + bq_1^2} + k_1 q_1^2 + k_2 \dot{q}_1 \dot{q}_2,$$

where a, b, k_1, and k_2 are constants. Find the equations of motion in the Hamiltonian formulation.

6. A Hamiltonian of one degree of freedom has the form

$$H = \frac{p^2}{2\alpha} - bqpe^{-\alpha t} + \frac{ba}{2}q^2 e^{-\alpha t}(\alpha + be^{-\alpha t}) + \frac{kq^2}{2},$$

where a, b, α, and k are constants.

a) Find a Lagrangian corresponding to this Hamiltonian.
b) Find an equivalent Lagrangian that is not explicitly dependent on time.
c) What is the Hamiltonian corresponding to this second Lagrangian, and what is the relationship between the two Hamiltonians?

7. Find the Hamiltonian for the system described in Exercise 13 of Chapter 5 and obtain Hamilton's equations of motion for the system. Use both the direct and the matrix approach in finding the Hamiltonian.

8. Repeat the preceding exercise except this time allow the *pendulum* to move in three dimensions, i.e., a spring-loaded spherical pendulum. Either the direct or the matrix approach may be used.

9. The point of suspension of a simple pendulum of length l and mass m is constrained to move on a parabola $z = ax^2$ in the vertical plane. Derive a Hamiltonian governing the motion of the pendulum and its point of suspension. Obtain the Hamilton's equations of motion.

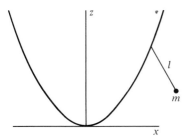

10. Obtain Hamilton's equations of motion for a plane pendulum of length l with mass point m whose radius of suspension rotates uniformly on the circumference of a vertical circle of radius a. Describe physically the nature of the canonical momentum and the Hamiltonian.

11. a) The point of suspension of a plane simple pendulum of mass m and length l is constrained to move along a horizontal track and is connected to a point on the circumference of a uniform fly wheel of mass M and radius a through a massless connecting rod also of length a, as shown in the figure. The fly wheel rotates about a center fixed on the track. Find a Hamiltonian for the combined system and the Hamilton's equations of motion.

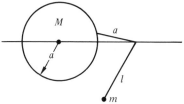

b) Suppose the point of suspension were moved along the track according to some function of time $x = f(t)$, where x reverses at $x = \pm 2a$ (relative to the center of the fly wheel). Again find a Hamiltonian and the Hamilton's equations of motion.

12. For the system described in Exercise 20 of Chapter 2 find the Hamiltonian of the system first in terms of coordinates in the laboratory system and then in terms of coordinates in the rotating systems. What are the conservation properties of the Hamiltonians, and how are they related to the energy of the system?

13. a) A particle of mass m and electric charge e moves in a plane under the influence of a central force potential $V(r)$ and a constant uniform magnetic field $\mathbf{B}$, perpendicular to the plane, generated by a static vector potential

$$\mathbf{A} = \tfrac{1}{2}\mathbf{B} \times \mathbf{r}.$$

Find the Hamiltonian using coordinates in the observer's inertial system.

b) Repeat part (a) using coordinates rotating relative to the previous coordinate system about an axis perpendicular to the plane with an angular rate:

$$\omega = -\frac{eB}{2mc}.$$

14. A uniform cylinder of radius a and density ρ is mounted so as to rotate freely around a vertical axis. On the outside of the cylinder is a rigidly fixed uniform spiral or helical track

along which a mass point m can slide without friction. Suppose a particle starts at rest at the top of the cylinder and slides down under the influence of gravity. Using any set of coordinates, arrive at a Hamiltonian for the combined system of particle and cylinder, and solve for the motion of the system.

15. Suppose that in the previous example the cylinder is constrained to rotate uniformly with angular frequency ω. Set up the Hamiltonian for the particle in an inertial system of coordinates and also in a system fixed in the rotating cylinder. Identify the physical nature of the Hamiltonian in each case and indicate whether or not the Hamiltonians are conserved.

16. A particle of mass m can move in one dimension under the influence of two springs connected to fixed points a distance a apart (see figure). The springs obey Hooke's law and have zero unstretched lengths and force constants k_1 and k_2, respectively.

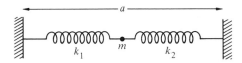

a) Using the position of the particle from one fixed point as the generalized coordinate, find the Lagrangian and the corresponding Hamiltonian. Is the energy conserved? Is the Hamiltonian conserved?
b) Introduce a new coordinate Q defined by

$$Q = q - b\sin\omega t, \qquad b = \frac{k_2 a}{k_1 + k_2}.$$

What is the Lagrangian in terms of Q? What is the corresponding Hamiltonian? Is the energy conserved? Is the Hamiltonian conserved?

17. a) The Lagrangian for a system of one degree of freedom can be written as

$$L = \frac{m}{2}(\dot{q}^2 \sin^2\omega t + \dot{q}q\omega \sin 2\omega t + q^2\omega^2).$$

What is the corresponding Hamiltonian? Is it conserved?
b) Introduce a new coordinate defined by

$$Q = q\sin\omega t$$

Find the Lagrangian in terms of the new coordinate and the corresponding Hamiltonian. Is H conserved?

18. Consider a system of particles interacting with each other through potentials depending only on the scalar distances between them and acted upon by conservative central forces from a fixed point. Obtain the Hamiltonian of the particle with respect to a set of axes, with origin at the center of force, which is rotating around some axis in an inertial system with angular velocity ω. What is the physical significance of the Hamiltonian in this case? Is it a constant of the motion?

19. It has been previously noted that the total time derivative of a function of q_i and t can be added to the Lagrangian without changing the equations of motion. What does such an addition do to the canonical momenta and the Hamiltonian? Show that the equations of motion in terms of the new Hamiltonian reduce to the original Hamilton's equations of motion.

20. Assume that the Lagrangian is a polynomial in $\dot{q}$ of no higher order than quadratic. Convert the $2n$ equations

$$p_i = \frac{\partial L}{\partial \dot{q}_i}, \qquad \dot{p}_i = \frac{\partial L}{\partial q_i},$$

into $2n$ equations for $\dot{q}_i$ and $\dot{p}_i$ in terms of q and p, using the matrix form of the Lagrangian. Show that these are the same equations as would be obtained from Hamilton's equations of motion.

21. A Hamiltonian-like formulation can be set up in which $\dot{q}_i$ and $\dot{p}_i$ are the independent variables with a "Hamiltonian" $G(\dot{q}_i, \dot{p}_i, t)$. [Here p_i is defined in terms of $q_i, \dot{q}_i$ in the usual manner.] Starting from the Lagrangian formulation show in detail how to construct $G(\dot{q}_i, \dot{p}_i, t)$, and derive the corresponding "Hamilton's equation of motion."

22. Obtain the Hamiltonian of a heavy symmetrical top with one point fixed, and from it the Hamilton's equations of motion. Relate these to the equations of motion discussed in Section 5–7 and, in particular, show how the solution may be reduced to quadratures. Also use the Routhian procedure to eliminate the cyclic coordinates.

23. In Exercise 13 of Chapter 1 there is given the velocity-dependent potential assumed in Weber's electrodynamics. What is the Hamiltonian for a single particle moving under the influence of such a potential?

24. Show that if λ_i are the eigenvalues of a square matrix, then if the reciprocal matrix exists it has the eigenvalues λ_i^{-1}.

25. Verify that the matrix $\mathbf{J}$ has the properties given in Eqs. (8–32) and (8–33) and that its determinant has the value $+1$.

26. Show that Hamilton's principle can be written as

$$\delta \int_1^2 [2H(\boldsymbol{\eta}, t) + \boldsymbol{\eta}\mathbf{J}\dot{\boldsymbol{\eta}}] \, dt = 0.$$

27. Verify that both Hamiltonians, Eq. (8–39) and Eq. (8–41), lead to the same motion as described by Eq. (8–38).

28. Treat the nutation of a "fast" top as an example of small oscillations about steady motion, here precession at constant θ. Find the frequency of nutation.

29. A symmetrical top is mounted so that it pivots about its center of mass. The pivot in turn is fixed a distance r from the center of a horizontal disc free to rotate about a vertical axis. The top is started with an initial rotation about its figure axis, which is initially at an angle θ_0 to the vertical. Analyze the possible nutation of the top as a case of small oscillations about steady motion.

30. Two mass points m_1 and m_2 are connected by a string that acts as a Hooke's-law spring with force constant k. One particle is free to move without friction on a smooth horizontal plane surface, the other hangs vertically down from the string through a hole in the surface. Find the condition for steady motion in which the mass point on the plane rotates uniformly at constant distance from the hole. Investigate the small oscillations in the radial distance from the hole, and in the vertical height of the second particle.

31. Show that the modified Hamilton's principle, in the form of Eq. (8–71), leads to Hamilton's equations of motion.

32. If the canonical variables are not all independent, but are connected by auxiliary conditions of the form

$$\psi_k(q_i, p_i, t) = 0,$$

show that the canonical equations of motion can be written

$$\frac{\partial H}{\partial p_i} + \sum_k \lambda_k \frac{\partial \psi_k}{\partial p_i} = \dot{q}_i, \qquad \frac{\partial H}{\partial q_i} + \sum_k \lambda_k \frac{\partial \psi_k}{\partial q_i} = -\dot{p}_i,$$

where the λ_k are the undetermined Lagrange multipliers. The formulation of the Hamiltonian equations in wich t is a canonical variable is a case in point, since a relation exists between p_{n+1} and the other canonical variables:

$$H(q_1, \ldots, q_{n+1}; p_1, \ldots, p_n) + p_{n+1} = 0.$$

Show that as a result of these circumstances the $2n + 2$ Hamilton's equations of this formulation can be reduced to the $2n$ ordinary Hamilton's equations plus Eq. (8–35) and the relation

$$\lambda = \frac{dt}{d\theta}.$$

Note that while these results are reminiscent of the relativistic covariant Hamiltonian formulation, they have been arrived at entirely within the framework of nonrelativistic mechanics.

33. A possible covariant Lagrangian for a system of one particle interacting with a field is

$$\Lambda = \tfrac{1}{2} m u_\lambda u_\lambda + D_{\lambda\nu}(x_\mu) m_{\lambda\nu},$$

where $D_{x\nu}(x_\mu)$ is an antisymmetric field tensor and $m_{\lambda\nu}$ is the antisymmetric angular momentum tensor,

$$m_{\lambda\nu} = m(x_\lambda u_\nu - x_\nu u_\lambda).$$

What are the canonical momenta? What is the corresponding covariant Hamiltonian?

CHAPTER 9
Canonical Transformations

When applied in a straightforward manner the Hamiltonian formulation usually does not materially decrease the difficulty of solving any given problem in mechanics. One winds up with practically the same differential equations to be solved as are provided by the Lagrangian procedure. The advantages of the Hamiltonian formulation lie not in its use as a calculational tool, but rather in the deeper insight it affords into the formal structure of mechanics. The equal status accorded to coordinates and momenta as independent variables encourages a greater freedom in selecting the physical quantities to be designated as "coordinates" and "momenta." As a result we are led to newer, more abstract ways of presenting the physical content of mechanics. While often of considerable help in practical applications to mechanical problems, these more abstract formulations are primarily of interest to us today because of their essential role in constructing the more modern theories of matter. Thus, one or another of these formulations of classical mechanics serves as a point of departure for both statistical mechanics and quantum theory. It is to such formulations, arising as outgrowths of the Hamiltonian procedure, that this and the next chapter are devoted.

9-1 THE EQUATIONS OF CANONICAL TRANSFORMATION

There is one type of problem for which the solution of the Hamilton's equations is trivial. Consider a situation in which the Hamiltonian is a constant of the motion, and where *all* coordinates q_i are cyclic. Under these conditions the conjugate momenta p_i are all constant:

$$p_i = \alpha_i,$$

and since the Hamiltonian cannot be an explicit function of either the time or the cyclic coordinates, it may be written as

$$H = H(\alpha_1 \ldots \alpha_n).$$

Consequently, the Hamilton's equations for $\dot{q}_i$ are simply

$$\dot{q}_i = \frac{\partial H}{\partial \alpha_i} = \omega_i, \tag{9-1}$$

where the ω_i's are functions of the α_i's only and therefore are also constant in time. Equations (9–1) have the immediate solutions

$$q_i = \omega_i t + \beta_i, \tag{9–2}$$

where the β_i's are constants of integration, determined by the initial conditions.

It would seem that the solution to this type of problem, easy as it is, can only be of academic interest, for it rarely occurs in practice that one finds that all the generalized coordinates are cyclic. But a given system can be described by more than one set of generalized coordinates. Thus, to discuss motion of a particle in a plane one may use as generalized coordinates either the Cartesian coordinates

$$q_1 = x, \qquad q_2 = y,$$

or the plane polar coordinates

$$q_1 = r, \qquad q_2 = \theta.$$

Both choices are equally valid, but one or the other set may be more convenient for the problem under consideration. It will be noticed that for central forces neither x nor y is cyclic, while the second set does contain a cyclic coordinate in the angle θ. The number of cyclic coordinates thus depends on the choice of generalized coordinates, and for each problem there may be one particular choice for which all coordinates are cyclic. If we can find this set the remainder of the job is trivial. Since the obvious generalized coordinates suggested by the problem will not normally be cyclic, we must first derive a specific procedure for *transforming* from one set of variables to some other set that may be more suitable.

The transformations considered in the previous chapters have involved going from one set of coordinates q_i to a new set Q_i by transformation equations of the form

$$Q_i = Q_i(q, t). \tag{9–3}$$

For example, the equations of an orthogonal transformation, or of the change from Cartesian to plane polar coordinates, have the general form of Eqs. (9–3). As has been previously noted in Exercise 15 of Chapter 1, such transformations are known as *point transformations*. But in the Hamiltonian formulation the momenta are also independent variables on the same level as the generalized coordinates. The concept of transformation of coordinates must therefore be widened to include the simultaneous transformation of the independent *coordinates* and *momenta*, q_i, p_i, to a new set Q_i, P_i, with (invertible) equations of transformation:

$$Q_i = Q_i(q, p, t),$$
$$P_i = P_i(q, p, t). \tag{9–4}$$

Thus the new coordinates will be defined not only in terms of the old coordinates but also in terms of the old momenta. Equations (9–3) may be said to define a transformation of *configuration space*; correspondingly Eqs. (9–4) define a transformation of *phase space*.

In developing Hamiltonian mechanics, only those transformations can be of interest for which the new Q, P are canonical coordinates. This requirement will be satisfied provided there exists some function $K(Q, P, t)$ such that the equations of motion in the new set are in the Hamiltonian form

$$\dot{Q}_i = \frac{\partial K}{\partial P_i}, \qquad \dot{P}_i = -\frac{\partial K}{\partial Q_i}. \tag{9-5}$$

The function K plays the role of the Hamiltonian in the new coordinate set.* It is important for future considerations that the transformations considered be problem-independent. That is to say, (Q, P) must be canonical coordinates not only for some specific mechanical systems, but for all systems of the same number of degrees of freedom. Equations (9–5) must be the form of the equations of motion in the new coordinates and momenta no matter what the particular initial form of H. We may indeed be incited to develop a particular transformation from (q, p) to (Q, P) to handle, say, a plane harmonic oscillator. But the same transformation must then also lead to Hamilton's equations of motion when applied, for example, to the two-dimensional Kepler problem.

As was seen in Section 8–5, if Q_i and P_i are to be canonical coordinates, they must satisfy a modified Hamilton's principle that can be put in the form

$$\delta \int_{t_1}^{t_2} (P_i \dot{Q}_i - K(Q, P, t)) \, dt = 0, \tag{9-6}$$

(where summation over the repeated index i is implied). At the same time the old canonical coordinates, of course, satisfy a similar principle:

$$\delta \int_{t_1}^{t_1} (p_i \dot{q}_i - H(q, p, t)) \, dt = 0. \tag{9-7}$$

The simultaneous validity of Eqs. (9–6) and (9–7) does not mean, of course, that the integrands in both expressions are equal. Since the general form of the modified Hamilton's principle has zero variation at the endpoints, both statements will be satisfied if the integrands are connected by a relation of the form

$$\lambda(p_i \dot{q}_i - H) = P_i \dot{Q}_i - K + \frac{dF}{dt}. \tag{9-8}$$

---

* It has been remarked in a jocular vein that if H stands for the Hamiltonian, K must stand for the Kamiltonian! Of course K is every bit as much a Hamiltonian as H, but the designation is occasionally a convenient substitute for the longer term "transformed Hamiltonian."

Here F is any function of the phase space coordinates with continuous second derivatives, and λ is a constant independent of the canonical coordinates and the time. The multiplicative constant λ is related to a particularly simple type of transformation of canonical coordinates known as a *scale transformation*. Suppose we change the size of the units used to measure the coordinates and momenta so that in effect we transform them to a set (Q', P') defined by

$$Q'_i = \mu q_i, \qquad P'_i = v p_i. \tag{9-9}$$

Then it is clear Hamilton's equations in the form of Eqs. (9–5) will be satisfied for a transformed Hamiltonian $K'(Q', P') = \mu v H(q, p)$. The integrands of the corresponding modified Hamilton's principles are, also obviously, related as

$$\mu v (p_i \dot{q}_i - H) = P'_i \dot{Q}'_i - K', \tag{9-10}$$

which is of the form of Eq. (9–8) with $\lambda = \mu v$. With the aid of a suitable scale transformation it will always be possible to confine our attention to transformations of canonical coordinates for which $\lambda = 1$. Thus, if we have a transformation of canonical coordinates $(q, p) \rightarrow (Q', P')$ for some $\lambda \neq 1$, then we can always find an intermediate set of canonical coordinates (Q, P) related to (Q', P') by a simple scale transformation of the form (9–9) such that μv also has the same value λ. The transformation between the two sets of canonical coordinates (q, p) and (Q, P) will satisfy Eq. (9–8), but now with $\lambda = 1$:

$$p_i \dot{q}_i - H = P_i \dot{Q}_i - K + \frac{dF}{dt}. \tag{9-11}$$

Since the scale transformation is basically trivial, as Landau and Lifshitz have characterized it,[*] the consequential transformations to be examined are those for which Eq. (9–11) holds.

A transformation of canonical coordinates for which $\lambda \neq 1$ will be called an *extended canonical transformation*. Where $\lambda = 1$, and Eq. (9–11) holds, we will speak simply of a *canonical transformation*. The conclusion of the previous paragraph may then be stated as saying that any extended canonical transformation can be made up of a canonical transformation followed by a scale transformation. Except where otherwise stated all future considerations of transformations between canonical coordinates will involve only canonical transformations. It is also convenient to give a specific name to canonical

---

[*] See Landau and Lifshitz, *Mechanics*, 1960, p. 144. The scale transformation as defined here is related to, but not identical with, the scale transformation involved in what is sometimes called *dilatation symmetry*. There the scale of time is also transformed and attention is focused on a particular class of Hamiltonians with the same functional dependence on the new and old canonical coordinates. See Exercise 2.

transformations for which the equations of transformation Eqs. (9–4) do not contain the time explicitly; they will be called *restricted canonical transformation.**

The last term on the right in Eq. (9–11) contributes to the variation of the action integral only at the endpoints and will therefore vanish if F is a function of (q, p, t) or (Q, P, t) or any mixture of the phase space coordinates since these have zero variation at the endpoints. Further, through the equations of transformation, Eqs. (9–4) and their inverses F can be expressed in terms partly of the old set of variables and partly of the new. Indeed, F is useful for specifying the exact form of the canonical transformation only when half of the variables (beside the time) are from the old set and half are from the new. It then acts, as it were, as a bridge between the two sets of canonical variables and is called the *generating function* of the transformation. To show how the generating function specifies the equations of transformation, suppose F were given as a function of the old and new generalized coordinates:

$$F = F_1(q, Q, t).$$ (9–12)

Equation (9–11) then takes the form

$$p_i \dot{q}_i - H = P_i \dot{Q}_i - K + \frac{dF_1}{dt},$$

$$= P_i \dot{Q}_i - K + \frac{\partial F_1}{\partial t} + \frac{\partial F_1}{\partial q_i} \dot{q}_i + \frac{\partial F_1}{\partial Q_i} \dot{Q}_i.$$ (9–13)

Since the old and the new coordinates, q_i and Q_i, are separately independent, Eq. (9–13) can hold identically only if the coefficients of $\dot{q}_i$ and $\dot{Q}_i$ each vanish:

$$p_i = \frac{\partial F_1}{\partial q_i},$$ (9–14a)

$$P_i = -\frac{\partial F_1}{\partial Q_i},$$ (9–14b)

leaving finally

$$K = H + \frac{\partial F_1}{\partial t}.$$ (9–14c)

---

* Most present day authors, but not all, use the term canonical transformation in ways equivalent to the definition given here. The rest of the terminology is not standard. No uniform practice exists in the literature and the confusion does not appear to be decreasing with time. The above terminology is offered as being adequate for physics needs. In much of the physics literature the term *contact transformation* is used as fully synonomous to canonical transformation, although it was introduced into mathematics in a different sense. Some authors use contact transformation only when there is also a transformation of the time coordinate. As we shall see this is equivalent to a restricted canonical transformation in an enlarged phase space in which time is one of the canonical coordinates, a formulation obviously particularly appropriate to special relativity theory. For a description of contact transformations in the sense of the mathematicians see C. Carathéodory, *Calculus of Variations* Part 1, 1965, Chapter 7; and H. Rund, *Hamiltonian–Jacobi Theory in the Calculus of Variations*, 1966, pp. 84–85.

Equations (9–14a) are n relations defining the p_i as functions of q_j, Q_j, and t. Assuming they can be inverted, they could then be solved for the n Q_j's in terms of q_j, p_j, and t, thus yielding the first half of the transformation equations (9–4). Once the relations between the Q_i's and the old canonical variables (q, p) have been established, they can be substituted into Eqs. (9–14b) so that they give the n P_i's as functions of q_j, p_j, and t, that is, the second half of the transformation equations (9–4). To complete the story, Eq. (9–14c) provides the connection between the new Hamiltonian, K, and the old one, H. One must be careful to read Eq. (9–14c) properly. First q and p in H are expressed as functions of Q and P through the inverses of Eqs. (9–4). Then the q_i in $\partial F_1 / \partial t$ are expressed in terms of Q, P in a similar manner and the two functions are added to yield $K(Q, P, t)$.

The procedure described shows how, starting from a given generating function F_1, the equations of the canonical transformation can be obtained. One can usually reverse the process: given the equations of transformation (9–4), an appropriate generating function F_1 may be derived. Equations (9–4) are first inverted to express p_i and P_i as functions of q, Q, and t. Equations (9–14a, b) then constitute a coupled set of partial differential equations that can be integrated, in principle, to find F_1 providing the transformation is indeed canonical. F_1 is always uncertain to within an additive arbitrary function of t alone (which doesn't affect the equations of transformation) and there may at times be other ambiguities.

It sometimes happens that it is not suitable to describe the canonical transformation by a generating function of the type $F_1(q, Q, t)$. For example, the transformation may be such that p_i cannot be written as functions of q, Q, and t, but rather will be functions of q, P, and t. One would then seek a generating function that is a function of the old coordinates q and the new momenta P. Clearly, Eq. (9–13) must then be replaced by an equivalent relation involving $\dot{P}_i$ rather than $\dot{Q}_1$. This can be accomplished by writing F in Eq. (9–11) as

$$F = \boxed{F_2(q, P, t)} - Q_i P_i. \qquad (9\text{–}15)$$

Substituting this F in Eq. (9–11) leads to

$$p_i \dot{q}_i - H = -\dot{Q}_i P_i - K + \frac{d}{dt} F_2(q, P, t). \qquad (9\text{–}16)$$

Again, the total derivative of F_2 is expanded and the coefficients of $\dot{q}_i$ and P_i collected, leading to the equations

$$p_i = \frac{\partial F_2}{\partial q_i}, \qquad (9\text{–}17a)$$

$$Q_i = \frac{\partial F_2}{\partial P_i}, \qquad (9\text{–}17b)$$

with

$$K = H + \frac{\partial F_2}{\partial t}. \qquad (9\text{–}17c)$$

As before, Eqs. (9–17a) are to be solved for P_i as functions of q_j, p_j, and t to correspond to the second half of the transformation equations (9–4). The remaining half of the transformation equations is then provided by Eqs. (9–17b).

The corresponding procedures for the remaining two general types of generating functions is by now obvious. A generating function F_3 of the old momenta p_i, the new coordinates Q_i, and time t is defined by setting

$$F = q_i p_i + \boxed{F_3(p, Q, t),}$$
(9–18)

in terms of which Eq. (9–11) becomes

$$-q_i \dot{p}_i - H = P_i \dot{Q}_i - K + \frac{d}{dt} F_3(p, Q, t).$$
(9–19)

The same process of equating coefficients leads to

$$q_i = -\frac{\partial F_3}{\partial p_i}, \qquad P_i = -\frac{\partial F_3}{\partial Q_i},$$
(9–20a)

and

$$K = H + \frac{\partial F_3}{\partial t},$$
(9–20b)

with the equations of transformation being derived from the $2n$ Eqs. (9–20a). If the generating function is of the form $F_4(p, P, t)$, then it is related to F by

$$F = q_i p_i - Q_i P_i + \boxed{F_4(p, P, t),}$$
(9–21)

so that Eq. (9–11) takes the form

$$-q_i \dot{p}_i - H = -Q_i \dot{P}_i - K + \frac{d}{dt} F_4(p, P, t).$$
(9–22)

The equations of transformation are then derived from the relations

$$q_i = -\frac{\partial F_4}{\partial p_i}, \qquad Q_i = \frac{\partial F_4}{\partial P_i},$$
(9–23a)

and again

$$K = H + \frac{\partial F_4}{\partial t}.$$
(9–23b)

It is tempting to look upon the four general types of generating functions as being related to each other through Legendre transformations. For example, the transition from F_1 to F_2 is equivalent to going from the variables q, Q to q, P with the relation

$$-P_i = \frac{\partial F_1}{\partial Q_i}.$$

This is just the form required for a Legendre transformation of the basis variables, as described in Section 8–1, and in analogy to Eq. (8–5) we would set

$$F_2(q, P, t) = F_1(q, Q, t) + P_i Q_i, \qquad (9\text{--}24)$$

which is equivalent to Eq. (9–15) combined with Eq. (9–12). All the other defining equations for the generating functions can similarly be looked on, in combination with Eq. (9–12) as Legendre transformations from F_1, with Eq. (9–21) describing a double Legendre transformation. The only drawback to this picture is that it might lead one to believe that for any given canonical transformation one can always find generating functions of all four types by means of these Legendre transformations. This is not always possible. There are transformations that are just not suitable for description in terms of certain forms of generating functions, as has been noted above and as will be illustrated in the next section with specific examples. If one tries to apply the Legendre transformation process, one is then led to generating functions that are identically zero or are indeterminate. For this reason we have preferred to define each type of generating function relative to F, which is some unspecified function of $2n$ independent coordinates and momenta.

Finally it should be emphasized that a suitable generating function doesn't have to conform to one of the four general types for *all* the degrees of freedom of the system. It is possible, and for some canonical transformations necessary, to use a generating function that is a mixture of the four types. To take a simple example, it may be desirable for a particular canonical transformation with two degrees of freedom to be defined by a generating function of the form

$$F'(q_1, p_2, P_1, Q_2, t).$$

This generating function would be related to F in Eq. (9–11) by the equation

$$F = F'(q_1, p_2, P_1, Q_2, t) - Q_1 P_1 + q_2 p_2,$$

and the equations of transformation would be obtained from the relations

$$p_1 = \frac{\partial F'}{\partial q_1}, \qquad Q_1 = \frac{\partial F'}{\partial P_1},$$

$$q_2 = -\frac{\partial F'}{\partial p_1}, \qquad P_2 = -\frac{\partial F'}{\partial Q_2},$$

with

$$K = H + \frac{\partial F'}{\partial t}.$$

Specific illustrations are given in the next section and in the exercises.

9-2 EXAMPLES OF CANONICAL TRANSFORMATIONS

The nature of canonical transformations and the role played by the generating function can best be illustrated by some simple yet important examples. Consider, first, a generating function of the second type with the particular form

$$F_2 = q_i P_i. \tag{9-25}$$

From Eqs. (9-17), the transformation equations are

$$p_i = \frac{\partial F_2}{\partial q_i} = P_i,$$

$$Q_i = \frac{\partial F_2}{\partial P_i} = q_i,$$

$$K = H.$$

The new and old coordinates are the same; hence F_2 merely generates the *identity transformation*.

A more general type of transformation is described by the generating function

$$F_2 = f_i(q_1, \ldots, q_n; t) P_i, \tag{9-26}$$

where the f_i may be any desired set of independent functions. By Eqs. (9-17b), the new coordinates Q_i are given by

$$Q_i = \frac{\partial F_2}{\partial P_i} = f_i(q, t). \tag{9-27}$$

Thus, with this generating function the new coordinates depend only on the old coordinates and the time and do not involve the old momenta. Such a transformation is therefore an example of the class of point transformations defined by Eqs. (9-3). In order to define a point transformation, the functions f_i must be independent and invertible, so that the q_j can be expressed in terms of the Q_i. Since the f_i are otherwise completely arbitrary, we may conclude that *all point transformations are canonical*. Equation (9-17c) furnishes the new Hamiltonian in terms of the old and of the time derivatives of the f_i functions.

It should be noted that F_2 as given by Eq. (9-26) is not the only generating function leading to the point transformation specified by the f_i. Clearly, the same point transformation is implicit in the more general form

$$F_2 = f_i(q_1, \ldots, q_n; t) P_i + g(q_1, \ldots, q_n, t), \tag{9-28}$$

where $g(q, t)$ is any (differentiable) function of the old coordinates and the time. Equations (9-27), the transformation equations for the coordinates, remain unaltered for this generating function. But the transformation equations for the momenta differ for the two forms. From Eqs. (9–17a) we have

$$p_j = \frac{\partial F_2}{\partial q_j} = \frac{\partial f_i}{\partial q_j} P_i + \frac{\partial g}{\partial q_j}, \tag{9–29}$$

using the form of F_2 given by Eq. (9–28). These equations may be inverted to give P as a function of (q, p) most easily by writing them in matrix notation:

$$\mathbf{p} = \mathbf{P}\frac{\partial \mathbf{f}}{\partial \mathbf{q}} + \frac{\partial g}{\partial \mathbf{q}}. \tag{9–29'}$$

Here $\mathbf{p}$, $\mathbf{P}$, and $\dfrac{\partial g}{\partial \mathbf{q}}$ are n-element one column (or row) matrices and $\dfrac{\partial \mathbf{f}}{\partial \mathbf{q}}$ is a square matrix whose ijth element is $\dfrac{\partial f_i}{\partial q_j}$. It follows then that $\mathbf{P}$ is a linear function of $\mathbf{p}$ given by

$$\mathbf{P} = \left(\mathbf{p} - \frac{\partial g}{\partial \mathbf{q}}\right)\left(\frac{\partial \mathbf{f}}{\partial \mathbf{q}}\right)^{-1}. \tag{9–30}$$

Thus, the transformation equations for Q are independent of g and depend only on the $f_i(q, t)$, but the transformation equations for P do depend on the form of g and are in general functions of both the old coordinates and momenta. The generating function given by Eq. (9–26) is only a special case of Eq. (9–28) for which $g = 0$, with correspondingly specialized transformation equations for P.

An instructive transformation is provided by the generating function of the first kind, $F_1(q, Q, t)$, of the form

$$F_1 = q_k Q_k. \tag{9–31}$$

The corresponding transformation equations, from (9–14a, b) are

$$p \tag{9–32a}$$

$$P_i = -\frac{\partial F_1}{\partial Q_i} = -q_i. \tag{9–32b}$$

In effect the transformation interchanges the momenta and coordinates; the new coordinates are the old momenta and the new momenta are essentially the old coordinates. This simple example should emphasize the independent status of generalized coordinates and momenta. They are both needed to describe the motion of the system in the Hamiltonian formulation, and the distinction between them is practically one of nomenclature. One can shift the names around with at most no more than a change in sign. There is no longer present in the theory any lingering remnant of the concept of q_i as a spatial coordinate and p_i as a mass times a velocity. Incidentally, one may see directly from Hamilton's equations,

$$\dot{p}_i = -\frac{\partial H}{\partial q_i}, \qquad \dot{q}_i = \frac{\partial H}{\partial p_i},$$

that this exchange transformation is canonical. If q_i is substituted for p_i the equations remain in the canonical form only if $-p_i$ is substituted for q_i.

It is also easy to see that the exchange transformation cannot be derived from a generating function of the type $F_2(q, P, t)$. With such a generating function the set of partial differential equations (9–17a),

$$p_i = \frac{\partial F_2(q, P, t)}{\partial q_i},$$

is required to give the p_i as functions of q, P, and t. But in the exchange transformation, p_i is not a function of P at all; it depends only on the new coordinates Q! Hence F_2 is just not suitable as a generating function for the exchange transformation.

By a similar argument, it is obvious an F_1 function cannot be used to generate the identity transformation. In that case the equations (9–14a),

$$p_i = \frac{\partial F_1(q, Q, t)}{\partial q_i},$$

would define p_i in terms of q, Q, and t; whereas in the identity transformation p_i depends only on P, being in fact equal to P_i. One cannot circumvent this difficulty by defining an F_1 function through a Legendre transformation (cf. Eq. 9–24):

$$F_1(q, Q, t) = F_2(q, P, t) - P_i Q_i.$$

If F_2 is substituted from Eq. (9–25) with q written as Q, then this recipe yields the result

$$F_1(q, Q, t) = Q_i P_i - P_i Q_i \equiv 0,$$

which is useless as a generating function. A similar dead end is obtained in attempting to construct an F_2 function to generate the exchange transformation.*

A transformation that leaves some of the (q, p) pairs unchanged, and interchanges the rest (with a sign change), is obviously a canonical transformation. But from the above considerations it cannot be derived from a generating function of one of the four "pure" forms discussed previously; the generating function must be of a "mixed" form. Thus in a system of two degrees of freedom, the transformation

$$Q_1 = q_1, \qquad P_1 = p_1,$$
$$Q_2 = p_2, \qquad P_2 = -q_2$$

---

* One can easily show, however, that an F_3 function can generate the identity transformation, and that the exchange transformation can be derived from an F_4 function.

is generated by the function

$$F = q_1 P_1 + q_2 Q_2,$$

which is a mixture of the F_1 and F_2 types.

As a final example consider a canonical transformation that can be used to solve the problem of the simple harmonic oscillator in one dimension. If the force constant is k, the Hamiltonian for this problem in terms of the usual coordinates is

$$H = \frac{p^2}{2m} + \frac{kq^2}{2}. \tag{9-33}$$

Designating the ratio k/m by ω^2, H can also be written as

$$H = \frac{1}{2m}(p^2 + m^2\omega^2 q^2). \tag{9-34}$$

This form of the Hamiltonian, as the sum of two squares, suggests a transformation in which H is cyclic in the new coordinate. If we could find a canonical transformation of the form

$$p = f(P)\cos Q, \tag{9-35a}$$

$$q = \frac{f(P)}{m\omega}\sin Q, \tag{9-35b}$$

then the Hamiltonian as a function of Q and P would be simply

$$K = H = \frac{f^2(P)}{2m}(\cos^2 Q + \sin^2 Q) = \frac{f^2(P)}{2m},$$

so that Q is cyclic. The problem is to find the form of the as yet unspecified function $f(P)$ in such a manner that the transformation be canonical. The ratio of the two equations (9–35) gives the relation

$$p = m\omega q \cot Q, \tag{9-36}$$

independent of $f(P)$. Equation (9–36) is of the form of Eq. (9–14a) for the F_1 type of generating function,

$$p = \frac{\partial F_1(q, Q)}{\partial q},$$

and the simplest solution for F_1 corresponding to Eq. (9–36) is clearly

$$F_1 = \frac{m\omega q^2}{2}\cot Q. \tag{9-37}$$

Equation (9–14b) then provides the other half of the equations of transformation,

$$P = -\frac{\partial F_1}{\partial Q} = \frac{m\omega q^2}{2\sin^2 Q}. \tag{9-38}$$

Solving for q we have

$$q = \sqrt{\frac{2P}{m\omega}} \sin Q, \qquad (9\text{--}39)$$

and comparison with Eq. (9–35a) shows that the only form for $f(P)$ leading to a canonical transformation is

$$f(P) = \sqrt{2m\omega P}. \qquad (9\text{--}40)$$

It follows then that the Hamiltonian in the transformed variables is

$$H = \omega P. \qquad (9\text{--}41)$$

Since the Hamiltonian is cyclic in Q, the conjugate momentum P is a constant. It is seen from Eq. (9–41) that P is in fact equal to the constant energy divided by ω:

$$P = \frac{E}{\omega}.$$

The equation of motion for Q reduces to the simple form

$$\dot{Q} = \frac{\partial H}{\partial P} = \omega,$$

with the immediate solution

$$Q = \omega t + \alpha, \qquad (9\text{--}42)$$

where α is a constant of integration fixed by the initial conditions. From Eq. (9–39) the solution for q is

$$q = \sqrt{\frac{2E}{m\omega^2}} \sin(\omega t + \alpha), \qquad (9\text{--}43)$$

which is the customary solution for a harmonic oscillator.*

It would seem that the use of contact transformations to solve the harmonic oscillator problem is similar to "cracking a peanut with a sledge hammer." We have here, however, a simple example of how the Hamiltonian can be reduced to a form cyclic in all coordinates by means of canonical transformations. Discussion of general schemes for the solution of mechanical problems by this technique will be reserved for the next chapter. For the present we shall continue to examine the formal properties of canonical transformations.

---

* It can be argued that F_1 does not unambiguously specify the canonical transformation, because in solving Eq. (9–38) for q we could have taken the negative square root instead of the positive root as (implied) in Eq. (9–39). However the two canonical transformations thus derived from F_1 differ only trivially; a shift in α by π corresponds to going from one transformation to the other. Nonetheless, it should be kept in mind that the transformations derived from a generating function may at times be double-valued or even have local singularities.

9–3 THE SYMPLECTIC APPROACH TO CANONICAL TRANSFORMATIONS

Another method of treating canonical transformations, seemingly unrelated to the generator formalism, can be derived in terms of the matrix or symplectic formulation of Hamilton's equations. By way of introduction to this approach let us consider a restricted canonical transformation, i.e., one in which time does not appear in the equations of transformation:

$$Q_i = Q_i(q, p),$$
$$P_i = P_i(q, p). \tag{9–44}$$

We know that the Hamiltonian function does not change in such a transformation. The time derivative of Q_i, on the basis of Eqs. (9–44), is to be found as

$$\dot{Q}_i = \frac{\partial Q_i}{\partial q_j} \dot{q}_j + \frac{\partial Q_i}{\partial p_j} \dot{p}_j = \frac{\partial Q_i}{\partial q_j} \frac{\partial H}{\partial p_j} - \frac{\partial Q_i}{\partial p_j} \frac{\partial H}{\partial q_j}. \tag{9–45}$$

On the other hand, the inverse of Eqs. (9–44),

$$q_j = q_j(Q, P),$$
$$p_j = p_j(Q, P), \tag{9–46}$$

enable us to consider $H(q, p, t)$ as a function of Q and P and to form the partial derivative

$$\frac{\partial H}{\partial P_i} = \frac{\partial H}{\partial p_j} \frac{\partial p_j}{\partial P_i} + \frac{\partial H}{\partial q_j} \frac{\partial q_j}{\partial P_i}. \tag{9–47}$$

Comparing Eqs. (9–45) and (9–47) it can be concluded that

$$\dot{Q}_i = \frac{\partial H}{\partial P_i},$$

that is, the transformation is canonical, only if

$$\left(\frac{\partial Q_i}{\partial q_j}\right)_{q,p} = \left(\frac{\partial p_j}{\partial P_i}\right)_{Q,P}, \qquad \left(\frac{\partial Q_i}{\partial p_j}\right)_{q,p} = -\left(\frac{\partial q_j}{\partial P_i}\right)_{Q,P}. \tag{9–48a}$$

The subscripts on the derivatives are to remind us that on the left-hand side of these equations Q_i is considered as a function of (q, p) (cf. Eqs. 9–44), while on the right-hand side the derivatives are for q_j and p_j as functions of (Q, P) (cf. Eqs. 9–46). A similar comparison of $\dot{P}_i$ with the partial of H with respect to Q_j leads to the conditions

$$\left(\frac{\partial P_i}{\partial q_j}\right)_{q,p} = -\left(\frac{\partial p_j}{\partial Q_i}\right)_{Q,P}, \qquad \left(\frac{\partial P_i}{\partial p_j}\right)_{q,p} = \left(\frac{\partial q_j}{\partial Q_i}\right)_{Q,P}. \tag{9–48b}$$

The sets of Eqs. (9–48) together are sometimes known as the "direct conditions" for a (restricted) canonical transformation.

The algebraic manipulation that leads to Eqs. (9–48) can be performed in a compact and elegant form if we make use of the symplectic notation for the Hamiltonian formulation introduced above at the end of Section 8–1. If $\boldsymbol{\eta}$ is a column matrix with the $2n$ elements q_i, p_i, then Hamilton's equations can be written, it will be remembered, as

$$\dot{\boldsymbol{\eta}} = \mathbf{J}\frac{\partial H}{\partial \boldsymbol{\eta}}, \tag{8–31}$$

where $\mathbf{J}$ is the antisymmetric matrix defined in Eq. (8–30). Similarly the new set of coordinates Q_i, P_i define a $2n$ element column matrix $\boldsymbol{\zeta}$, and for a restricted canonical transformation the equations of transformation (9–44) take the form

$$\boldsymbol{\zeta} = \boldsymbol{\zeta}(\boldsymbol{\eta}). \tag{9–49}$$

Analogously to Eq. (9–45) we can seek the equations of motion for the new variables by looking at the time derivative of a typical element of $\boldsymbol{\zeta}$:

$$\dot{\zeta}_i = \frac{\partial \zeta_i}{\partial \eta_j}\dot{\eta}_j, \qquad i, j = 1, \ldots, 2n.$$

In matrix notation this time derivative can be written as

$$\dot{\boldsymbol{\zeta}} = \mathbf{M}\dot{\boldsymbol{\eta}}, \tag{9–50}$$

where $\mathbf{M}$ is the Jacobian matrix of the transformation with elements

$$M_{ij} = \frac{\partial \zeta_i}{\partial \eta_j}. \tag{9–51}$$

Making use of the equations of motion for $\boldsymbol{\eta}$, Eq. (9–50) becomes

$$\dot{\boldsymbol{\zeta}} = \mathbf{MJ}\frac{\partial H}{\partial \boldsymbol{\eta}}. \tag{9–52}$$

Now, by the inverse transformation H can be considered as a function of $\boldsymbol{\zeta}$, and the derivative with respect to η_i evaluated as

$$\frac{\partial H}{\partial \eta_i} = \frac{\partial H}{\partial \zeta_j}\frac{\partial \zeta_j}{\partial \eta_i},$$

or, in matrix notation*

$$\frac{\partial H}{\partial \boldsymbol{\eta}} = \widetilde{\mathbf{M}}\frac{\partial H}{\partial \boldsymbol{\zeta}}. \tag{9–53}$$

---

* Readers of Section 7–3 will have recognized that Eq. (9–50) is the statement that $\boldsymbol{\eta}$ transforms contravariantly under the transformation, and Eq. (9–53) says that the partial derivative of H with respect to the elements of $\boldsymbol{\eta}$ transforms *covariantly* (cf Eqs. 7–48' and 7–49').

The combination of Eqs. (9–52) and (9–53) leads to the form of the equations of motion for any set of variables ζ transforming, independently of time, from the canonical set $\boldsymbol{\eta}$:

$$\dot{\zeta} = \mathbf{MJ\tilde{M}}\frac{\partial H}{\partial \zeta}. \qquad (9\text{–}54)$$

We have the advantage of knowing from the generator formalism that for a *restricted* canonical transformation the old Hamiltonian expressed in terms of the new variables serves as the new Hamiltonian:

$$\dot{\zeta} = \mathbf{J}\frac{\partial H}{\partial \zeta}.$$

The transformation, Eq. (9–49), will therefore be canonical if $\mathbf{M}$ satisfies the condition

$$\mathbf{MJ\tilde{M}} = \mathbf{J}. \qquad (9\text{–}55)$$

That Eq. (9–55) is also a necessary condition for a restricted canonical transformation is easily shown directly by reversing the order of the steps of the proof. Note that for an extended time-independent canonical transformation, where $K = \lambda H$, the condition of Eq. (9–55) would be replaced by

$$\mathbf{MJ\tilde{M}} = \lambda \mathbf{J}. \qquad (9\text{–}56)$$

Equation (9–55) may be expressed in various forms. Multiplying from the right by the matrix inverse to $\mathbf{\tilde{M}}$ leads to

$$\mathbf{MJ} = \mathbf{J\tilde{M}}^{-1}, \qquad (9\text{–}57)$$

(since the transpose of the inverse is the inverse of the transpose). The elements of the matrix equation (9–57) will be found to be identical with Eqs. (9–48a) and (9–48b). If Eq. (9–57) is multiplied by $\mathbf{J}$ from the left and $-\mathbf{J}$ from the right, then by virtue of Eq. (8–32) we have

$$\mathbf{JM} = \mathbf{\tilde{M}}^{-1}\mathbf{J},$$

or

$$\mathbf{\tilde{M}JM} = \mathbf{J}. \qquad (9\text{–}58)$$

Equation (9–55), or its equivalent version, Eq. (9–58), is spoken of as the *symplectic condition* for a canonical transformation, and the matrix $\mathbf{M}$ satisfying the condition is said to be a *symplectic matrix*. (Following the interpretation of Eq. (7–43) we can look on Eq. (9–58) also as saying that $\mathbf{M}$ is an orthogonal matrix in a phase space in which $\mathbf{J}$ is the metric tensor, but this is not a particularly fruitful approach.)

For a canonical transformation that contains the time as a parameter, the simple derivation given for the symplectic condition no longer holds. Nonetheless, the symplectic condition remains a necessary and sufficient condition for a canonical transformation even if it involves the time. It is possible to prove the general validity of the symplectic condition for all canonical transformations by straightforward, albeit lengthy, procedures resembling those employed for restricted canonical transformations.* Instead we shall take a different tack, one that takes advantage of the parametric form of the canonical transformations involving time. A canonical transformation of the form

$$\zeta = \zeta(\eta, t) \tag{9–59}$$

evolves continuously as time increases from some initial value t_0. It is a single-parameter instance of the family of continuous transformations first studied systematically by the mathematician Sophus Lie and as such plays a distinctive role in the transformation theory of classical mechanics. If the transformation

$$\eta \rightarrow \zeta(t) \tag{9–60a}$$

is canonical, then so obviously is the transformation

$$\eta \rightarrow \zeta(t_0). \tag{9–60b}$$

It follows then from the definition of canonical transformation that the transformation characterized by

$$\zeta(t_0) \rightarrow \zeta(t) \tag{9–60c}$$

is also canonical. Since t_0 in Eq. (9–60b) is a fixed constant, this canonical transformation satisfies the symplectic condition. If now the transformation of Eq. (9–60c) obeys the symplectic condition, it is easy to show (cf. Exercise 13) that the general transformation Eq. (9–60a) will also.

To demonstrate that the symplectic condition does indeed hold for canonical transformations of the type of Eq. (9–60c) we introduce the notion of an *infinitesimal canonical transformation* (abbreviated I.C.T.), a concept that will prove to be widely useful. As in the case of infinitesimal rotations such a transformation is one in which the new variables differ from the old only by infinitesimals. Only first-order terms in these infinitesimals are to be retained in all calculations. The transformation equations can then be written as

$$Q_i = q_i + \delta q_i, \tag{9–61a}$$

$$P_i = p_i + \delta p_i, \tag{9–61b}$$

or in matrix form

$$\zeta = \eta + \delta \eta. \tag{9–61c}$$

(Here δq_i and δp_i do *not* represent virtual displacements but are simply the infinitesimal changes in the coordinates and momenta.) An infinitesimal canonical transformation thus differs only infinitesimally from the identity

---

*For this approach see L. A. Pars, *A Treatice on Analytical Dynamics*, 1965, pp. 514–515; and G. S. S. Ludford and D. W. Yannitell, *Am. J. Phys.* **36**, 231 (1968).

transformation discussed in the previous section. In the generator formalism a suitable generating function for an I.C.T. would therefore be

$$F_2 = q_i P_i + \epsilon G(q, P, t),\tag{9–62}$$

where ϵ is some infinitesimal parameter of the transformation, and G is any (differentiable) function of its $2n + 1$ arguments. By Eq. (9–14a) the transformation equations for the momenta are to be found from

$$p_j = \frac{\partial F_2}{\partial q_j} = P_j + \epsilon \frac{\partial G}{\partial q_j}$$

or

$$\delta p_j \equiv P_j - p_j = -\epsilon \frac{\partial G}{\partial q_j}.\tag{9–63a}$$

Similarly, by Eq. (9–14b) the transformation equations for Q_j are determined by the relations

$$Q_j = \frac{\partial F_2}{\partial p_j} = q_j + \epsilon \frac{\partial G}{\partial P_j}.$$

Since the second term is already linear in ϵ, and P differs from p only by an infinitesimal, it is consistent to first order to replace P_j in the derivative function by p_j. We may then consider G as a function of q, p only (and possibly t). Following the usual practice we will refer to $G(q, p)$ as the *generating function of the infinitesimal canonical transformation*, although strictly speaking that designation belongs only to F. The transformation equation for Q_j can therefore be written as

$$\delta q_j = \epsilon \frac{\partial G}{\partial p_j}.\tag{9–63b}$$

Both transformation equations can be combined into one matrix equation

$$\delta \boldsymbol{\eta} = \epsilon \mathbf{J} \frac{\partial G}{\partial \boldsymbol{\eta}}.\tag{9–63c}$$

An obvious example of an infinitesimal canonical transformation would be the transformation of Eq. (9–60c) when t differs from t_0 by an infinitesimal t:

$$\boldsymbol{\zeta}(t_0) \rightarrow \boldsymbol{\zeta}(t_0 + dt),\tag{9–64}$$

with dt as the infinitesimal parameter ϵ. The continuous evolution of the transformation $\boldsymbol{\zeta}(\eta, t)$ from $\boldsymbol{\zeta}(\eta, t_0)$ means that the transformation $\boldsymbol{\zeta}(t_0) \rightarrow \boldsymbol{\zeta}(t)$ can be built up as a succession of such I.C.T.'s in dt steps. It will therefore suffice to show that the infinitesimal transformation, Eq. (9–64), satisfies the symplectic condition. But we can demonstrate simply from the transformation equations

(9–63) that the Jacobian matrix of any I.C.T. is a symplectic matrix. By definition the Jacobian matrix for an infinitesimal transformation is

$$\mathbf{M} \equiv \frac{\partial \boldsymbol{\zeta}}{\partial \boldsymbol{\eta}} = 1 + \frac{\partial \delta \boldsymbol{\eta}}{\partial \boldsymbol{\eta}},$$

or by Eq. (9–63c)

$$\mathbf{M} = 1 + \epsilon \mathbf{J} \frac{\partial^2 G}{\partial \boldsymbol{\eta} \partial \boldsymbol{\eta}}. \tag{9–65}$$

The second derivative in Eq. (9–65) is a square, symmetric matrix with elements

$$\left(\frac{\partial^2 G}{\partial \boldsymbol{\eta} \partial \boldsymbol{\eta}} \right)_{ij} = \frac{\partial^2 G}{\partial \eta_i \partial \eta_j}.$$

Because of the antisymmetrical property of $\mathbf{J}$, the transpose of $\mathbf{M}$ is

$$\widetilde{\mathbf{M}} = 1 - \epsilon \frac{\partial^2 G}{\partial \boldsymbol{\eta} \partial \boldsymbol{\eta}} \mathbf{J}. \tag{9–66}$$

The symplectic condition involves the value of the matrix product

$$\mathbf{M} \mathbf{J} \widetilde{\mathbf{M}} = \left(1 + \epsilon \mathbf{J} \frac{\partial^2 G}{\partial \boldsymbol{\eta} \partial \boldsymbol{\eta}} \right) \mathbf{J} \left(1 - \epsilon \frac{\partial^2 G}{\partial \boldsymbol{\eta} \partial \boldsymbol{\eta}} \mathbf{J} \right).$$

Consistent to first order in this product is

$$\mathbf{M} \mathbf{J} \widetilde{\mathbf{M}} = \mathbf{J} + \epsilon \mathbf{J} \frac{\partial^2 G}{\partial \boldsymbol{\eta} \partial \boldsymbol{\eta}} \mathbf{J} - \mathbf{J} \epsilon \frac{\partial^2 G}{\partial \boldsymbol{\eta} \partial \boldsymbol{\eta}} \mathbf{J}$$

$$= \mathbf{J},$$

thus demonstrating that the symplectic condition holds for any infinitesimal canonical transformation. By the chain of reasoning we have spun out, it therefore follows that *any* canonical transformation, involving time as a parameter or not, obeys the symplectic condition, Eq. (9–55) or Eq. (9–58).

The symplectic approach, for the most part, has been developed independently of the generating function method, except in the treatment of infinitesimal canonical transformations. They are, of course, connected. We shall sketch later, for example, a proof that the symplectic condition implies the existence of a generating function. But the connection is largely irrelevant. Both are valid ways of looking at canonical transformations and both encompass all of the needed properties of the transformations. For example, either the symplectic or the generator formalisms can be used to prove that canonical transformations have the four properties that characterize a group (cf. Exercise 13):

1. The identity transformation is canonical.
2. If a transformation is canonical so is its inverse.
3. Two successive canonical transformations (the "product" operation) define a transformation that is also canonical.
4. The product operation is associative.

We shall therefore be free to use either the generator or the symplectic approach at will, depending on which leads to the simplest treatment at the moment.

9-4 POISSON BRACKETS AND OTHER CANONICAL INVARIANTS

The *Poisson bracket* of two functions u, v with respect to the canonical variables (q, p) is defined as

$$[u, v]_{q,p} = \frac{\partial u}{\partial q_i} \frac{\partial v}{\partial p_i} - \frac{\partial u}{\partial p_i} \frac{\partial v}{\partial q_i}. \tag{9-67}$$

In this bilinear expression we have a typical symplectic structure, as in Hamilton's equations, where q is coupled with p, and p with $-q$. The Poisson bracket thus lends itself readily to being written in matrix form, where it appears as

$$[u, v]_\eta = \frac{\widetilde{\partial u}}{\partial \boldsymbol{\eta}} \mathbf{J} \frac{\partial v}{\partial \boldsymbol{\eta}}. \tag{9-68}$$

The transpose sign is used on the first matrix on the right-hand side to indicate explicitly that this matrix must be treated as a single row matrix in the multiplication. On most occasions this specific reminder will not be needed and the transpose sign may be omitted.

Suppose we choose the functions u, v out of the set of canonical variables (q, p) themselves. Then it follows trivially from the definition, either as Eq. (9-67) or (9-68), that these Poisson brackets have the values

$$[q_j, q_k]_{q,p} = 0 = [p_j, p_k]_{q,p},$$

and

$$[q_j, p_k]_{q,p} = \delta_{jk} = -[p_j, q_k]_{q,p}. \tag{9-69}$$

We can summarize the relations of Eqs. (9-69) in one equation by introducing a *square matrix Poisson bracket*, $[\boldsymbol{\eta}, \boldsymbol{\eta}]$, whose *lm* element is $[\eta_l, \eta_m]$. Equations (9-69) can then be written as

$$[\boldsymbol{\eta}, \boldsymbol{\eta}]_\eta = \mathbf{J}. \tag{9-70}$$

Now let us take for u, v the members of the transformed variables (Q, P), or $\boldsymbol{\zeta}$, defined in terms of (q, p) by the transformation equations (9-59). The set of all the Poisson brackets that can be formed out of (Q, P) comprise the matrix Poisson bracket defined as

$$[\boldsymbol{\zeta}, \boldsymbol{\zeta}]_\eta = \frac{\widetilde{\partial \boldsymbol{\zeta}}}{\partial \boldsymbol{\eta}} \mathbf{J} \frac{\partial \boldsymbol{\zeta}}{\partial \boldsymbol{\eta}}.$$

But we recognize the partial derivatives as defining the square Jacobian matrix of the transformation, so that the Poisson bracket relation is equivalent to

$$[\boldsymbol{\zeta}, \boldsymbol{\zeta}]_\eta = \widetilde{\mathbf{M}} \mathbf{J} \mathbf{M}. \tag{9-71}$$

If the transformation $\boldsymbol{\eta} \to \boldsymbol{\zeta}$ is canonical, then the symplectic condition holds and Eq. (9–71) reduces to

$$[\boldsymbol{\zeta}, \boldsymbol{\zeta}]_\eta = \mathbf{J}, \tag{9–72}$$

and conversely, if Eq. (9–72) is valid then the transformation is canonical.

Poisson brackets of the canonical variables themselves, such as Eqs. (9–70) or (9–72), are referred to as the *fundamental Poisson brackets*. Since we have from Eq. (9–70) that

$$[\boldsymbol{\zeta}, \boldsymbol{\zeta}]_\zeta = \mathbf{J}, \tag{9–73}$$

Eq. (9–72) states that the fundamental Poisson brackets of the ζ variables have the same value when evaluated with respect to *any* canonical coordinate set. In other words, the *fundamental Poisson brackets are invariant under canonical transformation*. We have seen from Eq. (9–71) that the invariance is a necessary and sufficient condition for the transformation matrix to be symplectic. The invariance of the fundamental Poisson brackets is thus in all ways equivalent to the symplectic condition for a canonical transformation.

It does not take many more steps to show that *all* Poisson brackets are invariant under canonical transformation. Consider the Poisson bracket of two functions u, v with respect to the η set of coordinates, Eq. (9–68). In analogy to Eq. (9–53) the partial derivative of v with respect to $\boldsymbol{\eta}$ can be expressed in terms of partial derivatives with respect to $\boldsymbol{\zeta}$ as

$$\frac{\partial v}{\partial \boldsymbol{\eta}} = \widetilde{\mathbf{M}} \frac{\partial v}{\partial \boldsymbol{\zeta}}$$

(that is, the partial derivative transforms covariantly). In a similar fashion,

$$\frac{\widetilde{\partial u}}{\partial \boldsymbol{\eta}} = \widetilde{\mathbf{M} \frac{\partial u}{\partial \boldsymbol{\zeta}}} = \frac{\widetilde{\partial u}}{\partial \boldsymbol{\zeta}} \mathbf{M}.$$

Hence the Poisson bracket Eq. (9–68) can be written

$$[u, v]_\eta = \frac{\widetilde{\partial u}}{\partial \boldsymbol{\eta}} \mathbf{J} \frac{\partial v}{\partial \boldsymbol{\eta}}$$

$$= \frac{\widetilde{\partial u}}{\partial \boldsymbol{\zeta}} \mathbf{M} \mathbf{J} \widetilde{\mathbf{M}} \frac{\partial v}{\partial \boldsymbol{\zeta}}.$$

If the transformation is canonical, the symplectic condition in the form of Eq. (9–55) holds, and we then have

$$[u, v]_\eta = \frac{\widetilde{\partial u}}{\partial \boldsymbol{\zeta}} \mathbf{J} \frac{\partial v}{\partial \boldsymbol{\zeta}} \equiv [u, v]_\zeta. \tag{9–74}$$

Thus, the Poisson bracket has the same value when evaluated with respect to any canonical set of variables—*all Poisson brackets are canonical invariants*. In writing the symbol for the Poisson bracket we have so far been careful to indicate by the subscript the set of variables in terms of which the brackets are defined. So long as we use only canonical variables that practice is now seen to be unnecessary, and we shall in general drop the subscript.*

The hallmark of the canonical transformation is that Hamilton's equations of motion are invariant in form under the transformation. Similarly, the canonical invariance of Poisson brackets implies that equations expressed in terms of Poisson brackets are invariant in form under canonical transformation. As we shall see, we can develop a structure of classical mechanics, paralleling the Hamiltonian formulation, expressed solely in terms of Poisson brackets. This Poisson bracket formulation, which has the same form in all canonical coordinates, is especially useful for carrying out the transition from classical to quantum mechanics. There is a simple "correspondence principle" that says that the classical Poisson bracket is to be replaced by a suitably defined commutator of the corresponding quantum operators.

The algebraic properties of the Poisson bracket are therefore of considerable interest. We have already used the obvious properties

$$[u, u] = 0 \tag{9–75a}$$

(antisymmetry):

$$[u, v] = -[v, u]. \tag{9–75b}$$

Almost equally obvious are the characteristics

(linearity):

$$[au + bv, w] = a[u, w] + b[v, w], \tag{9–75c}$$

where a and b are constants, and

$$[uv, w] = [u, w]v + u[v, w]. \tag{9–75d}$$

One other property is far from obvious, but is very important in defining the nature of the Poisson bracket. It is usually given in the form of *Jacobi's identity*, which states that if $u, v,$ and w are three functions with continuous second derivatives, then

$$[u, [v, w]] + [v, [w, u]] + [w, [u, v]] = 0, \tag{9–75e}$$

that is, the sum of the cyclic permutations of the double Poisson bracket of three functions is zero. There seems to be no simple way of proving Jacobi's identity for the Poisson bracket without lengthy algebra. However it is possible to mitigate the complexity of the manipulations by introducing a special nomenclature. We shall put subscripts on u, v, w (or functions of them) to denote partial derivatives by the corresponding canonical variable. Thus

$$u_i \equiv \frac{\partial u}{\partial \eta_i}, \qquad v_{ij} \equiv \frac{\partial v}{\partial \eta_i \partial \eta_j}.$$

---

* Note that for a scale transformation, or an extended canonical transformation, where the symplectic condition takes on the form of Eq. (9–56), then Poisson brackets do *not* have the same values in all coordinate systems. This is one of the reasons scale transformations are excluded from the class of canonical transformations that are useful to consider.

In this notation the Poisson bracket of u and v can be expressed as

$$[u, v] = u_i J_{ij} v_j.$$

Here J_{ij}, as usual, is simply the ijth element of $\mathbf{J}$. In the proof the only property of $\mathbf{J}$ that we shall need is its antisymmetry. Consider now the first double Poisson bracket in Eq. (9–75e):

$$[u, [v, w]] = u_i J_{ij} [v, w]_j = u_i J_{ij} (v_k J_{kl} w_l)_j.$$

Because the elements J_{kl} are constants, the derivative with respect to η doesn't act on them, and we have

$$[u, [v, w]] = u_i J_{ij} (v_k J_{kl} w_{lj} + v_{kj} J_{kl} w_l). \tag{9–76}$$

The other double Poisson brackets can be obtained from Eq. (9–76) by cyclic permutation of u, v, w. There are thus six terms in all, each being a four-fold sum over dummy indices $i, j, k,$ and l. Consider the term in Eq. (9–76) involving a second derivative of w:

$$J_{ij} J_{kl} u_i v_k w_{lj}.$$

The only other second derivative of w will appear in evaluating the second double Poisson bracket in (Eq. 9–75e):

$$[v, [w, u]] = v_k J_{kl} (w_j J_{ji} u_i)_l.$$

Here the term in the second derivative in w is

$$J_{ji} J_{kl} u_i v_k w_{jl}.$$

Since the order of differentiation is immaterial, $w_{lj} = w_{jl}$, and the sum of the two terms is given by

$$(J_{ij} + J_{ji}) J_{kl} u_i v_k w_{lj} = 0,$$

by virtue of the antisymmetry of J. The remaining four terms are cyclic permutations and can similarly be divided in two pairs, one involving second derivatives of u and the other of v. By the same reasoning each of these pairs sums to zero, and Jacobi's identity is thus verified.

If the Poisson bracket of u, v is looked on as defining a "product" operation of the two functions, then Jacobi's identity is the replacement for the associative law of multiplication. It will be remembered that the ordinary multiplication of arithmetic is associative, i.e., the order of a sequence of multiplications is immaterial:

$$a(bc) = (ab)c.$$

Jacobi's identity says that the bracket "product" is not associative and gives the effect of changing the sequence of "multiplications." Together with the properties (9–75b) and (9–75c), Jacobi's identity, Eq. (9–75e), defines a particular type of nonassociative algebra, called a *Lie algebra*, obeyed by the Poisson bracket. The

Poisson bracket operation is not the only type of "product" familiar to physicists that satisfies the conditions for a Lie algebra. It will be left to the exercises to show that that vector product of two vectors,

$$_V[A, B] \rightarrow A \times B,$$

and the commutator of two matrices,

$$_M[A, B] \rightarrow AB - BA,$$

satisfy the same Lie algebra conditions as the Poisson bracket. It is this last that makes it feasible to replace the classical Poisson bracket by the commutator of the quantum mechanical operators. In other words, the "correspondence principle" can work only because both the Poisson bracket and commutator are representations of a Lie algebra "product."*

There are other canonical invariants besides the Poisson bracket. One, mainly of historical interest now, is the *Lagrange bracket*. Suppose u and v are two functions out of a set of $2n$ independent functions of the canonical variables. By inversion the canonical variables can then be considered as functions of the set of $2n$ functions. On this basis, the Lagrange bracket of u and v with respect to the (q, p) variables is defined as

$$\{u, v\}_{q,p} = \frac{\partial q_i}{\partial u} \frac{\partial p_i}{\partial v} - \frac{\partial p_i}{\partial u} \frac{\partial q_i}{\partial v}, \tag{9–77}$$

or, in matrix notation,

$$\{u, v\}_\eta = \frac{\widetilde{\partial \eta}}{\partial u} J \frac{\partial \eta}{\partial v}. \tag{9–78}$$

Proof of the canonical invariance of the Lagrange bracket parallels that for the Poisson bracket. We form first the Lagrange bracket with respect to the ζ variables:

$$\{u, v\}_\zeta = \frac{\widetilde{\partial \zeta}}{\partial u} J \frac{\partial \zeta}{\partial v}. \tag{9–79}$$

However the derivatives can be expressed in terms of the η set as

$$\frac{\partial \zeta_i}{\partial v} = \frac{\partial \zeta_i}{\partial \eta_j} \frac{\partial \eta_j}{\partial v}, \qquad \text{or} \qquad \frac{\partial \zeta}{\partial v} = M \frac{\partial \eta}{\partial v}.$$

---

*Of course one must not mistake the mathematical acceptability of this version of the correspondence principle with its physical necessity. The introduction of the quantum commutation relations was a great act of physical discovery by the pioneers of quantum mechanics. All we show here is that there is a similarity in the mathematical structure of the Poisson bracket formulation of classical mechanics and the commutation relation version of quantum mechanics. The formal correspondence is that

$$[u, v] \rightarrow \frac{1}{ih}(uv - vu)$$

where on the left u, v are classical functions and on the right they are quantum operators.

Equation (9–79) can then be written as

$$\{u, v\}_\zeta = \frac{\widetilde{\partial \boldsymbol{\eta}}}{\partial u} \widetilde{\mathbf{M}} \mathbf{J} \mathbf{M} \frac{\partial \boldsymbol{\eta}}{\partial v},$$

which by virtue of the symplectic condition for a canonical transformation (9–58) becomes

$$\{u, v\}_\zeta = \frac{\partial \boldsymbol{\eta}}{\partial u} \mathbf{J} \frac{\partial \boldsymbol{\eta}}{\partial v} = \{u, v\}_\eta, \tag{9–80}$$

thus verifying the canonical invariance of the Lagrange bracket.

If for u and v we take two members of the set of canonical variables then we obtain the *fundamental Lagrange brackets*:

$$\{q_i, q_j\} = 0 = \{p_i, p_j\}; \qquad \{q_i, p_j\} = \delta_{ij}, \tag{9–81}$$

or, in matrix notation,

$$\{\boldsymbol{\eta}, \boldsymbol{\eta}\} = \mathbf{J}. \tag{9–82}$$

The Lagrange and Poisson brackets clearly stand in some kind of inverse relationship to each other, but the precise form of this relation is somewhat complicated to express. Let $u_i, i = 1, \ldots, 2n$, be a set of $2n$ independent functions of the canonical variables, to be represented by a column (or row) matrix $\mathbf{u}$. Then $\{\mathbf{u}, \mathbf{u}\}$ is the $2n \times 2n$ matrix whose ijth element is $\{u_i, u_j\}$, with a similar description for $[\mathbf{u}, \mathbf{u}]$. The reciprocal character of the two brackets manifests itself in the relation

$$\{\mathbf{u}, \mathbf{u}\}[\mathbf{u}, \mathbf{u}] = -\mathbf{1}. \tag{9–83}$$

If for u we choose the canonical set itself, $\boldsymbol{\eta}$, then Eq. (9–83) obviously follows from the fundamental bracket formulas, Eqs. (9–70) and (9–82), and the properties of $\mathbf{J}$. The proof for arbitrary u is not difficult if written in terms of the matrix definitions of the brackets and is reserved for the exercises. While the properties of the Lagrange and Poisson brackets parallel each other in many aspects it should be noted that the Lagrange brackets do *not* obey Jacobi's identity. Lagrange brackets therefore do not qualify as a "product" operation in a Lie algebra.

Another important canonical invariant is the magnitude of a volume element in phase space. A canonical transformation $\boldsymbol{\eta} \to \boldsymbol{\zeta}$ transforms the $2n$ dimensional phase space with coordinates η_i to another phase space with coordinates ζ_i. The volume element

$$(d\eta) = dq_1 dq_2 \ldots dq_n \, dp_1 \ldots dp_n$$

transforms to a new volume element

$$(d\zeta) = dQ_1 \, dQ_2 \ldots dQ_n \, dP_1 \ldots dP_n.$$

As is well known* the size of the two volume elements is related by the absolute value of the Jacobian determinant:

$$(d\zeta) = \|\mathbf{M}\|(d\eta).\tag{9-84}$$

But by taking the determinant of both sides of the symplectic condition, Eq. (9-55), we have

$$|\mathbf{M}|^2|\mathbf{J}| = |\mathbf{J}|.\tag{9-85}$$

Thus, in a real canonical transformation the Jacobian determinant is ± 1, and the absolute value is always unity, proving the canonical invariance of the volume element in phase space. It follows, also, that the volume of any arbitrary region in phase space,

$$J_n = \int \cdots \int (d\eta),\tag{9-86}$$

is a canonical invariant.

The volume integral in Eq. (9-86) is the final member of a sequence of canonical invariants known as the *integral invariants of Poincaré*, comprising integrals over subspaces of phase space of different dimensions. The other members of the sequence cannot be stated as simply as J_n and because they are not needed for the further development of the theory they will not be discussed here.†

Finally, the invariance of the fundamental Poisson brackets now enables us to outline a proof that the symplectic condition implies the existence of a generating function, as mentioned at the conclusion of the previous section. To simplify considerations we shall examine only a system with one degree of freedom; the general method of the proof can be directly extended to systems with many degrees of freedom.‡ We suppose that the first of the equations of transformation,

$$Q = Q(q,p), \qquad P = P(q,p),$$

is invertable so as to give p as a function q and Q, say

$$p = \phi(q,Q).\tag{9-87}$$

Substitution in the second equation of transformation gives P as some function of q and Q, say

$$P = \psi(q,Q).\tag{9-88}$$

---

* See, for example, W. Kaplan, *Advanced Calculus*, 2d. ed., 1973, p. 270.

† There are difficulties with the usual presentation of the lower order integral invariants. For an exposition of the problems see H. D. Block, *Quarterly Applied Math.* **12**, 201 (1954).

‡ For details of the proof for a general system see C. Carathèodory, *Calculus of Variations and Partial Differential Equations of the First Order*, English translation, Holden-Day, 1965, Vol. 1, Sect. 97, pp. 87–90. In the literature the connection between the symplectic approach and the generator formalism is sometimes referred to as the *Caratheodory theorem*.

In such case we would expect the transformation to be generated by a generating function of the first kind,* F_1, with Eqs. (9–87) and (9–88) appearing as

$$p = \frac{\partial F_1(q, Q)}{\partial q}, \qquad P = -\frac{\partial F_1}{\partial Q}(q, Q). \qquad (9\text{–}89)$$

If Eq. (9–89) holds, then it must be true that

$$\frac{\partial \phi}{\partial Q} = -\frac{\partial \psi}{\partial q}. \qquad (9\text{–}90)$$

Conversely, if we can show that Eq. (9–90) is valid, then there must exist a function F_1 such that p and P are given by Eqs. (9–89).

To demonstrate the validity of Eq. (9–90) we try to look on all quantities as functions of q and Q. Thus, we of course have the identity

$$\frac{\partial Q}{\partial Q} = 1,$$

but if Eq. (9–87) be substituted in the first transformation equation,

$$Q = Q(q, \phi(q, Q)), \qquad (9\text{–}91)$$

the partial derivative can also be written

$$\frac{\partial Q}{\partial Q} = \frac{\partial Q}{\partial p} \frac{\partial \phi}{\partial Q},$$

so that we have the relation

$$\frac{\partial Q}{\partial p} \frac{\partial \phi}{\partial Q} = 1. \qquad (9\text{–}92)$$

In the same spirit we evaluate the Poisson bracket

$$[Q, P] \equiv \frac{\partial Q}{\partial q} \frac{\partial P}{\partial p} - \frac{\partial P}{\partial q} \frac{\partial Q}{\partial p} = 1.$$

The derivatives of P are derivatives of ψ considered as a function of q and $Q(q, p)$. Hence the Poisson bracket can be written

$$[Q, P] = \frac{\partial Q}{\partial q} \frac{\partial \psi}{\partial Q} \frac{\partial Q}{\partial p} - \frac{\partial Q}{\partial p} \left(\frac{\partial \psi}{\partial q} + \frac{\partial \psi}{\partial Q} \frac{\partial Q}{\partial q} \right),$$

or, consolidating terms, as

$$[Q, P] = \frac{\partial \psi}{\partial Q} \left(\frac{\partial Q}{\partial q} \frac{\partial Q}{\partial p} - \frac{\partial Q}{\partial p} \frac{\partial Q}{\partial q} \right) - \frac{\partial Q}{\partial p} \frac{\partial \psi}{\partial q},$$

---

*Of course, if the Q transformation equation is not invertable, as in the identity transformation, then we would invert the P equation and be led to a generating function of the second kind.

and therefore

$$1 = -\frac{\partial Q}{\partial p}\frac{\partial \psi}{\partial q}. \tag{9-93}$$

Combining Eqs. (9-92) and (9-93) we have

$$\frac{\partial Q}{\partial p}\frac{\partial \phi}{\partial Q} = -\frac{\partial Q}{\partial p}\frac{\partial \psi}{\partial q}.$$

Since the partial derivative of Q with respect to p is the same on both sides of the equation, i.e., the other variable being held constant is q in both cases, and since the derivative doesn't vanish (else the Q equation could not be inverted), it follows that Eq. (9-90) must be true. Thus, from the value of the fundamental Poisson bracket $[Q, P]$, which we have seen is equivalent to the symplectic condition, we are led to the existence of a generating function. The two approaches to canonical transformations, though arrived at independently, are fully equivalent.

9-5 EQUATIONS OF MOTION, INFINITESIMAL CANONICAL TRANSFORMATIONS, AND CONSERVATION THEOREMS IN THE POISSON BRACKET FORMULATION

Almost the entire framework of Hamiltonian mechanics can be restated in terms of Poisson brackets. As a result of the canonical invariance of the Poisson brackets, the relations so obtained will also be invariant in form under a canonical transformation. Suppose for example we look for the total time derivative of some function of the canonical variables and time, $u(q, p, t)$, by use of Hamilton's equations of motion:

$$\frac{du}{dt} = \frac{\partial u}{\partial q_i}\dot{q}_i + \frac{\partial u}{\partial p_i}\dot{p}_i + \frac{\partial u}{\partial t} = \frac{\partial u}{\partial q_i}\frac{\partial H}{\partial p_i} - \frac{\partial u}{\partial p_i}\frac{\partial H}{\partial q_i} + \frac{\partial u}{\partial t},$$

or

$$\frac{du}{dt} = [u, H] + \frac{\partial u}{\partial t}. \tag{9-94}$$

In terms of the symplectic notation, the derivation of Eq. (9-94) would run

$$\frac{du}{dt} = \frac{\partial u}{\partial \boldsymbol{\eta}}\dot{\boldsymbol{\eta}} + \frac{\partial u}{\partial t} = \frac{\partial u}{\partial \boldsymbol{\eta}}\mathbf{J}\frac{\partial H}{\partial \boldsymbol{\eta}} + \frac{\partial u}{\partial t},$$

from whence Eq. (9-94) follows, by virtue of (9-68). Equation (9-94) may be looked on as the generalized equation of motion for an arbitrary function u in the Poisson bracket formulation. It contains Hamilton's equations as a special case when for u we substitute one of the canonical variables

$$\dot{q}_i = [q_i, H], \qquad \dot{p}_i = [p_i, H], \tag{9-95a}$$

or, in symplectic notation,

$$\dot{\boldsymbol{\eta}} = [\boldsymbol{\eta}, H].$$ (9–95b)

That Eq. (9–95) is identical with Hamilton's equations of motion may be seen directly from the observation that by the definition of the Poisson bracket, Eq. (9–68), we have

$$[\boldsymbol{\eta}, H] = \mathbf{J}\frac{\partial H}{\partial \boldsymbol{\eta}},$$ (9–96)

so that Eq. (9–95b) is simply another way of writing Eq. (8–31). Another familiar property may be obtained from Eq. (9–94) by taking u as H itself. Equation (9–94) then says that

$$\frac{dH}{dt} = \frac{\partial H}{\partial t},$$

as was obtained previously in Eq. (8–35). It should be noted that the generalized equation of motion is canonically invariant; it is valid in whatever set of canonical variables q, p is used to express the function u or to evaluate the Poisson bracket. However, the Hamiltonian used must be appropriate to the particular set of canonical variables. On transforming to another set of variables by a time-dependent canonical transformation one must also change to the transformed Hamiltonian K.

If u is a constant of the motion, then Eq. (9–94) says it must have the property

$$[H, u] = \frac{\partial u}{\partial t}.$$ (9–97)

All functions that obey Eq. (9–97) are constants of the motion, and conversely the Poisson bracket of H with any constant of the motion must be equal to the explicit time derivative of the constant function. We thus have a general test for seeking and identifying the constants of the system. For those constants of the motion not involving the time explicitly, the test of Eq. (9–97) reduces to requiring that their Poisson brackets with the Hamiltonian vanish.*

If two constants of the motion are known, the Jacobi identity provides a possible way for obtaining further constants. Suppose u and v are two constants of the motion not explicitly functions of time. Then if w in Eq. (9–75c) is taken to be H, the Jacobi identity says

$$[H, [u, v]] = 0;$$

that is, the Poisson bracket of u and v is also a constant in time. Even when the conserved quantities depend on time explicitly, it can be shown with a bit more

---

* In view of the "correspondence principle" between the classical Poisson bracket and the quantum commutator, it is seen that this statement corresponds to the well-known quantum theorem that conserved quantities commute with the Hamiltonian.

algebra (cf. Exercise 28) that *the Poisson bracket of any two constants of the motion is also a constant of the motion* (Poisson's theorem). Repeated application of the Jacobi identity in this manner can in principle lead to a complete sequence of constants of the motion. Quite often, however, the process is disappointing. The Poisson bracket of u and v frequently turns out to be a trivial function of u and v themselves, or even identically zero. Still, the possibility of generating new independent constants of motion by Poisson's theorem should be kept in mind.

The Poisson bracket notation can also be used to reformulate the basic equations of an infinitesimal canonical transformation. As discussed above (p. 394) such a transformation is a special case of a transformation that is a continuous function of a parameter, starting from the identity transformation at some initial value of the parameter (which may, for convenience, be set equal to zero). If the parameter is small enough to be treated as a first order infinitesimal, then the transformed canonical variables differ only infinitesimally from the initial coordinates:

$$\boldsymbol{\zeta} = \boldsymbol{\eta} + \delta\boldsymbol{\eta} \tag{9–98}$$

with the change being given in terms of the generator G through Eq. (9–63c):

$$\delta\boldsymbol{\eta} = \epsilon\mathbf{J}\frac{\partial G(\boldsymbol{\eta})}{\partial \boldsymbol{\eta}}.$$

Now, by the definition (9–68) of the Poisson bracket it follows that

$$[\boldsymbol{\eta}, u] = \mathbf{J}\frac{\partial u}{\partial \boldsymbol{\eta}} \tag{9–99}$$

(cf. Eq. 9–96), a relation that remains valid when the Poisson bracket is evaluated in terms of any other canonical variables. If u is taken to be G, it is seen that the equations of transformation for an infinitesimal canonical transformation can be written as

$$\delta\boldsymbol{\eta} = \epsilon[\boldsymbol{\eta}, G]. \tag{9–100}$$

Consider now an infinitesimal canonical transformation in which the continuous parameter is t (as was done in proving the symplectic condition) so that $\epsilon = dt$, and let the generating function G be the Hamiltonian. Then the equations of transformation for this I.C.T. become, by Eq. (9–100),

$$\delta\boldsymbol{\eta} = dt[\boldsymbol{\eta}, H] = \dot{\boldsymbol{\eta}}\, dt = d\boldsymbol{\eta}. \tag{9–101}$$

These equations state that the transformation changes the coordinates and momenta at the time t to the values they have at the time $t + dt$. Thus the motion of the system in a time interval dt can be described by an infinitesimal contact transformation generated by the Hamiltonian. Correspondingly, the system motion in a finite time interval from t_0 to t is represented by a succession of infinitesimal contact transformations, which, as we have seen, is equivalent to a single finite canonical transformation. Thus, the values of q and p at any time t can be obtained from their initial values by a canonical transformation that is a continuous function of time. According to this view the motion of a mechanical system corresponds to the continuous evolution or unfolding of a canonical transformation. In a very literal sense, the *Hamiltonian is the generator of the system motion with time.*

Conversely, there must exist a canonical transformation from the values of the coordinates and momenta at any time t to their constant initial values. Obtaining such a transformation is obviously equivalent to solving the problem of the system motion. At the beginning of the chapter it was pointed out that a mechanical problem could be reduced to finding the canonical transformation for which all momenta are constants of the motion. The present considerations indicate the possibility of an alternative solution by means of the canonical transformation for which *both* the momenta and coordinates are constants of the motion. These two suggestions will be elaborated in the next chapter in order to show how formal solutions may be obtained for any mechanical problem.

Implicit to this discussion has been an altered way of looking at a canonical transformation and the effect it produces. The notion of a canonical transformation was introduced as a change of the coordinates used to characterize phase space. In effect we switched from one phase space η with coordinates (q, p) to another, ζ, with coordinates (Q, P). If the state of the system at a given time was described by a point A in one system, it could also be described equally well by the transformed point A' (cf. Fig. 9–1). Any function of the system variables would have the same value for a given system configuration whether it was described by the (q, p) set or by the (Q, P) set. In other words, the function would have the same value at A' as at A. In analogy to the corresponding description of orthogonal transformations we may call this the *passive* view of a canonical transformation.

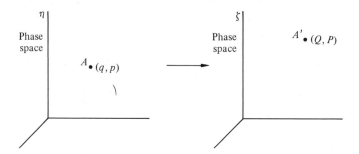

FIGURE 9–1
Illustration of the passive view of a canonical transformation.

In contrast, we have spoken of the canonical transformation generated by the Hamiltonian as relating the coordinates of one point in phase space to those of another point in the *same* phase space. From this viewpoint the canonical transformation accomplishes, in the mathematician's language, a mapping of the points of phase space onto themselves. In effect we have an *active* interpretation of the canonical transformation as "moving" the system point from one position, with coordinates (q, p), to another point, (Q, P), in phase space (cf. Fig. 9–2). Of course, the canonical transformation in itself cannot move or change the system configuration. What it does is express one configuration of the system in terms of another. With some classes of canonical transformation the active viewpoint is not helpful. For example, the point transformation from Cartesian coordinates to spherical polar coordinates is a canonical transformation, but an "active" interpretation of it would border on the ludicrous.

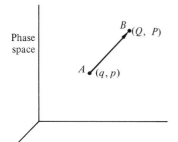

Phase space

B
$\bullet (Q, P)$

A
(q, p)

FIGURE 9–2
Illustration of the active view of a canonical transformation.

The active viewpoint is particularly useful for transformations depending continuously on a single parameter. On the active interpretation the effect of the transformation is to "move" the system point continuously on a curve in phase space as the parameter changes continuously. When the generator of the associated I.C.T. is the Hamiltonian, the curve on which the system point moves is the trajectory of the system in phase space.

If we consider the canonical transformation from the passive viewpoint, the phrase "change of a function u under canonical transformation" is meaningless. The function u doesn't change value whether it's evaluated at A or A'. It is true u will in general have a different functional dependence on (Q, P) than it has on (q, p) but its value will be the same at corresponding points; $u(A') = u(A)$. We can however give a meaning to the phrase if the canonical transformation is interpreted "actively." It can mean, for example, the change in the *value* of the function as the system point is moved from A to B. At point B in the same phase space u will have the same functional dependence on (Q, P) that it has on (q, p) at point A, but now the value of the function will in general be different. We shall use the symbol ∂ here to denote such a change in the value of a function under infinitesimal canonical transformation:

$$\partial u = u(B) - u(A), \qquad\qquad (9\text{–}102)$$

where, of course, A and B will be infinitesimally close. Using the matrix notation for the canonical variables the change in the function value under an I.C.T. would be defined as

$$\partial u = u(\boldsymbol{\eta} + \delta\boldsymbol{\eta}) - u(\boldsymbol{\eta}).$$

Expanding in a Taylor series and retaining terms in first order infinitesimals we have, by virtue of Eq. (9–63c),

$$\partial u = \frac{\partial u}{\partial \boldsymbol{\eta}}\, \delta \boldsymbol{\eta} = \epsilon \frac{\partial u}{\partial \boldsymbol{\eta}} \mathbf{J} \frac{\partial G}{\partial \boldsymbol{\eta}}.$$

Recalling the definition of the Poisson bracket, Eq. (9–68), we see the change can be written as

$$\partial u = \epsilon [u, G]. \tag{9–103}$$

An immediate application of Eq. (9–103) is to take for u one of the phase space coordinates themselves (or the matrix of the coordinates). We then have, by Eq. (9–100),

$$\partial \boldsymbol{\eta} = \epsilon [\boldsymbol{\eta}, G] = \delta \boldsymbol{\eta}.$$

Of course this result is obvious from the definition of the point B in relation to A; the "change" in the coordinates from A to B is just the infinitesimal difference between the old and new coordinate.

These considerations must be generalized somewhat in talking about the "change in the Hamiltonian." It must be remembered that the designation "Hamiltonian" does not mean a specific function, the same in all coordinate systems. Rather it refers to that function which in the given phase space defines the canonical equations of motion. Where the canonical transformation depends on the time the very meaning of "Hamiltonian" is also transformed. Thus $H(A)$ goes over not into $H(A')$ but into $K(A')$, and $H(A)$ will not necessarily have the same value as $K(A')$. In such case we shall mean by ∂H in effect the difference in the value of the Hamiltonian under the two interpretations:

$$\partial H = H(B) - K(A'). \tag{9–104}$$

Where the function itself does not change under the canonical transformation the two forms for the change, Eqs. (9–102) and (9–104), are identical since $u(A') = u(A)$. In general, K is related to H by the equation

$$K = H + \frac{\partial F}{\partial t},$$

where for an I.C.T. the generating function is given by Eq. (9–62) in terms of G. Since only G in that equation can be an explicit function of time, the value of the new Hamiltonian is given by

$$K(A') = H(A') + \epsilon \frac{\partial G}{\partial t} = H(A) + \epsilon \frac{\partial G}{\partial t},$$

and the change in the Hamiltonian is

$$\partial H = H(B) - H(A) - \epsilon \frac{\partial G}{\partial t}. \tag{9–105}$$

Following along the path that led from Eq. (9–103) we see that ∂H is given by

$$\partial H = \epsilon[H, G] - \epsilon\frac{\partial G}{\partial t}. \tag{9–106}$$

From the generalized equation of motion, Eq. (9–106), with G as u, it follows finally that the change in H is

$$\partial H = -\epsilon\frac{dG}{dt}. \tag{9–107}$$

If G is a constant of the motion, Eq. (9–107) says that it generates an infinitesimal canonical transformation that does not change the value of the Hamiltonian. Equivalently, *the constants of the motion are the generating functions of those infinitesimal canonical transformations that leave the Hamiltonian invariant.* Implied in this conclusion is a connection between the symmetry properties of the system and conserved quantities, a connection that is simplest to see for constants of the motion not explicitly depending on time. The change in the Hamiltonian under the transformation is then simply the change in the value of the Hamiltonian as the system is moved from configuration A to configuration B. If the system is symmetrical under the operation that produces this change of configuration, then the Hamiltonian will obviously remain unaffected under the corresponding transformation. Thus, to take a simple example, if the system is symmetrical about a given direction, then the Hamiltonian will not change in value if the system as a whole is rotated about that direction. It follows then that the quantity that generates (through an I.C.T.) such a rotation of the system must be conserved. The rotational symmetry of the system implies a particular constant of the motion. This is not the first instance of a connection between constants of the motion and symmetry characteristics. We encountered it previously (Sections 2–6, 8–2) in connection with the conservation of generalized momenta. Here, however, the theorem is more elegant, and more complete, for it embraces all independent constants of the motion and not merely the conserved generalized momenta.

The momentum conservation theorems appear now as a special case of the general statement. If a coordinate q_i is cyclic, the Hamiltonian is independent of q_i and will certainly be invariant under an infinitesimal transformation that involves a displacement of q_i alone. Consider, now, a transformation generated by the generalized momentum conjugate to q_i:

$$G(q, p) = p_i. \tag{9–108}$$

By Eqs. (9–63a and b) the resultant infinitesimal canonical transformation is

$$\delta q_j = \epsilon\delta_{ij},$$
$$\delta p_i = 0, \tag{9–109}$$

that is, exactly the required infinitesimal displacement of q_i and only q_i. We readily recognize this as the familiar momentum theorem: if a coordinate is cyclic its conjugate momentum is a constant of the motion. The observation that a displacement of one coordinate alone is generated by the conjugate momentum may be put in a slightly expanded form. If the generating function of an I.C.T. is given by

$$G = (\mathbf{J}\boldsymbol{\eta})_l = J_{lr}\eta_r, \tag{9-110}$$

then the equations of transformation is obtained from Eq. (9–63c) appear as

$$\delta\eta_k = \epsilon J_{ks}\frac{\partial G}{\partial \eta_s} = \epsilon J_{ks}J_{lr}\delta_{rs} = \epsilon J_{ks}J_{ls}.$$

By virtue of the orthogonality of **J** these reduce finally to

$$\delta\eta_k = \epsilon\delta_{kl}, \tag{9-111}$$

that is, a displacement of any canonical variable η_l alone is generated in terms of the conjugate variable in the form given by Eq. (9–110). Of course, if η_l is q_i, G from Eq. (9–110) is just p_i, and if η_l is p_i, G is then $-q_i$.

As a specific illustration of these concepts let us consider again the infinitesimal contact transformation of the dynamical variables that produces a rotation of the system as a whole by an angle $d\theta$. The physical significance of the corresponding generating function cannot depend upon the choice of initial canonical coordinates,* and it is convenient to use for this purpose the Cartesian coordinates of all particles in the system. Nor will there be any loss in generality if the axes are so oriented that the infinitesimal rotation is along the z axis. For an infinitesimal counter-clockwise rotation of each particle, the change in the position vectors is to be found from the infinitesimal rotation matrix of Eq. (4–105′) (p. 134). With a rotation only about the z axis the changes in the particle coordinates are

$$\delta x_i = -y_i\,d\theta, \qquad \delta y_i = x_i\,d\theta, \qquad \delta z_i = 0. \tag{9-112a}$$

The effect of the transformation on the components of the Cartesian vectors formed by the momenta conjugate to the particle coordinates is similarly given by

$$\delta p_{ix} = -p_{iy}\,d\theta, \qquad \delta p_{iy} = p_{ix}\,d\theta, \qquad \delta p_{iz} = 0. \tag{9-112b}$$

Comparing these transformation equations with Eqs. (9–63a and b) it is seen that the corresponding generating function is

$$G = x_i p_{iy} - y_i p_{ix}, \tag{9-113}$$

with $d\theta$ as the infinitesimal parameter ϵ. For a direct check note that

$$\delta x_i = d\theta\frac{\partial G}{\partial p_{ix}} = -y_i\,d\theta, \qquad \delta p_{ix} = -d\theta\frac{\partial G}{\partial x_i} = -p_{iy}\,d\theta,$$

$$\delta y_i = d\theta\frac{\partial G}{\partial p_{iy}} = x_i\,d\theta, \qquad \delta p_{iy} = -d\theta\frac{\partial G}{\partial y_i} = p_{ix}\,d\theta,$$

---

* This can most easily be seen from the canonically invariant Eq. 9–100. The change in the canonical variable η_i remains the same no matter in what set of canonical variables G is expressed.

agreeing with Eqs. (9–112). The generating function (9–113) in addition has the physical significance of being the z component of the total canonical angular momentum:

$$G = L_z \equiv (r_i \times p_i)_z. \tag{9-114}$$

Since the z axis was arbitrarily chosen, one can state that the generating function corresponding to an infinitesimal rotation about an axis denoted by the unit vector **n** is

$$G = \mathbf{L} \cdot \mathbf{n}. \tag{9-115}$$

It should be noted that the canonical angular momentum as defined here may differ from the mechanical angular momentum. If the forces on the system are derivable from velocity-dependent potentials, then the canonical momentum vectors p_i are not necessarily the same as the linear momentum vectors, and **L** in Eqs. (9–114) and (9–115) may not be the same as the mechanical angular momentum. The result obtained here is therefore a generalization of the conclusion given in Section 2–6 that the momentum conjugate to a rotation coordinate is the corresponding component of the total angular momentum. The proof presented there was restricted to systems with velocity-independent potentials. By virtue of Eqs. (9–108) and (9–109) we can now conclude that the momentum conjugate to a generalized coordinate that measures the rotation of the system as a whole about an axis **n** is the component of the total canonical angular momentum along the same axis. Just as the Hamiltonian is the generator of a displacement of the system in time so the angular momentum is the generator of the spatial rotations of the system.

It has already been noted that on the "active" interpretation a canonical transformation depending on a parameter "moves" the system point along a continuous trajectory in phase space. The finite transformation, we have said, could be looked on as the sum of an infinite succession of infinitesimal canonical transformations, each corresponding to an infinitesimal displacement along the curve. Formally, it should therefore be possible to obtain the finite transformation by integrating the expression for the infinitesimal displacements. We can do this by noting that each point on the trajectory in phase space corresponds to a particular value of the parameter, which we shall call α, starting from the initial system configuration denoted by $\alpha = 0$. If u is some function of the system configuration, then u will be a continuous function of α along the trajectory, $u(\alpha)$, with initial value $u_0 = u(0)$. (For simplicity we shall consider u as not depending explicitly on time.) Equation (9–103) for the infinitesimal change of u on the trajectory can be written as

$$\partial u = d\alpha [u, G],$$

or as a differential equation in the variable α:

$$\frac{du}{d\alpha} = [u, G]. \tag{9-116}$$

We can get $u(\alpha)$, and therefore the effect of the finite canonical transformation, by integrating this differential equation. A formal solution may be obtained by expanding $u(\alpha)$ in a Taylor series about the initial conditions:

$$u(\alpha) = u_0 + \alpha \frac{du}{d\alpha}\bigg|_0 + \frac{\alpha^2}{2!} \frac{d^2 u}{d\alpha^2}\bigg|_0 + \frac{\alpha^3}{3!} \frac{d^3 u}{d\alpha^3}\bigg|_0 + \cdots.$$

By Eq. (9–116) we have

$$\frac{du}{d\alpha}\bigg|_0 = [u, G]_0,$$

the zero subscript meaning that the value of the Poisson bracket is to be taken at the initial point, $\alpha = 0$. Repeated application of Eq. (9–116), taking $[u, G]$ itself as a function of the system configuration, gives

$$\frac{d^2 u}{d\alpha^2} = [[u, G], G],$$

and the process can be repeated to give the third derivative of u and so on. The Taylor series for $u(\alpha)$ thus leads to the formal series solution

$$u(\alpha) = u_0 + \alpha[u, G]_0 + \frac{\alpha^2}{2!}[[u, G], G]_0 + \frac{\alpha^3}{3!}[[[u, G], G], G]_0 + \cdots. \quad (9\text{--}117)$$

If for u we take any of the canonical variables ζ_i, with u_0 being then the starting set of variables η_i, then Eq. (9–115) is a prescription for finding the transformation equations of the finite canonical transformation generated by G.

It is not difficult to find specific examples showing that this procedure actually works. Suppose for G we take L_z, so that the final canonical transformation should correspond to a finite rotation about the z axis. The natural parameter to use for α is the rotation angle. For u let us take the x-coordinate of the ith particle in the system. Either by direct evaluation of the Poisson brackets or by inference from Eqs. (9–112a), it is easy to see that

$$[X_i, L_z] = -Y_i, \qquad [Y_i, L_z] = X_i, \qquad (9\text{--}118)$$

where capital letters have been used to denote the coordinates after some rotation θ, that is, the final coordinate. The initial coordinates, i.e., before rotation, are as usual represented by lower case letters. It follows then that

$$[X_i, L_z]_0 = -y_i,$$

$$[[X_i, L_z], L_z]_0 = -[Y_i, L_z]_0 = -x_i,$$

$$[[[X_i, L_z], L_z], L_z]_0 = -[X_i, L_z]_0 = y_i,$$

and so on. The series representation for X_i thus becomes

$$X_i = x_i - y_i\theta - x_i\frac{\theta^2}{2} + y_i\frac{\theta^3}{3!} + x_i\frac{\theta^4}{4!} - \cdots$$

$$= x_i\left(1 - \frac{\theta^2}{2!} + \frac{\theta^4}{4!} - \cdots\right) - y_i\left(\theta - \frac{\theta^3}{3!} + \cdots\right).$$

The two series will be recognized as the expansion for the cosine and sine respectively. Hence the equation for the finite transformation of X_i is

$$X_i = x_i \cos\theta - y_i \sin\theta,$$

which is exactly what we would expect for the finite rotation of a vector counterclockwise about the z axis.

For another example, consider the situation when $G = H$ and the parameter is the time. Equation (9–116) then reduces to the equation of motion for u:

$$\frac{du}{dt} = [u, H],$$

with the formal solution

$$u(t) = u_0 + t[u, H]_0 + \frac{t^2}{2!}[[u, H], H]_0 + \frac{t^3}{3!}[[[u, H], H], H]_0 + \cdots. \quad (9\text{–}119)$$

Here the subscript zero refers to the initial conditions at $t = 0$. Let us apply this prescription to the simple problem of one-dimensional motion with a constant acceleration a, for which the Hamiltonian is

$$H = \frac{p^2}{2m} - max,$$

with u as the position coordinate x. The Poisson brackets needed in Eq. (9–119) are easy to evaluate directly or from the fundamental brackets:

$$[x, H] = \frac{p}{m},$$

$$[[x, H], H] = \frac{1}{m}[p, H] = a.$$

Because this last Poisson bracket is a constant, all higher order brackets vanish identically and the series terminates, with the complete solution being given by

$$x = x_0 + \frac{p_0 t}{m} + \frac{at^2}{2}.$$

Remembering that $p_0/m = v_0$, this will be recognized as the familiar elementary solution to the problem.

It may be felt that what we have done here is a *tour de force*, a mere virtuoso performance. There is force to the objection. One would not propose the formal series solution, Eq. (9–119), as the preferred method for solving realistic problems in mechanics. It is surely one of the most recondite procedures one could conceive of for solving the easiest of freshman physics problems! Nonetheless, the technique provides insights into the structure of classical mechanics as based on canonical transformation theory. The series expansion shows directly that infinitesimal canonical transformations can generate finite canonical transformations, depending on a parameter, and thus lead to solutions to the equations of motion. Of particular interest for the relation between classical and quantum mechanics is the observation that the series in Eqs. (9–117) or (9–119) bear a family resemblance to the series for an exponential. The nest of Poisson brackets in the nth term can be considered as the nth repeated application (from the right!) of the operator $[\ ,G]$, or the nth power of the operator. Equation (9–119), for example, could symbolically be written as

$$u(t) = u e^{\hat{H}t}\Big|_0. \tag{9-120}$$

The exponential here means no more than its series representations and the symbol $\hat{H}$ is used to indicate the operator $[\ ,H]$. What we have here is very reminiscent of the Heisenberg picture in quantum mechanics* where the $u(t)$ become time-varying operators, whose time dependence is given in terms of $\exp[iHt/\hbar]$ in such a manner as to lead to the same equation of motion, Eq. (9–94). (The additional factor $i/\hbar$ arises out of the correspondence between the classical Poisson bracket and the quantum commutator.) The Poisson bracket formulation of mechanics is thus the classical analog of the Heisenberg picture of quantum mechanics.

9–6 THE ANGULAR MOMENTUM POISSON BRACKET RELATIONS

The identification of the canonical angular momentum as the generator of a rigid rotation of the system leads to a number of interesting and important Poisson bracket relations. Equation (9–103) for the change of a function u under an infinitesimal canonical transformation (on the "active" view) is also valid if u is taken as the component of a vector along a *fixed* axis in ordinary space. Thus, if $\mathbf{F}$ is a vector function of the system configuration, then

$$\partial F_i = [F_i, G].$$

It is important to note that the direction along which the component is taken must be fixed, i.e., not affected by the canonical transformation. If the direction itself is determined in terms of the system variables, then the transformation changes not only the value of the function but the nature of the function, just as with the Hamiltonian. With this understanding the change in a vector $\mathbf{F}$ under a

---

* See, for example, Schiff, *Quantum Mechanics*, 3d ed., 1968, pp. 170–171, or Merzbacher, *Quantum Mechanics*, 1961, pp. 343–349.

rotation of the system about a fixed axis $\mathbf{n}$, generated by $\mathbf{L} \cdot \mathbf{n}$, can be written in vector notation

$$\partial \mathbf{F} = d\theta[\mathbf{F}, \mathbf{L} \cdot \mathbf{n}]. \tag{9-121}$$

To put it in other words, Eq. (9–121) implies that the unit vectors $\mathbf{i}, \mathbf{j}, \mathbf{k}$ that form the basis set for $\mathbf{F}$ are not themselves rotated by $\mathbf{L} \cdot \mathbf{n}$.

The words describing what is meant by Eq. (9–121) must be chosen carefully for another reason. What is spoken of is the rotation of the system under the I.C.T., not necessarily the rotation of the vector $\mathbf{F}$. The generator $\mathbf{L} \cdot \mathbf{n}$ induces a spatial rotation of the system variables, not for example of some external vector such as a magnetic field or the vector of the acceleration of gravity. Under what conditions then does $\mathbf{L} \cdot \mathbf{n}$ generate a spatial rotation of $\mathbf{F}$? The answer is clear— when $\mathbf{F}$ is a function only of the system variables (q, p) and does not involve any external quantities or vectors not affected by the I.C.T. Only under these conditions does a spatial rotation imply a corresponding rotation of $\mathbf{F}$. We will designate such vectors as *system vectors*. The change in a vector under infinitesimal rotation about an axis $\mathbf{n}$ has been given several times before (cf. Eq. 2–50 and Eq. 4–111′, p. 173):

$$d\mathbf{F} = \mathbf{n} \, d\theta \times \mathbf{F}.$$

For a system vector $\mathbf{F}$, the change induced under an I.C.T. generated by $\mathbf{L} \cdot \mathbf{n}$ can therefore be written as

$$\partial \mathbf{F} = d\theta[\mathbf{F}, \mathbf{L} \cdot \mathbf{n}] = \mathbf{n} \, d\theta \times \mathbf{F}. \tag{9-122}$$

Equation (9–122) implies an important Poisson bracket identity obeyed by all system vectors:

$$[\mathbf{F}, \mathbf{L} \cdot \mathbf{n}] = \mathbf{n} \times \mathbf{F}. \tag{9-123}$$

Note that in Eq. (9–123) there is no longer any reference to a canonical transformation or even to a spatial rotation. It is simply a statement about the value of certain Poisson brackets for a specific class of vectors and, as such, can be verified by direct evaluation in any given case. Suppose, for example, we had a system of an unconstrained particle and used the Cartesian coordinates as the canonical space coordinates. Then the Cartesian vector $\mathbf{p}$ is certainly a suitable system vector. If $\mathbf{n}$ is taken as a unit vector in the z direction, then by direct evaluation we have

$$[p_x, xp_y - yp_x] = -p_y,$$
$$[p_y, xp_y - yp_x] = p_x,$$
$$[p_z, xp_y - yp_x] = 0.$$

The right-hand sides of these identities is clearly the same as the components of $\mathbf{k} \times \mathbf{p}$, as predicted by Eq. (9–123). On the other hand, suppose that in the same problem we tried to use for $\mathbf{F}$ the vector $\mathbf{A} = \frac{1}{2}(\mathbf{r} \times \mathbf{B})$ where $\mathbf{B} = B\mathbf{i}$ is a fixed vector along the x axis. The vector $\mathbf{A}$ will be recognized as the vector potential corresponding to a uniform magnetic field **B** in the x direction. As **A** depends on a vector external to the system we would expect it not to fit the characteristics of a system vector and Eq. (9–123) should not hold for it. Indeed, we see that the Poisson brackets involved are here

$$[0, xp_y - yp_x] = 0,$$

$$[\tfrac{1}{2}zB, xp_y - yp_x] = 0,$$

$$[-\tfrac{1}{2}yB, xp_y - yp_x] = -\tfrac{1}{2}Bx,$$

whereas the vector **k** × **A** has instead the components $(-\tfrac{1}{2}Bz, 0, 0)$.

The relation Eq. (9–123) may be expressed in various notations. Thus, in dyadic notation it appears as

$$[\mathbf{F}, L] = -\mathbf{1} \times \mathbf{F}, \tag{9–124}$$

where **1** is the unit dyadic **ii** + **jj** + **kk**. Equation (9–124) readily reduces to Eq. (9–123) by taking the dot product of both sides with **n** from the right. More useful is a form using the Levi–Civita density to express the cross product (cf. Eq. 4–113′, p. 172). The ith component of Eq. (9–123) for arbitrary **n** then can be written

$$[F_i, L_j n_j] = \epsilon_{ijk} n_j F_k,$$

which implies the simple result

$$[F_i, L_j] = \epsilon_{ijk} F_k. \tag{9–125}$$

An alternative statement of Eq. (9–125) is to note that if l, m, n are three indices in cyclic order, then

$$[F_l, L_m] = F_n, \qquad l, m, n \text{ in cyclic order.} \tag{9–125′}$$

Another consequence of Eq. (9–123) relates to the dot product of two system vectors: **F·G**. Being a scalar such a dot product should be invariant under rotation, and indeed the Poisson bracket of the dot product with **L·n** is easily shown to vanish:

$$[\mathbf{F \cdot G}, \mathbf{L \cdot n}] = \mathbf{F} \cdot [\mathbf{G}, \mathbf{L \cdot n}] + \mathbf{G} \cdot [\mathbf{F}, \mathbf{L \cdot n}]$$

$$= \mathbf{F \cdot n} \times \mathbf{G} + \mathbf{G \cdot n} \times \mathbf{F}$$

$$= \mathbf{F \cdot n} \times \mathbf{G} + \mathbf{F \cdot G} \times \mathbf{n}$$

$$= 0. \tag{126}$$

The magnitude of any system vector therefore has a vanishing Poisson bracket with any component of **L**.

Perhaps the most frequent application of these results arises from taking F to be the vector **L** itself. We then have

$$[\mathbf{L}, \mathbf{L \cdot n}] = \mathbf{n} \times \mathbf{L}, \tag{9–127}$$

$$[L_i, L_j] = \epsilon_{ijk} L_k, \tag{9–128}$$

and

$$[L^2, \mathbf{L} \cdot \mathbf{n}] = 0. \tag{9–129}$$

A number of interesting consequences follow from Eqs. (9–127) through (9–129). If L_x and L_y are constants of the motion, Poisson's theorem then states that $[L_x, L_y] = L_z$ is also a constant of the motion. Thus, if any two components of the angular momentum are constant, the total angular momentum vector is conserved. As a further instance, let us assume that in addition to L_x and L_y being conserved there is a Cartesian vector of canonical momentum $\mathbf{p}$ with p_z a constant of the motion. Not only then is L_z conserved but we have two further constants of the motion:

$$[p_z, L_x] = p_y$$

and

$$[p_z, L_y] = -p_x,$$

that is, both $\mathbf{L}$ and $\mathbf{p}$ are conserved. We have here an instance in which Poisson's theorem does yield new constants of the motion. Note, however, that if p_x, p_y, and L_z were the given constants of the motion, then their Poisson brackets are

$$[p_x, p_y] = 0,$$

$$[p_x, L_z] = p_y,$$

$$[p_y, L_z] = p_x.$$

Here no new constants can be obtained from Poisson's theorem.

It will be remembered from the fundamental Poisson brackets, Eqs. (9–69), that the Poisson bracket of any two canonical momenta must always be zero. But, from Eq. (9–128), L_i does not have a vanishing Poisson bracket with any of the other components of $\mathbf{L}$. Thus, while we have described $\mathbf{L}$ as the total canonical angular momentum by virtue of its definition as $\mathbf{r}_i \times \mathbf{p}_i$ (summed over all particles), no two components of $\mathbf{L}$ can simultaneously be canonical variables. However, Eq. (9–129) shows that any one of the components of $\mathbf{L}$, and its magnitude L, can be chosen to be canonical variables at the same time.*

---

* It has been remarked previously that the correspondence between quantum and classical mechanics is such that the quantum mechanical commutator goes over essentially into the classical Poisson bracket as $h \to 0$. Much of the formal structure of quantum mechanics appears as a close transcript of the Poisson bracket formulation of classical mechanics. All the results of this section therefore have close quantum analogs. For example, the fact that two components of $\mathbf{L}$ cannot be simultaneous canonical momenta appears as the well-known statement that L_i and L_j cannot have simultaneous eigenvalues. But L^2 and any L_i can be quantized together. Indeed, most of these relations are known far better in their quantum form than as classical theorems. Thus one of the earliest references to the classical Poisson brackets for angular momentum appears to be the 1930 treatise by Born and Jordan on *Elementare Quantenmechanik*. Again, while the general change of a vector function under rotation, Eq. (9–123), has long been familiar in quantum mechanics (cf. Condon and Shortley, *The Theory of Atomic Spectra*, p. 59), until very recently about the only reference to its classical version was in the famous thesis of H. B. G. Casimir, *Rotation of a Rigid Body in Quantum Mechanics*, 1931, p. 30.

9–7 SYMMETRY GROUPS OF MECHANICAL SYSTEMS*

It has already been pointed out that canonical transformations form a group. Canonical transformations that are analytic functions of continuous parameters form separate groups belonging to the class known as Lie groups. Thus the canonical transformations corresponding to spatial rotations of the system form a group with three parameters, e.g., the Euler angles of rotation. Rotations about a particular axis form a subgroup (actually the rotation group in two dimensions) with only one parameter. The group of finite transformations has the same properties as the group of the associated infinitesimal canonical transformations, and it is customary to work primarily with the I.C.T.'s as they are easier to handle. The Lie groups of I.C.T.'s whose generators are the constants of the motion of the system are known as the *symmetry groups* of the system for, as we have seen, such transformations leave the Hamiltonian invariant. Finding the symmetry groups of a system goes a long way to solving the problem of its classical motion and is even closer to a solution of the quantum-mechanical problem.

A system with spherical symmetry is invariant under rotation about any axis. One would expect therefore that one of its symmetry groups should be the rotation group in three dimensions $R(3) \equiv SO(3)$. The vector $\mathbf{L}$ is conserved in such a system in accord with our identification of the components of $\mathbf{L}$ as the generators of spatial rotations. The character of a Lie group is determined by the bracket (whether Poisson, commutators, etc.) relationships among the generators of the transformation making up the group. In general, if G_i are the generators of the individual one-parameter transformations, then in a Lie group the bracket relations are linear, of the form†

$$[G_i, G_j] = C_{ijk}G_k, \tag{9–130}$$

where the quantities C_{ijk} are called the *structure constants* of the Lie group (or of the associated Lie algebra involving the bracket "product"). For the group of transformations generated by L_i, Eq. (9–128) shows that the structure constants are $C_{ijk} = \epsilon_{ijk}$ and it is this relationship that stamps the group as being the rotation group in three dimensions. Thus, the matrix generators $\mathbf{M}_i$ of infinitesimal rotations, Eqs. (4–117), have been seen to obey the commutation relations, Eq. (4–118),

$$[\mathbf{M}_i, \mathbf{M}_j] = \epsilon_{ijk}\mathbf{M}_k, \tag{4–118}$$

that is, with the same structure constants as for L_i. The quantities L_i and $\mathbf{M}_i$ are different physically; the brackets in Eqs. (9–125) and (4–118) refer to different operations (although they share the same significant algebraic properties). But the identity of the structure constants for L_i and $\mathbf{M}_i$ show that they have the same group structure, that of $SO(3)$.

---

* This section may be omitted on first reading.

† See, for example, L. P. Eisenhart, *Continuous Groups of Transformations*, pp. 25–28.

For the bound Kepler problem we have seen (Section 3–9) that there exists in addition to **L** another conserved vector quantity, **A**, the Laplace vector defined as

$$\mathbf{A} = \mathbf{p} \times \mathbf{L} - \frac{mk\mathbf{r}}{r}. \tag{3-82}$$

The Poisson bracket relations of the components of **A** with themselves and with the components of **L** can be obtained in a straightforward manner. Since **A** clearly qualifies as a system vector, we immediately have the bracket relations

$$[A_i, L_j] = \epsilon_{ijk} A_k. \tag{9-131}$$

The Poisson brackets of the components of **A** among themselves cannot be obtained by any such simple stratagem, but after a fair amount of tedious manipulation it is found that*

$$[A_1, A_2] = -\left(p^2 - \frac{2mk}{r}\right) L_3. \tag{9-132}$$

The quantity on the right in the parentheses will be recognized as $2mH$, which has the conserved value $2mE$. If we therefore introduce a new constant vector **D** defined as

$$\mathbf{D} = \frac{\mathbf{A}}{\sqrt{-2mE}} \equiv \frac{\mathbf{A}}{\sqrt{2m|E|}} \tag{9-133}$$

(note that E is negative for bound motion!), then the components of **D** satisfy the Poisson bracket relation

$$[D_1, D_2] = L_3.$$

By cyclically permuting the indices the complete set of Poisson brackets follows immediately. Thus, the components of **L** and **D** together form a symmetry group for the bound Kepler problem, with structure constants to be obtained from the identities.

$$[L_i, L_j] = \epsilon_{ijk} L_k, \tag{9-128}$$

$$[D_i, L_j] = \epsilon_{ijk} D_k, \tag{9-134}$$

and

$$[D_i, D_j] = \epsilon_{ijk} L_k. \tag{9-135}$$

An examination of the fundamental matrices for rotation will show that the symmetry group for the bound Kepler problem is to be identified with the group of four-dimensional real rotations. Usually designated as SO(4) or R(4), such a transformation preserves the value of the quadratic form $x_\mu x_\mu$, where all the x_μ

---

* Some reduction in the length of the derivation is obtained by identifying $\mathbf{p} \times \mathbf{L}$ as a system vector **C**, and first evaluating the Poisson brackets $[C_1, (\mathbf{p} \times \mathbf{L})_2]$ and $[C_1, \mathbf{r}/r]$ making use of the fundamental Poisson brackets and Eqs. (9–125) to the utmost.

are real. An orthogonal transformation in four dimensions has ten conditions on the sixteen elements of the matrix, so only six are independent. By looking on the infinitesimal transformation as being made up of a sequence of rotations in the various planes, one can easily obtain the corresponding six generators. Three of them are rotations in the three distinct x_i-x_j planes and so correspond to the $\mathbf{M}_i$ generators of Eqs. (4–117), except that there are added zeros in the fourth row and column. The remaining three generate infinitesimal rotations in the x_i-x_4 planes. Thus the generator matrix for an infinitesimal rotation in the x_1-x_4 plane would be

$$
\mathbf{N}_1 = \begin{pmatrix} 0 & 0 & 0 & -1 \\ 0 & 0 & 0 & 0 \\ 0 & 0 & 0 & 0 \\ 1 & 0 & 0 & 0 \end{pmatrix},
\tag{9–136}
$$

with $\mathbf{N}_2$ and $\mathbf{N}_3$ given in corresponding fashion. Direct matrix multiplication shows that these six matrices satisfy the commutator (or Lie bracket) relations

$$
[\mathbf{M}_i, \mathbf{M}_j] = \epsilon_{ijk}\mathbf{M}_k,
$$
$$
[\mathbf{N}_i, \mathbf{M}_j] = \epsilon_{ijk}\mathbf{N}_k,
$$
$$
[\mathbf{N}_i, \mathbf{N}_j] = \epsilon_{ijk}\mathbf{M}_k.
$$

Since these are the same as the Poisson bracket relations, Eqs. (9–128), (9–134), and (9–135), the identification of the symmetry group of the bound Kepler problem with $R(4)$ is thus proven.

Note that for the Kepler problem with positive energy (i.e., scattering) $\mathbf{A}$ is still a constant of the motion,* but the appropriate reduced real vector, instead of $\mathbf{D}$, is $\mathbf{C}$ defined as

$$
\mathbf{C} = \frac{A}{\sqrt{2mE}},
\tag{9–137}
$$

and the Poisson bracket relations for $\mathbf{L}$ and $\mathbf{C}$ are now

$$
[L_i, L_j] = \epsilon_{ijk}L_k,
$$
$$
[C_i, L_j] = \epsilon_{ijk}C_k,
\tag{9–138}
$$
$$
[C_i, C_j] = -\epsilon_{ijk}L_k.
$$

Comparison with Eqs. (7–70) shows that these structure constants are the same as for the restricted Lorentz group, which must therefore be the symmetry group for the positive energy Kepler problem—in nonrelativistic mechanics. One must, of course, be careful not to read any kinship of physical ideas into this

---

* The arguments of Section 3–9 are independent of the sign of either E or the force constant k.

happenstance. The Kepler problem does *not* contain in it the seed of the basic conceptions of special relativity; it is purely a problem of nonrelativistic Newtonian mechanics. That the symmetry group may involve a space of higher dimension than ordinary space is connected with the fact that the symmetry we seek here is one in the six-dimensional phase space. The symmetry group consists of the canonical transformations in this space that leave the Hamiltonian unchanged. It should not be surprising, therefore, that the group can be interpreted in terms of transformations of spaces of more than three dimensions.

The two-dimensional isotropic harmonic oscillator is another mechanical system for which a symmetry group is easily identified. In Cartesian coordinates, the Hamiltonian for this system may be written as

$$H = \frac{1}{2m}(p_x^2 + m^2\omega^2 x^2) + \frac{1}{2m}(p_y^2 + m^2\omega^2 y^2). \tag{9-139}$$

As it doesn't depend on time explicitly the Hamiltonian is constant and is equal to the total energy of the system. The z axis is an axis of symmetry for the system, and hence the angular momentum along that axis (which is in fact the total angular momentum) is also a constant of motion:

$$L = xp_y - yp_x. \tag{9-140}$$

Further constants of the motion exist for this problem that can be written as components of a symmetrical two-dimensional tensor **A** defined as

$$A_{ij} = \frac{1}{2m}(p_i p_j + m^2\omega^2 x_i x_j). \tag{9-141}$$

Of the three distinct elements of the tensor the diagonal terms may be identified as the energies associated with the separate one-dimensional motions along the x and y axes, respectively. Physically, as there is no coupling between the two motions, the two energies must separately be constant. A little more formally, it is obvious from the way in which H has been written in Eq. (9–139) that A_{11} and A_{22} each have a vanishing Poisson bracket with H. The off-diagonal element of **A**,

$$A_{12} = A_{21} = \frac{1}{2m}(p_x p_y + m^2\omega^2 xy), \tag{9-142}$$

is a little more difficult to recognize. That it is a constant of motion may easily be seen by evaluating the Poisson bracket with H. In relation to the separate x and y motions, A_{11} and A_{22} are related to the amplitudes of the oscillations, whereas A_{12} is determined by the phase difference between the two vibrations. Thus, the solutions for the motion can be written as

$$x = \sqrt{\frac{2A_{11}}{m\omega^2}}\sin(\omega t + \theta_1),$$

$$y = \sqrt{\frac{2A_{22}}{m\omega^2}}\sin(\omega t + \theta_2),$$

and it then follows from Eq. (9–142) that

$$A_{12} = \sqrt{A_{11}A_{22}} \cos(\theta_2 - \theta_1). \tag{9–143}$$

The trace of the **A** tensor is the total energy of the harmonic oscillator. Out of the elements of the matrix we can form two other distinct constants of the motion, which it is convenient to write in the form

$$S_1 = \frac{A_{12} + A_{21}}{2\omega} = \frac{1}{2m\omega}(p_x p_y + m^2\omega^2 xy), \tag{9–144}$$

$$S_2 = \frac{A_{22} - A_{11}}{2\omega} = \frac{1}{4m\omega}(p_y^2 - p_x^2 + m^2\omega^2(y^2 - x^2)). \tag{9–145}$$

To these we may add a third constant of the motion:

$$S_3 = \frac{L}{2} = \tfrac{1}{2}(xp_y - yp_x). \tag{9–146}$$

The quantities S_i plus the total energy H form four algebraic constants of the motion not involving time explicitly. It is clear not all of them can be independent, because in a system of two degrees of freedom there can at most be only three such constants. We know that the orbit for the isotropic harmonic oscillator is an ellipse and three constants of the motion are needed to describe the parameters of the orbit in the plane—say, the semimajor axis, the eccentricity, and the orientation of the ellipse. The fourth constant of motion must relate to the passage of the particle through a specific point at a given time and would therefore be explicitly time-dependent. Hence there must exist a single relation connecting S_i and H. By direct evaluation it is easy to show that*

$$S_1^2 + S_2^2 + S_3^2 = \frac{H^2}{4\omega^2}. \tag{9–147}$$

By straight forward manipulation of the Poisson brackets one can verify that the S_i quantities satisfy the relations

$$[S_i, S_j] = \epsilon_{ijk} S_k. \tag{9–148}$$

These are the same relations as for the three-dimensional angular momentum vector, or for the generators of rotation in a three-dimensional space. The group of transformations generated by S_i may therefore be identified with R(3) or SO(3). Actually, there is some ambiguity in the identification. We have seen above (cf. p.

---

* An equivalent form of the condition Eq. (9–147) is that the determinant of **A** is $L^2\omega^2/4$. It will be recalled that similarly in the case of the Kepler problem, the components of the new vector constant of motion **A** were not all independent of the other constants of the motion. There exist indeed two relations linking **A**, **L**, and H, Eqs. (3–83) and (3–87).

151) that there is a homomorphism between R(3) and the unitary unimodular group SU(2). It turns out* that SU(2) is here more appropriate. To glimpse at the circumstances justifying this choice, it may be noted that Eq. (9–147) suggests there is a three-dimensional space each point of which corresponds to a particular set of orbital parameters. For a given system energy, Eq. (9–147) says the orbit "points" in this space lie on a sphere. The constants S_i generate three-dimensional rotations on this sphere, i.e., they change one orbit into another orbit having the same energy. It may be shown that S_1 generates a transformation that changes the eccentricity of the orbit and that for any given final eccentricity one can find *two* transformations leading to it. It is this double-valued quality of the transformation that indicates SU(2) rather than R(3) is the correct symmetry group for the two-dimensional harmonic oscillator. For higher dimensions the structure constants of the rotation groups and the SU(n) groups are no longer identical and a clear cut separation between the two can be made. For the three-dimensional isotropic harmonic oscillator there is again a tensor constant of the motion defined by Eq. (9–141), except that the indices now run from 1 to 3. The distinct components of this tensor, together with the components of **L** now satisfy Poisson bracket relations with the rather complicated structure constants that belong to SU(3). Indeed it is possible to show† that for the n-dimensional isotropic harmonic oscillator the symmetry group is SU(n).

It has previously been pointed out (cf. p. 104) that there exists a connection between the existence of additional algebraic constants of the motion—and therefore of higher symmetry groups—and degeneracy in the motions of the system. In the case of the Kepler and isotropic harmonic oscillator the additional constants of the motion are related to parameters of the orbit. Unless the orbit is closed, i.e., the motion is confined to a single curve, one can hardly talk of such orbital parameters. Only when the various components of the motion have commensurate periods, i.e. are degenerate, will the orbit be closed. The classic example is the two-dimensional anisotropic oscillator. When the frequencies in the x and y directions are integral multiples of each other, the particle traverses a closed Lissajous figure. But if the frequencies are incommensurate the motion of the particle is space-filling or ergodic, eventually coming as close as desired to any specific point in the rectangle defined by the energies of motion in the two directions. The exact nature of this connection between symmetry properties and degeneracy is still a subject for investigation. Attempts at finding complicated (and perhaps complex) symmetry groups for nondegenerate systems, applicable to all problems of the same number of degrees of freedom, have not yet proved fruitful. We shall have occasion in the next chapter to consider further the relation between symmetry and degeneracy.

---

*See H. V. McIntosh, Am. Jour. Phys. **27**, 620 (1959) for a fuller discussion.

†This was proved for the quantum-mechanical case by G. A. Baker, Jr., *Phys. Rev.* **103**, 1119 (1956), by arguments that can also be phrased in the language of classical mechanics.

9-8 LIOUVILLE'S THEOREM

As a final application of the Poisson bracket formalism we shall briefly discuss a fundamental theorem of statistical mechanics known as Liouville's theorem. While the exact motion of any system is completely determined in classical mechanics by the initial conditions, it is often impracticable to calculate an exact solution for complex systems. It would be obviously hopeless, for example, to calculate completely the motion of some 10^{23} molecules in a volume of gas. In addition, the initial conditions are often only incompletely known. We may be able to state that at t_0 a given mass of gas has a certain energy, but we cannot determine the initial coordinates and velocities of each molecule. Statistical mechanics therefore makes no attempt to obtain a complete solution for systems containing many particles. Its aim, instead, is to make predictions about certain average properties by examining the motion of a large number of identical systems. The values of the desired quantities are then computed by forming averages over all the systems in this *ensemble*. All the members of the ensemble are as like the actual system as our imperfect knowledge permits, but they may have any of the initial conditions that are consistent with this incomplete information. Since each system is represented by a single point in phase space the ensemble of systems corresponds to a swarm of points in phase space. Liouville's theorem states that the density of systems in the neighborhood of some given system in phase space remains constant in time.

The density, D, as defined above can vary with time through two separate mechanisms. Since it is the density in the neighborhood of a given system point, there will be an *implicit* dependence as the coordinates of the system (q_i, p_i) vary with time, and the system point wanders through phase space. There may also be an explicit dependence on time. The density may still vary with time even when evaluated at a fixed point in phase space. By Eq. (9–94) the total time derivative of D, due to both types of variation with time, can be written as

$$\frac{dD}{dt} = [D, H] + \frac{\partial D}{\partial t}, \qquad (9\text{–}149)$$

where the Poisson bracket arises from the implicit dependence, and the last term from the explicit dependence.

The ensemble of system points moving through phase space behaves much like a fluid in a multidimensional space, and there are numerous similarities between our discussion of the ensemble and the well-known notions of fluid dynamics. In Eq. (9–149) the total derivative is a derivative of the density as we follow the motion of a particular bit of the ensemble "fluid" in time. It is sometimes referred to as the *material* or *hydrodynamic* derivative. On the other hand, the partial derivative is at fixed (q, p); it is as if we station ourselves at a particular spot in phase space and measure the time variation of the density as the ensemble of system points flows by us. These two derivatives correspond to two viewpoints frequently used in considering fluid flow. The partial derivative at a fixed point in phase space is in line with the Eulerian viewpoint that looks on the coordinates solely as identifying a point in space. The total derivative fits in with the Lagrangian picture in which individual particles are followed in time; the coordinates in effect rather identify a particle than a point in space. Basically, our consideration of phase space has been more like the Lagrangian viewpoint; the collection of quanitites (q, p) identifies a system and its changing configuration with time.

Consider an infinitesimal volume in phase space surrounding a given system point, with the boundary of the volume formed by some surface of neighboring system points at the time $t = 0$. In the course of time the system points defining the volume move about in phase space and the volume contained by them will take on different shapes as time progresses. The dotted curve in Fig. 9–3 schematically indicates the evolution of the infinitesimal volume with time. It is clear that the number of systems within the volume remains constant, for a system initially inside can never get out. If some system point were to cross the border, it would occupy at some time the same position in phase space as one of the system points defining the boundary surface. Since the subsequent motion of a system is uniquely determined by its location in phase space at a particular time, the two systems would travel together from there on. Hence the system can never leave the volume. By the same token, a system initially outside can never enter the volume.

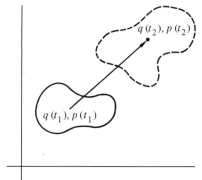

FIGURE 9–3
Motion of a volume in phase space.

It has been shown that on the active picture of a canonical transformation the motion of a system point in time is simply the evolution of a canonical transformation generated by the Hamiltonian. The canonical variables (q, p) at time t_2, as shown in Fig. 9–3, are related to the variables at time t_1 by a particular canonical transformation. The change in the infinitesimal volume element about the system point over the time interval is given by the same canonical transformation. Now, Poincaré's integral invariant, Eq. (9–87), says that a volume element in phase space is invariant under a canonical transformation. Therefore the size of the volume element about the system point cannot vary with time.

Thus, both the number of systems in the infinitesimal region, dN, and the volume, dV, are constants, and consequently the density

$$D = \frac{dN}{dV}$$

must also be constant in time, that is,

$$\frac{dD}{dt} = 0,$$

which proves Liouville's theorem. An alternative statement of the theorem follows from Eq. (9–149) as

$$\frac{\partial D}{\partial t} = -[D, H]. \tag{9–150}$$

When the ensemble of systems is in statistical equilibrium, the number of systems in a given state must be constant in time, which is to say that the density of system points at a given spot in phase space does not change with time. The variation of D with time at a fixed point corresponds to the partial derivative with respect to t, which therefore must vanish in statistical equilibrium. By Eq. (9–150) it follows that the equilibrium condition can be expressed as

$$[D, H] = 0.$$

We can ensure equilibrium, therefore, by choosing the density D to be a function of those constants of the motion of the system not involving time explicitly, for then the Poisson bracket with H must vanish. Thus, for conservative systems D can be any function of the energy, and the equilibrium condition is automatically satisfied. The characteristics of the ensemble will be determined by the choice of function for D. As an example, one well-known ensemble, the *microcanonical* ensemble, occurs if D is constant for systems having a given energy, zero otherwise.

These considerations have been presented here to illustrate the usefulness of the Poisson bracket formulation in classical statistical mechanics. Further discussion of these points would carry us far outside our field.

SUGGESTED REFERENCES

C. CARATHÉODORY, *Calculus of Variations and Partial Differential Equations.* Canonical or contact transformations were first introduced by mathematicians in the theory of partial differential equations. Many of the properties of the transformations that physicists use were first developed for these purposes. The most thorough exposition of these mathematical origins from which the physics applications came is in Carathéodory's masterful treatise. Fortunately it is now available in English translation. The pertinent references are in Chapters 4–7 of Vol. 1 on partial differential equations. They contain, among many other topics, a fuller proof of what is referred to above (p. 403) as

"Carathéodory's Theorem." One keeps coming back to Carathéodory time after time, to see how "it's done properly."

H. RUND, *Hamilton–Jacobi Theory in the Calculus of Variations*. This reference provides a somewhat different mathematical picture (in Chapter 2, Sections 12–14) with the physical applications kept clearly in mind. For much of the discussion, time is considered one of the canonical variables.

M. BORN, *The Mechanics of the Atom*. The subject of the canonical transformations of classical mechanics played an important role in the first formulations of both the older Bohr quantum theory and the newer quantum mechanics. Hence many treatises ostensibly devoted to one or the other of these forms of quantum mechanics often contain detailed expositions of the needed branches of classical mechanics. Outstanding among them is this 1924 volume of Born, written before the days of wave mechanics. The first chapter succinctly discusses canonical transformations and gives many interesting physical illustrations. There is no mention of Poisson brackets for they became of special interest to the modern physicist only with the advent of Heisenberg's and Dirac's formulation of quantum mechanics.

R. C. TOLMAN, *The Principles of Statistical Mechanics*. A veritable encyclopedia of theoretical physics, Chapter II of this bulky volume gives a brief but clear discussion of canonical transformations and similar topics in classical mechanics. The properties of Poisson brackets are included in the treatment. Section 19, Chapter III, is concerned with Liouville's theorem.

C. LANCZOS, *The Variational Principles of Mechanics*. Canonical transformations entered into mechanics first through perturbation theory in classical mechanics, indeed long before it was realized quite what they were. (References on these applications will be given below in Chapter 11 on perturbation theory.) By now any text on mechanics above the intermediate level devotes considerable attention to the subjects of Poisson bracket formulations and canonical transformations. Only a few references can be cited explicitly. Lanczos has a different viewpoint from others; he talks a good deal and writes relatively few equations. The subject is tied into its mathematical origins, but the use is physical.

L. PARS, *A Treatise on Analytical Dynamics*. Canonical transformations are made use of in this text a number of times before they are explicitly named and discussed (as contact transformations) in Chapter 24. Both the symplectic and the generator approaches are included. Particularly noteworthy is a proof in effect of Carathéodory's theorem that explicitly leads to the generating function.

J. L. SYNGE, *Classical Dynamics*. For the most part Synge works in a phase space that includes t and its conjugate momentum H, but otherwise the approach has considerable similarities with that adopted here. The symplectic method is given in matrix form, although the notation is quite different. A method is given for going from the symplectic to the generating function approach but the path is not clear or always convincing.

C. W. KILMISTER, *Hamiltonian Dynamics*. For those equipped with a modern mathematical background in tensor analysis and differential geometry, this little book presents an elegant and compact discussion with heavy emphasis on the symplectic approach. The author presents a "universal generator" in η and ζ from whence the four types may be derived.

E. J. SALETAN AND A. H. CROMER, *Theoretical Mechanics*. Considerable space is given to canonical transformations mostly from the symplectic viewpoint, but including the transition to the generating function approach. Canonical is used in the sense of what is here called extended canonical transformation. In addition there is added the highly unorthodox, if not downright dangerous, notion of a *canonoid* transformation—one that is canonical only for certain types of Hamiltonians. (Most applications of canonical transformations depend on the property that they be canonical for *all* Hamiltonians.) Canonical transformations depending on a continuous parameter are discussed explicitly.

E. C. G. SUDARSHAN AND N. MUKUNDA, *Classical Dynamics*. This is a treatment of mechanics permeated with a group-theoretical approach. It might be termed classical mechanics as viewed by theoretical-particle physicists. Much of the book is concerned indeed with canonical transformations and the implications of the symmetries of the system and the transformations.

H. V. McINTOSH, *Symmetry and Degeneracy*, in *Group Theory and its Applications*, vol. II, E. M. Loebl, ed. Reference is made again to this enthusiastic survey of the connections between symmetries of the system and the constants of the motion. While canonical transformations per se are rarely mentioned, the notion that the generators of symmetry operations provide the constants of motion appears quite frequently, and many examples are provided.

EXERCISES

1. One of the attempts at combining the two sets of Hamilton's equations into one tries to take q and p as forming a complex quantity. Show directly from Hamilton's equations of motion that for a system of one degree of freedom the transformation

$$Q = q + ip, \qquad P = Q^*$$

is not canonical if the Hamiltonian is left unaltered. Can you find another set of coordinates Q', P' that are related to Q, P by a change of scale only, and that are canonical?

2. a) For a one-dimensional system with the Hamiltonian

$$H = \frac{p^2}{2} - \frac{1}{2q^2},$$

show that there is a constant of the motion

$$D = \frac{pq}{2} - Ht.$$

b) As a generalization of part (a), for motion in a plane with the Hamiltonian

$$H = |\mathbf{p}|^n - ar^{-n},$$

where $\mathbf{p}$ is the vector of the momenta conjugate to the Cartesian coordinates, show that there is a constant of the motion

$$D = \frac{\mathbf{p} \cdot \mathbf{r}}{n} - Ht.$$

c) The transformation $Q = \lambda q, p = \lambda P$ is obviously canonical. However, the same transformation with t time dilatation, $Q = \lambda q, p = \lambda P, t' = \lambda^2 t$, is not. Show that, however, the equations of motion for q and p for the Hamiltonian in part (a) are invariant under the transformation. The constant of the motion D is said to be associated with this invariance.

3. In Section 8–4 some of the problems of treating time as one of the canonical variables are discussed. If we are able to sidestep these difficulties, show that the equations of transformation in which t is considered a canonical variable reduce to Eqs. (9–14) if in fact the transformation does not affect the time scale.

4. Show directly that the transformation

$$Q = \log\left(\frac{1}{q}\sin p\right), \qquad P = q\cot p$$

is canonical.

5. Show directly that for a system of one degree of freedom the transformation

$$Q = \arctan\frac{\alpha q}{p}, \qquad P = \frac{\alpha q^2}{2}\left(1 + \frac{p^2}{\alpha^2 q^2}\right)$$

is canonical, where α is an arbitrary constant of suitable dimensions.

6. The transformation equations between two sets of coordinates are

$$Q = \log(1 + q^{\frac{1}{2}}\cos p),$$

$$P = 2(1 + q^{\frac{1}{2}}\cos p)q^{\frac{1}{2}}\sin p.$$

a) Show directly from these transformation equations that Q, P are canonical variables if q and p are.

b) Show that the function that generates this transformation is

$$F_3 = -(e^Q - 1)^2 \tan p.$$

7. a) If each of the four types of generating functions exist for a given canonical transformation, use the Legendre transformation to derive relations between them.

b) Find a generating function of the F_4 type for the identity transformation and of the F_3 type for the exchange transformation.

c) For an orthogonal point transformation of q in a system of n degrees of freedom, show that the new momenta are likewise given by the orthogonal transformation of an n-dimensional vector whose components are the old momenta plus a gradient in configuration space.

8. Prove directly that the transformation

$$Q_1 = q_1, \qquad P_1 = p_1 - 2p_2,$$

$$Q_2 = p_2, \qquad P_2 = -2q_1 - q_2$$

is canonical and find a generating function.

9. For the point transformation in a system of two degrees of freedom,

$$Q_1 = q_1^2, \qquad Q_2 = q_1 + q_2,$$

find the most general transformation equations for P_1 and P_2 consistent with the overall transformation being canonical. Show that with a particular choice for P_1 and P_2 the Hamiltonian

$$H = \left(\frac{p_1 - p_2}{2q_1}\right)^2 + p_2 + (q_1 + q_2)^2$$

can be transformed to one in which both Q_1 and Q_2 are ignorable. By this means solve the problem and obtain expressions for q_1, q_2, p_1 and p_2 as functions of time and their initial values.

10. Show that the transformation for a system of one degree of freedom,

$$Q = q \cos \alpha - p \sin \alpha,$$

$$P = q \sin \alpha + p \cos \alpha,$$

satisfies the symplectic condition for any value of the parameter α. Find a generating function for the transformation. What is the physical significance of the transformation for $\alpha = 0$? For $\alpha = \pi/2$? Does your generating function work for both of these cases?

11. Determine whether the transformation

$$Q_1 = q_1 q_2, \qquad P_1 = \frac{p_1 - p_2}{q_2 - q_1} + 1,$$

$$Q_2 = q_1 + q_2, \qquad P_2 = \frac{q_2 p_2 - q_1 p_1}{q_2 - q_1} - (q_2 + q_1)$$

is canonical.

12. Show that the direct conditions for a canonical condition are given immediately by the symplectic condition expressed in the form

$$\mathbf{JM} = \mathbf{M}^{-1}\mathbf{J}.$$

13. The set of restricted canonical transformations has the group-property as defined by the conditions as set forth in Exercise 9 of Chapter 7. Verify this statement once using the invariance of Hamilton's principle under canonical transformation, c.f. Eq. (9–11), and again using the symplectic condition.

14. By any method you choose show that the following transformation is canonical:

$$x = \frac{1}{\alpha}(\sqrt{2P_1} \sin Q_1 + P_2), \qquad p_x = \frac{\alpha}{2}(\sqrt{2P_1} \cos Q_1 - Q_2),$$

$$y = \frac{1}{\alpha}(\sqrt{2P_1} \cos Q_1 + Q_2), \qquad p_y = -\frac{\alpha}{2}(\sqrt{2P_1} \sin Q_1 - P_2).$$

where α is some fixed parameter.

Apply this transformation to the problem of a particle of charge q moving in a plane that is perpendicular to a constant magnetic field $\mathbf{B}$. Express the Hamiltonian for this problem in the (Q_i, P_i) coordinates letting the parameter α take the form

$$\alpha^2 = \frac{qB}{c}.$$

From this Hamiltonian obtain the motion of the particle as a function of time.

15. Find under what conditions

$$Q = \frac{\alpha p}{x}, \qquad P = \beta x^2,$$

where α and β are constants, represents a canonical transformation for a system of one degree of freedom, and obtain a suitable generating function. Apply the transformation to the solution of the linear harmonic oscillator.

16. Prove that the transformation

$$Q_1 = q_1^2, \qquad\qquad Q_2 = q_2 \sec p_2,$$

$$P_1 = \frac{p_1 \cos p_2 - 2q_2}{2q_1 \cos p_2}, \qquad P_2 = \sin p_2 - 2q_1$$

is canonical, by any method you choose. Find a suitable generating function that will lead to this transformation.

17. a) Show that the transformation

$$Q = p + iaq, \qquad P = \frac{p - iaq}{2ia}$$

is canonical and find a generating function

b) Use the transformation to solve the linear harmonic oscillator problem.

18. a) The Hamiltonian for a system has the form

$$H = \frac{1}{2}\left(\frac{1}{q^2} + p^2 q^4\right).$$

Find the equation of motion for q.

b) Find a canonical transformation that reduces H to the form of a harmonic oscillator. Show that the solution for the transformed variables is such that the equation of motion found in part (a) is satisfied.

19. A system of n particles moves in a plane under the influence of interaction forces derived from potential terms depending only upon the scalar distances between particles.

a) Using plane polar coordinates for each particle (relative to a common origin), identify the form of the Hamiltonian for the system.

b) Find a generating function for the canonical transformation that corresponds to a transformation to coordinates rotating in the plane counterclockwise with a uniform angular rate ω (the same for all particles). What are the transformation equations for the momenta?

c) What is the new Hamiltonian? What physical significance can you give to the difference between the old and the new Hamiltonian?

20. a) In the problem of small oscillations about steady motion, show that at the point of steady motion all the Hamiltonian variables η are constant. If the values for steady motion are η_0 so that $\eta = \eta_0 + \xi$, show that to lowest nonvanishing approximation the effective Hamiltonian for small oscillation can be put as

$$H(\eta_0, \zeta) = \frac{1}{2}\xi S \xi,$$

where **S** is a square matrix with components functions of η_0 only.

b) Assuming all frequencies of small oscillation are distinct, let **M** be a square $2n \times 2n$ matrix formed by the components of a possible set of eigenvectors (for both positive and negative frequencies). Only the directions of the eigenvectors are fixed, not their magnitudes. Show that it is possible to apply conditions to the eigenvectors (in effect fixing their magnitudes) that make **M** the Jacobian matrix of a canonical transformation.

c) Show that the canonical transformation so found transforms the effective Hamiltonian to the form

$$H = i\omega_j q_j P_j,$$

where ω_j is the magnitude of the normal frequencies. What are the equations of motion in this set of canonical coordinates?

d) Finally, show that

$$F_2 = q_j P_j + \frac{i}{2}\frac{P_j^2}{\omega_j} - \frac{i}{4}\omega_j q_j^2$$

leads to a canonical transformation that decomposes H into the Hamiltonians for a set of uncoupled linear harmonic oscillators that oscillate in the normal modes.

21. a) Using the fundamental Poisson brackets find the values of α and β for which the equations

$$Q = q^\alpha \cos \beta p, \qquad P = q^\alpha \sin \beta p$$

represent a canonical transformation.

b) For what values of α and β do these equations represent an *extended* canonical transformation? Find a generating function of the F_3 form for the transformation.

c) On the basis of part (b), can the transformation equations be modified so that they describe a canonical transformation for all values of β?

22. For the symmetric rigid body obtain formulas for evaluating the Poisson brackets

$$[\dot{\phi}, f(\theta, \phi, \psi)], \qquad [\dot{\psi}, f(\theta, \phi, \psi)]$$

where θ, ϕ, and ψ are the Euler angles, and f is any arbitrary function of the Euler angles.

23. A charged particle moves in space with a constant magnetic field **B** such that

$$A = \tfrac{1}{2}(B \times r)$$

a) If v_j are the Cartesian components of the velocity of the particle, evaluate the Poisson brackets

$$[v_i, v_j], \qquad i \neq j = 1, 2, 3.$$

b) If p_i is the canonical momentum conjugate to x_i, also evaluate the Poisson brackets

$$[x_i, v_j], \qquad [p_i, v_j],$$
$$[x_i, \dot{p}_j], \qquad [p_i, \dot{p}_j].$$

24. The semimajor axis a of the elliptical Kepler orbit and the eccentricity e are functions of first integrals of the motion and therefore of the canonical variables. Similarly, the mean anomaly

$$\phi \equiv \omega(t - T) = \psi - e \sin \psi$$

is a function of r, θ, and the conjugate momenta. Here T is the time of periapsis passage and is a constant of motion. Evaluate the Poisson brackets that can be formed of a, e, ϕ, ω, and T. There are in fact only nine nonvanishing distinct Poisson brackets out of these quantities.

25. Show that the Jacobi identity is satisfied if the Poisson bracket sign stands for the commutator of two square matrices:

$$[\mathbf{A}, \mathbf{B}] = \mathbf{A}\mathbf{B} - \mathbf{B}\mathbf{A}.$$

Show also that for the same representation of the Poisson bracket that

$$[\mathbf{A}, \mathbf{B}\mathbf{C}] = [\mathbf{A}, \mathbf{B}]\mathbf{C} + \mathbf{B}[\mathbf{A}, \mathbf{C}].$$

26. Prove Eq. (9–83) using the symplectic matrix notation for the Lagrange and Poisson brackets.

27. Verify the analog of the Jacobi identity for Lagrange brackets,

$$\frac{\partial\{u, v\}}{\partial w} + \frac{\partial\{v, w\}}{\partial u} + \frac{\partial\{w, u\}}{\partial v} = 0,$$

where u, v, and w are three functions in terms of which the (q, p) set can be specified.

28. a) Prove that the Poisson bracket of two constants of the motion is itself a constant of the motion even when the constants depend on time explicitly.

b) Show that if the Hamiltonian and a quantity F are constants of the motion, then the nth partial derivative of F with respect to t must also be a constant of the motion.

c) As an illustration of this result, consider the uniform motion of a free particle of mass m. The Hamiltonian is certainly conserved, and there exists a constant of the motion

$$F = x - \frac{pt}{m}.$$

Show by direct computation that the partial derivative of F with t, which is a constant of the motion, agrees with $[H, F]$.

29. Show by the use of Poisson brackets that for a one-dimensional harmonic oscillator there is a constant of the motion u defined as

$$u(q, p, t) = \ln(p + im\,\omega q) - i\omega t, \qquad \omega = \sqrt{\frac{k}{m}}.$$

What is the physical significance of this constant of the motion?

30. A system of two degrees of freedom is described by the Hamiltonian

$$H = q_1 p_1 - q_2 p_2 - aq_1^2 + bq_2^2.$$

Show that

$$F_1 = \frac{p_1 - aq_1}{q_2} \qquad \text{and} \qquad F_2 = q_1 q_2$$

are constants of the motion. Are there any other independent algebraic constants of the motion? Can any be constructed from Jacobi's identity?

31. Set up the magnetic monopole described in Exercise 23 (Chapter 3) in Hamiltonian formulation (you may want to use spherical polar coordinates). By means of the Poisson bracket formulation show that the quantity D defined in that exercise is conserved.

32. Obtain the motion in time of a linear harmonic oscillator by means of the formal solution for the Poisson bracket version of the equation of motion as derived from Eq. (9–116). Assume that at time $t = 0$ the initial values are x_0 and p_0.

33. A particle moves in one dimension under a potential

$$V = \frac{mk}{x^2}.$$

Find x as a function of time, by using the symbolic solution of the Poisson bracket form for the equation of motion for the quantity $y = x^2$. Initial conditions are that at $t = 0$, $x = x_0$, and $v = 0$.

34. a) Verify that Eq. (9–123) can be written in dyadic form as

$$[\mathbf{F}, \mathbf{L}] = -\mathbf{1} \times \mathbf{F} = -\mathbf{F} \times \mathbf{1}.$$

b) Show that if $\mathbf{n}$ is a unit vector normal to the orbital plane of central force motion, then if $\mathbf{F}$ is any vector of position and momentum only, it follows that

$$[\mathbf{F}, \mathbf{n}] = -\frac{\mathbf{F} \times (\mathbf{1} - \mathbf{n}\mathbf{n})}{L},$$

where L is the magnitude of the orbital angular momentum.

35. a) For a single particle show directly, i.e., by direct evaluation of the Poisson Brackets, that if u is a scalar function only of r^2, p^2, and $\mathbf{r} \cdot \mathbf{p}$, then

$$[u, \mathbf{L}] = 0.$$

b) Similarly show directly that if $\mathbf{F}$ is a vector function,

$$\mathbf{F} = u\mathbf{r} + v\mathbf{p} + w(\mathbf{r} \times \mathbf{p}),$$

where u, v, and w are scalar functions of the same type as in part (a), then

$$[F_i, L_j] = \epsilon_{ijk} F_k.$$

36. a) Using the theorem concerning Poisson brackets of vector functions and components of the angular momentum, show that if $\mathbf{F}$ and $\mathbf{G}$ are two vector functions of the coordinates and momenta only, then

$$[\mathbf{F} \cdot \mathbf{L}, \mathbf{G} \cdot \mathbf{L}] = \mathbf{L} \cdot (\mathbf{G} \times \mathbf{F}) + L_i L_j [F_i, G_j].$$

b) Let $\mathbf{L}$ be the total angular momentum of a rigid body with one point fixed and let L_μ be its component along a set of Cartesian axes fixed in the rigid body. By means of part (a) find a general expression for

$$[L_\mu, L_\nu], \qquad \mu, \nu = 1, 2, 3.$$

(*Hint:* Choose for $\mathbf{F}$ and $\mathbf{G}$ unit vectors along the μ and ν axes.)

c) From the Poisson bracket equations of motion for L_μ derive Euler's equations of motion for a rigid body.

37. Set up the problem of the spherical pendulum in the Hamiltonian formulation, using spherical polar coordinates for the q_i. Evaluate directly in terms of these canonical variables the following Poisson brackets:

$$[L_x, L_y], \qquad [L_y, L_z], \qquad [L_z, L_x],$$

showing that they have the values predicted by Eq. (9–128). Why is it that p_θ and p_ψ can be used as canonical momenta, although they are perpendicular components of the angular momentum?

38. On page 419 above it is shown that if any two components of the angular momentum are conserved, then the total angular momentum is conserved. If two of the components are identically zero, the third must be conserved. From this it would appear to follow that in any motion confined to a plane, so that the components of the angular momentum in the plane are zero, the total angular momentum is constant. There appear to be a number of obvious contradictions to this prediction; e.g., the angular momentum of an oscillating spring in a watch, or the angular momentum of a plane disk rolling down an inclined plane in the vertical plane. Discuss the force of these objections and whether the statement of the theorem requires any restrictions.

39. a) Show from the Poisson bracket condition for conserved quantities that the Laplace–Runge–Lenz vector $\mathbf{A}$,

$$\mathbf{A} = \mathbf{p} \times \mathbf{L} - \frac{mk\mathbf{r}}{r},$$

is a constant of the motion for the Kepler problem.

b) Verify the Poisson bracket relations for the components of $\mathbf{A}$ as given by Eq. (9–131).

40. a) Verify that the components of the two dimensional matrix $\mathbf{A}$, defined by Eq. (9–141) are constants of the motion for the two dimensional isotropic harmonic oscillator problem.

b) Verify that the quantities S_i, $i = 1, 2, 3$, defined by Eqs. (9–144), (9–145), (9–146), have the properties stated in Eqs. (9–147) and (9–148).

CHAPTER 10
Hamilton–Jacobi Theory

It has already been mentioned that canonical transformations may be used to provide a general procedure for solving mechanical problems. Two methods have been suggested. If the Hamiltonian is conserved then a solution could be obtained by transforming to new canonical coordinates that are all cyclic, since the integration of the new equations of motion becomes trivial. An alternative technique is to seek a canonical transformation from the coordinates and momenta, (q, p), at the time t, to a new set of constant quantities, which may be the $2n$ initial values, (q_0, p_0), at $t = 0$. With such a transformation, the equations of transformation relating the old and new canonical variables are then exactly the desired solution of the mechanical problem:

$$q = q(q_0, p_0, t),$$

$$p = p(q_0, p_0, t),$$

for they give the coordinates and momenta as a function of their initial values and the time. This last procedure is the more general one, especially as it is applicable, in principle at least, even when the Hamiltonian involves the time. We shall therefore begin our discussion by considering how such a transformation may be found.

10–1 THE HAMILTON–JACOBI EQUATION
FOR HAMILTON'S PRINCIPAL FUNCTION

We can automatically ensure that the new variables are constant in time by requiring that the transformed Hamiltonian, K, shall be identically zero, for then the equations of motion are

$$\frac{\partial K}{\partial P_i} = \dot{Q}_i = 0,$$

$$-\frac{\partial K}{\partial Q_i} = \dot{P}_i = 0.$$

(10–1)

438

As we have seen, K must be related to the old Hamiltonian and to the generating function by the equation

$$K = H + \frac{\partial F}{\partial t},$$

and hence will be zero if F satisfies the equation

$$H(q, p, t) + \frac{\partial F}{\partial t} = 0. \tag{10–2}$$

It is convenient to take F as a function of the old coordinates q_i, the new constant momenta P_i, and the time; in the notation of the previous chapter we would designate the generating function as $F_2(q, P, t)$. To write the Hamiltonian in Eq. (10–2) as a function of the same variables, use may be made of the equations of transformation (cf. Eq. 9–17a),

$$p_i = \frac{\partial F_2}{\partial q_i},$$

so that Eq. (10–2) becomes

$$H\left(q_1, \ldots, q_n; \frac{\partial F_2}{\partial q_1}, \ldots, \frac{\partial F_2}{\partial q_n}; t\right) + \frac{\partial F_2}{\partial t} = 0. \tag{10–3}$$

Equation (10–3), known as the *Hamilton–Jacobi equation*, constitutes a partial differential equation in $(n + 1)$ variables, $q_1, \ldots, q_n; t$, for the desired generating function. It is customary to denote the solution of Eq. (10–3) by S and to call it *Hamilton's principal function*.

Of course, the integration of Eq. (10–3) only provides the dependence on the old coordinates and time; it would not appear to tell how the new momenta are contained in S. Indeed the new momenta have not yet been specified except that we know they must be constants. However, the nature of the solution indicates how the new P_i's are to be selected.

Mathematically Eq. (10–3) has the form of a first-order partial differential equation in $n + 1$ variables. Suppose that there exists a solution to Eq. (10–3) of the form

$$F_2 \equiv S = S(q_1, \ldots, q_n; \alpha_1, \ldots, \alpha_{n+1}; t), \tag{10–4}$$

where the quantities $\alpha_1, \ldots, \alpha_{n+1}$ are $n + 1$ *independent* constants of integration. Such solutions are known as *complete solutions* of the first-order partial differential equation.* One of the constants of integration, however, is in fact

---

* Equation (10–4) is not the only type of solution possible for Eq. (10–3). The most general form of the solution involves one or more arbitrary functions rather than arbitrary constants. See, for example, R. Courant and D. Hilbert: *Methods of Mathematical Physics*, Vol. II, 1962, pp. 24–28, and V. I. Smirnov: *A Course of Higher Mathematics*, Vol. IV, 1964, Section I11. Nor is there necessarily a unique solution of the form (10–4). There may be several complete solutions for the given equation. But all that is important for the subsequent argument is that there exist *a* complete solution.

irrelevant to the solution, for it will be noted that S itself does not appear in Eq. (10–3); only its partial derivatives with respect to q or t are involved. Hence, if S is some solution of the differential equation, then $S + \alpha$, where α is any constant, must also be a solution. One of the $n + 1$ constants of integration in Eq. (10–4) must therefore appear only as an additive constant tacked on to S. But by the same token an additive constant has no importance in a generating function, since only partial derivatives of the generating function occur in the transformation equations. Hence for our purposes a complete solution to Eq. (10–3) can be written in the form

$$S = S(q_1, \ldots, q_n; \alpha_1, \ldots, \alpha_n; t), \tag{10–5}$$

where none of the n independent constants is solely additive. In this mathematical garb S tallies exactly with the desired form for an F_2 type of generating function, for Eq. (10–5) presents S as a function of n coordinates, the time t, and n independent quantities α_i. We are therefore at liberty to take the n constants of integration to be the new (constant) momenta:

$$P_i = \alpha_i. \tag{10–6}$$

Such a choice does not contradict the original assertion that the new momenta are connected with the initial values of q and p at the time t_0. The n transformation equations (9–17a) can now be written as

$$p_i = \frac{\partial S(q, \alpha, t)}{\partial q_i}, \tag{10–7}$$

where q, α stand for the complete set of quantities. At the time t_0 these constitute n equations relating the n α's with the initial q and p values, thus enabling one to evaluate the constants of integration in terms of the specific initial conditions of the problem. The other half of the equations of transformation, which provide the new constant coordinates, appear as

$$Q_i = \beta_i = \frac{\partial S(q, \alpha, t)}{\partial \alpha_i}. \tag{10–8}$$

The constant β's can be similarly obtained from the initial conditions, simply by calculating the value of the right side of Eq. (10–8) at $t = t_0$ with the known initial values of q_i. Equations (10–8) can then be "turned inside out" to furnish q_j in terms of $\alpha, \beta,$ and t:

$$q_j = q_j(\alpha, \beta, t), \tag{10–9}$$

which solves the problem of giving the coordinates as functions of time and the initial conditions.* After the differentiation in Eqs. (10–7) has been performed, Eqs. (10–9) may be substituted for the q's, thus giving the momenta p_i as functions of the α, β, and t:

$$p_i = p_i(\alpha, \beta, t). \tag{10–10}$$

Equations (10–9) and (10–10) thus constitute the desired complete solution of Hamilton's equations of motion.

Hamilton's principal function is thus the generator of a canonical transformation to constant coordinates and momenta; *when solving the Hamilton–Jacobi equation we are at the same time obtaining a solution to the mechanical problem.* Mathematically speaking, we have established an equivalence between the $2n$ canonical equations of motion, which are first-order differential equations, to the first-order partial differential Hamilton–Jacobi equation. This correspondence is not restricted to equations governed by the Hamiltonian; indeed the general theory of first-order partial differential equations is largely concerned with the properties of the equivalent set of first-order ordinary differential equations. Essentially, the connection can be traced to the fact that both the partial differential equation and its canonical equations stem from a common variational principle, in this case Hamilton's modified principle.

To a certain extent the choice of the α_i's as the new momenta is arbitrary. One could just as well choose any n quantities, γ_i, which are independent functions of the α_i constants of integration:

$$\gamma_i = \gamma_i(\alpha_1, \ldots, \alpha_n). \tag{10–11}$$

By means of these defining relations Hamilton's principal function can be written as a function of q_i, γ_i, and t, and the rest of the derivation then goes through unchanged. It often proves convenient to take some particular set of γ_i's as the new momenta, rather than the constants of integration that appear naturally in integrating the Hamilton–Jacobi equation.

---

* As a mathematical point it may be questioned whether the process of "turning inside out" is feasible for Eqs (10–7) and (10–8), i.e., whether they can be solved for α_i and q_i respectively. The question hinges on whether the equations in each set are independent, for otherwise they are obviously not sufficient to determine the n independent quantities α_i or q_i as the case may be. To simplify the notation, let S_α symbolize members of the set of partial derivatives of S with respect to α_i, so that Eq. (10–8) is represented by $\beta = S_\alpha$. That the derivatives S_α in (10–8) form independent functions of the q's follows directly from the nature of a complete solution to the Hamilton–Jacobi equation; indeed this is what we mean by saying the n constants of integration are independent. Consequently the Jacobian of S_α with respect to q_i cannot vanish. Since the order of differentiation is immaterial, this is equivalent to saying that the Jacobian of S_q with respect to α_i cannot vanish, which proves the independence of Eqs. (10–7).

Further insight into the physical significance of S is furnished by an examination of its total time derivative, which can be computed from the formula

$$\frac{dS}{dt} = \frac{\partial S}{\partial q_i}\dot{q}_i + \frac{\partial S}{\partial t},$$

since the P_i's are constant in time. By Eqs. (10–7) and (10–3) this relation can also be written

$$\frac{dS}{dt} = p_i\dot{q}_i - H = L, \tag{10–12}$$

so that Hamilton's principal function differs at most from the indefinite time integral of the Lagrangian only by a constant:

$$S = \int L\,dt + \text{constant}. \tag{10–13}$$

Now, Hamilton's principle is a statement about the definite integral of L, and from it we obtained the solution of the problem via the Lagrange equations. Here the same action integral, in an indefinite form, furnishes another way of solving the problem. In actual calculations the result expressed by Eq. (10–13) is of no help, because one cannot integrate the Lagrangian with respect to time until q_i and p_i are known as functions of time, i.e., until the problem is solved.*

10–2 THE HARMONIC OSCILLATOR PROBLEM AS AN EXAMPLE OF THE HAMILTON–JACOBI METHOD

To illustrate the Hamilton–Jacobi technique for solving the motion of mechanical systems we shall work out in detail the simple problem of a one-dimensional harmonic oscillator. The Hamiltonian is

$$H = \frac{1}{2m}(p^2 + m^2\omega^2 q^2) \equiv E, \tag{10–14}$$

where

$$\omega = \sqrt{\frac{k}{m}}, \tag{10–15}$$

---

* Historically the recognition by Hamilton that the time integral of L is a special solution of a partial differential equation came before it was seen how the Hamilton–Jacobi equation can furnish the solution to a mechanical problem. It was Jacobi who realized that the converse was true, that by the techniques of canonical transformations any complete solution of the Hamiltonian–Jacobi equation could be used to describe the motion of the system.

k being the force constant. One obtains the Hamilton–Jacobi equation for S by setting p equal to $\partial S/\partial q$ and substituting in the Hamiltonian; the requirement that the new Hamiltonian vanishes becomes

$$\frac{1}{2m}\left[\left(\frac{\partial S}{\partial q}\right)^2 + m^2\omega^2 q^2\right] + \frac{\partial S}{\partial t} = 0. \tag{10–16}$$

Since the explicit dependence of S on t is involved only in the last term, a solution for Eq. (10–16) can be found in the form

$$S(q, \alpha, t) = W(q, \alpha) - \alpha t, \tag{10–17}$$

where α is a constant of integration (to be designated later as the transformed momentum). With this choice of solution the time can be eliminated from Eq. (10–16).

$$\frac{1}{2m}\left[\left(\frac{\partial W}{\partial q}\right)^2 + m^2\omega^2 q^2\right] = \alpha. \tag{10–18}$$

The integration constant α is thus to be identified with the total energy E. This can also be recognized directly from Eq. (10–17) and the relation (cf. Eq. (10–3))

$$\frac{\partial S}{\partial t} + H = 0,$$

which then reduces to

$$H = \alpha.$$

Equation (10–18) can be integrated immediately to

$$W = \sqrt{2m\alpha}\int dq \sqrt{1 - \frac{m\omega^2 q^2}{2\alpha}},$$

so that

$$S = \sqrt{2m\alpha}\int dq \sqrt{1 - \frac{m\omega^2 q^2}{2\alpha}} - \alpha t. \tag{10–19}$$

While the integration involved in Eq. (10–19) is not particularly difficult there is no reason to carry it out at this stage, for what is desired is not S but its partial derivatives. The solution for q arises out of the transformation equation (10–8):

$$\beta = \frac{\partial S}{\partial \alpha} = \sqrt{\frac{m}{2\alpha}}\int \frac{dq}{\sqrt{1 - \frac{m\omega^2 q^2}{2\alpha}}} - t,$$

which can be integrated without trouble as

$$t + \beta = \frac{1}{\omega}\arcsin q \sqrt{\frac{m\omega^2}{2\alpha}}. \tag{10–20}$$

Equation (10–20) can be immediately "turned inside out" to furnish q as a function of t and the two constants of integration α and β:

$$q = \sqrt{\frac{2\alpha}{m\omega^2}} \sin \omega(t + \beta), \qquad (10\text{–}21)$$

which is the familiar solution for a harmonic oscillator. Formally, the solution for the momentum comes from the transformation equation (10–7), which, using Eq. (10–18), can be written

$$p = \frac{\partial S}{\partial q} = \frac{\partial W}{\partial q} = \sqrt{2m\alpha - m\omega^2 q^2}.$$

In conjunction with the solution for q, Eq. (10–21), this becomes

$$p = \sqrt{2m\alpha(1 - \sin^2 \omega(t + \beta))},$$

or

$$p = \sqrt{2m\alpha} \cos \omega(t + \beta). \qquad (10\text{–}22)$$

Of course, this result checks with the simple identification of p as $m\dot{q}$.

To complete the story, the constants α and β must be connected with the initial conditions q_0 and p_0 at time $t = 0$. By squaring Eqs. (10–21) and (10–22) it is clearly seen that α is given in terms of q_0 and p_0 by the equation

$$2m\alpha = p_0^2 + m^2\omega^2 q_0^2. \qquad (10\text{–}23)$$

The same result follows immediately, of course, from the previous identification of α as the conserved total energy E. Finally, the phase constant β is related to q_0 and p_0 by

$$\tan \omega\beta = m\omega\frac{q_0}{p_0}. \qquad (10\text{–}24)$$

Thus, Hamilton's principal function is the generator of a canonical transformation to a new coordinate that measures the phase angle of the oscillation and to a new canonical momentum identified as the total energy.

If the solution for q is substituted into Eq. (10–19), Hamilton's principal function can be written as

$$S = 2\alpha \int \cos^2 \omega(t + \beta)\, dt - \alpha t = 2\alpha \int (\cos^2 \omega(t + \beta) - \tfrac{1}{2})\, dt.$$

Now, the Lagrangian is

$$L = \frac{1}{2m}(p^2 - m^2\omega^2 q^2)$$

$$= \alpha(\cos^2 \omega(t + \beta) - \sin^2 \omega(t + \beta))$$

$$= 2\alpha(\cos^2 \omega(t + \beta) - \tfrac{1}{2}),$$

so that S is the time integral of the Lagrangian, in agreement with the general relation (10–13). Note that the identity could not be proved until *after* the solution to the problem had been obtained.

10–3 THE HAMILTON–JACOBI EQUATION FOR HAMILTON'S CHARACTERISTIC FUNCTION

It was possible to integrate the Hamilton–Jacobi equation for the simple harmonic oscillator primarily because S could be separated into two parts, one involving q only and the other only time. Such a separation of variables is always possible *whenever the old Hamiltonian does not involve time explicitly.*

If H is not an explicit function of t, then the Hamilton–Jacobi equation for S becomes

$$\frac{\partial S}{\partial t} + H\left(q_i, \frac{\partial S}{\partial q_i}\right) = 0.$$

The first term involves only the dependence on t, whereas the second is concerned only with the dependence of S on the q_i. The time variable can therefore be separated by assuming a solution for S of the form

$$S(q_i, \alpha_i, t) = W(q_i, \alpha_i) - \alpha_1 t. \qquad (10\text{–}25)$$

Upon substituting this trial solution, the differential equation reduces to the expression

$$H\left(q_i, \frac{\partial W}{\partial q_i}\right) = \alpha_1, \qquad (10\text{–}26)$$

which no longer involves the time. One of the constants of integration appearing in S, namely α_1, is thus equal to the constant value of H. (Normally H will be the energy, but it should be remembered that this need not always be the case, cf. Section 8–2.)

The time-independent function W appears here merely as a part of the generating function S when H is constant. It can also be shown that W separately generates its own contact transformation with properties quite different from that generated by S. Consider a canonical transformation in which the new momenta are all constants of the motion α_i, and where α_1 in particular is the constant of motion H. If the generating function for this transformation be denoted by $W(q, P)$, then the equations of transformation are

$$p_i = \frac{\partial W}{\partial q_i}, \qquad Q_i = \frac{\partial W}{\partial P_i} = \frac{\partial W}{\partial \alpha_i}. \qquad (10\text{–}27)$$

While these equations resemble Eqs. (10–9) and (10–8) for Hamilton's principal function S, the condition now determining W is that H shall be equal to the new momentum α_1:

$$H(q_i, p_i) = \alpha_1.$$

Using Eqs. (10–27) this requirement becomes a partial differential equation for W:

$$H\left(q_i, \frac{\partial W}{\partial q_i}\right) = \alpha_1,$$

which is seen to be identical with Eq. (10–26). Since W does not involve the time, the new and old Hamiltonians are equal, and it follows that $K = \alpha_1$.

The function W, known as *Hamilton's characteristic function*, thus generates a canonical transformation in which all the new coordinates are cyclic. It was noted in the previous chapter that when H is a constant of the motion, a transformation of this nature in effect solves the mechanical problem involved, for the integration of the new equations of motion is then trivial. The canonical equations for P_i, in fact, merely repeat the statement that the momenta conjugate to the cyclic coordinates are all constant:

$$\dot{P}_i = -\frac{\partial K}{\partial Q_i} = 0, \qquad P_i = \alpha_i, \tag{10–28a}$$

Because the new Hamiltonian depends on only one of the momenta α_i, the equations of motion for $\dot{Q}_i$ are

$$\dot{Q}_i = \frac{\partial K}{\partial \alpha_i} = 1 \qquad i = 1,$$

$$= 0 \qquad i \neq 1,$$

with the immediate solutions

$$Q_1 = t + \beta_1 \equiv \frac{\partial W}{\partial \alpha_1},$$

$$\tag{10–28b}$$

$$Q_i = \beta_i \equiv \frac{\partial W}{\partial \alpha_i} \qquad i \neq 1.$$

The only coordinate that is not simply a constant of the motion is Q_1, which is equal to the time plus a constant. We have here another instance of the conjugate relationship between the time as a coordinate and the Hamiltonian as its conjugate momentum.

The dependence of W on the old coordinates q_i is determined by the partial differential equation (10–26), which, like Eq. (10–3), is also referred to as the Hamilton–Jacobi equation. There will now be n constants of integration in a complete solution, but again one of them must be merely an additive constant. The $n - 1$ remaining independent constants, $\alpha_2, \ldots, \alpha_n$, together with α_1 may then be taken as the new constant momenta. When evaluated at t_0 the first half of Eqs. (10–27) serve to relate the n constants α_i with the initial values of q_i and p_i. Finally, Eqs. (10–28b) can be solved for the q_i as a function of α_i, β_i, and the time t, thus completing the solution of the problem. It will be noted that $(n - 1)$ of the Eqs.

(10–28b) do not involve the time at all. One of the q_i's can be chosen as an independent variable, and the remaining coordinates can then be expressed in terms of it by solving only these time-independent equations. We are thus led directly to the *orbit equations* of the motion. In central force motion, for example, this technique would furnish r as a function of θ, without the need for separately finding r and θ as functions of time.

It is not necessary always to take α_1 and the constants of integration in W as the new constant momenta. Occasionally it is desirable rather to use some particular set of n independent functions of the α_i's as the transformed momenta. Designating these constants by γ_i the characteristic function W can then be expressed in terms of q_i and γ_i as the independent variables. The Hamiltonian will in general depend upon more than one of the γ_i's and the equations of motion for $\dot{Q}_i$ become

$$\dot{Q}_i = \frac{\partial K}{\partial \gamma_i} = v_i,$$

where the v_i's are functions of γ_i. In this case all the new coordinates are linear functions of the time:

$$Q_i = v_i t + \beta_i. \tag{10–29}$$

The characteristic function W possesses a physical significance similar to that for S. As W does not involve time explicitly, its total time derivative is

$$\frac{dW}{dt} = \frac{\partial W}{\partial q_i}\dot{q}_i = p_i \dot{q}_i,$$

and hence

$$W = \int p_i \dot{q}_i \, dt = \int p_i \, dq_i. \tag{10–13'}$$

The above integrals will be recognized as defining the abbreviated action, as used in Section 8–6. Again, this information is of little practical help; the form of W cannot be found a priori without obtaining a complete integral of the Hamilton–Jacobi equation. The procedures involved in solving a mechanical problem by either Hamilton's principal or characteristic function may now be summarized in the following tabular form:

The two methods of solution are applicable when the Hamiltonian

is any general function of q, p, t:	is conserved:
$H(q, p, t)$.	$H(q, p) = $ constant.

We seek canonical transformations to new variables such that

| all the coordinates and momenta Q_i, P_i are constants of the motion. | all the momenta P_i are constants. |

To meet these requirements it is sufficient to demand that the new Hamiltonian

shall vanish identically:

$$K = 0.$$

shall be cyclic in all the coordinates:

$$K = H(P_i) = \alpha_1.$$

Under these conditions the new equations of motion become

$$\dot{Q}_i = \frac{\partial K}{\partial P_i} = 0,$$

$$\dot{P}_i = -\frac{\partial K}{\partial Q_i} = 0,$$

$$\dot{Q}_i = \frac{\partial K}{\partial P_i} = v_i,$$

$$\dot{P}_i = -\frac{\partial K}{\partial Q_i} = 0,$$

with the immediate solutions

$$Q_i = \beta_i,$$

$$P_i = \gamma_i,$$

$$Q_i = v_i t + \beta_i$$

$$P_i = \gamma_i$$

which satisfy the stipulated requirements.

The generating function producing the desired transformation is Hamilton's

Principal Function:

$$S(q, P, t),$$

Characteristic Function:

$$W(q, P),$$

satisfying the Hamilton–Jacobi partial differential equation:

$$H\left(q, \frac{\partial S}{\partial q}, t\right) + \frac{\partial S}{\partial t} = 0.$$

$$H\left(q, \frac{\partial W}{\partial q}\right) - \alpha_1 = 0.$$

A complete solution to the equation contains

n nontrivial constants of integration $\alpha_1, \ldots, \alpha_n$.

$n - 1$ nontrivial constants of integration which together with α_1 form a set of n independent constants $\alpha_1, \ldots, \alpha_n$.

The new constant momenta, $P_i = \gamma_i$, can be chosen as any n independent functions of the n constants of integration:

$$P_i = \gamma_i(\alpha_1, \ldots, \alpha_n),$$

$$P_i = \gamma_i(\alpha_1, \ldots, \alpha_n),$$

so that the complete solutions to the Hamilton–Jacobi equation may be considered as functions of the new momenta:

$$S = S(q_i, \gamma_i, t).$$

$$W = W(q_i, \gamma_i).$$

In particular, the γ_i's may be chosen to be the α_i's themselves. One half of the transformation equations,

$$p_i = \frac{\partial S}{\partial q_i}, \qquad\qquad\qquad p_i = \frac{\partial W}{\partial q_i},$$

are fulfilled automatically, since they have been used in constructing the Hamilton–Jacobi equation. The other half,

$$Q_i = \frac{\partial S}{\partial \gamma_i} = \beta_i, \qquad\qquad Q_i = \frac{\partial W}{\partial \gamma_i} = v_i(\gamma_j)t + \beta_i,$$

can be solved for q_i in terms of t and the $2n$ constants β_i, γ_i. The solution to the problem is then completed by evaluating these $2n$ constants in terms of the initial values, (q_{i0}, p_{i0}), of the coordinates and momenta.

When the Hamiltonian does not involve time explicitly, both methods are suitable, and the generating functions are then related to each other according to the formula

$$S(q, P, t) = W(q, P) - \alpha_1 t.$$

10-4 SEPARATION OF VARIABLES IN THE HAMILTON–JACOBI EQUATION

It might appear from the preceding section that little practical advantage has been gained through the introduction of the Hamilton–Jacobi procedure. Instead of solving the $2n$ ordinary differential equations that make up the canonical equations of motion, one now must solve the partial differential Hamilton–Jacobi equation, and partial differential equations are notoriously complicated to solve. Under certain conditions, however, it is possible to separate the variables in the Hamilton–Jacobi equation, and the solution can then always be reduced to quadratures. In practice the Hamilton–Jacobi technique becomes a useful computational tool only when such a separation can be effected.

A coordinate q_j is said to be separable in the Hamilton–Jacobi equation when (say) Hamilton's principal function can be split into two additive parts, one of which depends only on the coordinate q_j and the other is entirely independent of q_j. Thus, if q_1 is taken as a separable coordinate, then the Hamiltonian must be such that one can write

$$S(q_1, \ldots, q_n; \alpha_1, \ldots, \alpha_n; t) = S_1(q_1; \alpha_1, \ldots, \alpha_n; t) + S'(q_2, \ldots, q_n; \alpha_1, \ldots, \alpha_n; t), \tag{10-30}$$

and the Hamilton–Jacobi equation can be split into two equations—one separately for S_1 and the other for S'. Similarly the Hamilton–Jacobi equation is described as *completely separable* (or simply, *separable*) if all the coordinates in the problem are separable. A solution for Hamilton's principal function of the form

$$S = \sum_i S_i(q_i; \alpha_1, \ldots, \alpha_n; t) \tag{10-31}$$

will then split the Hamilton–Jacobi equation into n equations of the type

$$H_i\left(q_i; \frac{\partial S_i}{\partial q_i}; \alpha_1, \ldots, \alpha_n\right) = \alpha_i \tag{10–32}$$

(no summation!). The constants α_i are referred to now as the *separation constants*. Each of the Eqs. (10–32) involves only one of the coordinates q_i and the corresponding partial derivative of S_i with respect to q_i. They are therefore a set of ordinary differential equations of a particularly simple form. Since the equations are only of first order, it is always possible to reduce them to quadratures; one has only to solve for the partial derivative of S_i with respect to q_i and then integrate over q_i.

The transition from Hamilton's principal function S to the characteristic function for conservative mechanical systems can be treated as an instance where t is a separable variable in the Hamilton–Jacobi equation. In such case one seeks a solution for S of a form corresponding to that prescribed by Eq. (10–30):

$$S(q, \alpha, t) = S_0(\alpha, t) + W(q, \alpha). \tag{10–32}$$

Since, under the assumption, H is not an explicit function of time, the Hamilton–Jacobi equation with this trial solution becomes

$$H\left(q, \frac{\partial W}{\partial q}\right) + \frac{\partial S_0}{\partial t} = 0. \tag{10–33}$$

The first term in Eq. (10–33) is independent of t and can depend only on the q_i's, while the second term can be a function at most of t. Hence the equation can hold only if the two terms are both constant, with equal and opposite values

$$\frac{\partial S_0}{\partial t} = -\alpha_1, \tag{10–34a}$$

$$H\left(q, \frac{\partial W}{\partial q}\right) = \alpha_1. \tag{10–34b}$$

The first equation is solved by $S_0 = -\alpha_1 t$, as in Eqs. (10–17) and (10–25), while the second is the Hamilton–Jacobi equation for W. The constant value of the Hamiltonian, α_1, thus appears in this procedure in the guise of the separation constant.

It is possible to find examples in which the Hamilton–Jacobi equation can be solved without separating the time variable (e.g., Exercise 8). Nonetheless, almost all useful applications of the Hamilton–Jacobi method involve Hamiltonians not explicitly dependent on the time, for which therefore t is a separable variable. The subsequent discussion on separability will therefore be restricted to such systems where H is a constant of the motion, and Hamilton's characteristic function W will be used exclusively.

We can easily show that any cyclic or ignorable coordinate is separable. Suppose that the cyclic coordinate is q_1; the conjugate momentum p_1 is a constant, say γ. The Hamilton–Jacobi equation for W is then

$$H\left(q_2, \ldots, q_n; \gamma; \frac{\partial W}{\partial q_2}, \ldots, \frac{\partial W}{\partial q_n}\right) = \alpha_1. \tag{10–35}$$

If we try a separated solution of the form

$$W = W_1(q_1, \alpha) + W'(q_2, \ldots, q_n; \alpha), \tag{10–36}$$

then it is obvious that Eq. (10–35) involves only the separate function W', while W_1 is the solution of the equation

$$P_1 = \gamma = \frac{\partial W_1}{\partial q_1}. \tag{10–37}$$

The constant γ is thus the separation constant, and the obvious solution for W_1 (to within a trivial additive constant) is

$$W_1 = \gamma q_1,$$

and W is given by

$$W = W' + \gamma q_1. \tag{10–38}$$

There is an obvious resemblance between Eq. (10–38) and the form S assumes when H is not an explicit function of time, Eq. (10–25). Indeed both equations can be considered as arising under similar circumstances. We have seen that t may be considered in some sense as a generalized coordinate with $-H$ as its canonical momentum, cf. Eq. (8–58). If H is conserved, then t may be treated as a cyclic coordinate and Eq. (10–34a) then directly corresponds to Eq. (10–37) leading to similar solutions.

If all but one of the coordinates are cyclic, then by repeated application of the procedure used above the Hamilton–Jacobi equation can be completely separated. To exhibit the separation specifically it is convenient this time to designate q_1 as the *noncyclic* coordinate so that all conjugate momenta p_i, $i > 1$, are constants $\alpha_2, \ldots, \alpha_n$. Following the same steps as for a single cyclic coordinate, the separated form for W then appears as

$$W = \sum_{i=1}^{n} W_i(q_i, \alpha) = W_1(q_1, \alpha) + \sum_{i=2}^{n} \alpha_i q_i, \tag{10–39}$$

(cf. Eq. (10–38)). Here W_1 is the solution of the reduced Hamilton–Jacobi equation:

$$H\left(q_1; \frac{\partial W_1}{\partial q_1}; \alpha_2, \ldots, \alpha_n\right) = \alpha_1. \tag{10–40}$$

Since this is an ordinary first-order differential equation in the independent variable q_1, it can be immediately reduced to quadratures, and the complete solution for W can be obtained.*

In general, a coordinate q_j can be separated if q_j and the conjugate momentum p_j can be segregated in the Hamiltonian into some function $f(q_j, p_j)$ that does not contain any of the other variables. If we then seek a trial solution of the form

$$W = W_j(q_j, \alpha) + W'(q_i, \alpha),$$

where q_i represents the set of all q's *except* q_j, then the Hamilton–Jacobi equation appears as

$$H\left(q_i, \frac{\partial W'}{\partial q_i}, f\left(q_j, \frac{\partial W_j}{\partial q_j}\right)\right) = \alpha_1. \tag{10-41}$$

In principle, at least, Eq. (10–41) can be inverted so as to solve for f:

$$f\left(q_j, \frac{\partial W_j}{\partial q_j}\right) = g\left(q_i, \frac{\partial W'}{\partial q_i}, \alpha_1\right). \tag{10-42}$$

The argument used previously in connection with Eq. (10–33) holds here in slightly varied guise: f is not a function of any of the q's except q_j; g on the other hand is independent of q_j. Hence Eq. (10–42) can hold only if both sides are equal to the same constant, independent of all q's:

$$f\left(q_j, \frac{\partial W_j}{\partial q_j}\right) = \alpha_j,$$

$$g\left(q_i, \frac{\partial W'}{\partial q_i}\right) = \alpha_j, \tag{10-43}$$

and the separation of the variable has been accomplished.

It must be emphasized that the separability of the Hamilton–Jacobi equation depends not only on the physical problem involved but also on the choice of the system of generalized coordinates employed. Thus, the one-body central force problem is separable in polar coordinates, but not in Cartesian coordinates. For

---

* The form of (10–39) may also be arrived at by the following considerations. It must be remembered that W is to be the generating function for a transformation to new coordinates that are all cyclic. But if $q_2, \ldots, q_n$ are already cyclic, no further transformation is needed for them. As far as they are concerned, W can be the identity transformation. Since the α_i's are the new momenta, the summation in (10–39) can be written as

$$\sum_{i=2}^{n} q_i P_i,$$

which will be recognized as the generator of the identity transformation (cf. Eq. (9–25)) for the coordinates $q_2, \ldots, q_n$.

some problems it is not at all possible to completely separate the Hamilton–Jacobi equation, the famous three-body problem being one illustration. On the other hand, in many of the basic problems of mechanics and atomic physics separation is possible in more than one set of coordinates. In general it is feasible to solve the Hamilton–Jacobi equation in closed form only when the variables are completely separable. Considerable ingenuity has therefore been devoted to finding the separable systems of coordinates appropriate to each problem.

No simple criterion can be given to indicate what coordinate systems lead to separable Hamilton–Jacobi equations for any particular problem. In the case of orthogonal coordinate systems the so-called Staeckel conditions have proved useful. They provide necessary and sufficient conditions for separability under certain circumstances. A proof of the sufficiency of the conditions will be found in Appendix D (which also lists further references to the literature on the subject). The Staeckel conditions themselves will be stated here, along with an illustration of their application.

The following restrictions are placed on the type of situation concerned:

1. The Hamiltonian is conserved.
2. The Lagrangian is no more than a quadratic function of the generalized velocities, so that the Hamiltonian takes the form (cf. Eq. (8–19)):

$$H = \tfrac{1}{2}(\mathbf{p} - \mathbf{a})\mathbf{T}^{-1}(\mathbf{p} - \mathbf{a}) + V(q). \tag{10–44}$$

3. The set of generalized coordinates q_i forms an orthogonal system of coordinates, so that the matrix $\mathbf{T}$ is diagonal. It follows that the inverse matrix $\mathbf{T}^{-1}$ is also diagonal with nonvanishing elements:

$$(\mathbf{T}^{-1})_{ii} = \frac{1}{T_{ii}} \qquad \text{(no summation)}. \tag{10–45}$$

For problems and coordinates satisfying this description, the Staeckel conditions state that the Hamilton–Jacobi equation will be completely separable if

a. the vector $\mathbf{a}$ has elements a_i that are functions only of the corresponding coordinate, that is, $a_i = a_i(q_i)$;
b. the potential function $V(q)$ can be written as a sum of the form

$$V(q) = \frac{V_i(q_i)}{T_{ii}}; \tag{10–46}$$

c. there exists an $n \times n$ matrix $\boldsymbol{\phi}$ with elements $\phi_{ij} = \phi_{ij}(q_i)$ such that

$$(\boldsymbol{\phi}^{-1})_{1j} = \frac{1}{T_{jj}} \qquad \text{(no summation)}. \tag{10–47}$$

If the Staeckel conditions are satisfied, then Hamilton's characteristic function is completely separable:

$$W(q) = \sum_i W_i(q_i),$$

with the W_i satisfying equations of the form

$$\left[\frac{\partial W_i}{\partial q_i} - a_i\right]^2 = -2V_i(q_i) + 2\phi_{ij}\gamma_j, \qquad (10\text{–}48)$$

where γ_j are constants of integration (and there is summation only over the index j).

While these conditions appear mysterious and complicated, their application usually is fairly straightforward. As an illustration of some of the ideas developed here about separability, the Hamilton–Jacobi equation for a particle moving in a central force will be discussed in polar coordinates. The problem will then be generalized to arbitrary potential laws, to furnish an application of the Staeckel conditions.

Let us first consider the central force problem in terms of the polar coordinates (r, ψ) in the plane of the orbit. The motion then involves only two degrees of freedom and the Hamiltonian has the form

$$H = \frac{1}{2m}\left(p_r^2 + \frac{p_\psi^2}{r^2}\right) + V(r), \qquad (10\text{–}49)$$

and is cyclic in ψ. Consequently Hamilton's characteristic function appears as

$$W = W_1(r) + \alpha_\psi \psi, \qquad (10\text{–}50)$$

where α_ψ is the constant angular momentum p_ψ conjugate to ψ. The Hamilton–Jacobi equation then becomes

$$\left(\frac{\partial W_1}{\partial r}\right)^2 + \frac{\alpha^2}{r^2} + 2mV(r) = 2m\alpha_1, \qquad (10\text{–}51)$$

where α_1 is the constant identified physically as the total energy of the system. Solving Eq. (10–51) for the partial derivative of W_1 we obtain

$$\frac{\partial W_1}{\partial r} = \sqrt{2m(\alpha_1 - V) - \frac{\alpha_\psi^2}{r^2}},$$

so that W is

$$W = \int dr \sqrt{2m(\alpha_1 - V) - \frac{\alpha_\psi^2}{r^2}} + \alpha_\psi \psi. \qquad (10\text{–}52)$$

With this form for the characteristic function the transformation equations (10–28b) appear as

$$t + \beta_1 = \frac{\partial W}{\partial \alpha_1} = \int \frac{m \, dr}{\sqrt{2m(\alpha_1 - V) - \dfrac{\alpha_\psi^2}{r^2}}}, \tag{10–53a}$$

and

$$\beta_2 = \frac{\partial W}{\partial \alpha_\psi} = -\int \frac{\alpha_\psi \, dr}{r^2 \sqrt{2m(\alpha_1 - V) - \dfrac{\alpha_\psi^2}{r^2}}} + \psi. \tag{10–53b}$$

Equation (10–53a) furnishes r as a function of t and agrees with the corresponding solution, Eq. (3–18), found in Chapter 3, with α_1 and α_ψ written explicitly as E and l, respectively. It has been remarked previously that the remaining transformation equations for Q_i, here only Eq. (10–53b), should provide the orbit equation. If the variable of integration in Eq. (10–53b) is changed to $u = 1/r$, the equation reduces to

$$\psi = \beta_2 - \int \frac{du}{\sqrt{\dfrac{2m}{\alpha_\psi^2}(\alpha_1 - V) - u^2}},$$

which agrees with Eq. (3–37) previously found for the orbit, identifying ψ as θ and β_2 as θ_0.

As a further example of separation of variables, we shall examine the same central force problem, but in spherical polar coordinates, that is, ignoring our a priori knowledge that the orbit lies in a plane. The appropriate Hamiltonian has been shown to be (cf. Eq. (8–21)):

$$H = \frac{1}{2m}\left(p_r^2 + \frac{p_\theta^2}{r^2} + \frac{p_\phi^2}{r^2 \sin^2\theta}\right) + V(r). \tag{10–54}$$

If the variables in the corresponding Hamilton–Jacobi equation are separable, then Hamilton's characteristic function must have the form

$$W = W_r(r) + W_\theta(\theta) + W_\phi(\phi). \tag{10–55}$$

The coordinate ϕ is cyclic in the Hamiltonian and hence

$$W_\phi = \alpha_\phi \phi \tag{10–56}$$

where α_ϕ is a constant of integration. In terms of this form for W, the Hamilton–Jacobi equation reduces to

$$\left(\frac{\partial W_r}{\partial r}\right)^2 + \frac{1}{r^2}\left[\left(\frac{\partial W_\theta}{\partial \theta}\right)^2 + \frac{\alpha_\phi^2}{\sin^2\theta}\right] + 2mV(r) = 2mE, \tag{10–57}$$

where we have explicitly identified the constant Hamiltonian with the total energy E. Note that all dependence on θ, and on θ alone, has been segregated into the expression within the square brackets. The Hamilton–Jacobi equation then conforms to the appearance of Eq. (10–41), and following the argument given there we see that the quantity in the square brackets must be a constant:

$$\left(\frac{\partial W_\theta}{\partial \theta}\right)^2 + \frac{\alpha_\phi^2}{\sin^2 \theta} = \alpha_\theta^2. \tag{10–58}$$

Finally the dependence of W on r is given by the remains of the Hamilton–Jacobi equation:

$$\left(\frac{\partial W_r}{\partial r}\right)^2 + \frac{\alpha_\theta^2}{r^2} = 2m(E - V(r)). \tag{10–59}$$

The variables in the Hamilton–Jacobi equation are thus completely separated. Equations (10–58) and (10–59) may be easily reduced to quadratures providing at least a formal solution for $W_\theta(\theta)$ and $W_r(r)$, respectively.

Note that the constants of integration α_ϕ, α_θ, α_1 all have directly recognizable physical meanings. The quantity α_ϕ is, of course, the constant value of the angular momentum about the polar axis:

$$\alpha_\phi = p_\phi = \frac{\partial W_\phi}{\partial \phi}. \tag{10–60}$$

To identify α_θ Eq. (10–58) can be rewritten as

$$p_\theta^2 + \frac{p_\phi^2}{\sin^2 \theta} = \alpha_\theta^2, \tag{10–58'}$$

so that the Hamiltonian, Eq. (10–54) appears as

$$H = \frac{1}{2m}\left(p_r^2 + \frac{\alpha_\theta^2}{r^2}\right) + V(r). \tag{10–54'}$$

Comparison with Eq. (10–49) for the Hamiltonian as expressed in terms of polar coordinates in the plane of the orbit shows that α_θ is the same as p_ψ, the magnitude of the total angular momentum:

$$\alpha_\theta = p_\psi \equiv l. \tag{10–61}$$

Lastly, α_1 is of course the total energy E. Indeed the three differential equations for the component parts of W can be looked on as statements of conservation theorems. Equation (10–60) says the z component of the angular momentum vector, $\mathbf{L}$, is conserved, while Eq. (10–58) states the conservation of the magnitude, l, of the angular momentum. And Eq. (10–59) is a form of the energy conservation theorem.

In this simple example some of the power and elegance of the Hamilton–Jacobi method begins to be apparent. A few short steps suffice to obtain the dependence of r on t and the orbit equation, Eqs. (10–53a and b), results derived earlier only with considerable labor. The conserved quantities of the central force problem also appear automatically. Separation of variables for the purely central force problem can also be performed in other coordinate systems, e.g., parabolic coordinates, and the conserved quantities appear there in forms appropriate to the particular coordinates.

Finally, we can employ the Staeckel conditions to find the most general form of a scalar potential V for a single particle for which the Hamilton–Jacobi equation is separable in spherical polar coordinates. The matrix $\boldsymbol{\phi}$ of the Staeckel conditions depends only on the coordinate system and not on the potential. Since the Hamilton–Jacobi equation is separable in spherical polar coordinates for at least one potential, i.e., the central force potential, it follows that the matrix $\boldsymbol{\phi}$ does exist. The specific form of $\boldsymbol{\phi}$ is not needed to answer our question.* Further, since **a** by hypothesis is zero, all we need do is apply Eq. (10–46) to find the most general separable form of V. From the kinetic energy (see p. 345), the diagonal elements of T are

$$T_{rr} = m, \qquad T_{\theta\theta} = mr^2, \qquad T_{\phi\phi} = mr^2 \sin^2 \theta.$$

By Eq. (10–40) it follows that the desired potential must have the form

$$V(q) = V_r(r) + \frac{V_\theta(\theta)}{r^2} + \frac{V_\phi(\phi)}{r^2 \sin^2 \theta}. \tag{10–62}$$

It is easy to verify directly that with this potential the Hamilton–Jacobi equation is still completely separable in spherical polar coordinates.

10–5 ACTION-ANGLE VARIABLES IN SYSTEMS OF ONE DEGREE OF FREEDOM

Of especial importance in many branches of physics are systems in which the motion is periodic. Very often we are interested not so much in the details of the orbit as in the frequencies of the motion. A very elegant and powerful method of handling such systems is provided by a variation of the Hamilton–Jacobi procedure. In this technique the integration constants α_i appearing directly in the solution of the Hamilton–Jacobi equation are not themselves chosen to be the new momenta. Instead we use suitably defined constants J_i, which form a set of n independent functions of the α_i's, and which are known as the *action variables*.

For simplicity we shall first consider in this section systems of one degree of freedom. It is assumed the system is conservative so that the Hamiltonian can be written as

$$H(q, p) = \alpha_1.$$

---

* For the actual form of $\boldsymbol{\phi}$ appropriate to spherical polar coordinates, see Appendix D.

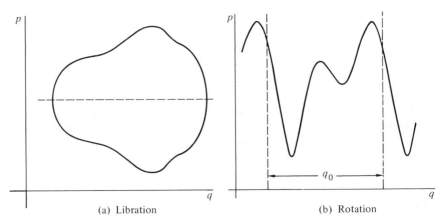

(a) Libration   (b) Rotation

FIGURE **10–1**
Orbit of the system point in phase space for periodic motion of one-dimensional systems.

Solving for the momentum, we have that

$$p = p(q, \alpha_1), \tag{10–63}$$

which can be looked on as the equation of the orbit traced out by the system point in the two-dimensional phase space when the Hamiltonian has the constant value α_1. What is meant by the term "periodic motion" is determined by the characteristics of the phase space orbit. Two types of periodic motion may be distinguished:

    1. In the first type the orbit is *closed*, as shown in Fig. 10–1(a), and the system point retraces its steps periodically. Both q and p are then periodic functions of the time with the same frequency. Periodic motion of this nature will be found when the initial position lies between two zeros of the kinetic energy. It is often designated by the astronomical name *libration*, although to a physicist it is more likely to call to mind the common oscillatory systems, such as the one-dimensional harmonic oscillator.

    2. In the second type of periodic motion the orbit in phase space is such that p is some periodic function of q, with period q_0, as illustrated in Fig. 10–1(b). Equivalently, this kind of motion implies that when q is increased by q_0, the configuration of the system remains essentially unchanged. The most familiar example is that of a rigid body constrained to rotate about a given axis, with q as the angle of rotation. Increasing q by 2π then produces no essential change in the state of the system. Indeed, the position coordinate in this type of periodicity is invariably an angle of rotation, and the motion will be referred to simply as *rotation*,[*] in contrast to libration. The values of q are no longer bounded but can increase indefinitely.

---

[*] Also sometimes designated as *circulation*, or *revolution*.

It may serve to clarify these ideas to note that both types of periodicity may occur in the same physical system. The classic example is the simple pendulum where q is the angle of deflection θ. If the length of the pendulum is l and the potential energy is taken as zero at the point of suspension, then the constant energy of the system is given by

$$E = \frac{p_\theta^2}{2ml^2} - mgl \cos \theta. \tag{10-64}$$

Solving Eq. (10-64) for p_θ, the equation of the path of the system point in phase space is

$$p_\theta = \pm\sqrt{2ml^2(E + mgl \cos \theta)}.$$

If E is less than mgl, then physical motion of the system can only occur for $|\theta|$ less than a bound, θ', defined by the equation

$$\cos \theta' = -\frac{E}{mgl}.$$

Under these conditions the pendulum oscillates between $-\theta'$ and $+\theta'$, which is a periodic motion of the libration type. The system point then traverses some such path in phase space as the curve 1 of Fig. 10-2. However, if $E > mgl$, all values of θ correspond to physical motion and θ can increase without limit to produce a periodic motion of the rotation type. What happens physically in this case is that the pendulum has so much energy that it can swing through the vertical position $\theta = \pi$ and therefore continues rotating. Curve 3 in Fig. 10-2 corresponds to the rotation motion of the pendulum. The limiting case when $E = mgl$ is illustrated by curves 2 and 2' in Fig. 10-2. At this energy the pendulum arrives at $\theta = \pi$, the vertical position, with zero kinetic energy, that is, $p_\theta = 0$. It is then in unstable equilibrium and could in principle remain there indefinitely. However, if there is the slightest perturbation it could continue its motion either along curve 2 or

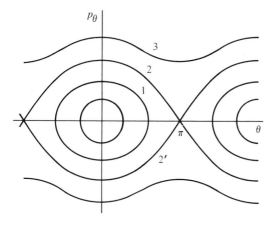

FIGURE 10-2
Phase space orbits for the simple pendulum.

switch to curve 2′—it could fall down either way. The point $\theta = \pi$, $p_\theta = 0$ is a saddle point of the Hamiltonian function $H = E(p_\theta, \theta)$ and there are two paths of constant E in phase space that intersect at the saddle point. We have here an instance (fortunately rare) of what has come to be called a *bifurcation*, a phenomenon of increasing interest in mathematics.

For either type of periodic motion we can introduce a new variable J designed to replace α_1 as the transformed (constant) momentum. The so-called action variable J is defined as

$$J = \oint p\, dq, \tag{10–65}$$

where the integration is to be carried over a complete period of libration or of rotation, as the case may be. (The designation as action variable stems from the resemblance of Eq. (10–65) to the abbreviated action of Section 8–6. Note that J always has the dimensions of an angular momentum.) From Eq. (10–63) it follows that J is always some function of α_1 alone:

$$\alpha_1 \equiv H = H(J). \tag{10–66}$$

Hence Hamilton's characteristic function can be written as

$$W = W(q, J). \tag{10–67}$$

The generalized coordinate conjugate to J, known as the *angle variable** w, is defined by the transformation equation:

$$w = \frac{\partial W}{\partial J}. \tag{10–68}$$

Correspondingly, the equation of motion for w is

$$\dot{w} = \frac{\partial H(J)}{\partial J} = v(J), \tag{10–69}$$

where v is a constant function of J only. Equation (10–69) has the immediate solution

$$w = vt + \beta, \tag{10–70}$$

so that w is a linear function of time, exactly as in Eq. (10–29).

So far the action-angle variables appear as no more than a particular set of the general class of transformed coordinates to which the Hamilton–Jacobi equation leads. Equation (10–68) could be solved for q as a function of w and J, which, in combination with Eq. (10–70), would give the desired solution for q as

---

* The name "action-angle variables" was first used by K. Schwarzschild in an epoch-making paper (Sitzungber. der Kgl. Akad. d. Wiss. 1916, p. 548), which unfortunately marked the end of a brilliant career, cut short tragically in World War I.

a function of time. But when employed in this fashion the variables have no significant advantage over any other set of coordinates generated by W. Their particular merit rises rather from the physical interpretation that can be given to v. Consider the change in w as q goes through a complete cycle of libration or rotation, as given by

$$\Delta w = \oint \frac{\partial w}{\partial q} dq. \tag{10–71}$$

By (Eq. (10–68) this can also be written

$$\Delta w = \oint \frac{\partial^2 W}{\partial q \, \partial J} dq. \tag{10–72}$$

Because J is a constant, the derivative with respect to J can be taken outside the integral sign:

$$\Delta w = \frac{d}{dJ} \oint \frac{\partial W}{\partial q} dq = \frac{d}{dJ} \oint p\,dq = 1, \tag{10–73}$$

where use has been made of the definition for J, Eq. (10–65).

Equation (10–73) states that w changes by unity as q goes through a complete period. But from Eq. (10–70) it follows that if τ is the period for a complete cycle of q, then

$$\Delta w = 1 = v\tau.$$

Hence the constant v can be identified as the reciprocal of the period,

$$v = \frac{1}{\tau} \tag{10–74}$$

and is therefore *the frequency associated with the periodic motion of q*. The use of action-angle variables thus provides a powerful technique for obtaining the frequency of periodic motion *without finding a complete solution to the motion of the system*. If it is known a priori that a system of one degree of freedom is periodic according to the definitions given above, then the frequency can be found once H is determined as a function of J. The derivative of H with respect to J, by Eq. (10–69), then directly gives the frequency v of the motion. The designation of w as an angle variable becomes obvious from the identification of v in Eq. (10–70) as a frequency. It also has been remarked that J has the dimensions of an angular momentum, and of course the coordinate conjugate to an angular momentum is an angle.*

---

* For some applications the action variable is defined in the literature of celestial mechanics as $(2\pi)^{-1}$ times the value given in Eq. (10–65). By Eq. (10–73) the corresponding angle variable is 2π times our definition and in place of v we have ω, the angular frequency. However, we shall stick throughout to the familiar definitions used in physics, as given above.

As an illustration of the application of action-angle variables to find frequencies, let us consider again the familiar linear harmonic oscillator problem. From Eqs. (10–18) and the defining equation (10–65), the constant action variable J is given by

$$J = \oint p\, dq = \oint \sqrt{2m\alpha - m^2\omega^2 q^2}\, dq,$$

where α is the constant total energy and ω is such that $\omega^2 = k/m$. The substitution

$$q = \sqrt{\frac{2\alpha}{m\omega^2}} \sin \theta$$

reduces the integral to

$$J = \frac{2\alpha}{\omega} \int_0^{2\pi} \cos^2 \theta\, d\theta, \tag{10–75}$$

where the limits are such as to correspond to a complete cycle in q. Because the average of $\cos^2 \theta$ over a complete cycle is $1/2$, Eq. (10–75) reduces simply to

$$J = \frac{2\pi\alpha}{\omega}$$

or, solving for α,

$$\alpha \equiv H = \frac{J\omega}{2\pi}. \tag{10–76}$$

The frequency of oscillation is therefore

$$\frac{\partial H}{\partial J} = \nu = \frac{\omega}{2\pi} = \frac{1}{2\pi}\sqrt{\frac{k}{m}}, \tag{10–77}$$

which is the customary formula for the frequency of a linear harmonic oscillator. Although it is entirely unnecessary for obtaining the frequencies, it is nevertheless instructive (and useful for future applications) to write the solutions, Eqs. (10–21) and (10–22), in terms of J and w. It will be recognized first that the combination $\omega(t + \beta)$ is by Eqs. (10–77) and (10–70) the same as $2\pi w$, with the constant of integration suitably redefined. Hence the solutions for q, Eq. (10–21), and p, Eq. (10–22), take on the form

$$q = \sqrt{\frac{J}{\pi m \omega}} \sin 2\pi w, \tag{10–78}$$

$$p = \sqrt{\frac{mJ\omega}{\pi}} \cos 2\pi w. \tag{10–79}$$

Note that Eqs. (10–78) and (10–79) can also be looked on as the transformation equations from the (w, J) set of canonical variables to the (q, p) canonical set.

10–6 ACTION-ANGLE VARIABLES FOR COMPLETELY SEPARABLE SYSTEMS*

Action-angle variables can also be introduced for certain types of motion of systems with many degrees of freedom, providing there exists one or more sets of coordinates in which the Hamilton–Jacobi equation is completely separable. As before, only conservative systems will be considered, so that Hamilton's characteristic function will be used. Complete separability means that the equations of canonical transformation have the form

$$p_i = \frac{\partial W_i(q_i; \alpha_1, \ldots, \alpha_n)}{\partial q_i}, \tag{10–80}$$

which provides each p_i as a function of the q_i and the n integration constants α_j:

$$p_i = p_i(q_i; \alpha_1, \ldots, \alpha_n). \tag{10–81}$$

Equation (10–81) is the counterpart of Eq. (10–63), which applied to systems of one degree of freedom. It will be recognized that Eq. (10–81) here represents the orbit equation of the projection of the system point on the (q_i, p_i) plane in phase space. We can define action-angle variables for the system when the orbit equations for *all* of the (q_i, p_i) pairs describe either closed orbits (libration, as in Fig. 10–1a) or periodic functions of q_i (rotation, as in Fig. 10–1b).

It should be emphasized that this characterization of the motion does not mean that each q_i and p_i will necessarily be periodic functions of the time, i.e., that they repeat their values at fixed time intervals. Even when each of the separated (q_i, p_i) sets are indeed periodic in this sense, the overall system motion need not be periodic. Thus, in a three-dimensional harmonic oscillator the frequencies of motion along the three Cartesian axes may all be different. In such an example it is clear the complete motion of the particle may not be periodic. If the separate frequencies are not rational fractions of each other, the particle will not traverse a closed curve in space but will describe an open "Lissajous figure." Such motion will be described as *multiply-periodic*. It is the advantage of the action-angle variables that they lead to an evaluation of all the frequencies involved in multiply-periodic motion without requiring a complete solution of the motion.

In analogy to Eq. (10–65) the action variables J_i are defined in terms of line integrals over complete periods of the orbit in the (q_i, p_i) plane:

$$J_i = \oint p_i \, dq_i. \tag{10–82}$$

If one of the separation coordinates is cyclic, its conjugate momentum is constant. The corresponding orbit in the q_i, p_i plane of phase space is then a horizontal straight line, which would not appear to be in the nature of a periodic

---

* Unless otherwise stated, the summation convention will *not* be used in this section.

motion. Actually the motion can be considered as a limiting case of the rotation type of periodicity, in which q_i may be assigned any arbitrary period. Since the coordinate in a rotation periodicity is invariably an angle, such a cyclic q_i always has a natural period of 2π. Accordingly, the integral in the definition of the action variable corresponding to a cyclic angle coordinate is to be evaluated from 0 to 2π, and hence

$$J_i = 2\pi p_i \tag{10-83}$$

for all cyclic variables.

By Eq. (10–80) J_i can also be written as

$$J_i = \oint \frac{\partial W_i(q_i; \alpha_1, \ldots, \alpha_n)}{\partial q_i} \, dq_i. \tag{10-84}$$

Since q_i is here merely a variable of integration, each action variable J_i is a function only of the n constants of integration appearing in the solution of the Hamilton–Jacobi equation. Further, it follows from the independence of the separate variable pairs (q_i, p_i) that the J_i's form n independent functions of the α_i's and hence are suitable for use as a set of new constant momenta. Expressing the α_i's as functions of the action variables, the characteristic function W can be written in the form

$$W = W(q_1, \ldots, q_n; J_1, \ldots, J_n) = \sum_j W_j(q_j; J_1, \ldots, J_n),$$

while the Hamiltonian appears as a function of the J_i's only:

$$H = \alpha_1 = H(J_1, \ldots, J_n). \tag{10-85}$$

As in the system of one degree of freedom, we can define conjugate angle variables w_i by the equations of transformation that here appear as

$$w_i = \frac{\partial W}{\partial J_i} = \sum_{j=1}^{n} \frac{\partial W_j(q_j; J_1, \ldots, J_n)}{\partial J_i}. \tag{10-86}$$

The w_i's satisfy equations of motion given by

$$\dot{w}_i = \frac{\partial H(J_1, \ldots, J_n)}{\partial J_i} = v_i(J_1, \ldots, J_n). \tag{10-87}$$

Because the v_i's are constants, functions of the action variables only, the angle variables are all linear functions of time

$$w_i = v_i t + \beta_i. \tag{10-88}$$

Note that in general the separate w_i's increase in time at different rates.

The constants v_i can be identified with the frequencies of the multiply-periodic motion, but the argument to demonstrate the relation is more subtle than for periodic systems of one degree of freedom. The transformation equations to the (w, J) set of variables implies that each q_j (and p_j) is a function of the constants J_i and the variables w_i. What we want to find is what sort of mathematical function the q's are of the w's. To do this we examine the change in a particular w_i when each of the variables q_j is taken through an integral number, m_j, of cycles of libration or rotation. In carrying out this purely mathematical procedure we are clearly *not* following the motion of the system in time. It is as if the flow of time were suspended and each of the q's were moved, manually as it were, independently through a number of cycles of their motion. In effect we are dealing with the virtual displacements of Chapter 1, and accordingly the infinitesimal change in w_i as the q_j's are changed infinitesimally will be denoted by δw_i and is given by

$$\delta w_i = \sum_j \frac{\partial w_i}{\partial q_j} dq_j = \sum \frac{\partial^2 W}{\partial J_i \, \partial q_j} dq_j,$$

where use has been made of Eq. (10–86). The derivative with respect to q_i vanishes except for the W_j constituent of W, so that by Eq. (10–80) δw_i reduces to

$$\delta w_i = \frac{\partial}{\partial J_i} \sum_j p_j(q_j, J) \, dq_j. \tag{10–89}$$

Equation (10–89) represents δw_i as the sum of independent contributions each involving only the q_j motion. The total change in w_i as a result of the specified maneuver is therefore

$$\Delta w_i = \sum_j \frac{\partial}{\partial J_i} \oint_{m_j} p_j(q_j, J) \, dq_j.$$

The differential operator with respect to J_i can be kept outside the integral signs because throughout the cyclic motion of q_i all the J's are of course constant. Below each integral sign the symbol m_j indicates the integration is over m_j cycles of q_j. But each of the integrals is, by the definition of the action variables, exactly $m_j J_j$. Since the J's are independent it follows that

$$\Delta w_i = m_i. \tag{10–90}$$

Further, it will be noted that if any q_j does not go through a complete number of cycles, then in the integration over q_j there will be a remainder of an integral over a fraction of a cycle and Δw_i will not have an integral value. If the sets of w's and m's are treated as vectors $\mathbf{w}$ and $\mathbf{m}$, respectively, Eq. (10–90) can be written as

$$\Delta \mathbf{w} = \mathbf{m}. \tag{10–90'}$$

Suppose, first, that the separable motions are all of the libration type so that each q_j, as well as p_j, returns to its initial value on completion of a complete cycle. The result described by Eq. (10–90') could now be expressed something as follows: $\boldsymbol{\eta}$ (the vector of q's and p's) is such a function of $\mathbf{w}$ that a change $\Delta \boldsymbol{\eta} = 0$ corresponds to a change $\Delta \mathbf{w} = \mathbf{m}$, a vector of integer values. Since the number of cycles in the chosen motions of q_j are arbitrary, $\mathbf{m}$ can be taken as zero except for $m_i = 1$, and all the components of $\boldsymbol{\eta}$ remain unchanged or return to their original value. Hence in the most general case the components of $\boldsymbol{\eta}$ must be periodic functions of *each* w_i with period unity; i.e., the q's and p's are multiply-periodic functions of the w's with unit periods. Such a multiply-periodic function can always be represented by a multiple Fourier expansion, which for q_k, say, would appear as

$$q_k = \sum_{j_1 = -\infty}^{\infty} \sum_{j_2 = -\infty}^{\infty} ,\ldots, \sum_{j_n = -\infty}^{\infty} a_{j_1,\ldots,j_n}^{(k)} \, e^{2\pi i (j_1 w_1 + j_2 w_2 + j_3 w_3 + \cdots + j_n w_n)}, \qquad \text{(libration)}$$

$$(10\text{–}91)$$

where the j's are n integer indices running from $-\infty$ to $+\infty$. By treating the set of j's also as a vector in the same n-dimensional space with $\mathbf{w}$, the expansion can be written more compactly as

$$q_k = \sum_{\mathbf{j}} a_{\mathbf{j}}^{(k)} \, e^{2\pi i \mathbf{j} \cdot \mathbf{w}}, \qquad \text{(libration).} \qquad (10\text{–}91')$$

If we similarly write Eq. (10–88) as a vector equation,

$$\mathbf{w} = \mathbf{v} t + \boldsymbol{\beta}, \qquad (10\text{–}88')$$

then the time dependence of q_k appears in the form

$$q_k(t) = \sum_{\mathbf{j}} a_{\mathbf{j}}^{(k)} \, e^{2\pi i \mathbf{j} \cdot (\mathbf{v} t + \boldsymbol{\beta})}, \qquad \text{(libration).} \qquad (10\text{–}92)$$

Note that in general $q_k(t)$ is *not* a periodic function of t. Unless the various v_i's are commensurate, i.e., rational multiples of each other, q_k will not repeat its values at regular intervals of time.* Finally it should be remembered that the coefficients $a_{\mathbf{j}}^{(k)}$ can be found by the standard procedure for Fourier coefficients, that is, they are given by the multiple integral over the unit cell in $\mathbf{w}$ space:

$$a_{\mathbf{j}}^{(k)} = \int_0^1 ,\ldots, \int_0^1 q_k(\mathbf{w}) e^{-2\pi i \mathbf{j} \cdot \mathbf{w}} (d\mathbf{w}). \qquad (10\text{–}93)$$

Here $(d\mathbf{w})$ stands for the volume element in the n-dimensional space of the w_i's.

When the motion is in the nature of a rotation, then in a complete cycle of the separated variable pair (q_k, p_k) the coordinate q_k does not return to its original value, but instead increases by the value of its period q_{0k}. Such a rotation coordinate is therefore not itself even multiply-periodic. However during the cycle we have seen that w_k increases by unity. Hence the function $q_k - w_k q_{0k}$ does return to its initial value and, like the librational coordinates, is a multiply-periodic

---

*As a function of t, q_k is described as a *quasi-periodic* function.

function of all the w's with unit periods. We can therefore expand the function in a multiple Fourier series analogous to Eq. (10–91):

$$q_k - w_k q_{0k} = \sum_j a_j^{(k)} e^{2\pi i j \cdot \mathbf{w}}, \qquad \text{(rotation)} \qquad (10\text{–}93)$$

or

$$q_k = q_{0k}(v_k t + \beta_k) + \sum_j a_j^{(k)} e^{2\pi i j \cdot (vt + \beta)}, \qquad \text{(rotation)}. \qquad (10\text{–}94)$$

Thus, it is always possible to derive a multiply-periodic function from a rotation coordinate, which can then be handled exactly like a libration coordinate. To simplify the further discussion we will therefore confine ourselves primarily to the libration type of motion.

The separable momentum coordinates, p_k, are by the nature of the assumed motion also multiply-periodic functions of the w's and can be expanded in a multiple Fourier series similar to Eq. (10–91). It follows then that any function of the several variable pairs (q_k, p_k) will also be multiply-periodic functions of the w's and can be written in the form

$$f(q, p) = \sum_j b_j e^{2\pi i j \cdot \mathbf{w}} = \sum_j b_j e^{2\pi i j \cdot (vt + \beta)}. \qquad (10\text{–}95)$$

For example, where the Cartesian coordinates of particles in the system are not themselves the separation coordinates, they can still be written as functions of time in the fashion of Eq. (10–95).

While Eqs. (10–91) and (10–92) represent the most general type of motion consistent with the assumed nature of the problem, not all systems will exhibit this full generality. In particular, for most problems simple enough to be used as illustrations of the application of action-angle variables, each separation coordinate q_k will be a function only of its corresponding w_k. When this happens q_k is then a periodic function of w_k (and therefore of time) and the multiple Fourier series reduces to a single Fourier series:

$$q_k = \sum_j a_j^{(k)} e^{2\pi i j w_k} = \sum_j a_j^{(k)} e^{2\pi i j (v_k t + \beta_k)}. \qquad (10\text{–}96)$$

In the language of Chapter 6, in such problems the q_k's are in effect the normal coordinates of the system. However, even when the motion in the q's can be so simplified, it frequently happens that functions of all the q's, such as Cartesian coordinates, remain multiply-periodic functions of the w's and must be represented as in Eq. (10–95). If the various frequencies v_k are incommensurate, then such functions are not periodic functions of time. The motion of a two-dimensional anisotropic harmonic oscillator provides a convenient and familiar example of these considerations. Suppose that in a particular set of Cartesian coordinates the Hamiltonian is given by

$$H = \frac{1}{2m}[(p_x^2 + 4\pi^2 m^2 v_x^2 x^2) + (p_y^2 + 4\pi^2 m^2 v_y^2 y^2)].$$

These Cartesian coordinates are therefore suitable separation variables, and each will exhibit simple harmonic motion with frequencies v_x and v_y, respectively. Thus the solutions for x and y are particularly simple forms of the single Fourier expansions of Eq. (10–96). Suppose now that the coordinates are rotated 45° about the z axis; the components of the motion along the new x', y' axes will be

$$x' = \frac{1}{\sqrt{2}} [x_0 \cos 2\pi(v_x t + \beta_x) + y_0 \cos 2\pi(v_y t + \beta_y)],$$

$$y' - \frac{1}{\sqrt{2}} [y_0 \cos 2\pi(v_y t + \beta_y) - x_0 \cos 2\pi(v_x t + \beta_x)].$$

(10–97)

If v_x/v_y is a rational number, these two expressions will be simply periodic, corresponding to a closed Lissajous figure. But if v_x and v_y are incommensurable, the Lissajous figure never exactly retraces its steps and Eqs. (10–97) provide simple examples of multiply-periodic series expansions of the form (10–95).

Even when q_k is a multiply-periodic function of all the w's, one intuitively feels there must be a special relationship between q_k and its corresponding w_k (and therefore v_k). After all, the argument culminating in Eq. (10–90) says that when q_k alone goes through its complete cycle, w_k is singled out as increasing by unity, while the other w's return to their initial value. It was only in 1961 that J. Vinti succeeded in expressing this intuitive feeling in a precise and rigorous statement.[*] Suppose that the time interval T contains m complete cycles of q_k plus a fraction of a cycle. In general the times required for each successive cycle will be different, since q_k will not be a periodic function of t. Then Vinti showed, on the basis of a theorem in number theory, that as T increases indefinitely,

$$\underset{T \to \infty}{L} \frac{m}{T} = v_k.$$

(10–98)

The *mean* frequency of the motion of q_k is therefore always given by v_k, even when the entire motion is more complicated than a simple periodic function with frequency v_k.

Barring commensurability of all the frequencies, a multiply-periodic function can always be formed from the generating function W. The defining equation for J_i, Eq. (10–84), in effect states that when q_i goes through a complete cycle, i.e., when w_i changes by unity, the characteristic function increases by J_i. It follows that the function

$$W' = W - \sum_k w_k J_k$$

(10–99)

remains unchanged when *each* w_k is increased by unity, all the other angle variables remaining constant. Equation (10–99) therefore represents a multiply-periodic function that can be expanded in terms of the w_i (or of the frequencies v_i),

---

[*] J. Vinti, *J. Res. Nat. Bur. Standards*, **65B**, 131 (1961). See also R. Garfinkel in *Space Mathematics*, Part 1, p. 57, 1966.

by a series of the form of Eq. (10–95). Since the transformation equations for the angle variables are

$$w_k = \frac{\partial W}{\partial J_k},$$

it will be recognized that Eq. (10–99) defines a Legendre transformation from the q, J basis to the q, w basis. Indeed, comparison with Eq. (9–15) in combination with Eq. (9–12) shows that if $W(q, J)$ is a generating function of the form $F_2(q, P)$, then $W'(q, w)$ is the corresponding generating function of the type $F_1(q, Q)$, transforming in both cases from the (q, p) variables to the (w, J) variables. While W' thus generates the same transformation as W it is of course *not* a solution of the Hamilton Jacobi equation.*

It has been emphasized that the system configuration is multiply-periodic only if the frequencies v_i are not rational fractions of each other. Otherwise the configuration repeats after a sufficiently long time and would therefore be simply periodic. The formal condition for the commensurability of all the frequencies is that there exist $n - 1$ relations of the form

$$\sum_{i=1}^{n} j_i v_i = 0, \tag{10–100}$$

where the j_i's are integers. By solving these equations we can then express any v_i as a rational fraction of any of the other frequencies. When there are only m relations of the form (10–100) between the fundamental frequencies, then the system is said to be *m-fold degenerate*. If m is equal to $n - 1$, so that the motion is simply periodic, then the system is said to be *completely degenerate*. Thus, whenever *the orbit of the system point is closed, the motion will be completely degenerate.*

There is an interesting connection between degeneracy and the coordinates in which the Hamiltonian–Jacobi equation is separable. It can be shown that the path of the system point for a nondegenerate system completely fills a limited region of both configuration and phase space (cf. Born, *The Mechanics of the Atom*, Appendix 1). Suppose the problem in such that the motion in any one of the separation coordinates is simply periodic and has therefore been shown to be independent of the motion of the other coordinates. Hence the path of the system point as a whole must be limited by the surfaces of constant q_i and p_i that mark the bounds of the oscillatory motion of the separation variables. (The argument is

---

* Action-angle variables have been defined here in terms of the simply periodic separable coordinates. It was then shown that the motion of the system as a whole was in general multiply-periodic. Mention should be made that it is possible to reverse the process. Starting from the recognition that the system motion is multiply-periodic, it is possible to introduce the action-angle variables such that the system configuration and the generating function $W'(q, w)$ are multiply-periodic in the w's with the period unity, and the Hamiltonian is cyclic in all w's. One can avoid in this manner the necessity of referring to the separation coordinates. For further details see Born, *The Mechanics of the Atom*, Section 15.

easily extended to rotation by limiting all angles to the region 0 to 2π.) These surfaces therefore define the volume in space that is densely filled by the system point orbit. It obviously follows that the separation of variables in nondegenerate systems must be unique; the Hamiltonian–Jacobi equation cannot be separated in two different coordinate systems (aside from trivial variations such as change of scale). The possibility of separating the motion in more than one set of coordinates thus normally provides evidence that the system is degenerate.*

The simplest examples of degeneracy occur when two or more of the frequencies are equal. If two of the force constants in a three-dimensional harmonic oscillator are equal, then the corresponding frequencies are identical and the system is single degenerate. In an isotropic linear oscillator the force constants are the same along all directions, all frequencies are equal, and the system is completely degenerate.

Whenever degeneracy is present the fundamental frequencies are no longer independent and the periodic motion of the system can be described by less than the full complement of n frequencies. Indeed, the m conditions of degeneracy can be used to reduce the number of frequencies to $n - m$, and the system motion is said to be $n - m$-fold periodic. The reduction of the frequencies may be most elegantly performed by means of a point transformation of the action-angle variables. The m degeneracy conditions may be written in summary form as

$$\sum_{i=1}^{n} j_{ki}v_i = 0, \qquad k = 1,\ldots,m. \tag{10–101}$$

Consider now a point transformation from (w, J) to (w', J') defined by the generating function (cf. Eq. 9–26) where the summation convention is used):

$$F_2 = \sum_{k=1}^{m}\sum_{i=1}^{n} J_k' j_{ki} w_i + \sum_{k=m+1}^{n} J_k' w_k. \tag{10–102}$$

The transformed coordinates are

$$\begin{aligned} w_k' &= \sum_{i=1}^{n} j_{ki} w_i, & k &= 1,\ldots,m, \\ w_k' &= w_k, & k &= m+1,\ldots,n. \end{aligned} \tag{10–103}$$

Correspondingly, the new frequencies are

$$\begin{aligned} v_k' = \dot{w}_k' &= \sum_{i=1}^{n} j_{ki} v_i = 0 & k &= 1,\ldots,m, \\ &= v_k & k &= m+1,\ldots,n. \end{aligned} \tag{10–104}$$

---

* Pathological cases exist in which a Hamiltonian may be separable in several coordinate systems, yet the orbits are closed, i.e., degenerate, only if they are contained within a particular subregion. Discontinuities may occur if the orbit crosses certain boundaries. See E. Onofri and M. Pauri, *Jour. Math. Phys.* **14**, 1106 (1973). Such restrictions to "local" degeneracy do not occur in the usual Hamiltonian that is differentiable and continuous.

Thus in the transformed coordinates m of the frequencies are zero and we are left with a set of $n - m$ independent frequencies. It is obvious that the new w'_k may also be termed as angle variables in the sense that the system configuration is multiply-periodic in the w'_k coordinates with the fundamental period unity. The corresponding constant action variables are given as the solution of the n equations of transformation

$$J_i = \sum_{k=1}^{m} J'_k j_{ki} + \sum_{k=m+1}^{n} J'_k \delta_{ki}. \tag{10–105}$$

The zero frequencies correspond to constant factors in the Fourier expansion. These are, of course, also present in the original Fourier series in terms of the v's, Eq. (10–91), occurring whenever the indices j_i are such that degeneracy conditions are satisfied. Since

$$v'_i = \frac{\partial H}{\partial J'_i},$$

the Hamiltonian must be independent of the action variables J'_i whose corresponding frequencies vanish. In a completely degenerate system the Hamiltonian can therefore be made to depend on only one of the action variables.

It should be noted that Hamilton's characteristic function W also serves as the generating function for the transformation from the (q, p) set to the (w', J') set. Since the J' quantities are n independent constants, the original constants of integration may be expressed in terms of the J' set, and W given as $W(q, J')$. In this form it is a generating function to a new set of canonical variables for which the J' quantities are the canonical momenta. But by virtue of the point transformation generated by the F_2 of Eq. (10–102) we know that w' is conjugate to J'. Hence it follows that the new coordinates generated by $W(q, J')$ must be the w' set, with equations of transformation given by

$$w'_i = \frac{\partial W}{\partial J'_i}. \tag{10–106}$$

(For a more formal proof of Eq. (10–106) based on the algebraic structure of Eq. (10–102), see Exercise 17.)

The problem of the bound motion of a particle in an inverse square law central force illustrates many of the phenomena involved in degeneracy. A discussion of this problem also affords an opportunity to show how the action-angle technique is applied to specific systems, and to indicate the connections with Bohr's quantum mechanics and with celestial mechanics. Accordingly, the next section is devoted to a detailed treatment of the Kepler problem in terms of action-angle variables.

10–7 THE KEPLER PROBLEM IN ACTION-ANGLE VARIABLES*

To exhibit all of the properties of the solution we shall examine the motion in space, rather than make use of our a priori knowledge that the orbit lies in a plane. In terms of spherical polar coordinates the Kepler problem becomes a special case of the general treatment given above in Section 10–4 for central force motion in space. Equations (10–54) through (10–61) can be taken over here immediately, replacing $V(r)$ wherever it occurs by its specific form

$$V(r) = -\frac{k}{r}. \tag{10–107}$$

The Hamilton–Jacobi equation with this potential has been demonstrated to be completely separable in spherical polar coordinates. We shall confine our discussion to the bound case, that is, $E < 0$. Hence the motion in each of the coordinates will be periodic—libration in r and θ, and rotation in ϕ. The conditions for the application of action-angle variables are thus satisfied, and we can proceed to construct the action variables on the basis of the defining equation (10–84). From Eq. (10–56) it follows that

$$J_\phi = \oint \frac{\partial W}{\partial \phi} d\phi = \oint \alpha_\phi \, d\phi. \tag{10–108a}$$

Similarly, on the basis of Eq. (10–58), J_θ is given by

$$J_\theta = \oint \frac{\partial W}{\partial \theta} d\theta = \oint \sqrt{\alpha_\theta^2 - \frac{\alpha_\phi^2}{\sin^2 \theta}} \, d\theta. \tag{10–108b}$$

Finally the integral for J_r from Eq. (10–59), is

$$J_r = \oint \frac{\partial W}{\partial r} dr = \oint \sqrt{2mE + \frac{2mk}{r} - \frac{\alpha_\theta^2}{r^2}} \, dr. \tag{10–108c}$$

The first integral is trivial; ϕ goes through 2π radians in a complete revolution and therefore

$$J_\phi = 2\pi\alpha_\phi = 2\pi p_\phi. \tag{10–109}$$

This result could have been predicted beforehand, for ϕ is a cyclic coordinate in H, and Eq. (10–109) is merely a special case of Eq. (10–83) for the action variables corresponding to cyclic coordinates. Integration of Eq. (10–108b) can be performed in various ways; a procedure involving only elementary rules of integration will be sketched here. If the polar angle of the total angular momentum vector is denoted by i, so that

$$\cos i = \frac{\alpha_\phi}{\alpha_\theta}, \tag{10–110}$$

---

* The summation convention will be resumed from here on.

then Eq. (10–108b) can also be written as

$$J_\theta = \alpha_\theta \oint \sqrt{1 - \cos^2 i \csc^2 \theta}\, d\theta.$$

The complete circuital path of integration is for θ going from a limit $-\theta_0$ to $+\theta_0$ and back again, where $\sin \theta_0 = \cos i$, or $\theta_0 = (\pi/2) - i$. Hence the circuital integral can be written as four times the integral over from 0 to θ_0, or after some manipulation,

$$J_\theta = 4\alpha_\theta \int_0^{\theta_0} \csc \theta \sqrt{\sin^2 i - \cos^2 \theta}\, d\theta.$$

The substitution

$$\cos \theta = \sin i \sin \psi$$

transforms the integral to

$$J_\theta = 4\alpha_\theta \sin^2 i \int_0^{\pi/2} \frac{\cos^2 \psi\, d\psi}{1 - \sin^2 i \sin^2 \psi}.$$

Finally, with the substitution

$$u = \tan \psi,$$

the integral becomes

$$J_\theta = 4\alpha_\theta \sin^2 i \int_0^\infty \frac{du}{(1 + u^2)(1 + u^2 \cos^2 i)} = 4\alpha_\theta \int_0^\infty du \left[\frac{1}{1 + u^2} - \frac{\cos^2 i}{1 + u^2 \cos^2 i} \right].$$

$$(10\text{–}111)$$

This last form involves only well-known integrals, and the final result* is

$$J_\theta = 2\pi\alpha_\theta[1 - \cos i] = 2\pi(\alpha_\theta - \alpha_\phi). \tag{10–112}$$

The last integral, for J_r, can now be written as

$$J_r = \oint \sqrt{2mE + \frac{2mk}{r} - \frac{(J_\theta + J_\phi)^2}{4\pi^2 r^2}}\, dr. \tag{10–113}$$

After performing the integration, this equation can be solved for the energy $E \equiv H$ in terms of the three action variables J_ϕ, J_θ, J_r. It will be noted that J_ϕ and J_θ can occur in E only in the combination $J_\theta + J_\phi$, and hence the corresponding

---

* In evaluating the integral of the second term in the final integrand of Eq. (10–111) it has been assumed $\cos i$ is positive. This is always possible, since there is no preferred direction for the z axis in the problem and it may be chosen at will. If $\cos i$ were negative, the sign of α_ϕ in Eq. (10–112) would be positive. For changes in the subsequent formulas, see Exercise 23.

frequencies v_ϕ and v_θ must be equal, indicating a degeneracy. This result has not involved the inverse square law nature of the central force; *any motion produced by a central force is at least singly degenerate.* The degeneracy is of course a consequence of the fact that the motion is confined to a plane normal to the constant angular momentum vector **L**. Motion in this plane implies that θ and ϕ are related to each other such that as ϕ goes through a complete 2π period θ varies through a complete cycle between the limits $(\pi/2) \pm i$. Hence the frequencies in θ and ϕ are necessarily equal.

The integral involved in Eq. (10–113) can be evaluated by elementary means, but the integration is most elegantly and quickly performed using the method of residues, a procedure first employed by Sommerfeld. For the benefit of those familiar with this technique we shall outline the steps involved in integrating Eq. (10–113). Bound motion can, of course, occur only when E is negative (cf. Section 3–3), and since the integrand is equal to $p_r = m\dot{r}$, the limits of the motion are defined by the roots r_1 and r_2 of the expression in the square root sign. If r_1 is the inner bound, as in Fig. 3–6, a complete cycle of r involves going from r_1 to r_2 and then back again to r_1. On the outward half of the journey, from r_1 to r_2, p_r is positive and we must take the positive square root. However, on the return trip to r_1, p_r is negative and the square root must likewise be negative. The integration thus involves both branches of a double-valued function, with r_1 and r_2 as the branch points. Consequently the complex plane can be represented as one of the sheets of a Riemann surface, slit along the real axis from r_1 to r_2 (as indicated in Fig. 10–3).

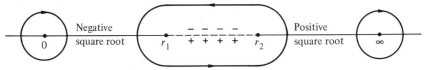

FIGURE 10–3
The complex r plane in the neighborhood of the real axis; showing the paths of integration occurring in the evaluation of J.

Since the path of integration encloses the line between the branch points, the method of residues cannot be applied directly. However, we may also consider the path as enclosing all the rest of the complex plane, the direction of integration now being in the reverse (clockwise) direction.* The integrand is single-valued in this region and there is now no bar to the application of the method of residues. Only two singular points are present, namely, the origin and infinity, and the integration path can be distorted into two clockwise-described circles enclosing

---

* To visualize this change in viewpoint it is convenient to think of the complex plane as projected stereographically onto the surface of a sphere, with the origin at the south pole, and the point ∞ at the north pole. The real axis becomes a meridian circle joining the two poles. Any closed integration path on the sphere divides the surface of the sphere into two areas. The path may then be considered as enclosing either of the two areas, depending upon the direction of the integration.

these two points. Now, the sign in front of the square root in the integrand must be negative for the region along the real axis below r_1, as can be seen by examining the behavior of the function in the neighborhood of r_1. If the integrand is represented as

$$-\sqrt{A + \frac{2B}{r} - \frac{C}{r^2}},$$

the residue at the origin is

$$R_0 = -\sqrt{-C}.$$

Above r_2 the sign of the square root on the real axis is found to be positive, and the residue is obtained by the standard technique of changing the variable of integration to $z = r^{-1}$:

$$-\oint \frac{1}{z^2}\sqrt{A + 2Bz - Cz^2}\,dz. \tag{10-114}$$

Expansion about $z = 0$ now furnishes the residue

$$R_\infty = -\frac{B}{\sqrt{A}}.$$

The total integral is $-2\pi i$ times the sum of the residues:

$$J_r = 2\pi i\left(\sqrt{-C} + \frac{B}{\sqrt{A}}\right), \tag{10-115}$$

or, upon substituting the coefficients A, B, and C:

$$J_r = -(J_\theta + J_\phi) + \pi k\sqrt{\frac{2m}{-E}}. \tag{10-116}$$

Equation (10–116) supplies the functional dependence of H upon the action variables; for solving for E we have

$$H \equiv E = -\frac{2\pi^2 mk^2}{(J_r + J_\theta + J_\phi)^2}. \tag{10-117}$$

It will be noted that, as predicted, J_θ and J_ϕ occur only in the combination $J_\theta + J_\phi$. More than that, all three of the action variables appear only in the form $J_r + J_\theta + J_\phi$. Hence all of the frequencies are equal; *the motion is completely degenerate*. This result could also have been predicted beforehand, for we know that with an inverse square law of force the orbit is closed for negative energies. With a closed orbit, the motion is simply periodic and therefore completely degenerate. If the central force contained an r^{-3} term, such as is provided by relativistic corrections, then the orbit is no longer closed but is in the form of a precessing ellipse. One of the degeneracies will be removed in this case, but the motion is still singly degenerate, since $v_\theta = v_\phi$ for all central forces. The one frequency for the motion here is given by

$$v = \frac{\partial H}{\partial J_r} = \frac{\partial H}{\partial J_\theta} = \frac{\partial H}{\partial J_\phi} = \frac{4\pi^2 mk^2}{(J_r + J_\theta + J_\phi)^3}. \tag{10-118}$$

If we evaluate the sum of the J's in terms of the energy from Eq. (10–117) the period of the orbit is

$$\tau = \pi k \sqrt{\frac{m}{-2E^3}}. \tag{10-119}$$

This formula for the period agrees with Kepler's third law, Eq. (3–71), if it is remembered that the semimajor axis a is equal to $-k/2E$.

The degenerate frequencies may be eliminated by canonical transformation to a new set of action-angle variables, following the procedure outlined in the previous section. Expressing the degeneracy conditions as

$$v_\phi - v_\theta = 0, \qquad v_\theta - v_r = 0,$$

the appropriate generating function is

$$F = (w_\phi - w_\theta)J_1 + (w_\theta - w_r)J_2 + w_r J_3. \tag{10-120}$$

The new angle variables are

$$w_1 = w_\phi - w_\theta,$$
$$w_2 = w_\theta - w_r, \tag{10-121}$$
$$w_3 = w_r,$$

and, as planned, two of the new frequencies, v_1 and v_2, are zero. We can obtain the new action variables from the transformation equations

$$J_\phi = J_1,$$
$$J_\theta = J_2 - J_1,$$
$$J_r = J_3 - J_2,$$

which yield the relations

$$J_1 = J_\phi,$$
$$J_2 = J_\phi + J_\theta, \tag{10-122}$$
$$J_3 = J_\phi + J_\theta + J_r.$$

In terms of these transformed variables the Hamiltonian appears as

$$H = -\frac{2\pi^2 mk^2}{J_3^2}, \tag{10-123}$$

a form involving only that action variable for which the corresponding frequency is different from zero.

If we are willing to use, from the start, our a priori knowledge that the motion for the bound Kepler problem is a particular closed orbit in a plane, then the integrals for J_θ and J_r can be evaluated very quickly and simply. For the J_θ integral we can apply a procedure suggested by J. H. Van Vleck. It will be recalled that when the defining equations for the generalized coordinates do not involve time explicitly, then

$$p_i \dot{q}_i = 2L_2 = 2T.$$

Knowing that the motion is confined to a plane, we can express the kinetic energy T either in spherical polar coordinates or in the plane polar coordinates (r, ψ). It follows, then, that

$$2T = p_r \dot{r} + p_\theta \dot{\theta} + p_\phi \dot{\phi} = p_r \dot{r} + p\dot{\psi}, \tag{10–124}$$

where $p(\equiv l)$ is the magnitude of the total angular momentum. Hence the definition for J_θ can also be written as

$$J_\theta \equiv \oint p_\theta \, d\theta = \oint p \, d\psi - \oint p_\phi \, d\phi.$$

Because the frequencies for θ and ϕ are equal, both ϕ and ψ vary by 2π as θ goes through a complete cycle of libration and the integrals defining J_θ reduce to

$$J_\theta = 2\pi(p - p_\phi) = 2\pi(\alpha_\theta - \alpha_\phi),$$

in agreement with Eq. (10–112).

The integral for J_r, Eq. (10–113), was evaluated in order to obtain $H \equiv E$ in terms of the three action variables. If we use the fact that the closed elliptical orbit in the bound Kepler problem is such that the frequency for r is the same as that for θ and ϕ, then the functional dependence of H on J can also be obtained from Eq. (10–124). In effect then we are evaluating J_r in a different way. The virial theorem for the bound orbits in the Kepler problem says that (cf. Eq. (3–30))

$$\bar{V} = -2\bar{T},$$

where the bar denotes an average over the single complete period of the motion. It follows that

$$H \equiv E = \bar{T} + \bar{V} = -\bar{T}. \tag{10–125}$$

Integrating Eq. (10–124) with respect to time over a complete period of motion we have

$$\frac{2\bar{T}}{\nu_3} = J_r + J_\theta + J_\phi = J_3, \tag{10–126}$$

where ν_3 is the frequency of the motion, i.e., the reciprocal of the period. Combining Eqs. (10–125) and (10–126) leads to the relation

$$-\frac{2}{J_3} = \frac{\nu_3}{H} = \frac{1}{H} \frac{dH}{dJ_3}, \tag{10–127}$$

where use has been made of Eq. (10–87). Equation (10–127) is in effect a differential equation for the functional behavior of H on J_3. Integration of the equation immediately leads to the solution

$$H = \frac{D}{J_3^2},\tag{10–128}$$

where D is a constant that cannot involve any of the J's, and must therefore depend only on m and k. Hence we can evaluate D by considering the elementary case of a circular orbit, of radius r_0, for which $J_r = 0$ and $J_3 = 2\pi p$. The total energy is here

$$H = -\frac{k}{2r_0}\tag{10–129}$$

(as can most immediately been seen from the virial theorem). Further, the condition for circularity, Eq. (3–41), can be written for the inverse square force law as

$$\frac{k}{r_0^2} = \frac{p^2}{mr_0^3} = \frac{J_3}{4\pi^2 mr_0^3}.\tag{10–130}$$

Eliminating r_0 between Eqs. (10–129) and (10–130) leads to

$$H = -\frac{2\pi^2 mk^2}{J_3^2}.$$

This result has been derived only for circular orbits. But Eq. (10–128) says it must also be correct for *all* bound orbits of the Kepler problem, and indeed it is identical with Eq. (10–123). Thus, if the existence of a single period for all coordinates is taken as known beforehand, it is possible to obtain $H(J)$ without direct evaluation of the circuital integrals.

In any problem with three degrees of freedom there must of course be six constants of motion. It has previously been pointed out that in the Kepler problem five of these are algebraic functions of the coordinates and momenta and describe the nature of the orbit in space, and only the last refers to the position of the particle in the orbit at a given time (cf. Section 3–9 above). It is easy to see that five parameters are needed to completely specify, say, the elliptical orbit of the bound Kepler problem in space. Since the motion is in a plane, two constants are needed to describe the orientation of that plane in space. One constant is required to give the *scale* of the ellipse, e.g., the semimajor axis a, and another the *shape* of the ellipse, say, through the eccentricity e. Finally, the fifth parameter must specify the *orientation* of the ellipse relative to some arbitrary direction in the orbital plane.

The classical astronomical *elements* of the orbit provide the orbital parameters almost directly in the form given above. Two of the angles appearing in these elements have unfamiliar but time-honored names. Their definitions, and functions as orbital parameters, can best be seen from a diagram, such as is given in Fig. 10–4. Here *xyz* defines the chosen set of axes fixed in space and the unit

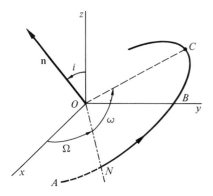

FIGURE 10–4
Angular elements of the orbit in the bound Kepler problem.

vector **n** characterizes the normal to the orbital plane. The intersection between the xy plane and the orbital plane is called the *line of nodes*. There are two points on the line of nodes at which the elliptical orbit intersects the xy plane; the point at which the particle enters from below into the upper hemisphere (or goes from the "southern" to the "northern" hemispheres) is known as the *ascending node*. In Fig. 10–4 the portion of the orbit shown that is in the southern hemisphere has been indicated as a dashed line for clarity. The dot-dashed line ON is a portion of the line of nodes containing the ascending node. We can measure the direction of ON in the xy plane by the angle xON, which is customarily denoted by Ω, and is known as the *longitude of the ascending node*. Finally, if C denotes the point of periapsis in the orbit, then the angle NOC in the orbital plane is denoted by ω and is called the *argument of the perihelion*.* The more familiar angle i, introduced above in Eq. (10–110) is in its astronomical usage known as the *inclination of the orbit*. One usual set of astronomical elements therefore consists of the six constants

$$i, \Omega, a, e, \omega, T,$$

where the last one, T, is the time of passage through the periapsis point. Of the remaining five, the first two define the orientation of the orbital plane in space, while a, e, and ω directly specify the scale, shape, and orientation of the elliptic orbit, respectively.

The action-angle variable treatment of the Kepler problem also leads to five algebraic constants of the motion. Three of them are obvious as the three constant action variables, J_1, J_2, and J_3. The remaining two are the angle variables w_1 and w_2, which are constants, because their corresponding frequencies are zero. It must therefore be possible to express the five constants J_1,

---

* This terminology appears to be commonly used even for orbits that are not around the sun.

J_2, J_3, w_1, and w_2 in terms of the classic orbital elements i, Ω, a, e, and ω, and vice versa. Some of these interrelations are immediately obvious. From Eqs. (10–122) and (10–112) it follows that

$$J_2 = 2\pi\alpha_\theta \equiv 2\pi l, \tag{10–131}$$

and hence, by Eq. (10–110),

$$\frac{J_1}{J_2} = \cos i. \tag{10–132}$$

As is well known, the semimajor axis a is a function only of the total energy E (cf. Eq. (3–61)) and therefore, by Eq. (10–123), a is given directly in terms of J_3:

$$a = -\frac{k}{2E} = \frac{J_3^2}{4\pi^2 mk}. \tag{10–133}$$

In terms of J_2 Eq. (3–62) for the eccentricity can be written as

$$e = \sqrt{1 - \frac{J_2^2}{4\pi^2 mka}},$$

or

$$e = \sqrt{1 - \left(\frac{J_2}{J_3}\right)^2}. \tag{10–134}$$

It remains only to relate the angle coordinates w_1 and w_2 to the classic orbit elements. Obviously, they must involve Ω and ω. In fact, it can be shown that for suitable choice of additive constants of integration they are indeed proportional to Ω and ω, respectively. This will be demonstrated for w_1; the identification of w_2 will be left as an exercise.

The equation of transformation defining w_1 is, by Eq. (10–106),

$$w_1 = \frac{\partial W}{\partial J_1}.$$

It can be seen either from Eq. (10–13′) (p. 447) or from the separated form of W, Eq. (10–55), that W can be written as the sum of indefinite integrals:

$$W = \int p_\phi \, d\phi + \int p_\theta \, d\theta + \int p_r \, dr. \tag{10–135}$$

As we have seen from the discussion on J_r, the radial momentum p_r does not involve J_1, but only J_3 (through E) and the combination $J_\theta + J_\phi = J_2$. Only the first two integrals are therefore involved in the derivative with respect to J_1. By Eq. (10–109),

$$p_\phi = \alpha_\phi = \frac{J_1}{2\pi}, \tag{10–136}$$

and by Eq. (10–58), with the help of Eqs. (10–131) and (10–136),

$$p_\theta = \pm\sqrt{\alpha_\theta^2 - \frac{\alpha_\phi^2}{\sin^2\theta}} = \pm\frac{1}{2\pi}\sqrt{J_2^2 - \frac{J_1^2}{\sin^2\theta}}.$$

It turns out that in order to relate w_1 to the ascending node, it is necessary to choose the negative sign of the square root.* The angle variable w_1 is therefore determined by

$$w_1 = \frac{\phi}{2\pi} + \frac{J_1}{2\pi}\int \frac{d\theta}{\sin^2\theta\sqrt{J_2^2 - J_1^2\csc^2\theta}},$$

or

$$2\pi w_1 = \phi + \cos i \int \frac{d\theta}{\sin^2\theta\sqrt{1 - \cos^2 i\csc^2\theta}}$$

$$= \phi + \int \frac{\cot i\csc^2\theta\, d\theta}{\sqrt{1 - \cot^2 i\cot^2\theta}}.$$

By a change of variable to u, defined through

$$\sin u = \cot i\cot\theta, \tag{10–137}$$

the integration can be performed trivially, and the expression for w_1 reduces to

$$2\pi w_1 = \phi - u. \tag{10–138}$$

The angle coordinate ϕ is the azimuthal angle of the projected on the xy plane measured relative to the x axis. Clearly, from Eq. (10–137) u is a function of the polar angle θ of the particle. But what is its geometrical significance? One can see what u is by reference to Napier's rules† as applied to the spherical triangle defined by the line of nodes, the radius vector, and the projection of the radius vector on the xy plane. However, it may be more satisfying to indulge in a little trigonometric manipulation and derive the relation *ab initio*. In Fig. 10–5 the line ON is the line of nodes, OR is the line of the radius vector at some time, and the dotted line OP is the projection of the radius vector on the xy plane. The angle that OP makes with the x axis is the azimuth angle ϕ. It is contended that u is the angle OP makes with the line of nodes. To prove this, imagine a plane normal both to the xy plane and to the line of nodes, which intersects the radius vector at unit distance from the origin O. The points of intersection with the three lines from the origin, A, B, and C, define with the origin four right triangles. Since OB

---

* Note that when the particle passes through the ascending node (cf. Fig. 10–4) θ is *decreasing* and the corresponding momentum is negative. In calculating J_θ it was not necessary to worry about the choice of sign because in going through a complete cycle both signs are encountered.

† See any book on spherical trigonometry, or such handbooks as the *Handbook of Mathematical Tables* (Chemical Rubber Publishing Co.) or *Handbook of Applied Mathematics* (Van Nostrand-Reinhold).

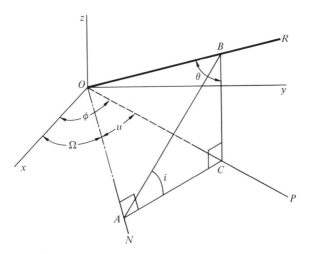

FIGURE 10–5
Diagram illustrating angles appearing in action-angle
treatment of the Kepler problem.

has unit length it follows that $BC = \cos\theta$ and therefore $AC = \cos\theta\cot i$. On the
other hand $OC = \sin\theta$ and therefore it is also true that $AC = \sin\theta\sin u$. Hence
$\sin u = \cot i$, which is identical with Eq. (10–137) and proves the stipulated
identification of the angle u. Figure 10–5 shows clearly that the difference
between ϕ and u must be Ω, so that

$$2\pi w_1 = \Omega. \tag{10–139}$$

In a similar fashion one can identify the physical nature of the constant w_2. Of
the integrals making up W, Eq. (10–135), the two over θ and r contain J_2 and are
therefore involved in finding w_2. After differentiation with respect to J_2, the
integral over θ can be performed by the same type of trigonometric substitution
as employed for w_1. The corresponding integral over r can be carried out in a
number of ways, most directly by using the orbit equation for r in terms of the
polar coordinate angle in the orbital plane. By suitable choice of the arbitrary
lower limit of integration it can thus be found that $2\pi w_2$ is the difference between
two angles in the orbital plane, one of which is the angle of the radius vector
relative to the line of nodes and the other is the same angle but relative to the line
of the periapsis. In other words, $2\pi w_2$ is the argument of the perihelion:

$$2\pi w_2 = \omega. \tag{10–140}$$

Detailed derivation is left to one of the exercises.

The method of action-angle variables does not strike one as the quickest way
to solve the Kepler problem, and the practical usefulness of the set of variables is
not obvious. However, their value has long been demonstrated in celestial

mechanics, where they appear under the guise of the *Delaunay variables*.* As will be seen in the next chapter they provide the natural elements of the orbit to be used in perturbation theory, e.g., when we seek to find the modifications of the nominal Kepler orbits produced by small deviations of the force from the inverse square law. Many of the basic studies on possible perturbations of satellite orbits were carried out in terms of the action-angle variables.

For a brief period action-angle variables, particularly for the Kepler problem, played a prominent role at the very frontiers of physics research. Shortly after the advent of Bohr's quantum theory of the atom in 1913, it was realized that the quantum conditions could be stated most simply in terms of the action variables. For a decade, starting about 1915, there was intense interest in the properties of action-angle variables, and much of the "old quantum theory" was built around them. In classical mechanics the action variables possess a continuous range of values, but this is no longer the case in quantum mechanics. The quantum conditions of Sommerfeld and Wilson required that the motion be limited to such orbits for which the "proper" action variables had discrete values that were integral multiples of h, the quantum of action. (By proper action variables are meant those J's whose frequencies are nondegenerate and different from zero. For example, J_3 is a proper action variable.) As Sommerfeld stated, the method of action-angle variables then provided "a royal road to quantization." One had only to solve the problem in classical mechanics using action-angle variables, and the motion could be immediately quantized by replacing the J's with integral multiples of Planck's constant h.

As an example of this procedure it may be noted that the quantized energy levels for a hydrogenic atom follow at once from Eq. (10–123) if k is set equal to Ze^2 and J_3 is replaced by nh:

$$E = -\frac{2\pi^2 m Z^2 e^4}{n^2 h^2}. \tag{10–141}$$

Here the integer n is known as the *principal quantum number* and is the sole quantum number for the completely degenerate system. The degeneracy will be partly removed if relativity corrections are introduced, producing a precession of the periapsis in the plane of the orbit. The angle variable w_2 that measures the position of the periapsis then varies with time, and the conjugate action variable becomes a "proper" variable and must also be quantized:

$$J_2 = kh,$$

where k is the azimuthal quantum number. Since both v_3 and v_2 are different from zero, the energy must depend on both J_3 and J_2, i.e., on n and k. We thus obtain the well-known relativistic fine structure of the hydrogen levels. The degeneracy can be completely removed by introducing a constant magnetic field along the

---

* As customarily defined, the Delaunay variables differ from the (J_i, w_i) set by multiplicative constants.

arbitrary polar axis. The plane of the orbit then executes a Larmor precession about the polar axis, producing a uniform increase of the angle variable w_1 with time. Therefore J_1 becomes a proper action variable in the presence of a magnetic field and must likewise obey quantum conditions:

$$J_1 = mh,$$

m being the *magnetic* quantum number. The energy now depends on all three quantum numbers, and removal of the degeneracy in this manner thus results in the Zeeman splitting of the atomic levels.*

Beyond the simple hydrogen atom in the periodic table it became progressively more complicated to apply the older quantum theory. The underlying classical problems could no longer be solved exactly, and it was necessary to treat many of the additional forces as small perturbing elements. While there are many points of resemblance between classical perturbation theory and the perturbation methods of wave mechanics, as we shall see, the classical techniques are far more involved than their quantum counterparts, especially where degeneracy occurs.

It soon became apparent, however, that the difficulties were not merely mathematical; the Bohr quantum theory was simply not an accurate picture of nature. As is well known, the impasse was broken with the almost simultaneous introduction of wave and matrix mechanics. Techniques for solving quantum problems were entirely different in these new theories, and interest in action-angle variables waned abruptly. Recent years, however, have seen something of a renaissance elsewhere in the use and application of action-angle variables, not only in celestial mechanics but in problems involving the motions of charged particles in electromagnetic fields. The so-called adiabatic invariance property of the action variables, to be discussed below in Section 11–7, has led to many fruitful applications of action-angle variables in plasma physics and in the design of particle accelerators.

Strangely enough, the root of the newer wave mechanics also arose out of Hamilton–Jacobi theory. If the Poisson bracket formulation of classical mechanics serves as a point of departure to matrix mechanics, the germ of wave mechanics is contained in the connection between Hamilton–Jacobi theory and geometrical optics. It is to the study of this connection that we now turn our attention.

10–8 HAMILTON–JACOBI THEORY, GEOMETRICAL OPTICS, AND WAVE MECHANICS

We shall consider only those systems for which the Hamiltonian is a constant of the motion and is identical with the total energy. Hamilton's principal and

---

* The splitting so obtained represents only the normal Zeeman effect. The correct abnormal Zeeman effect can, of course, be calculated only by including the effects of "spin."

characteristic functions are then related according to the equation

$$S(q, P, t) = W(q, P) - Et. \qquad (10\text{--}142)$$

Since the characteristic function is independent of time, the surfaces of constant W in configuration space have fixed locations. A surface characterized by a constant value of S must coincide at a given time with some particular surface of constant W. However, the value of W corresponding to a definite value of S changes with time in accordance with Eq. (10–12). Thus at $t = 0$ the surfaces $S = a$ and $S = b$ coincide with the surfaces for which $W = a$ and $W = b$, respectively (cf. Fig. 10–6). At a time dt later the surface $S = a$ now coincides with the surface for which $W = a + E\,dt$, and similarly $S = b$ is located at the surface $W = b + E\,dt$. In effect, in a time dt the surface $S = a$ has moved from $W = a$ to $W = a + E\,dt$. The motion of the surface in time is similar to the propagation of a wave front, such as, for example, that of a shock wave, across space. The surfaces of constant S may thus be considered as *wave fronts propagating in configuration space*.

Since the constant S surfaces in general change their shape in the course of time, the wave velocity, i.e., the velocity with which the surfaces move, will not be uniform for all points on the surfaces. However, it is possible to calculate the value of the wave velocity at any given point. For convenience we shall consider a system consisting of only one particle and take the Cartesian position coordinates as the generalized coordinates. Configuration space then reduces to ordinary three-dimensional space, which greatly simplifies the geometry of the problem. The wave velocity at a particular point on a surface of constant S is given by the perpendicular distance the wave front moves in an infinitesimal time dt, divided by the time interval dt. If the infinitesimal distance normal to the surface is denoted by ds, then the wave velocity is

$$u = \frac{ds}{dt}. \qquad (10\text{--}143)$$

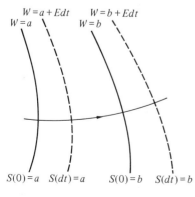

FIGURE **10–6**
The motion of the surfaces of constant S in configuration space.

Now in the time dt the S surface travels from a surface W to a new surface on which the value of the characteristic function is $W + dW$, where

$$dW = E\, dt.$$

The change dW is also related to the normal distance ds according to the formula

$$dW = |\nabla W|\, ds, \tag{10–144}$$

so that

$$u = \frac{ds}{dt} = \frac{E}{|\nabla W|}. \tag{10–145}$$

The magnitude of the gradient of W is furnished by the Hamilton–Jacobi equation which, in terms of Cartesian coordinates of a single particle, has the form

$$(\nabla W)^2 = 2m(E - V). \tag{10–146}$$

Hence the wave velocity is

$$u = \frac{E}{\sqrt{2m(E - V)}}. \tag{10–147}$$

Equation (10–147) may be expressed in a number of variant forms. The difference between E and V is simply the kinetic energy T, so that

$$u = \frac{E}{\sqrt{2mT}}. \tag{10–148}$$

For the one-particle system under consideration, $2mT = m^2v^2 = p^2$, and Eq. (10–148) can also be written as

$$u = \frac{E}{p} = \frac{E}{mv}. \tag{10–149}$$

Equation (10–149) states that the velocity of a point on a surface of constant S is inversely proportional to the spatial velocity of the particle whose motion is being described by S. It is but a simple step to show that the trajectories of the particle must always be normal to the surfaces of the constant S. The direction of the trajectory at any given point in space is determined by the direction of the momentum **p**. By Eqs. (10–27), however,

$$\mathbf{p} = \nabla W, \tag{10–150}$$

and the gradient of W determines the normal to the surfaces of constant S or W.

Any family of surfaces of constant W thus creates a set of trajectories of possible motion that are always normal to the surfaces. As a particle moves along one of the trajectories, the surfaces of S generating the motion will also travel through space, but the two motions do not keep step. In fact, when the particle slows down the surfaces move faster, and vice versa.

In these considerations we have specialized on a system of one particle for ease in discussion. But most of the results hold for many-particle systems if the kinetic energy is a quadratic function of the generalized velocities as in Eq. (8–85). As was seen at the end of Chapter 8, we can then construct a curvilinear configuration space in which the system point moves with velocity $\sqrt{2T}$ (cf. p. 370). Note that this is the same as the velocity of a single particle of unit mass. Thus, it is not surprising that it can be shown that the wave velocity of the S surfaces in the particular curvilinear configuration space is found to be*

$$u = \frac{E}{\sqrt{2(E - V)}} = \frac{E}{\sqrt{2T}}. \tag{10–147'}$$

Further, the reciprocal relation between the wave velocity, u, and the system point velocity, $\sqrt{2T}$, is preserved. Likewise, the possible system trajectories are again found to be normal to the surfaces of constant S. The transition to a many-particle system thus introduces no new physical results, and to simplify the mathematics we shall continue to confine the discussion to one-particle systems.

The surfaces of constant S have been characterized as wave fronts because they propagate in space in the same manner as wave surfaces of constant phase. We have even gone so far as to compute the wave velocity. But nothing has yet been said about the nature or origin of these waves whose fronts are surfaces of constant S. The most striking features of all wave motion result from their periodicity, and there has been no indication so far of the frequency and wavelength spectra of the waves associated with S. To throw light on these questions, let us examine some of the properties of a well-known wave motion— that of light waves.

The scalar wave equation of optics is

$$\nabla^2 \phi - \frac{n^2}{c^2} \frac{d^2 \phi}{dt^2} = 0, \tag{10–151}$$

where ϕ is a scalar quantity such as the scalar electromagnetic potential, c is the velocity of light in vacuo, and n is the index of refraction equal to the ratio of c to the velocity of light. In general, n depends upon the medium and will be a function

---

* For a discussion of the motion of the S surfaces in configuration space, see L. Brillouin. *Tensors in Mechanics and Elasticity*, trans. R. O. Brennan (New York: Academic, 1964), Chapter VIII.

of position in space. If n is constant, Eq. (10–151) is satisfied by a plane wave solution of the form

$$\phi = \phi_0\, e^{i(\mathbf{k}\cdot\mathbf{r} - \omega t)},\tag{10–152}$$

where the wave number k and the frequency ω are connected by the relation

$$k = \frac{2\pi}{\lambda} = \frac{n\omega}{c}.\tag{10–153}$$

Taking the direction of $\mathbf{k}$ for simplicity as being along the z axis, the plane wave solution can also be written

$$\phi = \phi_0 e^{ik_0(nz - ct)},\tag{10–154}$$

where k_0 is the wave number in the vacuum. We shall be interested, however, in the case of *geometrical optics*, where n is not exactly constant but varies slowly in space. The plane wave is then no longer a solution of the wave equation (10–151); the variation of the index of refraction with position will distort and bend the wave. Since n is assumed to change only gradually in space, we seek a solution resembling the plane wave as closely as possible:

$$\phi = e^{A(\mathbf{r}) + ik_0(L(\mathbf{r}) - ct)}.\tag{10–155}$$

The quantities A and L are taken as functions of position to be determined and are both considered as real. Therefore A is a measure of the amplitude of the wave. If n were constant, L would reduce to nz and in consequence is called the *optical path length* or phase of the wave. It is also frequently referred to as the *eikonal*. Successive applications of the gradient operator to the solution ϕ result in the relations

$$\nabla\phi = \phi\,\nabla(A + ik_0 L),$$

$$\nabla^2\phi = \phi[\nabla^2(A + ik_0 L) + (\nabla(A + ik_0 L))^2],$$

or

$$\nabla^2\phi = \phi[\nabla^2 A + ik_0\,\nabla^2 L + (\nabla A)^2 - k_0^2(\nabla L)^2 + 2ik_0\,\nabla A\cdot\nabla L].$$

The wave equation now becomes

$$ik_0[2\,\nabla A\cdot\nabla L + \nabla^2 L]\phi + [\nabla^2 A + (\nabla A)^2 - k_0^2(\nabla L)^2 + n^2 k_0^2]\phi = 0.\tag{10–156}$$

Since both A and L are real, the equation holds only if the two expressions in the square brackets separately vanish:

$$\nabla^2 A + (\nabla A)^2 + k_0^2(n^2 - (\nabla L)^2) = 0,\tag{10–157a}$$

$$\nabla^2 L + 2\,\nabla A\cdot\nabla L = 0.\tag{10–157b}$$

So far no approximation has been made; both equations are rigorous. We can now introduce the assumption that n varies only slowly with distance; in particular, that n does not change greatly over distances of the order of the wavelength. Effectively, this means that the wavelength is small compared to the dimension of any change in the medium, which is the assumption of geometrical optics. The term involving $k_0^2 = 4\pi^2/\lambda_0^2$ in Eq. (10–157a) is therefore the prominent one, and the equation reduced to the simple form

$$(\nabla L)^2 = n^2. \tag{10-158}$$

Equation (10–158) is known as the *eikonal equation of geometrical optics.** The surfaces of constant L determined by this equation are the surfaces of constant optical phase and thus define the wave fronts. The ray trajectories are everywhere perpendicular to the wave fronts and hence are also determined by Eq. (10–159).

We need not digress further into geometrical optics, for it will be seen that the eikonal equation (10–158) is identical in form with the mechanical Hamilton–Jacobi equation for W, (10–146). The characteristic function W plays the same role as the eikonal L and $[2m(E - V)]^{\frac{1}{2}}$ serves as the index of refraction. The Hamilton–Jacobi equation thus tells us that *classical mechanics corresponds to the geometrical optics limit of a wave motion* in which the rays orthogonal to the wave fronts correspond to the particle trajectories orthogonal to the surfaces of constant S. It is now clear why Huygens' wave theory and Newton's light corpuscles were able to account equally well for the phenomena of reflection and refraction, for both theories of geometrical optics are formally identical. The resemblance of the principle of least action to Fermat's principle of geometrical optics is also explained. In Jacobi's form of the least action principle, Eq. (8–89), we see that the integrand, $\sqrt{H - V}$, is to be replaced in geometrical optics by something proportional to the index of refraction or to the reciprocal of the wave velocity. Hence the principle of least action may also be written in the forms

$$\Delta \int n \, ds = \Delta \int \frac{ds}{u} = 0, \tag{10-159}$$

which are two well-known variations of Fermat's principle for the trajectories of light rays.

We have still not established the frequencies and wavelengths of the waves associated with classical motion. All that has been determined is that the wavelength must be very much smaller than the spatial extensions of the forces and potentials. Further than this we cannot go within the realm of classical mechanics. As a species of geometrical optics, classical mechanics is precisely the field in which phenomena depending on the wavelength (interference, diffraction, etc.) cannot occur. There is a duality of particle and wave even in classical mechanics, but the particle is the senior partner, and the wave aspect has no opportunity to display its unique characteristics.

We can speculate, nonetheless, on the form of the wave equation for which the Hamilton–Jacobi equation represents the shortwave length limit. The similarity of the eikonal equation (10–158) with the Hamilton–Jacobi equation (10–146) does not imply the equivalence of L with W: it is necessary merely that the two quantities be proportional to each other. We shall see that the constant of

---

* Although the eikonal equation has been derived here from the scalar wave equation, it can also be obtained from the wave equations for the vector field strengths of the electromagnetic field. See Born and Wolf, *Principles of Optics*.

proportionality is a measure of the magnitude of the wavelength. If W corresponds to L, then $S = W - Et$ must be proportional to the total phase of the light wave described by Eq. (10–155):

$$k_0(L - ct) = 2\pi\left(\frac{L}{\lambda_0} - vt\right).$$ (10–160)

Hence the particle energy E and the wave frequency v must be proportional, and we shall denote the constant ratio of the two quantities by the symbol h:

$$E = hv.$$ (10–161)

The wavelength and the frequency are in general connected to the wave velocity by the relation

$$\lambda v = u,$$

so that, by Eq. (10–149), λ is given by

$$\lambda = \frac{u}{v} = \frac{E/p}{E/h},$$

or

$$\lambda = \frac{h}{p}.$$ (10–162)

The expression for wave phase given in the right-hand side of Eq. (10–160) can also be written as

$$\frac{2\pi}{h}\left(\frac{Lh}{\lambda_0} - hvt\right).$$

Clearly the quantity in the parentheses is in the form of $W - Et = S$. This suggests then that the wave amplitude to be associated with the mechanical particle motion should have the form

$$\psi = \psi_0\, e^{iS/\hbar},$$ (10–163)

where, as customary, $\hbar \equiv h/2\pi$. If our picture is right, then the equation governing this wave amplitude should reduce in the limit of short λ; or equivalently, very small h, to the Hamilton–Jacobi equation. By hindsight, of course, we know that the equation satisfied by ψ is the Schrödinger (time-dependent) wave equation

$$\frac{\hbar^2}{2m}\nabla^2\psi - V\psi = \frac{\hbar}{i}\frac{\partial\psi}{\partial t}.$$ (10–164)

Is the Hamilton–Jacobi equation the shortwave length limit of the Schrödinger equation?

To investigate this question we can substitute Eq. (10–163) into the Schrödinger equation. We have

$$\frac{\partial\psi}{\partial t} = \frac{i}{\hbar}\psi\frac{\partial S}{\partial t}; \qquad \frac{\partial\psi}{\partial x} = \frac{i}{\hbar}\psi\frac{\partial S}{\partial x}.$$

Hence the Laplacian of ψ in Cartesian coordinates is equivalent to

$$\nabla^2\psi = \frac{i}{\hbar}\psi\nabla^2 S - \frac{\psi}{\hbar^2}(\nabla S)^2.$$

In terms of S the Schrödinger equation can therefore be written as

$$\left[\frac{1}{2m}(\nabla S)^2 + V\right] + \frac{\partial S}{\partial t} = \frac{i\hbar}{2m}\nabla^2 S. \qquad (10\text{--}165)$$

We recognize the quantity in brackets as the Hamiltonian in the Hamilton–Jacobi equation for a single particle as described in Cartesian coordinates. Indeed, Eq. (10–165) would be the classical Hamilton–Jacobi equation for Hamilton's principal function if only the right-hand side were zero. Equation (10–165) may be called the quantum-mechanical Hamilton–Jacobi equation; it reduces to the classical equation in the limit as $\hbar$ (and therefore λ) goes to zero.* The conditions for neglecting the $\nabla^2 S$ term can be stated in various ways. One is to note that $\nabla^2 S$ arises in association with $(\nabla S)^2$ in the evaluation of $\nabla^2 \phi$. Therefore one can drop the term if

$$\hbar \nabla^2 S \ll (\nabla S)^2$$

or, equivalently, if

$$\hbar \nabla \cdot \mathbf{p} \ll p^2.$$

In one dimension this would be the same as requiring that

$$\frac{\lambdabar}{p}\frac{dp}{dx} \ll 1, \qquad (10\text{--}165)$$

where $\lambdabar = \lambda/2\pi$, the reduced wavelength. Equation (10–166) in words says that classical mechanics, as the geometrical optics limit, is valid when the wavelength is so short that the momentum changes by a negligible fraction over a distance of $\lambdabar$. In turn this implies that the potential doesn't vary appreciably over a wavelength.†

It is seen that the difference between the quantum and classical versions of the Hamilton–Jacobi equation depends on the size of h. The smaller h, the smaller the wavelength and the better the approximation to geometrical optics. Now, the equivalence of the Hamilton–Jacobi and eikonal equations was first realized by Hamilton in 1834; the corresponding wave equation was first derived by de Broglie and Schrödinger in 1926. It has been stated that had Hamilton gone only a little further, he would have discovered the Schrödinger equation. This is not so; he lacked the experimental authority for the jump. In Hamilton's day classical

---

* If one expands S in Eq. (10–165) in a power series in $\hbar$, and collects coefficients of the same power of $\hbar$, the lowest approximation is obviously the classical Hamilton–Jacobi equation. The next approximation, linear in $\hbar$, leads to the WKB or semiclassical approximation.

† Incidentally, it can now be realized why the Schrödinger equation and the Hamilton–Jacobi equation have the same conditions for separability. The separated solutions have a different form, however. From Eq. (10–163) it is seen that when $W = S + Et$ is separated into a sum of functions each involving only one coordinate, then ψ is separated into a *product* of such functions.

mechanics was considered to be rigorously true, and there was no justification in experiment for considering it as an approximation to a broader theory. In other words, Hamilton had no reason to believe that the value of h was at all different from zero. The recognition that classical mechanics was only a geometrical optics *approximation* to a wave theory could come only when effects depending on the particle wavelength were discovered—as in the interference experiments of Davisson and Germer. Only then could physical reality be ascribed to h, which is, of course, the famous Planck's constant.*

Nevertheless, it can now be seen that classical mechanics contains within it the seeds of the quantum theory, and that the Hamilton–Jacobi formulation is particularly suited to show how to generalize from classical to wave mechanics. To go further into these subjects would take us beyond the scope of this book, which might well be titled "The Geometrical Optics of Wave Mechanics"!

SUGGESTED REFERENCES

R. Courant and D. Hilbert, *Methods of Mathematical Physics.* The Hamilton–Jacobi equation is an example of a first-order partial differential equation in many variables, and the mathematical characteristics of this type of equation are important (as has been noted) for the development of Hamilton–Jacobi theory. Volume 2 of Courant and Hilbert provide a painless (nearly) introduction into the relevant mathematics, first in an introductory section (§4) in Chapter 1 and then in a lengthy explicit discussion of Hamilton–Jacobi theory in Chapter 2.

C. Carathéodory, *Calculus of Variations and Partial Differential Equations of the First Order.* For many years the recognized authority on first-order partial differential equations, the exposition in Vol. I of Carathéodory goes into greater depth than Courant and Hilbert and is especially elaborate in developing the theory of what are called "characteristics." The Hamilton–Jacobi equation is not referred to by that name, but it, and its relation to mechanics, dominates most of the book.

C. Lanczos, *The Variational Principles of Mechanics.* The treatment here of Hamilton–Jacobi theory (in Chapter 8) emphasizes the foundations of the theory and its geometrical interpretations. There is considerable discussion of the necessary conditions for action-angle variables (referred to as "Delaunay's treatment") and their nature. Detailed applications are few.

D. Ter Haar, *Elements of Hamiltonian Mechanics.* The treatment of the Hamilton–Jacobi equation and its applications, including the action-angle variable approach to the Kepler problem, roughly parallels that given in the present chapter. Inevitably there are differences of emphasis and viewpoint, and the reader may benefit from comparing the two discussions.

M. Born, *The Mechanics of the Atom.* Physicists had a brief interest in action-angle variables in the heyday of the older quantum theory, when it provided the "royal road" to

---

* A similar situation occurred in the development of the wave theory of light. Until the phenomena of interference and diffraction were experimentally observed in light there was no reason to prefer Huygens' wave theory over Newton's corpuscular rays.

quantization. Born's book was a product of that period and remains one of the best discussions of Hamilton–Jacobi theory and action-angle variables accessible to physicists. It is outstanding in the wealth of the applications it presents. Born's discussion of action-angle variables and related perturbation theory is the source for the discussions on these areas as found in many text books in mechanics, and the present book is by no means a complete exception. The reader should be cautious, however, in accepting the statement's in Born's book about atomic stricture. Most of them are out of date.

A. SOMMERFELD, *Atomic Structure and Spectral Lines.* The exposition of Hamilton–Jacobi theory and action-angle variables to be found scattered through the text and appendices of this book is considerably less detailed than in Born. Probably for that reason it is often more readable. Especially noteworthy is the discussion of the connection between the number of systems of separation coordinates and the degeneracy of the motion. The evaluation of the integrals occurring in the Kepler problem by means of the theory of residues is explained in an appendix (and is also given in Born's book).

J. H. VAN VLECK, *Quantum Principles and Line Spectra.* The chapter of this work entiled "Mathematical Techniques" provides a quick survey of Hamilton–Jacobi theory and action-angle variables, with an introduction into perturbation theory. Most of the rest of the book is of historical interest only. The caution applied to Born's book is equally valid here and holds almost as well for Sommerfeld's volume.

B. GARFINKEL, *The Lagrange–Hamilton–Jacobi Mechanics* in *Space Mathematics, Part I.* Action-angle variables had been used in celestial mechanics (although not under that name) long before there was any physics interest in them, and their use remains today the elegant way of approaching perturbation theory. This reference shows how the subject is viewed from the standpoint of celestial mechanics, in an essay that's brief and concise but eminently readable (a characteristic notoriously lacking in many treatises on analytical mechanics). In a scant 36 pages it covers the field from Lagrangian mechanics through perturbation theory but packs an incredible amount of material in the brief compass. Examples are the Staeckel conditions for separability (there is an obvious factor of 2 missing in Eq. (82)), Vinti's theorem, and the Delaunay elements. Some of the notation and conventions differ from those customarily followed in physics.

L. A. PARS, *A Treatise on Analytical Dynamics.* Three chapters cover the area from the Hamilton–Jacobi equation to action-angle variables, which allows for a leisurely treatment and a plenitude of examples. Unusual is the discussion on separable systems as a property per se, independent of the application to the Hamilton–Jacobi equation. Staeckel's condition is treated at length. Action-angle variables are not mentioned explicitly (although angle variables are) but the sections on multiply-periodic motion are extensive.

H. V. MCINTOSH, *Symmetry and Degeneracy* in *Group Theory and its Applications*, Vol. 2, ed. E. M. Loebl. As might be expected the connection between separability and degeneracy (and the symmetry properties of the system) come in for considerable discussion here. The argument that the space-filling properties of the orbit in nondegenerate systems militates against separability is gone into in some detail, and the consequences for a wide variety of systems are considered (including such unusual ones as the magnetic monopole). All in all, this is probably the best available reference on the degeneracy–separability–symmetry connection available in the 1970s.

L. Brillouin, *Tensors in Mechanics and Elasticity*. This charmingly written book contains much information on a wide variety of topics, from differential geometry to the quantum mechanics of solids (of circa 1938). The motion of the surfaces of constant S in configuration space is presented in detail in Chapter VIII, and the connections linking classical mechanics, geometrical optics, and wave mechanics are thoroughly discussed in Chapter IX.

M. Born and E. Wolf, *Principles of Optics*. A standard reference on the application of Hamilton–Jacobi theory to geometrical optics is the rather formidable treatise by J. L. Synge, *Geometrical Optics*. Born and Wolf provide a more understandable introduction, with considerable attention paid to further extension to the Schrödinger wave equation. A chapter on the foundations of geometrical optics contains, among other material, a derivation of the eikonal equation for a vector field. The following chapter on the geometrical theory of optical imaging starts out, at least, with a Hamiltonian approach. Two appendices are of particular interest in this connection. The first is practically a short treatise on the calculus of variations, with emphasis on the Hamilton–Jacobi equation. The contents of the following appendix are clearly indicated by the title: "Light optics, electron optics and wave mechanics."

EXERCISES

1. For a conservative system show that by solving an appropriate partial differential equation one can construct a canonical transformation such that the new Hamiltonian is a function of the new *coordinates* only. (Do not use the exchange transformation.) Show how a formal solution to the motion of the system is given in terms of the new coordinates and momenta.

2. In the text the Hamilton–Jacobi equation for S was obtained by seeking a contact transformation from the canonical coordinates (q, p) to the constants (α, β). Conversely, if $S(q_i, \alpha_i, t)$ is any complete solution of the Hamilton–Jacobi equation (10–3), show that the set of variables (q_i, p_i) defined by Eqs. (10–7) and (10–8) are canonical variables, i.e., that they satisfy Hamilton's equations.

3. Solve the problem of the motion of a point projectile in a vertical plane, using the Hamilton–Jacobi method. Find both the equation of the trajectory and the dependence of the coordinates on time, assuming the projectile is fired off at time $t = 0$ from the origin with the velocity v_0, making an angle α with the horizontal.

4. Set up the problem of the heavy symmetrical top, with one point fixed, in the Hamilton–Jacobi method, and obtain the formal solution to the motion as given by Eq. (5–63).

5. Show that the function

$$S = \frac{m\omega}{2}(q^2 + \alpha^2)\cot \omega t - m\omega q\alpha \csc \omega t$$

is a solution of the Hamilton–Jacobi equation for Hamilton's principal function for the linear harmonic oscillator with

$$H = \frac{1}{2m}(p^2 + m^2\omega^2 q^2).$$

Show that this function generates a correct solution to the motion of the harmonic oscillator in time.

6. A charged particle is constrained to move in a plane under the influence of a central force potential (nonelectromagnetic) $V = \frac{1}{2}kr^2$, and a constant magnetic field **B** perpendicular to the plane, so that

$$(\mathbf{A} = \tfrac{1}{2}\mathbf{B} \times \mathbf{r}),$$

Set up the Hamilton–Jacobi equation for Hamilton's characteristic function in plane polar coordinates. Separate the equation and reduce it to quadratures. Discuss the motion if the canonical momentum p_θ is zero at time $t = 0$.

7. a) A single particle moves in space under a conservative potential. Set up the Hamilton–Jacobi equation in ellipsoidal coordinates u, v, ϕ defined in terms of the usual cylindrical coordinates r, z, ϕ by the equations

$$r = a\sinh v \sin u, \qquad z = a\cosh v \cos u.$$

For what forms of $V(u, v, \phi)$ is the equation separable?

b) Use the results of part (a) to reduce to quadratures the problem of a point particle of mass m moving in the gravitational field of two unequal mass points fixed on the z axis a distance $2a$ apart.

8. Suppose the potential in a problem of one degree of freedom is linearly dependent on time, such that the Hamiltonian has the form

$$H = \frac{p^2}{2m} - mAtx,$$

where A is a constant. Solve the dynamical problem by means of Hamilton's principal function, under the initial conditions $t = 0$, $x = 0$, $p = mv_0$.

9. Set up the plane Kepler problem in terms of the generalized coordinates

$$u = r + x,$$

$$v = r - x.$$

Obtain the Hamilton–Jacobi equation in terms of these coordinates, and reduce it to quadratures (at least).

10. One end of a uniform rod of length $2l$ and mass m rests against a smooth horizontal floor and the other against a smooth vertical surface. Assuming that the rod is constrained to move under gravity with its ends always in contact with the surfaces, use either of the Hamilton–Jacobi equations to reduce the solution of the problem to quadratures.

11. A particle is constrained to move on a roller coaster, the equation of whose curve is

$$z = A\cos^2 \frac{2\pi x}{\lambda}.$$

There is the usual constant downward force of gravity. Discuss the system trajectories in phase space under all possible initial conditions, describing the phase space orbits in as much detail as you can, paying special attention to turning points and transitions between different types of motion.

12. A particle of mass m moves in a plane in a square well potential:

$$V(r) = -V_0 \qquad 0 < r < r_0,$$
$$= 0 \qquad r > r_0.$$

a) Under what initial conditions can the method of action-angle variables be applied?

b) Assuming these conditions hold, use the method of action-angle variables to find the frequencies of the motion.

13. A particle moves in periodic motion in one dimension under the influence of a potential $V(x) = F|x|$, where F is a constant. Using action-angle variables find the period of the motion as a function of the particle's energy.

14. A particle of mass m moves in one dimension under a potential $V = -k/|x|$. For energies that are negative the motion is bounded and oscillatory. By the method of action-angle variables find an expression for the period of motion as a function of the particle's energy.

15. A particle of mass m moves in one dimension subject to the potential

$$V = \frac{a}{\sin^2\left(\dfrac{x}{x_0}\right)}.$$

Obtain an integral expression for Hamilton's characteristic function. Under what conditions can action-angle variables be used? Assuming these are met find the frequency of oscillation by the action-angle method. (The integral for J can be evaluated by manipulating the integrand so that the square root appears in the denominator.) Check your result in the limit of oscillations of small amplitude.

16. A particle of mass m is constrained to move on a curve in the vertical plane defined by the parametric equations

$$y = l(1 - \cos 2\phi),$$
$$x = l(2\phi + \sin 2\phi).$$

There is the usual constant gravitational force acting in the vertical y direction. By the method of action-angle variables find the frequency of oscillation for all initial conditions such that the maximum of ϕ is less than or equal to $\pi/4$.

17. In the action-angle formalism the arguments of Hamilton's characteristic function are the original coordinates q_k and the action variables J_k. In the case of degeneracy a subsequent canonical transformation is made to new variables (w_i', J_i') from (w_k, J_k), in order to replace the degeneracies by zero frequencies. By considering each J_k a function of the J_i' quantities as defined by Eq. (10–105), show that it remains true that

$$\frac{\partial W}{\partial J_i'} = w_i'.$$

18. For the system described in Exercise 12 of Chapter 6 find a linear point transformation to variables in which the Hamilton–Jacobi equation is separable. By use of the action-angle variables then find the eigenfrequencies of the system.

19. A three-dimensional harmonic oscillator has the force constant k_1 in the x and y directions and k_3 in the z direction. Using cylindrical coordinates (with the axis of the cylinder in the z direction) describe the motion in terms of the corresponding action-angle variables, showing how the frequencies can be obtained. Transform to the "proper" action-angle variables to eliminate degenerate frequencies.

20. Find the frequencies of a three-dimensional harmonic oscillator with *unequal* force constants using the method of action-angle variables. Obtain the solution for each Cartesian coordinate and conjugate momentum as functions of the action-angle variables.

21. a) In the harmonic oscillator of Exercise 20 allow all the frequencies to become equal (isotropic oscillator) so that the motion is completely degenerate. Transform to the "proper" action-angle variables, expressing the energy in terms of only one of the action variables.

b) Solve the problem of the isotropic oscillator in action-angle variables using spherical polar coordinates. Transform again to proper action-angle variables and compare with the result of part (a). Are the two sets of proper variables the same? What are their physical significances? This problem illustrates the feasibility of separating a degenerate motion in more than one set of coordinates. The nondegenerate oscillator can be separated only in Cartesian coordinates, not in polar coordinates.

22. The motion of a degenerate plane harmonic oscillator can be separated in any Cartesian coordinate system. Obtain the relations between the two sets of action-angle variables corresponding to two Cartesian systems of axes making an angle θ with each other. Note that the transformation between the two sets is *not* the orthogonal transformation of the rotation.

23. a) Evaluate the J_θ integral in the Kepler problem by the method of complex contour integration. To get the integral into a useful form it is suggested the substitution $\cos \theta = x \sin i$ might be made.

b) Verify the integration procedure used for J_θ in the text, carrying out the final integrations in Eq. (10–111).

c) Follow the consequences of the inclination being greater than 90°, i.e., $\cos i$ negative. In particular what are the changes in Eq. (10–112), in the canonical transformation to zero frequencies and therefore in Eqs. (10–122)? Can you write these equations in such a form that they are valid whether $\cos i$ is positive or negative?

24. Evaluate the integral for J_r in the Kepler problem by elementary means. This includes using tables of integrals, but if so, explicit and detailed references should be given to the tables used.

25. Show, by the method outlined in the text (or any other), that $2\pi w_2$ is ω, the argument of the periapsis, in the three-dimensional Kepler problem.

26. The so-called Poincaré elements of the Kepler orbits can be written as

$$w_1 + w_2 + w_3, \qquad\qquad J_\phi,$$

$$\frac{J_r}{\pi} \cos 2\pi(w_2 + w_1), \qquad\qquad \frac{J_r}{\pi} \sin 2\pi(w_2 + w_1),$$

$$\frac{J_\theta}{\pi} \cos 2\pi w_1, \qquad\qquad \frac{J_\theta}{\pi} \sin 2\pi w_1.$$

Show that they form a canonical set of coordinates, with the new coordinates forming the left-hand column, their conjugate momenta being given on the right-hand side.

27. Describe the phenomenon of small radial oscillations about steady circular motion in a central force potential as a one-dimensional problem in the action-angle formalism. With a suitable Taylor series expansion of the potential find the period of the small oscillations. Express the motion in terms of J and the conjugate angle variable.

28. Set up the problem of the relativistic Kepler motion in action-angle variables, using the Hamiltonian in the form given by Eq. (8–54). Show in particular that the total energy (including rest mass) is given by

$$\frac{E}{mc^2} = \frac{1}{\sqrt{1 + \dfrac{4\pi^2 k^2}{[(J_3' - J_2')c + \sqrt{J_2'^2 c^2 - 4\pi^2 k^2}]^2}}}.$$

Note that the degeneracy has been partly lifted, because the orbit is no longer closed, but is still confined to a plane. In the limit as c approaches infinity show that this reduces to Eq. (10–123).

CHAPTER 11

Canonical Perturbation Theory

11-1 INTRODUCTION

Almost all of the problems in classical mechanics discussed in this book so far, whether in the text or in the exercises, have been capable of exact solutions. It must not be thought, however, that complete solutions can be found for all, or even for most, problems in mechanics. Indeed it is probably the case that the vast majority of problems cannot be solved exactly. We have found solutions for the two-body Kepler problem, but the classical motion of three point-bodies acted on only by their mutual gravitational forces has proved intractable. Even for two bodies the solutions are implicit; no closed explicit formula can be found for the coordinates as a function of time (cf. p. 102), Section 3–8). There is thus considerable incentive for developing approximate methods of solution.

It often happens, fortunately, that in a physical problem that cannot be solved directly the Hamiltonian differs only slightly from the Hamiltonian for a problem that can be solved rigorously. The more complicated problem is then said to be a *perturbation* of the soluble problem, and the difference between the two Hamiltonians is called the *perturbation Hamiltonian*. Perturbation theory consists of techniques for obtaining approximate solutions based on the smallness of the perturbation Hamiltonian.

It should be noted that while the change in the Hamiltonian must be small, the eventual effect of the perturbation on the motion may be large. Thus, consider an isotropic plane harmonic oscillator, i.e., one with equal frequencies for motion in the x or y directions. The trajectory of the point oscillator is a simple closed curve; in general, an ellipse. Suppose that there is an infinitesimal perturbation of the force constant in the y direction so that the two frequencies are now unequal and incommensurate. As a result of this small perturbation in H, the trajectory is no longer closed and will in the course of time become space filling—in loose terms, pass through every point in a rectangle defined by the amplitudes in the x and y directions. We have seen that similarly large qualitative changes in the motion of a point satellite result from very small perturbations caused by the oblateness of the earth's gravitational field. In the unperturbed central field the satellite motion is confined to a plane; any gravitational quadrupole perturbation, no matter how small, causes motion out of the initial plane (so long as the axis of the quadrupole is not perpendicular to the plane).

The development of perturbation theory goes back to the earliest days of celestial mechanics. Newton realized, for example, that most of the oscillations in the moon's motion were the result of small changes in the attraction to the sun as the moon revolves about the earth. His initial attempts at a lunar theory including these effects corresponded roughly to a form of perturbation theory. Many of the subsequent developments in the formal structure of classical mechanics, such as Hamilton's canonical theory, stemmed in large measure from the desire to perfect perturbation techniques in celestial mechanics. The need for predicting highly accurate orbits for space vehicles and the enormously increased capacity for numerical computations have recently spurred further improvements in perturbation theory. (Most of these recent developments, however, lie outside the scope of our discussion.)

Classical perturbation theory can be divided into two approaches: time-dependent and time-independent perturbation. The terminology is chosen with an eye to perturbation theory as developed for quantum mechanics, and indeed there are many points of analogy between the classical perturbation techniques and their quantum counterparts. Generally speaking, classical perturbation theory is considerably more complicated than the corresponding quantum mechanical version. We shall treat time-dependent perturbation first as being the easier form to understand. While perturbation theory can be developed for all versions of classical mechanics, it is simplest to use the Hamilton–Jacobi formulation.

11–2 TIME-DEPENDENT PERTURBATION (VARIATION OF CONSTANTS)

Let $H_0(q, p, t)$ represent the Hamiltonian for the soluble, unperturbed problem. We imagine the solution has been obtained through Hamilton's principal function $S(q, \alpha, t)$, which generates a canonical transformation in which the new Hamiltonian, K_0, for the unperturbed problem is identically zero. The transformed canonical variables, (α, β), are then all constant in the unperturbed situation. Now consider the perturbed problem for which we write the Hamiltonian as

$$H(q, p, t) = H_0(q, p, t) + \Delta H(q, p, t). \tag{11–1}$$

As has been emphasized before, the canonical property of a given coordinate transformation is independent of the particular form of the Hamiltonian. Therefore the transformation

$$(p, q) \rightarrow (\alpha, \beta)$$

generated by $S(q, \alpha, t)$ *remains a canonical transformation for the perturbed problem.* Only now the new Hamiltonian will not vanish and the transformed variables may not be constant. For the perturbed problem the transformed Hamiltonian will be

$$K(\alpha, \beta, t) = H_0 + \Delta H + \frac{\partial S}{\partial t} = \Delta H(\alpha, \beta, t). \tag{11–2}$$

Hence the equations of motion satisfied by the transformed variables are now

$$\dot{\alpha}_i = -\frac{\partial \Delta H(\alpha, \beta, t)}{\partial \beta_i}, \qquad \dot{\beta}_i = \frac{\partial \Delta H(\alpha, \beta, t)}{\partial \alpha_i}. \tag{11–3}$$

Equations (11–3) are rigorous; no approximation has yet been made. If the set of $2n$ equations can be solved for α_i and β_i as functions of time, then the equations of transformation between (p, q) and (α, β) give q_j and p_j as functions of time, i.e., solve the problem. However, the exact solution of Eqs. (11–3) is usually no less difficult to obtain than for the original equations of motion. The use of Eqs. (11–3) as an alternative approach to the rigorous solution is therefore not particularly fruitful.

In the perturbation technique, however, advantage is taken of the fact that ΔH is small. The quantities (α, β) while no longer constant therefore do not change rapidly, at least compared to the explicit dependence of ΔH on time. A first-order approximation to the time variation of (α, β) is obtained by replacing α and β on the *right-hand side* of Eqs. (11–3) by their *constant* unperturbed values:

$$\dot{\alpha}_{1i} = \left.\frac{\partial \Delta H(\alpha, \beta, t)}{\partial \beta_i}\right|_0 ; \qquad \dot{\beta}_{1i} = -\left.\frac{\partial \Delta H(\alpha, \beta, t)}{\partial \alpha_i}\right|_0 . \tag{11–4}$$

Here α_{1i} and β_{1i} stand for the first-order perturbation solutions for α_i and β_i, respectively, and the vertical lines with subscript 0 indicate that after differentiation α and β are to be replaced by their unperturbed forms, i.e., the constants (α_0, β_0). Equations (11–4) can be put in matrix form by designating γ as the column matrix of the β and α canonical variables, so that

$$\dot{\gamma}_1 = \mathbf{J}\left.\frac{\partial \Delta H(\gamma, t)}{\partial \gamma}\right|_0 , \tag{11–5}$$

where $\mathbf{J}$ is the matrix given by Eq. (8–30). Equations (11–4) can now be integrated directly to yield the α_1 and β_1 as functions of time. Through the transformation equations one then obtained (q, p) as functions of time to first order in the perturbation. Clearly, the second-order perturbation is obtained by using the first-order dependence of α and β on time in the right-hand sides of Eqs. (11–4), and so on. In general, the nth order perturbation solution is obtained by integrating the equations (in matrix form) for γ_n given by

$$\dot{\gamma}_n = \mathbf{J}\left.\frac{\partial \Delta H(\gamma, t)}{\partial \gamma}\right|_{n-1} . \tag{11–6}$$

As a trivial example of these procedures, consider as the unperturbed system the force-free motion in one dimension of a particle of mass m. The unperturbed Hamiltonian is

$$H_0 = \frac{p^2}{2m}.$$

The momentum p is clearly conserved; call its constant value α. For this system the Hamilton–Jacobi equation is

$$\frac{1}{2m}\left(\frac{\partial S}{\partial x}\right)^2 + \frac{\partial S}{\partial t} = 0. \tag{11–7}$$

Because the system is conservative and x is cyclic, we know immediately that the solution for Hamilton's principal function is

$$S = \alpha x - \frac{\alpha^2 t}{2m}. \tag{11–8}$$

The transformed momentum is α; the transformed constant coordinate is

$$Q \equiv \beta = \frac{\partial S}{\partial \alpha} = x - \frac{\alpha t}{m}$$

or

$$x = \frac{\alpha t}{m} + \beta, \tag{11–9}$$

the expected solution for the force-free motion. While Eq. (11–9) is obvious a priori, this formal derivation via the Hamilton–Jacobi equation at least shows that α and β, so defined, form a canonical set.

Now suppose the perturbation Hamiltonian is

$$\Delta H = \frac{m\omega^2 x^2}{2}, \tag{11–10}$$

where ω is some constant. The total Hamiltonian is

$$H = H_0 + \Delta H = \frac{1}{2m}(p^2 + m^2\omega^2 x^2). \tag{11–11}$$

We are thus considering the harmonic oscillator potential as a perturbation on force-free motion! In terms of the α, β variables the perturbation Hamiltonian, by Eq. (11–9), is

$$\Delta H = \frac{m\omega^2}{2}\left(\frac{\alpha t}{m} + \beta\right)^2. \tag{11–12}$$

In the perturbed system the equations of motion for α, β are (cf. Eqs. (11–3))

$$\dot{\alpha} = -m\omega^2\left(\frac{\alpha t}{m} + \beta\right), \tag{11–13a}$$

$$\dot{\beta} = \omega^2 t\left(\frac{\alpha t}{m} + \beta\right). \tag{11–13b}$$

Note that

$$\dot{\beta} + \frac{t}{m}\dot{\alpha} = 0. \tag{11–14}$$

A rigorous solution of Eqs. (11–13) can be obtained by taking the time derivative of Eq. (11–13a):

$$\ddot{\alpha} = -\omega^2\alpha - m\omega^2\left(\dot{\beta} + \frac{\dot{\alpha}t}{m}\right) = -\omega^2\alpha. \tag{11-15}$$

Thus α in the perturbed system rigorously has a simple harmonic variation with time. From Eqs. (11–13a) and (11–9) it follows $x = -\dot{\alpha}/(m\omega^2)$ and hence the solution for x is also simple harmonic motion. Considered as rigorous equations of motion, Eqs. (11–13) therefore lead properly to the correct and well-known solution.

But now let's treat $m\omega^2$ ($\equiv k$, the force constant) as a small parameter and seek perturbation solutions. The first-order perturbation is obtained by replacing α and β on the right by their unperturbed values α_0 and β_0. For simplicity we'll take $x = 0$ initially, so that $\beta_0 = 0$; the initial value of p is then α_0. The first-order equations of motion are then

$$\dot{\alpha}_1 = -\omega^2\alpha_0 t, \qquad \dot{\beta}_1 = \alpha_0\frac{\omega^2 t^2}{m}, \tag{11-16}$$

with immediate solutions

$$\alpha_1 = \alpha_0 - \frac{\omega^2\alpha_0 t^2}{2}, \qquad \beta_1 = \frac{\alpha_0\omega^2 t^3}{3m}. \tag{11-17}$$

Solutions for x and p to first order are then

$$x = \frac{\alpha_1 t}{m} + \beta_1 = \frac{\alpha_0}{m\omega}\left[\omega t - \frac{\omega^3 t^3}{6}\right], \tag{11-18a}$$

and

$$p = \alpha_1 = \alpha_0\left[1 - \frac{\omega^2 t^2}{2}\right]. \tag{11-18b}$$

Substituting Eqs. (11–17) for α and β on the right-hand side of Eqs. (11–13), the second-order equations of motion become

$$\dot{\alpha}_2 = -\alpha_0\omega^2\left(t - \frac{\omega^2 t^3}{6}\right), \qquad \dot{\beta}_2 = \frac{\alpha_0\omega^2}{m}\left(t^2 - \frac{\omega^2 t}{b}\right), \tag{11-19}$$

with solutions

$$\alpha_2 = \alpha_0 - \frac{\omega^2\alpha_0 t^2}{2} + \frac{\omega^4\alpha_0 t^4}{24},$$

$$\beta_2 = \frac{\alpha_0\omega^2}{m}\left(\frac{t^3}{3} - \frac{\omega^2 t^5}{30}\right). \tag{11-20}$$

The corresponding second-order solutions for x and p are

$$x = \frac{\alpha_0}{m\omega}\left[\omega t - \frac{\omega^3 t^3}{3!} + \frac{\omega^2 t^5}{5!}\right],$$

$$p = \alpha_0\left[1 - \frac{\omega^2 t^2}{2!} + \frac{\omega^4 t^4}{4!}\right]. \tag{11-21}$$

By now we have enough to see where the nth order solution is going. The quantities in the brackets in Eqs. (11–21) are the first three terms in the expansion of the sine and cosine, respectively. In the limit of infinite order of perturbation clearly

$$x \to \frac{\alpha_0}{m\omega}\sin \omega t, \qquad p \to \alpha_0 \cos \omega t,$$

which are the standard solutions consistent with the initial conditions.

In the constant transformed variables (α, β) is contained information on the parameters of the unperturbed orbit. Thus, if the Kepler problem in three dimensions describes the unperturbed system, then a suitable set of (α, β) are the Delaunay variables, i.e., the constant action variables J_i and the constant terms in the corresponding angle variables w_i. We have seen in Section 10–7 that the Delaunay variables are simply related to the orbital parameters—semimajor axis, eccentricity, inclination, etc. The effect of the perturbation is to cause these parameters to vary with time. If the perturbation is small the variation of the parameters within one period of the unperturbed motion will also be small. Time-dependent perturbation theory—variation of constants—thus implies a picture in which the perturbed system moves during small intervals of time in an orbit of the same functional form as the unperturbed system, an orbit whose parameters, however, will be changing in time. The unperturbed orbit along which the system is momentarily traveling is sometimes described as the "osculating orbit." In position and tangent direction it matches instantaneously the true trajectory.

As determined by a perturbation treatment, the parameters of the osculating orbit may vary with time in two ways. There may be a periodic variation, in which the parameter comes back to an initial value in a time interval that to first order is usually the period of the unperturbed motion. Or there may remain a net increment in the value of the parameter at the end of each successive orbital period—and the perturbed parameters are said to exhibit *secular* change. Periodic effects of perturbation do not change the average parameters of the orbit; on the whole, the trajectory remains looking much like the unperturbed orbit. A secular change, no matter how small per orbital period, means that eventually, after many periods, the instantaneous perturbed parameters may be quite different from their unperturbed values. The major interest in a perturbation calculation will therefore often be in the secular terms only, and the periodic effects may be eliminated early in the game by averaging the perturbation over the unperturbed period. Effectively this is what was done in

Section 5-8 when the perturbing gravitational potential of the oblate earth was averaged over the satellite period (cf. Eq. 5–90).*

Often one would like to determine the time dependence of the orbital "constants"—e.g., eccentricity, or inclination—directly, rather than through the intermediary of the canonical set (α, β). This can be done easily through the Poisson bracket formalism. Let c_i be any set of $2n$ independent functions of the (α, β) constants of the unperturbed system:

$$c_i = c_i(\alpha, \beta). \tag{11-22}$$

One or more of the c_i may be the desired orbital parameters. Then in the perturbed system the time dependence of the c_i quantities is determined by the equations of motion

$$\dot{c}_i = [c_i, K] = [c_i, \Delta H]. \tag{11-23}$$

But $\Delta H(\alpha, \beta, t)$ may equally well, by the inverse of Eqs. (11–22), be considered a function of the c's and t, so that

$$[c_i, \Delta H] \equiv \frac{\partial c_i}{\partial \boldsymbol{\eta}} \mathbf{J} \frac{\partial \Delta H}{\partial \boldsymbol{\eta}} = \frac{\partial c_i}{\partial \boldsymbol{\eta}} \mathbf{J} \frac{\partial \Delta H}{\partial c_j} \frac{\partial c_j}{\partial \boldsymbol{\eta}}$$

$$= [c_i, c_j] \frac{\partial \Delta H}{\partial c_j}.$$

Hence,

$$\dot{c}_i = [c_i, c_j] \frac{\partial \Delta H}{\partial c_j}. \tag{11-24}$$

As with Eqs. (11–3), Eqs. (11–24) are rigorous equations of motion for the c_i's. They become first-order perturbation equations when the right-hand sides, including the Poisson brackets, are evaluated for the unperturbed motion. In general the nth-order perturbation is obtained when the right-hand sides are evaluated in terms of the $(n - 1)$st order of perturbation. Equations (11–24) thus correspond, in generalized form, to Eqs. (11–6).

---

* The circumstances may often be more complicated than as described in this paragraph. For example, the periodic variation of orbital parameters may exhibit more than one period. This would obviously occur when the perturbing potential has its own intrinsic periodicity, e.g., the varying perturbation of the sun's gravity on the earth–moon orbit as the earth revolves around the sun. Multiply-periodic behavior can also appear through interactions between perturbations. Thus the periodic perturbation of satellite parameters may show both short and long periods, and it is necessary to average over both kinds of periods to find the secular perturbation effects. Sometimes the dividing line between periodic and secular perturbations becomes a bit vague. What may appear as a secular perturbation in first order will at times on closer examination turn out to be a periodic perturbation with a very long period. Depending on the purpose of the calculation, it may still be advisable to treat it as a secular perturbation term. Nonetheless the distinction between periodic and secular terms remains useful and normally straightforward, especially in first-order perturbation theory.

A version of Eqs. (11–24) expressed in Lagrange brackets (cf. Eq. 9–77) is often found in the literature of celestial mechanics. Multiply the equation for c_i by the Lagrange bracket $\{c_k, c_i\}$ and sum over i:

$$\{c_k, c_i\}\dot{c}_i = \{c_k, c_i\}\{c_i, c_j\}\frac{\partial\,\Delta H}{\partial c_j}.$$

By the theorem expressed in Eq. (9–83) this reduces to

$$-\frac{\partial\,\Delta H}{\partial c_j} = \{c_j, c_i\}\dot{c}_i. \tag{11–25}$$

Historically, the perturbation equations of celestial mechanics are expressed in terms of the *disturbing function R*, defined as $-\Delta H$, so that Eqs. (11–25) appear as

$$\frac{\partial R}{\partial c_j} = \{c_j, c_i\}\dot{c}_i. \tag{11–25'}$$

Equations (11–24) or (11–25) are frequently denoted as the *Lagrange perturbation equations*. Further characteristics of time-dependent perturbation theory are best discussed as they occur in the course of applying the method in the next section to some typical examples.

11–3 ILLUSTRATIONS OF TIME-DEPENDENT PERTURBATION THEORY

A. *Period of the plane pendulum with finite amplitude.* In the limit of small oscillations a plane pendulum behaves like a harmonic oscillator and is isochronous, i.e., the frequency is independent of the amplitude. As the amplitude increases, however, the correct potential energy deviates from the harmonic oscillator form and the frequency shows a small dependence on the amplitude. The small difference between the potential energy and the harmonic oscillator limit can be considered as the perturbation Hamiltonian, and the shift in frequency derived from the time variation of the perturbed phase angle. In Section 11–4 below we shall find the same frequency shift in the time-independent version of perturbation theory.

The Hamiltonian for a plane pendulum, consisting of a mass point m at the end of a weightless rod of length l, is

$$H = \frac{p^2}{2ml^2} + mgl(1 - \cos\theta), \tag{11–26}$$

where, for simplicity, the momentum conjugate to θ is denoted by p. Expanding the $\cos\theta$ term in a Taylor series the Hamiltonian can be written as

$$H = \frac{p^2}{2ml^2} + \frac{mgl\theta^2}{2}\left(1 - \frac{\theta^2}{12} + \frac{\theta^4}{360} - \cdots\right). \tag{11–27}$$

The small amplitude limit consists of dropping all but the first term in the parentheses. We can get an idea of the magnitude of the correction terms by introducing artificially a parameter

$$\theta_1^2 = \frac{2E}{mgl} \tag{11–28}$$

and the related parameter

$$\lambda = \frac{\theta_1^2}{6} = \frac{E}{3mgl}.$$

The series in the parentheses then looks like

$$1 - \frac{\lambda}{2}\left(\frac{\theta}{\theta_1}\right)^2 + \frac{\lambda^2}{10}\left(\frac{\theta}{\theta_1}\right)^2 - \cdots.$$

Now, the ratio θ/θ_1 rises to the order of unity at the maximum amplitude. Indeed θ_1 is the maximum amplitude of oscillation when E, and therefore the amplitude, is small. Hence the rate of convergence of the expansion is determined by the magnitude of λ. If only one correction term is retained, first-order perturbation introduces terms of the order λ in the motion. Second-order perturbation with the same perturbation Hamiltonian introduces λ^2 terms. Thus, to obtain modifications of the motion consistently correct to λ^2, one would have to compute second-order perturbation on the λ term in the Hamiltonian, and first-order perturbation on the λ^2 term in the Hamiltonian. We shall here content ourselves with a consistent treatment to order λ, that is, retain only the first correction term in the Hamiltonian and carry out a first-order perturbation solution.

The unperturbed Hamiltonian derived from Eq. (11–27) can be put in the form of a harmonic oscillator by writing it as (cf. Eq. 10–14)

$$H = \frac{1}{2I}(p^2 + I^2\omega^2\theta^2), \tag{11–29}$$

where $I = ml^2$, the moment of inertia of the pendulum, and

$$\omega^2 = \frac{mgl}{I} = \frac{g}{l}. \tag{11–30}$$

A suitable set of canonical variables corresponding to a vanishing K for the unperturbed system are the action variable J and the phase angle β in the angle variable:

$$w = vt + \beta, \qquad v = \frac{\omega}{2\pi}. \tag{11–31}$$

The effect of the perturbation is to cause both J and β to vary with time. The equations of transformation relating p and θ to J and β, respectively, have already been given in Eqs. (10–78) and (10–79), which here take the form

$$\theta = \sqrt{\frac{J}{\pi I \omega}} \sin 2\pi(vt + \beta),$$

$$p = \sqrt{\frac{I J \omega}{\pi}} \cos 2\pi(vt + \beta). \tag{11–32}$$

In the unperturbed system J and β are constant and Eqs. (11–32) constitute the complete solutions for the motion. But the equations remain valid for the perturbed case, only J and β have time dependences to be determined.

The unperturbed Hamiltonian is $H_0 = Jv$, but the perturbation Hamiltonian takes the form

$$\Delta H = -\frac{mgl}{24} \theta^4 = -\frac{J^2}{24\pi^2 m p^2} \sin^4 2\pi(vt + \beta). \tag{11–33}$$

The first-order time dependence of β and J are to be obtained from

$$\dot{\beta} = \frac{\partial \Delta H}{\partial J}, \qquad \dot{J} = -\frac{\partial \Delta H}{\partial \beta}, \tag{11–34}$$

where on the right-hand side of each equation the unperturbed solution is to be used, that is, J and β are considered constant. Thus

$$\dot{\beta} = -\frac{J}{12\pi^2 m l^2} \sin^4 2\pi(vt + \beta). \tag{11–35}$$

Equation (11–35) says that to first order, $\dot{\beta}$ varies over the cycle of the unperturbed oscillation. But there is a net value for $\dot{\beta}$ when averaged over a complete cycle, for the average of $\sin^4$ is 3/8. Hence β exhibits a secular perturbation at a constant rate given by

$$\bar{\beta} = -\frac{J}{32\pi^2 m l^2}. \tag{11–36}$$

Viewed over times long compared to the unperturbed period, β has a time dependence

$$\beta = \bar{\beta} t + \beta_0. \tag{11–37}$$

Such a variation, when inserted in Eq. (11–32), says that on the average the first-order solution is still simple-harmonic with a frequency

$$v' = v + \bar{\beta}.$$

Now, in the unperturbed motion

$$J = \frac{2\pi E}{\omega} = 2\pi\omega\frac{El}{g},$$

so that $\bar{\beta}$, Eq. (11–36), becomes

$$\bar{\beta} = -\frac{\omega E}{16\pi mgl} = -\frac{v\theta_1^2}{16}. \tag{11–38}$$

The first-order fractional change in the frequency at a finite amplitude θ_1 is therefore

$$\frac{\Delta v}{v} = \frac{\bar{\beta}}{v} = -\frac{\theta_1^2}{16}, \tag{11–39}$$

a well-known result that can also be obtained by approximating the elliptic-function representation of the motion.*

From Eqs. (11–33) and (11–34) it is seen that to first order the time variation of J is

$$\dot{J} = -\frac{J^2}{3\pi ml^2} \sin^3 2\pi(vt + \beta) \cos 2\pi(vt + \beta).$$

The average of $\sin^3 \phi \cos \phi$ over even a half period of ϕ is zero; hence, J shows no secular perturbation. We would expect this result physically, as J is a measure of the amplitude of the oscillations (cf. Eqs. (11–32)), and the perturbation would not be such as to cause the amplitude to grow or decay with time.

B. *A central force perturbation of the bound Kepler problem.* In Exercise 14, Chapter 3, it was shown rigorously that if a potential with a $1/r^2$ form is added to the Coulomb potential, the orbit in the bound problem is an ellipse in a *rotating* coordinate system. In effect the ellipse rotates, and the periapsis appears to precess. Here we will find the precession rate by first-order perturbation theory, considering a somewhat more general form for the perturbing potential.

Suppose the total potential is

$$V = -\frac{k}{r} - \frac{h}{r^n}, \tag{11–40}$$

where n is an integer greater than or equal to $+2$. The constant h will be assumed to be such that the second term is a small perturbation on the first. The perturbation Hamiltonian is thus

$$\Delta H = -\frac{h}{r^n}, \qquad n \geqslant 2. \tag{11–41}$$

In the unperturbed problem the angular position of the periapsis in the plane of the orbit is given by the constant $\omega = 2\pi w_2$ (cf. Eq. 10–140). With the perturbation ω has a time dependence determined by

$$\dot{\omega} = 2\pi \frac{\partial \Delta H}{\partial J_2} = \frac{\partial \Delta H}{\partial l}, \tag{11–42}$$

---

* See for example, K. R. Symon, *Mechanics*, 3d. ed., p. 215, especially Eq. (5.35).

using the relation $J_2 = 2\pi l$ (Eq. 10–131). First-order perturbation results are obtained by evaluating ΔH, and the derivative, in terms of the unperturbed motion. Further, the instantaneous change in ω is rarely of interest. In most situations where the perturbation formalism is of value, $\dot{\omega}$ is so small the change in ω is difficult or impossible to perceive within a single orbital period; and it is sufficient to measure only the secular change in ω after many orbits. Therefore what is wanted is $\dot{\omega}$ averaged over a time interval τ, the period of the unperturbed orbit:

$$\overline{\dot{\omega}} \equiv \frac{1}{\tau} \int_0^\tau \frac{\partial \Delta H}{\partial l} \, dt.$$

The derivative can be taken outside the integral sign, since τ is a function of J_3 only (Eq. (10–119) combined with Eq. (10–123)), whereas the derivative is with respect to $l = J_2/2\pi$. Hence

$$\overline{\dot{\omega}} = \frac{\partial}{\partial l} \left(\frac{1}{\tau} \int_0^\tau \Delta H \, dt \right) = \frac{\partial \overline{\Delta H}}{\partial l}. \tag{11–43}$$

But the time average of the perturbation Hamiltonian is here

$$\Delta \overline{H} = -h \overline{\left(\frac{1}{r^n} \right)} = -\frac{h}{\tau} \int_0^\tau \frac{dt}{r^n}. \tag{11–44}$$

By using the conservation of angular momentum in the form $l \, dt = mr^2 \, d\psi$, the integral can be converted into one over ψ:

$$\Delta \overline{H} = -\frac{mh}{l\tau} \int_0^{2\pi} \frac{d\psi}{r^{n-2}} \tag{11–45}$$

$$= -\frac{mh}{l\tau} \left(\frac{mk}{l^2} \right)^{n-2} \int_0^{2\pi} (1 + e\cos(\psi - \psi'))^{n-2} \, d\psi, \tag{11–45'}$$

where r has been expressed in terms of ψ through the orbit equation, Eq. (3–51) (with ψ used in place of θ). In general, only terms involving even powers of the eccentricity e will give nonvanishing contributions to the integral. The derivative with respect to l also involves e and its powers, since, by Eq. (10–134) e is a function only of J_2 and J_3.

Two special cases are of particular interest. One occurs when $n = 2$, mentioned briefly at the start of this illustration. The average perturbation Hamiltonian is then simply

$$\overline{\Delta H} = -\frac{2\pi m h}{l\tau},$$

and the secular precession rate is

$$\overline{\dot{\omega}} = \frac{2\pi m h}{l^2 \tau}, \tag{11–46}$$

which agrees with Exercise 14 of Chapter 3.

The other case of interest is for $n = 3$ (a $1/r^3$ perturbation potential), for which Eq. (11–45') reduces to

$$\overline{\Delta H} = -\frac{2\pi m^2 hk}{l^3\tau}$$

and

$$\bar{\dot{\omega}} = \frac{6\pi m^2 hk}{l^4\tau}. \tag{11–47}$$

What make this choice of n of particular significance is that general relativity theory predicts a correction to Newtonian motion that can be construed as an r^{-3} potential. The so-called Schwarzschild spherically symmetric solution of the Einstein field equations corresponds to an additional Hamiltonian in the Kepler problem* of the form of Eq. (11–41), with $n = 3$ and

$$h = \frac{kl^2}{m^2c^2}, \tag{11–48}$$

so that Eq. (11–47) becomes

$$\bar{\dot{\omega}} = \frac{6\pi k^2}{\tau l^2 c^2}. \tag{11–49}$$

To apply Eq. (11–49) to the secular precession rate for the precession of a body revolving around the sun, k is set equal to GMm and Eq. (3–63), valid for the unperturbed ellipse, is used in the form

$$l^2 = mka(1 - e^2). \tag{11–50}$$

Equation (11–49) can then be put in the form

$$\bar{\dot{\omega}} = \frac{6\pi}{\tau(1 - e^2)}\left(\frac{R}{a}\right), \tag{11–51}$$

where R is the so-called gravitational radius of the sun:†

$$R = \frac{GM}{c^2} = 1.4766 \, \text{km}. \tag{11–52}$$

---

* See W. M. Smart, *Celestial Mechanics* (New York, Wiley, 1953) p. 243, where recognition of the equivalent perturbation potential is ascribed to Eddington. For a more recent reference see the massive monograph of Misner, Thorne, and Wheeler, *Gravitation* (San Francisco, California: Freeman, 1973). The r^{-3} correction potential can be derived from an equivalent one-dimensional energy equation given there on p. 668, Eq. (25.42), by carefully deciphering the array of elaborate symbols and remembering the near invisibility of c in most expressions.

† The numerical value is based on the 1968 JPL set of astronomical constants. See W. G. Melbourne et al., JPL Technical Report 32-1306, July 15, 1968.

For the planet Mercury $\tau = 0.2409$ sidereal years, $G = 0.2056$, and $a = 5.790$ $\times 10^7$ km; Eq. (11–51) then predicts a precession of the perihelion of Mercury arising from general relativity at an average rate of

$$\overline{\dot{\omega}} = 42.98''/\text{century}.$$

The observed secular precession of the perihelion of Mercury is over 100 times larger than this value, namely $5599.74 \pm 0.41''/\text{century}$. Most of this is due to the precession of the equinoxes, i.e., the motion of the reference point of longitude with respect to the galaxy (see Chapter 5, p. 225). Of the remainder, about $531.54''/\text{century}$ arises from perturbations of the orbit of Mercury by other planets. Only after these two sets of effects are subtracted from the observed precession does the small general relativity effect become visible. A 1973 evaluation* estimated this residual at $41.4'' \pm 0.9''/\text{century}$; the deviation from the theoretical prediction is not considered significant.

One point remains to be made. In the application to relativistic effects the constant h, Eq. (11–48), is a function of the value of l. It might be asked therefore that in finding $\dot{\omega}$, why doesn't the derivative with respect to l act also on h? The key here is that h is not functionally dependent on l as a canonical momentum; Equation (11–48) says only how the value of the constant h is determined in terms of the value of the orbit parameter l. In other words, the perturbation potential is a function of the dynamical variables only through r; it is not to be construed as velocity dependent.

C. *Precession of the equinoxes and of satellite orbits.*   The family of problems to be considered here was discussed previously in Section 5–8, which bears the same title. We wish to describe the relative motion of two bodies interacting through their gravitational attraction, one a spherically symmetric or point body, the other being slightly oblate with a resultant gravitational quadrupole moment. The effect of the slight oblate shape of the earth is physically that the torques exerted by the sun and moon on the equatorial bulge cause the earth's rotation axis to precess very slowly. Reciprocally, the effect on an object orbiting around the earth, such as the moon or an artificial satellite, is to cause the plane of the orbit to precess about the figure axis of the earth. The small magnitude of the gravitational quadrupole term, manifested by the very slow rate of precession, suggests that a perturbation treatment should be an extremely good approximation. We shall actually examine here only the case of the perturbation of a satellite's orbit; the reciprocal phenomenon of the precession of the equinoxes proceeds very similarly (though with different notation) from the same perturbation Hamiltonian and will be left for the exercises.

Since the emphasis here will be on a point satellite moving about a much more massive earth, the notation of Section (5–8) will be reversed here and m used

---

* C. W. Misner, K. S. Thorne, and J. A. Wheeler, *Gravitation* (1973) p. 1112.

to denote the mass of the satellite while M stands for the earth's mass. The total potential acting on the satellite, by Eq. (5–88), is then

$$V = -\frac{k}{r} + \frac{k}{M}\frac{(I_3 - I_1)}{r^3}P_2(\gamma), \tag{11-53}$$

where $k = GMm$ and γ is the cosine of the angle θ between the radius vector to the satellite and the earth's figure axis. For the perturbation Hamiltonian we therefore have

$$\Delta H = k\frac{I_3 - I_1}{2Mr^3}(3\cos^2\theta - 1). \tag{11-54}$$

The polar angle θ can be expressed in terms of the inclination angle of the orbit, i, and the angle of the radius vector in the orbital plane relative to the periapsis, ψ, (the so-called true anomaly) by the relation*

$$\cos\theta = \sin i \sin(\psi + \omega), \tag{11-55}$$

where ω is the argument of the periapsis. A small amount of manipulation enables us to rewrite the angular dependence of ΔH as

$$3\cos^2\theta - 1 = (\tfrac{1}{2} - \tfrac{3}{2}\cos^2 i) - \tfrac{3}{2}\sin^2 i\cos 2(\psi + \omega). \tag{11-56}$$

Now, because of the small size of the perturbation, the chief interest is in the cumulative effects of the secular perturbation. Thus, the precession of the orbital plane shows up as a secular change in Ω, the angle of the line of nodes (or longitude of the ascending node, see p. 479). By the same argument as used in the previous illustration we can obtain the secular effects by averaging ΔH prior to taking derivatives:

$$\overline{\Delta H} \equiv \frac{1}{\tau}\int_0^\tau \Delta H\, dt = \frac{m}{l\tau}\int_0^{2\pi} r^2\,\Delta H\, d\psi$$

$$= \frac{m^2 k^2 (I_3 - I_1)}{2Ml^3\tau}\int_0^{2\pi}(1 + e\cos\psi)(3\cos^2\theta - 1)\, d\psi. \tag{11-57}$$

The term in $\cos 2(\psi + \omega)$ in Eq. (11–56) gives zero contribution to the integral because it is orthogonal, in the interval of integration, to both 1 and $\cos\psi$. Hence the averaged perturbation Hamiltonian is

$$\overline{\Delta H} = \frac{\pi m^2 k^2 (I_3 - I_1)}{2Ml^3\tau}(1 - 3\cos^2 i). \tag{11-58}$$

---

* Equation (11–55) can be obtained in many ways, e.g., by matrix rotation of the plane of the orbit into the xy plane. It is given, most simply perhaps, by some old-fashioned trigonometric reasoning based on Fig. (10–5). As $OB = 1$, $BC = \cos\theta$, but $AB = \sin(\psi + \omega)$ and therefore BC is also $\sin i \sin(\psi + \omega)$.

In view of Eqs. (10–132) and (10–139) linking $\underline{\Omega}$ and i with the action-angle variables, the first-order perturbation value for $\dot{\underline{\Omega}}$ is to be found from

$$\overline{\dot{\Omega}} = 2\pi\overline{\dot{w}_1} = 2\pi\frac{\partial \overline{\Delta H}}{\partial J_1} = \frac{1}{l}\frac{\partial \overline{\Delta H}}{\partial \cos i}$$

or

$$\overline{\dot{\Omega}} = -\frac{3\pi m^2 k^2 (I_3 - I_1)\cos i}{Ml^4\tau}.$$

Finally, using Eq. (11–50) the average fractional change in Ω per unperturbed revolution is

$$\frac{\overline{\dot{\Omega}}\tau}{2\pi} = -\frac{3}{2}\frac{I_3 - I_1}{Ma^2}\frac{\cos i}{(1 - e^2)^2}, \tag{11–59}$$

which is the appropriate generalization of Eq. (5–96) to an elliptic satellite orbit.

Once the average perturbation Hamiltonian is known the effect of the perturbation on other average parameters of orbit can be found. Thus, the secular precession of the periapsis in the plane of the orbit is immediately given by

$$\overline{\dot{\omega}} = 2\pi\overline{\dot{w}_2} = 2\pi\frac{\partial \overline{\Delta H}}{\partial J_2} = \frac{\partial \overline{\Delta H}}{\partial l}.$$

The canonical variable J_2 occurs in ΔH as given by Eq. (11–58) in two forms: in the l^3 term in the denominator and in the term containing $\cos i = J_1/J_2$. Upon carrying out the derivative it is found that

$$\frac{\overline{\dot{\omega}}\tau}{2\pi} = \frac{3}{4}\frac{I_3 - I_1}{Ma^2(1 - e^2)}(5\cos^2 i - 1). \tag{11–60}$$

The maximum value of $\overline{\dot{\omega}}$ is thus about the same as that of $\overline{\dot{\Omega}}$, but the dependence on i is quite different. At critical inclinations of $63°26'$ and $116°34'$, the precession of the periapsis vanishes (at least to first order) and changes sign above and below these points. It is clear that, to first order, there is no secular change in either a or e, since $\overline{\Delta H}$ does not contain the constant parts of any of the angle variables. The shape and size of the osculating ellipse, when averaged over the orbital period, thus does not change with time.

It may be noted from the last two illustrations that the general relativity correction and the gravitational quadrupole field both give rise to a precession of the periapsis of an orbiting body. R. Dicke has raised the question, therefore, whether the observed precession of the perihelion of Mercury might not be explained by a small oblateness of the sun's gravitational potential. A partial answer can be seen from the perturbation results we have obtained. It will be remembered that Kepler's law says that the period of the planetary motion is proportional to the three-halves power of the semimajor axis. The precession rate predicted by general relativity for the different planets should then, by Eq.

(11–51), vary as $a^{-5/2}$. If the precession were due to the sun's oblateness, Eq. (11–60) would predict a dependence as $a^{-7/2}$. The data for the planets other than Mercury is rather crude, but it does come closer to the prediction of general relativity. Perhaps an even stronger factor is the precession of the ascending node of Mercury's orbit. If the sun's gravity field were oblate enough to account for the observed perihelion precession, then, by Eq. (11–51), there should be a significant accompanying precession of the ascending node. In contrast the general relativity perturbation is entirely central; the angular momentum vector, and therefore the normal to the orbital plane, remains unaffected. The observations do not suggest any significant precession of the node not accounted for by other known perturbations. While the observational errors are considerable, they are several times less than the predictions of the oblateness effect. It would therefore appear that the quadrupole component of the sun's gravity field cannot account for more than a minor fraction of the observed anomaly in the precession of the perihelion.

11–4 TIME-INDEPENDENT PERTURBATION THEORY IN FIRST ORDER WITH ONE DEGREE OF FREEDOM

Time-independent perturbation is not concerned with the time dependence of the erstwhile constants of the unperturbed system; rather it seeks to find the quantities that are constant in the perturbed system. It can be applied only to conservative systems that are separable and periodic in the unperturbed state and remain so in the perturbed situation. These limitations still admit to a large number of interesting problems. On the celestial scale, after all, the planetary motions are still periodic in all three coordinates even with full considerations of all conservative perturbations to the Kepler motion. And in a classical picture of the atom, a perturbing magnetic field, as in the Zeeman effect, does not qualitatively change the periodic nature of the electron's motion. In celestial mechanics, time-independent perturbation theory is usually known as von Zeipel's method;* we give here the version of it developed by Born for the needs of the old quantum theory.

For simplicity, the formalism will first be described for systems of only one degree of freedom, and in first order at that. In the next section we will remove these restrictions. Consider an unperturbed periodic system described by action-angle variables w_0, J_0, and a Hamiltonian $H_0(J_0)$. The unperturbed frequency is given by

$$v_0 = \frac{\partial H_0}{\partial J_0}, \quad \text{with } \omega_0 = v_0 t + \beta_0. \tag{11–61}$$

---

* Also known as Poincaré's method. The use of von Zeipel as the eponym is sometimes reserved for a procedure that treats perturbation terms with short periods separately from those with long periods (compare the discussion on degeneracy, p. 525 below). But the practice is by no means uniform. See the bibliography at the end of the chapter, also R. A. Howland, Jr., *Celestial Mechanics* **15**, 327 (1977), especially Section 1.

It follows from the property that the system is periodic in w_0 with period 1, that q can be written as a Fourier series in the form*

$$q = \sum_{k=-\infty}^{+\infty} A_k(J_0)e^{2\pi i k w_0} \tag{11-62}$$

(compare with Eq. (10–91)), with a similar expression for p. Now suppose we add a perturbation so that the Hamiltonian looks like

$$H = H_0 + \epsilon H_1, \tag{11-63}$$

where ϵ is some small parameter that can be varied continuously from 0. Since the perturbed system is still periodic there must be a new set of action-angle variables (w, J) appropriate to the system, with $H = H(J)$. The new J is constant and w is a linear function of time:

$$w = vt + \beta, \tag{11-64}$$

with the new frequency v to be obtained from $H(J)$. Nonetheless (w_0, J_0) remain canonical variables for the perturbed system, for they are related to the original set (q, p) by a canonical transformation. As has been repeatedly emphasized the canonical property of a transformation, and therefore of the set of variables, is independent of the specific form of the Hamiltonian. But now J_0 is not a constant and w_0 is not a linear function of time. However, q is still a periodic function of w_0, for Eq. (11–62) is nothing more than the canonical equation of transformation for q in terms of w_0 and J_0. We therefore have the situation that *both* w and w_0 advance by unity as q goes through a complete period of the motion.

The two sets of canonical variables, (w_0, J_0) and (w, J), must therefore be related by a canonical transformation, whose generator, $Y(w_0, J)$, can be found from the Hamilton–Jacobi equation for Hamilton's characteristic function. Since the perturbation Hamiltonian is small, the generating function must deviate from the identity function only by a small quantity. To first order in ϵ we can write therefore

$$Y(w_0, J) = w_0 J + \epsilon Y_1(w_0, J). \tag{11-65}$$

What we want to do is to find the functional dependence of H on J so that the perturbed frequency v can be found. Of course H is also a function of the parameter ϵ. If H as a function of J and ϵ is denoted by $\alpha(J, \epsilon)$, then to first order

$$\alpha(J, \epsilon) = \alpha_0(J) + \epsilon \alpha_1(J). \tag{11-66}$$

Hence the Hamilton–Jacobi equation for

$$H(w_0, J_0) = H\left(w_0, \frac{\partial Y}{\partial w_0}\right) = \alpha$$

---

* It will be convenient *not* to use the summation convention.

can be written to first order in ϵ as

$$H_0\left(\frac{\partial Y}{\partial w_0}\right) + \epsilon H_1\left(w_0, \frac{\partial Y}{\partial w_0}\right) = \alpha_0(J) + \epsilon\alpha_1(J). \qquad (11\text{–}67)$$

Here H_0 and H_1 are still functions of ϵ through Y. To obtain a right-hand side of Eq. (11–67) that is consistently to first order in ϵ only, we must clearly replace Y in H_1 by its zero approximation, namely $w_0 J$. Further, J_0 in H_0 can be expressed in terms of J through the transformation equation

$$J_0 = \frac{\partial Y}{\partial w_0} = J + \epsilon\frac{\partial Y_1}{\partial w_0}.$$

To express H_0 as a function of J to first order in ϵ, we expand $H_0(J_0)$ in a Taylor series about the point $J_0 = J$ and retain only the first term in ϵ. The derivatives in the Taylor expansion are, strictly speaking, derivatives with respect to J_0 evaluated at $J_0 = J$, but without loss of rigor they can be written as derivatives with respect to J, once J is substituted for J_0 in $H_0(J_0)$. Hence the H_0 term in Eq. (11–67) can be written

$$H_0\left(J + \epsilon\frac{\partial Y_1}{\partial w_0}\right) = H_0(J) + \epsilon\frac{\partial Y_1}{\partial w_0}\frac{\partial H_0(J)}{\partial J}. \qquad (11\text{–}68)$$

In view of Eq. (11–61) the Hamilton–Jacobi equation now becomes

$$H_0(J) + \epsilon\left[H_1(w_0, J) + v_0\frac{\partial Y_1}{\partial w_0}\right] = \alpha_0(J) + \epsilon\alpha_1(J). \qquad (11\text{–}69)$$

Equating terms of the same order in ϵ we have

$$\alpha_0(J) = H_0(J), \qquad (11\text{–}70)$$

as would have been expected from the limiting situations as $\epsilon \to 0$, and

$$\alpha_1(J) = H_1(w_0, J) + v_0\frac{\partial Y_1}{\partial w_0}. \qquad (11\text{–}71)$$

Equation (11–71) looks at first sight somewhat strange, for the right-hand side is a function of J only, while the left-hand side is nominally a function of both J and w_0. It can only be concluded that Y_1 must be such that the terms on the left-hand side depending on w_0 cancel, leaving only the constant terms independent of w_0. But it is easy to see that the derivative of Y_1 has no constant term. The equation of transformation determining w is

$$w = \frac{\partial Y}{\partial J} = w_0 + \epsilon\frac{\partial Y_1}{\partial J}. \qquad (11\text{–}72)$$

In order therefore that both w and w_0 advance by 1 as the system goes through a period we must require Y_1 to be a periodic function of w_0:

$$Y_1(w_0 J) = \sum_{k=-\infty}^{+\infty} y_k(J) e^{2\pi i k w_0}. \tag{11-73}$$

Indeed it is the condition that Y_1 have the form of Eq. (11–73), which guarantees that (w, J) are action-angle variables. Hence from Eq. (11–73) it follows that the term in the derivative of Y_1 with respect to w_0 for $k = 0$, i.e., the constant term, vanishes identically. Thus Eq. (11–71) can be written as

$$\alpha_1(J) = \overline{H_1(w_0, J)} - \left[(\overline{H}_1 - H_1) - v_0 \frac{\partial Y_1}{\partial w_0} \right], \tag{11-74}$$

where the bar denotes averaging over a complete period of w_0. The quantity in the bracket must cancel if α_1 is to be a function of J alone, so we have two conditions:

$$\alpha_1(J) = \overline{H_1(w_0, J)}, \tag{11-75}$$

and

$$\frac{\partial Y_1}{\partial w_0} = \frac{\overline{H}_1 - H_1}{v_0}. \tag{11-76}$$

The first of these equations tells us how to complete the functional dependence of H on J (to first order). We express the perturbation Hamiltonian in terms of the unperturbed motion through (w_0, J_0) and average it over a complete period of motion. The remaining dependence on J_0 is then the same (to this order) as the J dependence of α_1. In terms of α the new frequency is given as

$$v = \frac{\partial \alpha(J)}{\partial J} = v_0 + \epsilon \frac{\partial \alpha_1}{\partial J}. \tag{11-77}$$

The second equation, Eq. (11–76), is now a differential equation that can be solved for $Y_1(w_0, J)$ in terms of the behavior of H_1. Once Y_1 is found, the relations between (w_0, J_0) and (w, J) are determined to first order and then, by Eqs. (11–62) and (11–64), the solutions for (q, p) as functions of time are thereby found. Note, however, that if we want only the new frequency there is no need to find Y_1 at all; to this order Eq. (11–75) is then sufficient.

This classical formalism has a familiar appearance to those acquainted with time-independent perturbation theory in quantum mechanics. There, in first order, the shift of the energy eigenvalue (Hamiltonian) is given by the matrix element of the perturbation Hamiltonian (average over the unperturbed motion). Knowledge of the perturbed wave functions is not needed to obtain the energy shift, just as we don't need Y_1 here.

A brief example will illustrate the procedure. Consider the problem, treated in Section 11–3, of the plane pendulum with finite amplitude of oscillation. From Eq.

(11–33) the perturbation Hamiltonian can be written in terms of J_0 and w_0 as

$$\epsilon H_1 = -\frac{J_0^2}{24\pi^2 m l^2} \sin^4 2\pi w_0 \tag{11–78}$$

(where l, it will be remembered, is here only the length of the pendulum). It will be convenient to take as the parameter ϵ the small quantity θ_1^2, Eq. (11–28), the square of the amplitude of oscillation in the unperturbed situation. Since the average of $\sin^4$ over one period is 3/8, the functional form of α_1 is

$$\alpha_1(J) = \bar{H}_1 = \frac{-J^2}{64\pi^2 m l^2 \theta_1^2},$$

and

$$\frac{v - v_0}{\epsilon} = \frac{\partial \alpha_1}{\partial J} = -\frac{J}{32\pi^2 m l^2 \theta_1^2}. \tag{11–79}$$

To evaluate $v - v_0$ to first order in ϵ, it is permissible to replace J by J_0. Further, it follows from Eqs. (10–76) and (11–28) that

$$J_0 = \frac{2\pi E}{\omega_0} = \frac{\pi m g l \theta_1^2}{\omega_0} = 2\pi^2 m l^2 \theta_1^2 v_0. \tag{11–80}$$

Hence Eq. (11–79) reduces simply to

$$\frac{\partial \alpha_1}{\partial J} = -\frac{v_0}{16},$$

and the fractional change in v is

$$\frac{\Delta v}{v_0} \equiv \frac{v - v_0}{v_0} = -\frac{\theta_1^2}{16},$$

which is the same result as in Eq. (11–39).

There are instances in which H_1 vanishes and first-order perturbation gives no useful result. Thus in the so-called anharmonic oscillator the first-order term in the perturbation Hamiltonian is of the form ϵq^3. But since $\sin^3 2\pi w_0$ has a zero average, there is no resultant shift in the frequency to first order and one must seek higher approximations. The subject of second and higher perturbations, as well as the generalization to systems of many degrees of freedom, will be treated in the next section.

11–5   TIME-INDEPENDENT PERTURBATION THEORY TO HIGHER ORDER

We shall retrace the arguments of the previous section for conservative periodic separable systems of arbitrary number of degrees of freedom and to higher order in the perturbation parameter ϵ. For the unperturbed problem we assume a set of action-angle variables (w_{0i}, J_{0i}) such that the unperturbed Hamiltonian, H_0, is a function only of the action variables and correspondingly the w_{0i} are then linear functions of time. In the notation of Eq. (10–91') the relation between, say, q_k and the w_{0i} can be written compactly as

$$q_k = \sum_{\mathbf{j}} A_{\mathbf{j}}^{(k)}(\mathbf{J}_0) e^{2\pi i \mathbf{j} \cdot \mathbf{w}_0}, \tag{11–81}$$

where $\mathbf{j}, \mathbf{w}_0$ and $\mathbf{J}_0$ are n-dimensional vectors of the integer indices, angle variables, and action variables, respectively.

In the perturbed system $(\mathbf{w}_0, \mathbf{J}_0)$ remain a valid canonical set of variables. When expressed in terms of the set $(\mathbf{w}_0, \mathbf{J}_0)$ the perturbed Hamiltonian can be expanded in powers of a small perturbation parameter ϵ:

$$H(\mathbf{w}_0, \mathbf{J}_0, \epsilon) = H_0(\mathbf{J}_0) + \epsilon H_1(\mathbf{w}_0, \mathbf{J}_0) + \epsilon^2 H_2(\mathbf{w}_0, \mathbf{J}_0) + \cdots. \tag{11–82}$$

We seek a canonical transformation from $(\mathbf{w}_0, \mathbf{J}_0)$ to a new set $(\mathbf{w}, \mathbf{J})$ such that the $\mathbf{J}$ are all constants and the $\mathbf{w}$ therefore linear functions of time. In this set H is a function only of $\mathbf{J}$ (and ϵ) and, in its functional form with respect to $\mathbf{J}$, will be written as

$$\alpha(\mathbf{J}, \epsilon) = \alpha_0(\mathbf{J}) + \epsilon \alpha_1(\mathbf{J}) + \epsilon^2 \alpha_2(\mathbf{J}) + \cdots. \tag{11–83}$$

To obtain the perturbed frequencies through a given order in ϵ, it suffices to find the appropriate functions $\alpha_0, \alpha_1, \ldots$, for then the vector representing the frequencies is

$$\mathbf{v} = \mathbf{v}_0 + \epsilon \frac{\partial \alpha_1}{\partial \mathbf{J}} + \epsilon^2 \frac{\partial \alpha_2}{\partial \mathbf{J}} + \cdots. \tag{11–84}$$

The generator of the canonical transformation from $(\mathbf{w}_0, \mathbf{J}_0)$ to $(\mathbf{w}, \mathbf{J})$ is $Y(\mathbf{w}_0, \mathbf{J}, \epsilon)$, with a corresponding expansion in ϵ:

$$Y(\mathbf{w}_0, \mathbf{J}, \epsilon) = \mathbf{w}_0 \cdot \mathbf{J} + \epsilon Y_1(\mathbf{w}_0, \mathbf{J}) + \epsilon^2 Y_2(\mathbf{w}_0, \mathbf{J}) + \cdots. \tag{11–85}$$

We seek to find Y as the solution of the appropriate Hamilton–Jacobi equation:

$$H\left(\mathbf{w}_0, \frac{\partial Y}{\partial \mathbf{w}_0}, \epsilon\right) = \alpha(\mathbf{J}, \epsilon). \tag{11–86}$$

As before, the terms in α to a given order in ϵ are found by expanding both sides in powers of ϵ and collecting coefficients of the same order on both sides. We will illustrate the process for a second-order calculation, where the Hamilton–Jacobi equation reduces to

$$H_0\left(\frac{\partial Y}{\partial \mathbf{w}_0}\right) + \epsilon H_1\left(\mathbf{w}_0, \frac{\partial Y}{\partial \mathbf{w}_0}\right) + \epsilon^2 H_2\left(\mathbf{w}_0, \frac{\partial Y}{\partial \mathbf{w}_0}\right)$$

$$= \alpha_0(\mathbf{J}) + \epsilon \alpha_1(\mathbf{J}) + \epsilon^2 \alpha_2(\mathbf{J}). \tag{11–87}$$

Each of the terms on the left are functions of ϵ through the derivative of Y:

$$\mathbf{J}_0 = \frac{\partial Y}{\partial \mathbf{w}_0} = \mathbf{J} + \epsilon \frac{\partial Y_1}{\partial \mathbf{w}_0} + \epsilon^2 \frac{\partial Y_2}{\partial \mathbf{w}_0}. \qquad (11\text{-}88)$$

We again expand the terms H_i in a Taylor series around $\mathbf{J}_0 = \mathbf{J}$, retaining terms of order ϵ^2 in H_0 and of order ϵ in H_1, with $\mathbf{J}_0$ replaced directly by $\mathbf{J}$ in H_2. The expansions for H_0 and H_1, in matrix notation, are then

$$H_0\left(\frac{\partial Y}{\partial \mathbf{w}_0}\right) = H_0(\mathbf{J}) + \left(\epsilon \frac{\partial Y_1}{\partial \mathbf{w}_0} + \epsilon^2 \frac{\partial Y_2}{\partial \mathbf{w}_0}\right) \frac{\partial H_0}{\partial \mathbf{J}} + \frac{1}{2}\left(\epsilon \frac{\partial Y_1}{\partial \mathbf{w}_0}\right) \frac{\partial^2 H_0}{\partial \mathbf{J} \partial \mathbf{J}} \left(\epsilon \frac{\partial Y_1}{\partial \mathbf{w}_0}\right), \qquad (11\text{-}89)$$

$$H_1\left(\mathbf{w}_0, \frac{\partial Y}{\partial \mathbf{w}_0}\right) = H_1(\mathbf{w}_0, \mathbf{J}) + \epsilon \frac{\partial Y_1}{\partial \mathbf{w}_0} \frac{\partial H_1}{\partial \mathbf{J}}. \qquad (11\text{-}90)$$

Collecting powers of ϵ in Eq. (11-87) then leads to the following expressions for the first three terms in α:

$$\alpha_0 = H_0(\mathbf{J}), \qquad (11\text{-}91a)$$

$$\alpha_1 = \mathbf{v}_0 \frac{\partial Y_1}{\partial \mathbf{w}_0} + H_1(\mathbf{w}_0, \mathbf{J}), \qquad (11\text{-}91b)$$

$$\alpha_2 = \mathbf{v}_0 \frac{\partial Y_2}{\partial \mathbf{w}_0} + \Phi_2(\mathbf{w}_0, \mathbf{J}), \qquad (11\text{-}91c)$$

where

$$\Phi_2(\mathbf{w}_0, \mathbf{J}) = H_2(\mathbf{w}_0, \mathbf{J}) + \frac{\partial Y_1}{\partial \mathbf{w}_0} \frac{\partial H_1}{\partial \mathbf{J}} + \frac{1}{2} \frac{\partial Y_1}{\partial \mathbf{w}_0} \frac{\partial^2 H_0}{\partial \mathbf{J} \partial \mathbf{J}} \frac{\partial Y_1}{\partial \mathbf{w}_0}. \qquad (11\text{-}92)$$

Again, the equation of transformation linking $\mathbf{w}$ and $\mathbf{w}_0$ is given by

$$\mathbf{w} = \frac{\partial Y}{\partial \mathbf{J}} = \mathbf{w}_0 + \epsilon \frac{\partial Y_1}{\partial \mathbf{J}} + \epsilon^2 \frac{\partial Y_2}{\partial \mathbf{J}} + \cdots. \qquad (11\text{-}93)$$

In order for the (q, p) set to be periodic in both $\mathbf{w}_0$ and $\mathbf{w}$ with period 1, all of the Y_k terms must be periodic functions of $\mathbf{w}_0$, that is, of the form

$$Y_k(\mathbf{w}_0, \mathbf{J}) = \sum_{\mathbf{j}} B_{\mathbf{j}}^{(k)}(\mathbf{J}) e^{2\pi i \mathbf{j} \cdot \mathbf{w}_0}. \qquad (11\text{-}94)$$

Hence all derivatives of Y_k with respect to $\mathbf{w}_0$ have no constant term, and the first terms on the right of Eqs. (11-91b, c) do not contribute to the J dependence. Equations (11-91) can therefore also be written as

$$\alpha_0(\mathbf{J}) = H_0(\mathbf{J}), \qquad (11\text{-}95a)$$

$$\alpha_1(\mathbf{J}) = \overline{H_1(w_0, \mathbf{J})}, \qquad (11\text{-}95b)$$

$$\alpha_2(\mathbf{J}) = \overline{\Phi_2(\mathbf{w}_0, \mathbf{J})}, \qquad (11\text{-}95c)$$

where the bar denotes an average over the periods of all $\mathbf{w}_0$. We can conveniently express all of Eqs. (11–95) in a common format by

$$\alpha_i(\mathbf{J}) = \overline{\Phi_i(\mathbf{w}_0, \mathbf{J})}, \tag{11–95'}$$

where $\Phi_0 = H_0$ and $\Phi_1 = H_1$. In addition, Eqs. (11–91) have counterparts periodic in $\mathbf{w}_0$ with zero mean:

$$\mathbf{v}_0 \frac{\partial Y_i}{\partial \mathbf{w}_0} = \overline{\Phi}_i - \Phi_i. \tag{11–96}$$

For $i = 0$ and 1, Eqs. (11–91a), (11–95), and (11–96) reduce to the same form as found in the previous section, except they now involve all frequencies of the unperturbed motion. Note that in second-order perturbation the terms in Y_1 do not necessarily vanish in the mean. It is true that the derivatives of Y_1 themselves have zero mean, but they are multiplied by other functions that will be periodic in $\mathbf{w}_0$, and there is no guarantee that the average of the product vanishes. Hence to find the second-order correction to the frequencies, one needs to know the first-order canonical transformation. (Analogously in quantum mechanics, a second-order eigenvalue involves first-order corrections of the wave function.) In principle the coefficients $B_j^{(1)}$ defining Y_1 through Eq. (11–94) can be found directly from Eq. (11–96) for $i = 1$. Subtraction of the average means that $H_1 - \overline{H}_1$ can be expanded in a Fourier series analogous to Eq. (11–81) or (11–94) but without any constant term:

$$H_1 - \overline{H}_1 = \sum_{\mathbf{j} \neq 0} C_{\mathbf{j}}(\mathbf{J}) e^{2\pi i \mathbf{j} \cdot \mathbf{w}_0}. \tag{11–97}$$

Taking the derivative of Y_1 in Eq. (11–96) with respect to one of the $\mathbf{w}_0$, say $\dot{w}_{0k}$, will bring down a factor $2\pi i j_k$. Hence the matrix product on the left-hand side of Eq. (11–96) can be written

$$\mathbf{v}_0 \frac{\partial Y_1}{\partial \mathbf{w}_0} = \sum_{\mathbf{j} \neq 0} B_{\mathbf{j}}^{(1)}(\mathbf{J}) 2\pi i (\mathbf{j} \cdot \mathbf{v}_0) e^{2\pi i \mathbf{j} \cdot \mathbf{w}_0}. \tag{11–98}$$

From Eqs. (11–96) and (11–97) the coefficients in the series for Y_1 can be obtained as

$$B_{\mathbf{j}}^{(1)}(\mathbf{J}) = \frac{C_{\mathbf{j}}(\mathbf{J})}{2\pi i (\mathbf{j} \cdot \mathbf{v}_0)}, \qquad \mathbf{j} \neq 0. \tag{11–99}$$

It is true the constant terms in Y_1 are not determined in this way, but it is only the derivatives of Y_1 that enter into the expressions for α_i and these do not involve constant terms.

While we have carried out the procedure in detail only for second-order perturbation, it is easy to see that the general form of the higher order calculations must be similar; only the details of the algebra will be more complex. For the ith order perturbation we will again be able to write α_i in the form

$$\alpha_i(\mathbf{J}) = \mathbf{v}_0 \frac{\partial Y_i}{\partial \mathbf{w}_0} + \Phi_i(\mathbf{w}_0, \mathbf{J}). \tag{11–91d}$$

The first term on the right will come from the first-derivative term in the Taylor expansion of $H(\mathbf{J}_0)$ about $\mathbf{J}_0 = \mathbf{J}$, where all terms in the difference $\mathbf{J}_0 - \mathbf{J}$ are kept through order ϵ^i. Only in this term will Y_i appear; hence Φ_i can contain only the generators Y_k for order less than i. By virtue of the arguments already used for first- and second-order perturbations, the first term on the right in Eq. (11–91d) has zero mean when averaged over complete cycles in $\mathbf{w}_0$, and hence Eqs. (11–95) and (11–96) are valid in all orders. Of course, for $i > 2$, Φ_i becomes increasingly more complicated than Eq. (11–92) but it always contains only such functions as have already been found in lower order calculations. Thus, step by step, one could in principle work up to any order perturbation.

There are practical problems in such a series of calculations, of course, but the most serious and obvious conceptual difficulty occurs if the unperturbed system is degenerate, i.e., the set of frequencies $\mathbf{v}_0$ exhibit commensurabilities. As we see from Eq. (10–101) the existence of a degeneracy means there will be at least one vector of indices $\mathbf{j}$ such that $\mathbf{j} \cdot \mathbf{v}_0 = 0$. The corresponding coefficient $B_{\mathbf{j}}^{(1)}$ in the Fourier series for Y_1 will therefore, by Eq. (11–99), blow up. Indeed, something similar takes place even when the unperturbed system is *not* degenerate. Even if the frequencies are not exactly commensurate, as one goes to higher and higher values of the integer indices in $\mathbf{j}$ eventually there will be found a vector $\mathbf{j}$ for which $\mathbf{j} \cdot \mathbf{v}_0$ is very small even if not zero, and the corresponding coefficients B become very large (the so-called problem of "small divisors").* This crudely qualitative observation is the basis of the elegant proof by Poincaré at the end of the last century that the Fourier series for Y_1, and therefore for the motion, are only semiconvergent. Nonetheless, the series can be truncated at some reasonable values of the indices and still give extremely precise results, at least for times that are not too long.

We will discuss later what can be done in the presence of degeneracy, but at this point it may be well to illustrate a second-order calculation with a specific example of a system with one degree of freedom.

Consider a one-dimensional *anharmonic oscillator*, i.e., one with a term in q^3 in the potential energy. The Hamiltonian can be written as

$$H = \frac{1}{2m}\left[p^2 + m^2\omega_0^2 q^2\left(1 + \epsilon\frac{q}{q_0}\right)\right], \tag{11–100}$$

where ω_0 is the unperturbed angular frequency:

$$\omega_0 = 2\pi v_0 = 2\pi\sqrt{\frac{k}{m}},$$

---

* Similar phenomena, it will be recalled, are found in quantum mechanics, where degeneracy means that there are several states with the same energy E. Denominators of the form $E_i - E_j$ will then vanish, or become small even if there is no exact degeneracy.

q_0 is a reference amplitude that can be left unspecified for the moment, and ϵ is a small dimensionless parameter. Taken as an expansion in powers of ϵ, H consists of the terms

$$H_0 = \frac{1}{2m}[p^2 + m^2\omega_0^2q^2], \tag{11-101a}$$

$$H_1 = \frac{m\omega_0^2q^3}{2q_0}, \tag{11-101b}$$

and

$$H_i = 0, \qquad i \geqslant 2. \tag{11-101c}$$

Using the unperturbed action-angle variables (J_0, w_0) as canonical variables the nonvanishing parts of H can, by Eqs. (10–78) and (10–79) be written as

$$H_0 = J_0\nu_0 \tag{11-102a}$$

and

$$H_1 = \frac{m\omega_0^2}{2q_0}\left(\frac{J_0}{\pi m\omega_0}\right)^{3/2}\sin^3 2\pi w_0. \tag{11-102b}$$

The recipes of Eqs. (11–95a, b) then give as the lowest two terms in $\alpha(J)$

$$\alpha_0(J) = J\nu_0; \qquad \alpha_1(J) = 0.$$

To obtain the second-order term $\alpha_2(J)$, we note that since H_0 is linear in J, and H_2 vanishes, then Φ_2 reduces to

$$\Phi_2 = \frac{\partial Y_1}{\partial w_0}\frac{\partial H_1}{\partial J}.$$

But the vanishing of $\overline{H_1}$ means that Eq. (11–96) for $i = 1$ has the simple form

$$\frac{\partial Y_1}{\partial w_0} = -\frac{H_1}{\nu_0}.$$

Combining these two results leads to

$$\Phi_2 = -\frac{1}{2\nu_0}\frac{\partial H_1^2}{\partial J}. \tag{11-103}$$

Now from Eq. (11–102b),

$$H_1^2(w_0, J) = \frac{\nu_0 J^3}{2\pi^2 mq_0^2}\sin^6 2\pi w_0,$$

leading to

$$\Phi_2(w_0, J) = -\frac{3J^2}{4\pi^2 mq_0^2}\sin^6 2\pi w_0. \tag{11-104}$$

Since the average of $\sin^6$ over one period is 15/48, $\alpha_2(J)$ is simply

$$\alpha_2(J) = -\frac{15J^2}{64\pi^2 mq_0^2}, \tag{11-105}$$

and to second order in ϵ the perturbed frequency is

$$v = \frac{\partial \alpha}{\partial J} = v_0 - \epsilon^2 \frac{15J}{32\pi^2 m q_0^2}. \tag{11–106}$$

It is convenient to use for q_0 the maximum amplitude the oscillator would have for the given energy in its unperturbed form, so that

$$\frac{m\omega_0^2 q_0^2}{2} = E,$$

or, to lowest order,

$$m q_0^2 = \frac{J}{\pi \omega_0}. \tag{11–107}$$

In terms of this reference amplitude, Eq. (11–106) is equivalent to saying that the second-order fractional shift in the frequency is simply

$$\frac{\Delta v}{v_0} = -\frac{15}{16} \epsilon^2. \tag{11–108}$$

Other examples of perturbation theory in second order or for systems of many degrees of freedom are given in the exercises.

Mention has already been made of the difficulties that appear in perturbation theory arising out of the existence of degeneracy, e.g., the vanishing (or near vanishing) of $\mathbf{j} \cdot \mathbf{v}_0$ in the denominators of Eq. (11–99). Treatment of degeneracies in classical perturbation theory is much more complicated than in quantum mechanics. In some aspects, especially as relates to the handling of near-degeneracies, it is still a subject of expanding research. The mathematics that has been brought to bear on the problem is both subtle and complicated and a full exposition of current developments would be out of place here. Only some brief and introductory remarks can be made at this point.

We speak of exact (or "proper") degeneracy, as in Section 10–6 above, when the unperturbed frequencies $\mathbf{v}_0$ are such that there are one or more sets of integers $\mathbf{j}$ for which $\mathbf{j} \cdot \mathbf{v}_0 = 0$. As has been pointed out in Section 10–6, one can then transform to a new set of variables (w_0, J_0) for which the degeneracies appear as zero frequencies and the remaining nonzero unperturbed frequencies are not degenerate. The effect of the perturbation is to lift the degeneracy so that the corresponding frequencies are not exactly zero but have small values. In consequence there appear in the solution terms that have small frequencies, i.e., long periods. The corresponding angle variables are known as "slow" variables, in contrast to the angle variables with nondegenerate frequencies, which are therefore called the "fast" variables. Long-period terms may appear as secular terms over restricted time intervals; e.g., $\sin 2\pi v t$ can be taken as a linear function of t so long as $vt \ll 1$.

When there is exact degeneracy, a transformation is first made to the $(w_0, \mathbf{J}_0)$ set. The unperturbed Hamiltonian will be a function only of the nondegenerate J_0

variables; in all other respects Eq. (11–82) still represents the complete Hamiltonian. We now carry through the canonical transformation of the perturbation calculation, but only for the nonperturbed variables, leaving the degenerate variables unchanged. What is in effect the new Hamiltonian, Eq. (11–83), now has the form

$$\alpha(\mathbf{J}, \mathbf{J}'_0, \mathbf{w}'_0, \epsilon) = \alpha_0(\mathbf{J}) + \epsilon\alpha_1(\mathbf{J}, \mathbf{J}'_0, w'_0) + \epsilon^2\alpha_2(\mathbf{J}, \mathbf{J}'_0, \mathbf{w}'_0) + \cdots.$$

Here $\mathbf{w}'_0$ stands for the m (degenerate) variables that in the unperturbed problem have zero values and $\mathbf{J}'_0$ for their conjugate momenta. The transformed nondegenerate momenta are represented by $\mathbf{J}$. The result of the canonical transformation is thus to eliminate the "fast" variables, but to leave in terms with the "slow" variables. Note that since α is cyclic in w, the transformed $\mathbf{J}$ momenta are true constants of the motion, and $\alpha(\mathbf{J}, \mathbf{J}'_0, \mathbf{w}'_0, \epsilon)$ can be considered as a Hamiltonian of a system with m degrees of freedom. Further, since $\alpha_0(\mathbf{J})$ is a constant, independent of the remaining variables, it doesn't matter for the equations of motion of $(\mathbf{J}'_0, \mathbf{w}'_0)$ and can be dropped from α. Thus the new effective Hamiltonian is now of order ϵ; in effect the "unperturbed Hamiltonian" is $\epsilon\alpha_1(\mathbf{J}, \mathbf{J}'_0, \mathbf{w}'_0)$ and in *this* unperturbed problem $\mathbf{w}'_0$ no longer consists of zero values. If there is only one degeneracy condition, the effective problem is of only one degree of freedom and is in principle immediately integrable. With more degeneracy conditions, one can seek a second canonical transformation to eliminate the "slow" variable terms just as was done for the "fast" variables. In practice the procedure obviously becomes quite complicated.

It has already been pointed out, in connection with Eq. (11–99), that even with nondegenerate frequencies, small values of the divisor $\mathbf{j} \cdot \mathbf{v}_0$ will inevitably occur as the indices $\mathbf{j}$ become larger and larger. This phenomenon is referred to as *resonance*, implying that the amplitude of some particular term in the Fourier expansions becomes very large. It would seem therefore that the problems of degeneracy will always be with us, no matter what the unperturbed frequencies are! The situation is not all as bad as that, in part because of the nature of the perturbation Hamiltonians encountered in practice. From Eq. (11–99) it will be noted that what counts is not so much the value of $\mathbf{j} \cdot \mathbf{v}_0$ as the ratio

$$\frac{C_\mathbf{j}}{\mathbf{j} \cdot \mathbf{v}_0},$$

where $C_\mathbf{j}$ is the Fourier series expansion of the perturbation Hamiltonian H_1, cf. Eq. (11–97). It turns out that in celestial mechanics, at least, most perturbation Hamiltonians have what is called the *D'Alembert characteristic*. While the formal mathematical definition of the property is involved,* what it says, roughly, is that when the values of the integers in the $\mathbf{j}$ indices are larger than the exponent of ϵ in

---

*See G. E. O. Giacaglia, *Perturbation Methods in Non-Linear Systems* (New York: Springer-Verlag, 1972), pp. 279–280.

the Hamiltonian, the magnitudes of C_j fall rapidly, generally exponentially, with increasing values of the indices. The ratios in Eq. (11–99) then do not become too large, and the expansion process actually can be proved to converge when the frequencies v_0 satisfy an irrationality condition.

Resonant behavior in the presence of the D'Alembert characteristic, or generally when $C_j/(\mathbf{j} \cdot \mathbf{v}_0) < O(\epsilon^{\frac{1}{2}})$, is described as a *shallow resonance*. In principle, at least, shallow resonances do not upset the perturbation expansion process and can be tolerated without introducing new methods. There are situations where the ratio $C_j/(\mathbf{j} \cdot \mathbf{v}_0)$ becomes large, at least larger than order $\epsilon^{\frac{1}{2}}$, and these are referred to as *deep resonances*. Special methods have to be devised to handle deep resonances, particularly the so-called Bohlin expansion in powers of $\epsilon^{\frac{1}{2}}$ rather than in powers of ϵ. To go further would enter into the large and rapidly developing field of resonance phenomena in nonlinear oscillations, a field that would need a separate treatise for adequate treatment.

11–6 SPECIALIZED PERTURBATION TECHNIQUES IN CELESTIAL AND SPACE MECHANICS

As has been noted perturbation theory and celestial mechanics have evolved together since the time of Newton. A considerable number of specialized methods have been developed for the particular needs of celestial mechanics. The birth and rapid growth of space exploration and of the modern digital computer (almost simultaneously) has brought back to life a nearly dormant and hibernating field, with the creation of new approaches and fresh ways of looking at perturbation theory. No attempt will be made to discuss this long history, both ancient and modern, in any detail. All that is intended is to describe trends and to introduce the reader to the often peculiar terminology used in the literature.

Distinction is often made between *general* perturbation theories, which lead to analytic formulas, and *special* perturbation methods, involving numerical solutions of the equations for the perturbed system. (Mixed methods also exist, so the distinction is only rough.) The perturbation schemes discussed above are all examples of general perturbation theories. All of the early attempts at perturbation methods in celestial mechanics can be classified as "general." The first technique was that of the "variation of constants," which was developed, haltingly and with false starts, in the eighteenth century. Lagrange succeeded finally in putting the method on a firm basis in 1782 (not, of course, in the canonical version described above).

The construction of perturbation procedures in celestial mechanics was heavily influenced by the special nature of the problems to be treated. Until recently all the forces considered were gravitational, for the most part between point masses. The lunar problem, i.e., the motion of the moon around the earth, always loomed large, because the solar perturbations are considerably greater than in almost any other astronomical situation. Over the centuries, starting with Newton, repeated attacks were made on the "main problem" of lunar theory.

Here the earth and moon are treated as two mass points whose center-of-mass travels around the sun in a fixed Keplerian ellipse. Even in this simplified model, some idea of the complexity of the situation can be glimpsed by considering the form of the perturbation Hamiltonian. All that enters in ΔH are the components of the distances between the three bodies. But these distances are complicated functions of the radii vectors and the angles between them, especially when the motion is not confined to a plane. Usually the terms in ΔH are expressed as Fourier series of the angles involved. In turn, the angles are complicated series expressions in terms of the various Keplerian anomalies, and finally of the time. As a result the perturbation Hamiltonian, even when expressed in terms of the unperturbed motion, must be represented by nested trigonometric series. A good deal of mathematical effort in general perturbation theory is spent in finding the proper development of ΔH, usually in the form of the disturbing function $R = -\Delta H$.

Another general perturbation theory developed in the early nineteenth century is sometimes called "variation of coordinates." In its simplest form the equations of motion of a perturbed two-body problem are written in Cartesian coordinates, with a perturbation Hamiltonian $\Delta H = \epsilon V_1$. The equations then look like

$$\mu \ddot{x}_i = -\frac{kx_i}{r^3} - \epsilon \frac{\partial V_1}{\partial x_i},$$

where μ is the reduced mass of the two-body problem. One then looks for solutions as expansions in power of ϵ:

$$x_i = x_i^{(0)} + \epsilon x_i^{(1)} + \epsilon^2 x_i^{(2)} + \cdots,$$

collecting powers of ϵ when the expansion is substituted into the equations of motion. The method has the advantage of giving desired perturbed coordinates directly, but it fails to give a picture of how the perturbations distort the orbit and the parameters describing it.

A local maximum in the elaborateness of a general perturbation method was reached in the 1860s in Delaunay's analytic solution of the "main problem" of lunar theory. His procedure might be described as an early version of von Zeipel's method. In effect, he removed each periodic term in the disturbing function depending on an angle variable by a separate canonical transformation. After 20 years of effort, Delaunay published his results in two thick volumes containing little else than algebraic formulas for the perturbed motion of osculating elements, as expansions in powers of small quantities such as the eccentricities of the earth's and moon's orbits and the ratio of the mean angular speeds of the moon and the earth. Combined powers of these quantities were retained up to order 7, in some cases through order 9. The statistics of the algebraic manipulations required are stupefying—over 500 separate canonical transformations, involving in all over 10,000 individual terms—all done by hand, without assistance.

More sophisticated methods, some seminumerical, were devised by the end of the nineteenth century. They involve more complicated reference orbits, the so-called intermediaries, from which the perturbed motions are measured. Even a description of the underlying models would require too much space to give here; we can do no more than mention the names of some of them—e.g., Hansen's method or Hill's method.

Artificial satellites and digital computers brought both the need and the opportunity for a fresh look at perturbation methods. Radar and laser observations increased by several orders of magnitude the accuracy with which the coordinates of celestial bodies could be measured, and existing tables of the moon and the inferior planets were found to be inadequate. Hori and Deprit independently in the 1960s improved and reformulated the von Zeipel method so as to greatly simplify the calculations, at least conceptually. In the von Zeipel method, in order to obtain the motion as a function of time one must invert the equations of transformation, Eqs. (11–93) to find $\mathbf{w}_0$ as a function of $\mathbf{w}$. The inversion is often very difficult to carry out analytically in a systematic fashion. What Deprit proposed was to build up the transformation equations for finite values of ϵ out of those for the infinitesimal canonical transformation in the limit of very small ϵ. In basic principle, one uses the Poisson bracket solution for the differential equation governing the change in a function under an I.C.T., cf. Eqs. (9–114) and (9–115). The iterative method of constructing this solution, and the repetitive steps needed to expand the nested series of the disturbing function, lend themselves to computer algebraic manipulation. In this manner Deprit and his collaborators in 1970 repeated in short order Delaunay's program, carrying it to at least one higher order in small quantities. Comparison of the 1970 computer printouts with Delaunay's 1869 hand produced tables showed that out of the vast amounts of algebra he had committed errors in only seven coefficients!

Special perturbation techniques were not introduced much before the middle of the nineteenth century and could hardly have been practicable until at least manual desk calculators became available. The simplest (though not the earliest) is known as *Crowell's method*; it involves direct numerical integration of the Newtonian equations of motion for the n interacting bodies. Most often Cartesian coordinates are used, but the equations can also be expressed in polar coordinates. Notice that no advantage is taken of prior knowledge of the unperturbed motion. In *Encke's method*, instead, the equations of motion are expressed in terms of the difference of the coordinates from their values in unperturbed motion. Since these differences are small, at least at first, larger time steps can be used in the numerical integration. However, after a time the differences become large, especially if there are secular perturbation effects. After an interval one *rectifies* the reference orbit to, say, a new osculating orbit, with suitable additional rectifications in the course of time. Crowell's method, being more straightforward, lends itself better to computer operation than Encke's. For problems where only few orbital periods are involved as, e.g., in the passage of a space craft between the planets, it becomes the method of choice. Techniques of numerical integration are highly developed now, and it is often possible to put error bounds on the results, something that is very difficult to do in general perturbation theories.

On the other hand, numerical solutions of the equations of motion cannot settle questions of long-term stability of the motion. Ever since the beginning of the eighteenth century astronomers have debated the question of the stability of the solar system. Will the bounded planetary motion under the influence of gravitational forces continue indefinitely, or will the perturbations eventually lead to planetary collisions or cause one or another of the planets to leave the solar system? It was probably this concern that led to an almost obsessive interest in secular terms on the part of the early inventors of perturbation methods. Everything possible was done to identify them and, if possible, remove them. If, for example, a secular perturbation term could be demonstrated for the semimajor axis, then the conclusion would be that a planet would either leave the solar system or fall in the sun. It was early shown that no such term existed in first-order perturbation and all breathed easier. However, higher order perturbation calculations reopened the question. Poincaré's demonstration that the multiply-periodic Fourier series used in the expansions were only semiconvergent led regretfully to a verdict of "not proven."

Only in the last few decades has the stability question been freshly illuminated, by the application of new (and highly abstract) mathematical techniques. The methods of differential topology have been used to examine the global behavior of the possible orbits in phase space. A series of investigations, associated with the names C. L. Siegel, A. N. Kolmogorov, V. I. Arnold, and J. Moser, have shown that stable, bounded motion is possible for a system of n bodies interacting through gravitational forces only. That is to say, a nonnegligible fraction of the orbits (i.e., a group of finite measure) are confined to specific regions of phase space and remain so indefinitely in quasi-periodic motion.* The brilliance of the achievement and the power of the new methods are probably of greater significance than the specific result, for the ultimate fate of the solar system will likely be determined by dissipative and other nongravitational forces.

---

* It is curious that for the validity of statistical mechanics one would like almost the opposite result. For the notion of an ensemble as representative of the statistical behavior of a single system to be correct, one would prefer to show that a system point will eventually travel through all regions of phase space consistent with its initial macroscopic properties, e.g., energy. The phase space orbit should ergodically fill up all the phase space accessible to it. Remarkably, the topological analysis of possible motions winds up by making every one happy. As has been said, *some* orbits, a set of finite measure, maintain quasi-periodic motion indefinitely. But the overwhelming majority of initial conditions leads to motions that are ergodic, wandering over all available phase space and eventually coming as close as desired to any given point.

11-7 ADIABATIC INVARIANTS

At the first Solvay Conference in 1911, which grappled with the problems of introducing quantum notions into physics, a deceptively simple problem in classical mechanics was raised. Consider a bob on a string oscillating as a plane pendulum, with the string passing through a small hole in the ceiling. Now imagine that the string is either pulled up or let down slowly, so slowly that there is little change in the length of the pendulum during one period of oscillation. What happens to the frequency of oscillation during this process? Note that the energy of the pendulum is not conserved, for work is done on the system (or extracted from it) as the length of the string is altered. By elementary means it was demonstrated that for very slow change of the length E/v would be constant.* It will be recognized that this ratio is precisely the action variable J. The *adiabatic invariance* of the action variables under slow change of parameters was a very satisfying property to physicists developing quantum mechanics. As has been noted above (p. 483) the early recipes involved quantizing the values of the action variables to describe specific states of atomic systems. Since it was well known that slow variations of the atomic environment, e.g., of surrounding electromagnetic fields, did not induce transitions between states, the adiabatic invariance of the J's was comforting.

The original motivation for examining adiabatic invariance is no longer of concern, but in recent years there has been an intense revival of interest in the subject. Practical applications have been found in plasma physics, fusion technology, charged-particle accelerators, and even in galactic astronomy. Developments in the field are still going on, and the final word has not yet been said. All we can do here is present some basic considerations and describe briefly the trends in current research. For simplicity, only periodic systems of one degree of freedom will be examined, although the extension to many degrees of freedom normally is not difficult (i.e., in the absence of degeneracy). In broad outlines, the treatment is that given by Burgers in 1917.

We consider a system of one degree of freedom involving a parameter a. Implicit in the method is a picture of the system as initially conservative with a constant. Time dependence of a is then "switched on," and a varies slowly over a long time, eventually returning to a constant value. When a is constant the motion is periodic, and the slow change in the parameter does not alter the periodic nature of the motion. Although the changes in the motion are small in any one period, over a long interval of time the properties of the motion may

---

* According to at least one report of the 1911 Solvay Conference (Abhandlungen der Deutschen Bunsen-Gesellschaft, No. 7, 1914, p. 364) Lorentz remarked in the discussion that he had proposed the problem to Einstein some time previously. Einstein replied that he had demonstrated that the energy of the pendulum would remain proportional to v if the length were altered continuously and infinitesimally slowly.

undergo large quantitative changes. The switching on of the time dependence is thus in the nature of a small perturbation, and we are looking for secular changes in the motion.

When the parameter a is constant the system will be described by action-angle variables (w_0, J_0), such that the Hamiltonian is $H = H(J_0, a)$. It will be useful to consider these variables as derived from an original canonical set (q, p) via a generating function $W^*(q, w_0, a)$. The usual Hamilton–Jacobi equation of course leads to a generating function of the form $W(q, J_0, a)$., but the two generating functions are normally connected by a Legendre transformation (cf. Eq. 9–24):

$$W^*(q, w_0, a) = W(q, J_0, a) - J_0 w_0. \tag{11–109}$$

When a is allowed to vary with time, (w_0, J_0) of course remain as valid canonical variables, but the generating function is now an explicit function of time through the time dependence of a. Hence the appropriate Hamiltonian for the (w_0, J) set is now

$$K(\omega_0, J_0, a) = H(J_0, a) + \frac{\partial W^*}{\partial t}$$

$$= H(J_0, a) + \dot{a}\frac{\partial W^*}{\partial a}. \tag{11–110}$$

Now J_0 is no longer a constant and w_0 does not vary linearly with time. In effect, the second term in the Hamiltonian is a perturbation Hamiltonian and, as in the variation-of-constants method, the time dependence of J_0 is governed by the equation of motion

$$\dot{J}_0 = -\frac{\partial K}{\partial w_0} = -\dot{a}\frac{\partial}{\partial w_0}\left(\frac{\partial W^*}{\partial a}\right), \tag{11–111}$$

where, of course, the derivative in parenthesis is expressed, as is K, in terms of J_0, w_0, and a. In the spirit of a first-order perturbation theory, we look for a secular term, the average of $\dot{J}_0$ over the period of the unperturbed motion *for the appropriate a*. Since a varies slowly, a can be taken as constant during this time interval, and the average can be written as

$$\dot{J}_0 = -\frac{1}{\tau}\int_\tau \dot{a}\frac{\partial}{\partial w_0}\left(\frac{\partial W^*}{\partial a}\right) dt = -\frac{\dot{a}}{\tau}\int_\tau \frac{\partial}{\partial w_0}\left(\frac{\partial W^*}{\partial a}\right) dt + O(\dot{a}^2, \ddot{a}). \tag{11–112}$$

It will be remembered from Eq. (10–13') that W is given by the indefinite integral

$$W = \int p\, dq.$$

In one period of w_0 the generating function, W, therefore increases by J_0. At the same time $J_0 w_0$ also increases by J_0, since w_0 increases by unity. Hence, by Eq. (11–109) W^* is a periodic function of w_0 and both it and the derivative with respect to a can be expressed as a Fourier series:

$$\frac{\partial W^*}{\partial a} = \sum_k A_k(J_0, a) e^{2\pi i k w_0}. \tag{11–113}$$

The average, $\overline{\dot{J}_0}$, therefore has the form

$$\overline{\dot{J}_0} = -\frac{\dot{a}}{\tau} \int_\tau \sum_{k \neq 0} 2\pi i k A_k(J_0, a) e^{2\pi i k w_0} \, dt + O(\dot{a}^2, \ddot{a}).$$

Since the integrand has no constant term, the integral vanishes,

$$\overline{\dot{J}_0} = 0 + O(\dot{a}^2, \ddot{a}), \tag{11–114}$$

and $\overline{\dot{J}_0}$ has no secular variation to first order in $\dot{a}$, proving the desired property of adiabatic invariance.

Let us see how this derivation would work in detail for the problem of the harmonic oscillator:

$$H = \frac{1}{2m}(p^2 + m^2\omega^2 q^2)$$

where ω may be an explicit function of time. The equations of the canonical transformation from the (q, p) set to the (w_0, J_0) set are given by Eqs. (10–18) and (10–79), which can be written so as to facilitate the evaluation of W^*:

$$J_0 = \pi m \omega q^2 \csc^2 2\pi w_0 = -\frac{\partial W^*}{\partial w_0},$$

$$p = m\omega q \cot 2\pi w_0 = \frac{\partial W^*}{\partial q}. \tag{11–115}$$

To within constant (and therefore irrelevant) terms, W^* is found by integration of Eqs. (11–115) to be

$$W^*(q, w_0, \omega) = \frac{m\omega q^2}{2} \cot 2\pi w_0. \tag{11–116}$$

The derivative with respect to ω is

$$\frac{\partial W^*}{\partial \omega} = \frac{mq^2}{2} \cot 2\pi w_0,$$

or, using Eq. (10–78), as a function of w_0, J_0, and ω,

$$\frac{\partial W^*}{\partial \omega} = \frac{J_0}{4\pi\omega} \sin 4\pi w_0. \tag{11–117}$$

Thus $\dot{J}_0$ is given by the one-term Fourier expansion

$$\dot{J}_0 = -\frac{\dot{\omega}}{\omega} J_0 \cos 4\pi w_0, \tag{11–118}$$

which, as predicted, has no constant term. So far Eq. (11–118) is rigorous. Similarly the rigorous connection between w_0 and time is determined by the w_0 equation of motion

$$\dot{w}_0 = \frac{\partial K}{\partial J_0} = \frac{\partial H}{\partial J_0} + \dot{\omega}\frac{\partial}{\partial J_0}\left(\frac{\partial W^*}{\partial \omega}\right)$$

$$= \frac{\omega}{2\pi} + \frac{\dot{\omega}}{4\pi\omega}\sin 4\pi w_0. \tag{11–119}$$

In order to calculate an average of $\dot{J}_0$ over a period, including at least the first correction term, we begin to make approximations. First we will assume that over a particular period of the perturbed motion the ratio

$$\frac{\dot{\omega}}{\omega} \equiv \epsilon \tag{11–120}$$

is a constant, and one such that $\epsilon\tau \ll 1$. Equation (11–120) corresponds to a variation

$$\omega = \omega_0\,e^{\epsilon t} \approx \omega_0(1 + \epsilon t), \tag{11–121}$$

where t is measured from the start of the period interval, at which time $\omega(0) = \omega_0$. Equation (11–119) now looks like

$$\dot{w}_0 = \frac{\omega}{2\pi} + \frac{\epsilon}{4\pi}\sin 4\pi\,w_0. \tag{11–119'}$$

The zeroth-order solution is

$$2\pi w_0^{(0)} = \omega_0 t,$$

where the constant term has been set zero by suitable choice of the initial phase. To first order in ϵ, Eq. (11–119') becomes

$$\dot{w}_0^{(1)} = \frac{\omega_0(1 + \epsilon t)}{2\pi} + \frac{\epsilon}{4\pi}\sin 2\omega_0 t, \tag{11–122}$$

with the solution

$$2\pi w_0^{(1)} = \omega_0 t + \frac{\epsilon}{2}\left[\omega_0 t^2 + \frac{1 - \cos 2\omega_0 t}{2\omega_0}\right]. \tag{11–123}$$

Correspondingly the equation for $\dot{J}_0$ correct to second order in ϵ can be written as

$$\frac{d\ln J_0}{dt} = -\epsilon\cos\left(2\omega_0 t + \epsilon\left\{\omega_0 t^2 + \frac{1 - \cos 2\omega_0 t}{2\omega_0}\right\}\right).$$

Expanding the cosine, treating the term in ϵ as a small quantity to first order, the derivative reduces to

$$\frac{d\ln J_0}{dt} = -\epsilon\cos 2\omega_0 t + \epsilon^2\left\{\omega_0 t^2 + \frac{1 - \cos 2\omega_0 t}{2\omega_0}\right\}\sin 2\omega_0 t.$$

To find the secular behavior, this equation can be averaged over the period of the motion as it is at $t = 0$, i.e., over an interval $\tau = 2\pi/\omega_0$. In the averaging almost all terms on the right drop out, except the first inside the curly brackets, involving t^2. The final result is

$$\frac{\overline{d \ln J}}{dt} = \frac{\pi \epsilon^2}{\omega_0} = \frac{\omega_0 \delta^2}{4\pi}, \tag{11–124}$$

where $\delta = \epsilon\tau$, i.e., fractional change in ω over the period τ. Correspondingly, the fractional secular change in J over the period is

$$\frac{\Delta J}{J} = \frac{\delta^2}{2}. \tag{11–125}$$

As expected from the more general considerations, the secular change in the action variable shows no term in first order in ϵ. Only by retaining quantities of the order $\epsilon^2 = (\dot{\omega}/\omega)^2$ do we find any nonvanishing long-term change in J.

The adiabatic invariance of the action variables has proven to be especially useful in applications involving the motion of charged particles in electromagnetic fields. One of the simplest instances, and one with important practical consequences, concerns the motion of electrons in a uniform (or nearly uniform) constant magnetic field. As is well known, the charged particle in such a situation circles around the magnetic field lines. At the most basic level this can be shown from Newton's equations of motions. The Lorentz force in a constant magnetic field $\mathbf{B}$ is $(\mathbf{v} \times q\mathbf{B}/c)$; hence the equation of motion, Eq. (1–4), is

$$\frac{d\mathbf{v}}{dt} = \mathbf{v} \times \frac{q\mathbf{B}}{mc}. \tag{11–126}$$

Equation (11–126) says the velocity vector $\mathbf{v}$ rotates, without change of magnitude, about the direction of the magnetic field, with an angular frequency

$$\boldsymbol{\omega}_c = -\frac{q\mathbf{B}}{mc}. \tag{11–127}$$

An equivalent derivation can be formulated in terms of Lagrangian mechanics. It was shown, in Section 5–9, that the Lagrangian in this case can be written as

$$L = \frac{mv^2}{2} + \mathbf{M} \cdot \mathbf{B}, \tag{11–128}$$

where $\mathbf{M}$ is magnetic moment of the particle's motion defined in terms of its angular momentum $\mathbf{L}$ by

$$\mathbf{M} = \frac{q\mathbf{L}}{2mc}. \tag{11–129}$$

(Cf. Eq. 5–108.) In cylindrical coordinates with the z axis along the direction of B, the component of M parallel to B is

$$M_z = \frac{qr^2\dot{\theta}}{2c},$$
(11–130)

and the Lagrangian is

$$L = \frac{m}{2}(\dot{r}^2 + r^2\dot{\theta}^2 + \dot{z}^2) + \frac{q}{2c}Br^2\dot{\theta}.$$
(11–131)

Since θ is cyclic in the Lagrangian, the corresponding canonical momentum p_θ,

$$p_\theta = mr^2\dot{\theta} + \frac{qBr^2}{2c},$$
(11–132)

is a constant of the motion. Further, the radial equation of motion is

$$m\ddot{r} - r\dot{\theta}\left(m\dot{\theta} + \frac{qB}{c}\right) = 0.$$
(11–133)

A steady-motion solution to Eqs. (11–132) and (11–133) corresponds to r and $\dot{\theta}$ constant, with $\dot{\theta}$ having the value

$$\dot{\theta} = \omega_c \equiv -\frac{qB}{mc},$$
(11–134)

in agreement with Eq. (11–127).* In such case, $p_\theta = -(qBr^2/2c)$ and the action variable corresponding to θ is

$$J_\theta = \oint p_\theta\, d\theta = -\frac{\pi qBr^2}{c}.$$
(11–135)

By (11–130) we can write

$$\frac{qr^2}{c} = \frac{2M}{\omega_c}$$

(as M_z is equal to M for this motion), and therefore J_θ can also be written as

$$J_\theta = -\frac{2\pi MB}{\omega_c} = \frac{2\pi mc}{q}M.$$
(11–135')

The adiabatic invariance theorem implies that under sufficiently slow variation of the magnetic field J_θ remains constant. Equation (11–135') says that the magnetic

---

* The angular frequency ω_c is variously spoken of as cyclotron or gyration frequency and is twice as large as the Larmor frequency, Eq. (5–104). Of course, Larmor's theorem is not applicable here because the kinetic energy of the rotation around the lines of force is at least of the same magnitude as the remainder of the kinetic energy, and terms quadratic in B cannot be neglected.

moment is similarly invariant adiabatically. An alternative statement, on the basis of Eq. (11–135), is that B times the area of the orbit (i.e., the number of lines of force threading through the orbit) remains constant.

An adiabatic variation of B might arise if the magnetic field configuration remained static but was slightly nonuniform. If then the particle had a small z component of velocity, the resultant drift would move the particle slowly into regions of different B values. From Eqs. (11–134), (11–135), and (11–135′) it follows simply that the kinetic energy of motion around the lines of B is

$$T_{(\theta)} = \frac{mr^2\dot{\theta}^2}{2} = MB. \qquad (11\text{--}137)$$

Suppose a charged particle drifts in the direction of increasing B; by Eq. (11–137) the kinetic energy of rotation increases. As the total kinetic energy is conserved, the kinetic energy of longitudinal drift along the lines of force must decrease. Eventually the drift velocity goes to zero and the motion reverses in direction. If it can be arranged that B eventually increases in the other direction, the charged particle will remain confined, drifting back and forth between the two ends—the principle of the so-called mirror confinement. The complete story is of course more complicated, but the significance of the adiabatic invariance of M is clearly demonstrated.

It will be recalled that almost all phenomena of small oscillations about steady state or steady motion can be described in terms of harmonic oscillators. In consequence, there is a good deal of practical interest in questions of the invariance of J for a harmonic oscillator under slow, and not so slow, variations of a parameter. The study of oscillations in charged particle accelerators, for example, has led to a number of new insights. In addition, the mathematicians have used the problem to increase our knowledge about the solutions of differential equations with time-dependent coefficients. The results of these recent studies are too numerous to report in detail, much less to derive, here. But mention may be made of two properties of the harmonic oscillator problem of particular interest.

It is customary to introduce a time variable $t' = \epsilon t$, where ϵ is a small parameter. The time derivative of ω, which occurs in the "perturbation" Hamiltonian, then appears as

$$\dot{\omega} = \epsilon\frac{d\omega}{dt'}.$$

One then treats the derivative of ω with respect to t' as given by a fixed "program" of $\omega(t')$ variation, with the time scale on which this program is carried out being changed according to the value of ϵ. Thus a single parameter measures whether ω varies slowly or rapidly. Suppose the program involves starting (at $t' = -\infty$) at an initially constant value and winding up (at $t' = +\infty$) at some other constant value. We then look not at the change in J over a cycle of the motion, but at the change in J from its initial to its final value. The surprising result is that if $\omega(t')$ is sufficiently well-behaved, then the change is much less than Eq. (11–125) might lead one to expect. It has been proved* that, if $\omega(t')$ is real, bounded and analytic on and about the real axis, then the change in J is exponentially small, that is,

$$\frac{J(+\infty) - J(-\infty)}{J(+\infty)} \propto O(e^{-d/\epsilon}), \qquad (11\text{–}138)$$

where d is a real positive number having to do with the width of the strip of analyticity of ω in the complex t' plane.

An even more interesting result is that the harmonic oscillator with time-dependent frequency has an *exact* invariant, which in the limit of slow variation reduces to the action variable J. For a linear harmonic oscillator with Lagrangian and Hamiltonian given by

$$L = \frac{m}{2}(\dot{x}^2 - \omega^2(t)x^2); \qquad H = \frac{1}{2m}(p^2 + m^2\omega^2(t)x^2), \qquad (11\text{–}139)$$

the invariant is usually stated as

$$I = \frac{1}{2}\left[\frac{x^2}{r^2} + (r\dot{x} - x\dot{r})^2\right], \qquad (11\text{–}140)$$

where $r(t)$ is a function satisfying the differential equation

$$\ddot{r} + \omega^2(t)r - r^{-3} = 0. \qquad (11\text{–}141)$$

A formal mathematical derivation of the invariance of I is rather lengthy, but its physical meaning—and why it is constant—can be illuminated by some simple considerations. The motion of a linear harmonic oscillator with constant ω can be looked on as, say, the x component of a plane isotropic harmonic oscillator. This relationship is in no way altered if ω is time dependent. Thus, consider a central-force time-dependent potential

$$V = \frac{m}{2}\omega^2(t)r^2,$$

with Lagrangian

$$L = \frac{m}{2}(\dot{r}^2 + r^2\dot{\theta}^2 - \omega^2(t)r^2) = \frac{m}{2}[(\dot{x}^2 - \omega^2x^2) + (\dot{y}^2 - \omega^2y^2)]. \qquad (11\text{–}142)$$

In Cartesian coordinates the problem splits into two independent linear harmonic oscillators of the form of Eq. (11–139). In terms of plane polar coordinates we see that L is cyclic in θ and therefore the angular momentum is

---

* Several references can be given, but the easiest to follow is probably R. E. Meyer, *Jour. Applied Math. and Phys.* (ZAMP) **24**, 293 (1973).

still conserved, so that one equation of motion reduces to

$$r^2 \dot{\theta} = \frac{l}{m} \equiv h, \tag{11–143}$$

introducing a new constant h. Hence the linear harmonic oscillator can still be considered as the x projection of a plane isotropic harmonic oscillator with a constant angular momentum per unit mass measured by h. The radial equation of motion, from Eq. (11–142), is

$$m\ddot{r} + m\omega^2 r - mr\dot{\theta}^2 = 0,$$

or, in view of Eq. (11–143),

$$\ddot{r} + \omega^2 r - \frac{h^2}{r^3} = 0. \tag{11–144}$$

By using the conservation of h we can construct a constant of the motion defined as

$$I' = \frac{1}{2}\left[\frac{h^2 x^2}{r^2} + (r\dot{x} - x\dot{r})^2\right]. \tag{11–145}$$

To find the constant value of I', note that with $x = r\cos\theta$ we have

$$r\dot{x} - x\dot{r} = -h\sin\theta,$$

and therefore

$$I' = \frac{h^2}{2}.$$

Now, it is always possible to choose the arbitrary initial y amplitude, and the initial phase between the x and y motions, so that h has the numerical value of unity. Thus, suppose that for an initial period of time ω is constant at the value ω_0, so that x and y depend on time as $x = x_0 \sin\omega_0 t$, $y = y_0 \cos(\omega_0 t + \alpha)$. Then by the definition of h,

$$h = x\dot{y} - y\dot{x} = \omega_0 x_0 y_0 \cos\alpha.$$

If ω subsequently varies with time, h will of course preserve its initial value, which, by suitable choice of y_0 or α, can be set equal to unity no matter what the value of x_0. Thus it is always possible to associate with any time-dependent linear oscillator a corresponding plane isotropic oscillator for which $h = 1$. For this associated problem the radial equation of motion, Eq. (11–144), reduces to Eq. (11–141), and I' with $h = 1$ is precisely the invariant, Eq. (11–140). Physically speaking, the exact invariance of I, Eq. (11–140), is nothing more than a statement of the conservation of the angular momentum of the associated plane isotropic oscillator problem.* It remains only to consider the relation of I to J. If ω is constant, then a possible solution to the radial equation is r constant with value $\omega^{-1/2}$. (The motion of the associated plane problem is in a circle.) As $\dot{r}$ is then zero, the quantity I, Eq. (11–140), is

$$I = \frac{1}{2\omega}(\dot{x}^2 + \omega^2 x^2) = \frac{E}{m\omega} = \frac{J}{2\pi m}.$$

Thus in the zero-order approximation the exact invariant, as expected, is proportional to the action variable J.

It has been possible to sketch here only the highlights of the subject of adiabatic invariants. The ramifications of the field go into many areas of classical and quantum physics and of modern mathematics; details can be followed up in the references to this chapter. Further developments, both of the underlying theory and of the applications of adiabatic invariants, can be confidently expected in the future.

SUGGESTED REFERENCES

M. BORN, *Mechanics of the Atom.* Perturbation theory for classical mechanics wears somewhat different faces, depending on whether one's interest is in physics, celestial mechanics, the flight of space vehicles, or modern mathematics. Until recently physics textbooks tended to follow closely Born's treatment, itself the product of the development of the "old" quantum mechanics. Born's discussion, mainly of time-independent perturbation theory, is to be found principally in his Chapter 4, but there are various bits and pieces scattered through the book.

E. J. SALETAN AND A. H. CROMER, *Theoretical Mechanics.* This reference is cited as one of the better treatments along the line of Born. Only time-independent perturbation is described. Some examples, mainly referring to the harmonic oscillator, are considered in detail. There is also a short section on adiabatic invariants.

J. M. A. DANBY, *Fundamentals of Celestial Mechanics.* Once past the level of Kepler and most of Newton, celestial mechanics consists almost entirely of perturbation theory and the three-body problem. The literature on perturbation theory in celestial mechanics is therefore practically coextensive with that on celestial mechanics itself, and any citations must be highly selective. Danby's text is a relatively recent (1962) exposition of what might be called the classical version of perturbation theory, with up-to-date applications. It is unusually lucid for a field that generally runs to solid pages of formulas, and it has an extensive annotated bibliography. The von Zeipel method is not mentioned, nor are more modern developments such as the use of Lie series.

---

* The exact invariance of I for the linear time-dependent harmonic oscillator was first enunciated explicitly by H. R. Lewis, Jr., cf. *Jour. Math. Phys.* **9**, 1976 (1973), although the constancy of related quantities had been observed before in accelerator theory. The connection to the associated plane oscillator was apparently first noted by Eliezer and Gray, *SIAM J. Appl. Math.* **30**, 463 (1976).

B. Garfinkel, *Lagrange–Hamilton–Jacobi Mechanics*, in *Space Mechanics, Part 1*, ed. by J. B. Rosser. Ten concise and densely packed pages in this article describe perturbation theory as viewed by an expert in celestial mechanics. Assuming a background in canonical transformation theory and the Hamilton–Jacobi equation, it sweeps breathlessly from variation of constants to von Zeipel's method. It may be all you want.

Y. Hagihara, *Celestial Mechanics, Vol. 2: Perturbation Theory* (in two parts). In contrast, this reference covers the applications of perturbation theory to celestial mechanics in exhaustive detail, requiring some 900 pages. References to the current and historical literature appear to be nearly complete. If you want to find out what has actually been done in using perturbation techniques to solve the problems of celestial mechanics, this seems to be the place to look, although the text in words (what there is of it) is occasionally hard to follow. The Lie-series reformulation of the von Zeipel method is here, but the modern approach to stability theory is reserved for another volume.

R. Deutsch, *Orbital Dynamics of Space Vehicles*. It may be contested whether space technology provides an area of classical mechanics distinct from celestial mechanics. Perhaps the separation lies in the observation that "space mechanics" was born with a computer in its mouth. This reference for the most part reads like a textbook in celestial mechanics, and a good one at that. Various methods of perturbation theory are described in detail, including the specialized ones such as Hansen's method. The applications, however, mostly arise from space technology, as, for example, perturbation of artificial satellite orbits.

G. E. O. Giacaglia, *Perturbation Methods in Non-Linear Systems*. This is probably the best reference for a survey of modern developments in perturbation theory—from Poincaré and Lindstedt through Arnold and Moser—in a reasonably understandable form. The viewpoint appears basically to be that of an applied mathematician. The text, a grayish reproduction of typescript, is physically hard to read.

R. Abraham and E. Marsden, *Foundations of Mechanics*. There is a new language being used in the development and exposition of mechanics—that of differential topology. The physicist newcomer needs to take an intensive course in the language before its pronouncements become intelligible. It seems likely that in the area of global stability of perturbed motion the new language has scored notable successes not accessible by other means. However the expository advantages for the more conventional areas of mechanics appear highly doubtful. For those who wish to swim in these new waters, this newly revised text provides nearly encyclopedic coverage. Some 156 pages offer preliminaries on differential topology and the calculus on manifolds, but they require an orientation towards the methods of abstract mathematics. The applications in celestial mechanics form Part IV (pp. 619–740).

J. Moser, *Stable and Random Motions in Dynamical Systems*. This short book reproduces the text of five lectures given in 1972. Moser has himself been responsible for many of the advances in the modern treatment of stability problems. He gives here a survey of the developments in this century with emphasis on celestial mechanics. Theorems are often stated without proof, and considerable mathematical sophistication is expected on the part of the reader. Nevertheless it succeeds better than Abraham and Marsden in conveying both the flavor and the successes of the newer techniques.

T. G. Northrop, *The Adiabatic Motion of Charged Particles*. Although developments since the early 1960s are naturally not to be found here, this brief monograph provides a good introduction to the complexities of calculations based on adiabatic invariants. The applications are to plasma "devices," e.g., mirror machines.

B. LEHNERT, *Dynamics of Charged Particles*. This reference is nearly contemporaneous with the preceding one and provides a somewhat more voluminous discussion of the same area. Some problems associated with plasma devices are discussed that do not bear on adiabatic invariants, e.g., radiation from the charged particles.

EXERCISES

1. By the method of time-dependent perturbation theory carry the solution for the linear harmonic oscillator (in which the potential is considered a perturbation on the free particle motion) out through *third*-order terms, assuming the initial condition $\beta_0 = 0$. Find expressions for both x and p as functions of time and show that they agree with the corresponding terms in the expansion of the usual harmonic solutions.

2. A mass point m hangs at one end of a vertically hung Hook's-law spring of force constant k. The other end of the spring is oscillated up and down according to $z_1 = a \cos \omega_1 t$. By treating a as a small quantity, obtain a first-order solution to the motion of m in time, using the method of variation of constants. What happens as ω_1 approaches the unperturbed frequency ω_0?

3. a) A linear harmonic oscillator of force constant k has its mass suddenly increased by a fractional amount ϵ. By first-order time-independent perturbation theory find the resultant shift in the frequency of the oscillator to first order in ϵ. Show that to the same order in ϵ your result agrees with the rigorous prediction for the shift.

b) Repeat part (a), for the effect of increasing k by a fractional amount ϵ.

4. In Section 11–3, first-order perturbation in the time-independent format is used to find the effect of finite amplitude on the period of a plane pendulum. With the same formulation use Eq. (11–72) to find the first-order corrections in the dependence of θ on time. (Remember that J_0 is now a function of time.)

5. Carry out a consistent second-order perturbation calculation (using whichever method you choose) of the correction to the frequency of a plane pendulum as the result of finite amplitude of oscillation. All terms of order λ^2 should be retained in the Hamiltonian and in the perturbation treatment.

6. A mass particle is constrained to move in a straight line and is attached to the ends of two ideal springs of equal force constants, as shown in the diagram. The unstretched length

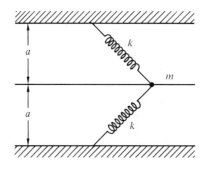

of each spring is $b \le a$. By first-order perturbation theory find the lowest order correction to the frequency of oscillation for finite amplitude of oscillation. What happens as a approaches b in magnitude?

7. a) Show that to lowest order in correction terms the relativistic (but noncovariant) Hamiltonian for the one-dimensional harmonic oscillator has the form

$$H = \frac{1}{2m}(p^2 + m^2\omega^2 q^2) - \frac{1}{8}\frac{p^4}{m^3 c^2}.$$

b) By first-order perturbation theory calculate the lowest order relativistic correction to the frequency of the harmonic oscillator. Express your result as the fractional change in the frequency.

8. A plane isotropic harmonic oscillator is perturbed by a change in the Hamiltonian of the form

$$\epsilon H_1 = p_x^2 p_y^2.$$

Using time-independent first-order perturbation theory find the shift in the frequencies.

9. A model of the atomic Stark effect can be made by taking the Kepler elliptic orbit in a plane and perturbing it by a potential $\Delta V = -Kx$. According to first-order perturbation theory, what happens to the frequencies of motion? This model can also be used as a first approximation to the effect of the light pressure of solar radiation on the orbit of an earth satellite.

10. By considering the work done to alter adiabatically the length l of a plane pendulum, prove by elementary means the adiabatic invariance of J for the plane pendulum in the limit of vanishing amplitude.

11. A plane pendulum of small amplitude is constrained to move on an inclined plane, as shown in the accompanying figure. How does its amplitude change when the inclination angle α of the plane is changed slowly?

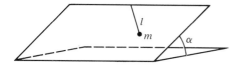

12. Consider the system described in Exercise 13 of Chapter 10. Suppose the parameter F is slowly varied from an initial value. What happens to the energy of the particle? The amplitude of oscillation? The period?

13. a) Show that the following transformation is canonical:

$$Q = -\arctan\left[\frac{r}{x}\phi(p, x)\right],$$

$$P = \frac{m}{2}\left[\frac{x^2}{r^2} + \phi^2(p, x)\right] = \frac{mI}{2},$$

where r is (to this point) an arbitrary function of t and

$$\phi(p, x) = \frac{rp}{m} - x\dot{r}.$$

b) Find a generating function of the $F_1(q, Q, t)$ type.

c) Show that if this canonical transformation is applied to the time-dependent harmonic oscillator of Eq. (11–139) and $r(t)$ satisfies Eq. (11–141), then the transformed Hamiltonian is cyclic in Q so that P is a constant of the motion. This constitutes an independent proof of the exact invariance of I defined by Eq. (11–140).

CHAPTER 12

Introduction to the Lagrangian and Hamiltonian Formulations for Continuous Systems and Fields

All the formulations of mechanics discussed up to this point have been devised for treating systems with a finite or at most a denumerably infinite number of degrees of freedom. There are some mechanical problems, however, that involve continuous systems, as, for example, the problem of a vibrating elastic solid. Here each point of the continuous solid partakes in the oscillations, and the complete motion can only be described by specifying the position coordinates of *all* points. It is not difficult to modify the previous formulations of mechanics so as to handle such problems. The most direct method is to approximate the continuous system by one containing discrete particles and then examine the change in the equations describing the motion as the continuous limit is approached.

12–1 THE TRANSITION FROM A DISCRETE TO A CONTINUOUS SYSTEM

We shall apply this procedure to an infinitely long elastic rod that can undergo small longitudinal vibrations, i.e., oscillatory displacements of the particles of the rod parallel to the axis of the rod. A system composed of discrete particles that approximates the continuous rod is an infinite chain of equal mass points spaced a distance a apart and connected by uniform massless springs having force constants k (cf. Fig. 12–1).* It will be assumed that the mass points can move only along the length of the chain. The discrete system will be recognized as an extension of the linear polyatomic molecule discussed in Chapter 6. We can therefore obtain the equations describing the motion by the customary techniques for small oscillations. Denoting the displacement of the ith particle from its equilibrium position by η_i, the kinetic energy is

$$T = \frac{1}{2}\sum_i m\dot{\eta}_i^2, \qquad (12\text{–}1)$$

---

* We use an infinite chain rather than a finite one to avoid at this point a discussion of the exceptional end mass points (in the discrete case) or boundary conditions (for the continuum).

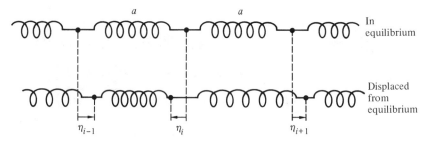

FIGURE 12–1

A discrete system of equal mass points connected by springs, as an approximation to a continuous elastic rod.

where m is the mass of each particle. The corresponding potential energy is the sum of the potential energies of each spring as the result of being stretched or compressed from its equilibrium length (cf. Section 6–4):

$$V = \frac{1}{2}\sum_i k(\eta_{i+1} - \eta_i)^2. \tag{12–2}$$

Combining Eqs. (12–1) and (12–2), the Lagrangian for the system is

$$L = T - V = \frac{1}{2}\sum_i (m\dot{\eta}_i^2 - k(\eta_{i+1} - \eta_i)^2), \tag{12–3}$$

which can also be written as

$$L = \frac{1}{2}\sum_i a\left[\frac{m}{a}\dot{\eta}_i^2 - ka\left(\frac{\eta_{i+1} - \eta_i}{a}\right)^2\right] = \sum_i aL_i, \tag{12–4}$$

where a is the equilibrium separation between the points (cf. Fig. 12–1). The resulting Lagrange equations of motion for the coordinates η_i are

$$\frac{m}{a}\ddot{\eta}_i - ka\left(\frac{\eta_{i+1} - \eta_i}{a^2}\right) + ka\left(\frac{\eta_i - \eta_{i-1}}{a^2}\right) = 0. \tag{12–5}$$

The particular form of L in Eq. (12–4), and of the corresponding equations of motion, has been chosen for convenience in going to the limit of a continuous rod as a approaches zero. It is clear that m/a reduces to μ, the mass per unit length of the continuous system, but the limiting value of ka may not be so obvious. For an elastic rod obeying Hooke's law it will be remembered that the extension of the rod *per unit length* is directly proportional to the force or tension exerted on the rod, a relation that can be written as

$$F = Y\xi,$$

where ξ is the elongation per unit length and Y is Young's modulus. Now the extension of a length a of a discrete system, per unit length, will be $\xi = (\eta_{i+1} - \eta_i)/a$. The force necessary to stretch the spring by this amount is

$$F = k(\eta_{i+1} - \eta_i) = ka\left(\frac{\eta_{i+1} - \eta_i}{a}\right),$$

so that ka must correspond to the Young's modulus of the continuous rod. In going from the discrete to the continuous case, the integer index i identifying the particular mass point becomes the continuous position coordinate x; instead of the variable η_i we have $\eta(x)$. Further, the quantity

$$\frac{\eta_{i+1} - \eta_i}{a} = \frac{\eta(x + a) - \eta(x)}{a}$$

occurring in L_i obviously approaches the limit

$$\frac{d\eta}{dx},$$

as a, playing the role of dx, approaches zero. Finally, the summation over a discrete number of particles becomes an integral over x, the length of the rod, and the Lagrangian (12–4) appears as

$$L = \frac{1}{2} \int \left(\mu\dot{\eta}^2 - Y\left(\frac{d\eta}{dx}\right)^2 \right) dx. \tag{12–6}$$

In the limit as a goes to zero, the last two terms in the equation of motion (12–5) become

$$\operatorname*{Lim}_{a \to 0} -\frac{Y}{a}\left\{ \left(\frac{d\eta}{dx}\right)_x - \left(\frac{d\eta}{dx}\right)_{x-a} \right\},$$

which clearly defines a second derivative of η. Hence the equation of motion for the continuous elastic rod is

$$\mu\frac{d^2\eta}{dt^2} - Y\frac{d^2\eta}{dx^2} = 0, \tag{12–7}$$

the familiar wave equation in one dimension with the propagation velocity

$$v = \sqrt{\frac{Y}{\mu}}. \tag{12–8}$$

Equation (12–8) is the well-known formula for the velocity of longitudinal elastic waves.

This simple example is sufficient to illustrate the salient features of the transition from a discrete to a continuous system. The most important fact to grasp is the role played by the position coordinate x. It is *not* a generalized coordinate; it serves merely as a continuous index replacing the discrete i. Just as each value of i corresponded to a different one of the generalized coordinates, η_i, of the system, so here for each value of x there is a generalized coordinate $\eta(x)$. Since η depends also on the continuous variable t, we should perhaps write more accurately $\eta(x, t)$, indicating that x, like t, can be considered as a parameter entering into the Lagrangian. If the continuous system were three-dimensional, rather than one-dimensional as here, the generalized coordinates would be distinguished by three continuous indices x, y, z, and would be written as $\eta(x, y, z, t)$. Note that the quantities x, y, z, and t are completely independent of each other and appear only as explicit variables in η. Derivatives of η with respect to any of them can therefore always be written as total derivatives without any ambiguity. Equation (12–6) also shows that the Lagrangian appears as an integral over the continuous index x; in the corresponding three-dimensional case the Lagrangian would have the form

$$L = \int \int \int \mathscr{L}\, dx\, dy\, dz, \tag{12–9}$$

where $\mathscr{L}$ is known as the *Lagrangian density*. For the longitudinal vibrations of the continuous rod the Lagrangian density is

$$\mathscr{L} = \frac{1}{2}\left\{\mu\left(\frac{d\eta}{dt}\right)^2 - Y\left(\frac{d\eta}{dx}\right)^2\right\}, \tag{12–10}$$

corresponding to the continuous limit of the quantity L_i appearing in Eq. (12–4). It is the Lagrangian density, rather than the Lagrangian itself, that will be used to describe the motion of the system.

12–2 THE LAGRANGIAN FORMULATION FOR CONTINUOUS SYSTEMS

It will be noted from Eq. (12–9) that $\mathscr{L}$ for the elastic rod, besides being a function of $\dot{\eta} \equiv \partial\eta/\partial t$, also involves a spatial derivative of η, namely, $\partial\eta/\partial x$; x and t thus play a similar role as parameters of the Lagrangian density. If there were local forces present in addition to the nearest neighbor interactions, then $\mathscr{L}$ would be a function of η itself as well as of the spatial gradient of η. Of course, in the general case $\mathscr{L}$ might well be an explicit function of x and t also. So the Lagrangian density for any one-dimensional continuous system would appear as a function of the form

$$\mathscr{L} = \mathscr{L}\left(\eta, \frac{d\eta}{dx}, \frac{d\eta}{dt}, x, t\right). \tag{12–11}$$

The total Lagrangian, following Eq. (12–10), is then the integral of $\mathscr{L}$ over the range of x defining the system, and Hamilton's principle in the limit of the continuous system appears as

$$\delta I = \delta \int_1^2 \int \mathscr{L}\, dx\, dt = 0. \tag{12–12}$$

If Hamilton's principle for the continuous system is to have any usefulness, it must be possible to derive the continuous limit of the equation of motion, e.g., Eq. (12–7), directly by variation of the double integral of $\mathscr{L}$ in Eq. (12–12). We can carry out this variation by methods that differ only slightly from those used in Chapter 2 for a discrete system. The variation is only on η and its derivatives; the parameters x and t are not affected by the variation either directly or in the ranges of integration. Just as the variation of η is taken to be zero at the endpoints t_1 and t_2, so the variation of η at the limits x_1 and x_2 of the integration in x is also to be zero. As in Section 2–2 a suitable varied path of integration in the η space can be obtained, for example, by choosing η from a one-parameter family of possible η functions:

$$\eta(x, t; \alpha) = \eta(x, t; 0) + \alpha\zeta(x, t). \tag{12–13}$$

Here $\eta(x, t, 0)$ stands for the correct function that will satisfy Hamilton's principle, and ζ is any well-behaved function that vanishes at the endpoints in t and x. If I is considered as a function of α, to be an extremum for $\eta(x, t, 0)$ the derivative of I with respect to α vanishes at $\alpha = 0$. By straightforward differentiation,

$$\frac{dI}{da} = \int_{t_1}^{t_2} \int_{x_1}^{x_2} dx\, dt \left\{ \frac{\partial \mathscr{L}}{\partial \eta} \frac{\partial \eta}{\partial \alpha} + \frac{\partial \mathscr{L}}{\partial \frac{d\eta}{dt}} \frac{\partial}{\partial \alpha}\left(\frac{d\eta}{dt}\right) + \frac{\partial \mathscr{L}}{\partial \frac{d\eta}{dx}} \frac{\partial}{\partial \alpha}\left(\frac{d\eta}{dx}\right) \right\}. \tag{12–14}$$

Because the variation of η, that is, $\alpha\zeta$, vanishes at the endpoints, integration by parts in x and t yields the relations

$$\int_{t_1}^{t_2} \frac{\partial \mathscr{L}}{\partial \frac{d\eta}{dt}} \frac{\partial}{\partial \alpha}\left(\frac{d\eta}{dt}\right) dt = -\int_{t_1}^{t_2} \frac{d}{dt}\left(\frac{\partial \mathscr{L}}{\partial \frac{d\eta}{dt}}\right) \frac{\partial \eta}{\partial \alpha} dt,$$

and

$$\int_{x_1}^{x_2} \frac{\partial \mathscr{L}}{\partial \frac{d\eta}{dx}} \frac{\partial}{\partial \alpha}\left(\frac{d\eta}{dx}\right) dx = -\int_{x_1}^{x_2} \frac{d}{dx}\left(\frac{\partial \mathscr{L}}{\partial \frac{d\eta}{dx}}\right) \frac{\partial \eta}{\partial \alpha} dx.$$

Hamilton's principle can therefore be written as

$$\int_{t_1}^{t_2} \int_{x_1}^{x_2} dx\, dt \left\{ \frac{\partial \mathscr{L}}{\partial \eta} - \frac{d}{dt}\left(\frac{\partial \mathscr{L}}{\partial \frac{d\eta}{dt}}\right) - \frac{d}{dx}\left(\frac{\partial \mathscr{L}}{\partial \frac{d\eta}{dx}}\right) \right\} \left(\frac{\partial \eta}{\partial \alpha}\right)_0 = 0, \tag{12–15}$$

and by the same arguments as in Section (2–2) the arbitrary nature of the varied path implies the vanishing of the expression in the curly brackets:

$$\frac{d}{dt}\left(\frac{\partial \mathscr{L}}{\partial \frac{d\eta}{dt}}\right) + \frac{d}{dx}\left(\frac{\partial \mathscr{L}}{\partial \frac{d\eta}{dx}}\right) - \frac{\partial \mathscr{L}}{\partial \eta} = 0. \tag{12–16}$$

Equation (12–16) is the appropriate form of the equation of motion as derived from Hamilton's principle, Eq. (12–12).

A system of n discrete degrees of freedom will have n Lagrange equations of motion; for the continuous system with an infinite number of degrees of freedom we seem to obtain only one Lagrange equation! It must be remembered, however, that the equation of motion for η is a differential equation involving the time only, and in that sense Eq. (12–15) furnishes a separate equation of motion for each value of x. The continuous nature of the indices x appears in that Eq. (12–15) is a partial differential equation in the two variables x and t, yielding η as $\eta(x, t)$.

For the specific instance of longitudinal vibrations in an elastic rod, it is seen from the form of the Lagrangian density, Eq. (12–10), that

$$\frac{\partial \mathscr{L}}{\partial \dfrac{d\eta}{dt}} = \mu \frac{d\eta}{dt}, \qquad \frac{\partial \mathscr{L}}{\partial \dfrac{d\eta}{dx}} = -Y\frac{d\eta}{dx}, \qquad \frac{\partial \mathscr{L}}{\partial \eta} = 0.$$

Thus, as desired, the Euler–Lagrange equation, Eq. (12–16), reduces properly to the equation of motion, Eq. (12–7).

The Lagrangian formulation developed here for one-dimensional continuous systems needs obviously to be extended to two- and three-dimensional situations, e.g., a general elastic solid. Further, instead of one field quantity η there may be several, e.g., displacement from an equilibrium position would be described by a spatial vector $\boldsymbol{\eta}$ with three components. There is no difficulty in carrying out the mathematical steps for the more general situation in close parallelism to the one-component one-dimensional case. However, the formulas become lengthy and cumbersome if written in the same manner, especially in view of the two tiers of derivatives. Considerable gain in notational simplicity can be achieved by noticing that time t and the spatial coordinates x, y, z play the same type of mathematical role in Hamilton's principle. The field quantities are functions of the coordinates of both time and space that are to be treated as independent variables. No variation of the field quantities occurs at the limits of integration in Hamilton's principle over both time and space.

It is mathematically convenient to think in terms of a four-dimensional space with coordinates $x_0 = t$, $x_1 = x$, $x_2 = y$, $x_3 = z$. No physical significance is implied for this space. As in Chapter 7 a Roman letter subscript refers only to the three coordinates of the physical space, a Greek letter subscript refers to all four coordinates. Use of the summation convention with respect to repeated indices will be resumed for the rest of the chapter. The various components of the field quantities will be symbolized by a subscript ρ. It should be emphasized that the subscript may cover a multitude of forms. At times it will stand for a single index having two, three, four, or more values. Or it may stand for multiple indices. Thus if the field quantity is a spatial tensor of second rank, then ρ really refers to two subscript indices. Finally, a derivative of the field quantities with respect to any one of the four coordinates x_ν will be denoted by the subscript ν separated from ρ

by a comma. Where there is only one field quantity a blank space precedes the comma. Examples are

$$\eta_{\rho,v} \equiv \frac{d\eta_\rho}{dx_v}; \qquad \eta_{,j} \equiv \frac{d\eta}{dx_j}; \qquad \eta_{i,\mu v} = \frac{d^2\eta_i}{dx_\mu\,dx_v}. \tag{12-17}$$

Only the derivatives of the field quantities will be symbolized in this manner.

In this notation, the most general form of the Lagrangian density to be considered here is written as

$$\mathscr{L} = \mathscr{L}(\eta_\rho, \eta_{\rho,v}, x_v). \tag{12-18}$$

The total Lagrangian is then an integral over three space:

$$L = \int \mathscr{L}(dx_i), \tag{12-19}$$

but rarely occurs explicitly. Hamilton's principle appears an integral over a region in four-space:

$$\delta I = \delta \int \mathscr{L}(dx_\mu) = 0, \tag{12-20}$$

where the variation of the η_ρ vanishes at the bounding surface S of the region of integration. The derivation of the corresponding Euler–Lagrange equations of motion proceeds symbolically as before. We consider a one-parameter set of varied functions that reduce to $\eta_\rho(x_v)$ as the parameter α goes to zero. As previously, a possible suitable set can be constructed, for example, by adding to η_ρ the product $\alpha\zeta_\rho$, where $\zeta_\rho(x_v)$ are convenient arbitrary functions vanishing on the bounding surface. The vanishing of the variation of I is equivalent to setting the derivative of I with respect to α equal to zero:*

$$\frac{dI}{d\alpha} = \int \left(\frac{\partial\mathscr{L}}{\partial\eta_\rho} \frac{\partial\eta_\rho}{\partial\alpha} + \frac{\partial\mathscr{L}}{\partial\eta_{\rho,v}} \frac{\partial\eta_{\rho,v}}{\partial\alpha} \right) (dx_\mu).$$

Integration by parts yields

$$\frac{dI}{d\alpha} = \int \left[\frac{\partial\mathscr{L}}{\partial\eta_\rho} - \frac{d}{dx_v} \left(\frac{\partial\mathscr{L}}{\partial\eta_{\rho,v}} \right) \right] \frac{\partial\eta_\rho}{\partial\alpha} (dx_\mu) + \int (dx_\mu) \frac{d}{dx_v} \left(\frac{\partial\mathscr{L}}{\partial\eta_{\rho,v}} \frac{\partial\eta_\rho}{\partial\alpha} \right). \tag{12-21}$$

The second integral vanishes in the limit as α goes to zero, as can be seen in various ways. One can examine it term by term: carrying out the integration for the particular x_v of each derivative term, which then vanishes because the derivative with respect to α is zero at the endpoints. Or the integral can be transformed by a four-dimensional divergence theorem into an integral over the surface bounding the region of integration in four-space. The surface integral

---

* Unless otherwise noted, the summation convention will be used in the rest of this chapter, for all types of subscripts.

again vanishes because the variation of η_ρ in the vicinity of the correct field functions is zero on the surface. Equation (12–21) in the limit as α goes to zero therefore reduces to

$$\left(\frac{dI}{d\alpha}\right)_0 = \int (dx_\mu) \left[\frac{\partial \mathscr{L}}{\partial \eta_\rho} - \frac{d}{dx_\nu}\left(\frac{\partial \mathscr{L}}{\partial \eta_{\rho,\nu}}\right)\right]\left(\frac{\partial \eta_\rho}{\partial \alpha}\right)_0. \tag{12–22}$$

Again, the arbitrary nature of the variation of each η_ρ means that Eq. (12–22) is satisfied only when each of the square brackets vanishes:

$$\frac{d}{dx_\nu}\left(\frac{\partial \mathscr{L}}{\partial \eta_{\rho,\nu}}\right) - \frac{\partial \mathscr{L}}{\partial \eta_\rho} = 0. \tag{12–23}$$

Equations (12–23) represent a set of partial differential equations for the field quantities, with as many equations as there are different values of ρ. It may be worth repeating that since the space coordinates x_i are in effect indices for the field quantities, each of Eqs. (12–23) in effect corresponds to an entire set of Lagrange differential equations of motion in the discrete case.

For a one-dimensional continuous system, where ν takes on only the values 0 and 1, Eq. (12–23) expands to the same form as Eq. (12–16). The compactness of the notation is evident even in so simple an example. Of course, the use of a four-dimensional space for symbolic convenience in no way requires covariant behavior (in the physicist's sense of the word) of any of the quantities in that space. A good illustration is provided by the case of the acoustic field in a perfect gas. Here the field quantities are the components of the vector $\boldsymbol{\eta}$ representing the small displacement of the gas particles from their positions in the absence of the sound vibrations. What we seek are the equations of motion for the longitudinal vibrations in the gas, i.e., the vector wave equation for the propagation of sound. The appropriate Lagrangian density for an acoustic field is derived in Appendix E and can be written as

$$\mathscr{L} = \tfrac{1}{2}(\mu_0 \dot{\boldsymbol{\eta}}^2 + 2P_0 \nabla \cdot \boldsymbol{\eta} - \gamma P_0 (\nabla \cdot \boldsymbol{\eta})^2). \tag{12–24}$$

Here μ_0 is the equilibrium mass density, and P_0 the equilibrium pressure, of the gas. The first term in $\mathscr{L}$ is clearly a kinetic energy density, while the remaining terms represent the change in the potential energy of the gas per unit volume as the result of the work done on or by the gas in the course of the contractions and expansions that are the mark of acoustic vibrations. Appearing in the potential energy we have the constant γ, the ratio of the specific heats at constant pressure and volume, which enters because the compression and rarefaction of the gas by the sound waves is done adiabatically and not isothermally. In the four-dimensional notation, the Lagrangian density becomes

$$\mathscr{L} = \tfrac{1}{2}(\mu_0 \eta_{i,0} \eta_{i,0} + 2P_0 \eta_{i,i} - \gamma P_0 \eta_{i,i} \eta_{j,j}). \tag{12–25}$$

The middle term in $\mathscr{L}$ clearly does not contribute to the equation of motion, because the partial derivative of the term with respect to $\eta_{i,j}$ is zero or a constant.

Hence the equations of motion, Eq. (12–23), takes the form

$$\mu_0 \eta_{j,00} - \gamma P_0 \eta_{i,ij} = 0, \quad j = 1, 2, 3. \tag{12–26}$$

Converted back into spatial vector notation, Eqs. (12–26) appear as the single vector equation

$$\mu_0 \frac{d^2\boldsymbol{\eta}}{dt^2} - \gamma P_0 \nabla\nabla \cdot \boldsymbol{\eta} = 0. \tag{12–27}$$

Two points may be made about the equation of motion, Eqs. (12–26) or (12–27), and the Lagrangian density from which it was derived. One is that Eq. (12–27) may be put in a more recognizable form by making use of the fact that for small amplitude vibrations the fractional change in the gas density, σ, is related to $\boldsymbol{\eta}$ by the equation

$$\sigma = -\nabla \cdot \boldsymbol{\eta}$$

(cf. Appendix E). Operating on Eq. (12–27) with the divergence operator then gives the scalar equation

$$\nabla^2\sigma - \frac{\mu_0}{\gamma P_0} \frac{d^2\sigma}{dt^2} = 0, \tag{12–28}$$

which is readily recognized as the three-dimensional wave equation, with the customary expression

$$v = \sqrt{\frac{\gamma P_0}{\mu_0}} \tag{12–29}$$

for the velocity of sound in gases. If, as in point mechanics, the main goal of the Lagrangian formulation is to derive the equations of motion from a Lagrangian, then a suitable Lagrangian density corresponding to Eq. (12–28) is

$$\mathscr{L} = \tfrac{1}{2}(\mu_0\sigma^2_{,0} - \gamma P_0\sigma_{,i}\sigma_{,i}). \tag{12–30}$$

Equation (12–30) of course does not have the same physical content as the Lagrangian density (12–25), and one cannot be derived directly from the other, but (12–30) *does* imply the scalar wave equation, (12–28).

The other point comes from the observation that the $\nabla \cdot \boldsymbol{\eta}$ term in Eq. (12–24) does not contribute to the equations of motion. This corresponds to the property of discrete systems that the Lagrangian is uncertain to a total time derivative of an arbitrary function of the generalized coordinates and time. With continuous systems the corresponding statement is that $\mathscr{L}$ is uncertain to any "four-divergence," i.e., to a term of the form

$$\frac{dF_v(\eta_\rho, x_\mu)}{dx_v}$$

where the F_v are any four (differentiable) functions of the field quantities η_ρ and the coordinates x_μ. That such a term makes no contribution to the variation of the action integral is obvious. Application of the divergence theorem in four-space converts the volume integral into an integral over the bounding surface where the variation of F_v is zero. In symbols, the relevant variation can be written

$$\delta \int (dx_\mu) \frac{dF_v(\eta_\rho, x_\mu)}{dx_v} = \delta \int F_v(\eta_\rho, x_\mu)\, d\sigma_v = 0, \qquad (12\text{--}31)$$

where $d\sigma_v$ represents the components of an element of surface (in four-space) oriented along the direction of the outward normal.* It will be seen that the middle term in Eq. (12–24) falls into this category with the functions F_v given by

$$F_0 = 0, \qquad F_i = 2P_0\eta_i.$$

The Lagrangian formulation for a continuous set of generalized coordinates has been developed in order to treat continuous mechanical systems such as an elastic solid in longitudinal oscillation, or a gas vibrating in such a manner as to set up acoustic waves. As has been implied the formulation may also be used, even in the absence of a mechanical system, to describe the equations governing a *field*. Mathematically, a field is no more than a set of one or more independent functions of space and time, and the generalized coordinates fit this definition accurately. There is no necessary requirement that the field be related to some underlying mechanical system. In thus breaking the connection between the Lagrangian field description and purely mechanical motion we are but recapitulating the history of physics. For example, the electromagnetic field was long thought of in terms of the elastic vibrations of a mysterious ether. Only in recent times was it realized that the ether had no other role than being the subject of the verb "to undulate."† We recognize equally well that the variational procedures developed here also stand independent of the notion of a continuous mechanical system, and that they serve to furnish the equations describing any space–time field. Hamilton's principle then becomes in effect a convenient and compact description of the field, one which upon expansion leads to the field equations.

Within this larger context the Lagrangian density need not be given as the difference of a kinetic and potential energy density. Instead we may use any expression for $\mathscr{L}$ that leads to the desired field equations. Thus we have seen that a Lagrangian density for the sound field, Eq. (12–24), is given naturally in terms of the vector displacement $\boldsymbol{\eta}$. But we also noted that the field can also be described in terms of a scalar σ, the fractional change in gas density. The wave equation in terms of σ can be derived from a Lagrangian density given by Eq. (12–30), which

---

* If, for example, the surface in four-space is a surface of constant time, then the only nonvanishing component of $d\sigma_v$ is $d\sigma_0 = dV \equiv dx\,dy\,dz$. The direction of the outward normal is thus along the time axis.

† In his 1894 Presidential Address to the British Association for the Advancement of Science, the Earl of Salisbury said that "the main, if not the only, function of the word aether has been to furnish a nominative case to the verb 'to undulate'." (See p. 8 of the 1894 Report of the BAAS.) It is possible the same thought had been expressed earlier.

is not the same as Eq. (12–24), and which cannot be deduced by reference to any underlying mechanical system. It leads to the correct field equation, and that is all that is desired.

But in addition to implying the field equations, the Lagrangian density has more to tell about the physical nature of the field. As with systems of a discrete number of degrees of freedom, the structure of the Lagrangian also contains information on conserved properties of the system. One such set of conservation theorems is discussed in the next section.*

12–3 THE STRESS-ENERGY TENSOR AND CONSERVATION THEOREMS

An analog to the conservation of Jacobi's integral in point mechanics, Section 2–6, can be derived here, and in much the same manner. All one has to remember is that the treatment of time must be extended in parallel fashion to the x_i since they are all independent parameters in $\mathscr{L}$. Thus, instead of the time derivative of L, we seek to evaluate the total derivative of $\mathscr{L}$ with respect to x_μ:

$$\frac{d\mathscr{L}}{dx_\mu} = \frac{\partial\mathscr{L}}{\partial\eta_\rho}\eta_{\rho,\mu} + \frac{\partial\mathscr{L}}{\partial\eta_{\rho,\nu}}\eta_{\rho,\mu\nu} + \frac{\partial\mathscr{L}}{\partial x_\mu}.$$

By the equations of motion, Eq. (12–23), this becomes (with a slight change in notation in addition)

$$\frac{d\mathscr{L}}{dx_\mu} = \frac{d}{dx_\nu}\left(\frac{\partial\mathscr{L}}{\partial\eta_{\rho,\nu}}\right)\eta_{\rho,\mu} + \frac{\partial\mathscr{L}}{\partial\eta_{\rho,\nu}}\frac{d\eta_{\rho,\mu}}{dx_\nu} + \frac{\partial\mathscr{L}}{\partial x_\mu}$$

$$= \frac{d}{dx_\nu}\left(\frac{\partial\mathscr{L}}{\partial\eta_{\rho,\nu}}\eta_{\rho,\mu}\right) + \frac{\partial\mathscr{L}}{\partial x_\mu}.$$

Combining total derivatives, this can be written

$$\frac{d}{dx_\nu}\left\{\frac{\partial\mathscr{L}}{\partial\eta_{\rho,\nu}}\eta_{\rho,\mu} - \mathscr{L}\delta_{\mu\nu}\right\} = -\frac{\partial\mathscr{L}}{\partial x_\mu}. \tag{12–32}$$

Suppose, now, that $\mathscr{L}$ does not depend explicitly on x_μ. This usually means that $\mathscr{L}$ represents a free field, i.e., contains no external driving sources or sinks that interact with the field at explicit space points and with given time dependence. In effect this means no interaction between the field and point particles moving in space and time through the field. Under this condition, Eq. (12–32) takes on the form of a set of divergence conditions,

$$\frac{dT_{\mu\nu}}{dx_\nu} = 0, \tag{12–33}$$

---

* A more general attack on the conservation properties inherent in the Lagrangian will be found in Section 12–7 on Noether's theorem.

on a quantity with the form of a four-tensor of the second rank:

$$T_{\mu\nu} = \frac{\partial \mathscr{L}}{\partial \eta_{\rho,\nu}} \eta_{\rho,\mu} - \mathscr{L}\delta_{\mu\nu}. \tag{12-34}$$

That these equations have only the *form* of tensor equations in four-space is emphasized because as yet the four-space has no transformation properties—space and time are still distinct—and there is no transformation requirement on $T_{\mu\nu}$. However, the space portions of these quantities do behave like vectors and tensors in ordinary space, that is, T_{ij} are the components of a three-dimensional tensor of the second rank.

The similarity between $T_{\mu\nu}$ and Jacobi's integral, Eq. (2–53), is obvious. It becomes especially clear for the component T_{00}:

$$T_{00} = \frac{\partial \mathscr{L}}{\partial \dot{\eta}_\rho} \dot{\eta}_\rho - \mathscr{L}. \tag{12-34'}$$

In mechanical systems the Lagrangian density often has the form $\mathscr{L} = \mathscr{T} - \mathscr{V}$, the difference between a kinetic energy density and a potential energy density. This is the case, for example, with the Lagrangian densities for the elastic rod, Eq. (12–10), and for sound vibrations, Eq. (12–24), with the kinetic energy density having the form of one half the mass density times a square of the displacement velocity:

$$\mathscr{T} = \tfrac{1}{2}\mu\dot{\eta}_\rho\dot{\eta}_\rho.$$

By the same arguments as used in point mechanics, T_{00} can then be identified as a total energy density.

The corresponding identification tags to be put on the other elements of $T_{\mu\nu}$ can be suggested by writing the set of Eqs. (12–33) as

$$\frac{dT_{\mu 0}}{dt} + \frac{dT_{\mu j}}{dx_j} = 0, \tag{12-35}$$

or

$$\frac{dT_{\mu 0}}{dt} + \nabla \cdot \mathbf{T}_\mu = 0, \tag{12-35'}$$

where $\mathbf{T}_\mu$ are a set of four space-vectors. In either form, Eqs. (12–35) or (12–35') appear as equations of continuity, which say that the time rate of change of some density plus the divergence of some corresponding flux or current density vanishes (cf. Eq. 7–76). In turn, the equations of continuity imply the conservation of some integral quantities providing the field volume is finite, i.e., the field can be contained within a volume beyond which the field quantities are zero. Define, in such case, integral quantities R_μ by

$$R_\mu = \int T_{\mu 0} \, dV, \tag{12-36}$$

where the volume integral extends beyond the region containing the field. Then, by Eqs. (12–35′),

$$\frac{dR_\mu}{dt} = \int \nabla \cdot \mathbf{T}_\mu \, dV = \int \mathbf{T}_\mu \cdot d\mathbf{A} = 0. \tag{12–37}$$

It is because of these conservation theorems, derived from Eq. (12–33), that the four arrays $T_{\mu\nu}$, $\mu = 1, 2, 3, 4$, are known in the parlance of modern physics as *conserved currents*.

We should therefore expect T_{0i} to play the role of the components of an energy current density. That this is reasonable can be seen again from considerations of the longitudinal vibration field in an elastic rod. Imagine the rod divided by an imaginary cut at point x (cf. Fig. 12–2). From the considerations that led to the Lagrangian, Eq. (12–6), the force exerted by the part of the rod on the right to extend the part that is to the left of the cut is

$$Y \frac{d\eta}{dx}.$$

Hence there is a tension at x in the left-hand portion of equal magnitude but of opposite direction. Further, the left-hand portion is being stretched by an amount that at x is η, and the rate at which this extension changes in time is $\dot{\eta}$. Hence the rate of work being done by the tension at the cut is

$$-\dot{\eta} Y \frac{d\eta}{dx}, \tag{12–38}$$

which is thus the rate at which energy is being transferred to the right per unit time. Comparison shows that this is exactly T_{01} for the appropriate Lagrangian density of Eq. (12–10). If T_{00} is an energy density then the quantity R_0 of Eq. (12–36) can be identified as the total energy in the field. The fourth component of the conservation equation (12–37) therefore says that the total field energy is

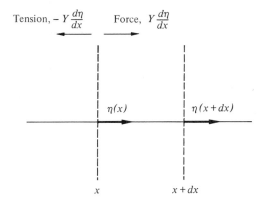

Tension, $-Y \frac{d\eta}{dx}$ Force, $Y \frac{d\eta}{dx}$

$\eta(x)$ $\eta(x+dx)$

x $x+dx$

FIGURE 12–2
Diagram illustrating calculation of energy current density in elastic rod.

conserved if T_{0i} vanishes on the bounding surface, i.e., if the system does not radiate energy to the outside.

Physical meaning for the T_{i0} components can be suggested similarly by turning once more to the vibrations of the elastic rod. If the particles in the rod moved by the same amount all along the rod, the motion would be that of a rigid body, i.e., no oscillatory disturbances. The net change of mass in a length dx of the rod as a result of the motion would clearly be zero, since as much mass moves past $x + dx$ as past x. There would still be a net momentum density $\mu\dot{\eta}$ for this case of rigid body motion. When wave motion takes place, a net mass change in the length dx exists, amounting at any given time to (cf. Fig. 12–2)

$$\mu(\eta(x) - \eta(x + dx)) = -\mu\frac{d\eta}{dx}dx.$$

The additional momentum in the interval resulting from the wave motion is therefore

$$-\mu\dot{\eta}\frac{d\eta}{dx}dx.$$

Thus, there is an additional momentum density, above and beyond that of the steady state motion, that may be identified as the wave or field momentum density:

$$-\mu\dot{\eta}\frac{d\eta}{dx}. \tag{12–39}$$

This quantity is just $-T_{10}$ for the Lagrangian density given by Eq. (12–10). Thus we are led to identify $-T_{i0}$ as the components of field momentum density and $-R_i$ as the total (linear) momentum of the field, at least in this four-dimensional convention.

The equations of continuity, Eqs. (12–35′), then suggest that $-\mathbf{T}_i$ must represent the vector flux density for the ith component of the field momentum density. We ascribe a vector property to $\mathbf{T}_i$ because there can be, for example, a flow in the y direction of the x component the momentum density, as measured by $-T_{12}$. An alternative interpertation of T_{ij} comes from considering the displacement field of an elastic solid. It is well known that in such a solid besides the compression forces normal to a surface there are also shear forces, along a surface element. The entire assemblage of forces can be described by saying that the force $d\mathbf{F}$ acting on an element of area $d\mathbf{A}$ is expressed in terms of a *stress tensor* $\mathbf{T}$ such that*

$$d\mathbf{F} = \mathbf{T} \cdot d\mathbf{A}.$$

---

* See, for example, K. R. Symon, *Mechanics*, 3rd ed., pp. 431–439.

Hence the net force, say in the x direction, on a rectangular volume element $dx\,dy\,dz$ has a contribution from the forces on the surfaces in yz planes given by (cf. Fig. 12–3)

$$[T_{11}(x + dx) - T_{11}(x)]\,dy\,dz = \frac{dT_{11}}{dx}\,dx\,dy\,dz,$$

but there is also a contribution from the surfaces in the xz plane;

$$[T_{12}(y + dy) - T_{12}(y)]\,dx\,dz = \frac{dT_{12}}{dy}\,dx\,dy\,dz,$$

and similarly from the xy planes. Newton's equations of motion here correspond to saying that the time rate of change of the momentum density in the x direction, $-T_{10}$, is equal to the x component of the force on a unit volume element:

$$-\frac{dT_{10}}{dt} = \frac{dT_{11}}{dx} + \frac{dT_{12}}{dy} + \frac{dT_{13}}{dz},$$

which is precisely the x component of Eq. (12–35′). For this particular field T_{ij} can be identified as the elements of the three-dimensional stress tensor; hence the origin of the name "stress-energy tensor" for $T_{\mu\nu}$.

By considerations of a continuous mechanical system we have thus been able to attach physical identifications, or associations, with each of the components of

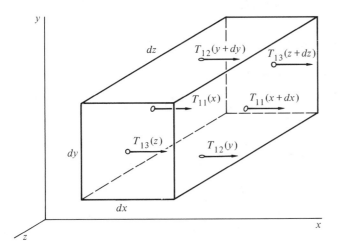

FIGURE 12–3
Force in x direction on a volume element $dx\,dy\,dz$ of an elastic solid.

the stress-energy tensor. In summary form, these labels are

T_{00}	field energy density,
$\mathbf{T}_0$, with components T_{0j}	field energy current density,
$-T_{i0}$	field momentum density, ith component,
$-\mathbf{T}_i$, with components T_{ij}	current density for the ith component of the field momentum density,
T_{ij}	three-dimensional stress tensor.

It must be remembered that although the example of mechanical systems gave birth to the procedures and nomenclature, the formalism can be applied to any field irrespective of its nature or origin. A classical theory of fields can be constructed not only for vibrations of an elastic solid, but for the electromagnetic field, for the "field" of the Schrödinger wave function, or for the relativistic field describing a "scalar" meson, among others. We shall examine some of these examples in more detail later on. For present purposes of illustration we can consider a two component field, η_1 and η_2, whose properties are deliberately chosen not to correspond to any existing theory. Suppose the field equations were

$$a\nabla^2\eta_1 + b\frac{d\eta_1}{dt} - c\frac{d^2\eta_1}{dt^2} = f\eta_2,$$

$$a\nabla^2\eta_2 - b\frac{d\eta_2}{dt} - c\frac{d^2\eta_2}{dt} = g\eta_1. \tag{12-40}$$

It's easily seen that these field equations can be derived from a Lagrangian density

$$\mathscr{L} = c\dot{\eta}_1\dot{\eta}_2 + \frac{b}{2}(\dot{\eta}_1\eta_2 - \eta_1\dot{\eta}_2) - a\nabla\eta_1 \cdot \nabla\eta_2 - \frac{1}{2}(f\eta_2^2 + g\eta_1^2). \tag{12-41}$$

The components of the stress-energy tensor for this Lagrangian density are

$$T_{00} = c\dot{\eta}_1\dot{\eta}_2 + a\nabla\eta_1 \cdot \nabla\eta_2 + \frac{1}{2}(f\eta_2^2 + g\eta_1^2), \tag{12-42}$$

(as with the discrete system energy, the terms linear in velocity drop out)

$$T_{0i} = -a\left(\dot{\eta}_1\frac{d\eta_2}{dx_i} + \dot{\eta}_2\frac{d\eta_1}{dx_i}\right), \tag{12-43}$$

$$T_{i0} = c\left(\dot{\eta}_1\frac{d\eta_2}{dx_i} + \dot{\eta}_2\frac{d\eta_1}{dx_i}\right) + \frac{b}{2}\left(\eta_2\frac{d\eta_1}{dx_i} - \eta_1\frac{d\eta_2}{dx_i}\right), \tag{12-44}$$

$$T_{ij} = -a\left(\frac{d\eta_1}{dx_i}\frac{d\eta_2}{dx_j} + \frac{d\eta_1}{dx_j}\frac{d\eta_2}{dx_i}\right) - \mathscr{L}\delta_{ij}. \tag{12-45}$$

It will be noticed that in this example, the three-dimensional tensor $\mathbf{T}$ is symmetric. This is a physically desirable, one might almost say necessary, characteristic for the spatial part of the stress-energy tensor. Recalling the identification of R_i, the conservation equations, Eq. (12–37), say that for a closed noninteracting system the total linear momentum of the field is conserved. We would expect no less. But there should be a corresponding conservation theorem for the total angular momentum of the field. It is simple to construct a quantity that should act as an angular momentum density. Since angular momentum is an axial vector, it would be expected that the components of the angular momentum density are the elements of an antisymmetric tensor of the second rank. A suitable form for this angular momentum (i.e., "moment of momentum") density tensor is

$$\mathcal{M}_{ij} = -(x_i T_{j0} - x_j T_{i0}), \tag{12–46}$$

with the total angular momentum of the field given by

$$M_{ij} = \int \mathcal{M}_{ij} \, dV.$$

In as much as t and x_i are completely independent variables, the time rate of change of M_{ij} is

$$\frac{dM_{ij}}{dt} = -\int \left(x_i \frac{dT_{j0}}{dt} - x_j \frac{dT_{i0}}{dt} \right) dV,$$

or, from the continuity conditions, Eqs. (12–35),

$$\frac{dM_{ij}}{dt} = -\int \left(x_i \frac{dT_{jk}}{dx_k} - x_j \frac{dT_{ik}}{dx_k} \right) dV.$$

Integration by parts converts this expression to

$$\frac{dM_{ij}}{dt} = -\int \frac{d}{dx_k} (x_i T_{jk} - x_j T_{ik}) \, dV + \int (T_{ji} - T_{ij}) \, dV. \tag{12–47}$$

The first integral on the right is in the form of a volume integral of a divergence. It is therefore equal to an integral over the bounding surface, which vanishes for a closed nonradiating system. Finally, if $T_{ij} = T_{ji}$, the second integral is also zero. Thus the total angular momentum of the field is conserved if **T** is symmetric.

For the example chosen, the stress tensor is indeed symmetric. There is no guarantee that this will always be so, and there are well-known fields for which the stress tensor as obtained directly is not symmetric. However the desirable conservation property can often be rescued by noting that just like $\mathcal{L}$, the stress-energy tensor $T_{\mu\nu}$ is not uniquely defined.* The form of $T_{\mu\nu}$, Eq. (12–34), was originally chosen because it satisfied the divergence conditions, Eq. (12–33).

---

* This should not be unexpected. In electromagnetic theory, for example, it has long been understood (and has been a source of controversy) that expressions for densities of the energy or for energy current (Poynting's vector) are uncertain to within quantities that are space divergences. Added quantities of this nature contribute nothing to the observable integrals such as the total energy or the total energy current.

Therefore $T_{\mu\nu}$ is indeterminate to any function whose 4-divergence vanishes. A quite general candidate is a quantity of the form

$$\frac{d\psi_{\mu\nu\lambda}}{dx_\lambda},$$

where $\psi_{\mu\nu\lambda}$ is an arbitrary set of functions of the field variables, but with the antisymmetry property that

$$\psi_{\mu\nu\lambda} = -\psi_{\mu\lambda\nu}.$$

In evaluating the 4-divergence, ν and λ will be dummy indices, and the 4-divergence will consist of pairs such as (no summation convention!)

$$\frac{d^2\psi_{\mu\nu\lambda}}{dx_\nu\,dx_\lambda} + \frac{d^2\psi_{\mu\lambda\nu}}{dx_\lambda\,dx_\nu},$$

which vanishes identically because of the antisymmetry condition. This ambiguity in $T_{\mu\nu}$ thus often makes it possible to "symmetrize" the stress-energy tensor, a process that is almost always carried out in constructing classical fields. Examples will be given later.

12-4 HAMILTONIAN FORMULATION, POISSON BRACKETS AND THE MOMENTUM REPRESENTATION

It is possible to obtain a Hamiltonian formulation for systems with a continuous set of coordinates much as was done in Chapter 8 for discrete systems. To indicate the method of approach, we return briefly to the linear chain of mass points discussed in Section 12-1. Conjugate to each η_i there is a canonical momentum

$$p_i = \frac{\partial L}{\partial \dot{\eta}_i} = a\frac{\partial L_i}{\partial \dot{\eta}_i}. \tag{12-48}$$

The Hamiltonian for the system is therefore

$$H \equiv p_i\dot{\eta}_i - L = a\frac{\partial L_i}{\partial \dot{\eta}_i}\dot{\eta}_i - L,$$

or

$$H = a\left(\frac{\partial L_i}{\partial \dot{\eta}_i}\dot{\eta}_i - L_i\right). \tag{12-49}$$

It will be remembered that in the limit of the continuous rod, when a goes to zero, $L_i \to \mathscr{L}$ and the summation in Eq. (12-49) becomes an integral:

$$H = \int dx\left(\frac{\partial \mathscr{L}}{\partial \dot{\eta}}\dot{\eta} - \mathscr{L}\right). \tag{12-50}$$

$$\pi_i = \frac{\partial \mathscr{L}}{\partial \dot{\eta}_i}$$

The individual canonical momenta p_i, as given by Eq. (12–48), vanish in the continuous limit, but we can define a *momentum density*, π, that remains finite:

$$\lim_{a \to 0} \frac{p_i}{a} \equiv \pi = \frac{\partial \mathscr{L}}{\partial \dot{\eta}}. \tag{12–51}$$

Equation (12–50) is in the form of a space integral over a *Hamiltonian density*, $\mathscr{H}$, defined by

not　　3.14159...

$$H = \int \mathscr{H} \, d\mathscr{V}$$

$$\mathscr{H} = \pi \dot{\eta} - \mathscr{L}. \tag{12–52}$$

While a Hamiltonian formulation can thus be introduced in a straightforward manner for classical fields, it will be noticed that the procedure singles out the time variable for special treatment. It is therefore in contrast to the development we have given for the Lagrangian formulation where the independent variables of time and space were handled symmetrically. For this reason the Hamiltonian approach, at least as introduced here, lends itself less easily to incorporation in a relativistically covariant description of fields. The Hamiltonian way of looking at fields has therefore not proved as useful as the Lagrangian method, and a rather brief description should suffice here.

The obvious route for generalizing to a three-dimensional field described by field quantities η_ρ is to define, analogously to Eq. (12–51), the canonical momentum densities

$$\pi_\rho(x_\mu) = \frac{\partial \mathscr{L}}{\partial \dot{\eta}_\rho}. \tag{12–53}$$

The quantities $\eta_\rho(x_i, t)$, $\pi_\rho(x_i, t)$ together define the infinite-dimensional *phase space* describing the classical field and its time development. A conservation theorem can be found for π_ρ that is roughly similar to that for the canonical momentum in discrete systems. If a given field quantity η_ρ is cyclic in the sense that $\mathscr{L}$ does not contain η_ρ explicitly (as in the case of Eq. 12–10) then the Lagrange field equation looks like an existence statement for a conserved current:

$$\frac{d}{dx_\mu} \frac{\partial \mathscr{L}}{\partial \eta_{\rho,\mu}} = 0,$$

or

$$\frac{d\pi_\rho}{dt} + \frac{d}{dx_i} \frac{\partial \mathscr{L}}{\partial \eta_{\rho,i}} = 0. \tag{12–54}$$

It follows that if η_ρ is cyclic, there is an integral conserved quantity

$$\Pi_\rho = \int dV \, \pi_\rho(x_i, t).$$

The obvious generalization of Eq. (12–52) for a Hamiltonian density is

$$\mathscr{H}(\eta_\rho, \eta_{\rho,i}, \pi_\rho, x_\mu) = \pi_\rho \dot{\eta}_\rho - \mathscr{L}, \tag{12–55}$$

where it is assumed that functional dependence on $\dot{\eta}_\rho$ can be eliminated by inversion of the defining equations (12–53). From this definition it follows that

$$\frac{\partial \mathscr{H}}{\partial \pi_\rho} = \dot{\eta}_\rho + \pi_\lambda \frac{\partial \dot{\eta}_\lambda}{\partial \pi_\rho} - \frac{\partial \mathscr{L}}{\partial \dot{\eta}_\lambda} \frac{\partial \dot{\eta}_\lambda}{\partial \pi_\rho} = \dot{\eta}_\rho \tag{12–56}$$

by Eq. (12–53). The other half of the canonical field equation is more cumbersome. When expressed in terms of the canonical variables, $\mathscr{H}$ is a function of η_ρ through the explicit dependence of $\mathscr{L}$, and through $\dot{\eta}_\rho$. Hence

$$\frac{\partial \mathscr{H}}{\partial \eta_\rho} = \pi_\lambda \frac{\partial \dot{\eta}_\lambda}{\partial \eta_\rho} - \frac{\partial \mathscr{L}}{\partial \dot{\eta}_\lambda} \frac{\partial \dot{\eta}_\lambda}{\partial \eta_\rho} - \frac{\partial \mathscr{L}}{\partial \eta_\rho} = - \frac{\partial \mathscr{L}}{\partial \eta_\rho}. \tag{12–57}$$

Using the Lagrange equations this can be written

$$\frac{\partial \mathscr{H}}{\partial \eta_\rho} = -\frac{d}{dx_\mu}\left(\frac{\partial \mathscr{L}}{\partial \eta_{\rho,\mu}}\right) = -\dot{\pi}_\rho - \frac{d}{dx_i}\left(\frac{\partial \mathscr{L}}{\partial \eta_{\rho,i}}\right). \tag{12–58}$$

Because of the appearance of $\mathscr{L}$ we still don't have a useful form. By an exactly similar derivation, however, we find that

$$\frac{\partial \mathscr{H}}{\partial \eta_{\rho,i}} = \pi_\lambda \frac{\partial \dot{\eta}_\lambda}{\partial \eta_{\rho,i}} - \frac{\partial \mathscr{L}}{\partial \dot{\eta}_\lambda} \frac{\partial \dot{\eta}_\lambda}{\partial \eta_{\rho,i}} - \frac{\partial \mathscr{L}}{\partial \eta_{\rho,i}} = - \frac{\partial \mathscr{L}}{\partial \eta_{\rho,i}}. \tag{12–59}$$

Hence we can write as the second half of the canonical equations

$$\frac{\partial \mathscr{H}}{\partial \eta_\rho} - \frac{d}{dx_i}\left(\frac{\partial \mathscr{H}}{\partial \eta_{\rho,i}}\right) = -\dot{\pi}_\rho. \tag{12–60}$$

Equations (12–56) and (12–60) can be put in a notation more closely approaching Hamilton's equations for a discrete system by introducing the notion of a *functional derivative* defined as

$$\frac{\delta}{\delta \psi} = \frac{\partial}{\partial \psi} - \frac{d}{dx_i} \frac{\partial}{\partial \psi_{,i}}. \tag{12–61}$$

Since $\mathscr{H}$ is not a function of $\pi_{\rho,i}$, Eqs. (12–56) and (12–60) can be written as

$$\dot{\eta}_\rho = \frac{\delta \mathscr{H}}{\delta \pi_\rho}, \qquad \dot{\pi}_\rho = -\frac{\delta \mathscr{H}}{\delta \eta_\rho}. \tag{12–62}$$

Note that in the same symbolism the Lagrange equations, Eq. (12–23), take the form

$$\frac{d}{dt}\left(\frac{\partial \mathscr{L}}{\partial \dot{\eta}_\rho}\right) - \frac{\delta \mathscr{L}}{\delta \eta_\rho} = 0. \tag{12–63}$$

About the only advantage of the functional derivative, however, is that of the resultant similarity with discrete system. It supresses, on the other hand, the parallel treatment of time and space variables.

$$dH = \frac{\partial \mathcal{H}}{\partial \eta_k} d\eta_k + \frac{\partial \mathcal{H}}{\partial \pi_k} d\pi_k + \frac{\partial \mathcal{H}}{\partial \eta_{k,j}} d\eta_{k,j} + \frac{\partial \mathcal{H}}{\partial t} dt$$

Other properties of $\mathcal{H}$ can be obtained by expanding the total time derivative of Eq. (12–55), recalling that $\dot{\eta}_\rho$ is to be considered a function of η_ρ, $\eta_{\rho,i}$, π_ρ, and x_μ. We therefore have that

$$\frac{d\mathcal{H}}{dt} = \dot{\pi}_\rho \dot{\eta}_\rho + \pi_\rho \frac{d\dot{\eta}_\rho}{dt} - \frac{\partial \mathcal{L}}{\partial \eta_\rho} \dot{\eta}_\rho - \frac{\partial \mathcal{L}}{\partial \dot{\eta}_\rho} \frac{d\dot{\eta}_\rho}{dt} - \frac{\partial \mathcal{L}}{\partial \eta_{\rho,i}} \frac{d\eta_{\rho,i}}{dt} - \frac{\partial \mathcal{L}}{\partial t}.$$

The second and fourth terms on the right cancel by virtue of the defining equations (12–53), so the derivative simplifies to

$$\frac{d\mathcal{H}}{dt} = \dot{\pi}_\rho \dot{\eta}_\rho - \frac{\partial \mathcal{L}}{\partial \eta_\rho} \dot{\eta}_\rho - \frac{\partial \mathcal{L}}{\partial \eta_{\rho,i}} \frac{d\eta_{\rho,i}}{dt} - \frac{\partial \mathcal{L}}{\partial t}. \tag{12-64}$$

On the other hand, considering $\mathcal{H}$ as a function of η_ρ, $\eta_{\rho,i}$, π_ρ, and x_μ, the total time derivative is

$$\frac{d\mathcal{H}}{dt} = \dot{\pi}_\rho \frac{\partial \mathcal{H}}{\partial \pi_\rho} + \frac{\partial \mathcal{H}}{\partial \eta_\rho} \dot{\eta}_\rho + \frac{\partial \mathcal{H}}{\partial \eta_{\rho,i}} \frac{d\eta_{\rho,i}}{dt} + \frac{\partial \mathcal{H}}{\partial t}. \tag{12-65}$$

The expression on the right has been written so as to facilitate comparison with the right-hand side of Eq. (12–64). Thus the first terms of both expressions are the same, by virtue of Eq. (12–56). The second pair similarly match by Eq. (12–57), and Eq. (12–59) shows the equivalence of the third pair. Hence the final pair of terms must be equal:

$$\frac{\partial \mathcal{H}}{\partial t} = -\frac{\partial \mathcal{L}}{\partial t}, \tag{12-66}$$

which corresponds to Eq. (8–13) for discrete systems.

On the other hand, the analog of Eq. (8–35) does *not* hold, i.e., the total and partial time derivatives of $\mathcal{H}$ are not in general the same. By use of Hamilton's equations of motion, Eqs. (12–56) and (12–60), and with an interchange in the order of differentiations, Eq. (12–65) can be written

$$\frac{d\mathcal{H}}{dt} = \frac{\partial \mathcal{H}}{\partial \pi_\rho} \frac{d}{dx_i} \left(\frac{\partial \mathcal{H}}{\partial \eta_{\rho,i}} \right) + \frac{\partial \mathcal{H}}{\partial \eta_{\rho,i}} \frac{d\dot{\eta}_\rho}{dx_i} + \frac{\partial \mathcal{H}}{\partial t}.$$

Using Eq. (12–56) and combining terms we have finally

$$\frac{d\mathcal{H}}{dt} = \frac{d}{dx_i} \left(\dot{\eta}_\rho \frac{\partial \mathcal{H}}{\partial \eta_{\rho,i}} \right) + \frac{\partial \mathcal{H}}{\partial t}, \tag{12-67}$$

which is as close as one can get to Eq. (8–35).

However, Eq. (12–67) is really an old friend in slight disguise. It surely occasions no surprise that $\mathcal{H}$, Eq. (12–55), is identical with T_{00}, Eq. (12–35), which had already been identified with the energy density. Further, it follows by Eq. (12–59) that

$$\dot{\eta}_\rho \frac{\partial \mathcal{H}}{\partial \eta_{\rho,i}} = -\dot{\eta}_\rho \frac{\partial \mathcal{L}}{\partial \eta_{\rho,i}} = -T_{0i}.$$

When $\mathscr{L}$ does not contain t explicitly, by Eq. (12–66) neither does $\mathscr{H}$, and in that case, Eq. (12–67) reduces to

$$\frac{dT_{00}}{dt} + \frac{dT_{0i}}{dx_i} = 0,$$

which is the first of the divergence conservation equations, Eq. (12–35). We have seen that the existence of a conserved current implies the conservation of an integral quantity, here

$$P_0 \equiv H = \int \mathscr{H} \, dV. \tag{12–68}$$

Thus, if $\mathscr{H}$ is not an explicit function of time, the conserved quantity is not $\mathscr{H}$, but the integral quantity H.

The total Hamiltonian H is but one example of functions that are volume integrals of densities. A general formalism for the time derivative of such integral quantities can be formulated directly. Consider some density $\mathscr{U}$ that is a function of the phase space coordinates (η_ρ, π_ρ), their spatial gradients, and possibly of x_μ:

$$\mathscr{U} = \mathscr{U}(\eta_\rho, \pi_\rho, \eta_{\rho,i}, \pi_{\rho,i}, x_\mu). \tag{12–69}$$

The corresponding integral quantity is

$$U(t) = \int \mathscr{U} \, dV, \tag{12–70}$$

where the volume integral extends over all space encompassed by the bounding surface on which η_ρ and π_ρ vanish. Differentiating U with respect to time we have, in general,

$$\frac{dU}{dt} = \int dV \left\{ \frac{\partial \mathscr{U}}{\partial \eta_\rho} \dot{\eta}_\rho + \frac{\partial \mathscr{U}}{\partial \eta_{\rho,i}} \dot{\eta}_{\rho,i} + \frac{\partial \mathscr{U}}{\partial \pi_\rho} \dot{\pi}_\rho + \frac{\partial \mathscr{U}}{\partial \pi_{\rho,i}} \dot{\pi}_{\rho,i} + \frac{\partial \mathscr{U}}{\partial t} \right\}. \tag{12–71}$$

Consider a term such as

$$\int dV \frac{\partial \mathscr{U}}{\partial \eta_{\rho,i}} \dot{\eta}_{\rho,i} = \int dV \frac{\partial \mathscr{U}}{\partial \eta_{\rho,i}} \frac{d\dot{\eta}_\rho}{dx_i}.$$

An integration by parts, remembering that η_ρ and derivatives vanish on the bounding surfaces, yields

$$\int dV \frac{\partial \mathscr{U}}{\partial \eta_{\rho,i}} \dot{\eta}_{\rho,i} = -\int dV \dot{\eta}_\rho \frac{d}{dx_i} \left(\frac{\partial \mathscr{U}}{\partial \eta_{\rho,i}} \right).$$

A similar reduction holds for the term in $\dot{\pi}_{\rho,i}$. Collecting coefficients of $\dot{\eta}_\rho$ and $\dot{\pi}_\rho$ respectively, we see then that in terms of the δ notation (Eq. 12–61), Eq. (12–71) reduces to

$$\frac{dU}{dt} = \int dV \left\{ \frac{\delta \mathscr{U}}{\delta \eta_\rho} \dot{\eta}_\rho + \frac{\delta \mathscr{U}}{\delta \pi_\rho} \dot{\pi}_\rho + \frac{\partial \mathscr{U}}{\partial t} \right\}. \tag{12–72}$$

Finally, inserting the canonical equations of motion (12–62) for $\dot{\eta}_\rho$ and $\dot{\pi}_\rho$ we have

$$\frac{dU}{dt} = \int dV \left\{ \frac{\delta \mathscr{U}}{\delta \eta_\rho} \frac{\delta \mathscr{H}}{\delta \pi_\rho} - \frac{\delta \mathscr{H}}{\delta \eta_\rho} \frac{\delta \mathscr{U}}{\delta \pi_\rho} \right\} + \int dV \frac{\partial \mathscr{U}}{\partial t}. \qquad (12\text{–}73)$$

The first integral on the right clearly corresponds to the Poisson bracket form. If $\mathscr{U}$ and $\mathscr{W}$ are two density functions, then these considerations suggest defining the Poisson bracket of the integral quantities by

$$[U, W] = \int dV \left\{ \frac{\delta \mathscr{U}}{\delta \eta_\rho} \frac{\delta \mathscr{W}}{\delta \pi_\rho} - \frac{\delta \mathscr{W}}{\delta \eta_\rho} \frac{\delta \mathscr{U}}{\delta \pi_\rho} \right\}. \qquad (12\text{–}74)$$

Let us also define what is to be meant by the partial derivative of U with respect to t in the obvious fashion as

$$\frac{\partial U}{\partial t} = \int dV \frac{\partial \mathscr{U}}{\partial t}. \qquad (12\text{–}75)$$

Equation (12–73) can then be written as

$$\frac{dU}{dt} = [U, H] + \frac{\partial U}{\partial t}, \qquad (12\text{–}76)$$

which corresponds precisely, in this notation, to Eq. (9–94) for discrete systems. Since by definition the Poisson bracket of H with itself vanishes, Eq. (12–76) then specializes to

$$\frac{dH}{dt} = \frac{\partial H}{\partial t}, \qquad (12\text{–}77)$$

which is the integral form of Eq. (12–67) and the field-theory version of Eq. (8–35).

A Poisson bracket formalism for classical fields thus appears as a natural outgrowth of the Hamiltonian formulation. But one cannot carry out a Poisson bracket description of field theory in step-by-step correspondence with that for discrete systems. Notice, for example, that the Poisson brackets are defined here only in terms of a pair of densities. One cannot therefore easily set up Poisson brackets corresponding to the fundamental Poisson brackets in discrete mechanics. It is true that π_ρ is a density, but η_ρ is not. Further, if x_i plays the role of continuous indices on the mechanical variables, then fundamental Poisson brackets should involve functions at different values of x_i, which is not easily brought into the present formulation. For this reason there has been little exploration of canonical transformations for classical fields, a subject that for discrete systems proved to be so rich and consequential. It is also difficult to carry through the steps for quantization, which usually involve the replacement of the fundamental Poisson brackets by the quantum commutators.

There is, however, one way of treating classical fields that provides for almost all of the Hamiltonian and Poisson bracket formulation of discrete mechanics.

Indeed the main idea behind this treatment is replacing the continuous space variable or index by a denumerable discrete index. We can see how to do this by referring again to the longitudinal oscillations of the elastic rod. Suppose the rod is of finite length $L = x_2 - x_1$. The requirement that η vanish at the extremities is a boundary condition that could be provided physically by placing the rod between two perfectly rigid walls. Then the amplitude of oscillation can be represented by a Fourier series:

$$\eta(x) = \sum_{n=0}^{\infty} q_n \sin \frac{2\pi n(x - x_1)}{2L}. \tag{12–78}$$

Instead of the continuous index x we have the discrete index n. We are allowed to use this representation for all x only when $\eta(x)$ is a well-behaved function, which most physical field quantities are.

For simplicity in illustrating how the scheme may be carried out it will be assumed there is only one real field quantity η that can be expanded in a three-dimensional Fourier series of the form

$$\eta(\mathbf{r}, t) = \frac{1}{V^{\frac{1}{2}}} \sum_{k} q_k(t) e^{i\mathbf{k} \cdot \mathbf{r}}. \tag{12–79}$$

Here $\mathbf{k}$ is a wave vector that can take on only discrete magnitudes and directions, such that only an integral (or sometimes, half-integral) number of wavelengths fit into a given linear dimension. We say that $\mathbf{k}$ has a discrete spectrum. The scalar index k stands for some ordering of the set of integer indices used to denumerate the discrete values of $\mathbf{k}$, and V is the volume of the system, appearing in a normalization factor. Because η is real we must have $q_k^* = q_{-k}$.

The orthogonality of the exponentials over the volume can be stated as the relation

$$\frac{1}{V} \int e^{i(\mathbf{k} - \mathbf{k}') \cdot \mathbf{r}} \, dV = \delta_{k,k'}. \tag{12–80}$$

In effect, the allowed values of $\mathbf{k}$ are those for which the condition (12–80) is satisfied (as can be seen by looking at the one-dimensional Fourier series). It follows that the coefficients of expansion, $q_k(t)$, are given by

$$q_k(t) = \frac{1}{V^{\frac{1}{2}}} \int e^{-i\mathbf{k} \cdot \mathbf{r}} \eta(\mathbf{r}, t) \, dV. \tag{12–81}$$

In similar fashion the canonical momentum density can be expanded as

$$\pi(\mathbf{r}, t) = \frac{1}{V^{\frac{1}{2}}} \sum_{k} p_k(t) e^{-i\mathbf{k} \cdot \mathbf{r}}, \tag{12–82}$$

again with $p_k^* = p_{-k}$. Correspondingly, the expansion coefficients, $p_k(t)$, are to be found from

$$p_k(t) = \frac{1}{V^{\frac{1}{2}}} \int e^{i\mathbf{k} \cdot \mathbf{r}} \pi(\mathbf{r}, t) \, dV. \tag{12–83}$$

Both q_k and p_k are integral quantities in the pattern of Eq. (12–70). We can therefore ask for the Poisson brackets of such quantities. Since the exponentials do not involve the field quantities we have, by Eq. (12–74),

$$[q_k, p_{k'}] = \frac{1}{V} \int dV\, e^{i(\mathbf{k'}-\mathbf{k})\cdot\mathbf{r}} \left\{ \frac{\delta\eta}{\delta\eta}\frac{\delta\pi}{\delta\pi} - \frac{\delta\pi}{\delta\eta}\frac{\delta\eta}{\delta\pi} \right\}$$

$$= \frac{1}{V} \int dV\, e^{i(\mathbf{k'}-\mathbf{k})\cdot\mathbf{r}},$$

or, by Eq. (12–80),

$$[q_k, p_{k'}] = \delta_{k,k'}. \tag{12–84}$$

It is further obvious from the definition of the Poisson bracket, Eq. (12–74), that

$$[q_k, q_{k'}] = 0 = [p_k, p_{k'}]. \tag{12–85}$$

Thus the Poisson brackets for q_k, p_k form a set of fundamental Poisson brackets, which suggests that we look upon them as canonical coordinates. The form of the equations of motion they obey thus becomes of considerable interest.

By Eq. (12–76) the time dependence of q_k is to be found from

$$\dot{q}_k(t) = [q_k, H] = \frac{1}{V^{\frac{1}{2}}} \int dV\, e^{-i\mathbf{k}\cdot\mathbf{r}} \left\{ \frac{\delta\eta}{\delta\eta}\frac{\delta\mathcal{H}}{\delta\pi} - \frac{\delta\mathcal{H}}{\delta\eta}\frac{\delta\eta}{\delta\pi} \right\}$$

or

$$\dot{q}_k(t) = \frac{1}{V^{\frac{1}{2}}} \int dV\, e^{-i\mathbf{k}\cdot\mathbf{r}} \frac{\delta\mathcal{H}}{\delta\pi}. \tag{12–86}$$

On the other hand we have that

$$\frac{\partial H}{\partial p_k} = \int dV \frac{\partial\mathcal{H}}{\partial\pi}\frac{\partial\pi}{\partial p_k}. \tag{12–87}$$

Inasmuch as $\mathcal{H}$ is not a function of the gradient of π, the partial derivative is the same as the functional derivative. Further, from Eq. (12–82) we have

$$\frac{\partial\pi}{\partial p_k} = \frac{1}{V^{\frac{1}{2}}} e^{-i\mathbf{k}\cdot\mathbf{r}}. \tag{12–88}$$

Then (12–87) is identical with (12–86) and we have

$$\dot{q}_k = \frac{\partial H}{\partial p_k}. \tag{12–89}$$

The equation of motion for p_k can be obtained similarly, with but one extra step. We again have

$$\dot{p}_k = [p_k, H] = -\frac{1}{V^{\frac{1}{2}}} \int dV\, e^{i\mathbf{k}\cdot\mathbf{r}} \frac{\delta\mathcal{H}}{\delta\eta},$$

but now

$$\frac{\partial H}{\partial q_k} = \int dV \left\{ \frac{\partial \mathcal{H}}{\partial \eta} \frac{\partial \eta}{\partial q_k} + \frac{\partial \mathcal{H}}{\partial \eta_{,j}} \frac{\partial \eta_{,j}}{\partial q_k} \right\}.$$

However, by carrying out an integration by parts on the term involving the components of the gradient of η, the integral can be reduced to one involving the functional derivative,

$$\frac{\partial H}{\partial q_k} = \int dV \frac{\delta \mathcal{H}}{\delta \eta} \frac{\partial \eta}{\partial q_k},$$

and then by Eq. (12–77) it follows that

$$\dot{p}_k = -\frac{\partial H}{\partial q_k}. \tag{12–90}$$

The quantities q_k and p_k thus obey Hamilton's equations of motion.

In a sense we have come about full circle. We started off this chapter with a discrete system employing a denumerable number of generalized coordinates. By then going to the limit of a continuous set of variables we could see how to treat continuous systems. Finally, we have introduced a description of the continuous system in terms of a denumerable, discrete, set of coordinates that obey the same type of mechanics as the discrete system we started with. Because of the formal correspondence with the variables of discrete systems, the q_k and p_k quantities are the obvious candidates for quantization when we go from classical to quantum field theory. Indeed the q_k correspond to what are spoken of as the "occupation numbers" for the field.

We could describe the field in terms of discrete coordinates because the finite size of the system, and the boundary conditions, permitted a discrete Fourier expansion. Equivalently we can say that the expansion is made over a discrete spectrum of plane waves. Since the wave vector **k** is in quantum mechanics directly proportional to the momentum of the particle associated with the plane wave, the expansions used here are often spoken of as the *momentum representation*. We need not be restricted to plane wave expansions. A denumerable set of coordinates can be found whenever the field functions can be expanded in terms of a discrete set of orthonormal eigenfunctions. The development for such general expansions parallels the steps followed here for plane wave functions and is discussed in detail in some of the literature referenced at the end of the chapter.

12–5 RELATIVISTIC FIELD THEORY

We saw in Chapter 7 that there is considerable difficulty in constructing relativistically covariant Lagrangian and Hamiltonian descriptions of particle mechanics. Part of the problem can be traced to the separate roles played by space and time coordinates. For point particles the space coordinates are mechanical variables while time is a monotonic parameter. But in classical field theory there is a natural similarity in handling space and time coordinates. They are all parameters, together defining a point in space-time continuum at which the field variables are to be determined. While the four-dimensional space-time system has been used so far only for reasons of notational simplicity, the easy and natural way it fitted into the formulation suggests a relativistically covariant description is quite feasible for classical fields. Indeed only relatively minor tinkering has to be done to the formulation already presented so that it can handle relativistic fields in a manner that is manifestly Lorentz covariant.

Three points require specific attention: (1) the nature (and metric) of the four-dimensional space used, (2) the Lorentz transformation properties of the field quantities, Lagrangian densities, and related functions, and (3) the covariant description of the limits of integration. The simple Cartesian four-space with coordinates t, x, y, z that we have implicitly used so far is not convenient for exhibiting Lorentz invariance. Of the various metrics and spaces that appear in theories of special relativity (cf. Sections 7–2, 7–3) we shall prefer to use Minkowski space. Admittedly it results in the occasional appearance, in otherwise real physical quantities, of factors involving i, but it has the virtues of familiarity and transparent notational simplicity. (Translation of the formulation into other metrics is considered in some of the exercises.) Accordingly, the Greek letter indices will now be considered to run from 1 to 4, with $x_4 = ict$. It will be noted that the Lagrange equations (12–23) are unaffected by this change. Indeed, the term

$$\frac{d}{dx_v}\left(\frac{\partial \mathscr{L}}{\partial \eta_{\rho,v}}\right)$$

remains unaltered by a scale change of any of the x_v, and the other term in the Lagrange equation does not involve the coordinates at all. Further, the change in space does not affect the formulation of Hamilton's principle in Eq. (12–20), since it only introduces a multiplicative constant.

All of the quantities related to the field and associated equations must now have some definite Lorentz covariant properties. The field quantities must therefore consist of world tensors of some given rank — world scalar, world vector, etc. In principle, η_ρ need not be restricted to any one of these categories but may stand for a set of such; e.g., two world scalars. The Lagrangian and Hamiltonian densities must also be covariant. In Hamilton's principle the volume element (dx_v) of four-space is invariant under Lorentz transformation. Since we usually think of the action I as a scalar this means that the Lagrangian density (and therefore $\mathscr{H}$) should be world scalars. That is to say, they must be functions of the field quantities (possibly along with external covariant quantities) in such manner as to form world scalars. It then follows that the stress-energy tensor, $T_{\mu v}$, as defined by Eq. (12–34) is automatically a world tensor of the second rank. The change in the four-space, however, means that the components of $T_{\mu v}$ may be altered in value. Of course, T_{ij} is unaffected by the switch from (t, x, y, z) to (x, y, z, ict) —providing

Lagrangian density is the same. For the other components we can see from Eqs. (12–34) and (12–34′) that under the same conditions the correspondences are

$$T_{44} \rightarrow T_{00},$$

$$T_{j4} \rightarrow ic\,T_{j0}, \tag{12–91}$$

$$T_{4j} \rightarrow -\frac{i}{c}T_{0j}.$$

The Lagrangian density is of course uncertain to a multiplicative constant factor. It is customary to choose the factor such that T_{44} (or its symmetrized form) directly represents the energy density in the field. In the chosen four-space the quantities R_μ, Eq. (12–36), are now defined as

$$R_\mu = \int T_{\mu 4}\,dV. \tag{12–92}$$

Consider a related set P_μ defined as

$$P_\mu = \frac{i}{c}R_\mu. \tag{12–93}$$

It follows then, from Eq. (12–91) and the interpretation given above (cf. p. 560) for T_{i0}, that P_i represents the components of the total linear momentum of the field, and P_4 is $(iE)/c$, where E is the total energy in the field. This suggests that P_μ forms the relativistic world vector of energy—momentum for the field (cf. p. 307). However, one has to show still that R_μ and P_μ transform like world vectors under a Lorentz transformation. To prove this property we have to examine what is meant by an integration over three-space in a covariant formulation and indeed how the integration limits are to be treated in general.

The first instance where the covariance of the limits of integration may be questioned is in Hamilton's principle. In Eq. (12–20) the integral appears manifestly covariant, but the limits of integration derived from Eq. (12–12) are not. The spatial integration is over some fixed volume in three-space followed by an integration over time between t_1 and t_2. But an integration over V for fixed t is not a covariant concept, for simultaneity ("constant time") is not necessarily preserved under Lorentz transformation. A suitable covariant description is to say the integration is conducted over a hypersurface of three dimensions that is *space-like*. By a space-like surface we mean one in which all world vectors lying in it are space-like (cf. p. 301). The vectors normal to such a surface are time-like. Now, any vector connecting two world points on a surface of constant time is certainly space-like, for its x_4 component vanishes. Hence a surface at constant time is a particular example of a space-like surface. But such a surface retains its character in all Lorentz frames, because the space-like or time-like quality of a vector is not affected by the Lorentz transformation. In a similar fashion what is in one frame an integration over t at a fixed point can be described covariantly as an integration over a time-like surface. With a system of one dimension (in physical space), the integration in Hamilton's principle as given in Eq. (12–12) is over the rectangle shown in Fig. 12–4. A Lorentz transformation is a rotation in Minkowski space, and the sides of the rectangle will not be parallel to the axes in the transformed space. But we can describe the integration in all Lorentz frames as being over a region in four-space contained between two space-like hypersurface surfaces and bounded by intersecting time-like surfaces.

The appropriate covariant description of integral quantities such as P_μ is then given as

$$P_\mu = \frac{i}{c} \int_S T_{\mu\nu} \, dS_\nu, \tag{12–94}$$

where the integration is over a region on a space-like hypersurface for which the vector elements of surface, in the direction of the surface normal, are dS_ν. As $T_{\mu\nu}$ is a world tensor of the second rank, it is obvious that P_μ so defined is a world vector. But now we can show that the components of P_μ given by (12–94) reduce to a volume integral in ordinary three-space, *providing* it is divergenceless, i.e., satisfies Eq. (12–33). Imagine a region in four-space defined by three surfaces: S_1 and S_2 that are space-like, and S_3 that is time-like (cf. Fig. 12–5). By a four-dimensional divergence theorem a volume integral of a divergence can be replaced by a surface integral:

$$\int_{V_4} \frac{dT_{\mu\nu}}{dx_\nu}(dx_\nu) = \int_{S_1 + S_2 + S_3} T_{\mu\nu} \, dS_\nu. \tag{12–95}$$

The integration over S_3 corresponds to an integration over t at constant $\mathbf{r}$. By allowing the volume to expand sufficiently, the integral over this surface will involve $\mathbf{r}$ outside the system, where all field quantities vanish. Because of the assumed divergenceless property of $T_{\mu\nu}$, the integral on the left-hand side also vanishes. Therefore, if the normals to the space-like surfaces are taken in the same sense,

$$\int_{S_1} T_{\mu\nu} \, dS_\nu = \int_{S_2} T_{\mu\nu} \, dS_\nu. \tag{12–96}$$

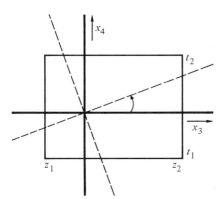

FIGURE 12–4
Regions of integration in Hamilton's principle for a system extending in only one dimension.

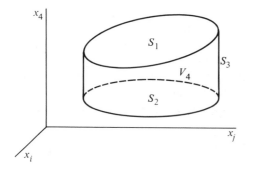

FIGURE **12–5**
Schematic integration volume in four-space

If S_1 is any arbitrary space-like surface, and S_2 is a particular surface for which x_4, or t, is constant, then by Eq. (12–96)

$$\int_{S_1} T_{\mu\nu}\, dS_\nu = \int T_{\mu 4}\, dV. \tag{12–97}$$

The four-vector transformation property of the left-hand side is obvious; hence, the right-hand side, i.e., R_μ according to Eq. (12–92) also transforms as a four-vector. Further, if *both* S_1 and S_2 are surfaces at constant t, say t_1 and t_2, respectively, then Eq. (12–96) is equivalent to

$$R_\mu(t_1) = R_\mu(t_2),$$

which is thus the covariant way proving that R_μ is conserved in time.

With some care, therefore, the conserved integral quantities can still be used within the framework of a relativistic theory of classical fields. We shall not always carry through the detailed correspondence but will let it suffice in most instances that the volume integration refers to a particular Lorentz frame in which the space-like hypersurface is a region in three-space at constant t. For the angular momentum density, it may be noted that the covariant analog of $\mathcal{M}_{ij}$, Eq. (12–46), is a four-tensor of third rank:

$$\mathcal{M}_{\mu\nu\lambda} = \frac{i}{c}(x_\mu T_{\nu\lambda} - x_\nu T_{\mu\lambda}), \tag{12–98}$$

which is antisymmetric in μ and ν. The corresponding global or integral quantity is

$$M_{\mu\nu} = \int \mathcal{M}_{\mu\nu\lambda}\, dS_\lambda, \tag{12–99}$$

where the integration is over a space-like hypersurface. If the Lorentz frame is chosen such that the surface is one at constant t, then

$$M_{\mu\nu} \rightarrow \int \mathcal{M}_{\mu\nu 4}\, dV,$$

which corresponds to the previous definition. The rest of the argument on the conservation of M_{ij} for symmetrical stress-energy tensors then can be carried out as before by considering this particular Lorentz frame.

As constructed in the previous section, the Hamiltonian formulation sharply distinguishes between the time coordinate and the space coordinates. This is not to say that it is necessarily nonrelativistic, merely that the formulation is not manifestly covariant. We must imagine the Hamiltonian framework as constructed in terms of the time as seen by each particular observer. Providing the field quantities and derived functions have suitable transformation properties, this construction for each Lorentz frame is not in violation of special relativity.

One further point needs to be made here. By allowing η_ρ to stand for a set of covariant field quantities, we allow for the possibility that the system consists of two or more fields that interact with each other. The complete Lagrangian density may consist of a sum of Lagrangian densities representing the free fields plus terms that describe the interactions between the fields. It will be remembered that one of the difficulties of relativistic point mechanics was the problem of considering interactions between particles that necessarily implied action-at-a-distance. However, interactions between fields can be at a point and, therefore, consistent with special relativity. One can often go further and treat the interaction between a field and a particle at a given point in space-time. There is thus the possibility of considering relativistically a system consisting of a continuous field, a discrete particle, and the interaction between them. How this can be done in a specific case will be shown in the next section, which provides illustrations of relativistic field theories.

12–6 EXAMPLES OF RELATIVISTIC FIELD THEORIES

Three examples, of increasing complexity, will be discussed.

A. *Complex scalar field.* Any complex field will be described by two independent parts, which can be expressed either as the real and imaginary part of the field or as the complex field itself and its complex conjugate. We shall follow the latter alternative. Accordingly, the Lagrangian density and associated functions will here be given in terms of two independent field variables, ϕ and ϕ^*, each of which are world scalars.* For this particular example we choose the Lagrangian density

$$\mathscr{L} = -c^2 \phi_{,\lambda} \phi^*_{,\lambda} - \mu_0^2 c^2 \phi \phi^*, \tag{12–100}$$

where μ_0 is a constant. Notice, that as required, $\mathscr{L}$ is a world scalar. Expressed in terms of space and time variables, $\mathscr{L}$ is written as

$$\mathscr{L} = \dot{\phi}\dot{\phi}^* - c^2 \nabla\phi \cdot \nabla\phi^* - \mu_0^2 c^2 \phi \phi^*. \tag{12–100'}$$

---

* As will be seen in the next section, complex fields lead naturally to an associated charge and current density, and this is the main reason for their introduction in physical theories.

To obtain the field equation for which $\eta_\rho = \phi^*$ note that

$$\frac{\partial \mathscr{L}}{\partial \phi^*_{,\nu}} = -c^2 \phi_{,\nu}, \qquad \frac{\partial \mathscr{L}}{\partial \phi^*} = -\mu_0^2 c^2 \phi.$$

Hence the Lagrange–Euler field equation is

$$\phi_{,\nu\nu} - \mu_0^2 \phi = 0, \tag{12–101}$$

or, in equivalent forms,

$$\sum_\nu \frac{d^2 \phi}{dx_\nu^2} - \mu_0^2 \phi = 0 \tag{12–102}$$

and

$$\nabla^2 \phi - \frac{1}{c^2} \frac{d^2 \phi}{dt^2} - \mu_0^2 \phi = 0. \tag{12–102'}$$

In terms of the D'Alembertian (defined above, p. 303), the field equation can also be written covariantly as

$$(\Box^2 - \mu_0^2)\phi = 0. \tag{12–103}$$

Similarly, from the symmetry of $\mathscr{L}$, the field equation obtained when $\eta_\rho = \phi^*$ is

$$(\Box^2 - \mu_0^2)\phi^* = 0. \tag{12–103'}$$

The basic field equation satisfied by both ϕ and ϕ^* is known as the Klein–Gordon equation* and, as given here, represents the relativistic analog of the Schrödinger equation for a charged zero-spin particle of rest mass energy μ_0.

The stress-energy tensor defined by Eq. (12–34) has components

$$T_{\mu\nu} = -c^2 \phi_{,\mu} \phi^*_{,\nu} - c^2 \phi^*_{,\mu} \phi_{,\nu} + c^2 (\phi_{,\lambda} \phi^*_{,\lambda} + \mu_0^2 \phi \phi^*) \delta_{\mu\nu} \tag{12–104}$$

and is clearly symmetrical. As the Lagrangian density describes a free field, without interactions to the outside world, $\mathscr{L}$ does not contain x_ν explicitly and the conservation theorem (12–33) holds for $T_{\mu\nu}$, as can be verified directly. To introduce the Hamiltonian formulation we have to distinguish between the time and space coordinates in some particular Lorentz frame. The conjugate momenta, according to Eq. (12–53), are then

$$\pi = \frac{\partial \mathscr{L}}{\partial \dot\phi} = \dot\phi^*, \qquad \pi^* = \frac{\partial \mathscr{L}}{\partial \dot\phi} = \dot\phi. \tag{12–105}$$

---

* In some of the modern literature the Klein–Gordon equation in the form of Eq. (12–103) is given with a plus sign for the $\mu_0^2 \phi$ term. The reason stems from the use of the metric given by Eq. (7–53) with a corresponding change of sign in the definition of the D'Alembertian. See Exercise 7.

It follows that the Hamiltonian density (which has the same magnitude as T_{44}) take the form

$$\mathcal{H} \equiv \pi\dot{\phi} + \pi^*\dot{\phi}^* - \mathcal{L},$$

$$= \pi\pi^* + c^2\nabla\phi \cdot \nabla\phi^* + \mu_0^2 c^2 \phi\phi^*. \tag{12-106}$$

We leave to the exercises the verification that the Hamilton equations of motion reduce to Eqs. (12–105) and the Klein–Gordon equations. For the moment, all that will be done here is to illustrate the transformation to the momentum representation. The expansions (12–79) and (12–82) may be introduced into the Hamiltonian density. Since the field is not real, we do *not* have that $q_k^* = q_{-k}$. In effect, q_k and q_k^* now stand for two independent sets of discrete coordinates, one representing ϕ and the other ϕ^*. The total Hamiltonian is a sum of volume integrals over the three terms in Eq. (12–106). As a typical example, consider

$$\mu_0^2 \int \phi\phi^* \, dV = \frac{\mu_0^2}{V} \sum_{k,k'} \int q_k q_{k'}^* \, e^{i(\mathbf{k}-\mathbf{k}')\cdot\mathbf{r}} \, dV,$$

which by Eq. (12–80) reduces to

$$\mu_0^2 q_k q_k^*.$$

The only other term requiring any special note at all is that involving the divergences, which introduce a factor $(i\mathbf{k}) \cdot (-i\mathbf{k}')$ in the integrand. The final form for H can be written as

$$H = p_k p_k^* + \omega_k^2 q_k q_k^*, \tag{12-107}$$

where

$$\omega_k^2 = c^2 k^2 + \mu_0^2. \tag{12-108}$$

Each of the terms of the summation in Eq. (12–107) is in the form of a harmonic oscillator of unit mass with frequency ω_k. This can be seen explicitly by evaluating the Hamilton's equations of motion, (12–89) and (12–90). In the momentum or plane wave representations, the fields ϕ and ϕ^* are thus replaced by discrete systems of harmonic oscillators, much in the same manner that the sound field in a finite solid is looked on as a collection of "phonons." The discrete spectrum of "vibrations" of our scalar charged field is given by Eq. (12–108). Quantization of the field (i.e., the so-called second quantization) is done most simply via the momentum representation. In effect, the motion of each of the harmonic oscillators is quantized as would be done for an actual harmonic oscillator. But this subject certainly lies outside our province.

B. *The Sine–Gordon equation and associated field.* If the scalar field in the previous example were taken as real (i.e., $\phi^* = \phi$) and to exist in only one spatial dimension, then the obvious corresponding Lagrangian density along the model of Eq. (12–100′) would be

$$\mathcal{L} = \frac{c^2}{2}\left(\frac{\dot{\phi}^2}{c^2} - \left(\frac{\partial\phi}{\partial x}\right)^2 - \mu_0^2\phi^2\right). \tag{12–109}$$

(The factor of $\frac{1}{2}$ is introduced for convenience; it clearly does not affect the form of the equations of motion.) The associated field equation is

$$\frac{\partial^2\phi}{\partial x^2} - \frac{1}{c^2}\frac{\partial^2\phi}{\partial t^2} = \mu_0^2\phi, \tag{12–110}$$

which is the one-dimensional Klein–Gordon equation. Note that it is linear in the field $\phi(x, t)$.

We can look upon the Lagrangian density of Eq. (12–109) as a small-field approximation to a Lagrangian density of the form

$$\mathcal{L} = \frac{c^2}{2}\left(\frac{\dot{\phi}^2}{c^2} - \left(\frac{\partial\phi}{\partial x}\right)^2\right) - \mu_0^2 c^2(1 - \cos\phi), \tag{12–111}$$

which has the corresponding field equation

$$\frac{\partial^2\phi}{\partial x^2} - \frac{1}{c^2}\frac{\partial^2\phi}{\partial t^2} = \mu_0^2 \sin\phi. \tag{12–112}$$

Inevitably, if perhaps frivolously, Eq. (12–112) has come to be known as the sine–Gordon equation. If the Klein–Gordon equation, Eq. (12–110), is reminiscent of the harmonic oscillator, then the "potential" term in the Lagrangian Eq. (12–111) recalls the potential term of the linear pendulum. Indeed, Eq. (12–112) has also been called, perhaps more appropriately, the pendulum equation.

In this one-dimensional world the stress-energy tensor has only four components. As x and t again do not appear explicitly in $\mathcal{L}$, the elements of the tensor satisfy conservation equations, which are here two in number. Details will be left to the exercises, but of particular interest is the energy density T_{44}:

$$T_{44} = \frac{1}{2}\left(\dot{\phi}^2 + c^2\left(\frac{\partial\phi}{\partial x}\right)^2\right) + \mu_0^2 c^2(1 - \cos\phi),$$

which is of course the same in magnitude as the Hamiltonian density

$$\mathcal{H} = \frac{1}{2}\left(\pi^2 + c^2\left(\frac{\partial\phi}{\partial x}\right)^2\right) + \mu_0^2 c^2(1 - \cos\phi), \tag{12–113}$$

where the conjugate momentum is

$$\pi(x, t) = \dot{\phi}.$$

The momentum representation for the Klein–Gordon field as the sum over harmonic oscillators means that in the one-dimensional case the field can be built up as a superposition of plane waves of the form

$$q_k(t)\,e^{ikr} = A_0(k)\,e^{i(kr - \omega_k t)},$$

where k and ω_k are related by the *dispersion relation*, Eq. (12–108). For the field obeying the sine–Gordon equation, it is much more difficult to apply a momentum representation, because of the presence of the $\cos\phi$ term in $\mathscr{H}$. But one can still solve the sine–Gordon equation by something resembling a traveling wave. A solution for ϕ in Eq. (12–112) that has the form of a disturbance traveling with a speed v, but otherwise keeping its shape, must be a function only of $\tau = t - x/v$. In that case Eq. (12–112) reduces to

$$\frac{d^2\phi}{d\tau^2} - A \sin\phi = 0,$$

where

$$A = \frac{\mu_0^2 c^2 v^2}{c^2 - v^2}. \qquad (12\text{–}114)$$

In terms of the variable τ the equation of motion is indeed that for a simple pendulum of finite amplitude. For very small amplitude, we know that ϕ is a simple harmonic motion in τ with ω given by Eq. (12–108) for a wave number $k = \omega/v$, independent of the amplitude. With finite amplitude, we also know from our study of the pendulum, that while ϕ will still be periodic, the frequency ω will also depend on the amplitude. That is to say, the dispersion relation will be amplitude dependent. This is a characteristic, of course, of nonlinear equations, of which the sine–Gordon equation is one example. The Klein–Gordon equation is linear, but the dispersion equation, Eq. (12–108), is said to be nonlinear, i.e., ω_k is not a linear function of k. It becomes linear only when $\mu_0 \to 0$, when the Klein–Gordon equation reduces the usual linear wave equation.

We can thus describe the sine–Gordon equation as being nonlinear, with a nonlinear amplitude-dependent dispersion relation. Further examination reveals that it can have solutions with properties shared by only a few other nonlinear equations. These solutions are traveling wave disturbances that can interact with each other—pass through each other—and emerge with unchanged shape except perhaps for a phase shift. Such solutions are also found, for example, for the nonlinear Korteweg–deVries equation,

$$\frac{\partial\phi}{\partial t} + \alpha\phi\frac{\partial\phi}{\partial x} + v\frac{\partial^3\phi}{\partial x^3} = 0, \qquad (12\text{–}115)$$

where α and v are constants. These solitary waves that preserve their shape even through interactions have been termed "solitons" and are finding an expanding area of application throughout physics, from elementary particles through solid-state physics. The pendulum sine–Gordon equation, for example, has been used to describe families of elementary particles and also shows up in connection with the theory of the Josephson junction.

C. *Classical field of a Dirac particle.* Here the field consists of four complex scalar quantities appearing in two arrays, ψ and $\psi^\dagger$. For present purposes, ψ can be considered as a four-element column matrix and $\psi^\dagger$ as the adjoint matrix. A suitable Lagrangian density is

$$\mathscr{L} = i\psi^\dagger \gamma_\mu \psi_{,\mu} + m\psi^\dagger \psi. \tag{12–16}$$

Here m is a constant equal to the mass of the particle being represented (in certain units) and γ_μ is a set of four 4×4 Dirac matrices that are generalizations of the Pauli 2×2 matrices σ_i, used in Chapter 4 (cf. Eq. 4–74).* The field thus has eight components, four for ψ and four for $\psi^\dagger$. If ψ_λ represents an element of ψ, and $\psi_v^\dagger$ one of the elements of $\psi^\dagger$, then $\mathscr{L}$ in Eq. (12–116) could be written in expanded form as

$$\mathscr{L} = i\psi_v^\dagger (\gamma_\mu)_{v\lambda} \psi_{\lambda,\mu} + m\psi_v^\dagger \psi_v.$$

Here $(\gamma_\mu)_{v\lambda}$ is the $v\lambda$ element of γ_μ. It is much more convenient however, and equally unambiguous, to retain the matrix notation for ψ and $\psi^\dagger$ throughout.

Since $\mathscr{L}$ does not contain $\psi^\dagger_{,\mu}$, the Euler–Lagrange equations for the $\psi^\dagger$ variables are particularly simple to obtain:

$$\frac{\partial \mathscr{L}}{\partial \psi^\dagger} = i\gamma_\mu \psi_{,\mu} + m\psi = 0. \tag{12–117}$$

For the ψ variables the corresponding Euler–Lagrange equations have almost as simple a form:

$$\frac{d}{dx_\mu}\left(\frac{\partial \mathscr{L}}{\partial \psi_{,\mu}}\right) - \frac{\partial \mathscr{L}}{\partial \psi} = i\psi^\dagger_{,\mu}\gamma_\mu - m\psi^\dagger = 0. \tag{12–118}$$

Equations (12–117) constitute the well-known Dirac wave equation, and Eqs. (12–118) correspond to the adjoint form.

The formal stress-energy tensor is easily found to be

$$T_{\mu v} = i\psi^\dagger \gamma_v \psi_{,\mu} - \mathscr{L}\delta_{\mu v},$$

as can be verified directly by explicit representation of ψ and $\psi^\dagger$ as matrices. However $T_{\mu v}$ is not symmetric as it stands. Further the Hamiltonian formulation of Section 12–4 cannot be carried out because $\mathscr{L}$ is at most linear in the time

---

* Represented as matrices of 2×2 matrices, the Dirac matrices can be defined as

$$\gamma_i = \begin{pmatrix} 0 & \sigma_i \\ -\sigma_i & 0 \end{pmatrix}, \qquad \gamma_4 = \begin{pmatrix} 1 & 0 \\ 0 & -1 \end{pmatrix}. \tag{12–116}$$

The explicit representations are not needed for the illustration of $\psi, \psi^\dagger$ as an example of a classical field.

$$\frac{d}{dt}\left(\frac{\partial \mathscr{L}}{\partial \dot{\phi}}\right) + \frac{d}{dx_k}\left(\frac{\partial \mathscr{L}}{\partial \phi_k}\right) - \frac{\partial \mathscr{L}}{\partial \phi} = 0$$

$$\mathcal{L} = \frac{E^2 - B^2}{8\pi} - \rho\phi + \frac{1}{c}\,\mathbf{J}\cdot\mathbf{A}$$

derivatives of the field quantities. Thus π does not involve either $\dot{\psi}$ or $\dot{\psi}^\dagger$ and, as in particle mechanics, one cannot then invert the defining equations to eliminate the time derivatives. The Dirac field does have other interesting aspects, which will appear toward the end of this section and in the next section.

D. *The electromagnetic field.* Probably the most familiar example of a classical field not built upon a mechanical system is the electromagnetic field. Yet its formulation is considerably more complicated than the illustrations already considered and is at times ambiguous. We cannot expect to go into all the ramifications here but will touch on some of the highlights.

Maxwell's equations, Eqs. (1–60), in microscopic form (i.e., not in a macroscopic medium) consist of two homogeneous equations,

$$\nabla \cdot \mathbf{B} = 0, \qquad \nabla \times \mathbf{E} + \frac{1}{c}\frac{\partial \mathbf{B}}{\partial t} = 0, \qquad (12\text{–}119)$$

$$\boxed{\mathbf{E} = -\nabla\phi - \frac{1}{c}\frac{\partial \mathbf{A}}{\partial t}}$$

and two inhomogeneous equations,

$$\boxed{\mathbf{B} = \nabla \times \mathbf{A}}$$

$$\nabla \cdot \mathbf{E} = 4\pi\rho, \qquad \nabla \times \mathbf{B} - \frac{1}{c}\frac{\partial \mathbf{E}}{\partial t} = \frac{4\pi\mathbf{j}}{c}. \qquad (12\text{–}120)$$

The homogeneous equations imply that $\mathbf{E}$ and $\mathbf{B}$ can be expressed in terms of a scalar ϕ and vector potential $\mathbf{A}$, which together form a four-vector A_μ. However, the equations (1–62) and (1–63), which define the relations between A_μ and $(\mathbf{E}, \mathbf{B})$, do not completely fix the values of the potentials. Indeed, A_μ is undetermined to within the four-gradient of any scalar function. An additional relation, the *gauge condition*, must be added. In most of our considerations it will not be necessary to fix the gauge explicitly. Whenever a specific gauge condition is required we shall use the Lorentz gauge (cf. p. 303), which leads to particularly simple forms for the wave equations and can be stated in the obviously covariant form

$$\frac{dA_\mu}{dx_\mu} = 0. \qquad (12\text{–}121)$$

Equations (1–62) and (1–63) can also be expressed covariantly through an antisymmetric world tensor of the second rank, the *field tensor* $F_{\mu\nu}$, defined as

$$F_{\mu\nu} = \frac{\partial A_\nu}{\partial x_\mu} - \frac{\partial A_\mu}{\partial x_\nu} \equiv A_{\nu,\mu} - A_{\mu,\nu}. \qquad (12\text{–}122)$$

The gauge uncertainty in A_μ appears in the fact that if Λ is any scalar function, then

$$\frac{d\Lambda}{dx_\mu} \equiv \Lambda_{,\mu} \qquad (12\text{–}123)$$

can be added to A_μ without affecting the value of $F_{\mu\nu}$. Clearly $F_{\mu\nu}$ is a sort of four-dimensional curl of the vector A_μ. The purely spatial elements, F_{ij}, are indeed the components of the curl of $\mathbf{A}$ and therefore are given in terms of $\mathbf{B}$. For the elements F_{4j} we have

$$F_{4j} = \frac{\partial A_j}{\partial x_4} - \frac{\partial A_4}{\partial x_j} = -\frac{i}{c}\frac{\partial A_j}{\partial t} - i\frac{\partial \phi}{\partial x_j} = iE_j,$$

by Eq. (1–63). Hence the full form of the field tensor is

$$\mathbf{F} = \begin{pmatrix} 0 & B_3 & -B_2 & -iE_1 \\ -B_3 & 0 & B_1 & -iE_2 \\ B_2 & -B_1 & 0 & -iE_3 \\ iE_1 & iE_2 & iE_3 & 0 \end{pmatrix}. \tag{12-124}$$

It would be expected that the homogeneous equations, (12–119), would be satisfied identically in terms of $F_{\mu\nu}$, as it is these equations that led to the potentials. Indeed, it is not difficult to see that equations (12–119) can be written as the four equations

$$\frac{\partial F_{\mu\nu}}{\partial x_\lambda} + \frac{\partial F_{\lambda\mu}}{\partial x_\nu} + \frac{\partial F_{\nu\lambda}}{\partial x_\mu} = 0, \qquad \text{(no summation!)} \tag{12-125}$$

where μ, ν, λ are any three cyclic set out of the four indices.* (When μ, ν, λ are chosen as 1, 2, 3, we obviously have the $\nabla \cdot \mathbf{B}$ equation.) But in terms of A_μ Eqs. (12–125) are identically true, for they can then be written as

$$A_{\nu,\mu\lambda} - A_{\mu,\nu\lambda} + A_{\mu,\lambda\nu} - A_{\lambda,\mu\nu} + A_{\lambda,\nu\mu} - A_{\nu,\lambda\mu} = 0,$$

which vanishes by cancellation of pairs. It is therefore only the inhomogeneous equations that define the field in terms of its sources, and that are to be considered as the field equations. In terms of the field tensor, Eqs. (12–120) can be written as

$$\frac{dF_{\mu\nu}}{dx_\nu} = \frac{4\pi j_\mu}{c} \tag{12-126}$$

where, as before, j_μ is the four-vector $(\mathbf{j}, i\rho c)$. For example, when $\mu = 4$, Eq. (12–123) is obviously the $\nabla \cdot \mathbf{E}$ equation. Equations (12–125) and (12–126) together are the covariant form of Maxwell's equations. When we seek a Lagrangian formulation of the electromagnetic field we need concern ourselves only with Eqs. (12–126), as the others are satisfied by definition.

---

* Equation (12–125) can be written more economically (with the summation convention) as

$$\epsilon_{\mu\nu\lambda\rho}\frac{\partial F_{\lambda\rho}}{\partial x_\nu} = 0, \tag{12-125'}$$

where $\epsilon_{\mu\nu\lambda\rho}$ is the four-dimensional permutation symbol (cf. p. 172).

If the potential components A_μ are treated as the field quantities* then a suitable Lagrangian density for the electromagnetic field is

$$\mathscr{L} = -\frac{F_{\lambda\rho}F_{\lambda\rho}}{16\pi} + \frac{j_\lambda A_\lambda}{c}. \tag{12-127}$$

To obtain the Euler–Lagrange equations we note that

$$\frac{\partial\mathscr{L}}{\partial A_\mu} = \frac{j_\mu}{c}; \qquad \frac{\partial\mathscr{L}}{\partial A_{\mu,\nu}} = -\frac{F_{\lambda\rho}}{8\pi}\frac{\partial F_{\lambda\rho}}{\partial A_{\mu,\nu}}.$$

Now, from the defining equations (12–122), the derivative of $F_{\lambda\rho}$ vanishes except when $\lambda = \mu, \rho = \nu$ and $\lambda = \nu, \rho = \mu$. Hence,

$$\frac{\partial\mathscr{L}}{\partial A_{\mu,\nu}} = \frac{F_{\mu\nu}}{8\pi} - \frac{F_{\nu\mu}}{8\pi} = \frac{F_{\mu\nu}}{4\pi}, \tag{12-128}$$

and the Euler–Lagrange equations are

$$\frac{1}{4\pi}\frac{dF_{\mu\nu}}{dx_\nu} - \frac{j_\mu}{c} = 0,$$

identical with Eqn. (12–126).†

The Lagrangian density in Eq. (12–127) is clearly in the form of a free-field term plus a term describing the interaction with the outside world in terms of the four-vector of current density. In general $\mathscr{L}$ will therefore be an explicit function of x_μ through the spatial and time dependence of the charges and currents. Only for the free-field case will there be conserved currents in terms of the stress-energy tensor. Discussion of the stress-energy tensor will therefore be confined to free fields. From Eq. (12–128) it follows that $T_{\mu\nu}$ then becomes

$$T_{\mu\nu} \equiv A_{\lambda,\mu}\frac{\partial\mathscr{L}}{\partial A_{\lambda,\nu}} - \mathscr{L}\delta_{\mu\nu} = \frac{A_{\lambda,\mu}F_{\lambda\nu}}{4\pi} - \mathscr{L}\delta_{\mu\nu}.$$

In this form $T_{\mu\nu}$ is not symmetric. But if we subtract from it a term involving the sum $A_{\mu,\lambda}F_{\lambda\nu}$ then we could obtain a symmetrized form:

$$\hat{T}_{\mu\nu} = T_{\mu\nu} - \frac{A_{\mu,\lambda}F_{\lambda\nu}}{4\pi} = -\frac{F_{\lambda\mu}F_{\lambda\nu}}{4\pi} - \mathscr{L}\delta_{\mu\nu}. \tag{12-129}$$

---

* Part of the difficulty in handling the electromagnetic field arises from the fact that the components A_μ are not entirely independent—to be unique they must be connected through some gauge condition, such as Eq. (12–121). However, it will be sufficient for present purposes if we treat the gauge condition as a "weak" constraint in the sense of p. 329.

† The second term in $\mathscr{L}$, Eq. (12–127), often appears in the literature with a minus sign, which is a consequence of using a different metric for the four-space. See Exercise 14 below.

As noted before this is a legitimate procedure, without effect on conservation laws or integral quantities, only if the added term has a specific form, as described on p. 562. To see that the term does conform to the prescription, note that

$$A_{\mu,\lambda}F_{\lambda v} = \frac{d}{dx_\lambda}(A_\mu F_{\lambda v}) - A_\mu \frac{dF_{\lambda v}}{dx_\lambda}.$$

But, in the absence of external currents,

$$\frac{dF_{\lambda v}}{dx_\lambda} = -\frac{dF_{v\lambda}}{dx_\lambda} = 0$$

by Eq. (12–126). Further, $A_\mu F_{\lambda v} = -A_\mu F_{v\lambda}$, so the symmetrizing term is indeed the four-divergence of a quantity with the required antisymmetry properties.

Consider now $\hat{T}_{44}$, which should be an energy density. By Eq. (12–129)

$$\hat{T}_{44} = -\frac{F_{\lambda 4}F_{\lambda 4}}{4\pi} + \frac{F_{\mu v}F_{\mu v}}{16\pi}.$$

From the explicit form of the field tensor we see that $F_{\lambda 4}F_{\lambda 4} = -E^2$, and $F_{\mu v}F_{\mu v}$, which is just the sum of the squares of all terms in the tensor, is $2(B^2 - E^2)$. Hence,

$$\hat{T}_{44} = \frac{E^2}{4\pi} + \frac{B^2 - E^2}{8\pi} = \frac{E^2 + B^2}{8\pi}, \tag{12–130}$$

the usual expression for the energy density in the electromagnetic field. From Eq. (12–91) it should also be expected that $ic\hat{T}_{4j}$ should be the components of an energy flux density vector. Consider, for example,

$$ic\hat{T}_{41} = -ic\frac{F_{\lambda 4}F_{\lambda 1}}{4\pi} = -\frac{ic}{4\pi}(iE_2 B_3 - iE_3 B_2)$$

or

$$ic\hat{T}_{41} = \frac{c}{4\pi}(\mathbf{E} \times \mathbf{B})_1,$$

which is the 1-component of the customary Poynting's vector for energy flow of the electromagnetic field. The remaining components of $\hat{T}_{\mu v}$ also conform to the familiar interpretation of the electromagnetic field properties, as will be shown in the exercises at the end of the chapter. Note that if $T_{\mu v}$ had been used instead of the symmetrical form neither the energy density nor the energy flux density would have the usual form. But integral quantities over volumes or enclosing surfaces would remain the same, and these after all are the observables.

A few further points can be made. The Lagrangian density can be expressed directly in terms of A_μ by expanding $F_{\lambda \rho}$. We have from Eq. (12–122) that

$$F_{\lambda \rho}F_{\lambda \rho} = 2(A_{\rho,\lambda}A_{\rho,\lambda} - A_{\lambda,\rho}A_{\rho,\lambda}).$$

But the second term in the parentheses can be further manipulated:

$$A_{\lambda,\rho}A_{\rho,\lambda} \equiv \frac{dA_\lambda}{dx_\rho}\frac{dA_\rho}{dx_\lambda} = \frac{d}{dx_\rho}\left(A_\lambda\frac{dA_\rho}{dx_\lambda}\right) - A_\lambda\frac{d}{dx_\lambda}\left(\frac{dA_\rho}{dx_\rho}\right).$$

Of the terms on the extreme right the first contributes nothing to the action integral because it is a four-divergence leading to an integral over a surface on which A_λ vanishes. If the Lorentz gauge Eq. (12–121) is assumed, then the remaining term vanishes identically. Hence, *for the Lorentz gauge*, the Lagrangian density is equivalent to

$$\mathscr{L}' = -\frac{A_{\mu,\nu}A_{\mu,\nu}}{8\pi} + \frac{j_\mu A_\mu}{c}. \tag{12–131}$$

For this Lagrangian density the terms entering the Euler–Lagrange equations are

$$\frac{\partial\mathscr{L}'}{\partial A_{\mu,\nu}} = -\frac{A_{\mu,\nu}}{4\pi}, \qquad \frac{\partial\mathscr{L}'}{\partial A_\mu} = \frac{j_\mu}{c}.$$

Hence, Eq. (12–131) implies the field equations

$$\Box^2 A_\mu + \frac{4\pi j_\mu}{c} = 0,$$

which are the well-known wave equations for the four-vector potential, Eq. (7–81), when the Lorentz gauge is used.

Finally, it has already been noted that $\mathscr{L}$ for an electromagnetic field, Eq. (12–131), consists of a free-field Lagrangian density plus a term describing the interaction of a continuous charge and current density with the field. It is tempting to see how far we can go toward introducing field-particle interactions, by localizing the charge to a point. This is most easily done by considering the physical situation in some particular Lorentz frame, i.e., as seen by a particular observer. Manifest covariance is thereby abandoned, but the result still conforms to special relativity, as it derives from a clearly relativistic theory. The current density is a measure of the motion of the charges, and in any given system $\mathbf{j}$ is defined in terms of the charge density ρ by the relation

$$\mathbf{j}(\mathbf{r}, t) = \rho(\mathbf{r}, t)\mathbf{v}(\mathbf{r}, t).$$

Here $\mathbf{v}$ is the velocity "field" of the continuous charge distribution. The localization can be carried out through the use of the well-known Dirac δ-function. In three-dimensional form the δ-function has the property that if $f(\mathbf{r})$ is any function of space, then

$$\int dV f(\mathbf{r})\delta(\mathbf{r} - \mathbf{s}(t)) = f(\mathbf{s}),$$

where $\mathbf{s}(t)$ is the spatial position, say, of a particle at time t (so long as $\mathbf{s}$ is inside the volume of integration). Thus, the spatial charge and current density corresponding to a particle of charge q at point $\mathbf{s}$ is

$$\rho = q\,\delta(\mathbf{r} - \mathbf{s})$$

and

$$\mathbf{j} = q\delta(\mathbf{r} - \mathbf{s})\mathbf{v}(\mathbf{r}).$$

If we write $\mathscr{L}$ of Eq. (12–127) as the sum of a free-field term $\mathscr{L}_0$ and an interaction term, the Lagrangian as seen in the given Lorentz frame is

$$L = \int dV \mathscr{L}_0 - \int dV \rho\phi + \frac{1}{c}\int dV \mathbf{A}\cdot\mathbf{j}$$

$$= \int dV \mathscr{L}_0 - q\phi + \frac{q}{c}\mathbf{A}\cdot\mathbf{v}. \tag{12–132}$$

The interaction terms in Eq. (12–132) are exactly the same as those in Eq. (7–141) for the Lagrangian of a single particle in an electromagnetic field. This suggests that a single Lagrangian can be formed for the complete system of particle and field that, analogous to Eq. (7–141), would look like

$$L = -mc^2\sqrt{1 - \beta^2} - q\phi + \frac{q}{c}\mathbf{v}\cdot\mathbf{A} + \int dV \mathscr{L}_0. \tag{12–133}$$

Considered as a function of the field tensor or potentials, this Lagrangian implies the field equations; considered as a function of the particle coordinates, L leads to the particle equations of motion. The mechanical descriptions of the continuous field and the discrete particle have in effect been put under one wing, expressed in a common formalism!

To describe the field-particle interaction covariantly runs into much the same difficulties experienced in Section 7–9 when seeking a covariant Lagrangian formulation for the free particle. It is much simpler if the particle itself is described by a field (as in relativistic quantum mechanics) for field–field interactions naturally fall into a covariant picture. Thus the complex Dirac field, illustrated in the previous example, is the relativistic quantum representation of an electron with spin $\frac{1}{2}$. In Dirac theory there is a four-vector of charge current density given by

$$j_\mu = -q\psi^\dagger\gamma_\mu\psi, \tag{12–134}$$

so that the interaction Lagrangian density is

$$\frac{j_\mu A_\mu}{c} = -\frac{q}{c}\psi^\dagger\gamma_\mu A_\mu\psi. \tag{12–135}$$

A complete Lagrangian density for the two fields and their mutual interactions would, for Lorentz gauge, have the form

$$\mathscr{L} = -\frac{A_{\mu\nu}A_{\mu\nu}}{8\pi} - \frac{q}{c}\psi^\dagger\gamma_\mu A_\mu\psi + i\psi^\dagger\gamma_\mu\psi_{,\mu} + m\psi^\dagger\psi. \qquad (12\text{–}136)$$

With $\psi^\dagger$ as the field variables, the resultant field equation can be written

$$i\gamma_\mu\psi_{,\mu} - \frac{q}{c}\gamma_\mu A_\mu\psi + m\psi = 0 \qquad (12\text{–}137)$$

or

$$\gamma_\mu\left[i\frac{d}{dx_\mu} - \frac{q}{c}A_\mu\right]\psi + m\psi = 0, \qquad (12\text{–}137')$$

which is the Dirac equation with electromagnetic interaction. The expression in the bracket in Eq. (12–137′) has a familiar form, for i times the derivative operator is the wave-mechanical representation of the momentum operator (in the notation used here, which sets $\hbar = 1$). Hence, the bracket is the Dirac quantum analog of the expression $p_\mu - q/cA_\mu$, which we have encountered so many times before. With A_μ as the field variables the field equations are

$$\Box^2 A_\mu = \frac{4\pi q}{c}\psi^\dagger\gamma_\mu\psi, \qquad (12\text{–}138)$$

which is the electromagnetic vector potential wave equation interacting with the Dirac field.

    A considerable branch of modern physics is concerned with the construction of fields to represent various types of elementary particles. Of course, all such theories are quantum-mechanical, but many features of quantum field theories will have concomitant or nearly corresponding classical analogs. There is little a priori physical guidance in the construction of possible Lagrangian densities and interaction terms for the various particles. Some constraint on the form of these functions comes from covariance limitations. For example, the terms in $\mathscr{L}$ must be combinations of field and other quantities in such a manner as to produce a world scalar. Usually, $\mathscr{L}$ is also restricted to the field quantities or their first derivatives, although Lagrangian densities with higher derivatives (cf. Exercise 9) have also been explored. Additional requirements on the form of the terms are also provided, or suggested, by conservation and invariance properties, implicit in the Lagrangians. These properties go beyond the conservation conditions contained in the stress-energy tensor and are usually to be found by the application of a powerful procedure known as Noether's theorem, which forms the subject of the next section.

12–7 NOETHER'S THEOREM*

A recurring theme throughout has been that symmetry properties of the Lagrangian (or Hamiltonian) imply the existence of conserved quantities. Thus, if the Lagrangian does not contain explicitly a particular coordinate of displacement, then the corresponding canonical momentum is conserved. The absence of explicit dependence on the coordinate means the Lagrangian is unaffected by a transformation that alters the value of that coordinate; it is said to be invariant, or symmetric under the given transformation. Similarly, invariance of the Lagrangian under time displacement implies conservation of energy. The formal description of the connection between invariance or symmetry properties and conserved quantities is contained in Noether's theorem. It is in the four-space of classical field theory that the theorem attains its most sophisticated and fertile form. For that reason explicit discussion of the theorem has been reserved for the treatment of fields, although a discrete-system version can also be derived.

Symmetry under coordinate transformation refers to the effects of an infinitesimal transformation of the form

$$x_\mu \rightarrow x'_\mu = x_\mu + \delta x_\mu, \tag{12–139}$$

where the infinitesimal change δx_μ may be a function of all the other x_ν. Noether's theorem also considers the effect of a transformation in the field quantities themselves, which may be described by

$$\eta_\rho(x_\mu) \rightarrow \eta'_\rho(x'_\mu) = \eta_\rho(x_\mu) + \delta\eta_\rho(x_\mu). \tag{12–140}$$

Here $\delta\eta_\rho(x_\mu)$ measures the effect of both the changes in x_μ and in η_ρ and may be a function of all the other field quantities η_λ. Note that the change in one of the field variables at a particular point in x_ν space is a different quantity $\bar{\delta}\eta_\rho$:

$$\eta'_\rho(x_\mu) = \eta_\rho(x_\mu) + \bar{\delta}\eta_\rho(x_\mu). \tag{12–141}$$

The description of the transformations in terms of infinitesimal changes from the untransformed quantities indicates we are dealing only with *continuous* transformations. Thus, symmetry under inversion in three dimensions (parity symmetry) is not one of the symmetries for which Noether's theorem can be applied. As a consequence of the transformations of both the coordinates and the field quantities the Lagrangian appears, in general, as a different function of both the field variables and the space-time coordinates:

$$\mathscr{L}(\eta_\rho(x_\mu), \eta_{\rho,\nu}(x_\mu), x_\mu) \rightarrow \mathscr{L}'(\eta'_\mu(x'_\mu), \eta'_{\rho,\nu}(x'_\mu), x'_\mu). \tag{12–142}$$

The version of Noether's theorem that will be presented now is not the most general form possible but is such as to facilitate the derivation without

---

* Emmy Noether, 1882–1935, one of the leading mathematicians of this century, has been properly described as "the greatest of women mathematicians." The original publication of the theorem was in the *Nachrichten Gesell. Wissenschaft. Gottingen* **2**, 235 (1918). See the article on Noether by C. H. Kimberling, *Am. Math. Monthly* **79**, 136 (1972).

significantly restricting the scope of the theorem or the usefulness of the conclusions. Three conditions will be assumed to hold:

1. The four-space is euclidean. This requirement is dispensible but will be assumed here for simplicity. It restricts the relativistic space-time to Minkowski space, which is complex but euclidean.

2. The Lagrangian density displays the same functional form in terms of the transformed quantities as it does of the original quantities, that is,

$$\mathscr{L}'(\eta'_\rho(x'_\mu), \eta'_{\rho,\nu}(x'_\mu), x'_\mu) = \mathscr{L}(\eta'_\rho(x'_\mu), \eta'_{\rho,\nu}(x'_\mu), x'_\mu). \tag{12–143}$$

This type of condition has not previously entered our discussions of conserved quantities, mainly because it has been automatically satisfied under the transformations considered. When cyclic coordinates are transformed by displacement the functional dependence of the Lagrangian on the variables is unaltered by the implied shift in origin. But in our present extended types of transformation it becomes a symmetry property that needs study. Thus the *free-field* version of the Lagrangian density for the electromagnetic field, Eq. (12–127), retains its functional form when A_μ is subject to a gauge transformation, while the corresponding version of Eq. (12–131) obviously does not, even though both are cyclic in A_μ. The requirement expressed by Eq. (12–143) is known as *form-invariance*. Note also that Eq. (12–143) ensures that the equations of motion have the same form whether expressed in terms of the old or the new variables. The condition of form-invariance is not the most general circumstance under which this is true; the original and transformed Lagrangian densities may also differ by a four-divergence without modifying the equations of motion. Indeed, it is possible to carry out the derivation of Noether's theorem with such an extended version of form-invariance because the volume integral of the four-divergence term vanishes. But for simplicity we shall restrict ourselves to Eq. (12–143).

3. The magnitude of the action integral is invariant under the transformation, that is to say,

$$I' \equiv \int_{\Omega'} (dx'_\mu) \mathscr{L}'(\eta'_\rho(x'_\mu), \eta'_{\rho,\nu}(x'_\mu), x'_\mu) = \int_{\Omega} (dx_\mu) \mathscr{L}(\eta_\rho(x_\mu), \eta_{\rho,\nu}(x_\mu), x_\mu). \tag{12–144}$$

Again, this represents an extension of, and includes, our previous symmetry properties for cyclic coordinates. The Lagrangian does not change numerically under translation of a cyclic coordinate, nor does the value of the action integral. Equation (12–144) will be called the condition of *scale-invariance*. Our second and third conditions thus represent generalizations of the symmetry or invariance conditions that led to the existence of conserved quantities for discrete systems.

Combination of Eqs. (12–143) and (12–144) gives the requirement

$$\int_{\Omega'} \mathscr{L}(\eta'_\rho(x'_\mu), \eta'_{\rho,\nu}(x'_\mu), x'_\mu)(dx'_\mu) - \int_{\Omega} \mathscr{L}(\eta_\rho(x_\mu), \eta_{\rho,\nu}(x_\mu), x_\mu)(dx_\mu) = 0.$$

In the first integral x'_μ now represents merely a dummy variable of integration and can therefore be relabeled x_μ. But of course there remains a change in the domain of integration, so the condition becomes

$$\int_{\Omega'} \mathscr{L}(\eta'_\rho(x_\mu), \eta'_{\rho,\nu}(x_\mu), x_\mu)(dx_\mu) - \int_{\Omega} \mathscr{L}(\eta_\rho(x_\mu), \eta_{\rho,\nu}(x_\mu), x_\mu)(dx_\mu) = 0. \qquad (12\text{–}145)$$

The sequence of transformations of space and of integration region is illustrated in Fig. 12–6 for a space of two dimensions. Equation (12–145) says that if in the action integral over (x_μ) space we replace the original field variables by the transformed quantities, and transform the region of integration, then the action integral remains unaltered.

Under the infinitesimal transformations of Eqs. (12–139) and (12–140) the first-order difference between the integrals in Eq. (12–145) thus consists of two parts, one being an integral over Ω and the other an integral over the difference volume $\Omega' - \Omega$. An example in one-dimension will show how the terms are to be formed. Consider the difference of two integrals:

$$\int_{a+\delta a}^{b+\delta b} [f(x) + \delta f(x)]\, dx - \int_a^b f(x)\, dx$$

$$= \int_a^b \delta f(x)\, dx + \int_b^{b+\delta b} [f(x) + \delta f(x)]\, dx - \int_a^{a+\delta a} [f(x) + \delta f(x)]\, dx. \qquad (12\text{–}146)$$

To first order in small quantities the last two terms on the right can be written as

$$\int_b^{b+\delta b} f(x)\, dx - \int_a^{a+\delta a} f(x)\, dx = \delta b f(b) - \delta a f(a).$$

To this approximation, Eq. (12–146) becomes

$$\int_{a+\delta a}^{b+\delta b} [f(x) + \delta f(x)]\, dx - \int_a^b f(x)\, dx = \int_a^b \delta f(x)\, dx + f(x)\, \delta x \Big|_a^b, \qquad (12\text{–}147)$$

or

$$= \int_a^b \left[\delta f(x) + \frac{d}{dx}(\delta x f(x)) \right] dx. \qquad (12\text{–}148)$$

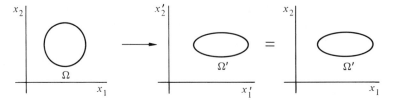

FIGURE 12–6
Schematic illustration of the transformation of the invariant action integral.

The multidimensional analog of Eq. (12–147) then says that the invariance condition of Eq. (12–145) takes the form

$$\int_{\Omega'} \mathscr{L}(\eta', x_\mu)\, d(x_\mu) - \int_{\Omega} \mathscr{L}(\eta, x_\mu)\, dx_\mu$$

$$= \int_{\Omega} [\mathscr{L}(\eta', x_\mu) - \mathscr{L}(\eta, x_\mu)]\,(dx_\mu) + \int_{S} \mathscr{L}(\eta)\,\delta x_\mu\, dS_\mu = 0. \quad (12\text{–}149)$$

Here, $\mathscr{L}(\eta, x_\mu)$ is shorthand for the full functional dependence, S is the three-dimensional surface of the region Ω (corresponding to the endpoints a and b in the one-dimensional case), and δx_μ is in effect the difference vector between points on S and corresponding points on the transformed surface S' (cf. Fig. 12–7). Corresponding to Eq. (12–148), the last integral can be transformed by the four-dimensional divergence theorem so for the invariance condition we have

$$0 = \int_{\Omega} (dx_\mu) \left\{ [\mathscr{L}(\eta', x_\mu) - \mathscr{L}(\eta, x_\mu)] + \frac{d}{dx_\nu}(\mathscr{L}(\eta, x_\mu)\,\delta x_\nu) \right\}. \quad (12\text{–}150)$$

Now, by Eq. (12–141), the difference term in the square brackets can be written to first order as

$$\mathscr{L}(\eta'_\rho(x_\mu), \eta'_{\rho,\nu}(x_\mu), x_\mu) - \mathscr{L}(\eta(x_\mu), \eta_{\rho,\nu}(x_\mu), x_\mu) = \frac{\partial \mathscr{L}}{\partial \eta_\rho}\,\bar{\delta}\eta_\rho + \frac{\partial \mathscr{L}}{\partial \eta_{\rho,\nu}}\,\bar{\delta}\eta_{\rho,\nu}.$$

The important property of the $\bar{\delta}$ change is that it is a change of η at a fixed point in x_μ space (unlike the δ variation, Eq. (12–140)). Hence, it commutes with the spatial differentiation operator, i.e., the order of

$$\bar{\delta} \quad \text{and} \quad \frac{d}{dx_\nu}$$

can be interchanged. Symbolically,

$$\mathscr{L}(\eta', x_\mu) - \mathscr{L}(\eta, x) = \frac{\partial \mathscr{L}}{\partial \eta_\rho}\,\bar{\delta}\eta_\rho + \frac{\partial \mathscr{L}}{\partial \eta_{\rho,\nu}}\,\frac{d\bar{\delta}\eta_\rho}{dx_\nu},$$

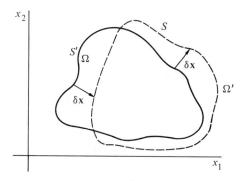

FIGURE 12–7

Illustration, in two dimensions, of the integration regions involved in the transformation of the action integral.

or, using the Lagrange field equations,

$$\mathcal{L}(\eta', x_\mu) - \mathcal{L}(\eta, x) = \frac{d}{dx_\nu}\left(\frac{\partial \mathcal{L}}{\partial \eta_{\rho,\nu}}\bar{\delta}\eta_\rho\right).$$

Hence the invariance condition, Eq. (12–150), appears as

$$\int (dx_\mu)\frac{d}{dx_\nu}\left\{\frac{\partial \mathcal{L}}{\partial \eta_{\rho,\nu}}\bar{\delta}\eta_\rho + \mathcal{L}\,\delta x_\nu\right\} = 0, \tag{12–151}$$

which is already in the form of a conserved current equation.

It is helpful, however, to develop the condition further by specifying the form of the infinitesimal transformation in terms of R infinitesimal parameters ϵ_r, $r = 1, 2, \ldots, R$, such that the change in x_ν and η_ρ is linear in the ϵ_r:

$$\delta x_\nu = \epsilon_r X_{r\nu}, \qquad \delta\eta_\rho = \epsilon_r \Psi_{r\rho}. \tag{12–152}$$

The functions $X_{r\nu}$ and $\Psi_{r\rho}$ may depend on the other coordinates and field variables, respectively. If the transformation symmetry related to the coordinates only, and corresponded to a displacement of a single coordinate x_λ, then these functions are simply

$$X_{r\nu} = \delta_{\nu\lambda}, \qquad \Psi_{r\rho} = 0. \tag{12–153}$$

Thus, the transformations contained in the form of Eq. (12–152) constitute a far more extensive test for symmetries than we have so far used. From Eqs. (12–140) and (12–141) it follows that to first order $\delta\eta$ and $\bar{\delta}\eta$ are related by

$$\delta\eta_\rho = \bar{\delta}\eta_\rho + \frac{\partial\eta_\rho}{\partial x_\sigma}\delta x_\sigma. \tag{12–154}$$

Hence,

$$\bar{\delta}\eta_\rho = \epsilon_r(\Psi_{r\rho} - \eta_{\rho,\sigma}X_{r\sigma}). \tag{12–155}$$

Substituting Eqs. (12–152) and (12–155) in the invariance condition, Eq. (12–151), we have

$$\int \epsilon_r \frac{d}{dx_\nu}\left\{\left(\frac{\partial \mathcal{L}}{\partial \eta_{\rho,\nu}}\eta_{\rho,\sigma} - \mathcal{L}\,\delta_{\nu\sigma}\right)X_{r\sigma} - \frac{\partial \mathcal{L}}{\partial \eta_{\rho,\nu}}\Psi_{r\rho}\right\}(dx_\mu) = 0.$$

Since the ϵ_r parameters are arbitrary, there exist r conserved currents with differential conservation theorems:

$$\frac{d}{dx_\nu}\left\{\left(\frac{\partial \mathcal{L}}{\partial \eta_{\rho,\nu}}\eta_{\rho,\sigma} - \mathcal{L}\,\delta_{\nu\sigma}\right)X_{r\sigma} - \frac{\partial \mathcal{L}}{\partial \eta_{\rho,\nu}}\Psi_{r\rho}\right\} = 0. \tag{12–156}$$

Equations (12–156) form the main conclusion of Noether's theorem, which thus says that if the system (or the Lagrangian density) has symmetry properties such that conditions (2) and (3) above hold for transformations of the type of Eqs. (12–152), then there exist r conserved quantities.

The conservation of the stress-energy tensor is easily recovered as a special case of Eq. (12–151). If $\mathscr{L}$ does not contain any of the x_μ, then it and therefore the action integral will be invariant under transformations such as Eq. (12–153), where λ takes on all the values μ. Equation (12–156) then reduces to

$$\frac{d}{dx_\nu}\left\{\left(\frac{\partial\mathscr{L}}{\partial\eta_{\rho,\nu}}\eta_{\rho,\sigma} - \mathscr{L}\,\delta_{\nu\sigma}\right)\delta_{\sigma\mu}\right\} = \frac{d}{dx_\nu}\left(\frac{\partial\mathscr{L}}{\partial\eta_{\rho,\nu}}\eta_{\rho,\mu} - \mathscr{L}\,\delta_{\mu\nu}\right),$$

which is identical with Eqs. (12–33) with $T_{\mu\nu}$ given by Eq. (12–34).

A large number of other symmetries are covered by transformations of the form of Eq. (12–152). One of the most interesting is a family of transformations of the field variables only, called *gauge transformations of the first kind,** and such that

$$\delta x_\nu = 0, \qquad \delta\eta_\rho = \epsilon c_\rho\eta_\rho \qquad \text{(no summation on } \rho\text{)}, \qquad (12\text{–}157)$$

where the c_ρ are constants. If the Lagrangian density, and therefore the action integral, is invariant under this transformation, then there is a conservation equation of the form

$$\frac{d\Theta_\nu}{dx_\nu} = 0, \qquad (12\text{–}158)$$

where

$$\Theta_\nu = c_\rho\frac{\partial\mathscr{L}}{\partial\eta_{\rho,\nu}}\eta_\rho. \qquad (12\text{–}159)$$

Equation (12–158) is in the form of an equation of continuity with Θ_ν in the role of a current density j_ν. Hence, invariance under a gauge transformation of the first kind leads to the identification of a conserved current that would be appropriate for an electric charge and current density to be associated with the field.

As an illustration consider the first example of Section (12–6), the complex scalar field. A transformation of the type

$$\phi' = \phi\,e^{i\epsilon}, \qquad \phi^{*'} = \phi^*e^{-i\epsilon} \qquad (12\text{–}160)$$

corresponds in infinitesimal form to a gauge transformation of the first type, Eq. (12–157), with

$$c = i, \qquad c^* = -i.$$

It is obvious that the Lagrangian density of Eq. (12–100) is invariant under the transformation (12–160). Hence, there is an associated current density for the Klein–Gordon field that can be given as

$$j_\mu = iq\left(\frac{d\phi}{dx_\mu}\phi^* - \phi\frac{d\phi^*}{dx_\mu}\right), \qquad (12\text{–}161)$$

---

* The familiar gauge transformation of the electromagnetic field, which adds a four-gradient $\Lambda_{,\mu}$ to A_μ, is part of a gauge transformation of the second kind and is not considered here.

which is in agreement with the conventional quantum mechanical current density. Note that the entire derivation of the conserved charge current density depends on the fact that the field is complex. Thus, as mentioned above, a real field does not lead to a charge or current density associated with the field. To describe fields associated with charged particles, one must use a pair of complex fields, ϕ and ϕ^* for the (spin-less) Klein–Gordon particle, ψ and $\psi^\dagger$ for the spin $\frac{1}{2}$ Dirac electron.

The Lagrangian density for the Dirac field, Eq. (12–116), is invariant under the same transformation if we replace ϕ and ϕ^* in Eqs. (12–160) by the four-element field variables ψ and ψ^*. Accordingly, there is a conserved current associated with the gauge invariance given by

$$\Theta_\mu = i \frac{\partial \mathscr{L}}{\partial \psi_{,\mu}} \psi,$$

inasmuch as $\mathscr{L}$ does not contain $\psi^\dagger{}_{,\mu}$. Following the pattern of the Klein–Gordon field, we would expect then that the Dirac field has an electric current density

$$j_\mu = iq(i\psi^\dagger\gamma_\mu)\psi = -q\psi^\dagger\gamma_\mu\psi.$$

Indeed, this is exactly the form for j_μ stated in Eq. (12–134) and used in exhibiting the interaction between the Dirac and the electromagnetic fields. The choice of Eq. (12–134), apparently pulled out of the air at that time, is now seen to be a consequence, via Noether's theorem, of the gauge invariance of the Lagrangian density.

It should be remarked that while Noether's theorem proves that a continuous symmetry property of the Lagrangian density leads to a conservation condition, the converse is not true. There appear to be conservation conditions that cannot correspond to any symmetry property. The most prominent examples at the moment are the fields that have soliton solutions, e.g., are described by the sine-Gordon equation or the Korteweg–deVries equation (cf. p. 577).

Consider, for example, the Lagrangian density for the sine-Gordon equation, Eq. (12–109). As x and t do not appear explicitly, the Lagrangian density is invariant under translations of space and time in the manner fulfilling the conditions of Noether's theorem. In addition, there is a symmetry under a Lorentz transformation (in x, t space). No other symmetry is apparent. One would therefore expect no more than three conserved quantities from application of Noether's theorem. Yet it has been demonstrated, by methods lying outside the Lagrangian description of fields, that there exists an infinite number of conserved quantities. That is to say, an infinite number of distinct functions F_i and G_i that are polynomials of ϕ and its derivatives can be found for which

$$\frac{dF_i}{dt} + \frac{dG_i}{dx} = 0,$$

so that the volume integrals of the F_i are constant in time. It appears that the presence of such an infinite set of conserved quantities is a necessary condition in order for the field to describe solitons. Research on solitons is in a vigorous state of development so that the last word has probably not yet been said on the relation between conserved quantities and the nature of the field.

Finally, we can easily deduce the version of Noether's theorem that should apply to discrete systems. Here the four coordinates of space time are no longer parametric variables on equal footing—the space coordinates revert to their status as mechanical variables (or functions thereof) and only time remains to fill the role of a parameter. The action integral instead of being a four-dimensional volume integral,

$$I = \int \mathscr{L}(dx_\mu),$$

is a one-dimensional integral in t:

$$I = \int L \, dt.$$

Instead of the continuously indexed field variables $\eta_\rho(x_\mu)$ we have the discrete generalized coordinates $q_k(t)$. It is straightforward enough to recapitulate with these translations the steps that led to Noether's theorem. One could repeat in this manner the arguments contained in Eqs. (12–139) through (12–156) as applied to discrete systems. But the effect of the conversion is sufficiently obvious and clear, that one can readily see the translation need be done directly only on the final result, Eq. (12–156).

The rules for the translation can be summarized as

$$\mathscr{L} \to L,$$
$$x_\mu \text{ or } x_\nu \to t,$$
$$\eta_\rho \to q_k,$$
$$\eta_{\rho,\nu} \to \dot{q}_k.$$

$$(12\text{–}162)$$

Further, all sums over four-valued Greek indices reduce to one term, in t. As a result, the transformations, Eq. (12–152), under which the Lagrangian is to exhibit form and scale invariance become

$$\delta t = \epsilon_r X_r, \qquad \delta q_k = \epsilon_r \Psi_{rk}. \qquad (12\text{–}163)$$

Equation (12–156), the statement of the conservation theorems resulting from the invariance, now becomes

$$\frac{d}{dt} \left\{ \left(\frac{\partial L}{\partial \dot{q}_k} \dot{q}_k - L \right) X_r - \frac{\partial L}{\partial \dot{q}_k} \Psi_{rk} \right\} = 0. \qquad (12\text{–}164)$$

Equation (12–164) is the statement of the conclusions of Noether's theorem for a discrete mechanical system.

The expression in the parentheses in Eq. (12–164) is our old friend the Jacobi integral h of Eq. (2–53), or equivalently in terms of (q, p), the Hamiltonian. Indeed, we can recover the conservation of h by considering a transformation that involves a displacement of time only:

$$X_r = \delta_{r1}, \qquad \Psi_{rk} = 0. \tag{12–165}$$

If the Lagrangian is not an explicit function of time, then clearly the form of the Lagrangian and the value of the action integral are unaffected by this transformation. But Noether's theorem, Eq. (12–64), then says that as a result there is a conservation theorem

$$\frac{d}{dt}\left(\frac{\partial L}{\partial \dot{q}_k}\dot{q}_k - L\right) = 0,$$

which is identical with the familiar conclusion of Section (2–6).

Suppose further that a particular coordinate q_l is cyclic. Then the Lagrangian and the action is invariant under a transformation for which

$$X_r = 0, \qquad \Psi_{rk} = \delta_{kl}\delta_{r1}. \tag{12–166}$$

Equation (12–164) then immediately implies the single conservation statement

$$\frac{d}{dt}\left(\frac{\partial L}{\partial \dot{q}_l}\right) = 0,$$

or

$$\dot{p}_l = 0.$$

Thus, the theorems on the conservation both of Jacobi's integral and of the generalized momentum conjugate to a cyclic coordinate are subsumed under Noether's theorem as stated in Eq. (12–164). The connection between symmetry properties of a mechanical system and conserved quantities has run as a thread throughout formulations of mechanics as presented here. Having come full circle, as it were, and rederived by sophisticated techniques symmetry theorems found in the first chapters, it seems an appropriate point at which to end our discussions.

SUGGESTED REFERENCES

J. C. SLATER AND N. H. FRANK, *Mechanics.* This somewhat elderly text still provides a prime reference for a readable, if at times elementary, discussion of the transition from discrete to continuous systems. In particular the path from the discrete chain to the continuous string is examined in Chapter VII with respect to transverse vibrations.

LORD RAYLEIGH, *The Theory of Sound.* This treatise naturally contains much material on the vibrations of continuous bodies. A discussion of the wave equation for the propagation of sound in gases will be found in Chapter XI, Volume 2, where the question of adiabatic versus isothermal motion of the gas is examined in great detail.

G. WENTZEL, *Introduction to the Quantum Theory of Fields.* Most monographs on the quantum theory of fields start with a discussion of classical fields and often provide the best references for the classical theory. Wentzel's book was one of the first. It is still worthy of reference because of the extent and lucidity of its treatment of the classical aspects. What the book has to say about the quantum formulation has long since been superseded however and is only of historical interest now. Minkowski space is used throughout.

A. O. BARUT, *Electrodynamics and Classical Theory of Fields and Particles.* Perhaps the best single reference to the classical formulation of fields, Barut's book packs a tremendous amount of information in relatively small compass—Lagrangians, field equations, stress-energy tensors, and conserved quantities for a variety of fields, etc. It attempts to give a covariant treatment of the Hamiltonian formulation, marked by considerable complexity. A number of oddities from the literature are collected here; e.g., spinor form of Maxwell's equations, which may be safely ignored at this introductory level. Noether's theorem is discussed, but not by name. (Parenthetically it may be noted that the treatment of Noether's theorem in Section 12–7 above was inspired in good part by an article by T. H. Boyer, *American Journal of Physics* 34, 475, June 1966.) It should be remembered that Barut uses the four-space with trace -2.

E. J. SALETAN AND A. H. CROMER, *Theoretical Mechanics.* Most recent texts on classical mechanics for physicists include a treatment of the classical theory of fields, and this is one of the best (their Chapter VIII). Four-space with trace -2 is used throughout so the formulas abound with metric tensors and raised or lowered indices (although the reasons for their particular version of the formalism are not always clear). Particularly noteworthy is the generalization of the momentum representation by a general expansion of the field in orthonormal functions.

J. D. JACKSON, *Classical Electrodynamics.* As has been mentioned, the second edition has converted from Minkowski space to one with trace -2. However the procedures for handling quantities in such a space are described in detail in Chapter 11 on special relativity. The classical field formalism is treated in Chapter 12, naturally with great emphasis on the electromagnetic field, which is discussed in all aspects. Especially noteworthy is a description of the Proca Lagrangian, which is a suggested form of the Lagrangian density for the electromagnetic field if the photon has a mass. The overall discussion is painstaking and distinguished by great clarity.

EXERCISES

1. a) The transverse vibrations of a stretched string can be approximated by a discrete system consisting of equally spaced mass points located on a weightless string. Show that if the spacing is allowed to go to zero, the Lagrangian approaches the limit

$$L = \frac{1}{2} \int \left[\mu \dot{\eta}^2 - T \left(\frac{\partial \eta}{\partial x} \right)^2 \right] dx$$

for the continuous string, where T is the fixed tension. What is the equation of motion if the density μ is a function of position?

b) Obtain the Lagrangian for the continuous string by finding the kinetic and potential energies corresponding to transverse motion. The potential energy can be obtained from the work done by the tension force in stretching the string in the course of the transverse vibration.

2. a) Describe the field of sound vibrations in a gas in the Hamiltonian formalism and obtain the corresponding Hamilton equations of motion.

b) Generalizing the momentum expansion to a vector field, express the Hamiltonian for the acoustic modes of a gas in the momentum representation.

3. Obtain Hamilton's equations of motion for a continuous system from the modified Hamilton's principle, following the procedure of Section 8–5.

4. Show that if ψ and ψ^* are taken as two independent field variables, the Lagrangian density

$$\mathscr{L} = \frac{h^2}{8\pi^2 m} \nabla\psi \cdot \nabla\psi^* + V\psi^*\psi + \frac{h}{4\pi i}(\psi^*\dot{\psi} - \psi\dot{\psi}^*)$$

leads to the Schrödinger equation

$$-\frac{h^2}{8\pi^2 m}\nabla^2\psi + V\psi = \frac{ih}{2\pi}\frac{\partial\psi}{\partial t},$$

and its complex conjugate. What are the canonical momenta? Obtain the Hamiltonian density corresponding to $\mathscr{L}$.

5. Show that

$$G_i = -\int \pi_k \frac{\partial\eta_k}{\partial x_i}\,dV$$

is a constant of the motion if the Hamiltonian density is not an explicit function of position. The quantity G_i can be identified as the total linear momentum of the field along the x_i direction. The similarity of this theorem with the usual conservation theorem for linear momentum of discrete systems is obvious.

6. a) In a four-space that is not Euclidean, the D'Alembertian is defined as

$$\Box^2 = g^{\mu\nu}\frac{\partial^2}{\partial x^\mu \partial x^\nu}.$$

Here $g^{\mu\nu}$ is the contravariant metric tensor, which in the flat space of special relativity is indeed the same as $g_{\mu\nu}$ (c.f. Section 7–3). For the metric tensor of trace -2, Eq. (7–53), find the explicit form of the D'Alembertian so defined.

b) A suitable Lagrangian for the charged scalar meson field in this metric is

$$\mathscr{L} = \frac{1}{2}\left(g^{\mu\nu}\frac{\partial\phi}{\partial x^\mu}\frac{\partial\phi^*}{\partial x^\nu} - \mu_0^2\phi\phi^*\right).$$

Show that one of the corresponding field equations is

$$(\Box^2 + \mu_0^2)\phi = 0.$$

Show also that in light of part (a) this equation is actually identical with Eq. (12–103).

7. To the Lagrangian density for the scalar charged meson, Eq. (12–100), add the following term to represent the interaction with an electromagnetic field:

$$\frac{j_\lambda A_\lambda}{c}$$

where

$$j_\lambda = i(\phi\phi^*_{,\lambda} - \phi_{,\lambda}\phi^*).$$

What are the field equations for ϕ and ϕ^*? What happens to the conserved currents and associated conservation theorems?

8. Suppose the Lagrangian density in Hamilton's principle is a function of higher derivatives of the field quantities η_ρ:

$$\mathscr{L} = \mathscr{L}(\eta_\rho, \eta_{\rho,\mu}, \eta_{\rho,\mu\nu}, x_\lambda).$$

Assuming the vanishing of the variation at the endpoints, what is the form of the field equations corresponding to such a Lagrangian density?

9. Consider a scalar field quantity η that, for simplicity, is a function only of x and t. Suppose now that the Hamiltonian density is a function of higher spatial derivatives of η and π, that is,

$$\mathscr{H} = \mathscr{H}(\eta, \eta_{,x}, \pi, \pi_{,x}, \pi_{,xx}).$$

What are the corresponding Hamilton equations of motion?

10. Show that the Korteweg–deVries equation corresponds to the field equation for a scalar field ψ with Lagrangian density

$$\mathscr{L} = \frac{1}{2}\psi_x\psi_t + \frac{\alpha}{6}\psi_x^3 - \frac{\nu}{2}\psi_{xx}^2,$$

where the subscripts indicate derivatives with respect to the variables indicated, provided ψ is a potential function for the quantity ϕ of Eq. (12–115):

$$\phi = \frac{\partial\psi}{\partial x}.$$

11. Consider a Hamiltonian density in (x, t) space:

$$\mathscr{H} = \eta^3 + \frac{1}{2}\eta^2_{,x} + \pi^3_{,x} + \frac{1}{2}\pi^2_{,xx}.$$

Show that the Hamilton equations of motion correspond to a form of the Korteweg–deVries equation, Eq. (12–115), if

$$\eta = \phi(x, t)$$

$$\pi = \int_{-\infty}^{x} \phi(x', t)\, dx'.$$

12. Evaluate explicitly iT_{j4}/c and T_{ij} for the symmetrized stress-energy tensor of the free electromagnetic field as given by Eq. (12–129). What can be said about the physical meaning of these components?

13. a) In a four-space with metric $g_{\mu\nu}$ of trace -2 as defined by Eq. (7–53) in Section 7–3, evaluate explicitly the elements of the covariant (mathematically speaking) tensor $F_{\mu\nu}$ of the electromagnetic field. Also give the elements of the matrix with one index lifted and with two indices lifted:

$$F^\lambda_{\ \rho} = g^{\lambda\mu}F_{\mu\nu}; \qquad F^{\lambda\rho} = g^{\lambda\mu}F_{\mu\nu}g^{\rho\nu}.$$

b) Show that in this metric a Lagrangian density leading to Maxwell's equations with external currents is given by

$$\mathscr{L} = -\frac{F^{\lambda\rho}F_{\lambda\rho}}{16\pi} - \frac{j^{\lambda}A_{\lambda}}{c}.$$

c) Find the elements of the symmetric stress-energy tensor in the same metric.

14. Equation (12–131) presents an alternative Lagrangian density for the electromagnetic field. Find the corresponding stress-energy tensor. Can you give physical meaning to the components? For the free field what are the conservation equations?

Proof of
Bertrand's Theorem*

The orbit equation under a conservative central force, Eq. (3–34), may be written

$$\frac{d^2u}{d\theta^2} + u = J(u),$$

(A–1)

where

$$J(u) = -\frac{m}{l^2}\frac{d}{du}V\left(\frac{1}{u}\right) = -\frac{m}{l^2u^2}f\left(\frac{1}{u}\right).$$

(A–2)

The condition for a circular orbit of radius $r_0 = u_0^{-1}$, Eq. (3–41), now takes the form

$$u_0 = J(u_0).$$

(A–3)

In addition, of course, the energy must satisfy the condition of Eq. (3–42). If the energy is slightly above that needed for circularity, and the potential is such that the motion is stable, then u will remain bounded and vary only slightly from u_0 and $J(u)$ can be expressed in terms of the first term in a Taylor series expansion about $J(u_0)$:

$$J(u) = u_0 + (u - u_0)\frac{dJ}{du_0} + O((u - u_0)^2).$$

(A–4)

As is customary, the derivative appearing in Eq. (A–4) is a shorthand symbol for the derivative of J with respect to u evaluated at $u = u_0$. If the difference $u - u_0$ is represented by x, the orbit equation for motion in the vicinity of the circularity conditions is then

$$\frac{d^2x}{d\theta^2} + x = x\frac{dJ}{du_0},$$

or

$$\frac{d^2x}{d\theta^2} + \beta^2x = 0,$$

(A–5)

---

* See Section 3–6.

where

$$\beta^2 = 1 - \frac{dJ}{du_0}. \tag{A-6}$$

In order for x to describe a bounded stable oscillation, β^2 must be positive definite. From the definition, Eq. (A–2), we have

$$\frac{dJ}{du} = \frac{2m}{l^2 u^3} f\left(\frac{1}{u}\right) - \frac{m}{l^2 u^2} \frac{d}{du'} f\left(\frac{1}{u}\right) = -\frac{2J}{u} - \frac{m}{l^2 u^2} \frac{d}{du'} f\left(\frac{1}{u}\right).$$

In view of the circularity conditions, Eq. (3–41) or Eq. (A–3), it then follows that

$$\frac{dJ}{du_0} = -2 + \frac{u_0}{f_0} \frac{df}{du_0},$$

where, in addition to the convention employed for the derivatives, f_0 stands for $f(1/u_0)$. Hence β^2 is given by

$$\beta^2 = 3 - \frac{u_0}{f_0} \frac{df}{du_0} = 3 + \frac{r}{f} \frac{df}{dr}\bigg|_{r=r_0}, \tag{A-7}$$

which is the same as Eq. (3–46), and the stability condition, $\beta^2 > 0$, thus reduces to Eq. (3–43).

By a suitable choice of origin of θ, the solution to Eq. (A–5) for β^2 positive definite can be written

$$x = a \cos \beta\theta \tag{A-8}$$

(cf. Eq. (3–45)). In order for the orbit to remain closed when the energy and angular momentum are thus slightly disturbed from circularity, the quantity β must be a rational number. We are concerned with finding force laws such that for a wide range of initial conditions, i.e., for a wide range of u_0, the orbits deviating slightly from circularity remain closed. Under these circumstances, as argued in the main text, β must have the same value over the entire domain of u_0, and Eq. (A–7) can be looked on as a differential equation for $f(1/u)$ or $f(r)$. The desired force law must therefore conform to a dependence on r given in Eq. (3–48):

$$f(r) = -\frac{k}{r^{3-\beta^2}}, \tag{A-8}$$

where k is some constant and β is a rational number.

The force law of Eq. (A–8) still permits a wide variety of behavior for the force. However we seek more stringent conditions on the force law, by requiring that even when the deviations from circularity are considerable the orbit remain closed. We must therefore at least deal with deviations of u from u_0 so large that we must keep more terms than the first in the Taylor series expansion of $J(u)$. Equation (A–4) may therefore be replaced by

$$J(u) = u_0 + xJ' + \frac{x^2}{2}J'' + \frac{x^3}{6}J''' + O(x^4), \tag{A-9}$$

where it is understood the derivatives are evaluated at $u = u_0$. In terms of this expansion of $J(u)$ the orbit equation becomes

$$\frac{d^2x}{d\theta^2} + \beta^2 x = \frac{x^2 J''}{2} + \frac{x^3 J'''}{6}. \qquad \text{(A–10)}$$

We seek to find the nature of the source law such that even when the deviation from the circular orbit, x, is large enough that the terms on the right cannot be neglected, the solution to Eq. (A–10) still represents a closed orbit. For small perturbations from circularity we know x has the behavior described by Eq. (A–8), which represents the fundamental term in a Fourier expansion in terms of $\beta\theta$. We seek therefore a closed-orbit solution by including a few more terms in the Fourier expansion:

$$x = a_0 + a_1 \cos \beta\theta + a_2 \cos 2\beta\theta + a_3 \cos 3\beta\theta. \qquad \text{(A–11)}$$

The amplitudes a_0 and a_2 must be of smaller magnitude than a_1 because they vanish faster than a_1 as circularity is approached. As will be seen a_3 must be of even lower order of magnitude than a_0 or a_2, which is why terms in $\cos 4\beta\theta$ and so on can be neglected. Consequently in the x^2 term on the right of Eq. (A–10) the factors in $\cos 3\beta\theta$ are dropped, and in the x^3 term only factors in $\cos \beta\theta$ are kept. In evaluating the right-hand side powers and products of the cosine functions are reduced by means of such identities as

$$\cos \beta\theta \cos 2\beta\theta = \tfrac{1}{2}(\cos \beta\theta + \cos 3\beta\theta)$$

and

$$\cos {}^3\beta\theta = \tfrac{1}{4}(3 \cos \beta\theta + \cos 3\beta\theta).$$

Consistently keeping terms through the order of a_1^3 in this manner, Eq. (A–10) with the solution (A–11) can be reduced to

$$\beta^2 a_0 - 3\beta^2 a_2 \cos 2\beta\theta - 8\beta^3 a_3 \cos 3\beta\theta$$

$$= \frac{a_1^2}{4} J'' + \left[\frac{2a_1 a_0 + a_1 a_2}{2} J'' + \frac{J''' a_1^3}{8} \right] \cos \beta\theta \qquad \text{(A–12)}$$

$$+ \frac{a_1^2}{4} J'' \cos 2\beta\theta + \left[\frac{a_1 a_2}{2} J'' + \frac{J''' a_1^3}{24} \right] \cos 3\beta\theta.$$

In order for the solution to be valid, the coefficient of each cosine term must separately vanish, leading to four conditions on the amplitude and the derivatives of J:

$$a_0 = \frac{a_1^2 J''}{4\beta^2}, \qquad \text{(A–13a)}$$

$$a_2 = -\frac{a_1^2 J''}{12\beta^2}, \qquad \text{(A–13b)}$$

$$0 = \frac{2a_1 a_0 + a_1 a_2}{2} J'' + \frac{J''' a_1^3}{8}, \tag{A-13c}$$

$$a_3 = -\frac{1}{8\beta^3}\left[\frac{a_1 a_2}{2} J'' + \frac{J''' a_1^3}{24}\right]. \tag{A-13d}$$

It should be remembered that we already have shown, on the basis of slight deviations from circularity, that for closed orbits the force law must have the form of Eq. (A–8) or that $J(u)$ is given by

$$J = +\frac{mk}{l^2}u^{1-\beta^2}. \tag{A-14}$$

Keeping in mind the circularity condition, Eq. (A–3), the various derivatives at u_0 can be evaluated as

$$J'' = \frac{\beta^2(1-\beta^2)}{u_0} \tag{A-15a}$$

and

$$J''' = \frac{-\beta^2(1-\beta^2)(1+\beta^2)}{u_0^2}. \tag{A-15b}$$

Equations (A–13a, b) thus say that a_0/a_1 and a_2/a_1 are of the order of a_1/u_0, which is by supposition a small number. Further, Eq. (A–13d) shows that a_3/a_1 is of the order of $(a_1/u_0)^2$, which justifies the earlier statement that a_3 is of lower order of magnitude than a_0 or a_2.

Equation (A–13c) is a condition on β only, the condition that in fact is the principal conclusion of Bertrand's theorem. Substituting Eqs. (A–13a, b) and (A–15) into Eq. (A–13c) yields the condition

$$\beta^2(1-\beta^2)(4-\beta^2) = 0. \tag{A-16}$$

For deviations from a circular orbit, that is, $\beta \neq 0$, the only solutions are

$$\beta^2 = 1, \qquad f(r) = -\frac{k}{r^2} \tag{A-17a}$$

and

$$\beta^2 = 4, \qquad f(r) = -kr. \tag{A-17b}$$

Thus the only two possible force laws consistent with the solution are either the gravitational inverse-square law or Hooke's law!

We started out with orbits that were circular. These are possible for all attractive force laws over a wide range of l and E, whose values in turn fix the orbital radius. The requirement that the circular orbit be *stable* for all radii already restricts the form of the force law through the inequality condition of $\beta^2 > 0$ (Eq. (3–48)). If we further seek force laws such that orbits that deviate only slightly from a circular orbit are still closed, no matter what the radius of the reference orbit, then the force laws are restricted to the discrete set given by Eq. (A–8) with rational values of β. In order for the orbits to remain closed for larger deviations from circularity, no matter what the initial conditions of the reference orbit, only two of these rational values are permitted: $|\beta| = 1$ and $|\beta| = 2$. Since we know that these attractive force laws do in fact give closed orbits for all E and l leading to bounded motion, they must be the *only* force laws leading to closed orbits for all bounded motion.

APPENDIX B

Euler Angles in Alternate Conventions*

The Euler angles as defined in the text are specified by an initial rotation about the original z axis through an angle ϕ, a second rotation about the intermediate x axis through an angle θ, and a third rotation about the final z axis through an angle ψ. This sequence is here denoted as the "x convention," referring to the choice of the second rotation. Other conventions are possible, and two in particular have found frequent application in particular fields. Formulas will be given here for properties of a general rotation in terms of the Euler angles of these two alternate conventions.

y CONVENTION

As mentioned in the text, this convention has become almost standard practice in quantum mechanics and related fields. It differs from the x convention only in that the second rotation is about the intermediate y axis. Transcription from the x to the y convention is particularly simple because θ retains its meaning in both conventions and the changes for the other angles are easily obtained. In the x convention, ϕ is the angle between the line of nodes and the x axis; in the y convention, it is the same angle measure to the y axis. Similarly in the x convention, ψ is the angle between the line of nodes and the x' axis; while in the y convention, it is the same angle relative to the y' axis. Temporarily using subscripts to indicate the convention used, these relations imply the connection (cf. Fig. 4–7)

$$\phi_x = \phi_y + \frac{\pi}{2}$$

$$\psi_x = \psi_y - \frac{\pi}{2}$$

(B–1y)

or

$$\sin \phi_x = \cos \phi_y, \qquad \sin \psi_x = -\cos \psi_y,$$

$$\cos \phi_x = -\sin \phi_y, \qquad \cos \psi_x = \sin \psi_y.$$

(B–2y)

---

* See Section 4–4.

With this recipe one obtains the following formulas in terms of the Euler angles in the y convention:

Rotation matrix:

$$A = \begin{pmatrix} -\sin\psi\sin\phi + \cos\theta\cos\phi\cos\psi & \sin\psi\cos\phi + \cos\theta\sin\phi\cos\psi & -\cos\psi\sin\theta \\ -\cos\psi\sin\phi - \cos\theta\cos\phi\sin\psi & \cos\psi\cos\phi - \cos\theta\sin\phi\sin\psi & \sin\psi\sin\theta \\ \sin\theta\cos\psi & \sin\theta\sin\psi & \cos\theta \end{pmatrix}$$

$$\text{(B-3y)}$$

The same result can be obtained by noting that the exchange of y for x corresponds to a rotation of the reference frames about the z axis through an angle of $-\pi/2$ or $3\pi/2$. One can therefore translate the A matrix from x convention to y convention by a similarity transformation by the orthogonal matrix G:

$$G = \begin{pmatrix} 0 & -1 & 0 \\ 1 & 0 & 0 \\ 0 & 0 & 1 \end{pmatrix},$$

again leading to Eq. (B–3).

Cayley–Klein parameters. As before, the Q matrix can be obtained as the product of the Q matrices for each rotation, the only difference being in the Q_θ matrix, which is now

$$Q_\theta = e^{i\sigma_2\frac{\theta}{2}} = 1\cos\frac{\theta}{2} + i\sigma_2\sin\frac{\theta}{2}$$

$$= \begin{pmatrix} \cos\frac{\theta}{2} & \sin\frac{\theta}{2} \\ -\sin\frac{\theta}{2} & \cos\frac{\theta}{2} \end{pmatrix}. \qquad \text{(B-4y)}$$

From the matrix product $Q = Q_\psi Q_\theta Q_\phi$ (or by the translation equations (B–3)) one then obtains the Cayley–Klein parameters as

$$\alpha = e^{i\left(\frac{\psi + \phi}{2}\right)}\cos\frac{\theta}{2}, \qquad \beta = e^{i\left(\frac{\psi - \phi}{2}\right)}\sin\frac{\theta}{2},$$

$$\text{(B-5y)}$$

$$\gamma = -e^{-i\left(\frac{\psi - \phi}{2}\right)}\sin\frac{\theta}{2}, \qquad \delta = e^{-i\left(\frac{\psi + \phi}{2}\right)}\cos\frac{\theta}{2}.$$

Euler Parameters. It immediately follows from Eq. (4–65) and Eq. (B–5y) that in the y convention the Euler parameters are given by

$$e_0 = \cos\frac{\psi + \phi}{2}\cos\frac{\theta}{2}, \qquad e_2 = \cos\frac{\psi - \phi}{2}\sin\frac{\theta}{2},$$

$$e_1 = \sin\frac{\psi - \phi}{2}\sin\frac{\theta}{2}, \qquad e_3 = \sin\frac{\psi + \phi}{2}\cos\frac{\theta}{2}. \tag{B–6y}$$

Components of angular velocity. Either by direct use of the translation equations, (B–2), or by following through the physical meanings of the component parts of $\boldsymbol{\omega}$, one can obtain the following components of $\boldsymbol{\omega}$ along the body axes in the y convention:

$$\omega_{x'} = -\dot{\phi}\sin\theta\cos\psi + \dot{\theta}\sin\psi,$$

$$\omega_{y'} = \dot{\phi}\sin\theta\sin\psi + \dot{\theta}\cos\psi, \tag{B–7y}$$

$$\omega_{z'} = \dot{\phi}\cos\theta + \dot{\psi}.$$

Similarly the components of $\boldsymbol{\omega}$ along the space axes are

$$\omega_x = -\dot{\theta}\sin\phi + \dot{\psi}\sin\theta\cos\phi,$$

$$\omega_y = \dot{\theta}\cos\phi + \dot{\psi}\sin\theta\sin\phi, \tag{B–8y}$$

$$\omega_z = \dot{\psi}\cos\theta + \dot{\phi}.$$

Finally, note that

$$\cos\left(\frac{\Phi}{2}\right) = e_0 = \cos\frac{\psi + \phi}{2}\cos\frac{\theta}{2}, \tag{B–9y}$$

which is the same as Eq. (4–97) for x convention.

xyz CONVENTION

In this convention each rotation is about a differently labeled axis. Obviously, various sequences of rotations are still possible. It appears that most U.S. and British aerodynamicists prefer* the sequence in which the first rotation is the *yaw* angle ϕ about a z axis, the second is the *pitch* angle θ about an intermediary y axis, and the third is a *bank* or *roll* angle ψ about the final x axis (or figure axis of the vehicle).† Of the three elementary rotation matrices **D** remains the same as Eq. (4–43), **C** appears as

$$\mathbf{C} = \begin{pmatrix} \cos\theta & 0 & -\sin\theta \\ 0 & 1 & 0 \\ \sin\theta & 0 & \cos\theta \end{pmatrix}, \tag{B–10xyz}$$

---

* See R. L. Pio, "Euler Angle Transformations," *IEEE Transactions on Automatic Control* AC11, 707 (1966).

† In the engineering literature this is sometimes referred to as the 321 sequence, i.e., the first rotation is about the 3 axis, etc.

and **B** is the same as Eq. (4–44) (with ψ in place of θ, of course). The product **BCD** gives the

Rotation matrix

$$
\mathbf{A} = \begin{vmatrix}
\cos\theta\cos\phi & \cos\theta\sin\phi & -\sin\theta \\
\sin\psi\sin\theta\cos\phi - \cos\psi\sin\phi & \sin\psi\sin\theta\sin\phi + \cos\psi\cos\phi & \cos\theta\sin\psi \\
\cos\psi\sin\theta\cos\phi + \sin\psi\sin\phi & \cos\psi\sin\theta\sin\phi - \sin\psi\cos\phi & \cos\theta\cos\psi
\end{vmatrix}
$$

(B–11xyż)

Cayley–Klein parameters. The $\mathbf{Q}$ matrix is given by

$$
\mathbf{Q} = \mathbf{Q}_\psi \mathbf{Q}_\theta \mathbf{Q}_\phi = \begin{vmatrix} \cos\dfrac{\psi}{2} & i\sin\dfrac{\psi}{2} \\ i\sin\dfrac{\psi}{2} & \cos\dfrac{\psi}{2} \end{vmatrix} \begin{vmatrix} \cos\dfrac{\theta}{2} & \sin\dfrac{\theta}{2} \\ -\sin\dfrac{\theta}{2} & \cos\dfrac{\theta}{2} \end{vmatrix} \begin{vmatrix} e^{i\phi/2} & 0 \\ 0 & e^{-i\phi/2} \end{vmatrix}.
$$

On carrying out the multiplication it is found that the Cayley–Klein parameters have the form

$$
\alpha = \delta^* = \left(\cos\frac{\psi}{2}\cos\frac{\theta}{2} - i\sin\frac{\psi}{2}\sin\frac{\theta}{2} \right) e^{i\phi/2}
$$

(B–12 xyz)

$$
\beta = -\gamma^* = \left(\cos\frac{\psi}{2}\sin\frac{\theta}{2} + i\sin\frac{\psi}{2}\cos\frac{\theta}{2} \right) e^{-i\phi/2}.
$$

Euler parameters. From Eqs. (B–11xyz) it follows that the Euler parameters are

$$
\cos\frac{\Phi}{2} = e_0 = \cos\frac{\psi}{2}\cos\frac{\theta}{2}\cos\frac{\phi}{2} + \sin\frac{\psi}{2}\sin\frac{\theta}{2}\sin\frac{\phi}{2},
$$

$$
e_1 = \sin\frac{\psi}{2}\cos\frac{\theta}{2}\cos\frac{\phi}{2} - \cos\frac{\psi}{2}\sin\frac{\theta}{2}\sin\frac{\phi}{2},
$$

(B–13 xyz)

$$
e_2 = \cos\frac{\psi}{2}\sin\frac{\theta}{2}\cos\frac{\phi}{2} + \sin\frac{\psi}{2}\cos\frac{\theta}{2}\sin\frac{\phi}{2},
$$

$$
e_3 = -\sin\frac{\psi}{2}\sin\frac{\theta}{2}\cos\frac{\phi}{2} + \cos\frac{\psi}{2}\cos\frac{\theta}{2}\sin\frac{\phi}{2}.
$$

Note that the cosine of the total angle of rotation now has a different form from either the x or the y convention.

Components of angular velocity. Clearly ω_ψ lies along the body x axis, ω_ϕ along the space z axis, and ω_θ along the intermediate y axis, and therefore in the final yz plane. The resulting components along body axes are

$$
\omega_{x'} = \dot{\psi} - \dot{\phi}\sin\theta,
$$

$$
\omega_{y'} = \dot{\theta}\cos\psi + \dot{\phi}\cos\theta\sin\psi,
$$

(B–14xyz)

$$
\omega_{z'} = -\dot{\theta}\sin\psi + \dot{\phi}\cos\theta\cos\psi.
$$

Similarly the components of $\boldsymbol{\omega}$ along the space axes are

$$\omega_x = \dot{\psi} \cos \theta \cos \phi - \dot{\theta} \sin \phi,$$

$$\omega_y = \dot{\psi} \cos \theta \sin \phi + \dot{\theta} \cos \phi, \qquad \text{(B–15xyz)}$$

$$\omega_z = \dot{\phi} - \dot{\psi} \sin \theta.$$

APPENDIX C

Transformation Properties of $d\Omega$*

The components $d\Omega_j$ are the elements of the 3×3 antisymmetric matrix ϵ, Eq. (4–105), and are formally given in terms of the permutation symbol by the relation

$$\epsilon_{mn} = \epsilon_{mnj} \, d\Omega_j. \tag{C–1}$$

Under an orthogonal transformation whose matrix is **B**, the matrix ϵ transforms by a similarity transformation

$$\boldsymbol{\epsilon}' = \mathbf{B}\boldsymbol{\epsilon}\mathbf{B}^{-1} = \mathbf{B}\boldsymbol{\epsilon}\tilde{\mathbf{B}}, \tag{C–2}$$

with typical component

$$\epsilon'_{kl} = b_{km} b_{ln} \epsilon_{mn}. \tag{C–3}$$

The antisymmetry property is preserved under a similarity transformation; hence the elements of ϵ' can also be written as

$$\epsilon'_{kl} = \epsilon_{kli} \, d\Omega'_i. \tag{C–4}$$

With these representations of the antisymmetric matrices the transformation properties of $d\Omega_j$ are thus given by

$$\epsilon_{kli} \, d\Omega'_i = b_{km} b_{ln} \epsilon_{mnj} \, d\Omega_j. \tag{C–5}$$

If **A** is any 3×3 matrix, then the determinant of **A** is given by†

$$|\mathbf{A}| = \epsilon_{pqi} a_{pm} a_{qn} a_{ij}, \tag{C–6}$$

---

* See Section 4–8.

† Most books on linear algebra or "college algebra" give the equivalent of this expansion for the determinant; e.g., R. R. Stoll, *Linear Algebra and Matrix Theory*, p. 92; S. Lipschutz, *Theory and Problems of Linear Algebra*, p. 172; and G. Strang, *Linear Algebra and its Applications*, p. 157. Indeed, many of the older books use Eq. (C–6) as the definition of a determinant, e.g., the comprehensive monograph by H. W. Turnbull, *The Theory of Determinants, Matrices and Invariants*, p. 12. (The more modern approach is to define a determinant in terms of its manipulative properties.) See also Chapter 4 in G. Arfken, *Mathematical Methods for Physicists*; T. C. Bradbury, Theoretical Mechanics, pp. 26–28; and G. Goertzel and N. Tralli, *Some Mathematical Methods of Physics*, Appendix 1A.

providing *mnj* is a cyclic permutation of 1, 2, 3. Interchange of *m* with *n*, so that the permutation is not cyclic, clearly changes the sign of the expression, for the operation is equivalent to interchanging two columns in a determinant. Further, the expression vanishes if any two of the indices *mnj* are equal, for a determinant with two equal columns is zero. These properties with respect to the indices are exactly those of the permutation symbol. Hence Eq. (C–6) can be written more generally as

$$\epsilon_{mnj}|\mathbf{A}| = \epsilon_{pqi}a_{pm}a_{qn}a_{ij}, \tag{C–7}$$

without restriction on the indices *mnj*. We can apply Eq. (C–7) to the orthogonal matrix **B**, and remembering that the square of the determinant of an orthogonal matrix is always +1, we obtain

$$\epsilon_{mnj} = \epsilon_{pqi}b_{pm}b_{qn}b_{ij}|\mathbf{B}|. \tag{C–8}$$

Equation (C–5) can now be rewritten as

$$\epsilon_{kli}\,d\Omega'_i = \epsilon_{pqi}b_{km}b_{pm}b_{ln}b_{qn}b_{ij}|\mathbf{B}|. \tag{C–9}$$

By the orthogonality property of **B** we have

$$b_{km}b_{pm} = \delta_{kp}, \qquad b_{ln}b_{qn} = \delta_{lq},$$

so that Eq. (C–9) reduces to

$$\epsilon_{kli}\,d\Omega'_i = \epsilon_{kli}|\mathbf{B}|b_{ij}\,d\Omega_j. \tag{C–10}$$

It follows that

$$d\Omega'_i = |\mathbf{B}|b_{ij}\,d\Omega_j, \tag{C–11}$$

which is the result, Eq. (4–110), that we wanted to prove.

The Staeckel Conditions for Separability of the Hamilton–Jacobi Equation

We show here that the Staeckel conditions (see above p. 453) provide sufficient conditions for the separability of the Hamilton–Jacobi equation, i.e., if they are satisfied the variables in the Hamilton–Jacobi equation can be separated. A proof of the necessity of the Staeckel conditions (within certain limits) will be found in Pars.* The proof given here roughly follows the procedure outlines by Garfinkel.†

From Eqs. (10–44) and (10–45) it follows that the Hamiltonian for the system can be written in the form

$$H = \frac{1}{2} \sum_i \frac{(p_i - a_i)^2}{T_{ii}} + V(q). \tag{D-1}$$

As the system is conservative, both forms of the Hamilton–Jacobi equation, with the help of Eq. (10–46), reduce to

$$\frac{1}{2} \sum_i \frac{1}{T_{ii}} \left[\left(\frac{\partial W}{\partial q_i} - a_i \right)^2 + 2V_i(q_i) \right] = \alpha_1, \tag{D-2}$$

where α_1 is the first of n constants of integration. Equation (D–2) can be written compactly by defining a number of n-dimensional vectors:

$$\mathbf{b} \quad \text{with elements} \quad \frac{1}{T_{ii}},$$

$$\mathbf{c} \quad \text{with elements} \quad \left(\frac{\partial W}{\partial q_i} - a_i \right)^2,$$

$$\mathbf{V} \quad \text{with elements} \quad V_i(q_i).$$

Equation (D–2) now appears as the scalar equation (in matrix notation)

$$\mathbf{b}(\mathbf{c} + 2\mathbf{V}) = 2\alpha_1. \tag{D-3}$$

---

* L. A. Pars, *A Treatise on Analytical Dynamics*, 1965, pp. 321–323.

† B. Garfinkel in *Space Mathematics, Part I*, (J. B. Rosser, ed.) 1966, pp. 52–45. (There are some obvious misprints: a factor of 2 in Eq. (82) and a minus sign in Eq. (87).)

The last of the Staeckel conditions, Eq. (10–47), can also be written in matrix form by defining a vector $\boldsymbol{\delta}_1$ with elements

$$(\boldsymbol{\delta}_1)_j = \delta_{1j}, \tag{D–4}$$

that is, the first element is 1 and all the rest are zero. Then Eq. (10–47) appears in our present notation as

$$\mathbf{b} = \boldsymbol{\delta}_1 \boldsymbol{\phi}^{-1}. \tag{D–5}$$

It is now claimed that a solution to (D–3) is contained in the form

$$\mathbf{c} + 2\mathbf{V} = 2\boldsymbol{\phi}\boldsymbol{\gamma}, \tag{D–6}$$

where $\boldsymbol{\gamma}$ is a vector with constant, but otherwise unspecified, elements. For Eq. (D–3) to hold, it must then follow that

$$\mathbf{b}\boldsymbol{\phi}\boldsymbol{\gamma} = \alpha_1,$$

or by Eq. (D–5)

$$\boldsymbol{\delta}_1\boldsymbol{\gamma} = \alpha_1. \tag{D–7}$$

The nature of the $\boldsymbol{\delta}_1$ vector, as defined by Eq. (D–4), is such that Eq. (D–7) reduces to

$$\gamma_1 = \alpha_1, \tag{D–7$'$}$$

that is, the first element of $\boldsymbol{\gamma}$ must be the integration constant α_1.

Equation (D–6) is equivalent to the set of equations

$$\left(\frac{\partial W}{\partial q_i} - a_i(q_i)\right)^2 = -2V_i(q_i) + 2\sum_j \phi_{ij}(q_i)\gamma_j. \tag{D–8}$$

As all other terms in the ith equation depend only on q_i, a complete solution for W can be found in a separated form,

$$W(q_1,\ldots,q_n) = \sum_i W_i(q_i), \tag{D–9}$$

so that we finally have

$$\left(\frac{\partial W_i(q_i)}{\partial q_i} - a_i(q_i)\right)^2 = -2V_i(q_i) + 2\sum_i \phi_{ij}(q_i)\gamma_j, \tag{D–10}$$

which is equivalent to Eq. (10–48).

Note that the matrix $\boldsymbol{\phi}$ depends only on the diagonal elements of the $\mathbf{T}$ matrix and on the nature of the coordinate system, and not at all on the "potential vector" $\mathbf{V}$. As an example, consider a suitable form for $\boldsymbol{\phi}$ to be used in problems with a single particle in spherical polar coordinates. We can deduce the elements of $\boldsymbol{\phi}$ by considering a simple problem of this nature in which the Hamilton–Jacobi equation is directly separable. Motion in space under the influence of a central force provides such a problem. If we designate the r, θ, and ϕ

coordinates by the subscript indices 1, 2, and 3, respectively, then the separated equations (10–59), (10–58), and (10–56) can be written as

$$\left(\frac{\partial W_1}{\partial r}\right)^2 = -2mV(r) + 2m\alpha_1 - \frac{\alpha_2}{r^2}, \tag{D–11a}$$

$$\left(\frac{\partial W_2}{\partial \theta}\right)^2 = \alpha_2^2 - \frac{\alpha_1}{\sin^2 \theta}, \tag{D–11b}$$

$$\left(\frac{\partial W_3}{\partial \phi}\right)^2 = \alpha_3^2. \tag{D–11c}$$

Because we know that each element of the first row of $\boldsymbol{\phi}^{-1}$ is to have the factor $1/m$, then we would expect that each element of $\boldsymbol{\phi}$ should have the factor m. Comparison of Eqs. (D–11) with Eqs. (D–10) suggest then the following elements for γ and $\boldsymbol{\phi}$:

$$\gamma_1 = \alpha_1, \qquad \gamma_2 = \frac{\alpha_2}{2m}, \qquad \gamma_3 = \frac{\alpha_3^2}{2m}, \tag{D–12}$$

and

$$\boldsymbol{\phi} = m \begin{pmatrix} 1 & -\dfrac{1}{r^2} & 0 \\ 0 & 1 & -\csc^2 \theta \\ 0 & 0 & 1 \end{pmatrix}. \tag{D–13}$$

Direct evaluation then shows that the inverse matrix $\boldsymbol{\phi}^{-1}$ is

$$\boldsymbol{\phi}^{-1} = \frac{1}{m} \begin{pmatrix} 1 & \dfrac{1}{r^2} & \dfrac{1}{r^2 \sin^2 \theta} \\ 0 & 1 & \csc^2 \theta \\ 0 & 0 & 1 \end{pmatrix}, \tag{D–14}$$

and it is clear that both $\boldsymbol{\phi}$ and $\boldsymbol{\phi}^{-1}$ have the properties required of them.

APPENDIX E

Lagrangian Formulation of the Acoustic Field in Gases

To study the field equations describing sound vibrations in gases, the displacement of each particle of the gas from its normal position will be denoted by the vector $\boldsymbol{\eta}$, with components η_i, $i = 1, 2, 3$. Each point xyz in space will thus have three generalized coordinates associated with it. It will be assumed that the disturbance is always small, so that the pressure P and density μ differ only slightly from their equilibrium values P_0 and μ_0, respectively.

In a discrete system the problem is set up in the Lagrangian formulation by finding the kinetic and potential energies and writing the Lagrangian as the difference of these quantities. Here the Lagrangian we seek is the volume integral of a density $\mathscr{L}$. The kinetic and potential energies can similarly be obtained as volume integrals of densities $\mathscr{T}$ and $\mathscr{V}$ respectively, with the relation

$$\mathscr{L} = \mathscr{T} - \mathscr{V}. \tag{E–1}$$

The kinetic energy density presents no problem; bearing in mind that the displacements from equilibrium are small, we have

$$\mathscr{T} = \frac{\mu_0}{2}\dot{\boldsymbol{\eta}}^2 = \frac{\mu_0}{2}(\dot{\eta}_1^2 + \dot{\eta}_2^2 + \dot{\eta}_3^2). \tag{E–2}$$

To obtain the potential energy density is a more difficult task. The potential energy of the gas is a measure of the work the gas can do in expanding against the pressure. Essentially, it arises from what the seventeenth-century scientists were fond of calling the "spring" of the gas. Consider a mass of gas M with equilibrium volume

$$V_0 = \frac{M}{\mu_0} \tag{E–3}$$

sufficiently small that $\mathscr{V}$ is constant over the volume. Then $\mathscr{V}V_0$ represents the potential energy of the quantity of gas. As a result of the sound disturbance, the volume changes from V_0 to $V_0 + \Delta V$. Now, in a change in volume dV the work

performed on the system, i.e., the increase in the potential energy, is $-P\,dV$.*
Hence the potential energy corresponding to a volume change from V_0 to
$V_0 + \Delta V$ is

$$\mathscr{V} V_0 = -\int_{V_0}^{V_0+\Delta V} P\,dV.$$

It might be thought that since ΔV is small, the integral can be approximated by
$P_0\,\Delta V$. As we shall see, this term actually does not contribute to the equations of
motion. It is therefore necessary to go to the next approximation, in which the
curve of P vs. V is replaced by a straight line in the region from V_0 to $V_0 + \Delta V$ (cf.
Fig. E–1):

$$\int_{V_0}^{V_0+\Delta V} P\,dV = P_0\,\Delta V + \frac{1}{2}\left(\frac{\partial P}{\partial V}\right)_0 (\Delta V)^2. \tag{E-4}$$

To evaluate the derivative of P with respect to V, we must digress for a
moment into thermodynamics. The first inclination might be to use Boyle's law,

$$PV = C, \tag{E-5}$$

for the relation between the pressure and the volume, and this was the procedure
followed by Newton. It leads to the wrong result, however, because Eq. (E–5)
assumes that the changes in pressure and volume occur *isothermally*. Actually, the
vibrations of sound are almost always so rapid that there is no time for conduction
to remove the heat developed and equalize the temperatures. The contractions and
expansions instead take place *adiabatically*, i.e., without loss of heat. Under these
conditions the relation between P and V is

$$PV^\gamma = C, \tag{E-6}$$

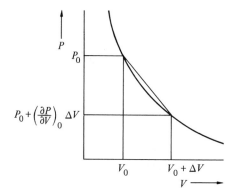

FIGURE E–1
Pressure-volume diagram for a gas.

---

* The customary elementary derivation is as follows. The force exerted on an element of
surface dA by the external system is $P\,dA$, pointing inwards. In expanding, the surface moves
a distance dx outward along the normal, and the external work done is $-P\,dA\,dx = -P\,dV$.

where γ is the constant ratio of the specific heats at constant pressure and volume.* Hence the desired derivative is

$$\left(\frac{\partial P}{\partial V}\right)_0 = -\frac{\gamma P_0}{V_0}. \tag{E-7}$$

It is convenient to express the change in volume in terms of the associated density change. Since $V = M/\mu$, the change in V is given by

$$\Delta V = -\frac{M}{\mu_0^2}\Delta\mu = -V_0\sigma, \tag{E-8}$$

where the fractional change in the density has been denoted by σ:

$$\mu = \mu_0(1 + \sigma). \tag{E-9}$$

Combining Eqs. (E–3, 4, 7, and 9) the potential energy density appears as

$$\mathcal{V} = P_0\sigma + \frac{\gamma P_0}{2}\sigma^2. \tag{E-10}$$

This is still not in the form useful for the Lagrangian; we have yet to express σ in terms of $\boldsymbol{\eta}$. Consider any finite volume V in space. The mass flowing out of this volume due to the small disturbance from equilibrium is given by

$$\mu_0 \int \boldsymbol{\eta} \cdot d\mathbf{A},$$

evaluated over the surface of the volume. The volume integral of the change in density must be exactly equal to this mass transport:

$$-\mu_0 \int \sigma \, dV = \mu_0 \int \boldsymbol{\eta} \cdot d\mathbf{A}. \tag{E-11}$$

By the divergence theorem the relation (E–11) can be written

$$-\int \sigma \, dV = \int \nabla \cdot \boldsymbol{\eta} \, dV,$$

and since the equality holds for any arbitrary volume, we must have†

$$\sigma = -\nabla \cdot \boldsymbol{\eta}. \tag{E-12}$$

---

* For derivation see M. W. Zemansky, *Heat and Thermodynamics*, 5th ed. (New York: McGraw-Hill, 1968), Section 5–5.

† Equation (E–12) may be recognized from its more familiar form,

$$\dot{\mu} = -\nabla \cdot \mu\dot{\boldsymbol{\eta}},$$

as the *equation of continuity* for the gas flow.

With this connection, the final form of the potential energy density is

$$\mathscr{V} = -P_0 \nabla \cdot \boldsymbol{\eta} + \frac{\gamma P_0}{2} (\nabla \cdot \boldsymbol{\eta})^2. \tag{E–13}$$

It can now be seen that the term in $\mathscr{V}$ linear in σ cannot contribute to the total potential energy. By Eq. (E–11) the volume integral of σ is minus the surface integral of $\boldsymbol{\eta}$, and if the surface completely encloses the system this must be zero, i.e., there is no transport of mass out of the system. That this term has a vanishing contribution to L is not yet sufficient reason to omit it from $\mathscr{V}$. Conceivably, the functional behavior of the term might still have an effect on the equation of motion. (It will be remembered that the covariant Hamiltonian of a system may be zero, but the equations of motion, of course, do not vanish.) The term will therefore be retained for the moment. The complete Lagrangian density can therefore be written as:

$$\mathscr{L} = \frac{1}{2} (\mu_0 \dot{\boldsymbol{\eta}}^2 + 2P_0 \nabla \cdot \boldsymbol{\eta} - \gamma P_0 (\nabla \cdot \boldsymbol{\eta})^2), \tag{E–14}$$

which is Eq. (12–24).

Bibliography

Editions listed are (for the most part) those consulted in the preparation of this book. Where later editions or reprints are known they are noted after the entry. No claim to bibliographic completeness is made. If a given work is listed in the Suggested References section of a chapter, this fact is indicated by the appropriate chapter number in parenthesis after the reference.

General Treatises on Classical Mechanics

AHARONI, J., *Lectures on mechanics*. Oxford: Oxford University Press, 1972. (Chapter 5)

AMES, JOSEPH SWEETMAN, AND FRANCIS D. MURNHAGHAN, *Theoretical mechanics*. Boston: Ginn and Company, 1929. New York: Dover, 1958.

APPEL, PAUL, *Traité de mécanique rationelle, Tome 2: Dynamique des systèmes. Mécanique analytique*, 6th ed. Paris: Gauthier-Villars, 1953.

BARGER, VERNON D., AND MARTIN G. OLSSON, *Classical mechanics, a modern perspective*. New York: McGraw-Hill, 1973. (Chapter 5)

BARTLETT, JAMES H., *Classical and modern mechanics*. University, Alabama: University of Alabama Press, 1975.

BRADBURY, T. C., *Theoretical mechanics*. New York: Wiley, 1968. (Chapter 4)

CORBEN, H. C., AND PHILIP STEHLE, *Classical mechanics*, 2nd ed. New York: Wiley, 1960. New York: R. E. Krieger, 1974. (Chapter 6)

FINKELSTEIN, ROBERT J., *Nonrelativistic mechanics*. Reading, Mass.: W. A. Benjamin, 1973.

FOWLES, GRANT R., *Analytical mechanics*. New York: Holt, Rinehart and Winston, 1962.

GANTMACHER, F., *Lectures in analytical mechanics*. Translated from Russian. Moscow: Mir Publishers, 1970. Reprint, 1975.

GREENWOOD, DONALD T., *Principles of dynamics*. Englewood Cliffs, N.J.: Prentice-Hall, 1965.

———. *Classical dynamics*. Englewood Cliffs, N.J.: Prentice-Hall, 1977.

GROESBERG, SANFORD WALTON, *Advanced mechanics*. New York: Wiley, 1968. (Chapter 5)

HAMEL, GEORG, *Theoretische Mechanik, eine einhertliche einführung in die gesamte Mechanik*. Berlin: Springer-Verlag, 1949. Corrected reprint, 1967. (Chapter 4)

HAUSER, WALTER, *Introduction to the principles of mechanics*. Reading, Mass.: Addison-Wesley, 1965. (Chapter 1)

KANE, THOMAS R., *Dynamics*. New York: Holt, Rinehart and Winston, 1968.

KILMISTER, C. W., *Hamiltonian dynamics*. New York: Wiley, 1964. (Chapter 9)

———. *Lagrangian dynamics*. New York: Plenum Press, 1967.

KILMISTER, C. W., AND J. E. REEVE, *Rational mechanics*. New York: American Elsevier, 1966. (Chapter 1)

KONOPINSKI, EMIL JAN, *Classical descriptions of motion*. San Francisco: Freeman, 1969.

LANCZOS, CORNELIUS, *The variational principles of mechanics*, 4th ed. Toronto: University of Toronto Press, 1970. (Chapters 1, 2, 8, 9, 10)

LANDAU, L. D., AND E. M. LIFSHITZ, *Mechanics*, 3rd ed. *Course of Theoretical Physics*, vol. 1. Translated from Russian. Oxford: Pergamon Press, 1976. (Chapters, 1, 3, 6)

MACMILLAN, WILLIAM DUNCAN, *Theoretical mechanics*.
    Vol. 1: *Statics and the dynamics of a particle*. New York: McGraw-Hill, 1927. (Chapter 3)
    Vol. 3: *Dynamics of rigid bodies*. New York: McGraw-Hill, 1936. New York: Dover, 1960. (Chapter 5)

MARION, JERRY B., *Classical dynamics of particles and systems*, 2nd ed. New York: Academic Press, 1970. (Chapter 3)

MEIROVITCH, LEONARD, *Methods of analytical dynamics*. New York: McGraw-Hill, 1970. (Chapters 5, 6)

MILNE, E. A., *Vectorial mechanics*. New York: Interscience Publishers, 1948.

MORGENSTERN, D., AND I. SZABO, *Vorlesungen über theoretische Mechanik*. Berlin: Springer-Verlag, 1961.

OSGOOD, WILLIAM F., *Mechanics*. New York: Macmillan, 1937. (Chapter 1)

PARS, L. A., *A treatise on analytical dynamics*. London: Heinemann, 1965.

SALETAN, EUGENE J., AND ALAN H. CROMER, *Theoretical mechanics*. New York: Wiley, 1971. (Chapters 4, 9, 11, 12)

SLATER, JOHN C., AND NATHANIEL H. FRANK, *Mechanics*. New York: McGraw-Hill, 1947. (Chapter 12)

SOMMERFELD, ARNOLD, *Mechanics. Lectures on theoretical physics*, vol. 1. New York: Academic Press, 1952. (Chapter 5)

SPOSITO, GARRISON, *An introduction to classical dynamics*. New York: Wiley, 1976.

SUDARSHAN, E. C. G., AND N. MUKUNDA, *Classical dynamics: A modern perspective*. New York: Wiley, 1974. (Chapter 9)

SYMON, KEITH R., *Mechanics*, 3rd ed. Reading, Mass.: Addison-Wesley, 1971. (Chapters 1, 5, 7, 8)

SYNGE, JOHN L., *Classical dynamics*, in vol. 3, part 1 of *Encyclopedia of physics*. Berlin: Springer-Verlag, 1960. (Chapters 8, 9)

SYNGE, JOHN L., AND BYRON A. GRIFFITH, *Principles of mechanics,* 3rd ed. New York: McGraw-Hill, 1959.

TER HARR, D., *Elements of Hamiltonian mechanics.* Amsterdam: North-Holland, 1961. Second ed., Oxford; Pergamon Press, 1971. (Chapters 6, 8, 10)

THIRRING, WALTER, *A course in mathematical physics I: Classical dynamical systems.* Translated from German. New York: Springer-Verlag, 1978.

THOMSON, WILLIAM (LORD KELVIN), AND PETER GUTHRIE TAIT, *Treatise on natural philosophy.* Cambridge: Cambridge University Press, 1879. Slightly revised, 1896. Reprinted as *Principles of mechanics and dynamics,* New York: Dover, 1962.

WEBSTER, ARTHUR GORDON, *The dynamics of particles and of rigid, elastic, and fluid bodies.* Leipzig: B. G. Teubner, 1904. New York: Stechert-Hafner, 1920.

WELLS, DALE A., *Theory and problems of Lagrangian dynamics.* New York: McGraw-Hill (Schaum), 1967. (Chapters 2, 8)

WHITTAKER, E. T., *A treatise on the analytical dynamics of particles and rigid bodies,* 4th ed. Cambridge: Cambridge University Press, 1937. (Chapters 1, 2, 3, 4, 6, 8)

ZAJAC, ALFRED, *Basic principles and laws of mechanics.* Boston: D. C. Heath, 1966.

————. *Principles of classical mechanics and field theory,* in vol. 3, part 1 of *Encyclopedia of physics.* Berlin: Springer-Verlag, 1960. (Chapters 1, 8) (*See also* J. L. Synge; and C. Truesdell and R. A. Toupin.)

Works on Special Aspects of Classical Mechanics

ABRAHAM, RALPH, AND JERROLD E. MARSDEN, *Foundations of mechanics,* 2nd ed. Reading, Mass.: Benjamin/Cummings, 1978. (Chapter 11)

ANDERSON, JAMES L., *Principles of relativity physics.* New York: Academic Press, 1967.

ARNOLD, V. I., *Mathematical methods of classical mechanics.* Translated from the Russian edition of 1974. New York: Springer-Verlag, 1978.

ARNOLD, V. I., AND A. AVEZ, *Ergodic problems of classical mechanics.* New York: W. A. Benjamin, 1968.

BERGMANN, PETER GABRIEL, *Introduction to the theory of relativity.* New York: Prentice-Hall, 1942.

BORN, MAX, *The mechanics of the atom.* Translated by J. W. Fisher. London: G. Bell and Sons, 1927. Reprint, New York: Ungar, 1967. (Chapters 9, 10, 11)

BROWN, ERNEST W., *An introductory treatise on the lunar theory.* Cambridge: Cambridge University Press, 1896. Reprint, New York: Dover, 1960.

BROUWER, DIRK, AND GERALD M. CLEMENCE, *Methods of celestial mechanics.* New York: Academic Press, 1961.

BRUNET, PIERRE, *Etude historique sur le principe de la moindre action.* Paris: Herrmann et Cie, 1938. (Chapter 8)

BYLERLY, WILLIAM ELWOOD, *An introduction to the use of generalized coordinates in mechanics and physics.* Boston: Ginn, 1913. New York: Dover, 1965.

CHEN, YU, *Vibrations: Theoretical methods.* Reading, Mass.: Addison-Wesley, 1966. (Chapter 6)

CRANDALL, STEPHEN H., et al., *Dynamics of mechanical and electromechanical systems.* New York: McGraw-Hill, 1968.

DANBY, J. M. A., *Fundamentals of celestial mechanics.* New York: Macmillan, 1962. (Chapters 3, 11)

DEUTSCH, RALPH, *Orbital dynamics of space vehicles.* Englewood Cliffs, N.J., Prentice-Hall. 1963. (Chapter 11)

DZIOBEK, OTTO, *Mathematical theories of planetary motions.* Register Publishing Co., 1892. Reprint, New York: Dover, 1962.

EINSTEIN, ALBERT, *The meaning of relativity,* 5th ed. Princeton: Princeton University Press, 1956. (Chapter 7)

FINLAY-FREUNDLICH, E., *Celestial mechanics.* New York: Pergamon Press, 1958.

FOCK, V., *The theory of space, time and gravitation,* 2nd English ed. New York: Pergamon (Macmillan), 1964. (Chapter 7)

FRENCH, A. P., *Special relativity.* (The M.I.T. Introductory Physics Series.) New York: W. W. Norton, 1968. (Chapter 7)

GARFINKEL, BORIS, *The Lagrange–Hamilton–Jacobi mechanics,* in *Space mathematics, Part 1,* J. Barkley Rosser, ed., *Lectures in applied mathematics,* vol. 5. Providence, R. I.: American Mathematical Society, 1966. (Chapters 10, 11)

GEYLING, FRANZ T., AND H. ROBERT WESTERMAN, *Introduction to orbital mechanics.* Reading, Mass.: Addison-Wesley, 1971.

GIACAGLIA, G. E. O., *Perturbation methods in non-linear systems.* New York: Springer-Verlag, 1972. (Chapter 11)

GOSSICK, B. R., *Hamilton's principle and physical systems.* New York: Academic Press, 1967. (Chapter 2)

GRAY, ANDREW, *A treatise on gyrostatics and rotational motion.* London: Macmillan, 1918. New York: Dover, 1959. (Chapter 5)

GRAMMEL, R., *Der Kreisel.* Berlin: Springer-Verlag, 1950.

HAGEDORN, R., *Relativistic kinematics.* New York: W. A. Benjamin, 1963. Reprint, 1973. (Chapter 7)

HAGIHARA, YUSUKE, *Celestial mechanics,* vol. 2, parts 1 & 2, *Perturbation theory.* Cambridge, Mass.: M.I.T. Press, 1972. (Chapter 11)

KERNER, EDWARD H., ed., *The theory of action-at-a-distance in relativistic particle dynamics.* A reprint collection. New York: Gordon and Breach, 1972.

KLEIN, FELIX, *The mathematical theory of the top.* New York: Scribners, 1897. Reprinted in *Congruence of sets and other monographs,* Bronx, N.Y.: Chelsea, 1967.

KLEIN, FELIX, AND ARNOLD SOMMERFELD, *Über die Theorie des Kreisels* (4 vols.). Leipzig: B. G. Teubner, 1897–1910. Reprint, New York: Johnson, 1965. (Chapter 5)

KOTKIN, G. L., AND V. G. SERBO, *Collection of problems in classical mechanics.* Translated from Russian. Oxford: Pergamon Press, 1971.

KURTH, RUDOLF, *Introduction to the mechanics of the solar system.* New York: Pergamon Press, 1969.

LEHNERT, BO, *Dynamics of charged particles.* Amsterdam: North-Holland, 1964. (Chapter 11)

LEIMANIS, EUGENE, *The general problem of motion of coupled rigid bodies about a fixed point.* New York: Springer-Verlag, 1965. (Chapter 5)

MACH, ERNST, *The science of mechanics,* 5th English ed. LaSalle, Ill.: Open Court, 1942. (Chapter 1)

MAGNUS, KURT, *Kreisel Theorie und Anwendungen.* Berlin: Springer-Verlag, 1971.

MANN, RONALD A., *The classical dynamics of particles: Galilean and Lorentz Relativity.* New York: Academic Press, 1974. (Chapter 7)

MCCUSKEY, S. W., *Introduction to celestial mechanics.* Reading, Mass.: Addison-Wesley, 1963. (Chapter 3)

MERCIER, ANDRÉ, *Analytical and canonical formalism in physics.* Amsterdam: North-Holland, 1959.

MISNER, CHARLES W.; KIP S. THORNE; JOHN ARCHIBALD WHEELER, *Gravitation.* San Francisco: Freeman, 1973. (Chapter 7)

MØLLER, C., *The theory of relativity,* 2nd ed. Oxford: Oxford University Press, 1972.

MOSER, JURGEN, *Stable and random motions in dynamical systems, with special emphasis on celestial mechanics.* Princeton: Princeton University Press, 1973. (Chapter 11)

MOULTON, FOREST RAY, *An introduction to celestial mechanics,* 2nd ed. New York: Macmillan, 1914.

NORTHROP, THEODORE G., *The adiabatic motion of charged particles.* New York: Interscience, 1963. (Chapter 11)

OLSON, HARRY F., *Solution of engineering problems by dynamical analogies,* 2nd ed. Princeton, N.J.: D. Van Nostrand, 1966 (corrected reprint of *Dynamical analogies,* 2nd ed., 1958). (Chapter 2)

PLUMMER, H. C., *An introductory treatise on dynamical astronomy.* Cambridge: Cambridge University Press, 1918. Reprint, New York: Dover, 1960. (Chapter 3)

POINCARÉ, HENRI, *Les méthodes nouvelles de la mécanique céleste* (3 vols.). Paris: Gauthier-Villars, 1892–99. Reprint, New York: Dover, 1957.

ROUTH, EDWARD JOHN, *Dynamics of a system of rigid bodies.* Part I, *Elementary Part,* 5th ed. London: Macmillan, 1891. Part II, *Advanced Part,* 5th ed. London, Macmillan, 1892. (Chapter 5)

SANTILLI, RUGGERO MARIO, *Foundations of theoretical mechanics I.* New York: Springer-Verlag, 1978.

SARD, R. D., *Relativistic mechanics: Special relativity and classical particle dynamics.* New York: W. A. Benjamin, 1970. (Chapter 7)

SCHWARTZ, HERMAN M., *Introduction to special relativity.* New York: McGraw-Hill, 1968. (Chapter 7)

SEIGEL, CARL LUDVIG, AND J. K. MOSER, *Lectures on celestial mechanics*. Translated from German. New York: Springer-Verlag, 1971.

SMART, W. M., *Celestial mechanics*. New York:Wiley, 1953. Reprint, 1961.

SOPER, DAVISON EUGENE, *Classical field theory*. New York: Wiley, 1976.

STERNBERG, SHLOMO, *Celestial mechanics*. New York: W. A. Benjamin, 1969.

STERNE, THEODORE E., *An introduction to celestial mechanics*. New York: Interscience, 1960.

TISSERAND, F., *Traité de mécanique céleste, Tome I, Perturbation des planètes* . . . . Paris: Gauthier-Villars, 1889. New printing, 1960.

TRUESDELL, C., *Essays in the history of mechanics*. New York: Springer-Verlag, 1968. (Chapter 1)

TRUESDELL, C., AND R. A. TOUPIN, *The classical field theories*, in vol. 3, part 1 of *Encyclopedia of Physics*. Berlin: Springer-Verlag, 1960. (Chapter 1)

WINTNER, AUREL, *The analytical foundations of celestial mechanics*. Princeton: Princeton University Press, 1941.

WRIGLEY, WALTER; WALTER M. HOLLISTER; AND WILLIAM G. DENHARD, *Gyroscopic theory, design and instrumentation*. Cambridge, Mass.: M.I.T. Press, 1969. (Chapter 5)

YOURGRAU, WOLFGANG, AND STANLEY MANDELSTAM, *Variational principles in dynamics and quantum theory,* 3rd ed. Philadelphia: Saunders, 1968.

Works in Other Branches of Physics and Mathematics, Containing Material of Interest for Classical Mechanics

ARFKEN, GEORGE, *Mathematical methods for physicists*. New York: Academic Press, 1966. Second ed., 1970. (Chapter 4)

BADGER, PARKER H., *Equilibrium thermodynamics*. Boston: Allyn and Bacon, 1967. (Chapter 8)

BARUT, A. O., *Electrodynamics and classical theory of fields and particles*. New York: Macmillan, 1964. (Chapters 7, 12)

BLISS, GILBERT AMES, *Calculus of variations*. Carus Mathematical Monographs, 1. LaSalle, Ill.: Open Court, 1925. (Chapter 2)

BOCHER, MAXIME, *Introduction to higher algebra*. New York: Macmillan, 1907.

BORN, MAX, AND PASCUAL JORDAN, *Elementare Quantenmechanik*. Berlin: Julius Springer, 1930. Ann Arbor: J. W. Edwards, 1946.

BORN, MAX, AND EMIL WOLF, *Principles of optics*. London: Pergamon Press, 1959. Fifth ed., Oxford: Pergamon Press, 1975. (Chapter 10)

BRAND, LOUIS, *Vector and tensor analysis*. New York: Wiley, 1947. (Chapter 5)

BRILLOUIN, LEON, *Tensors in mechanics and elasticity*. Translated from the 1938 French edition. New York: Academic Press, 1964. (Chapter 10)

BORISENKO, A. I., AND I. E. TARAPOV, *Vector and tensor analysis with applications*. Translated and revised by R. A. Silverman. Englewood Cliffs, N.J.: Prentice-Hall, 1968. (Chapter 4)

CARATHÉODORY, C., *Calculus of variations and partial differential equations*. San Francisco: Holden-Day, vol. 1, 1965; vol. 2, 1967. Translated from German edition, 1935; revised, 1956. (Chapters 9, 10)

CASIMIR, H. B. G., *Rotation of a rigid body in quantum mechanics*. Groningen: J. B. Wolters, 1931.

CONDON, E. U., AND G. H. SHORTLEY, *The theory of atomic spectra*. Cambridge: Cambridge University Press, 1935.

COURANT, R., AND D. HILBERT, *Methods of mathematical physics*, 2 vols. New York: Interscience, vol. 1, 1953; vol. 2, 1962. (Chapters 2, 8, 10)

CRONIN, JEREMIAH A.; DAVID F. GREENBERG; AND VALENTINE L. TELEGDI, *University of Chicago graduate problems in physics, with solutions*. Chicago: University of Chicago Press, 1977.

DAVYDOV, A. S., *Quantum mechanics*. Oxford: Pergamon Press, 1965. Second ed., 1976.

DIRAC, PAUL A. M., *Lectures on quantum mechanics*. New York: Yeshiva University, 1964.

EISELE, JOHN A., AND ROBERT M. MASON, *Applied matrix and tensor analysis*. New York: Wiley-Interscience, 1970.

EISENHART, LUTHER PFAHLER, *Continuous groups of transformation*. Princeton: Princeton University Press, 1933. Reprint, New York: Dover, 1961.

ELSGOLC, L. E., *Calculus of variations*. Translated from Russian. Reading, Mass.: Addison-Wesley, 1962.

GARDNER, MARTIN, *New mathematical diversions from Scientific American*. New York: Simon & Schuster, 1971.

GELFAND, I. M., AND S. V. FOMIN, *Calculus of variations*. Translated and edited by Richard A. Silverman. Englewood Cliffs, N.J.: Prentice-Hall, 1963.

GIBBS, J. WILLARD, *Vector analysis*. Edited by E. B. Wilson. New York: Scribner, 1901. New Haven: Yale University Press, 1931.

GOERTZEL, GERALD, AND NUNZIO TRALLI, *Some mathematical methods of physics*. New York: McGraw-Hill, 1960.

GUILLEMIN, ERNST A., *The mathematics of circuit analysis*. New York: Wiley, 1949. (Chapter 6)

HERZBERG, GERHARD, *Infrared and Raman spectra of polyatomic molecules*. New York: D. Van Nostrand, 1945. (Chapters 4, 6)

HESS, SEYMOUR L., *Introduction to theoretical meterology*. New York: Holt, Rinehart and Winston, 1959. (Chapter 4)

HILDEBRAND, FRANCIS B., *Methods of applied mathematics*, 2nd ed. Englewood Cliffs, N.J.: Prentice-Hall, 1965.

HIRSCHFELDER, JOSEPH O.; CHARLES F. CURTISS; AND R. BYRON BIRD, *Molecular theory of gases and liquids*. New York: Wiley, 1954. Slightly revised edition, 1964. (Chapter 3)

JACKSON, JOHN DAVID, *Classical electrodynamics*. New York: Wiley, 1962. Second ed., 1975.

JEFFREYS, H., AND BERTHA S. JEFFREYS, *Methods of mathematical physics,* 2nd ed. Cambridge: Cambridge University Press, 1950. Third ed., 1972. (Chapters 4, 6)

KAPLAN, WILFRED, *Advanced calculus,* 2nd ed. Reading, Mass.: Addison-Wesley, 1973.

LANDAU, L. D., AND E. M. LIFSHITZ, *Quantum mechanics, non-relativistic theory,* 2nd ed., in vol. 3 of *Course of theoretical physics.* Oxford: Pergamon Press, 1965. Third ed., 1977.

LANG, SERGE, *A second course in calculus,* 3rd ed. Reading, Mass.: Addison-Wesley, 1973.

LINDSAY, ROBERT BRUCE, AND HENRY MARGENAU, *Foundations of physics.* New York: Wiley, 1936. (Chapters 1, 8)

LIPSCHUTZ, SEYMOUR, *Theory and problems of linear algebra.* Schaum's Outline Series. New York: McGraw-Hill, 1968.

LOEBL, ERNEST M., ed., *Group theory and its applications,* vol. II. New York: Academic Press, 1971. (Chapters 3, 9, 10)

LOOMIS, LYNN H., AND SHLOMO STERNBERG, *Advanced calculus.* Reading, Mass.: Addison-Wesley, 1968.

MARGENAU, HENRY, AND GEORGE MOSELEY MURPHY, *The mathematics of physics and chemistry.* New York: D. Van Nostrand, 1943. New York: Krieger, 1976.

MARION, JERRY B., *Principles of vector analysis.* New York: Academic Press, 1965. Second ed., 1970. (Chapter 4)

MATHEWS, JON, AND R. L. WALKER, *Mathematical methods of physics.* New York: W. A. Benjamin, 1965. Second ed., 1970.

MCINTOSH, HAROLD. *See* Loebl, Ernest M.

MERZBACHER, EUGEN, *Quantum mechanics.* New York: Wiley, 1961. Second ed., 1970.

MUNK, WALTER H., AND GORDON J. F. MACDONALD, *The rotation of the earth, a geophysical discussion.* Cambridge: Cambridge University Press, 1960. (Chapter 5)

NERING, EVAR D., *Linear algebra and matrix theory.* New York: Wiley, 1963.

NEWTON, ROGER G., *Scattering theory of waves and particles.* New York: McGraw-Hill, 1966. (Chapter 3)

PANOFSKY, WOLFGANG K. H., AND MELBA PHILLIPS, *Classical electricity and magnetism,* 2nd ed. Reading, Mass.: Addison-Wesley, 1962.

PARS, L. A., *An introduction to the calculus of variations.* New York: Wiley, 1962. (Chapter 2)

PEARSON, CARL E., ed., *Handbook of applied mathematics.* New York: Van Nostrand Reinhold, 1974.

PEASE, MARSHAL C., III, *Methods of matrix algebra.* New York: Academic Press, 1965.

LORD RAYLEIGH, *The theory of sound* (2 vols.), 2nd ed. London: Macmillan, 1894–1896. New York: Dover, 1945. (Chapters 1, 12)

ROSE, M. E., *Elementary theory of angular momentum.* New York: Wiley, 1957.

ROZENTAL, S., *Niels Bohr: His life and work.* Amsterdam: North-Holland, 1967. New York: Wiley, 1967.

RUND, HANNO, *The Hamilton–Jacobi theory in the calculus of variations.* London: D. Van Nostrand, Ltd., 1966. (Chapters 2, 7, 8, 9)

SCHIFF, LEONARD I., *Quantum mechanics,* 3rd ed. New York: McGraw-Hill, 1968.

SMIRNOV, V. I., *A course of higher mathematics* (5 vols.). Translated from Russian. Oxford: Pergamon Press, 1964.

SOMMERFELD, ARNOLD, *Atomic structure and spectral lines.* Translated by H. L. Brose from the 5th German edition of 1931. New York: Dutton, 1934. (Chapter 10)

STACEY, FRANK D., *Physics of the earth.* New York: Wiley, 1969. Second ed., 1977. (Chapters 4, 5)

STOLL, ROBERT R., *Linear algebra and matrix theory.* New York: McGraw-Hill, 1952. Reprint, New York: Dover, 1969. (Chapter 4)

STRANG, GILBERT, *Linear algebra and its applications.* New York: Academic Press, 1976.

SYNGE, J. L., *Geometrical optics.* Cambridge: Cambridge University Press, 1937.

SYNGE, J. L., AND A. SCHILD, *Tensor calculus.* Toronto: University of Toronto Press, 1949. (Chapters 4, 6, 7)

TOLMAN, RICHARD C., *The principles of statistical mechanics.* Oxford: Oxford University Press, 1938. (Chapter 9)

TURNBULL, H. W., *The theory of determinants, matrices and invariants,* 2nd ed. London: Blackie, 1945.

VAN VLECK, J. H., *Quantum principles and line spectra.* Bulletin No. 54 of the National Research Council. Washington D.C.: National Research Council, 1926. (Chapter 10)

WEINSTOCK, ROBERT, *Calculus of variations.* New York: McGraw-Hill, 1952. (Chapter 2)

WENTZEL, GREGOR, *Quantum theory of fields.* Translated by C. Houtermans and J. W. Jauch. New York: Interscience, 1949. (Chapter 12)

WEYL, HERMAN, *The classical groups.* Princeton: Princeton University Press, 1939. Second ed., reprint of first, with additions, 1946.

WHITHAM, G. B., *Linear and nonlinear waves.* New York: Wiley, 1974.

WIGNER, EUGENE P., *Group theory, and its application to the quantum mechanics of atomic spectra.* Translated from German by J. J. Griffin. New York: Academic Press, 1959.

WILLS, A. P., *Vector analysis, with an introduction to tensor analysis.* New York: Prentice-Hall, 1931. New York: Dover, 1931. (Chapter 5)

WILSON, E. BRIGHT, JR.; J. C. DECIUS; AND PAUL C. CROSS, *Molecular vibrations, the theory of infrared and Raman vibrational spectra.* New York: McGraw-Hill, 1955. (Chapter 6)

ZEMANSKY, MARK W., *Heat and thermodynamics,* 5th ed. New York: McGraw-Hill, 1968.

Tables of Integrals and Mathematical Functions, and Other Works of Reference

ABRAMOWITZ, MILTON, AND IRENE A. STEGUN, eds., *Handbook of mathematical functions, with formulas, graphs and mathematical tables*. NBS Applied Math. Series 55. Washington, D.C.: Government Printing Office, 1964. New York: Dover, 1965.

GRADSHTEYN, I. S., AND I. M. RYZHIK, *Table of integrals, series and products*, 4th ed. Prepared by Yu V. Geronimus and M. Yu Tseytlin. Translated from Russian. New York: Academic Press, 1965.

PEIRCE, B. O., *A short table of integrals*, 3rd ed. Boston: Ginn, 1929. Fourth ed. (by B. O. Peirce and R. M. Foster), Boston: Ginn, 1956.

SELBY, SAMUEL M., ed., *Handbook of mathematical tables*. Cleveland, Ohio: Chemical Rubber Publishing Co. (CRC), 1962.

SPIEGEL, MURRAY R., *Mathematical handbook of formulas and tables*. New York: McGraw-Hill, 1968.

INDEX OF SYMBOLS

In choosing the various symbols, certain general principles have been followed whenever possible. Vectors have been denoted by bold-faced Roman characters, while tensors of second rank or higher, and matrices, are represented by bold-faced sans serif letters. When a vector is specifically treated as a tensor of the first rank, or represented by a column (or row) matrix it is occasionally symbolized in bold-faced Roman letters, but most often the sans serif type face is used. For Greek symbols, the same bold face is used for vectors, matrices, and tensors.

A dot above a letter invariably denotes differentiation with respect to time. Primes are frequently used to denote quantities that have been subjected to a transformation of some kind. In Chapter 4 primes on coordinates refer to body sets of axes, as distinguished from the unprimed space sets of axes, but not subsequently. Generally, primes are also used to mark symbols involving some quantity relative to the center of mass. There are in addition many other exceptional usages of the prime symbol, which will be clear within the context of the given argument. Two instances may however be noted specifically. Scattering cross sections in terms of the angle in the laboratory system (*not* the angle in the center of mass system) are marked by a prime (see p. 117). And in Section 7–9 a prime on a character denotes differentiation with respect to a parameter θ used in place of time.

To simplify the appearance of many of the formulas, various subscript notations are used from time to time to indicate differentiation with respect to the subscripted variable (or its index). See, for example, p. 399f, the footnote on p. 441, and all of Chapter 12 after p. 551.

In discussing canonical transformations, lower case letters are often used for the original variables, and capitals for the transformed variables. Subscripts 0 frequently denote initial or equilibrium values. As is customary, complex conjugates are denoted by an asterisk. Contrary usage of the asterisk is always specifically noted.

This index of symbols is not intended to be complete. Indexing is hardly necessary for commonly used symbols, e.g., Cartesian coordinates (x, y, z); plane polar coordinates (r, θ) (or variants: (r, ϕ), (r, ψ) etc.);

spherical polar coordinates (r, θ, ϕ), or the notation of vector analysis. Generally speaking, the index omits (except where there may be ambiguity) "scratch" symbols, which are used only within a page or two of their definition, and not reused again elsewhere in the same meaning.

Symbols defined in exercises are not normally indexed except when they are of wider usefulness or may give rise to ambiguity; such references are indicated by page numbers in italics. When two or more page numbers are listed the first is to an initial, often incidental, reference. Subsequent page numbers indicate fuller definitions or descriptions.

V_{ij} potential energy coefficients about equilibrium, 245

V, V velocity, and speed, of the center of mass, 115

V   square matrix of V_{ij} coefficients, 247

$\mathcal{V}$ potential energy density, 556

v   velocity vector, 1

v speed, 3

W work, 3

W Hamilton's characteristic function, 443, 445f

W^* Hamilton's characteristic function in terms of (q, w), 532

W_i separated Hamilton's characteristic function, 451, 454

W   Jacobian matrix of canonical momenta with respect to coordinates, 359

w_i angle variables, 460, 464

w_1, w_2, w_3 transformed angle variables of the Kepler problem, 476

w_0, J_0 unperturbed action-angle variables, 515f

w   vector (column or row matrix) of angle variables, 465

$\mathbf{w}_0$ vector (column or row matrix) of unperturbed angle variables, 520

X, Y, Z; or X_1, X_2, X_3 Components of eigenvectors, 158f

$X_{r\nu}$ function describing transformation of x_ν in Noether's theorem, 592

X   square matrix of eigenvector components, 160

x, y variables in problem of heavy symmetrical top, 219

x deviation of orbit from circularity, $u - u_0$, 601

$x_i = x_1, x_2, x_3$ Cartesian coordinates in ordinary space, 132

$x_\mu = x_1, x_2, x_3, x_4$ Cartesian coordinates in Minkowski space, 280

$x_\mu = x_1, x_2, x_3, x_0$ coordinates in real four-space, 288

x   column or row matrix of a vector, 138

Y generating function in perturbation theory, 516

Y Young's modules, 546

Y_i expansion functions of perturbation generating function, 516, 520

Z, Z' atomic number, 108

INDEX OF GREEK SYMBOLS

Ω vector of precession, 210

$d\Omega$ differential vector of infinitesimal rotation, 169

ω, ω_i angular velocity of rotation, angular frequency, 29, 101

ω, ω_k normal frequencies of small oscillation, 246

ω angular frequency of the linear harmonic oscillator, 389

ω argument of the perihelion, 479

$\boldsymbol{\omega}$ angular velocity vector, 175

$\boldsymbol{\omega}$ frequency of Thomas precession, 288

$\boldsymbol{\omega}_c$ cyclotron or gyration frequency, 535

$\boldsymbol{\omega}_l, \omega_l$ Larmor frequency, 234

MISCELLANEOUS SYMBOLS

0   matrix with zero elements, 347

1   unit matrix, 139

1   unit dyadic, 194

$'$ marks derivative with respect to θ (q.v.), 327

∇ gradient operator, 4

$\Box$ four-dimensional gradient operator, 303

$\Box^2$ D'Alembertian, 303

$[,]$ commutator, Lie bracket, 173, 401

$[,]$ Poisson bracket, 397

$\{ , \}$ Lagrange bracket, 401

$\dfrac{\delta}{\delta\psi}$ functional derivative (with respect to ψ), 564

INDEX

(Page numbers in italics refer to exercises.)

4-vectors:

$$J_\alpha = (c\rho, \vec{J}) \qquad X_\alpha = (ict, \vec{x})$$

$$\partial_\alpha = \left(\frac{1}{c}\frac{\partial}{\partial t}, \vec{\nabla}\right)$$

$$A_\alpha = (\phi, \vec{A})$$

Anti-symmetric

$$F_{\alpha\beta} = \partial_\alpha A_\beta - \partial_\beta A_\alpha = \begin{pmatrix} 0 & & & \\ E_x & 0 & & \\ E_y & B_z & 0 & \\ E_z & -B_y & B_z & 0 \end{pmatrix}$$

Lorentz transform of $F_{\mu\nu}$:

$$F'_{\alpha\beta} = \frac{\partial X'_\alpha}{\partial X_\mu} \frac{\partial X'_\beta}{\partial X_\nu} F_{\mu\nu} = A F \tilde{A}$$

Lorentz transformation matrix

Ex: $v = v_x$ $\quad E'_x = E_x, \quad B'_x = B_x$

$$E'_y = \gamma(E_y - \beta B_z)$$

$$E'_z = \gamma(E_z + \beta B_y)$$

$$B'_y = \gamma(B_y + \beta E_z)$$

$$B'_z = \gamma(B_z - \beta E_y)$$

Learn this!!

$$\boxed{L = -mc^2\sqrt{1-\beta^2} - V}$$